InterActions© IN PHYSICAL SCIENCE™

Fred Goldberg

Sharon Bendall

Patricia Heller

Robert Poel

84 Business Park Drive, Armonk, NY 10504 Phone (914) 273-2233
Fax (914) 273-2227 Toll Free (888) 698-TIME (8463) www.its-about-time.com

It's About Time, President
Tom Laster

Director of Product Development
Barbara Zahm, Ph.D.

Creative/Art Director
John Nordland

Project Editor
Ruta Demery

Design/Production
Emerson Cooke
Joan Harten
Marie Killoran
Kadi Sarv

Illustrations
Dennis Falcon

Technical Illustrations
Carlos Lopez

Project Coordinators
Loretta Steeves
Monica T. Rodriguez

Editorial Staff
Andrea Maywhort
Elisabeth McGrath
Minal Singh

Photo Research
Lisa Marmelstein
Bernardo Saravia

Printed and bound in the United States of America.

ISBN-10: 1-58591-360-X
ISBN-13: 978-1-58591-360-2

2 3 4 5 6 QW 10 09 08 07 06

This project was supported, in part, by the National Science Foundation.
Opinions expressed are those of the authors and not necessarily those of the National Science Foundation.

InterActions©

IN PHYSICAL SCIENCE™

Fred Goldberg, Ph.D.

Fred Goldberg is Professor of Physics at San Diego State University. Since the 1980s he has been involved in physics education. Initially his group studied student understanding in topical areas of physics, and later studied students' beliefs about physics knowledge and learning. They then focused on developing strategies that addressed student difficulties. Many strategies involved the use of computer technology: videodisks, animations, graphics programs, and simulations. Since the late 1990s, his group has focused on studying how students learn in a technology-rich, collaborative learning environment. He has directed or co-directed many large National Science Foundation grants on research on learning, on development of curriculum materials for middle school, high school and college, and on preservice teacher education. He has served on the editorial boards of the *American Journal of Physics*, *The Physics Teacher*, and the *International Journal of Science Education*. In 2003 he was the recipient of the Robert A. Millikan Award from the American Association of Physics Teachers for notable and creative contributions to the teaching of physics.

Sharon Bendall

After earning degrees in physics from Memphis State University and Arizona State University, Sharon Bendall began her career as a professional physicist at the IBM Thomas J. Watson Research Center. During this period she developed an appreciation for what people need to learn in school to prepare them for a career in the 'real world.' Later she turned from being a scientist to being a science educator. Since then, she has taught physics at San Diego State University and the University of San Diego, conducted research in how students learn physics, developed physics instructional materials for university and middle school students, and developed materials to help teachers teach. She has directed or co-directed physics education projects funded by the National Science Foundation, has served as a member of the Research in Physics Education Committee of the American Association of Physics Teachers, has served as a parent representative to her children's school district, has published journal articles, and given numerous professional talks and teacher workshops.

Patricia Heller, Ph.D.

Patricia Heller is Professor (Emeritus) of Science Education in the Department of Curriculum and Instruction at the University of Minnesota. She received graduate degrees in physics and science education from the University of Washington and the University of Michigan, respectively. She has a wide range of teaching experiences, including general science at the elementary school level, physics, chemistry and physical science at the high school level, and science education for elementary and secondary teachers at the college level. Her research focus has been in two areas: student difficulties with the conceptual and mathematical aspects of problem solving, and the design and evaluation of an instructional approach to help students overcome their difficulties with these two aspects of problem solving. She has directed or co-directed physics education projects funded by the National Science Foundation, and has published many journal articles and given myriad professional talks and workshops.

Robert Poel, Ph.D.

Robert Poel is Professor of Physics and Science Education (Emeritus) at Western Michigan University in Kalamazoo, Michigan. He began his professional career as a middle and high school physics and mathematics teacher in Battle Creek, Michigan. After realizing that helping students learn how the universe works was a challenging and rewarding profession, he returned to graduate school to fill in some of the large gaps that remained in his own science and pedagogical background. This was the beginning of a long journey that continues to the present in the areas of teacher preparation, professional development, and inquiry oriented science curriculum. Since 1980, he has worked extensively with elementary and secondary teachers in several national professional development projects that include Operation Physics (OP), Powerful Ideas in Physical Science (PIPS), and Constructing Physics Understanding (CPU). He has served on the Committee on Physics in Pre-High School Education of the American Association of Physics Teachers and has taught many content workshops and given numerous professional talks for K-12 science teachers and science educators.

InterActions in Physical Science™ formerly known as the (CIPS Project) was supported, in part, by grants ESI-9812299 and ESI-0138900 from the National Science Foundation. Additional support was provided by San Diego State University and Western Michigan University.

NSF Program Officer
Gerhard Salinger

Project Directors
Fred Goldberg
San Diego State University
Sharon Bendall
San Diego State University
Patricia Heller
University of Minnesota
Robert Poel
Western Michigan University

Other Development Staff
Gulcin Cirik
William Doerge
Heide Doss-Hammel
Michael McKean

Editor
Judith Leggett

Graphics and Illustrations
Carlos Lopez

Fiction Writer
Kelly McCullough

Video Support
Michael Noon
Pat Walker, WalkerVision

Computer Simulation Software
OpenTeach Software, Inc.

Teacher Contributors
Kathleen Blair
John Bohn
Holly Eaton
Frank Forrester
Brad Lappin
Leanne Larson
Jeff Major
Geof Martin
Andreanna Murphy
Abigail Paulsen
Holly Pennix
Mary Roobol
Sally Sondreal
Brian Vedder
Chad Wagner

Other Contributors
Clarisa Bercovich
Gary Blakmer
Judith Bransky
Cynthia D'Angelo
Tony DiMauro
Matzi Eliahu
April Maskiewicz
Graham Oberem
Cody Sandifer
Giovanni Stephens-Robledo

Content Reviewers
John Hubisz
North Carolina
State University
Neil Wolf
Dickinson College
(Emeritus)

Safety Reviewer
Edward Robeck
Salisbury University,
Salisbury, MD

Equity Reviewer
Marcia Fetters
Western Michigan University.
Kalamazoo, MI

External Evaluators
Sean Smith and
Eric Banilower
Horizons Research
Incorporated, Chapel Hill, NC

Welcome Letter to Students

Dear Students,

Welcome to *InterActions in Physical Science*, which we call *InterActions* for short. A great team of scientists and educators developed *InterActions*. We come from a wide range of backgrounds. Some of us started as scientists and others started as science teachers. We all love science. However, more importantly, we all care about how students, like you, think and learn about science. We care about what interests and excites you and about what causes you problems. We have poured more than five years of labor and love into this program. *InterActions* started with our original ideas and grew to where thousands of students across the USA were using the program and telling us what works for them and what doesn't. And, we listened and kept improving the program.

In *InterActions* you will be very busy learning about the world around you. You will do experiments, collect evidence, think, talk with students in your team, think, talk with students in your class, think, talk with your teacher, read, think some more, develop ideas, and answer questions. Sometimes, your teacher will give you Scientists' Consensus Ideas forms. These will give you a chance to see that all the thinking you did and the ideas you developed are the same as those of the scientists.

We also created an *InterActions Kids* website. There's a lot of great stuff going on there. Check it out!

We hope you have a great *InterActions* year!

The *InterActions* Team

Can You Think Like a Scientist?

The answer is **Yes!** Scientists are no different from you or me, except in one important way – *scientists think and develop their ideas using carefully collected evidence.* This helps make sure that the ideas the scientists develop are valid or true.

The results of thinking like a scientist have huge effects on our everyday lives. Imagine how different the world was 100 years ago – no TV's, phones, airplanes, or computers. Thinking like a scientist is a very powerful way to think.

And you are about to learn how to think like a scientist for yourself.

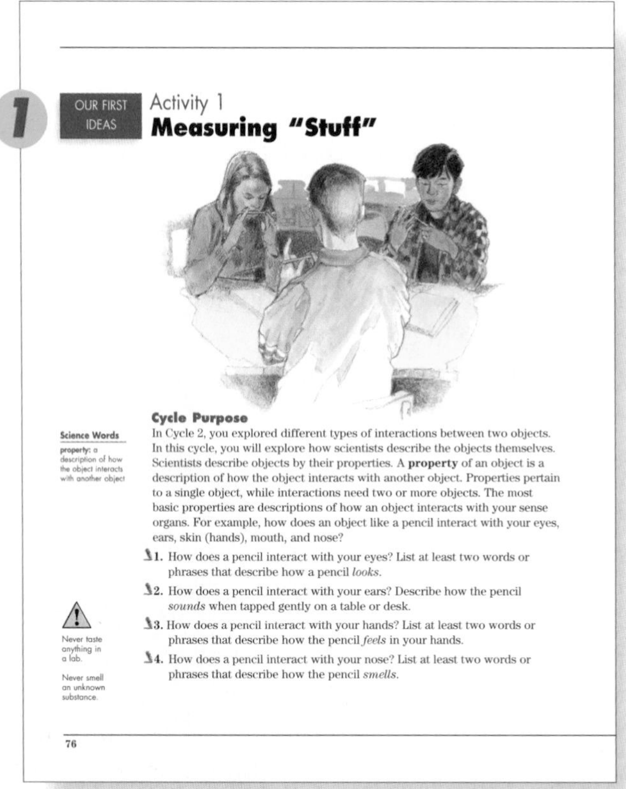

1 OUR FIRST IDEAS

Activity 1

Measuring "Stuff"

Cycle Purpose

In Cycle 2, you explored different types of interactions between two objects. In this cycle, you will explore how scientists describe the objects themselves. Scientists describe objects by their properties. A **property** of an object is a description of how the object interacts with another object. Properties pertain to a single object, while interactions need two or more objects. The most basic properties are descriptions of how an object interacts with your sense organs. For example, how does an object like a pencil interact with your eyes, ears, skin (hands), mouth, and nose?

Science Words

property: a description of how the object interacts with another object

1. How does a pencil interact with your eyes? List at least two words or phrases that describe how a pencil *looks.*
2. How does a pencil interact with your ears? Describe how the pencil *sounds* when tapped gently on a table or desk.
3. How does a pencil interact with your hands? List at least two words or phrases that describe how the pencil *feels* in your hands.
4. How does a pencil interact with your nose? List at least two words or phrases that describe how the pencil *smells.*

Never taste anything in a lab.

Never smell an unknown substance.

76

1 Our First Ideas

You will begin by thinking about what you already know about how something in the world around you happens. You'll share these first ideas with members of your team and with the rest of your class. This sharing will give you a chance to learn about other students' first ideas.

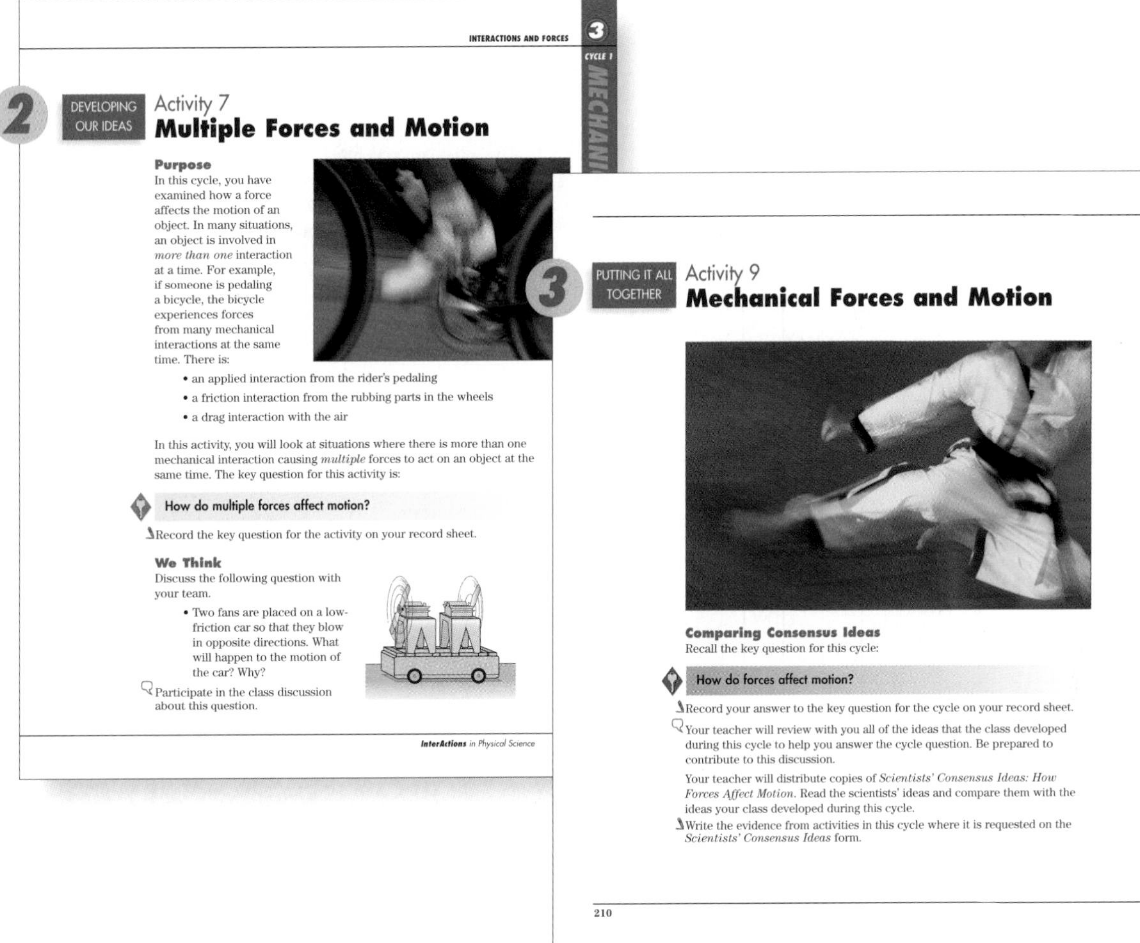

INTERACTIONS AND FORCES 3

CYCLE 1 MECHANI

2 DEVELOPING OUR IDEAS

Activity 7

Multiple Forces and Motion

Purpose

In this cycle, you have examined how a force affects the motion of an object. In many situations, an object is involved in *more than one* interaction at a time. For example, if someone is pedaling a bicycle, the bicycle experiences forces from many mechanical interactions at the same time. There is:

- an applied interaction from the rider's pedaling
- a friction interaction from the rubbing parts in the wheels
- a drag interaction with the air

In this activity, you will look at situations where there is more than one mechanical interaction causing *multiple* forces to act on an object at the same time. The key question for this activity is:

How do multiple forces affect motion?

Record the key question for the activity on your record sheet.

We Think

Discuss the following question with your team.

- Two fans are placed on a low-friction car so that they blow in opposite directions. What will happen to the motion of the car? Why?

Participate in the class discussion about this question.

InterActions in Physical Science

3 PUTTING IT ALL TOGETHER

Activity 9

Mechanical Forces and Motion

Comparing Consensus Ideas

Recall the key question for this cycle:

How do forces affect motion?

Record your answer to the key question for the cycle on your record sheet.

Your teacher will review with you all of the ideas that the class developed during this cycle to help you answer the cycle question. Be prepared to contribute to this discussion.

Your teacher will distribute copies of *Scientists' Consensus Ideas: How Forces Affect Motion*. Read the scientists' ideas and compare them with the ideas your class developed during this cycle.

Write the evidence from activities in this cycle where it is requested on the *Scientists' Consensus Ideas* form.

210

2 Developing Our Ideas

Now that you have identified your first ideas, you need a chance to find out more. Are your first ideas on the mark? Are they in some ways problematic? Just like a scientist, you will do some exploring to find out. You will collect evidence by doing an exploration, watching a teacher demonstration or video, or by running a computer simulation. You will receive the guidance you need to confirm or change the ideas you already had and to develop new ideas.

3 Putting It All Together

How do your ideas compare to scientists' ideas? In this activity, you will tie together all of the ideas you developed in your explorations and compare them to the ideas of actual scientists. You may be surprised at how good the ideas that you developed are!

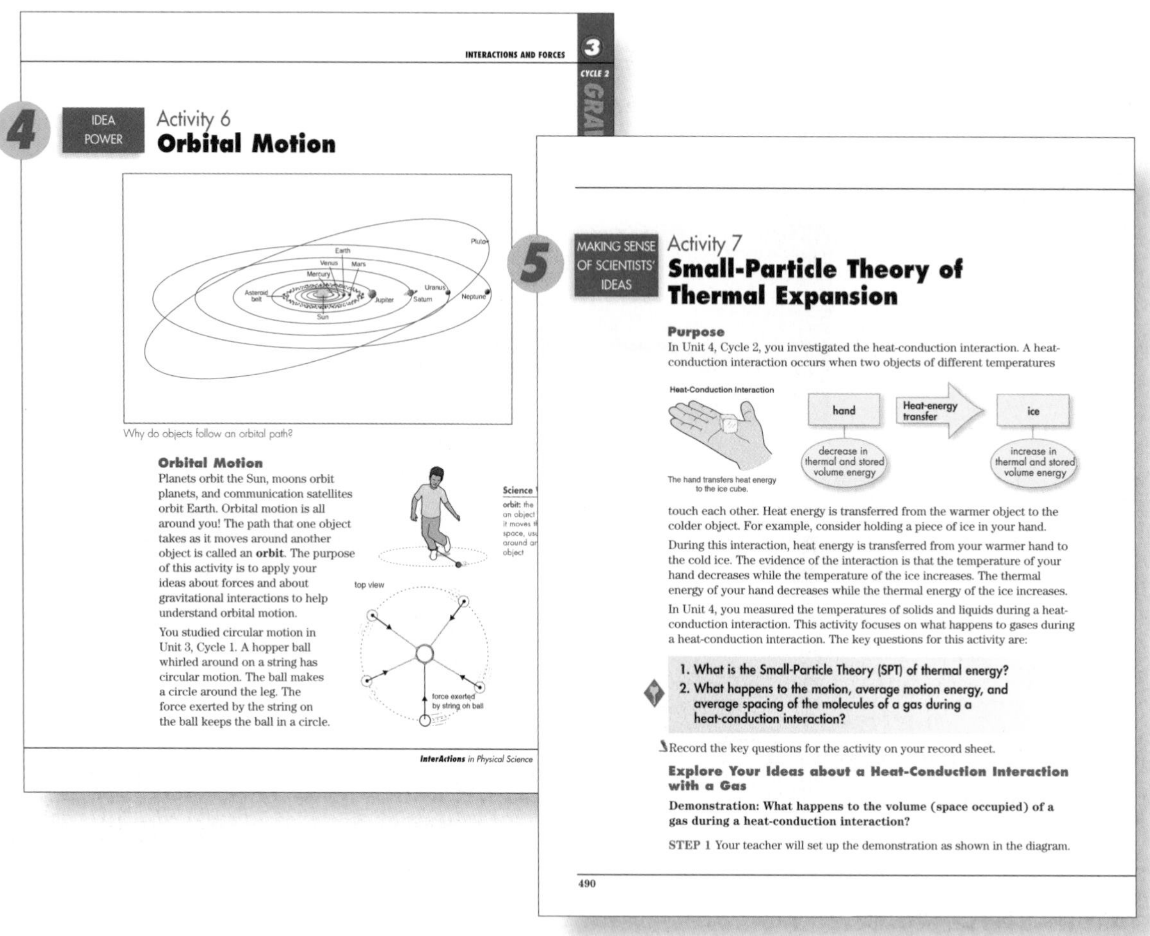
INTERACTIONS AND FORCES 3 CYCLE 2

4 IDEA POWER Activity 6

Orbital Motion

Why do objects follow an orbital path?

Orbital Motion

Planets orbit the Sun, moons orbit planets, and communication satellites orbit Earth. Orbital motion is all around you! The path that one object takes as it moves around another object is called an **orbit**. The purpose of this activity is to apply your ideas about forces and about gravitational interactions to help understand orbital motion.

You studied circular motion in Unit 3, Cycle 1. A hopper ball whirled around on a string has circular motion. The ball makes a circle around the leg. The force exerted by the string on the ball keeps the ball in a circle.

InterActions in Physical Science

5 MAKING SENSE OF SCIENTISTS' IDEAS Activity 7

Small-Particle Theory of Thermal Expansion

Purpose

In Unit 4, Cycle 2, you investigated the heat-conduction interaction. A heat-conduction interaction occurs when two objects of different temperatures touch each other. Heat energy is transferred from the warmer object to the colder object. For example, consider holding a piece of ice in your hand.

During this interaction, heat energy is transferred from your warmer hand to the cold ice. The evidence of the interaction is that the temperature of your hand decreases while the temperature of the ice increases. The thermal energy of your hand decreases while the thermal energy of the ice increases.

In Unit 4, you measured the temperatures of solids and liquids during a heat-conduction interaction. This activity focuses on what happens to gases during a heat-conduction interaction. The key questions for this activity are:

1. What is the Small-Particle Theory (SPT) of thermal energy?
2. What happens to the motion, average motion energy, and average spacing of the molecules of a gas during a heat-conduction interaction?

Record the key questions for the activity on your record sheet.

Explore Your Ideas about a Heat-Conduction Interaction with a Gas

Demonstration: What happens to the volume (space occupied) of a gas during a heat-conduction interaction?

STEP 1 Your teacher will set up the demonstration as shown in the diagram.

490

4 Idea Power

Now it's time to apply your ideas. Each learning cycle has one or two **Idea Power** activities. In the **Idea Power** activities you will apply the ideas you developed to new situations in the world around you. You'll have the chance to evaluate analyses and explanations written by others, and you'll even get to analyze and explain some new situations on your own. Using ideas is part of the job of any real scientist.

5 Making Sense of Scientists' Ideas

Sometimes, the experiments needed to develop a scientific idea are very complex and require expensive equipment. In these cases, you will not be able to develop the ideas on your own. Instead, you will be told about ideas that scientists have already developed. Then you'll do some exploring to help you make sense of these ideas. In these activities, you will also do more reading and answering questions than perhaps in other types of activities. But as always, it's your job to see how these ideas make sense to the scientist in you.

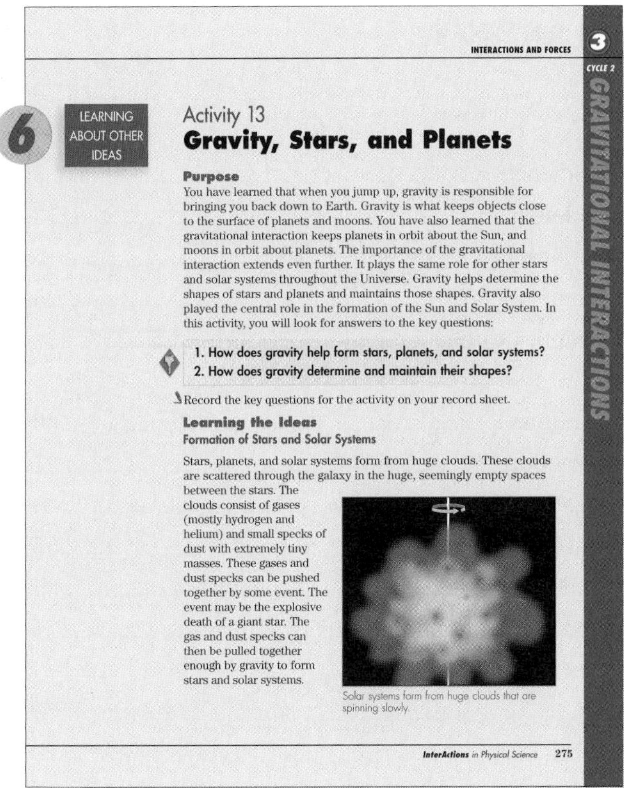

INTERACTIONS AND FORCES 3

CYCLE 2 GRAVITATIONAL INTERACTIONS

6 LEARNING ABOUT OTHER IDEAS

Activity 13

Gravity, Stars, and Planets

Purpose

You have learned that when you jump up, gravity is responsible for bringing you back down to Earth. Gravity is what keeps objects close to the surface of planets and moons. You have also learned that the gravitational interaction keeps planets in orbit about the Sun, and moons in orbit about planets. The importance of the gravitational interaction extends even further. It plays the same role for other stars and solar systems throughout the Universe. Gravity helps determine the shapes of stars and planets and maintains those shapes. Gravity also played the central role in the formation of the Sun and Solar System. In this activity, you will look for answers to the key questions:

1. How does gravity help form stars, planets, and solar systems?
2. How does gravity determine and maintain their shapes?

Record the key questions for the activity on your record sheet.

Learning the Ideas

Formation of Stars and Solar Systems

Stars, planets, and solar systems form from huge clouds. These clouds are scattered through the galaxy in the huge, seemingly empty spaces between the stars. The clouds consist of gases (mostly hydrogen and helium) and small specks of dust with extremely tiny masses. These gases and dust specks can be pushed together by some event. The event may be the explosive death of a giant star. The gas and dust specks can then be pulled together enough by gravity to form stars and solar systems.

Solar systems form from huge clouds that are spinning slowly.

InterActions in Physical Science 275

6 Learning About Other Ideas

Any scientist is always looking for new ideas. In these activities, you'll get to learn about a variety of other interesting ideas. Your teacher will decide which of the **Learning About Other Ideas** activities your class will do. The evidence for these ideas is presented through readings, demonstrations, movies, or computer simulations, rather than hands-on explorations. If your class doesn't do one of these activities in your classroom, you can still use the activity to find out about the ideas on your own.

Thinking like a scientist is exciting... and you're about to learn how.

Look for these features to help you as you go along.

There are seven Units in the *InterActions* book and each Unit has one to three Cycles of Learning.

Each Cycle of Learning builds from the previous Cycles and is made up from several Activities. Each Activity begins with a **Purpose** giving you a reason for what you will be doing in that Activity.

Each cycle has a Cycle Key Question that drives that cycle.

Each activity has an Activity Key Question that drives that activity.

Scientists often record their results in lab books. When you see this symbol, you should record the information you have found during your explorations in your workbook.

Like scientists, you will be gathering data as you complete your explorations. When you see this symbol, you should record your data in a table in your workbook.

All scientists need to share their ideas, questions, and discoveries with others. An important part of *InterActions* involves sharing your ideas with others. When you see this symbol, you will be sharing your ideas with your class.

SECTION A: INTERACTIONS AND ENERGY

SECTION C: INTERACTIONS OF MATERIALS

InterActions

As you learn science by using *InterActions in Physical Science*, you will discover that the term interactions has many different meanings.

In the real world objects can act on and influence each other. In these cases we say that the objects are interacting with each other. For example, when a soccer player kicks a ball, his foot and the ball are interacting with each other. When light from the Sun warms up your soda, the Sun and your soda are interacting. When a magnet picks up a piece of metal, the magnet and metal are interacting with each other. Scientists have classified many different types of interactions, and in this class you will study many of them.

When you learn science in the *InterActions* classroom you will be performing experiments, discussing ideas with your fellow classmates, reading material, and participating in discussions led by your teacher. When you perform experiments you are interacting with the apparatus. When you discuss and share ideas with other students, you are interacting with them. When you are listening or talking to your teacher, you are interacting with him or her. When you read your science book, you are interacting with it.

Finally, one of the main purposes of this course is to help you use science to explain many interesting things in the real world. When you apply the ideas you learn to explain these interesting things, you are interacting with the real world.

As you use *InterActions*, think about all these meanings of the term interactions. It can refer to the special ways that scientists see the world. It can refer to what will be happening in your *InterActions* classroom. It can also refer to the connections you will make between what you learn in this class and the world around you.

SECTION A

UNIT 1:
BUILDING A FOUNDATION

UNIT 2:
INTERACTIONS AND ENERGY

INTERACTIONS AND ENERGY

SECTION A

Interactions and Energy

Interactions and Energy introduces the foundations that you will use throughout your *InterActions in Physical Science* course. You will learn some of the fundamentals of science, like how to conduct a fair experiment and how to decide if your conclusion from an experiment is valid. You will also learn how to describe the world around you using interactions and energy ideas. While you are learning this, you will be developing skills that will help you interact with the materials in your laboratory and the other students in your classroom. By the end of Section A, you will already be thinking like a real scientist and working with other students as a team!

Section A has two units.

What will you learn about in Unit 1?

In Unit 1, you will learn about how to conduct experimental investigations. Then you will specifically investigate magnetic, electric-charge, and electric-circuit interactions. Finally, you will learn various ways of measuring the amount of substances.

What type of activities are in Unit 1?

Almost all activities in Unit 1 are *Developing Our Ideas*. In these activities, you do experiments to develop new science ideas. Doing the experiments will help you develop skills like "Follow Directions and Stay on Task." You will also have many group and class discussions. Then you can work on skills like "Respect Your Team and Their Ideas." There are, of course, other skills you will be developing. In Cycles 2 and 3, there are some activities called *Learning About Other Ideas*. These have less discussion.

What are the sections in each activity?

Each activity in Unit 1 is written in a similar way. The first activity in each cycle introduces the *Cycle Purpose* and ◆*Key Question*. The activity then continues with an activity *Purpose* and a ◆*Key Question* for the activity. These sections help you understand why you are doing the activity. Next is an *I Think* or *We Think* section. Here you will think about what you know before you do the experiments. You will begin by thinking about things on your own. Then you will learn skills to help you to share your first ideas with others. In the *Explore Your Ideas* section, you will do experiments and make observations. Then there is the *Make Sense of Your Ideas* section. This is when you work on making sense of what you observed. Finally, the activity ends with a section called *My Ideas* or *Our Consensus Ideas*. In this section, you answer the *Key Question* for the activity.

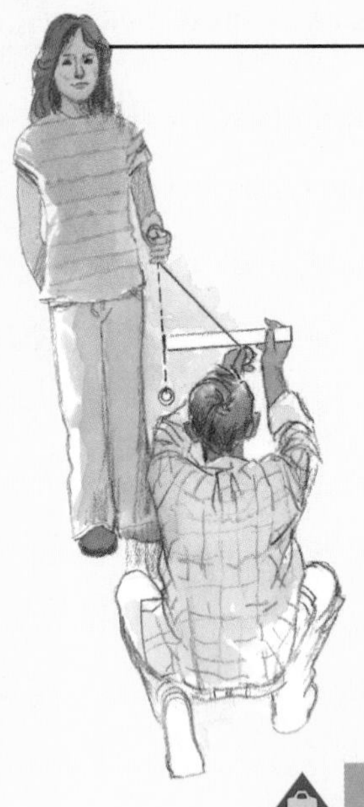

All InterActions in Physical Science units are divided into sections called "cycles of learning." In Unit 1 there are three cycles.

Cycle 1: Science Experiments

In the first cycle, you will investigate pendulums and magnets. You will learn how to make and interpret measurements, and how to decide if an experiment is well-designed. You will also learn how to judge the conclusions you draw from your experiments.

You will decide if the conclusions are logical and well-reasoned. Each cycle has a *Key Question.* Here is the key question for Cycle 1.

How can you tell if the conclusion from an experiment is valid?

Cycle 2: Introducing Interactions

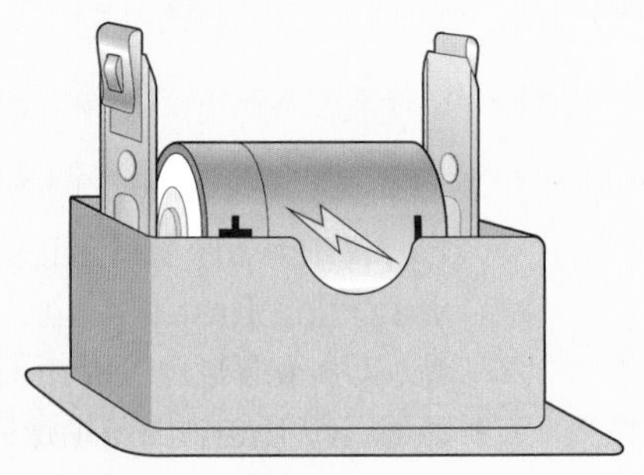

In this cycle you will learn how scientists describe our world in terms of interactions between objects. You will investigate three types of interactions: magnetic, electric charge, and electric circuit. You will also construct and analyze an electromagnet and study how a buzzer works. Here is the key question for Cycle 2.

How can you describe interactions?

Cycle 3: Interactions and Properties

In this cycle you learn what a property of an object is, and what are some properties that would help you identify objects. You will also be introduced to the meaning of mass, volume, and density, all of which are important properties of objects. Here are the key questions for Cycle 3.

1. **How do scientists measure the amount of stuff in an object?**
2. **What are some properties of objects that help you decide what kind of material they are made of?**

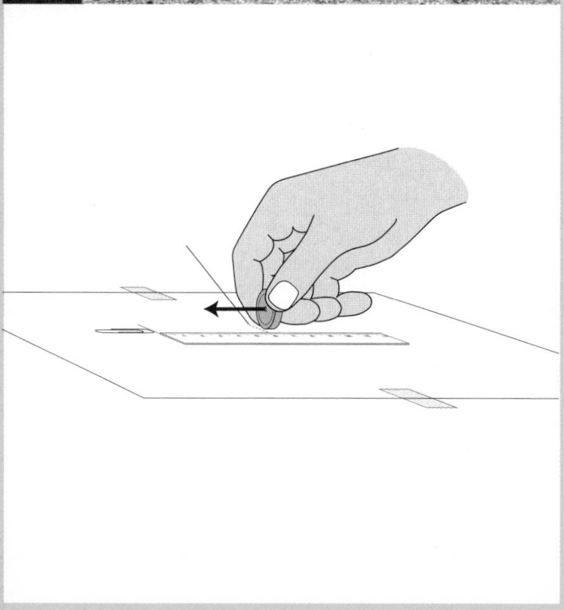

CYCLE 1:
SCIENCE EXPERIMENTS

Cycle Key Question

How can you tell if the conclusion from an experiment is valid?

Activity 1: Measurements in Science

Activity 2: Relationships in Science

Activity 3: Good and Poor Experiment Designs

Activity 4: Evaluating Experiment Designs

Activity 5: Evaluating Experiment Conclusions

Activity 1
Measurements in Science

Cycle Purpose

Scientists try to understand how the world works. First, they may develop an idea about how something works. Then they do experiments to test their idea. Finally, they use the results of the experiments to draw conclusions about their idea. In this cycle, you will learn how to determine if a conclusion is good. The key question for the cycle is:

How can you tell if the conclusion from an experiment is valid?

Record the cycle key question on your record sheet.

Activity Purpose

When scientists do experiments, they make lots of careful measurements. Therefore, it is important that their measurements are as accurate as possible. The purpose of this activity is for you to think about making measurements and how to interpret the values you obtain. Here is the key question for this activity:

When you measure something, can you obtain an exact value?

Record the key question for the activity on your record sheet.

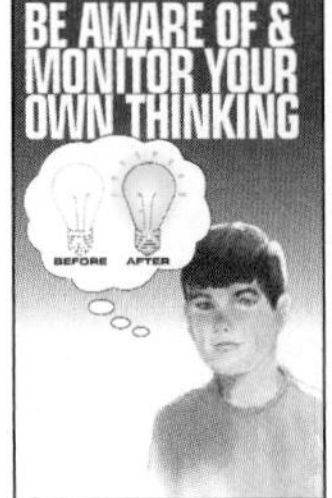

The skill you will be working on in this activity is *Be Aware of & Monitor Your Own Thinking.* To practice this skill, you will write down your ideas on your record sheet.

I Think

A pendulum is something hanging from a fixed point. It could be a ball or washer tied to the end of a string. When the pendulum is pulled back and let go, it swings back and forth. Many large clocks have pendulums.

Playground swings are pendulums.

Obviously, the pendulum takes a certain amount of time to swing back and forth. If you are very, very careful, you will be able to measure that exact time. That's the "true" value.
No, I don't think so. Although there may be an exact value for the time, you can never measure that value perfectly. You can never measure the "true" value.
Carlos
Chantel

Two students, Carlos and Chantel, wanted to measure how long it took for a pendulum to swing back and forth. They used a washer tied to a string as a pendulum. They disagreed on whether or not they could obtain the "true" (exact) value for their measurement.

Do you agree with Carlos, with Chantel, or with neither?

Without talking to your group members, record your own answer and your own reason on your record sheet.

Participate in a class discussion.

Explore Your Ideas

To find out whether Carlos or Chantel makes more sense, you and your partner will perform an experiment involving a pendulum.

Experiment 1: How long does it take for a pendulum to make 10 back and forth swings?

In performing this experiment, you will gather evidence that will help you answer the key question for this activity.

STEP 1 With your partner, decide how to perform the experiment. Be prepared to describe your method if asked.

Record how long it takes for your pendulum to make 10 complete back and forth swings.

STEP 2 Post your measurement in a class data table.

Have your teacher check your plan for safety.

You and your partner will need:

- clock with a second hand
- pendulum (washer attached to a string)
- metric ruler

Make Sense of Your Ideas

After everyone has entered data into the class table, examine all the data in the table. Discuss the answers to the next two questions with your partner. Then record your answers.

1. Why do you think different students in the class obtained different values for their measurement?

2. What suggestions do you have for improving the method used to make the measurements?

Participate in a class discussion to talk about the answers to the questions.

The class will decide on the "best" method of measuring the time for 10 back and forth swings.

3. Write down the method the class decides is the best.

Explore Your Ideas

Experiment 2: Repeat Experiment 1.

By repeating the experiment, you will have a chance to gather further evidence that will help you answer the key question for this activity.

STEP 1 Repeat the experiment with your partner, using the "best" method.

Record how long it takes for your pendulum to make 10 complete back and forth swings.

STEP 2 Post your measurement in a new class table.

Make Sense of Your Ideas

Look over the data in the new class table.

Think about the answers to the next questions and discuss them with your partner. Then record your answers on your record sheet.

1. Are all the posted values exactly the same, or is there some variation in the values?

2. If there is some variation in posted values, do you think it is possible for your class to repeat this experiment so that everyone would measure the exact same value? Why or why not?

3. Suppose your class needed to decide on a single *best* value for the time it takes the pendulum to swing back and forth 10 times. What value should be used? Why do you think so?

4. Last year, eight teams of students measured how long it took for a pendulum to swing back and forth 10 times. They all used the same procedure for their measurements. Below is a table listing their values.

Table: Time for 10 Back and Forth Swings of the Pendulum

Team	Time (s)	Team	Time (s)
A	15	E	14
B	14	F	16
C	9	G	15
D	16	H	15

The class then tried to decide on a single *best* value for its measurement. They wondered what to do about Team C's value because it was so very different from all the other measurements. Team C admitted they were careless in how they made their measurement.

Do you think Team C's measured value should be included in determining the best value for the class? Why or why not?

Participate in a class discussion to talk about the answers to the questions.

My Ideas

The key question for this activity is:

When you measure something, can you obtain an exact value?

1. Based on what you learned in this activity, answer the key question.
2. Suppose you repeated the experiment in this activity. What would you do differently, so you could report the best value for the time?

Scientists have agreed on ways to answer questions like these, based on performing many experiments. The ways that scientists make and interpret measurements are given in *How To Make and Interpret Experiment Measurements* in the Appendix.

Follow along as your teacher reviews the information with you.

DEVELOPING OUR IDEAS

Activity 2
Relationships in Science

Purpose

Sometimes, the ideas scientists want to test are about the relationship between two things. Scientists call the things in an experiment that can be changed or are different **variables.** Some variables of the pendulum are:

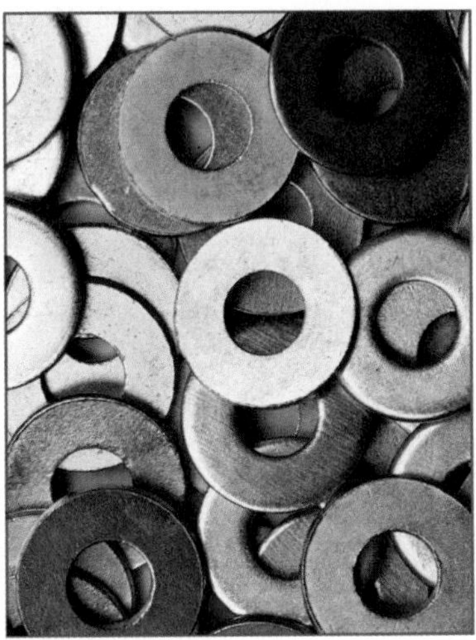

- its length
- how many washers you use
- what kind of string you use
- how far back you pull the pendulum at the start before letting go

Science Words

variable: something in an experiment that changes or can be changed

hypothesis: a statement that can be proved or disproved by experimental or observational evidence

In the experiment in this activity, you will change the length of the pendulum. Here is the key question for this activity:

What is the relationship between the length of a pendulum and how long it takes to swing back and forth 10 times?

Record the key question on your record sheet.

In this activity, you will continue working on the skill *Be Aware of & Monitor Your Own Thinking.*

I Think

Scientists usually have an idea about what they think will happen before they do an experiment. Their idea is called a **hypothesis.** (The plural is hypotheses.) A hypothesis is an educated guess, and is based on past experience.

Answer the following questions on your record sheet.

1. The following statements describe hypotheses about the relationship between two variables. One variable is the length of the pendulum, and the other variable is the time it takes to swing back and forth 10 times.

Which of the following is your hypothesis about the relationship between the two variables?

a) I think that as the length of a pendulum *increases*, it will take *more* time to swing back and forth 10 times.

b) I think that as the length of the pendulum *increases*, it will take *less* time to swing back and forth 10 times.

c) I think that as the length of a pendulum *increases*, the time it takes to swing back and forth 10 times will *stay the same*. This means there is no relationship between the two variables.

2. What reason(s) do you have for your hypothesis?

Explore Your Ideas

To answer the key question for the activity, you must deliberately change the length of a pendulum. Each set of partners in your class will use a different length pendulum.

You and your partner will need:

- clock with a second hand
- pendulum (washer attached to a string, knot at top of string)
- meter stick
- access to a calculator

Experiment: If the length of a pendulum increases, what happens to the time it takes to swing back and forth 10 times?

Before you do any experiment, you have to know how you will do it! As you learned in the last activity, the procedure you use is very important. Everyone will follow the exact same procedure, except for using a different length for the pendulum. Your teacher will demonstrate the procedure for you.

Read and follow the directions below very carefully. Because you don't expect to measure an exact value, you will make four measurements of the time for 10 swings. You will then calculate the average as your best value. Take turns being the *Timer* and the *Counter*.

STEP 1 *Timer:* Use a meter stick to measure the length of the string in centimeters. Measure from the knot to where the washer is tied to the string.

On your record sheet, record your measurement in a data table like the one shown.

Table: Time for 10 Swings of the Pendulum

Length of Pendulum = ____ cm	Time for 10 Swings (s)
Trial 1	
Trial 2	
Trial 3	
Trial 4	
Best Value	
Uncertainty	

STEP 2 *Timer:* Pinch the string between your fingers at the knot. Hold the pendulum so it can swing gently back and forth without hitting anything. Don't swing the pendulum—keep your hand still. Be sure you can see the second hand of the clock, so you can tell your partner when to start. Read the time after 10 full swings.

STEP 3 *Counter:* Hold the meter stick so one end lines up with the pendulum string. Pull the washer to the side 15 cm (no more).

STEP 4 *Counter:* When the Timer says "Go!", let go of the pendulum washer. Keep track of the number of back and forth pendulum swings and say "Stop!" just when the pendulum completes its 10th back-and-forth swing.

Record your measurement in your data table for trial 1.

STEP 5 Repeat Steps 2 through 4 for trial 2.

Record your measurement in your data table for trial 2.

STEP 6 Exchange roles and repeat Steps 2 through 4 for trials 3 and 4.

Record your measurement in your data table for trials 3 and 4.

STEP 7 Determine the best value and the uncertainty for your data. Refer to *How To Make and Interpret Experiment Measurements* found in the Appendix.

Record your best value and the uncertainty in the table on your record sheet.

Complete the following sentences on your record sheet:

Our team's best value for the time for 10 swings is _____ s (seconds) with an uncertainty of _____ s.
This means that the true value is probably within the range between _____ s and _____ s.

STEP 8 Your teacher will have drawn or posted a thick horizontal line with markings for the number of seconds for 10 back and forth swings. Tape the knot of your pendulum string along the horizontal line, using your best value.

Complete the class graph on your record sheet.

Graph: Time for 10 Swings versus Pendulum Length

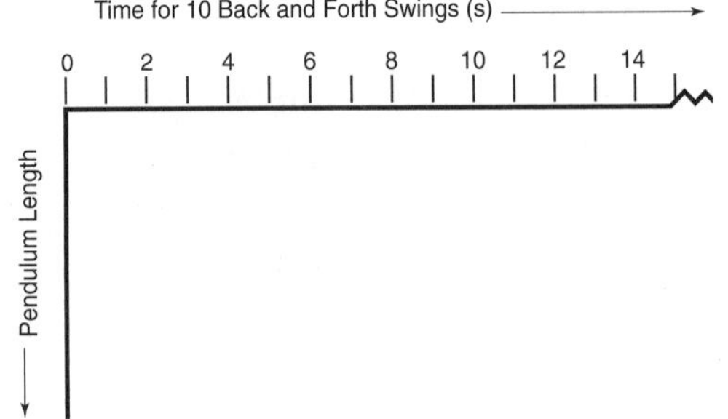

Make Sense of Your Ideas

Examine the class graph. To see if there is a pattern, try drawing the best smooth line through the points representing the washers. (The line does not need to be straight; it can be curved.) As the length of a pendulum increases, what happens to the number of seconds it takes to make 10 back and forth swings?

1. Write your conclusion on your record sheet.
2. In looking at your class graph, you may have noticed that some points (positions of washers) were above or below the smooth line you drew. Why do you think not all of the class data points ended up on a smooth line?

Participate in a discussion about what can be concluded from the class graph and the answer to Question 2.

3. Look back at the hypothesis you made at the beginning of the activity. How has your thinking changed after collecting data in the experiment?

Comparing your thinking now with what it was before is part of the process of being aware of and monitoring your own thinking. That is the skill you have been practicing in this activity.

4. Imagine that two teams each measured the time for 10 swings using pendulums. Each team calculated its best value and uncertainty.

   Team 1: Best value was 20 s with an uncertainty of 1 s.

   Team 2: Best value was 21 s with an uncertainty of 1 s.

Unfortunately, the teams forgot to write down the lengths of their pendulums. Since the best values were different, can you conclude that the pendulums of the two teams were different lengths? Explain your thinking.

Participate in a class discussion about the answer to Question 4.

My Ideas

The key question for this activity is:

What is the relationship between the length of a pendulum and how long it takes to swing back and forth 10 times?

Are you satisfied that you now know the answer to the key question? Explain.

Activity 3
Good and Poor Experiment Designs

Purpose

In Activity 2, you did an experiment to find the relationship between the length of a pendulum and how long it takes to make 10 back and forth swings. Suppose you wanted to convince someone that your results and conclusions were right. First, you would need to convince them that your experiment was a good experiment.

I think my experiment conclusion was right!
I'm not sure. What if there was something wrong with the experiment that influenced the results?
Carlos
Chantel

In this activity, you will first consider whether there is a relationship between the size of a magnet and its strength. Then you will perform an experiment to test your idea. After discussing the results of the experiment with the class, you will consider what makes the design of the experiment good or poor. Here is the key question for this activity:

What makes the design of an experiment good or poor?

Record the key question on your record sheet.

I Think

Imagine you have three magnets of different sizes and a way to measure how strong they are.

Write your answers to the following questions on your record sheet.

1. *My Hypothesis:* Which of the following is your hypothesis about the relationship between the size of magnets and their strength?
   - **a)** I think that as the size of the magnets *increases*, their strength will *increase*.
   - **b)** I think that as the size of the magnets *increases*, their strength will *decrease*.
   - **c)** I think that as the size of the magnets *increases*, their strength *will not change*. This means there is no relationship between the two variables.
2. *My Reason:* What reason would you give to *convince* someone of your hypothesis?

Explore Your Ideas

Working as an Efficient Research Team

Most modern scientists and engineers do not conduct experiments alone. They work in research teams. The teams share responsibilities. One member of the team may set up the experiment. Another may make the observations and take measurements. The members often have specific tasks that they do.

Scientists often have to follow complicated directions to get pieces of equipment to work. They read the directions to each other as they set up the equipment and try to make it work. They may need to collect special measuring instruments from a storeroom. Then, at the end of each day, they must return the instruments.

Once they start collecting data, all trials must be recorded in a journal. To accomplish all of these tasks, one scientist often keeps track of the time they have to set up and complete the experiment.

To Do

Team Member 1
Supply Master: gathers materials.

Team Member 2
Procedure Specialist: reads instructions and steps aloud.

Team Member 3
Team Manager: makes sure team stays on task and all team members are participating.

Team Member 4
Recycling Engineer: returns materials.

All Team Members
take turns measuring the distance when the magnet first attracts the paper clip.

In this course, you will also carry out specific tasks when you do experiments. Read the *To Do* list on the previous page. Your teacher will assign each member of your team a task. In this activity, you will practice the skill *Follow Directions and Stay on Task.* This skill will help your team complete the experiment in the limited time you have.

Your teacher will also be observing how each team is working. At the end of the experiment, you will fill out a form to evaluate how well the members worked together and carried out their tasks. (See the *Team Skills Chart.*) Be prepared to participate in a discussion about how teams can work effectively.

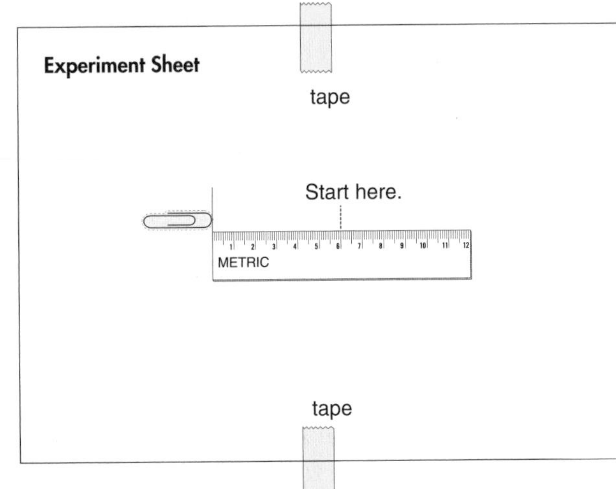

Experiment: If the size of a magnet increases, what happens to its strength?

Your team will need:

- 3 magnets of different sizes
- copy of the Experiment Sheet
- paper clip
- 2 pieces of tape
- access to a calculator

STEP 1 Tape the experiment sheet to the table so it won't move.

STEP 2 Lay the paper clip directly on top of its outline. Make sure the end of the paper clip lines up with the zero on the printed ruler.

STEP 3 Hold the front edge of the first magnet at the 6 cm (60 mm) mark as shown. (Don't lay the magnet flat on the paper.)

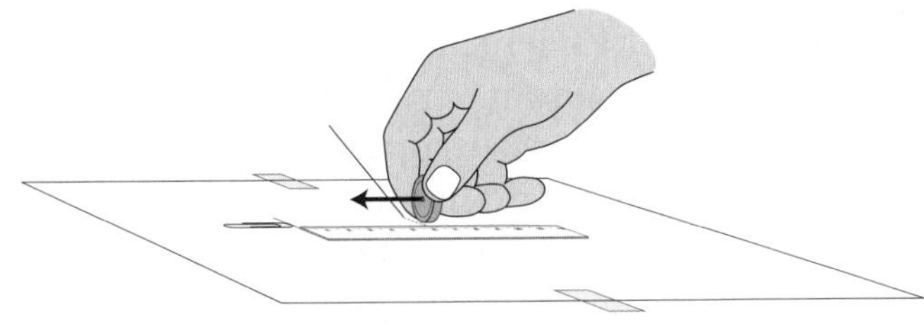

STEP 4 Slowly slide the edge of the magnet along the ruler's edge toward the paper clip. When the paper clip is attracted to the magnet, stop sliding the magnet.

Record the distance between the magnet and the zero position in Table 1 as shown on the following page.

STEP 5 *Take turns* repeating Steps 2 through 4 for two more trials.

STEP 6 Calculate the best value and uncertainty for the distance the magnet was from the zero position when it attracted the paper clip.

Record the best value and uncertainty in the table.

STEP 7 Repeat Steps 2 through 6 for the second and third magnets.

Table 1: Strength of Magnets and Their Sizes			
	Size of Magnets		
Distance When Magnet Attracted Paper Clip (mm)	Large Magnet	Medium Magnet	Small Magnet
Trial 1			
Trial 2			
Trial 3			
Best Value			
Uncertainty			
Our Ranking			

1. Complete the following statements on your record sheet. Take into account your calculated values for the uncertainties.

For the large magnet, the best value is _____ mm and the uncertainty is _____ mm. This means that the true value for the large magnet is probably within the range between _____ mm and _____ mm.

For the medium magnet, the best value is_____ mm and the uncertainty is _____ mm. This means that the true value for the medium magnet is probably within the range between _____ mm and _____ mm.

For the small magnet, the best value is _____ mm and the uncertainty is _____ mm. This means that the true value for the small magnet is probably within the range between _____ mm and _____ mm.

STEP 8 In deciding whether one magnet is stronger or weaker than another magnet, you need to look at the distances of the small, medium, and large magnets. The largest distance corresponds to the strongest magnet; the smallest distance corresponds to the weakest magnet. To rank the three magnets in terms of strongest, middle, or weakest, you need to decide whether the distances for the magnets are the same or different. Two distances are considered different only if there is no overlap in their range of possible values when taking into account their best values and uncertainties. If there is an overlap, the two values are considered to be the same.

For example, imagine another team (using completely different magnets) had the following best values and uncertainties for their magnets.

Large magnet: The best value is 48 mm with an uncertainty of 2 mm. Thus, the true value is probably within the range between 46 mm and 50 mm.

Medium magnet: The best value is 34 mm and the uncertainty is 3 mm. Thus, the true value is probably within the range between 31 mm and 37 mm.

Small magnet: The best value is 31 mm and the uncertainty is 2 mm. Thus, the true value is probably within the range between 29 mm and 33 mm.

2. What claim can the team make regarding how the strength of the large magnet compares to the strength of the medium magnet?

3. What claim can the team make about how the strength of the medium magnet compares to the strength of the small magnet?

Participate in a class discussion about the answers to these questions.

STEP 9 Now look over your team's own data. Rank the magnet strengths by using the terms *strongest*, *middle*, or *weakest*. If your data does not support the claim that one of the magnets is stronger than another, then give them both the same ranking. (In that case, you would not use the middle ranking.)

Record the rankings in Table 1 on your record sheet.

STEP 10 Post your team's rankings for the magnets in a class data table.

Make Sense of Your Ideas

Make Sense of the Experiment

Examine the class data table.

1. Write your conclusion for the experiment by completing the following sentence on your record sheet:

> As the size of the magnets used in this experiment increases, the strength of the magnets ____________ (*increases, decreases, does not seem to depend on size*).

Complete a Team Skills Chart like the one shown below. For each team member write "yes" or "no" in each cell depending on whether she/he did what is stated. Do this on your own. Then compare your ratings with other team members.

Participate in a brief class discussion about the conclusion for the experiment. Your teacher will also lead a class discussion about how teams can work together effectively.

Team Skills Chart					
Place your team members' names here.					
Followed the directions for how to do the experiment.					
Did his/her specific task.					
Took turns measuring the distance when the magnet pulled the paper clip.					

Make Sense of Good and Poor Experiment Designs

In your test of the relationship between magnet size and strength, you used at least two magnets made of the same material. One-half of your class used a third magnet also made of the *same* material. The other half of your class used a third magnet made of a *different* material (silver). Would your conclusion for the experiment be any different if those teams who had used magnets all made of the same material had been considered separately from those teams who used a magnet made of a different material? To consider that, divide the magnet experiment data into two parts: one where the three magnets were made of the same material, and another where one magnet was made of a different material.

Complete Table 2 on your record sheet. Use the rankings of those teams who used magnets all made of the same material. Record *strongest, middle,* or *weakest* in each column.

Table 2: Strength Rankings for All Magnets Made of the Same Material

Ranking of Magnets	Size of Magnets		
	Large Magnet	Medium Magnet	Small Magnet
Team 1			
Team 2			
Team 3			
Team 4			
Team 5			

Complete Table 3 on your record sheet. Use the rankings of those teams who used magnets including one made of a different material. Record *strongest, middle,* or *weakest* in each column.

Table 3: Strength Rankings for One Magnet Made of a Different Material

Ranking of Magnets	Size of Magnets		
	Large Magnet	Medium Magnet	Small Magnet*
Team 6			
Team 7			
Team 8			
Team 9			
Team 10			

**Magnet made of a different material*

2. Write your conclusion for the experiment where teams used magnets made of the same material.
3. Write your conclusion for the experiment where teams used magnets where two were made of the same material and one was made from a different material.

When scientists form a conclusion about a relationship, they need to be sure that their conclusion is *valid* or correct. Other scientists will accept the results of an experiment only if the conclusion is valid.

So, how can scientists decide whether their conclusions are valid? One way is to look at the design of the experiment.

First, consider the experiment done by teams who reported data in Table 2. In that case, only one variable, the size of the magnet, was changed. Any change in magnet strength could only be caused by the change in the size of the magnet. The experiment done by the Table 2 teams was a fair test. The conclusion drawn from their data was valid.

Next, consider the experiment done by teams who reported their data in Table 3. Those teams also changed the size of the magnet. But they changed another variable too. They changed the material from which the magnet was made.

So, the Table 3 teams could not be sure whether any effect they saw was caused by the different *sizes* of the magnets or was caused by the different *materials* from which the magnets were made or was caused by *both* variables. The experiment they performed was not a fair test, and the conclusion drawn from their data was not valid.

For an experiment to be a fair test, the only two variables that can change are those for which you are trying to determine a relationship. All other variables must be kept the same. The procedures for the experiment must also stay the same. If this is not the case, the experiment is not a fair test and you cannot form a valid conclusion.

Participate in a class discussion to review the conclusions for the teams who recorded their data in Table 2 or Table 3.

Now let's consider another example. In Activity 2, you did an experiment to answer the question: "If the length of a pendulum increases, what happens to the time it takes to make 10 back and forth swings?"

To answer this question, you deliberately changed the pendulum length. Each pair of partners had a pendulum with a different length. To find out what happens when the pendulum length changes, you measured how much time it took for the pendulum to make 10 back and forth swings. Was the design of this experiment a fair test? That is, were all pendulums treated in the same way?

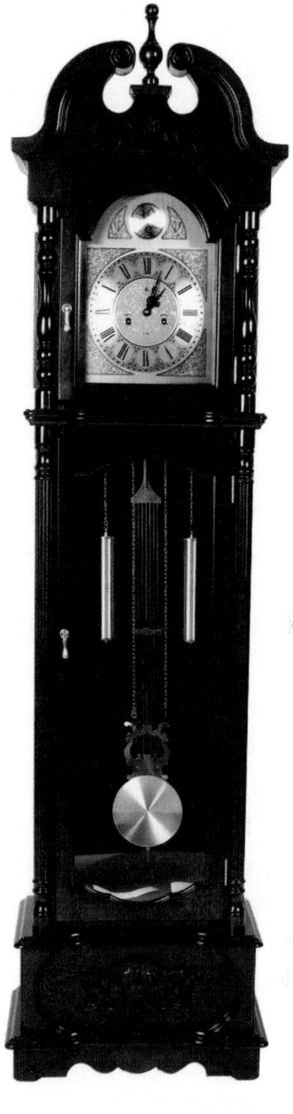

To figure out if the design of the pendulum experiment was a fair test, review the pendulum activity (Activity 2) and then answer these questions.

4. In the experiment, the length of the pendulum was changed and the time to make 10 back and forth swings was measured. Were any other variables deliberately changed?
5. What variables were kept the same in the pendulum activity?
6. Was the design of the pendulum experiment a fair test? State your reasons.

Participate in a class discussion of these questions.

Make Sense of Science Words

Consider the key question for this activity: "What makes the design of an experiment good or poor?" To understand how scientists might answer this question, you need to know the meaning of some terms:

- variables
- relationship
- fair test
- manipulated variable
- hypothesis
- responding variable

A **variable** is something about an object or event that can change or be different. For example, the length of someone's hair can change over time. The color of someone's eyes does not change over time, but one person may have green eyes and another person may have brown eyes. Hair length and eye color are variables. Variable comes from the verb "to vary," which means to change.

Science Words

variable: something in an experiment that changes or can be changed

Science Words

hypothesis: a statement that can be proved or disproved by experimental or observational evidence

relationship: an idea about what happens to one variable when a second variable changes

manipulated variable (also called independent variable): in an experiment, a variable that can be deliberately changed by the scientist and that determines the values of other variables (called responding variables)

responding variable (also called dependent variable): in an experiment, a variable that responds to the change in the values of the manipulated variables. The scientist cannot set the values of the manipulated variables directly.

fair test: an experiment in which only the manipulated and responding variables are allowed to change and all other variables and conditions are kept the same

Sometimes, the **hypothesis** a scientist wants to test concerns the **relationship** between two variables. A relationship is an idea about what happens to one variable when a second variable changes. The first variable may increase, decrease, or change in some more complicated way. Or it may not change at all, in which case there is no relationship between the two variables. To test their hypothesis about a relationship between two variables, scientists *deliberately change* one variable and *measure what happens* to another variable.

- The variable that is deliberately changed in an experiment is called the **manipulated variable** (also known as the independent variable).
- The variable that *responds to the change* (the variable that you measure) is called the **responding variable** (also known as the dependent variable).

An experiment is a **fair test** when both of the following are true:

- Only the manipulated and responding variables are allowed to change.
- All the other variables and conditions in the experiment are kept the same. (They are *controlled.*)

Discuss the following questions with your team and write your best answers on your record sheet.

7. In the experiment involving three magnets of the same material, what was the *manipulated variable*? What was the *responding variable*?

8. In the pendulum experiment, what was the manipulated variable? What was the *responding variable*?

My Ideas

The key question for this activity is:

What makes the design of an experiment good or poor?

Based on what you learned in this activity, write your answer to the key question.

Participate in a class discussion to review the answer to this question.

DEVELOPING OUR IDEAS

Activity 4
Evaluating Experiment Designs

Purpose

In the last activity, you learned about good and poor experiment designs. A good design involves a fair test of the experiment question. A poor design does not. Scientists have developed criteria (standards) for evaluating an experiment. The criteria are used to decide if the experiment is a fair test. The purpose of this activity is for you to practice using these criteria. Here is the key question for this activity.

What are the criteria for evaluating an experiment design for a fair test?

Record the key question on your record sheet.

Explore Your Ideas

Imagine that you do the following experiment for a science project:

Experiment: If different Earth's materials are heated by sunlight, is the final temperature the same or different for each?

You decide to work with two of the Earth's materials, dirt and water. You fill one bucket with dirt and half-fill an identical bucket with water. You place them so each bucket receives the same amount of sunlight. You place identical thermometers at the same depth and location in each bucket. You make sure each thermometer reads the same temperature at the beginning. Two hours later, you measure the temperature of each bucket. You compare the two temperatures. Is this experimental design good? Is the experiment a fair test?

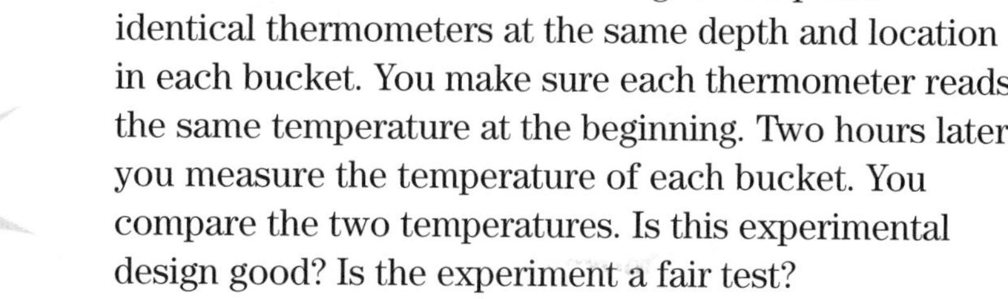

bucket filled with dirt

bucket half-filled with water

Analyze the Experiment

1. What is the question the experiment is designed to answer?
2. Identify the *manipulated variable* (the variable that was deliberately changed) and the *responding variable* (the variable that was measured).

3. Sometimes, the variable that is deliberately changed (the manipulated variable) can have many different values, like the many lengths of the pendulums in Activity 2. Sometimes, the manipulated variable has only a few values, like the three sizes of magnets (small, medium, large) in Activity 3. In this experiment, how many different types of Earth's materials are being used? What are the values of the manipulated variable?
4. What method is used to measure the responding variable?
5. What other variables are kept the same (controlled) during the experiment?

Now you are ready to make a decision about whether the experiment is a fair test, and if its design is good. You only have one more question to answer.

6. Is there any other variable (besides the manipulated and responding variables) that changes in the experiment? If the answer is "no," then the experiment is a fair test. If the answer is "yes," then the experiment is not a fair test.
7. Is the experiment a fair test? Explain your answer.

Use the questions above to analyze the following two experiment designs. Decide whether each experiment is or is not a fair test.

STEP 1 Analyze the following freezing-water experiment.

Imagine you are asked to help an elementary student with her science project. The student wants to find out if hot or cold water freezes faster.

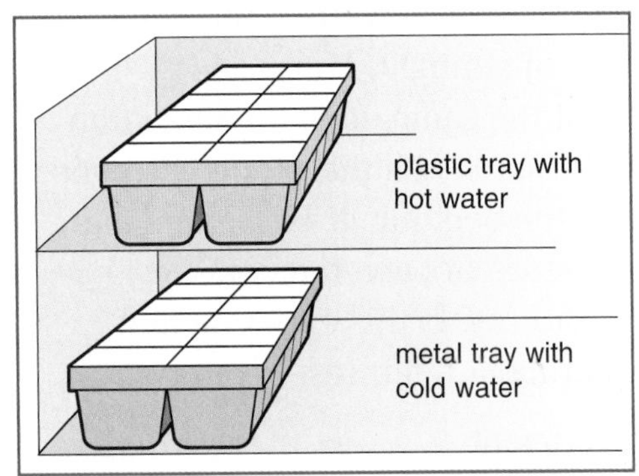

Experiment 2: If you start with two different temperatures of water (one hot and one cold), how does the time it takes for the water to freeze change?

The student started the experiment at home. She filled two ice trays full of water. One ice tray was plastic and was filled with hot water. The other ice tray was metal and was filled with cold water. Both trays were the same size.

She placed both ice trays in the freezer. She placed one tray on the top shelf and one on the bottom shelf. She then checked the ice trays every 15 minutes to see if the water had frozen. In order to help the student, you first have to decide if her experiment is a fair test.

Refer to Questions 1-7, this time answering them for the freezing-water experiment.

STEP 2 Analyze the following corn experiment.

Imagine that you are a botanist, a scientist who studies plants. You want to breed a new kind of corn plant to help countries that do not have enough land for growing food. You have bred four different types of new corn seeds. You are ready to do an experiment to find out which type of corn seed will grow best in crowded conditions.

Experiment 3: If the type of corn seed is changed, then what happens to its health and growth rate in crowded conditions?

You have four identical pots. You plant 20 corn seeds in each pot. Each pot has a different type of corn seed.

All of the pots have the same amount of soil, and all the seeds are planted at a depth of 4.0 cm. Pot #1 has soil from your back yard. Pot #2 has clay soil. Pot #3 has sandy soil. Pot #4 has topsoil with added fertilizer. You water each pot every day with 500 mL of water.

Pot 1

Backyard soil

Pot 2

Clay soil

Pot 3

Sandy soil

Pot 4

Topsoil and fertilizer

After 3 weeks, *all* of the corn seeds in Pot #4 have sprouted and have green, healthy leaves. Many of the seeds in the other pots did not sprout, and those that did were not as tall and healthy.

You conclude that the type of corn seeds in Pot #4 is the best. You decide to produce a lot of these seeds to sell to people who have a limited amount of land for growing crops.

Was your experiment a fair test?

Return to Questions 1-7, this time answering them for the corn experiment.

Participate in a class discussion to review your answers.

My Ideas

The key question for this activity is:

What are the criteria for evaluating an experiment design for a fair test?

Based on what you learned in this activity, write your answer to the key question.

Your teacher will go over *How To Analyze an Experiment Design and Determine if the Experiment is a Fair Test.* Compare your ideas to those on the form. You can use this form in the future when you need to analyze an experiment to decide whether it is or is not a fair test.

Activity 5

Evaluating Experiment Conclusions

Purpose

In the previous activity, you learned that an experiment to determine a relationship between variables must be a fair test. If it is not a fair test, then any conclusion drawn from the data is not valid.

Suppose, however, that the experiment *is* a fair test. Then someone draws a conclusion from the data. Would you accept the conclusion as valid? You might accept it if you decided that the person supported the conclusion with *good* reasons. You might not accept it if you decided the supporting reasons were not good. For a conclusion to be valid, it must be supported with *good* reasons. The purpose of this activity is to find out how you can tell if the reason for a conclusion is good or poor. The key question for this activity is:

How can you tell when a supporting reason for a conclusion is good or poor?

Record the key question on your record sheet.

During this activity, you will also work on the skills *Contribute Your Ideas and Reasons* and *Respect Your Team and Their Ideas*.

Explore Your Ideas

A class did an experiment to find the relationship between two variables. One variable (the manipulated variable) was pendulum length. The other variable (the responding variable) was the amount of time for 10 back and forth swings.

Five teams took part in the experiment. They did a different experiment from the one you did earlier. Each team was given the same set of four pendulums of different lengths (50 cm, 60 cm, 70 cm, and 80 cm). They were asked to measure the time for each pendulum to make 10 back and forth swings. The table on the next page shows the results from the five teams, and includes the class best values and uncertainties. All measurements were made to the nearest 0.1 s. You can assume the experiment was a fair test.

Five students in the class were asked to draw a conclusion from the class data table and to support their conclusions with reasons. The conclusions and reasons are presented on this page and the next page. The first student supported her conclusion with a *good* reason, so her conclusion is valid. The next four students, however, supported their conclusions with *poor* reasons. Their conclusions are not valid.

On your own, read all the conclusions and supporting reasons. Think about why the reason provided by the first student is considered good. Then think about why the reasons provided by the last four students are considered to be poor.

Class Data Table: Time for 10 Swings versus Length of Pendulum

Time for 10 Swings (s)	Length of Pendulum			
	50 cm	60 cm	70 cm	80 cm
Team 1	13.8	15.5	17.2	18.2
Team 2	14.0	15.2	17.1	18.2
Team 3	14.6	14.5	17.2	17.2
Team 4	13.8	15.5	16.4	18.0
Team 5	14.0	15.5	16.8	18.2
Class Best Value	14.0	15.2	16.9	18.0
Class Uncertainty	0.4	0.5	0.4	0.5

Student providing good supporting reason (Conclusion is valid.)

Student A

Conclusion – I conclude that as the length of a pendulum increases, the time it takes to swing back and forth 10 times also increases.

Reason – As can be seen from the class data table best values, as the length of the pendulum increases from 50 cm to 60 cm to 70 cm to 80 cm, the time to make 10 swings increases from 14.0 s to 15.2 s to 16.9 s to 18.0 s. Taking into account the uncertainties, all of these values are true increases.

Students providing poor supporting reasons (Conclusions are not valid.)

Student B

Conclusion – I conclude there is no relationship between the length of a pendulum and the time it takes to swing back and forth 10 times.

Reason – The data for Team 3 shows that as the pendulum length increases from 50 cm all the way to 80 cm, the time to make 10 swings first decreases, then increases, then remains the same.

Student C

Conclusion – I conclude that as the length of a pendulum increases, the time it takes to swing back and forth 10 times decreases.

Reason – The longer the pendulum, the closer it is to the ground, where gravity is stronger. Stronger gravity pulls the pendulum more quickly.

Student D

Conclusion – I conclude that as the length of a pendulum increases, the time it takes to swing back and forth 10 times also increases.

Reason – Team 5 showed that as the pendulum length increased from 50 cm to 60 cm to 70 cm to 80 cm, the time for 10 swings increased from 14.0 s to 15.5 s to 16.8 s to 18.2 s.

Student E

Conclusion – I conclude that as the length of a pendulum increases, the time it takes to swing back and forth 10 times also increases.

Reason – Longer things move more slowly than shorter things.

What makes the reasoning used to support a conclusion poor? To decide this, use the following procedure with your partner. This procedure will help you practice these important skills: *Contribute Your Ideas and Reasons* and *Respect Your Team and Their Ideas.*

STEP 1 *On your own*, think about what makes the reasons poor. Base your answer on the examples provided by the students.

STEP 2 *Share* your ideas and your reasons with your partner.

STEP 3 *Listen* carefully to your partner's ideas and reasons.

STEP 4 *Together*, decide what makes the reasons used to support a conclusion poor. This response might be a mixture of your own and your partner's ideas and reasons, or just one person's ideas and reasons.

What makes the reasoning used to support a conclusion poor? Write your answer on your record sheet.

Participate in a class discussion.

Make Sense of Your Ideas

In the previous activities, you learned that an experiment must be a fair test. If it is not, then any conclusion you draw is not valid.

If it is a fair test, you must evaluate the reasons to decide whether the conclusion is valid. The reasons tell how the data (observations) from the experiment actually support the conclusion. Scientists call the data **evidence**. Data can be collected using any of the five senses or instruments that extend your senses. Examples of instruments include meter sticks, clocks, and thermometers.

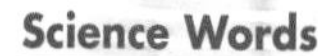

Science Words

evidence: in an experiment, the data collected by the researcher

In order for their conclusions to be valid, scientists support their conclusions with good reasons. To be *good*, each supporting reason must meet the following two criteria:

- It must *include evidence* from the experiment, not an opinion.
- It must use *all the available evidence*, not just part of the evidence.

If a reason includes an opinion and/or uses only part of the evidence, then the reason is poor, and the conclusion is not valid.

Look over the reasons used by students B, C, D, and E on the previous pages.

1. For which student(s) is the reason poor because it includes an opinion, rather than evidence?
2. For which student(s) is the reason poor because it only uses part of the data, rather than all the available data?

Look at *How To Evaluate an Experiment Conclusion*, located in the Appendix. This lists all the criteria that you use to decide whether a conclusion is valid or not valid.

Your teacher will review the criteria with you and will go over your decisions about why the students' reasons are poor.

Read the following conclusion and reason that a student named Nguyen wrote. It is about the experiment to determine the relationship between magnet size and strength for magnets made of the same material. (You can assume the experiment was a fair test.)

Nguyen

Nguyen

Conclusion – I conclude that when the magnet size increases, its strength increases.

Reason – For the three magnets used in our experiment, as the size of the magnets increased from small to medium to large, the class's strength rankings of the magnets increased from weakest to middle to strongest.

Isabel

Is this conclusion valid or not valid?

Isabel used *How To Evaluate an Experiment Conclusion* to help decide. She assumed that the experiment was a fair test because she was told that it was. Then she determined that Nguyen supported his conclusion with a good reason because it was based on evidence and used all the class data. Therefore, Isabel decided that Nguyen's conclusion was valid.

Consider another example of a conclusion and reason written for the same experiment by Rebecca.

Rebecca

Conclusion – I conclude that when magnet size increases, its strength increases.

Reason – I think we did our experiment correctly.

Is Rebecca's conclusion valid or not valid?

Chantel used *How To Evaluate an Experiment Conclusion* to help decide. She assumed that the experiment was a fair test because she was told that it was. Then she determined that the reason Rebecca used to support her conclusion was poor because her reason was an opinion. Even though Rebecca's conclusion was a correct statement, her supporting reason was poor. Chantel decided that Rebecca's conclusion was not valid. (Scientists place high value on providing good supporting reasons.)

Chantel

Rebecca

Our Consensus Ideas

The key question for this activity is:

How can you tell when a supporting reason for a conclusion is good or poor?

1. Based on what you learned in this activity, work with your partner to write your answer to the key question.

Use what you learned in this activity and *How To Evaluate an Experiment Conclusion* to decide if the two conclusions below are *valid* or *not valid*. You can assume that the experiment is a fair test and that these two students had access to the class data table.

Record your answers and reasons after discussing them with your partner. Follow the same approach used by Rebecca and Chantel.

2. Carlos wrote the following conclusion and reason:

   Conclusion: We conclude that there is no relationship between magnet size and strength.

   Reason: Our medium-sized magnet and our largest magnet both pulled the paper clip from the same distance away.

   Is the conclusion *valid* or *not valid*? Explain.

3. Nguyen wrote the following conclusion and reason:

   Conclusion: We conclude that when magnet size increases, its strength increases.

   Reason: We know that larger objects are stronger than smaller objects.

   Is the conclusion *valid* or *not valid*? Explain.

Participate in a class discussion about your answers.

CYCLE 2:
INTRODUCING INTERACTIONS

Cycle Key Question

How can you describe interactions?

Activity 1: Evidence of Interactions

Activity 2: The Magnetic Interaction

Activity 3: The Electric-Charge Interaction

Activity 4: The Electric-Circuit Interaction

Activity 5: Electromagnets and Buzzers

Activity 6: Interaction between a Magnet and an Electric Current

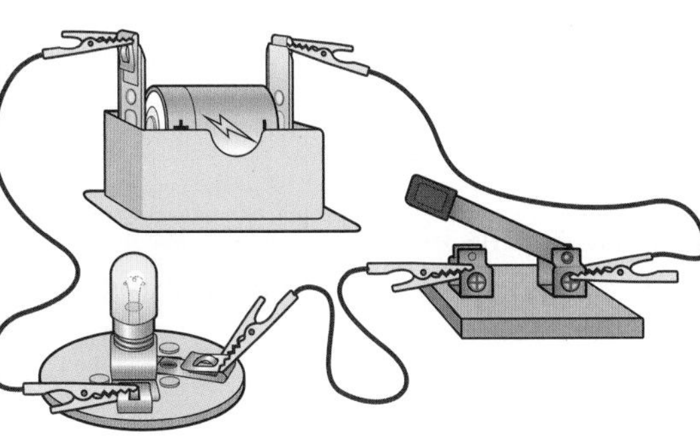

DEVELOPING OUR IDEAS

Activity 1
Evidence of Interactions

Cycle Purpose

Scientists make an important assumption about the natural world. They assume that everything they observe can be explained in terms of interactions. When two objects interact, they act on or influence each other. Jointly, their actions cause an effect. In this cycle, you will investigate interactions. The key question for the cycle is:

How can you describe interactions?

Record the key question for the cycle on your record sheet.

At the end of Cycle 1, you learned that a scientist always tries to support an experiment conclusion with evidence. In fact, scientists try to support any claim they make with good evidence. A claim is a statement that you believe to be true.

To convince someone else that your claim is valid (true), you need to support it with evidence. In the case of an interaction, the evidence is usually *what happens* to the two objects during the time they are interacting.

A soccer player kicks a ball to move it down the field. A scientist would explain why the ball moves away from her in terms of an *interaction* between the player and the ball.

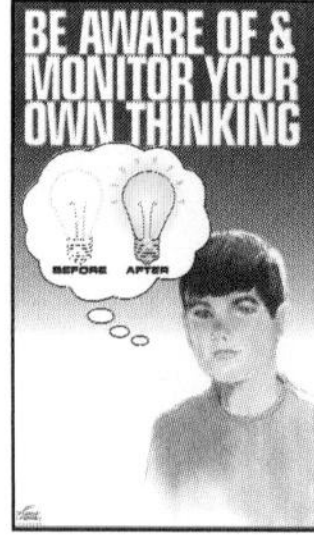

Activity Purpose

In this activity, you will practice making your own claims about whether two objects interact with each other. You will also practice supporting your claims with evidence.

The key question for this activity is:

What is the evidence that an interaction has occurred?

Record the key question for the activity on your record sheet.

I Think

Your teacher will do three demonstrations of different interactions. Observe what happens in each case. Figure out the evidence that would support the claim that the two objects act on or influence each other (interact).

Demonstration A: Striking a Table with a Ruler

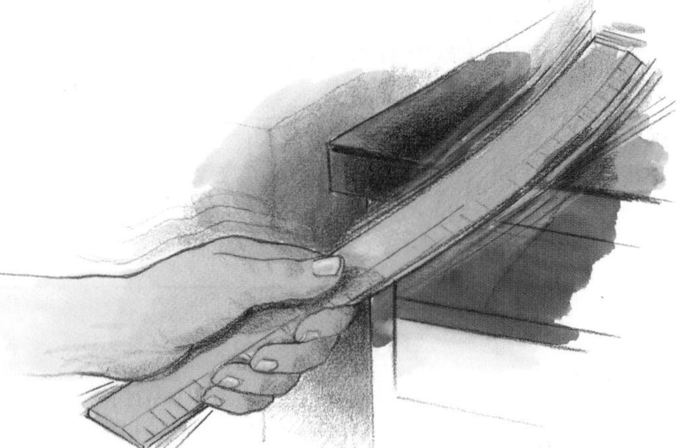

Your teacher will strike a table with a ruler. Consider what happens during the brief time interval while the ruler and table are touching.

When the ruler hits the table, there is a sudden *change in sound.* At first, there is no sound, then there is a very loud crack. There is also a sudden *change in motion* of the ruler as the table brings it to a stop during the interaction. The time interval during which this interaction occurs is very, very short.

1. With your teacher's guidance, complete the following.

Demonstration A

Time Interval: while the ruler and table are touching

Claim:	Object 1: ruler	interacts with	Object 2: table

1. Evidence: There is a change (are changes) in ____________________.

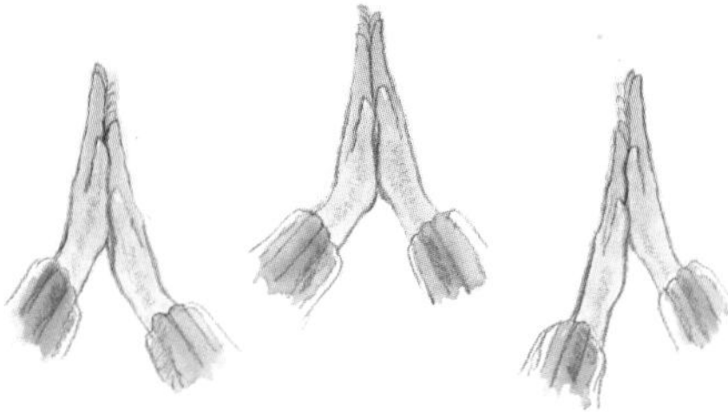

Demonstration B: Rubbing Your Hands Together

Rub your hands together firmly ten times.
What happens? As you rubbed your hands together ten times, you probably felt your hands get warmer. There was a *change in temperature* of both hands.

2. On your record sheet, fill in the evidence that your hands interacted. Be prepared to share your answers with the class.

Demonstration B

Time Interval: while rubbing right hand against left hand

Claim:	Object 1: hand	interacts with	Object 2: other hand

2. Evidence: There is a change (are changes) in________________.

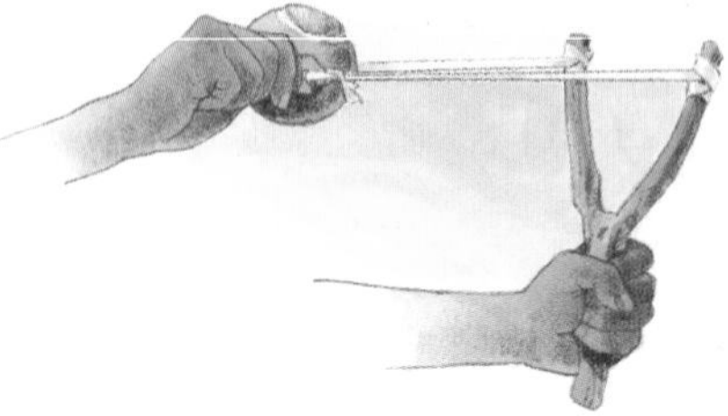

Demonstration C: Shooting a Ball with a Slingshot

Shooting a ball with a slingshot is actually two interactions in a sequence. This is called an "interaction chain." Your teacher will pull back on the elastic band of the slingshot. The first interaction is between your teacher's hand and the elastic band. Consider what happens during this time interval.

3. Fill in the evidence for the first interaction.

Your teacher will release the elastic band. The second interaction is between the stretched elastic band and the ball. What happens during the time interval when the rubber band is pushing the ball?

Demonstration C

Time Interval: while shooting a slingshot

Claim:	Object 1: hand	interacts with	Object 2: elastic band	which interacts with	Object 3: ball

3. Evidence: There is a change (are changes) in ________________.

4. Evidence: There is a change (are changes) in ________________.

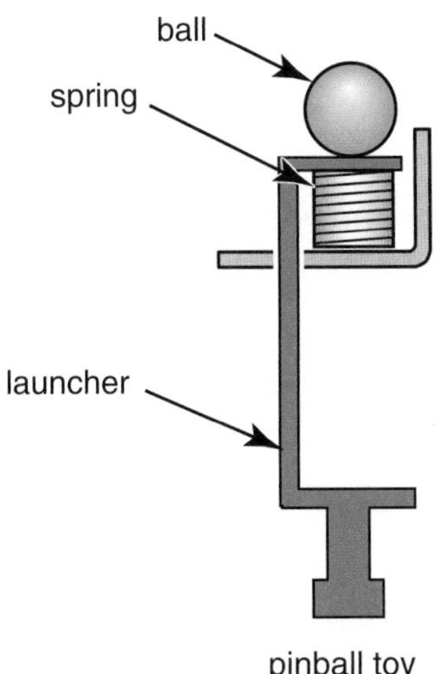

pinball toy

4. Fill in the evidence for the second interaction on your record sheet.

Your team will need:

- container with materials listed below
- scissors
- pencil or crayon
- file cards
- ribbon
- washer
- flashlight
- small pinball toy

Explore Your Ideas

Your tasks are to:

- make two single interactions.
- make one interaction chain (involving two separate interactions).
- record the time interval and the evidence for each interaction that you observe.

During this activity, you will practice the skill *Respect Your Team and Their Ideas.* You will learn one way you can disagree with someone's idea while communicating respect for the person.

To Do

Team Member 1
Supply Master: gathers materials.

Team Member 2
Procedure Specialist: reads instructions and steps aloud.

Team Member 3
Team Manager: makes sure team stays on task and all team members are participating.

Team Member 4
Recycling Engineer: returns materials.

STEP 1 Have one team member select one or two objects from the items listed. Your hand can be one of the two objects in an interaction.

Make the two objects interact (act on or influence each other).

Tell your team *when* you think the two objects act on or influence each other. This is the *time interval* of interest.

Then state your *evidence* (reasons) for why you think they are interacting.

For example, suppose you want to support the claim that when you drop a lump of clay on the floor, the clay interacts with the floor. You might say something like this, "I think the clay and the floor interact, because there is a sound when the clay hits the floor."

STEP 2 The other team members *take turns* either agreeing with your claim and your evidence (reasons), or suggesting changes or additions. They do this by stating their own idea and reason. Here are some possible ways to respond to the example:

"I *agree* that the clay and the floor are interacting and I *agree* with your evidence."

"Yes, I *agree* that the clay and the floor are interacting. I think additional evidence is that the shape of the clay changes when it hits the floor. The lump of clay gets squashed."

"I *don't agree* that the clay and the floor are interacting. Nothing is really changing. It is still the same clay."

STEP 3 After everyone has had a chance to give his or her ideas and reasons, choose the evidence that *you all agree on* as a team.

This might be a mixture of one or more team members' ideas, or your team might decide that one member's idea is the best.

1. Complete an interaction form, similar to the ones in the demonstration, on your record sheet. Be sure to include the time interval, the objects that are interacting, and the evidence.

STEP 4 Take turns repeating Steps 1 through 3 until your team has completed *two single interactions* and *one interaction chain.*

2. Complete an interaction form for the second *single interaction.*

3. Complete an interaction form for the *interaction chain.*

Have your teacher check your planned interaction for safety.

STEP 5 Select your team's favorite interaction (or interaction chain). Choose one member of your team to demonstrate this interaction to the class. That member will also need to describe the evidence for the interaction(s).

Make Sense of Your Ideas

Participate in a class discussion.

As you learned in this activity, two objects interact when they act on or influence each other to cause an effect. Changes that provide evidence for an interaction might include:

- changes in the *motion* of one or more objects.
- changes in the *sound* produced by one or more objects.
- changes in the *temperature* of one or more objects.
- changes in the *shape or size* of one or more objects.
- changes in the *illumination* on one or more objects.
- changes in the *color or pattern* of one or more objects.

When you make a claim that two objects are interacting, you need to support that claim with evidence. Any of the above changes would provide that evidence.

Our Consensus Ideas

The key question for this activity is:

What is the evidence that an interaction has occurred?

Write your answer to the key question based on your class discussion.

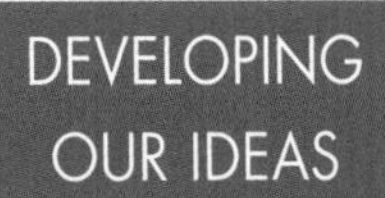
DEVELOPING OUR IDEAS

Activity 2
The Magnetic Interaction

Purpose

In the last activity, you learned that scientists assume that everything they observe can be explained in terms of interactions between objects. There are millions and millions of interactions occurring all the time on Earth. Suppose every interaction were unique. That is, suppose every interaction were different from every other interaction in *every* way. Then it would be impossible to make sense of the world.

Scientists group interactions together. They believe there is only a limited number of types of interactions. But how can you know scientifically that one type of interaction is similar to or different from another?

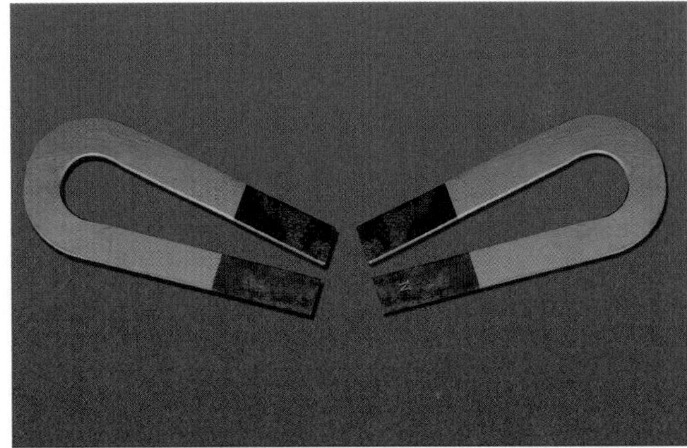
If you place a magnet near a second magnet, you can cause the second magnet to move. If you just push a magnet with your hand, you can also cause it to move. To claim that these interactions are different, you need evidence that supports your claim.

Scientists try to answer that question by identifying the characteristics of a particular type of interaction. These characteristics are called the **defining characteristics** of the interaction. They make one type of interaction different from all other types.

Science Words

defining characteristics: the characteristics that are specific to a given type of interaction

In this activity, you will perform experiments with magnets in order to determine the defining characteristics of the interaction between them. The key question is:

What are the defining characteristics of a magnetic interaction?

Record the key question on your record sheet.

We Think

Discuss the following two questions with your team. Make sure each of you has a chance to share your ideas. You do not need to write down your answers.

- Suppose someone gave you two objects and claimed they were magnets. What evidence would you need to be sure that the claim was valid?
- Are there any other objects (besides magnets) that interact with magnets?

Your team will need:
- 2 magnets
- bag with 4 different kinds of metals

Participate in a class discussion.

Explore Your Ideas

You will now gather evidence to answer the questions raised in the previous section. First, you will perform an exploration to look for the evidence for a magnet–magnet interaction. You will then perform an exploration to determine what other objects might interact with a magnet. Finally, you will consider interactions with a magnetic compass.

Remember to share responsibilities among your team members.

To Do

Team Member 1 *Supply Master:* gathers materials.

Team Member 2 *Procedure Specialist:* reads instructions and steps aloud.

Team Member 3 *Team Manager:* makes sure team stays on task and all team members are participating.

Team Member 4 *Recycling Engineer:* returns materials.

All Team Members take turns sliding the magnet.

Exploration 1: If a magnet is brought near another magnet, what happens to the other magnet?

In Cycle 1, you performed several experiments. Each involved making measurements and recording the data in a table. You then made a conclusion to answer the experiment question. From now on, we will use the term *exploration*, rather than *experiment*, to refer to any situation where you make observations and collect data. Sometimes an exploration will involve making measurements and recording the data in a table. Other times, an exploration will involve only making one or a few observations and writing a sentence or two about what actually happened. In both cases, however, your observations will provide evidence to support your ideas.

STEP 1 Place the two magnets flat on the table, far enough apart so they do not seem to influence each other. Slowly slide the first magnet closer and closer to the second magnet.

1. What happens to the second magnet?

STEP 2 Move the magnets far apart again. Turn over the first magnet. Repeat Step 1.

2. What happens to the second magnet? How does this compare with the previous step?

Scientists use the word attract when an object moves toward another nearby object. They use the word repel when an object moves away from another nearby object.

3. Complete the following statement on your record sheet.

When two magnets interact, they ________ (*always attract each other, always repel each other, can either attract or repel each other*). My evidence is _______.

Temporarily place one of the magnets far away from all other materials on the table.

Exploration 2: If a magnet is brought near a metal, what happens to the metal?

STEP 1 Place a magnet and a piece of steel, such as a steel nail, on the table. Place them far enough apart so they do not seem to influence each other. Slowly slide the magnet closer and closer to the piece of steel.

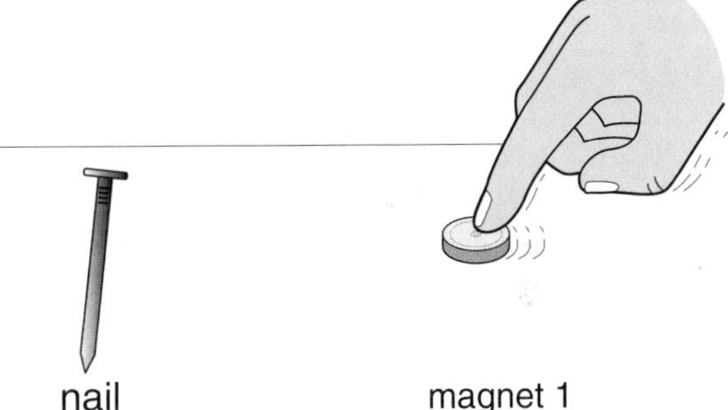

Use a data table like the one shown below to write your observations on your record sheet. Under Step 1, write what happens when the magnet is brought closer and closer to the piece of steel (your observations).

Table: Observations of How Metals Interact with a Magnet

Material	Step 1 What happens when the magnet is brought closer and closer to the piece of material?	Step 2 What happens when the magnet is turned over and brought closer and closer to the piece of material? How does this differ, if at all, from the previous step?
Steel		
Copper		
Aluminum		
Nickel		

STEP 2 Move the objects far apart again. Turn over the magnet. Repeat Step 1.

Under Step 2 in your table, write what happens when the magnet is turned over and brought closer and closer to the piece of steel. How does this differ, if at all, from the previous step?

STEP 3 Replace the steel with a piece of *copper*. Repeat Steps 1 and 2.

STEP 4 Replace the copper with a piece of *aluminum*. Repeat Steps 1 and 2.

STEP 5 Replace the aluminum with a piece of *nickel*. Repeat Steps 1 and 2.

4. Did the magnet interact *with all the metals*, *with none of the metals*, or *with only some of the metals*?

Science Words

magnetic materials: metals that interact with magnets

Metals that interact with a magnet are called **magnetic materials**.

5. Complete the following statement on your record sheet.

When a magnet and magnetic material interact, they _____. (*always attract each other, always repel each other, can either attract or repel each other*). My evidence is ______.

Exploration 3: Is a compass needle a magnet?

STEP 1 Read the discussion between Isabel and Otis about how a compass works.

I think the compass needle is a magnet because it always points in the north-south direction.

Isabel

I do not think the compass needle is a magnet. It is a magnetic material, like steel. It points in the north-south direction because the Earth is a giant magnet and attracts it.

Otis

6. Which student do you think is right, Isabel or Otis? Or do you have a different answer?

STEP 2 Your teacher will show you either a demonstration or a movie of a magnet being brought close to the *east* label on the compass.

7. What happens to the end of the compass needle that was originally pointing north? Does it move *toward the magnet*, *away from the magnet*, or does it *remain motionless*?

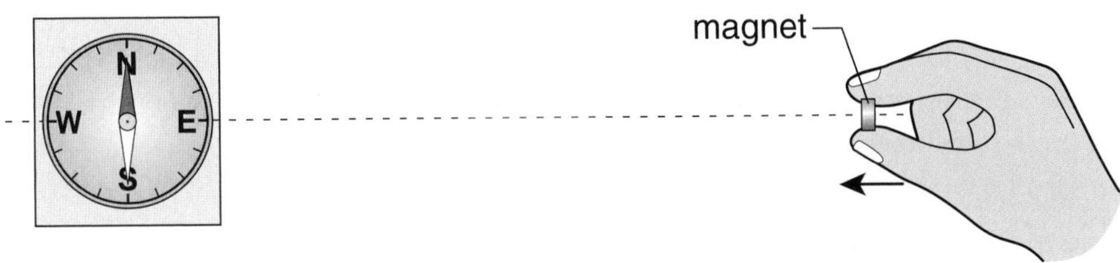

STEP 3 The previous step will be repeated, but this time, with the magnet turned around so the other flat face is aimed toward the compass.

8. What happens to the end of the compass needle that was originally pointing north? Does it move *toward the magnet*, *away from the magnet*, or does it *remain motionless*?

9. Is this the same observation you made in Step 2 or is the observation different?

10. Complete the following statement on your record sheet:

A compass needle _____ (*is*, *is not*) a magnet. My evidence is _______.

Your teacher will review the observations with the class.

Make Sense of Your Ideas

Steps for Discussing Ideas within Your Team

1. One team member states their answer to a question. The member should state both the answer and the reasons for their thinking. Reasons should include the *evidence* from an experiment or exploration.

   "I think that ______________ because ______________."

2. The other team members *take turns* either agreeing with the answer, or suggesting any additions or changes to the answer. They should do this by stating their own idea and reason.

   "I agree with your answer *because* you covered everything" or
   "I agree with your answer and I think we could add some more evidence from ______________" or

   "I disagree with your answer because I think we ______________."

3. After everyone has had a chance to give their answers and reasons, the team should choose the answer it likes best.

4. Take turns repeating Steps 1 through 3 for each question. The team manager decides who will go first, second, third, and fourth.

Your teacher will examine with the class the steps for discussing ideas within your team.

Discuss each of the four questions with your team members. Practice your skill *Respect Your Team and Their Ideas* (*answers* and *reasons*) by following the step-by-step procedure described above.

1. Do magnets interact with each other?
2. Suppose you had two objects and you knew that one of them was a magnet. What test(s) do you need to do to convince yourself that the other object was also a magnet?
3. What test(s) do you need to do to convince yourself that the other object was a magnetic material, but not a magnet?

4. Look back over the discussion between Isabel and Otis. Which of them do you agree with now?

Some of you may have been surprised to find that a magnet attracts some metals, but not others. What about nonmetals like paper, water, glass, etc.? Do these interact with a magnet? As you probably know from previous experience, nonmetals do not interact with a magnet.

Participate in a class discussion about the questions.

Our Consensus Ideas

In this activity, you and your classmates gathered evidence to answer the key question:

What are the defining characteristics of a magnetic interaction?

1. On your record sheet, write the team answer to the question, along with the supporting reasons (*evidence*).

Scientists have also done experiments and explorations similar to the ones you have. They have developed a set of ideas that describe the defining characteristics of the magnetic interaction. Your teacher will hand out a form titled *Scientists' Consensus Ideas: The Magnetic Interaction*. As you read through this form, think about the evidence from this activity that would support each of the ideas.

2. Indicate the experiments or explorations in this activity that provide the evidence to support the scientists' ideas.

Go over the evidence with your teacher.

DEVELOPING OUR IDEAS

Activity 3
The Electric-Charge Interaction

Purpose

In the previous activity, you studied the defining characteristics of the magnetic interaction. It turns out there is another type of interaction that is similar and yet different. You are probably familiar with some of its effects. If you drag your feet across a carpet, you sometimes get a shock when you try to open a door. This is the effect of the electric-charge interaction.

Clothes often stick together when you take them out of a dryer because of the electric-charge interaction.

When some objects rub together, like your shoes and the carpet, they become electrically charged. They can then interact with each other according to the electric-charge interaction. In this activity, you will gather evidence to answer the questions below.

1. What are some defining characteristics of the electric-charge interaction?

2. What are some of the similarities and differences between the magnetic and electric-charge interactions?

Record the key questions for the activity on your record sheet.

We Think

In the previous activity, you learned that one magnet can either attract or repel another magnet. It depends on which ends (or faces) of the two magnets are brought near each other.

- Do you think that electrically charged objects behave the same way as two magnets, or do you think they behave differently? Why do you think so?

Discuss these questions with your team. Write your team's best answer on your record sheet.

Participate in a whole-class discussion.

Your team will need:

- small balloon
- piece of wool
- window cleaner
- small piece of paper (about 10 cm by 10 cm)
- small piece of aluminum foil (about 10 cm by 10 cm) or tinsel
- paper towels
- piece of tape (about 2 cm long)

Explore Your Ideas

You will do two explorations to determine some of the defining characteristics of the electric-charge interaction. Your observations will also help you to decide on some similarities and differences with the magnetic interaction.

Remember to share responsibilities among your team members.

Exploration 1: What happens when a charged object is brought near metallic and nonmetallic objects that are not charged?

STEP 1 *Procedure Specialist:* Tear the small piece of paper into tiny bits. Make a small pile on the table. Next, tear the small piece of aluminum foil into tiny pieces. Make a separate pile on the table.

STEP 2 *Supply Master:* Blow up the balloon. Put a small piece of tape on one side. The tape is just to mark the side on which you can hold the balloon. After this, touch the balloon with your hands *only on the side with the tape.* Clean the balloon with the window cleaner. Set it aside for a few minutes.

Electric-charge effects are often difficult to produce. This is especially true when surfaces are dirty with grease from hands and fingers.

To get accurate results with this exploration, be sure to:

- Clean the balloon with window cleaner.
- Only touch the side of the balloon with the tape.

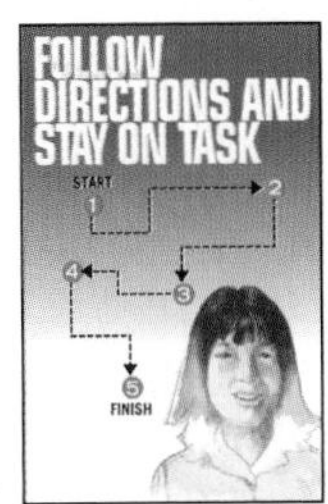

To Do

Team Member 1
Supply Master: gathers materials.

Team Member 2
Procedure Specialist: reads instructions and steps aloud.

Team Member 3
Team Manager: makes sure team stays on task and all team members are participating.

Team Member 4
Recycling Engineer: returns materials.

When the balloon is dry, pick it up *on the side with the tape.* Rub the opposite side of the balloon vigorously with the piece of wool.

STEP 3 Move the charged balloon *slowly* from above down toward the small bits of paper. Observe what happens to the paper.

Record what happens to the paper in Table 1 on your record sheet.

STEP 4 *Manager:* Hold the side of the balloon with the tape. Clean the opposite side of the balloon with window cleaner. When the balloon is dry, rub the opposite side of the balloon vigorously with the piece of wool. Move this charged side of the balloon slowly from above down toward the pieces of aluminum foil. Observe what happens to the aluminum foil.

Record what happens to the aluminum foil in Table 1.

Use the terms attract or repel in answering the following question. Use both terms if appropriate.

Table 1: What Happens to Uncharged Objects near Charged Objects

	Uncharged Objects	
	Paper (Nonmetal)	**Aluminum Foil** (Metal)
Charged Balloon		

1. What happens when a charged object (balloon) is brought near nonmetallic and metallic objects that are not charged?

Participate in a brief class discussion about your observations and answer to the previous question.

Exploration 2: What happens when charged objects are brought near each other?

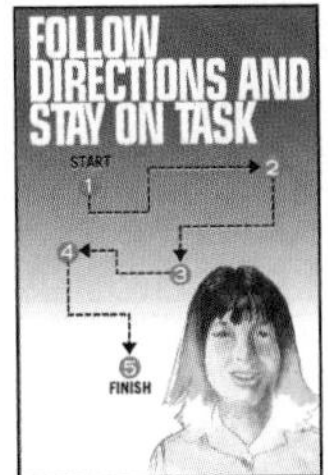

Your teacher will explain how you will work through this exploration's directions. You must read and follow the directions very carefully as you work. To get accurate results with this exploration, do not let the charged tape rub against your hand or clothing.

Your team will need:

- roll of Scotch® Magic™ tape
- paper or index card
- scissors
- meter stick
- pencil
- support for meter stick (heavy book or clamp)

Table 2: What Happens When Charged Objects Are Brought near Each Other

	Suspended Charge T1	Suspended Charge B1
Non-sticky side of charge T2		
Sticky side of charge T2		
Non-sticky side of charge B2		
Sticky side of charge B2		

STEP 1 *Manager:* Cut the paper or index card into 8 paper tabs, 2 cm by 2 cm. Lay two paper tabs on the table 15 cm apart. Tear off a piece of Magic™ tape about 15 cm long. Place the tape over the paper tabs. Press the tape onto the tabs and the table. Using a pencil, label the piece of tape B1 (bottom piece one).

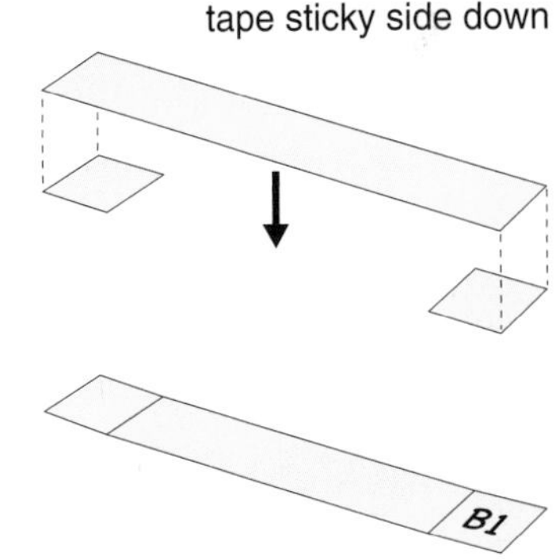

STEP 2 Lay two paper tabs on top of the bottom piece of tape B1. Tear off another piece of tape, the same length as before. Place it directly on top of the B1 piece of tape. Rub the top piece of tape firmly. Use a pencil to label the top piece of tape T1 (top piece 1).

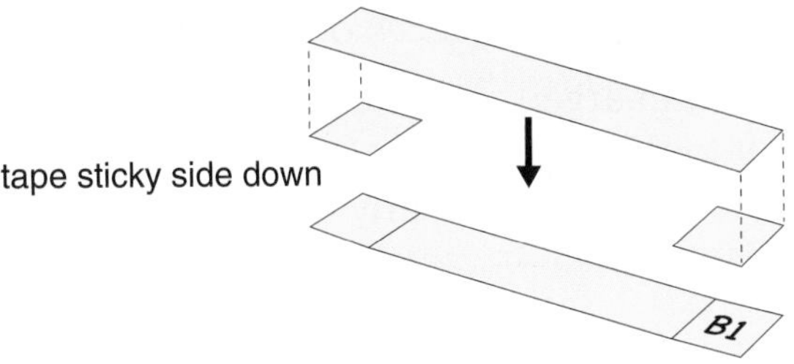

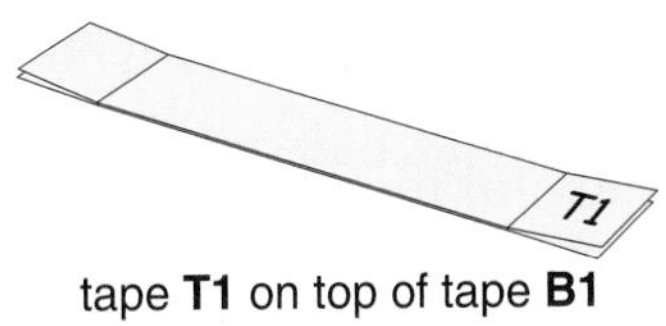

tape **T1** on top of tape **B1**

STEP 3 *Recycling Engineer:* Repeat Steps 1 and 2 with two more pieces of tape. This time label the bottom piece of tape B2 and the top piece of tape T2.

STEP 4 Carefully lift up the combination T1 and B1, still attached, off the table. Pull both pieces of tape between your fingers. This will help provide better results.

STEP 5 Using the paper tabs, rip apart the top piece from the bottom piece.

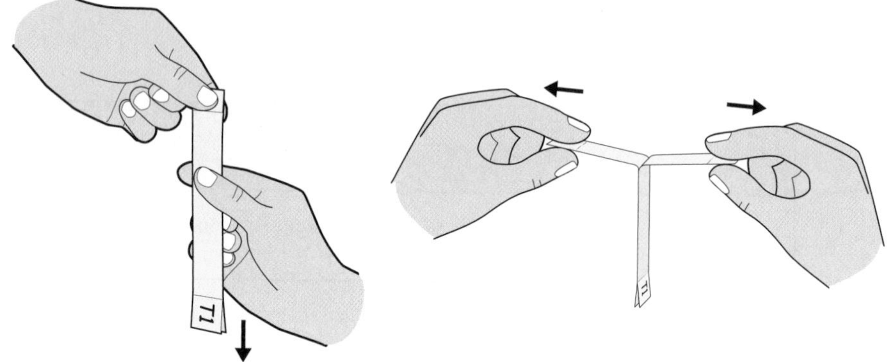

This ripping motion will cause each of the two pieces of tape (T1 and B1) to become electrically charged.

STEP 6 Attach each piece of tape (sticky side away from you) to the meter stick. The pieces should be about 20 cm from each other as shown in the diagram.

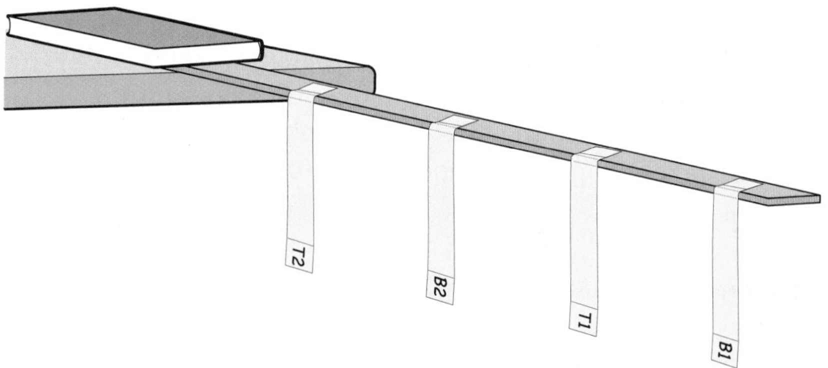

STEP 7 Repeat Steps 4, 5, and 6 for the other two pieces of tape (T2 and B2).

At this point, your team should have two charged top pieces of tape (T1 and T2) and two charged bottom pieces of tape (B1 and B2). For the remainder of this exploration, you will investigate how these differently charged pieces of tape may interact with each other.

STEP 8 *Supply Master:* Hold T2 by the two end tabs and move the *non-sticky* side slowly toward the suspended T1. Do not let them touch. Do the charged tapes *move toward each other (attract), move away from each other (repel),* or *is there no effect (no change in motion)*?

Record what happens in Table 2.

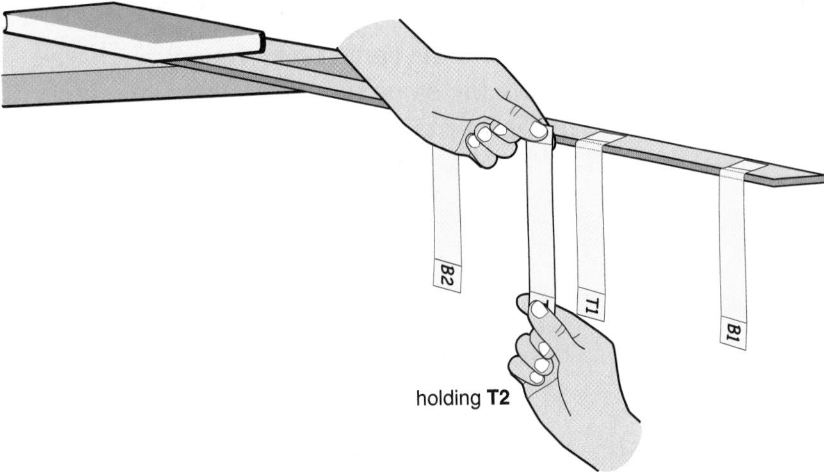

STEP 9 Turn T2 around and move the *sticky* side slowly toward the suspended T1. Do not let them touch.

Record what happens (*attract, repel,* or *no effect*) in Table 2.

2. Was the effect on suspended charge T1 the same or different when you turned around charge T2 and brought it near?

STEP 10 Move the non-sticky side of T2 toward the suspended B1.

Record what happens (*attract, repel,* or *no effect*) in Table 2.

STEP 11 Turn T2 around and move the sticky side slowly toward the suspended B1. Do not let them touch. Attach T2 to the edge of the meter stick.

Record what happens (*attract, repel,* or *no effect*) in Table 2.

3. Was the effect on suspended charge B1 the same or different when you turned around charge T2 and brought it near?

STEP 12 *Procedure Specialist:* Remove B2 from the edge of the meter stick and move its non-sticky side slowly toward the suspended T1. Turn B2 around and again bring it near T1.

Record what happens *(attract, repel,* or *no effect)* for both cases.

4. Was the effect on suspended charge T1 the same or different when you turned around charge B2 and brought it near?

STEP 13 *Procedure Specialist:* Take B2 and move its non-sticky side toward the suspended B1. Turn B2 around and again bring it near B1.

Record what happens *(attract, repel,* or *no effect)* for both cases.

5. Was the effect on suspended charge B1 the same or different when you turned around charge B2 and brought it near?

Your teacher will help you summarize the class's results for Table 2.

We will assume that T1 and T2 have the same type of electric charge. We will also assume that B1 and B2 have the same type of electric charge. However, the B1 and B2 charges are different from the T1 and T2 charges.

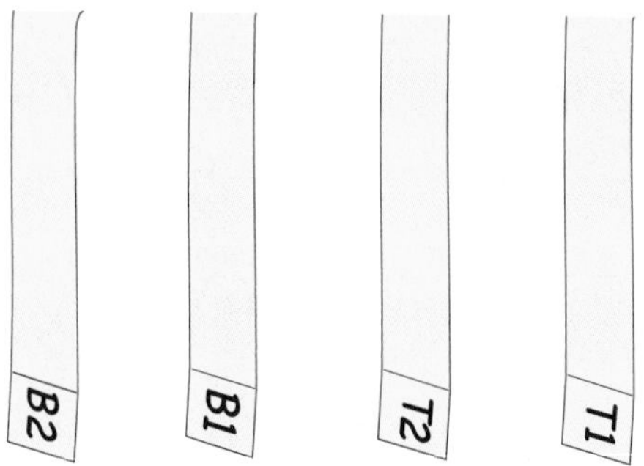

Use this information, along with your class's observations in Table 2, to answer the following questions:

6. What happens when two charged objects with the *same* type of charge (the two top charges or the two bottom charges) are brought near each other?

7. What happens when two charged objects with a *different* type of charge (one top charge and one bottom charge) are brought near each other?

8. When two charged objects are interacting (either attracting or repelling) and one is then turned around, is the interaction effect the *same* or *different?* What is your evidence?

Participate in a class discussion about your answers to the questions.

Make Sense of Your Ideas

Read through the first four ideas listed in *Scientists' Consensus Ideas: The Magnetic Interaction* that you examined in Activity 2. Compare these ideas of magnetic interaction to your observations of *electric-charge interaction* in this activity.

Write the defining characteristics of the electric-charge interaction. Use evidence from this activity to support your ideas.

Our Consensus Ideas

The key questions for this activity are:

1. What are some defining characteristics of the electric-charge interaction?

2. What are some of the similarities and differences between the magnetic and electric-charge interactions?

1. You answered the first key question in the Make Sense of Your Ideas section. Discuss the second key question with your team. Write your best answer on your record sheet.

Participate in a class discussion about your answers to the two key questions.

Scientists have also developed a set of ideas that describe the defining characteristics of the electric-charge interaction. Your teacher will hand out a form entitled *Scientists' Consensus Ideas: The Electric-Charge Interaction.* As you read through this form, think about the evidence from this activity that would support each of the ideas.

2. Indicate the explorations in this activity that provide the evidence to support the scientists' ideas.

Go over the form with your teacher.

DEVELOPING OUR IDEAS

Activity 4
The Electric-Circuit Interaction

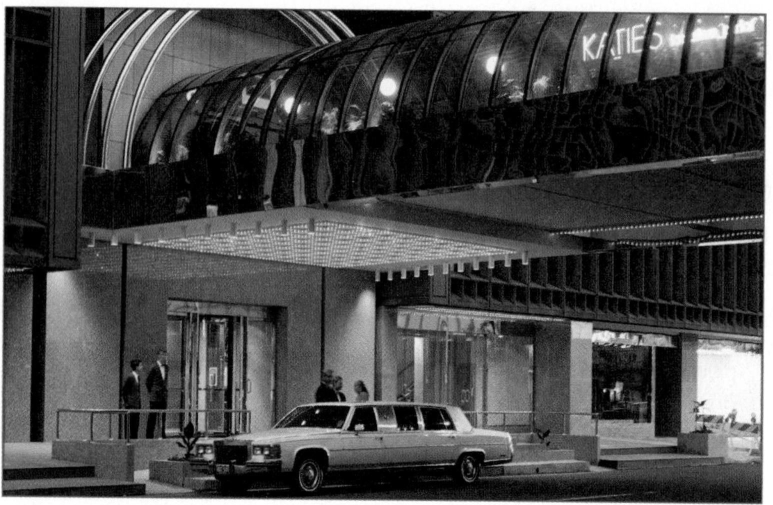

If you want to operate a number of devices that use electricity, you need to know what variables affect the interaction.

Purpose

In the previous two activities, you studied the defining characteristics of magnetic and electric-charge interactions. In other units, you will study other types of interactions. It is important to know the defining characteristics of each type of interaction, but you need to know more.

For every interaction, it is important to know *what* variables influence the interaction. It is also important to know *how* they influence the interaction.

In this activity, you will study the *electric-circuit interaction.* This interaction occurs when a source of electricity is connected to a device that uses electricity to operate. Here are the key questions for the activity:

1. What are the defining characteristics of an electric-circuit interaction?
2. What are some variables that influence the electric-circuit interaction?

Record the key questions for the activity on your record sheet.

We Think

Imagine you have an electric circuit with two bulbs and each bulb is glowing.

- If you unscrew one bulb and remove it from its socket, what do you think would happen to the other bulb? Why?

Discuss these questions with your team.

Participate in a whole-class discussion.

Do not leave circuits connected. Wires can get hot. Do not set up on any material that can burn or melt.

Explore Your Ideas

You will perform a series of five explorations to help answer the key questions for the activity.

We use the term *cell* to refer to one D-cell. This cell is used in many flashlights. Each cell provides a voltage of 1.5 V (volts). You may call it a battery. However, scientists call a battery a device that combines two or more cells.

Remember to share responsibilities among your team members.

Science Words

electric circuit: the path followed by an electric current from a power source through devices that use electricity and back to the source

Exploration 1: When does an electric-circuit interaction occur?

STEP 1 Mount the cell in the cell holder and screw the bulb into the bulb holder.

Use two hook-up wires and connect the cell to the bulb to make the bulb light up. The closed loop is called an **electric circuit**. When the bulb is lit, there is an *electric-circuit interaction* between the cell and the bulb.

STEP 2 Add a switch to the circuit that can be used to turn the bulb on and off. You will need a third hook-up wire.

Your team will need:

- 2 cells
- 2 cell holders
- 2 bare bulbs
- 2 bulb holders
- switch
- hook-up wires
- metal and non-metal objects

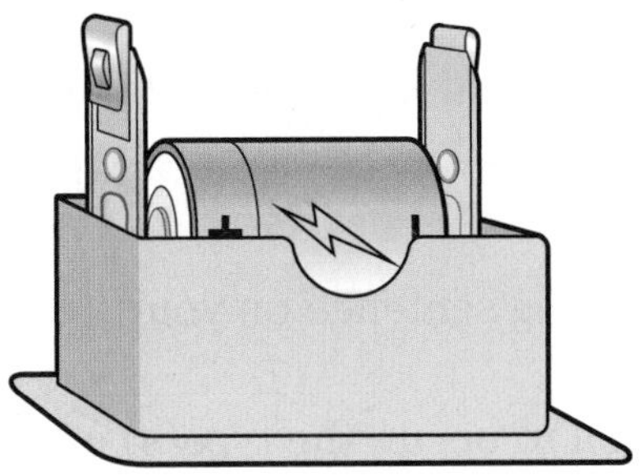

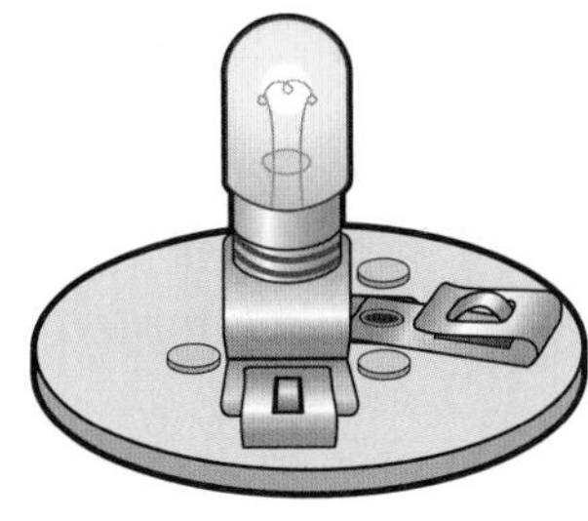

1. Draw a picture of the circuit when the bulb is lit.

2. Look carefully at how the wires are connected to each of the circuit elements—the cell (in its holder), the bulb (in its holder), and the switch. How many connections are there to each element in the circuit (one connection, two connections, or more than two)?

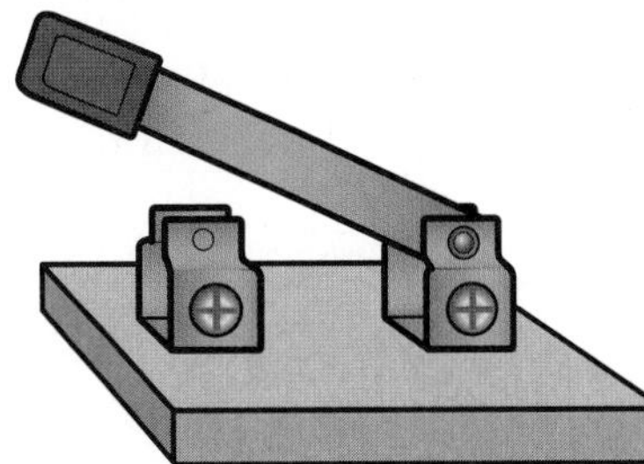

3. When does an electric-circuit interaction occur?

To Do

Team Member 1
Supply Master: gathers materials.

Team Member 2
Procedure Specialist: reads instructions and steps aloud.

Team Member 3
Team Manager: makes sure team stays on task and all team members are participating.

Team Member 4
Recycling Engineer: returns materials.

Exploration 2: What types of materials are necessary for an electric-circuit interaction to occur?

In the previous exploration, you discovered that if you hook up a circuit similar to the one shown on the right and close the switch, the bulb will glow. That provides evidence that there is an electric-circuit interaction occurring between the cell and bulb.

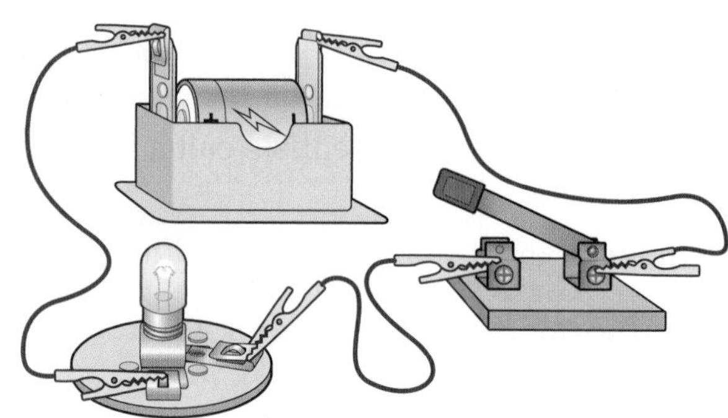

STEP 1 Add another hook-up wire and attach the ends of two hook-up wires to an iron nail, as shown in the diagram below.

When you close the switch, does the bulb glow? Record your observation in the table.

STEP 2 Open the switch and replace the iron nail with the piece of paper.

When you close the switch, does the bulb glow? Record your observation in the table.

STEP 3 Repeat Step 2 for at least three other metals and three other nonmetals.

In each case, record in the table the type of material and your observation about whether the bulb glows or not.

4. Examine your table. Complete the following sentence on your record sheet:

In order for an electric-circuit interaction to occur, the type(s) of material that must be included in the circuit is (are) _________ .
The evidence is ________.

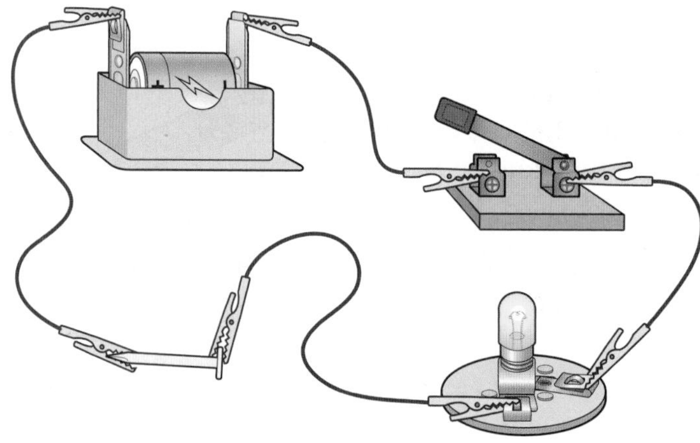

Materials placed in a circuit that allow the bulb to glow are called **conductors.** Materials that do not allow the bulb to glow are called **non-conductors.**

Science Words

electrical conductor: a material that allows electric current to exist in it

electrical non-conductor: a material that does not allow electric current to exist in it

single-loop circuit (series circuit): a circuit that has all its parts connected in a single loop

multi-loop circuit (parallel circuit): a circuit in which two or more single loops connect to the same cells

Table 1: Materials That Allow the Bulb to Glow

Material	Does the Bulb Glow? (Yes or No)
iron nail	
paper	

Exploration 3: How can you hook up more than one bulb to a cell?

STEP 1 Hook up one cell and two bulbs in a single loop. Use as many hook-up wires as needed. This circuit is called a **single-loop circuit** (or **series circuit**).

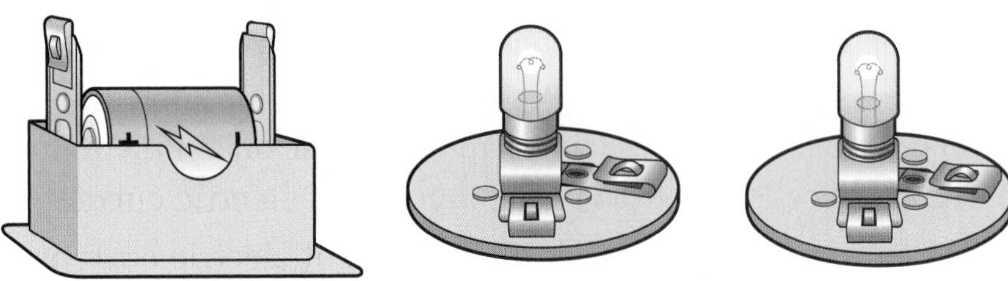

5. Draw a picture of this circuit and label it *series circuit.*
6. Unscrew one of the two bulbs from its socket. Leave the other bulb alone. What happens to the other bulb? Does it remain lit or does it go out?

STEP 2 There is a different way of hooking up two bulbs to the same cell. Hook up each bulb in its own separate loop to the cell. This circuit arrangement is called a **multi-loop circuit** (or **parallel circuit**).

7. Draw a picture of this circuit and label it *parallel circuit.*
8. Unscrew one of the two bulbs from its socket. Leave the other bulb alone. What happens to the other bulb? Does it remain lit or does it go out?
9. Why do you think the result for Question 8 was different from the result for Question 6?

Exploration 4: If the number of cells in the circuit increases, what happens to the brightness of the bulb?

STEP 1 Hook up two cells and one bulb in a single-loop (series) circuit. Make sure you connect the plus side of one cell to the minus side of the other cell.

10. Is the bulb in this two-cell and one-bulb circuit *brighter than*, *dimmer than*, or *equally as bright* as the bulb in the one-cell, one-bulb circuit in Exploration 1?

STEP 2 Borrow *one* additional cell and hook-up wire from another group. Hook up three cells and one bulb in a series circuit.

11. What happens to the brightness of the bulb when an additional cell is added to the circuit?

12. Write your conclusion for this exploration by answering the following question: If the number of cells in the circuit increases, what happens to the brightness of the bulb?

Exploration 5: If the number of cells in the circuit increases, what happens to the amount of electric current in the circuit?

Science Words

simulator: a machine or computer program that models a given environment or situation for the purpose of training or research

ammeter: a device that measures the amount of electric current in a circuit

When the bulb is turned on, something is flowing in the circuit. It is called an electric current. Suppose you add cells to a single-loop circuit with one bulb.

13. What is your hypothesis about the relationship between the number of cells and the amount of electric current? Choose one of the following.

a) The more cells, the *greater* the amount of electric current.

b) The more cells, the *smaller* the amount of electric current.

c) There is *no relationship* between the number of cells and the amount of electric current.

Discuss the question above with your partner and record your answer, including your reason.

A special computer program called a **simulator** can be used to test your hypothesis. The simulator was programmed with real data from many explorations.

A device that measures the amount of electric current in a circuit is called an **ammeter**. The unit commonly used for electric current is a milliampere (mA).

There is one important difference between using a computer simulator and using real cells, bulbs, switches, and ammeters to collect data. All real instruments (like ammeters) have uncertainties associated with their measurements. If you performed the exploration many times with the same circuit, you would get somewhat different values each time. You would then report the average as your best value and include the uncertainty in your measurements.

The simulator you will be using, however, was programmed so that it *always* shows the same amount for each measurement, no matter how many times you repeat it. Therefore, each single measurement provides the best value.

When you are collecting a lot of data, the simulator has an advantage over real equipment because it is very easy to use.

The four circuits shown in the diagram will be tested.

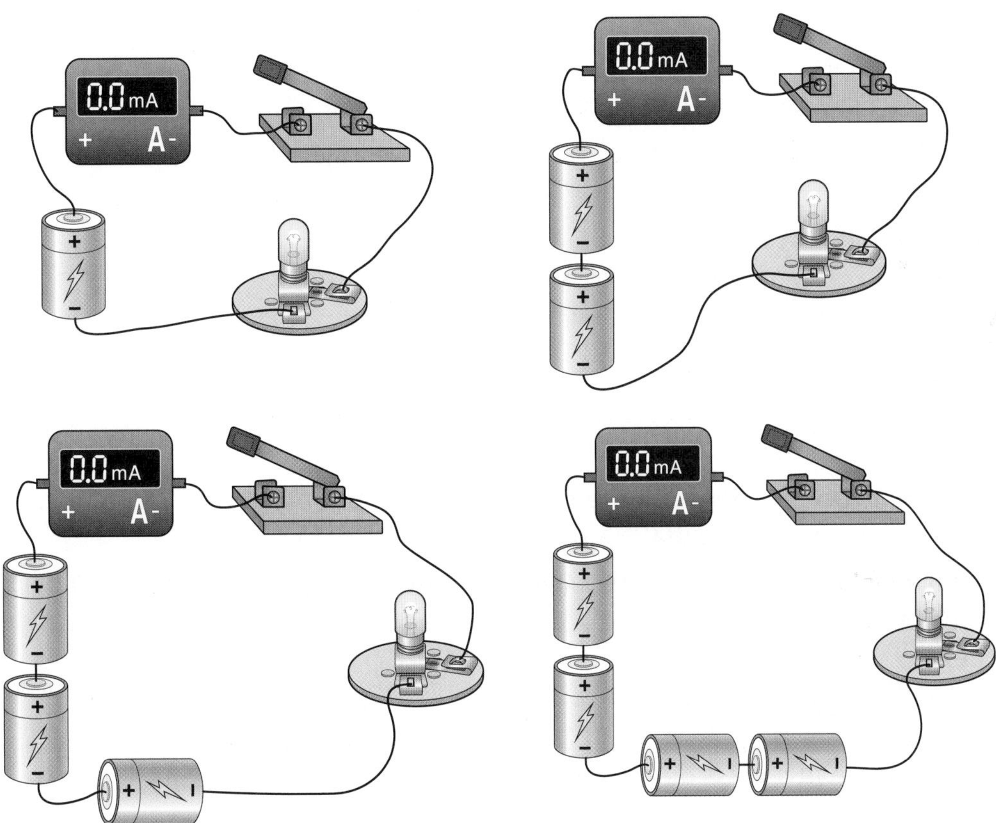

Table 2: Amount of Electric Current versus Number of Cells	
Number of Cells	Amount of Electric Current (mA)
1	
2	
3	
4	

In the second column of Table 2 on your record sheet, copy the value of the ammeter readings for each of the four circuits when the simulator is run and the switch is closed.

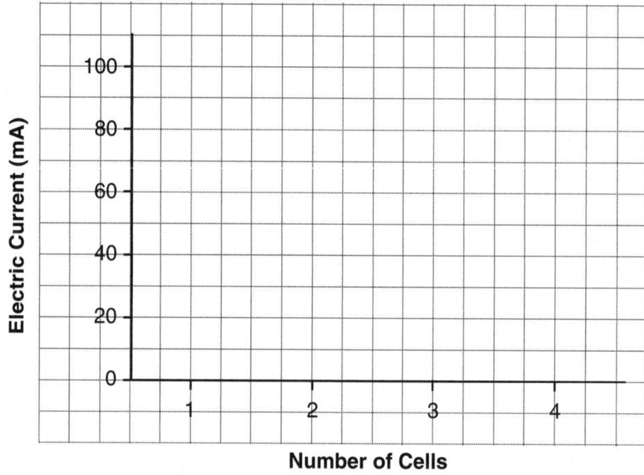

14. Sometimes, people like to see data displayed on a graph. Your teacher will show you how to make a bar graph of your data.

15. Write your conclusion by completing the following on your record sheet.

If the number of cells in the circuit increases, the amount of electric current ______ (*increases, decreases, remains the same*). The evidence is ______.

Participate in a class discussion to review the five explorations.

Make Sense of Your Ideas

In the previous activity, you learned about the electric-charge interaction. It can help you understand what happens in an electric circuit. One end of a cell is positively charged (the end with the knob, usually with a plus symbol). The other end is negatively charged. The wires of the circuit and the bulb contain both positive and negative charges. Only the negative charges (electrons) can move through the conducting wires and bulb. The positive charges (protons) are fixed in position and do not move. When the circuit is connected together, the positive end of the battery attracts the negative charges in the wires and the negative end of the battery repels (pushes away) the negative charges. This causes the negative charges to move through the circuit in the direction from the negative to the positive side of the battery. As negative charges enter the positive side of the battery, equal numbers of negative charges leave the negative side of the battery. The negative charges, then, continuously flow around the circuit. This flow of charges is called **electric current.**

Science Words

electric current: a flow of negative charges around a circuit

The first key question for this activity was: *What are the defining characteristics of an electric-circuit interaction?* Answering this question involves knowing three things: the objects involved in the interaction, how they are connected to each other, and the evidence that the interaction has occurred.

1. What kind of objects are involved in an electric-circuit interaction?

2. How are the objects connected together?

3. What is the evidence that an electric-circuit interaction has occurred?

In Exploration 3, you observed what happens when a bulb is removed from its socket.

4. What happens to the electric-circuit interaction when there is a break in an electric-circuit loop?

The second key question for this activity was: *What are some variables that influence the electric-circuit interaction?* A variable is something that might make the interaction effects stronger or weaker. Variables might include the number of cells, the number of bulbs, or the number of wires used. An interaction effect could be the brightness of a bulb or the amount of electric current in the circuit. Think about what variables you changed in the explorations you performed.

5. What variable or variables can influence the electric-circuit interaction? What is your evidence?

The following two questions are more challenging. They will help you make connections between what you learned in this activity and the real world.

6. You can buy two types of holiday lights.

   In type A, when you plug in the string of lights and one bulb burns out, all the bulbs go out.

   In type B, when you plug in the string and one bulb burns out, the other bulbs stay on.

   In one type, the bulbs are all connected together in a series circuit with the electrical source. In the other type, the bulbs are connected in a parallel circuit with the electrical source. Which type, A or B, is connected in series? Which type is connected in parallel?

7. Consider how your electrical devices are connected together at home. Suppose you have a room with two different lamps. Do you think the lamps are connected together in series or in parallel? How do you know?

Participate in a class discussion about the answers to the questions.

Our Consensus Ideas

The key questions for this activity are:

1. What are the defining characteristics of an electric-circuit interaction?

2. What are some variables that influence the electric-circuit interaction?

1. Think about your answers to the questions in Make Sense of Your Ideas. Write your best answers to the key questions on your record sheet.

Scientists have performed explorations with electric circuits and have determined the *defining characteristics* and the *variables that influence* the electric-circuit interaction. Your teacher will distribute a copy of *Scientists' Consensus Ideas: The Electric-Circuit Interaction.* As you read through this document, think about the evidence from this activity that would support each of the ideas.

2. Indicate the explorations in this activity that provide the evidence to support the scientists' ideas.

Go over the form with your teacher.

DEVELOPING OUR IDEAS

Activity 5
Electromagnets and Buzzers

Small electromagnets are used inside buzzers, larger ones are used inside television sets, and huge ones are used to lift scrap metal.

Purpose

You have been studying magnetic and electric-circuit interactions. One device that depends on both the magnetic interaction and the electric-circuit interaction for its operation is the **electromagnet**. An electromagnet consists of a coil of wire connected to a source of electricity. The coil is usually wrapped around a magnetic material.

In this activity, you will first gather evidence to see how a coil of wire connected to a battery can behave like a magnet. You will then construct your own electromagnet. Finally, you will investigate how a buzzer uses an electromagnet to work.

The key question for this activity is:

How do electromagnets and buzzers work?

Record the key question on your record sheet.

We Think

Think about what you know about electromagnets.

- What is an electromagnet? How does it work?

Discuss these questions with your team.

Participate in a whole-class discussion.

Science Words

electromagnet: An electromagnet consists of a coil of wire connected to a source of electricity. The coil is usually wrapped around a magnetic material.

Explore Your Ideas

Explore the Electromagnet

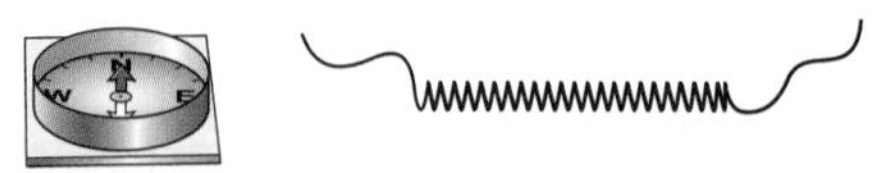

Exploration 1: Under what circumstances will a coil of wire interact with a magnet?

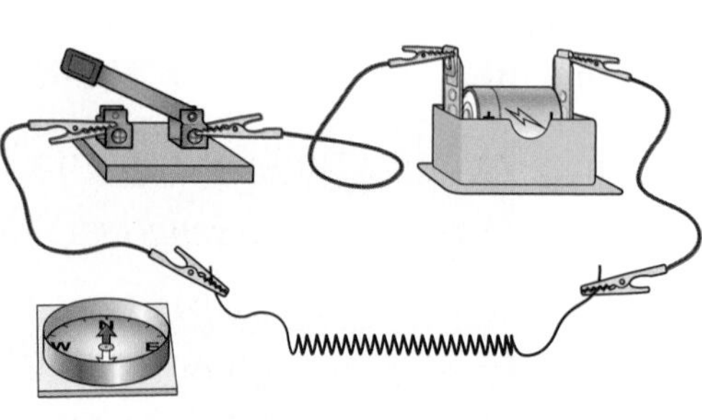

Your instructor will show you either a demonstration or a movie of the following steps.

STEP 1 A coil of wire is brought near the magnetic compass needle as shown.

1. Is the colored part of the compass needle *attracted to the coil, repelled from the coil,* or *is there no effect?*

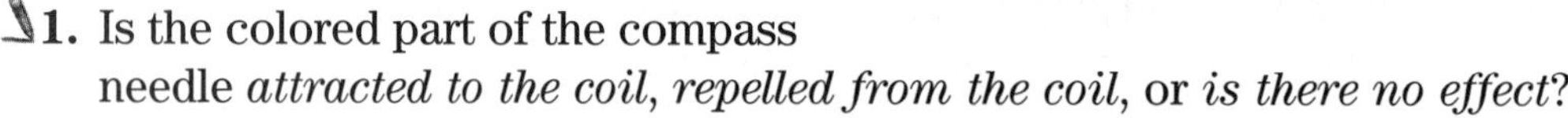

STEP 2 The coil of wire is now connected to a cell and the switch is closed.

2. Is the colored part of the compass needle *attracted to the coil, repelled from the coil,* or *is there no effect?*

STEP 3 An iron nail is inserted inside the coil of wire and the switch is closed.

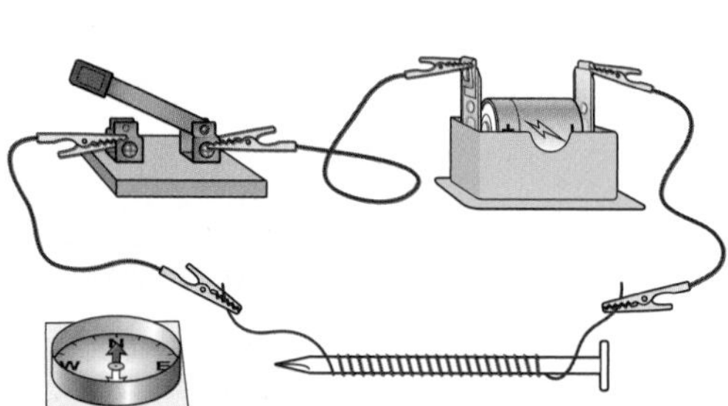

3. Is the colored part of the compass needle *attracted to the coil/nail, repelled from it,* or *is there no effect?*

4. How does the deflection (rotation) of the colored part of the compass needle compare to when there was no iron nail inside the coil?

STEP 4 The compass is moved to the other side of the coil/nail.

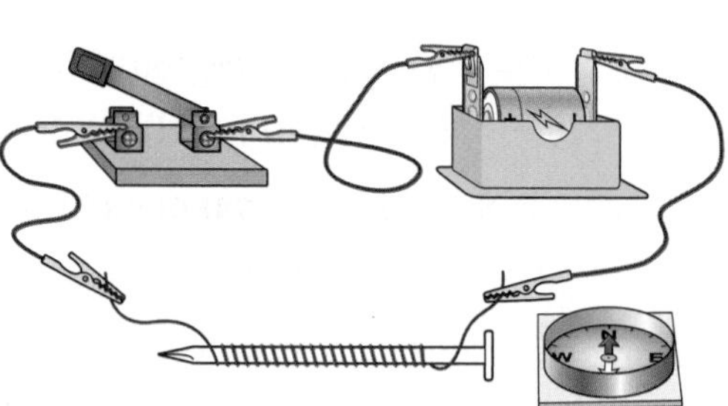

5. Is the colored part of the compass needle *attracted to the coil/nail, repelled from it,* or *is there no effect?*

6. How does the deflection (rotation) of the colored part of the compass needle compare to when the compass was near the other side of the coil/nail?

The wire may get hot. Do not touch it until after the switch has been open for several minutes.

7. What is the evidence that the electromagnet interacts with a magnet?

8. Does an electromagnet behave like a magnetic material or like a magnet? What is your evidence?

Participate in the class discussion to go over the answers to the last two questions.

Exploration 2: How can you make an electromagnet stronger?

Remember to share responsibilities among your team members.

Your team will need:

- 2 or 3 cells in holders
- switch
- 4 hook-up wires
- large nail
- long piece of enameled wire with insulation rubbed off the ends
- 2 or 3 large washers
- ruler

STEP 1 Wrap about *30 turns* of wire around the large nail. Spread the coils over the length of the nail. Connect a hook-up wire to each end of the wire and assemble the circuit as shown in the diagram. (The enamel should be removed from the ends of the wire. Make sure you connect the hook-up wires to these parts.) Keep the switch off (handle up). The nail and coil of wire is an electromagnet.

STEP 2 Close the switch. Pick up the electromagnet and bring the flat head end of the nail to one of the washers. Try to lift the washer up into the air and hold it there. If the electromagnet is not strong enough to lift the washer, figure out how to make it stronger so it can lift the washer.

To Do

Team Member 1
Supply Master: gathers materials.

Team Member 2
Procedure Specialist: reads instructions and steps aloud.

Team Member 3
Team Manager: makes sure team stays on task and all team members are participating.

Team Member 4
Recycling Engineer: returns materials.

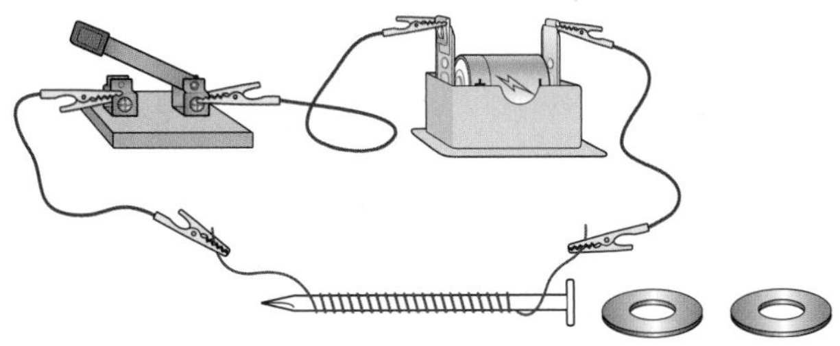

9. How did you make your electromagnet strong enough to lift the washer? (If it was already strong enough, record that.)

STEP 3 While the washer is lifted in the air, open the switch.

10. What happens to the washer?

11. Why do you think this happened?

STEP 4 Find two different ways to make your electromagnet stronger so it can lift two washers at the same time.

12. What two different ways did you find that work?

STEP 5 When you are finished, carefully remove the coil of wire from the nail and straighten the wire out as best you can.

Make Sense of Your Ideas

Make Sense of the Electromagnet

Discuss the following questions with your team and record your answers.

1. What are *two* variables that can influence the strength of the magnetic interaction between an electromagnet and a magnetic material (steel washer)?

2. How does each variable influence the strength of the electromagnetic interaction? For each variable, complete the following statements on your record sheet:

> As the ____ (*write the variable*) increases, the strength of the magnetic interaction between the electromagnet and the magnetic material ____ (*increases, decreases*).

Participate in the class discussion.

Your team will need:

- wooden or heavy plastic ruler (about 30 cm long)
- rubber band
- cell in holder
- 2 hook-up wires
- small buzzer

Explore Your Ideas

Explore the Buzzer

Buzzers produce a buzzing sound. That's what they do, but how do they work? There are really two basic questions. What actually produces the buzzing sound and how does the buzzer do it? After working through the remainder of this activity, you should be able to answer these questions.

Exploration: How is a sound produced?

STEP 1 Use the palm of your hand to hold a ruler tightly against the table. Leave about half of the ruler sticking out over the edge, as shown in the diagram.

STEP 2 Use your finger to flick the end of the ruler. Listen to the sound produced.

1. What is the ruler doing while the sound is produced?

STEP 3 Stretch a rubber band between your fingers and pluck it at its middle. Listen to the sound produced.

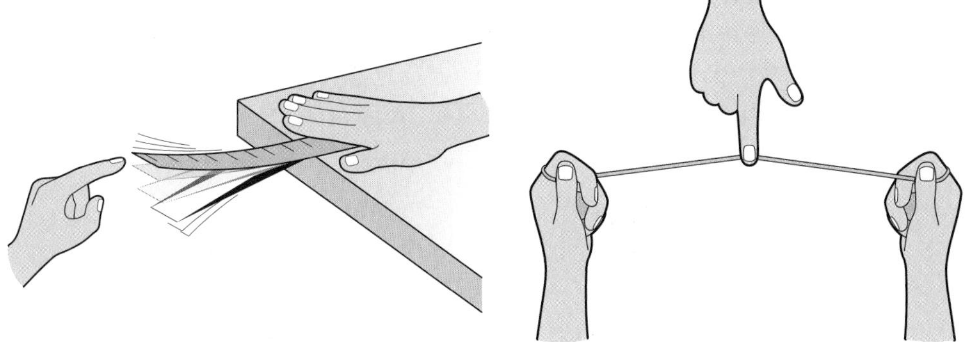

2. What is the rubber band doing while the sound is produced?

STEP 4 Hook up the battery cell to the small buzzer and listen to the sound. If you do not hear any sound, switch the hook-up wires to the positive and negative ends of the battery cell.

3. What do you think is happening inside the buzzer to make that sound?

STEP 5 Your teacher will show you a movie of a large buzzer. Watch what happens and listen to the sound produced when the buzzer is connected to cells. The movie also shows what is inside a small buzzer, similar to the one you used above.

Here is a simplified picture of the inside of the large buzzer.

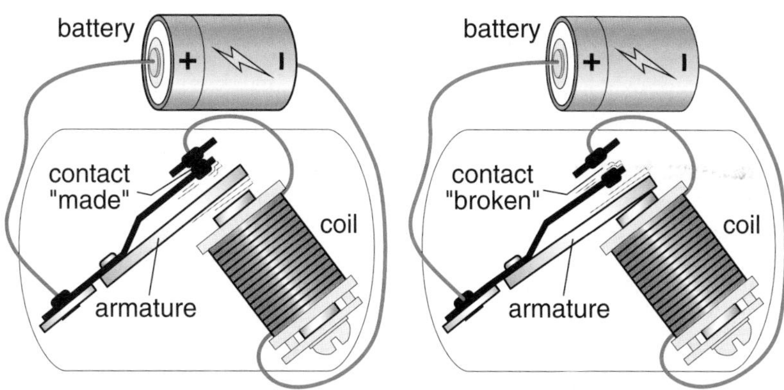

Make Sense of Your Ideas

Make Sense of the Buzzer

Below is a "story" that tells how the buzzer works. Some words or phrases are left out. Here is a list of possible words or phrases that can fit. Each word or phrase is used only once.

- electromagnet
- vibrate
- armature
- electric-circuit
- over and over again
- open
- magnetic

Complete the story on your record sheet.

All sounds are produced by vibrating objects. In the buzzer, the _______________ moves rapidly back and forth to produce the "buzzing" sound. What makes it _______________? When the contact is "made" (see the diagram on the previous page), there is an electric-circuit interaction between the cell and the coil. This _______________ interaction creates an electric current in the coil and the coil becomes a(n) _______________. There is then a(n) _______________ interaction between the electromagnet and the armature (made of iron). This interaction pulls the armature towards the coil. When this happens, however, the circuit becomes _______________ (contact "broken," no electric current), the electromagnet loses its strength, and the armature returns to its original position (contact "made"). The circuit is again closed and the process repeats itself _______________, causing the armature to vibrate.

Your teacher will review the story with the class.

Our Consensus Ideas

The key question for this activity is:

How do electromagnets and buzzers work?

Look over your answers to the Make Sense of Your Ideas sections. Are you now confident that you know the answer to the question?

LEARNING ABOUT OTHER IDEAS

Activity 6
Interaction between a Magnet and an Electric Current

Purpose

Life wouldn't be the same without motors! CD players, fridges, hair dryers, and starters in cars all use motors.

Remember in Activity 2, you observed magnetic interactions with other magnets, different metals, and the compass needle? In Activity 4, you set up an electric circuit. You explored what kinds of things made up a circuit and you learned about different types of circuits. Magnetism and electricity may seem like completely different things, but actually there are lots of devices in our everyday world that depend on the interaction *between* them.

The interaction between a magnet and an electric current is called the *electromagnetic interaction*. In this activity, you will learn how this new type of interaction can help explain how motors and meters work. The key question for the activity is:

How does the electromagnetic interaction help explain how motors and meters work?

Record the key question on your record sheet.

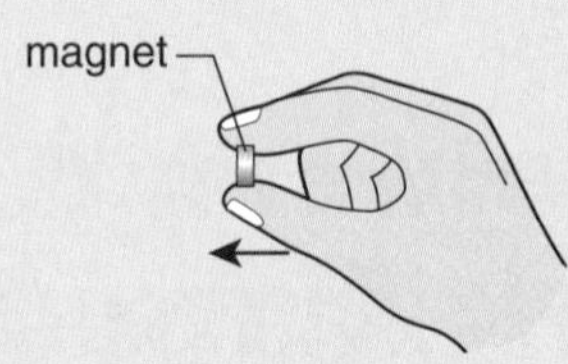

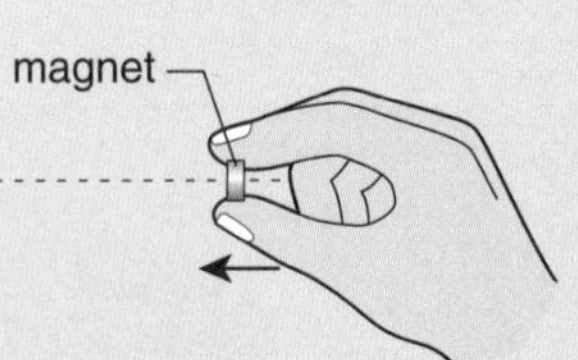

Learning the Ideas

Electromagnetic Interactions

In Activity 2, you learned that the needle on a magnetic compass is actually a magnet. You can prove this by first holding a magnet nearby and observing that the compass needle turns.

You can then turn the magnet around and observe that the compass needle turns the opposite way.

You observe no effect when a magnet is held near a piece of copper. If you wrap a coil of copper wire around a magnetic compass, you will observe no effect on the compass needle. These observations suggest there is no interaction between the copper wire and the compass-needle magnet.

Now imagine that you connect the copper wire wrapped around the compass to a circuit with a battery, bulb, and switch.

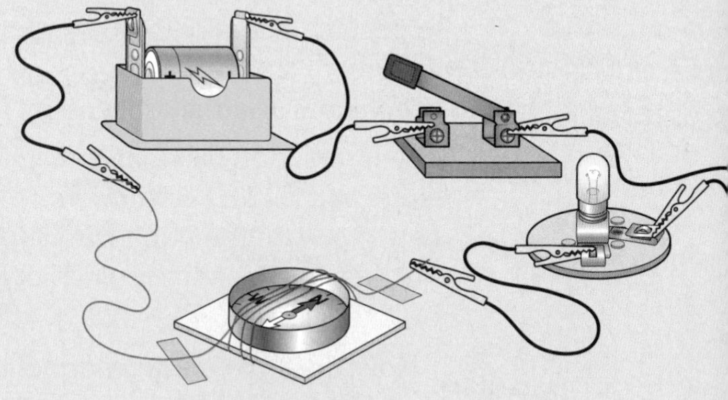

Your teacher will show a demonstration or movie of this setup.

1. What happens to the magnetic compass needle when the switch is closed?
2. What evidence is there that there is an electric current in the circuit? (After all, you cannot *see* the electric current in the wires.)

Your teacher will review answers to these questions with the class.

The observation of what happened in the circuit provides evidence for an interaction between an electric current (in the wires) and a magnet (the compass needle). This interaction is called the *electromagnetic interaction.*

Motors

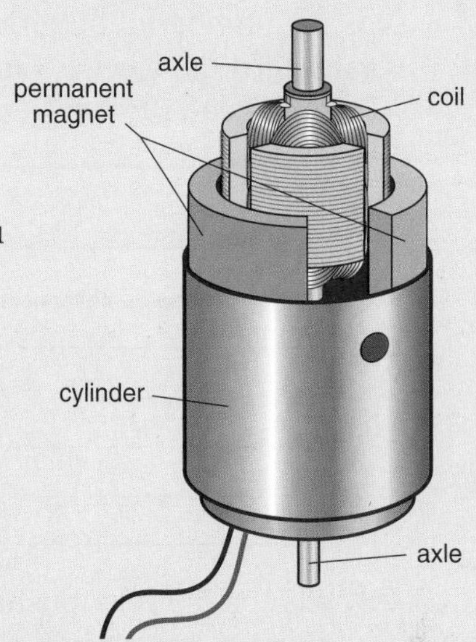

The electromagnetic interaction between the electric current in the coil of wire and the compass-needle magnet causes the needle to rotate. A magnet can also cause a coil of wire with an electric current to rotate. That helps you to understand how an *electric motor* works.

A motor involves both an electric-circuit interaction and an electromagnetic interaction. In a motor, a coil of wire is mounted on an axle (rotor shaft) and attached to a battery or another source of electrical energy. This creates an electric current in the wires of the coil (the electric-circuit interaction). A set of permanent magnets is on the inside of the cylinder that holds the coil and axle. When there is an electric current in the coil of wire, the coil behaves like an electromagnet, and the current in the coil interacts with the permanent magnets like a magnet. The interaction between the electric current and the magnet is the electromagnetic interaction. This interaction causes the coil of wire on the axle to rotate, usually at a high speed. Any other device attached to the axle, like fan blades, will also rotate.

Meters

When there is an electric current in wires wrapped around a magnetic compass, the compass needle turns. Are the amount of electric current in the circuit and the amount the compass needle turns related?

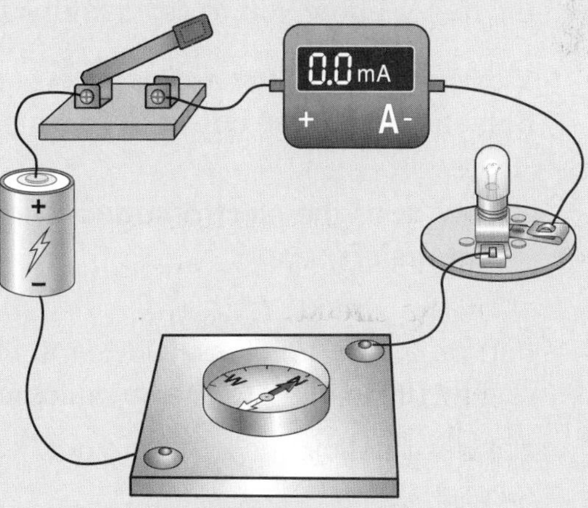

Your teacher will run a computer simulation of a circuit that has one or more cells, a switch, ammeter, bulb, and compass. As you learned in Activity 4, when the number of batteries is increased in a circuit, the value of the electric current in the circuit increases.

As the number of batteries in the circuit is changed, record the corresponding values for the electric current and the compass deflection in the table on your record sheet.

Table: Compass Deflection versus Amount of Current		
Number of Batteries	Electric Current (mA)	Compass Deflection (Number of Degrees)
1		
2		
3		
4		

3. What is the relationship between the compass deflection and the amount of electric current in the circuit?

Your teacher will review answers to this question with the class.

Many ammeters are made with a thin needle that is attached to a magnet surrounded by loops of wire, just like the compass in the simulation. Like the compass needle, the magnet-needle combination is free to rotate. When the meter is connected to a circuit and electric current is in the loops of wire, the current causes the meter needle to rotate. As with the compass, the amount of rotation depends on the amount of current. Numbers printed on the meter allow you to read values for the electric current.

What We Have Learned

Remember the key question for this activity:

How does the electromagnetic interaction help explain how motors and meters work?

Participate in the class discussion to review the answer to the key question.

Answer the key question on your record sheet.

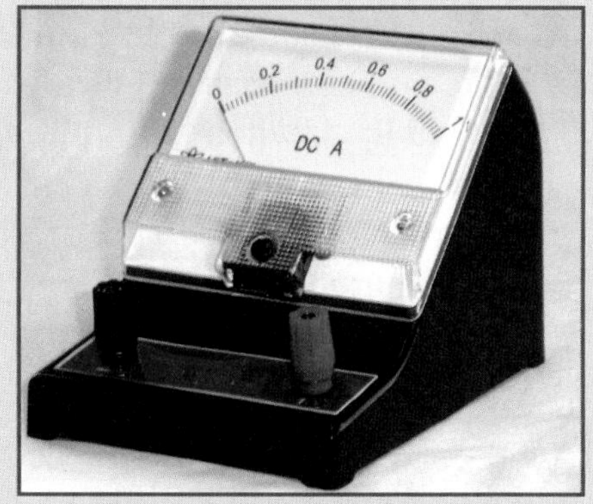

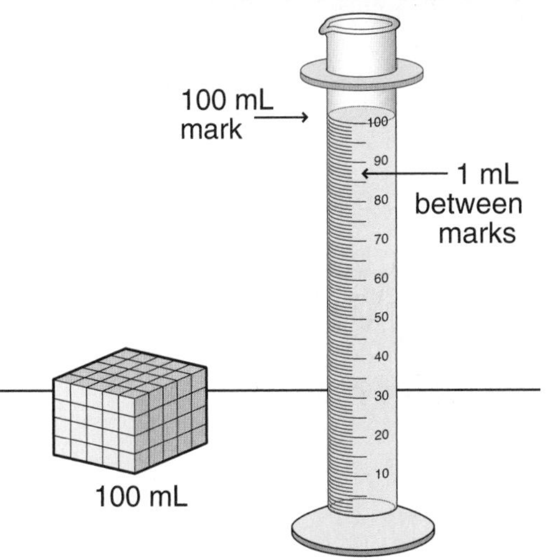

CYCLE 3:
INTERACTIONS AND PROPERTIES

Cycle Key Questions

1. How do scientists measure the amount of stuff in an object?

2. What are some properties of objects that help you decide what kind of material they are made of?

Activity 1: Measuring "Stuff"

Activity 2: Volume of Solids

Activity 3: Volume of Liquids

Activity 4: Measuring Mass

Activity 5: Density

Activity 6: Characteristic Properties

Activity 7: Calculating Density

OUR FIRST IDEAS

Activity 1
Measuring "Stuff"

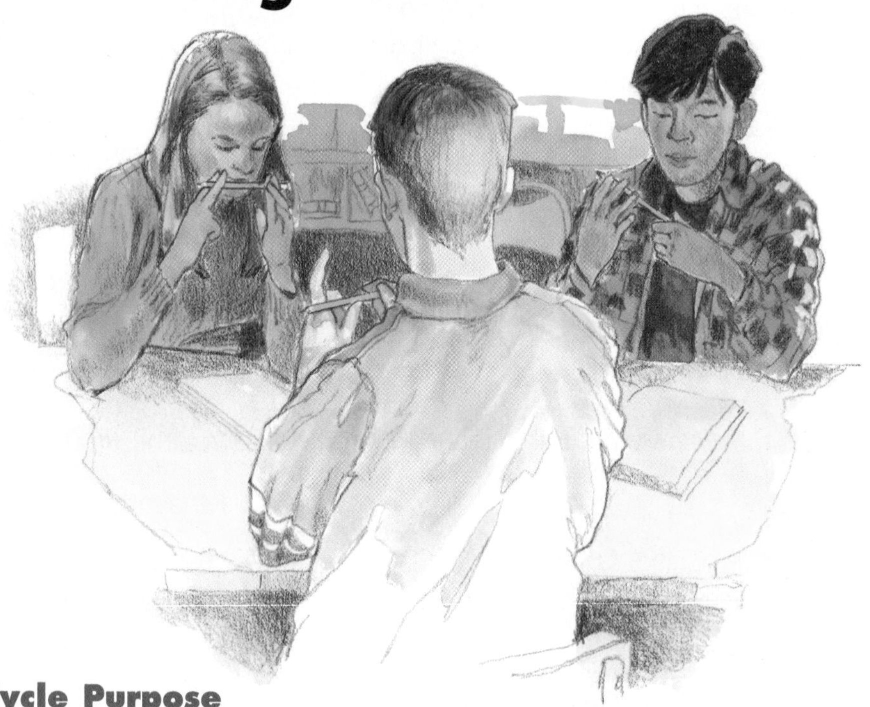

Cycle Purpose

Science Words

property: a description of how the object interacts with another object

In Cycle 2, you explored different types of interactions between two objects. In this cycle, you will explore how scientists describe the objects themselves. Scientists describe objects by their properties. A **property** of an object is a description of how the object interacts with another object. Properties pertain to a single object, while interactions need two or more objects. The most basic properties are descriptions of how an object interacts with your sense organs. For example, how does an object like a pencil interact with your eyes, ears, skin (hands), mouth, and nose?

1. How does a pencil interact with your eyes? List at least two words or phrases that describe how a pencil *looks*.

2. How does a pencil interact with your ears? Describe how the pencil *sounds* when tapped gently on a table or desk.

Never taste anything in a lab.

Never smell an unknown substance.

3. How does a pencil interact with your hands? List at least two words or phrases that describe how the pencil *feels* in your hands.

4. How does a pencil interact with your nose? List at least two words or phrases that describe how the pencil *smells*.

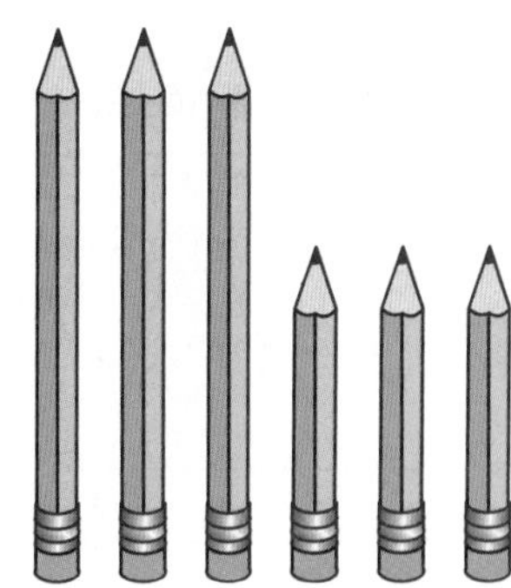

Consider the six pencils shown in the diagram. You could use your sense of sight to describe the pencils as short and long. However, the most accurate way to describe the lengths of the pencils is to *measure* their lengths with a ruler. Your "description" of each pencil would thus consist of a number and a unit of measurement of length.

All measurements are descriptions of how an object interacts with a measuring instrument. Measuring instruments, like a ruler, extend and refine your senses. They allow you to describe objects more accurately. Sometimes, you need to know more about an object than how long it is. You need to know *how much* stuff you have. The first key question you will explore in this cycle is:

1. How do scientists measure the amount of stuff in an object?

In some situations, you also need to know the *kind of material* an object is made of. You have already encountered some properties of materials. For example, you learned that some materials are magnetic and others are nonmagnetic. This property describes how the materials interact with a magnet.

Another property of materials is how they interact in an electric circuit. You learned that some materials are conductors and others are nonconductors. This property is determined by whether or not the materials allow an electric current in a circuit (and make a bulb glow).

However, suppose you have two metal blocks that look silvery. Neither material is magnetic. Both materials are conductors. In this case, the interaction of the blocks with a magnet and with an electric circuit are not useful properties for figuring out the kind of metal the blocks are made of. The second key question you will explore in this cycle is:

2. What are some properties of objects that help you decide what kind of material they are made of?

Record the key questions for the cycle on your record sheet.

Measurement and Standard Units

All measurements involve *counting* units. For example, to measure the **length** of a pencil, you would count the number of "unit" lengths that would fit along the pencil.

Science Words

length: a measure of distance

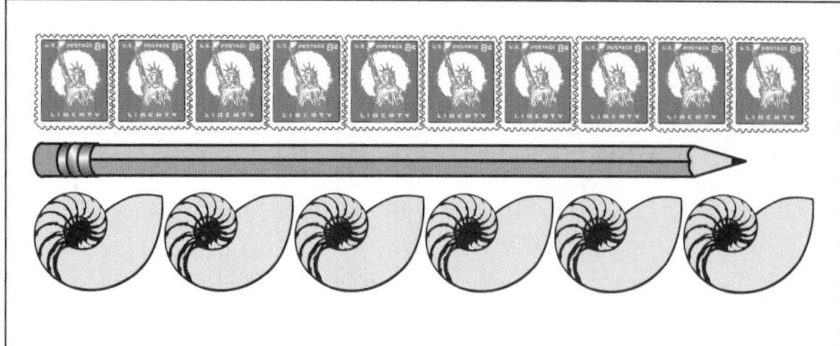

Any objects that are all the same length could be used for a unit of length. For example, the pencil in the diagram is 5.5 unit-shells long.

5. How many unit-stamps long is the object?

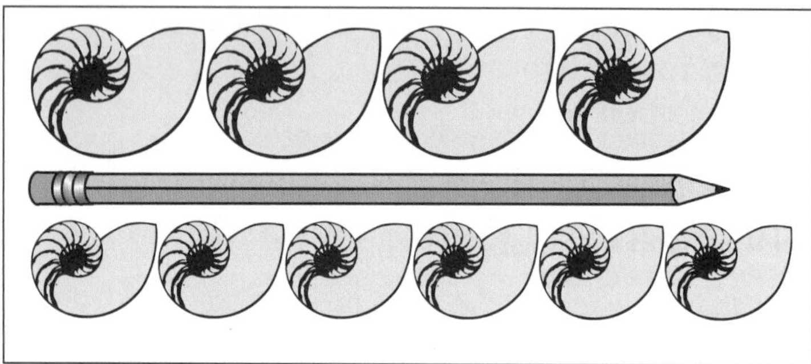

You would have a problem, however, if two people measure the length using different unit shells. For example, the pencil in the diagram at left is 4.0 large-shell units long, but 5.5 small-shell units long.

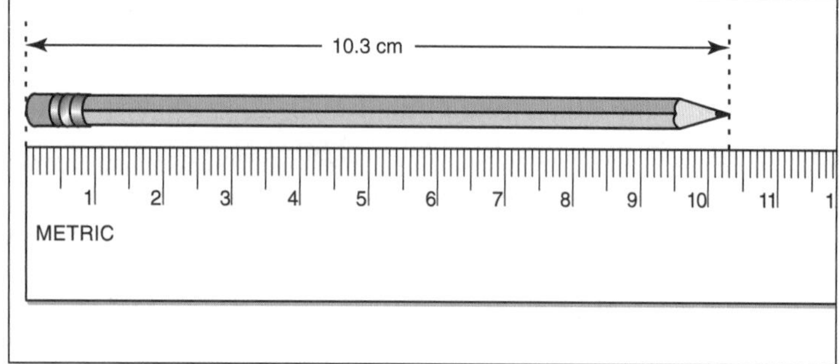

To get the same measurement for the length of the pencil, everyone has to agree on a standard unit of length. The *standard unit* of length is the meter (m). To measure smaller objects, you can use a centimeter (cm), which is one-hundredth of a meter.

We Think

Imagine that someone gave you a jar of homemade jelly. Discuss the following questions with your team, and then record your team's answer.

How could you measure how much jelly there is in the jar? Is there more than one way to measure how much jelly there is in the jar?

Our Class Ideas

Participate in a class discussion about your answers.

Write the class ideas about how you can measure how much jelly there is in the jar.

DEVELOPING OUR IDEAS

Activity 2
Volume of Solids

Purpose

Imagine that you had a solid block of material. How could you describe how much material there is in this block?

One way to measure the amount of material in the block is to measure its size. You need to find how much *room* or *space* it takes up. The property called **volume** is the measurement of how much space something occupies. The key question for this activity is:

Science Words

volume: the measurement of how much space something occupies

How are the volumes of cubes and rectangular solids measured?

Record the key question on your record sheet.

We Think

What method could you use to determine the volume of the solid block pictured above?

Participate in a class discussion.

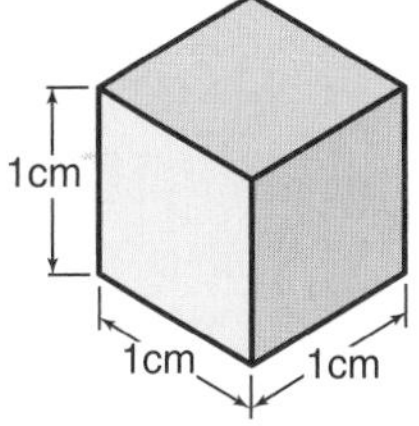

This cube has a volume of $1\text{ cm} \times 1\text{ cm} \times 1\text{ cm} = 1\text{ cm}^3$.

Standard Units of Volume

The volume (space occupied) of an object is the number of standard-unit cubes that fit inside the object.

For small objects, the standard unit of volume is a cube one centimeter (cm) on each side. The space occupied by the 1-cm unit cube is one cubic centimeter (abbreviated 1 cm^3).

To get a feeling for how much space a 1-cm cube occupies, your teacher will pass out some standard-unit cubes. Each cube is 1 cm on each side.

For regular solids like cubes, rectangles, cones, spheres, and cylinders, there are mathematical equations for calculating volume.

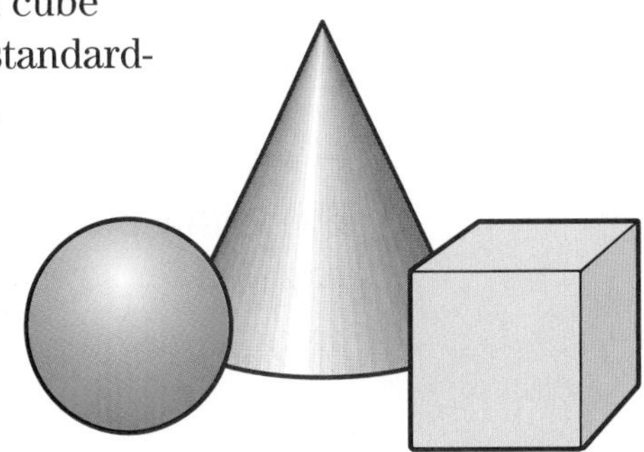

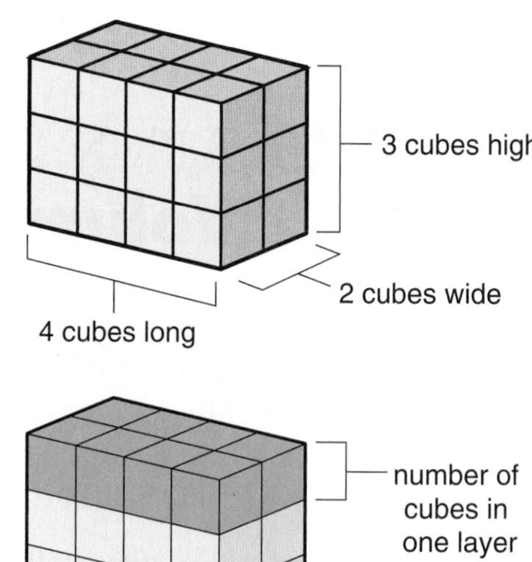

Explore Your Ideas

Your team will need:

- set of standard-unit cubes (1 cm on each side)
- 2 rulers
- access to a calculator

Exploration 1: How is using an equation to calculate the volume of a rectangular solid *the same* as counting the number of standard-unit cubes that fit inside the solid?

STEP 1 Use the standard-unit cubes to build a solid that is 4 cubes long, 2 cubes wide, and 3 cubes high, as shown in the diagram.

STEP 2 Count the number of cubes in one layer.

Record this number on your record sheet in Table 1.

Table 1: Counting and Calculating

Counting 1-cm cubes	Cubes in One Layer		Number of Layers	Volume (cm^3)
				____ cm^3
Using an Equation	**Width (cm)**	**Length (cm)**	**Height (cm)**	**Volume (cm^3)**
	____ cm^3	____ cm^3	____ cm^3	____ cm^3

STEP 3 Count the number of layers of unit cubes.

Record this number in Table 1 on your record sheet.

STEP 4 The volume of the rectangular solid is the sum of the cubes in each layer.

In this case, each layer contains the same number of cubes. So instead of adding, you can multiply to get the total number of unit cubes.

$$\text{Volume of rectangular solid} = \text{length} \times \text{width} \times \text{height}$$

Record the volume in Table 1 on your record sheet.

STEP 5 A mathematical equation for a volume is just a shorthand way of counting the number of unit cubes that fit inside a regular solid. For example, for a rectangular solid:

$$\begin{aligned}\text{Volume of rectangular solid} &= (\text{unit cubes in one layer}) \times (\text{number of layers}) \\ &= [(\text{number of cubes long}) \times (\text{number of cubes wide})] \times (\text{number of layers}) \\ &= \text{length} \times \text{width} \times \text{height}\end{aligned}$$

Use your ruler to measure the length, width, and height of your rectangular solid. Record the measurements to the nearest tenth of a centimeter; do not round off to the nearest centimeter. For example, you would record 5.1 cm, not just 5 cm. Have your partner check your measurements.

Record your measurements in Table 1 on your record sheet.

STEP 6 Use the equation to calculate the volume of the rectangle. Have your partner check your calculation.

Record the volume of the solid in Table 1 on your record sheet.

Participate in a brief class discussion about how using the volume formula for a rectangular solid is the same as counting the number of unit cubes that fit inside the solid.

Your team will need:

- 2 metric rulers
- 2 solids, 1 cube and 1 rectangular solid
- access to a calculator

Exploration 2: Using an Equation to Calculate the Volume of a Solid

STEP 1 So that you can make measurements at the same time, one team member should start with the cube and another member should start with the rectangular solid. Measure the length, width, and height of your object to the nearest tenth of a centimeter. Make sure you don't measure the same side of your object *twice*.

On your record sheet, record your measurements in a table similar to Table 2.

STEP 2 Use the equation to calculate the volume of your object. Have your partner check your calculation. (For a cube, the length, width, and height are the same.)

Record the volume of your object in Table 2.

STEP 3 Have each team member repeat Steps 1 and 2 for each solid.

Record your team's measurements and calculations in Table 2.

Table 2: Measured Volumes of Solids

Solid Cube	Width (cm)	Length (cm)	Height (cm)	Volume (cm^3)
Team Member 1	____ cm	____ cm	____ cm	____ cm^3
Team Member 2	____ cm	____ cm	____ cm	____ cm^3
Team Member 3	____ cm	____ cm	____ cm	____ cm^3
Team Member 4	____ cm	____ cm	____ cm	____ cm^3
			Volume Best Value:	____ cm^3
			Uncertainty:	____ cm^3
Rectangular Solid	**Width (cm)**	**Length (cm)**	**Height (cm)**	**Volume (cm^3)**
Team Member 1	____ cm	____ cm	____ cm	____ cm^3
Team Member 2	____ cm	____ cm	____ cm	____ cm^3
Team Member 3	____ cm	____ cm	____ cm	____ cm^3
Team Member 4	____ cm	____ cm	____ cm	____ cm^3
			Volume Best Value:	____ cm^3
			Uncertainty:	____ cm^3

STEP 4 Calculate the best value and uncertainty for the volume of each solid. If necessary, refer to *How To Make and Interpret Experiment Measurements* in the Appendix.

Record your volume best values and uncertainties in Table 2 on your record sheet.

Participate in a brief class discussion about your measurements and calculations to see if your volume and uncertainty values are reasonable.

Some of you may remember that the area of a rectangle is its length times its width.

Area of rectangle = length × width

Since the base of a rectangular solid is a rectangle, you can write the equation for the volume of a rectangular solid in a slightly different way.

Volume of rectangular solid = (length × width) × height
= area of base × height

This relationship for measuring volume can also be used for other solids. The base of a cylindrical solid is a circle, so the volume = area of circle × height.

Make Sense of Your Ideas

Think about what you did in the two explorations in this activity. How did you measure the volume of cubes and rectangular solids?

Participate in a class discussion about your answer.

Our Consensus Ideas

The key question for this activity is:

How are the volumes of cubes and rectangular solids measured?

Record the class consensus ideas on your record sheet.

DEVELOPING OUR IDEAS

Activity 3
Volume of Liquids

Purpose

In Activity 2 you learned how to measure the volume of rectangular solids. If you have a rectangular aquarium, you can calculate the volume of water you need to fill it. Suppose instead, the container was shaped like a cylinder.

What problem will you have when trying to use the same method of fitting standard-unit cubes that you used for the rectangular container?

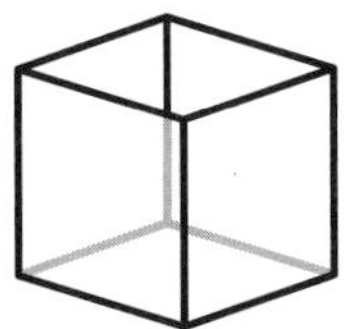

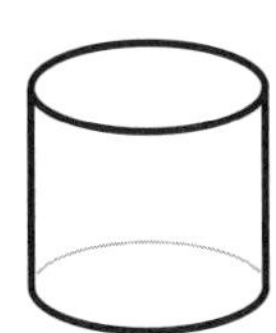

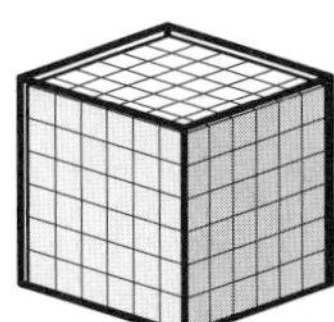

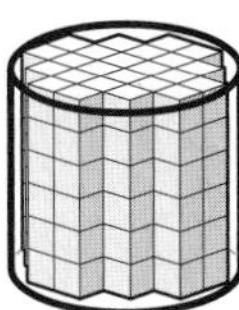

In this activity, you will explore a new method for measuring the volume of liquids that is accurate for measuring the capacity of containers of any shape.

The key question for this activity is:

How is the volume of liquids measured?

Record the key question on your record sheet.

We Think

Look at the two containers in the diagram.

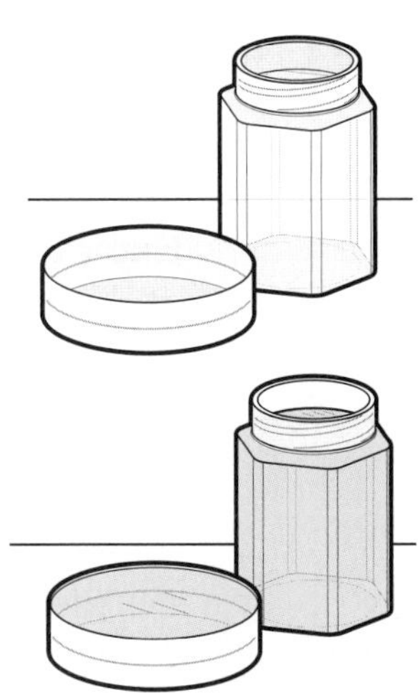

1. Imagine that the containers were filled with water. How do you think the volumes of water in the two containers compare? Do you think one container has a larger volume than the other? Write your reasons.
2. How could you measure the volume of water in each container?

Participate in a class discussion.

Explore Your Ideas

Graduated Cylinders

If you piled 1000 one-centimeter cubes to form a larger unit you would have 1000 cm^3, a unit called a cubic decimeter. This unit is also called a liter. So, 1 cm^3 is one-thousandth of a liter. Another name for 1 cm^3 is one milliliter (1 mL). The prefix *milli* means one-thousandth.

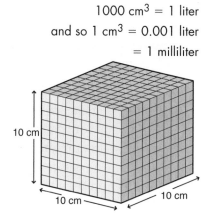

1000 cm^3 = 1 liter
and so 1 cm^3 = 0.001 liter
= 1 milliliter

10 cm
10 cm
10 cm

Follow standard safety rules for using glassware and handling spills.

Imagine you have a plastic container that holds exactly 100 standard 1-cm cubes. You fill the container with water and pour the water into a glass cylinder. The cylinder now contains 100 cm^3 or 100 mL of water. You could mark the 100 mL level. With a ruler, you could divide the height into 100 equal lengths. Each mark would represent 1 mL. See the diagram below.

Science Words

graduated cylinder: a tall, thin tube, marked off in units, used to measure volume of a liquid

A marked cylinder is called a **graduated cylinder**. Graduated means "marked off" or "divided up." You can use a graduated cylinder to measure the volume of water in any container. You pour the water from the container into the graduated cylinder, then read the mark next to the top of the water.

How To Use a Graduated Cylinder describes how to read the top of the water in a graduated cylinder. Be sure you have read this *How To...* and answered the questions before you continue.

Your teacher may also review some safety rules in *How To Follow Safety Rules during an Experiment.*

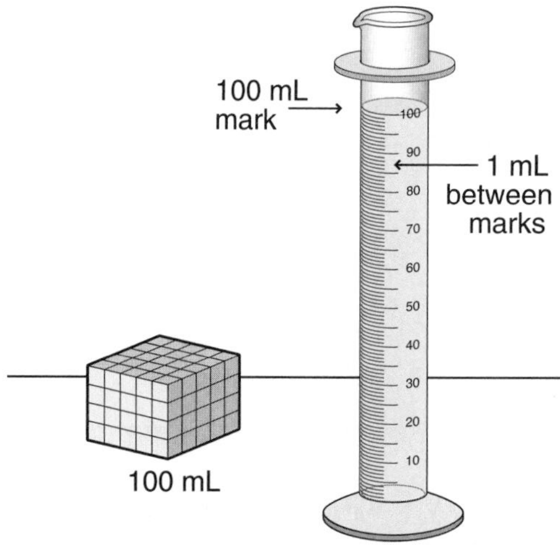

Your team will need:

- small hexagonal jar
- flat metal tin
- 100-mL graduated cylinder
- 50-mL graduated cylinder
- tub
- large beaker or bottle of water
- funnel
- eyedropper

Exploration 1: Volume and Height of Water in Graduated Cylinders

STEP 1 Place the large graduated cylinder in the tub. Pour water through the funnel to just below the 50-mL mark of the graduated cylinder. Use an eyedropper to fill the graduated cylinder to exactly 50 mL, measuring the volume. When measuring, move your head down so that your eye is at the same level as the water level. See the diagram on the next page. The bottom of the curve should line up at 50 mL.

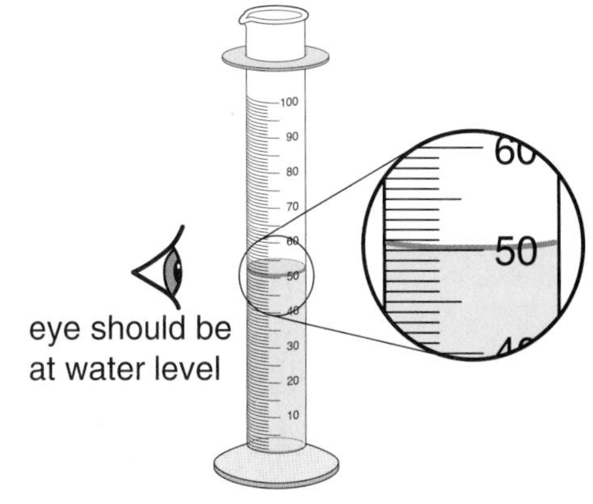

STEP 2 Using the funnel to prevent spills, pour the water from the large graduated cylinder into the small graduated cylinder. Measure the volume of water in the small graduated cylinder.

1. How is it that the small graduated cylinder holds the same volume of water as the larger cylinder does, even though the water level in the small cylinder is much higher than it was in the large cylinder?

Exploration 2: What is the volume of liquid in two containers?

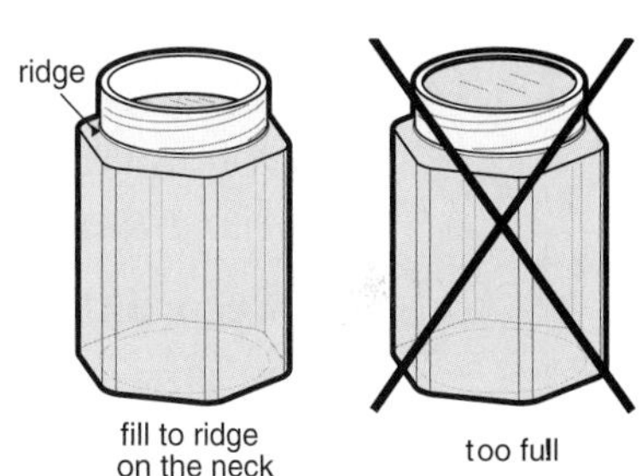

STEP 1 Place the hexagonal jar in the tub. Using a funnel, fill the jar with water to just below the neck. Use the eyedropper to fill the jar to the middle of the ridge on the neck.

STEP 2 Place a graduated cylinder in the tub. Place the funnel in the cylinder. *Carefully* lift the jar so that no water spills. Pour the water into the graduated cylinder through the funnel. Measure the volume of the water.

Table: Measured Volumes of Containers		
	Volume of Hexagonal Jar (mL)	Volume of Flat Metal Tin (mL)
Team Member 1		
Team Member 2		
Team Member 3		
Team Member 4		
Team Average		
Uncertainty		

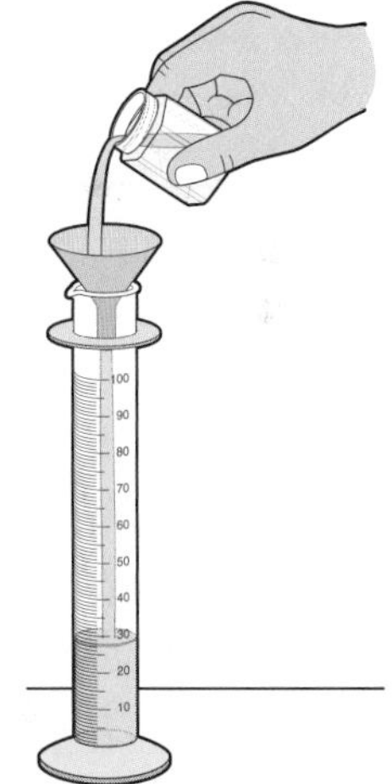

Record your measurement in the correct column of your data table on your record sheet.

STEP 3 Pour the water from the cylinder back into the large beaker.

STEP 4 Each team member then repeats Steps 1 to 3 to measure the volume of the hexagonal jar.

STEP 5 Repeat Steps 1 to 4 to measure the volume of the flat metal tin. To do this, fill the tin all the way to the rim. Lift it very carefully and pour the water through the funnel into the graduated cylinder.

Record your team's measurements in your data table.

Science Words

outliers: values far from most others in a set of data

STEP 6 After all team members have made careful measurements of the liquid volume for each container, *check for obvious mistakes* that cause **outliers**. (Outliers are values far from most others in a set of data.) The volume measurements may be slightly different, but if there is an obvious outlier, that volume measurement should be redone.

STEP 7 To find the best value of the liquid volumes, calculate the average liquid volume for each container. Each team member should compute the average volume of liquid in each container.

Record your team's best values in your data table.

STEP 8 Each team member should also compute the uncertainty in the four volume measurements for each container. Use the formula in *How To Make and Interpret Experiment Measurements.*

Record the *uncertainty* for each container in your data table.

2. How close or far apart are the measured volumes of the two containers? Compare this result to what you expected.

Make Sense of Your Ideas

In this activity and in Activity 2, you:

- measured the volume of a cube and a rectangular solid by measuring the lengths of three sides and using a mathematical formula
- measured the volume of water in two different containers by pouring the water into a graduated cylinder.

Discuss the following questions with your team and record their answer.

1. How would you decide whether to use the calculation method or the liquid-pouring method to measure volume?

2. What could lead to errors in volume measurements made using the formula method? The liquid-pouring method?

Participate in the class discussion about your answers.

Our Consensus Ideas

The key question for this activity is:

How is the volume of liquids measured?

1. Write your answer to the key question.

Participate in the class discussion about your volume estimates.

2. Record the class consensus ideas on your record sheet.

Activity 4
Measuring Mass

Purpose

You learned in the last two activities that one way to measure the amount of material is to measure the volume of an object. Another way to measure the amount of material is to measure the **mass** of an object. The more mass an object has, the more matter it has and the heavier it is. This activity has three key questions. In the first part of the activity, you will learn about mass, and answer the first two key questions:

Science Words

mass: the amount of matter that a body contains

1. How is mass measured?
2. How is mass different from volume?

There are two properties related to the amount of material of any *solid* or *liquid* object – its volume (the space it occupies) and its mass. What about *gases*, like the air you breathe?

In the second part of this activity, you will find out whether mass is also a property of gases, and you will answer the third key question:

3. Do gases have mass?

Record the key questions on your record sheet.

You use gases every day. The air you breathe is a gas. Party balloons are sometimes filled with helium gas. Some appliances for the home, like hot water heaters, furnaces, and kitchen stoves, burn methane gas. Firefighters wear gas masks to avoid breathing the gases produced when wood and other solids burn.

Part A: Developing Ideas about Mass

The mass of an object is measured by using an equal-arm balance. The mass of an object is the number of standard-unit masses that balance the object.

The standard unit of mass you will be using in *InterActions in Physical Science* is the gram (g). A gram is one-thousandth of a kilogram (kg), which is the unit of mass scientists and many countries use to measure the mass of large objects such as people.

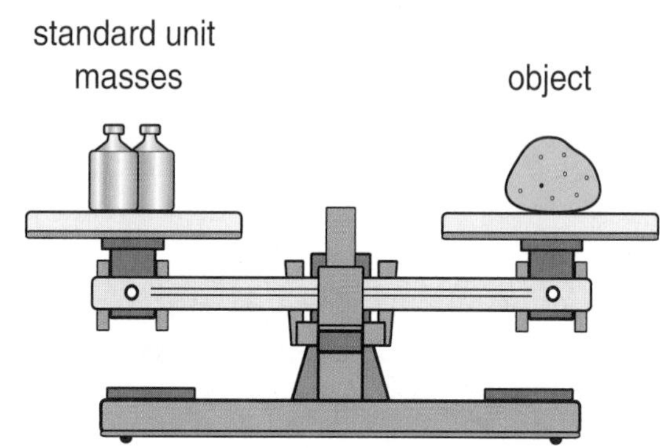

Many commercial balances have standard masses built into the balance. For example, the large balance in the doctor's office for measuring your mass has standard masses built into it. Some of these balances actually measure weight. Weight is not the same as mass although they are related. If you know one, you can figure out the other. In Unit 3, you will study weight. In this cycle, however, we will use the term mass.

To give you a feeling for how much mass a gram is, your teacher will pass out standard-unit cubes. The standard-unit cubes are specially designed not only to have a volume of *exactly* 1 cm^3 (or 1 mL), but also to have a mass of *exactly* 1 g.

We Think

Discuss the question below with your team.

The standard-unit cubes are examples of objects that all have the same volume and the same mass. Do you think that objects with the same volume always have the same mass? Write your reasoning.

Participate in a class discussion about your answer.

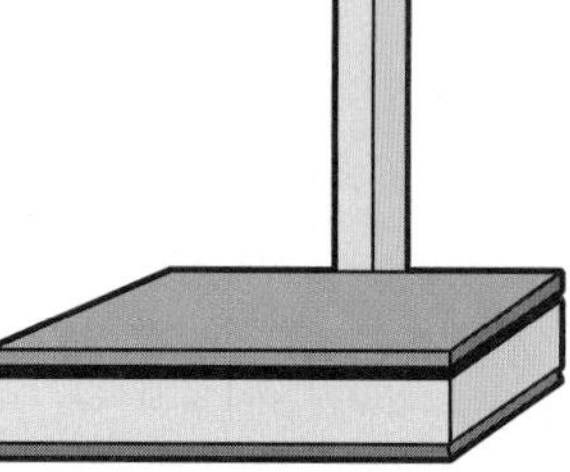

Explore Your Ideas

STEP 1 Each team will measure the mass of three cubes made of different types of material.

Fill in the material (such as wood or copper) that the objects are made of in the top row of Table 1 on your record sheet.

Your team will need:

- 3 cubes about the same size (volume), each one made of a different substance
- mass balance

Table 1: Measured Masses of Solid Objects			
	Mass of ____ Cube (g)	Mass of ____ Cube (g)	Mass of ____ Cube (g)
Team Member 1			
Team Member 2			
Team Member 3			
Team Member 4			
Best Value (average)			
Uncertainty			
Class Average Uncertainty:			

To use the mass balances, follow the instructions on *How To Use Mass Balances.* Treat the balances with care to avoid damaging them!

STEP 2 Each team member in sequence should use the balance to measure the mass of each of the three cubes. Measure as *carefully* and *accurately* as you can.

If possible, measure the mass precisely to the nearest tenth of a gram. Do not round off to the nearest gram. For example, you would record 37.2 g, not just 37 g.

After measuring the mass amounts, take the cube off the balance and reset the balance to zero.

Record both your *individual* measurements and those of your team members in rows 2 through 5 of Table 1 on your record sheet.

STEP 3 For the mass balance, an unavoidable measurement error may come from rubbing between its moving parts. Even though the pointer may be precisely lined up on the balance, the parts might have "stuck" in a slightly different position for each mass measurement.

Each team member should compute the best value and the uncertainty in the four mass measurements for each cube. If necessary, refer to *How To Make and Interpret Experiment Measurements.*

In your data table, record the volume best value and uncertainty for each cube.

STEP 4 Post the uncertainty for each cube in a class data table.

Use the class data table to calculate your class's average uncertainty in the mass measurement. You will use this uncertainty in mass for later activities.

Record the class average uncertainty at the bottom of your table.

Make Sense of Your Ideas

With your team, examine the results. Then discuss and answer the following:

1. What are some mistakes that are avoidable when you measure mass?
2. If you make no mistakes, what are some unavoidable sources of uncertainty in measuring the mass of an object?
3. The three objects you measured had approximately the same *volume.* Why do you think they differ so much in mass?
4. Can you think of two objects that have the same mass but different volumes?
5. Which has more mass – a kilogram of sand or a kilogram of Styrofoam®? Why?

Participate in a class discussion about your answers.

Part B: Developing Ideas about Gases

We Think

Suppose you measure the mass of a soft soccer ball (start mass). Then you pump air, which is a gas, into the ball until the ball is firm and measure the mass again (end mass).

What do you think will happen to the mass of the ball? Why?

Participate in a class discussion of your prediction and reason.

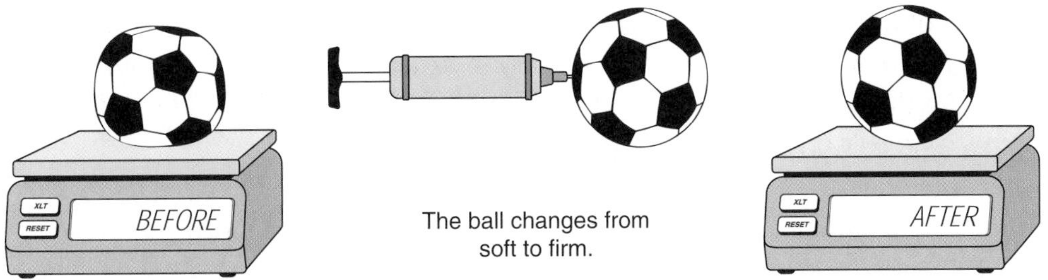

BEFORE
The ball changes from soft to firm.
AFTER

Explore Your Ideas

Your teacher will do the soccer-ball exploration.

Record the measurements in Table 2 on your record sheet.

Table 2: Pumping Air Into a Soccer Ball		
Before Air Pumped into Soccer Ball (g)	After Air Pumped into Soccer Ball (g)	Change in Mass (g) (After — Before)

Did the mass of the soccer ball *increase*, *decrease*, or *stay the same* after your teacher put air into the soccer ball (taking into account possible uncertainty)? Include your evidence.

Make Sense of Your Ideas

The third key question for this activity is: Do gases have mass? Does your conclusion provide supporting evidence that gases have mass or provide evidence that gases do not have mass? Explain.

Participate in a class discussion about your answers.

Our Consensus Ideas

The first key question in this activity is:

1. How is mass measured?

1. Write your best answer for this key question on your record sheet.

Participate in a class discussion about your answers.

2. Write the class consensus ideas about the first key question: How is mass measured?

Mass and volume are similar because they are both used to describe the *amount of material* of an object. The second key question in this activity is:

2. How is mass different from volume?

3. Write your best answer for this key question on your record sheet. In answering this question, look back over the definitions of mass and volume. Think about how these two definitions are different from each other.

Participate in a class discussion about your answers.

4. With the guidance of your teacher, write the class consensus ideas about the second key question: How is mass different from volume?

The third key question in this activity is:

3. Do gases have mass?

5. Write your answer for this key question on your record sheet.

Participate in the class discussion about your answers.

6. Write the class consensus ideas about the third key question: Do gases have mass?

Activity 5
Density

Purpose

In the previous activities, you measured two properties related to the amount of material of a solid or liquid object – its volume (the space it occupies) and its mass. However, the volume and mass of objects do not tell you about the *kind of materials* you have. The key question you will explore in this activity is:

What property can help you decide what kind of material an object is made of?

Record the key question on your record sheet.

We Think

Two students want to construct a model airplane that will fly when launched. They have two different wooden materials available. They need to decide which one is better. For this purpose, they take a *small* cube of the first material and a *bigger* cube of the second material. They measure the masses of the cubes. They find that the mass of the first cube is 2 g and the mass of the second one is 7 g.

When trying to decide which material is best, the two students have different opinions.

Since lighter airplanes will fly better, the first material is better because it has less mass.

Carlos

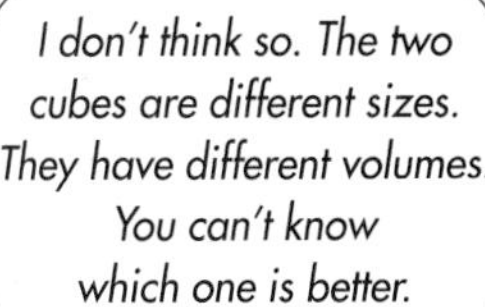

I don't think so. The two
cubes are different sizes.
They have different volumes.
You can't know
which one is better.

Nadia

1. Do you agree with Carlos, Nadia, or neither of them? Why? Discuss with your group. Record your answers on your record sheet.

Participate in a class discussion about your answers.

Table 1: Mass of Cubes			
Cube Material*	**Mass † (g)**	**Cube Material***	**Mass † (g)**
Aluminum	44.9 g	Milky Plastic (Nylon)	19.7 g
Brass	139.9 g	Oak Wood	12.7 g
Clear Plastic (acrylic)	16.2 g	Pine Wood	10.1 g
Copper	147.2 g	Poplar Wood	7.9 g
Gray Plastic (PVC)	24.2 g	Steel	127.8 g

* *Cubes have approximately the same volume (15 - 16 cm³).*
† *Uncertainty is 0.1 g.*

Equal Volumes of Different Materials

In Activity 4, each team measured the mass of three cubes of approximately the same volume. Another class used a similar set of cubes. The measurements of the mass of these cubes are shown in Table 1.

Use Table 1 to answer the following question.

2. Do equal volumes of different materials have the same mass or different masses? Be sure to include your evidence.

Science Words

density: the mass of per standard unit of volume

Scientists use the relationship between mass and volume to define a property that can help you decide what kind of material an object is made of. Suppose you carefully measured the mass of a standard unit of volume of different materials. This property is called **density**. Equal volumes of the same material have the same mass, and equal volumes of different materials have different masses. The standard unit of volume commonly used in determining density is a one-centimeter cube (1 cm^3).

This *Table of Densities* shows the mass (in grams) of 1 cm^3 of different solids, liquids, and gases.

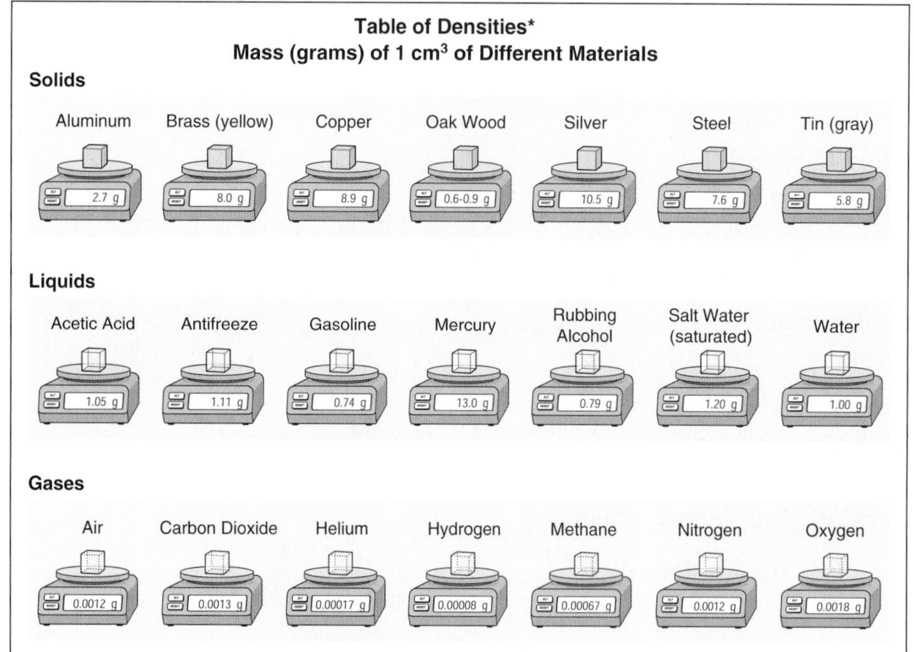

1 cm^3 of aluminum

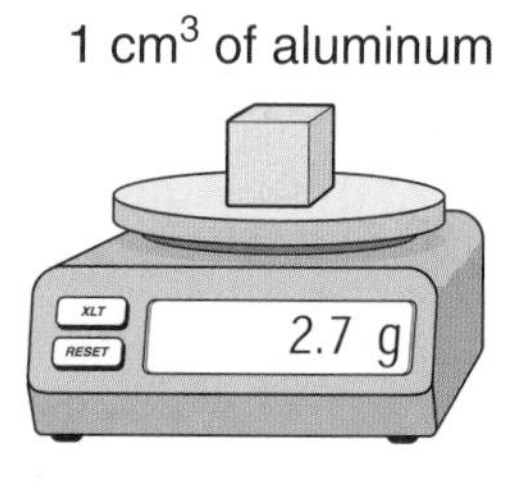

* *Approximate values at sea level and 20°C.*

Explore Your Ideas

The density of an object – the mass per standard unit of volume – helps you decide what kind of material an object is made of. Density is called a **characteristic property**. All characteristic properties of materials are measurements (numbers) that are different for different kinds of materials.

Science Words

characteristic property: a measurement (numbers) that is different for different kinds of materials

Table 2: Mass of Board-Game Cubes

Cube	Mass of 1cm³ Cube (g)	Type of Material
# 1	8.8	
# 2	5.8	
# 3	2.8	

STEP 1 Suppose a board game comes with three cubes that are one centimeter on each side. The cubes have a thin coat of paint, so you cannot tell what materials the cubes are made of. You carefully measure the mass of each cube. You record the mass in a table similar to Table 2. The uncertainty in your measurements is 0.2 g.

What material is each cube made of? Use the Table of Densities on the previous page. Record your answers in the table on your record sheet.

Participate in a class discussion about the answer to this question.

STEP 2 Jen has four painted solid (not hollow) cylinders that are the same size. She carefully measures the mass of each cylinder, as shown in the diagram. The uncertainty in her measurements is 0.2 g.

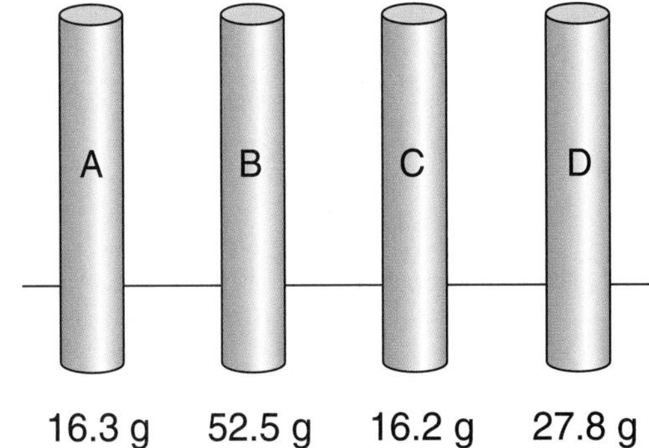

Discuss the following questions with a partner. Then record your best answers.

1. Are any cylinders made of the same materials? Explain your reasoning.
2. Which cylinders are made of different materials? Explain your reasoning.

Participate in a class discussion about the answers to these questions.

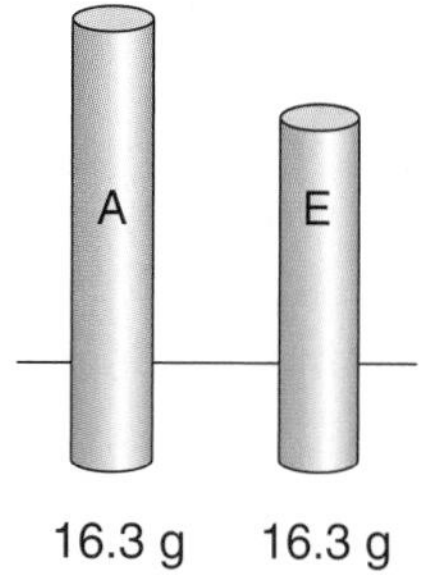

STEP 3 Jen has a fifth painted cylinder E that is the same mass as cylinder A, but a smaller volume.

Discuss the following question with a partner. Then record your best answer.

3. Are cylinders A and E made of the same material or different materials? Explain your reasoning.

Make Sense of Your Ideas

Imagine that you have two metal blocks, both the same silvery color. How could you decide what kind of material each object is made of? To tell what kind of material an object is made of, you need to test for its properties.

Most properties of objects (like mass, volume, magnetic or nonmagnetic, or electrical conductor or nonconductor) do not help you decide what kind of material an object is made of.

1. Why do the two properties, magnetic or nonmagnetic and electrical conductor or nonconductor, *not* help you decide if a metal object is made of aluminum, tin, silver, or titanium?

2. Why does the mass of a metal object *not* help you decide if it is made of aluminum, tin, silver, or titanium?

3. Describe what you would do to decide what the metal blocks were made of.

Our Consensus Ideas

The key question for this activity is:

What property can help you decide what kind of material an object is made of?

1. Answer the key question for this activity.

Participate in a class discussion about the answer to this question.

2. Write the class consensus on your record sheet.

Activity 6
Characteristic Properties

Purpose

In Activity 5 you learned about density. Density is the mass of a standard unit of volume of a material. Density is called a characteristic property because it helps you decide what kind of material an object is made of.

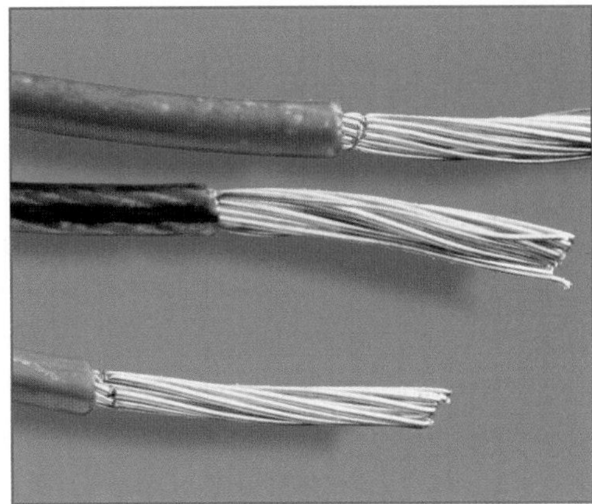

Some materials are electrical conductors, while others are nonconductors.

The property of electrical conductor or nonconductor you learned about in Cycle 2 is not a characteristic property of materials. For example, suppose you have two metal wires that look alike. They have the same silvery color. In a circuit, both wires allow a bulb to light. The simple test of placing the wire in the circuit does not help you decide what kind of metal a wire is made of.

In this activity, you will explore the following key question:

What is a characteristic property of wires?

Record the key question on your record sheet.

We Think

Imagine you put together a circuit with a cell, a bulb, and a piece of aluminum, as shown in the diagram. You observed the brightness of the bulb.

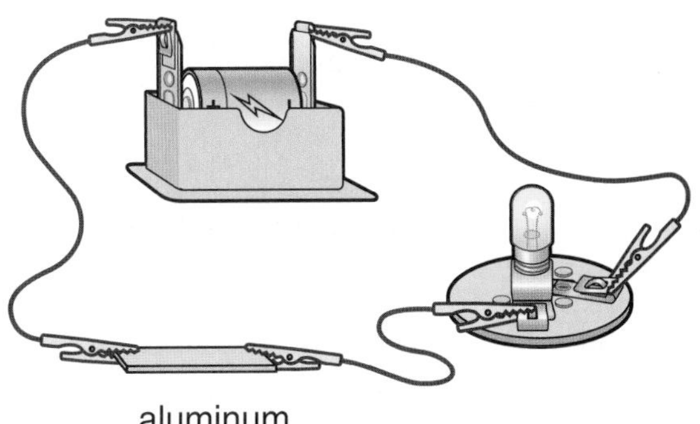

If you replaced the piece of aluminum with a piece of a different metal that had the exact same size, would the bulb glow more brightly, less brightly, or exactly the same? Why do you think so?

Participate in a class discussion.

Explore Your Ideas

Exploration 1: What is the relationship between the length of a wire and the brightness of a bulb in a circuit?

Your team will need:

- 2 cells in holders
- round-shaped bulb in socket
- switch
- hook-up wires
- meter stick
- bare nichrome wire
- bare copper wire
- tape

You will test two different wires, a nichrome wire and a copper wire. The manipulated variable is the length of the wire between the two clips. The responding variable is the brightness of the bulb.

Your teacher will assign roles for each team member. Review your tasks as a Supply Master, Procedure Specialist, Team Manager, or Recycling Engineer.

STEP 1 Set up the circuit shown in the diagram at the right with the nichrome wire taped down. Make sure the cells are connected properly as shown.

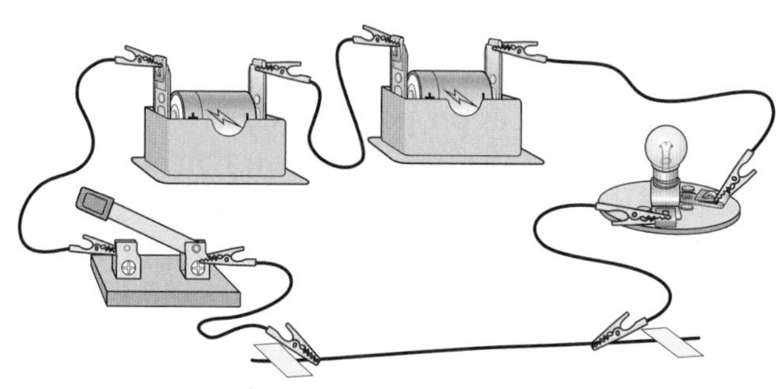

40 cm of nichrome wire

STEP 2 Use the meter stick and move the clips so they are 40 cm apart. Then close the switch.

STEP 3 Now move the right clip slowly closer and closer to the left clip until they are almost touching. Observe what happens (if anything) to the brightness of the bulb. Take turns moving the clip and observing the bulb.

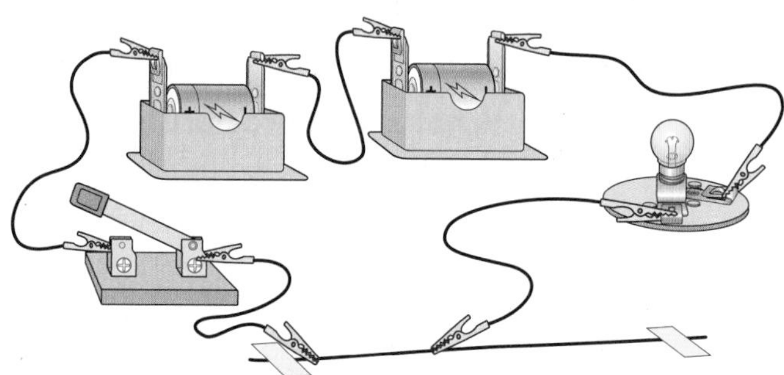

The length of wire between the clips is now shorter.

To Do

Team Member 1
Supply Master: gathers materials.

Team Member 2
Procedure Specialist: reads instructions and steps aloud.

Team Member 3
Team Manager: makes sure team stays on task and all team members are participating.

Team Member 4
Recycling Engineer: returns materials.

Record your observation in Table 1 on your record sheet.

Table 1: Length of Wire and Bulb Brightness

Wire Material	Brightness of the Bulb When the Length of Wire Decreases
Nichrome	
Copper	

STEP 4 Repeat Steps 1 through 3 with the copper wire.

Record your observation in Table 1 on your record sheet.

1. What is the relationship between the length of a nichrome wire between the connecting leads and the brightness of the bulb in the circuit? Be sure to include your evidence.
2. What is the relationship between the length of a copper wire and the brightness of the bulb in the circuit? Be sure to include your evidence.

Participate in a class discussion about Exploration 1.

Exploration 2: What is the relationship between the length of a wire and the electric current in a circuit?

In Cycle 2, Activity 4, you used a computer simulation to determine the effect of the number of cells in a circuit on the electric current. The simulator uses an ammeter to measure the electric current in a circuit.

When you recorded your data in Exploration 1, you may have used values like dim, bright, and really bright. These terms are not precise enough when determining the relationships between variables. This is why scientists develop measurement instruments.

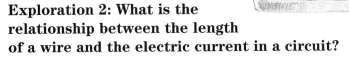

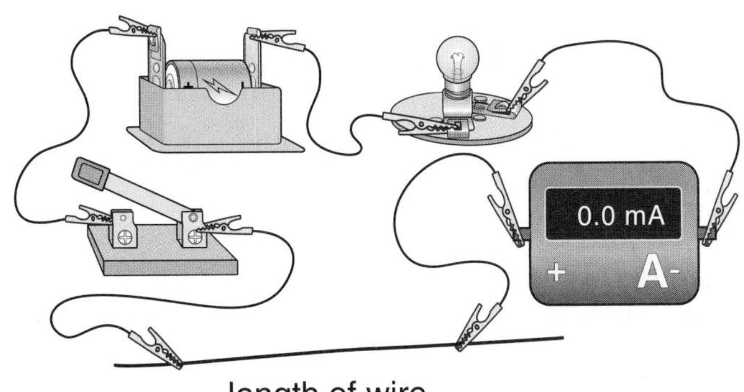

Table 2: Length of Wire and Electric Current*		
Wire Material	Wire Length (cm)	Electric Current (mA)
Nichrome	40	419
	20	591
	10	743
	5	853
Copper	40	979
	20	990
	10	995
	5	997

**All other variables that influence the interaction are kept constant.*

Assume an uncertainty of 1mA.

Table 2 shows the results of an exploration when an ammeter is used in the circuit to measure electric current. Discuss the following three questions with your team and record your team's answers.

3. What is the relationship between the length of a nichrome wire and the amount of electric current in a circuit? Include your evidence from Table 2.

4. What is the relationship between the length of a copper wire and the amount of electric current in a circuit? Include your evidence.

5. Does the amount of electric current in the circuit depend on the type of material of the wire? What is the evidence?

Participate in a class discussion about your answers to these questions.

Make Sense of Your Ideas

You did not see any difference in the brightness of the bulb when you changed the length of the copper wire in Exploration 1. This is because copper is a much better electrical conductor than nichrome. The bulb was so bright that you could not see the effect of the small changes in the electric current as you changed the wire length.

Characteristic properties are measurements (numbers) that are different for different kinds of materials. They can help you decide what kind of material an object is made of. In this activity, you explored how the kind of wire material (nichrome and copper) influences the amount of electric current in a circuit.

The electric-circuit interaction is influenced by many variables, not just the kind of material. For example, you found that three variables influence the electric-circuit interaction:

- the wire material
- the length of the wire in the circuit
- the number of cells in the circuit (Cycle 2, Activity 4).

A fourth variable, the thickness of the wire, also influences the strength of the electric-circuit interaction.

Suppose you design a circuit to determine a characteristic property of wires.

You want to compare the electric current in the circuit when you use different kinds of wire material. The ammeter measures the electric current. In order to ensure a fair test, you need to keep constant the length and thickness of the wire in the circuit and the number of cells in the circuit.

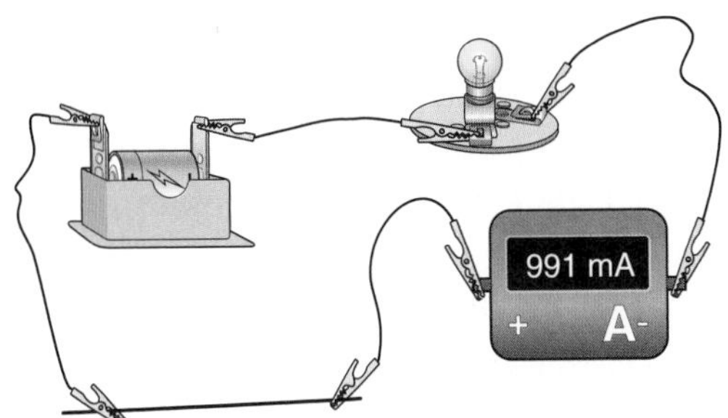

characteristic property experiment

Table 3: Electrical Conductivity of Different Wires*	
Wire Material	**Electric Current (mA)**
Aluminum	991
Brass	979
Copper	995
Nichrome	743
Steel	801
Tin	996

** All other variables that influence the interaction are kept constant.*

The results of this exploration are shown in Table 3. The characteristic property of wires being measured in this exploration is called electrical conductivity.

1. Which kind of wire material is the best conductor of an electric current? Justify your answer using information in Table 3.

2. Which kind of wire material is the worst conductor of an electric current? Justify your answer using information in Table 3.

Our Consensus Ideas

The key question for this activity is:

What is a characteristic property of wires?

1. Write your answer for this key question on your record sheet.

Participate in a class discussion about your answer.

2. Write the class consensus idea about the key question on your record sheet.

Your teacher will distribute a copy of *Scientists' Consensus Ideas: Properties of Objects and Materials.*

3. In the spaces provided on the form, write examples of activities and explorations in this cycle that provided practice in applying the ideas.

Go over the form with your teacher.

LEARNING ABOUT OTHER IDEAS

Activity 7
Calculating Density

Steel forms the "backbone" of much of our country. Bridges, skyscrapers, and some car frames are made of steel. How can you tell if a metal is steel?

Purpose

The Table of Densities in Activity 5 lists the masses of a unit volume of some solids, liquids, and gases (in picture form). For example, a unit volume (one cubic centimeter) of steel has a mass of 7.6 g. That means the density of steel is 7.6 g for every cubic centimeter.

Density is important because it is a characteristic property of a substance. In this activity, you will learn ways of determining the density of a sample of a substance, whatever its volume or mass. In this activity, the key question is:

How can you determine the density of an object?

Record the key question on your record sheet.

Learning the Ideas

The density of an object can be calculated from its mass and volume by using the following relationship:

$$\text{Density} = \frac{\text{Mass}}{\text{Volume}}$$

For small samples of solids, mass is often measured in grams (g) and volume in cubic centimeters (cm^3). The unit of density is then grams per cubic centimeter (g/cm^3). For example, steel has a density of 7.6 g/cm^3.

For liquids, the unit of volume most often used is the milliliter (mL), which is identical to the cubic centimeter. For example, water has a density of 1.00 g/mL or 1.00 g/cm^3.

The Table of Densities in this activity lists in a chart the densities in units of g/cm^3 or g/mL for the same solids, liquids, and gases shown in the table in Activity 5.

Calculating Density

As an example of calculating density, consider the following problem:

A small piece of shiny gray metal has a mass of 6.1 g and a volume of 0.58 cm^3. What is its density? What is the object made of?

$$\text{Density} = \frac{\text{Mass}}{\text{Volume}} = \frac{6.1\text{ g}}{0.58\text{ cm}^3} = 10.5\text{ g/cm}^3$$

The density is 10.5 g/cm^3 (grams per cubic centimeter). Look at the Table of Densities in this activity and find a material with that density. With the additional evidence that the material has a gray color, you can determine that this material is probably silver.

Table of Densities* (Remember that 1 mL = 1 cm^3)

Material	Density
Solids	
Aluminum	2.7 g/cm^3
Brass (yellow)	8.0 g/cm^3
Copper	8.9 g/cm^3
Oak Wood	0.6–0.9 g/cm^3
Steel	7.6 g/cm^3
Tin (gray)	5.8 g/cm^3
Silver	10.5 g/cm^3
Liquids	
Acetic Acid	1.05 g/mL
Antifreeze	1.11 g/mL
Gasoline	0.74 g/mL
Mercury	13.0 g/mL
Rubbing Alcohol	0.79 g/mL
Salt Water (saturated)	1.20 g/mL
Water	1.00 g/mL
Gases	
Air	0.0012 g/cm^3
Carbon Dioxide	0.0013 g/cm^3
Helium	0.00017 g/cm^3
Hydrogen	0.00008 g/cm^3
Methane	0.00067 g/cm^3
Nitrogen	0.0012 g/cm^3
Oxygen	0.0018 g/cm^3

* *Approximate values at sea level and 20°C.*

1. A large block of gray material has a mass of 270 g and a volume of 100 cm^3. What is its density? What is the object made of?

Calculating Mass

Sometimes you know what an object is made of and either its mass or volume, but you need to determine the other quantity. Consider this example.

Suppose you have a one-liter (1000 mL) container of gasoline. What is the mass of the gasoline?

To solve this problem, you first need the density of gasoline. From the Table of Densities in this activity you can determine that the density of gasoline is 0.74 g/cm^3 or 0.74 g/mL. You can then calculate the mass as follows:

Since *Density = Mass / Volume*, you can rearrange the equation to give you the mass as:

$$\text{Mass} = \text{Density} \times \text{Volume}$$

So in this case:

$$\text{Mass} = \text{Density} \times \text{Volume} = 0.74\text{ g/mL} \times 1000\text{ mL} = 740\text{ g}$$

2. What is the mass of the alcohol in a 500-mL bottle of alcohol?

Your teacher will review answers to this question with the class.

Calculating Volume

Now let's consider another example. Suppose you purchased a 45.0-g chunk of copper. What is its volume?

You can rearrange *Density* = *Mass* / *Volume* to solve for the volume as:

$$\text{Volume} = \frac{\text{Mass}}{\text{Density}}$$

So in this case, you would look up the density of copper. It is 8.9 g/cm^3. Then:

$$\text{Volume} = \frac{\text{Mass}}{\text{Density}} = \frac{45.0\text{ g}}{8.9\text{ g/cm}^3} = 5.06\text{ cm}^3$$

3. What is the volume of a 120-g piece of aluminum?

Your teacher will review answers to this question with the class.

Calculating Density of a Liquid from Measurements of Mass and Volume

You often need to measure both the mass and volume of an unknown material in order to determine its density. Consider the following situation. You have a container with a liquid in it and you want to identify the liquid. You can measure its mass and volume as follows:

Place the container with the liquid on an electronic mass balance and read its mass in grams.

Pour all the liquid into a 100-mL graduated cylinder. The diagram at the left shows the level of the liquid.

4. What is the volume of the liquid?

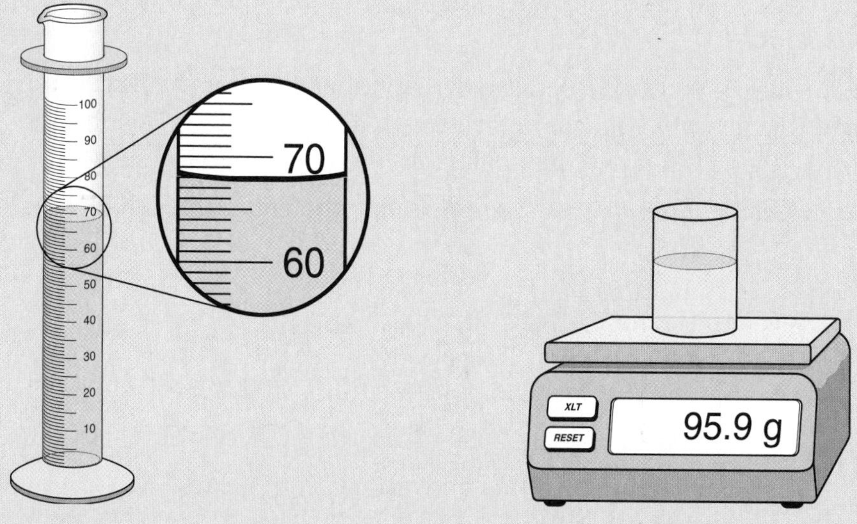

Finally, place the empty container on the mass balance.

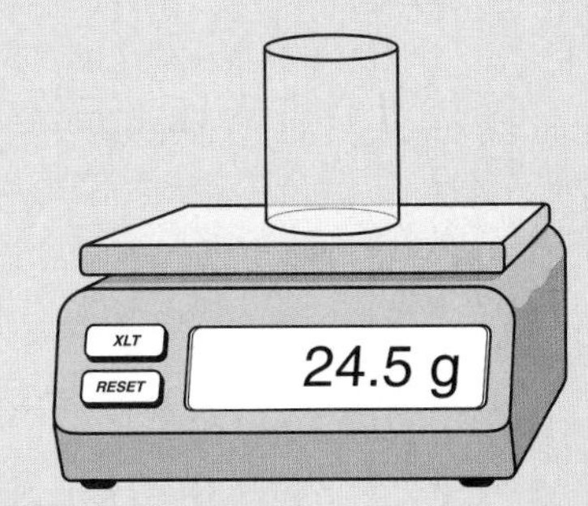

5. What is the mass of the liquid?
6. What is the density of the liquid?
7. What might the liquid actually be?

Your teacher will review answers to these questions with the class.

Calculating Density of an Irregular Solid

You learned in Activity 3 that if you are given a rectangular solid, you could measure its width, length, and height and calculate its volume by multiplying the three measurements together.

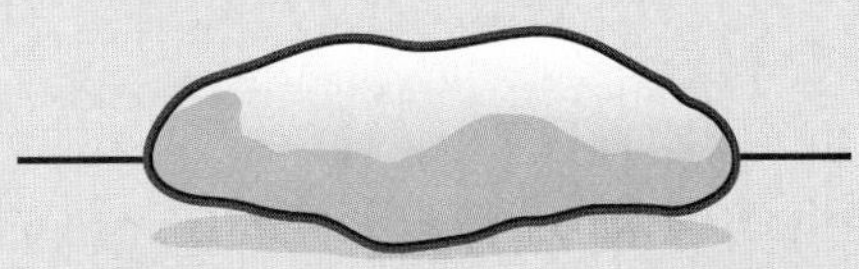

However, how do you measure the volume of a solid that is irregular in shape like the one shown in the diagram at the right?

A common way of doing this if the object is not too large is to use the method of **water displacement**. To use this method, you would choose a graduated cylinder large enough for the object to fit in without taking up more than about one-third of the inside volume of the cylinder. Then you would fill the graduated cylinder about halfway up with water and measure the volume of the water precisely. See the diagram below.

Science Words

water displacement: a method used to measure the volume of an object by measuring the amount of water that it displaces

8. What is the volume of the water?

You would then place the solid into the graduated cylinder as shown on the next page. It must completely sink in the water. If it does not sink, the water displacement method will not work. When the solid sinks to the bottom, the water level in the cylinder will rise. The solid is displacing, or pushing away, the water. Next you would measure the combined volume of the water and the solid.

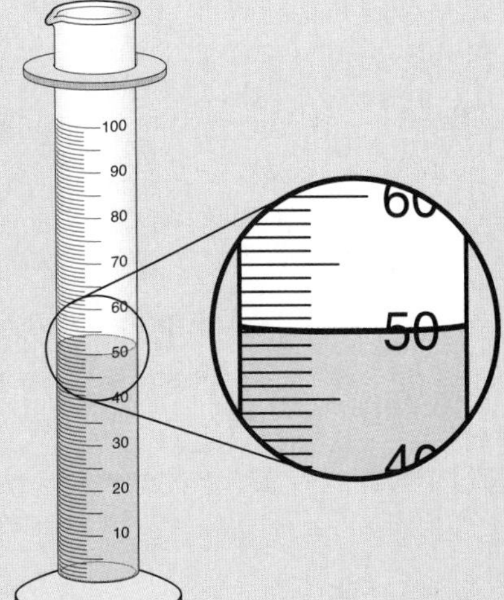

If you try this, be careful as you put the solid object into the graduated cylinder.

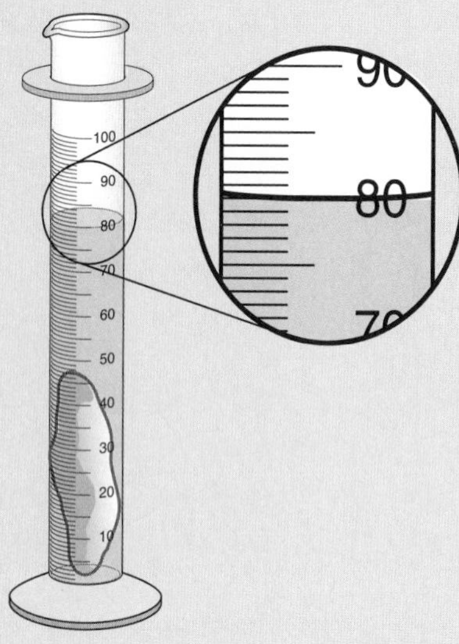

9. What is the volume of the water and the solid?

The difference between this combined volume and the volume of water by itself equals the volume of the solid.

Volume of solid = Volume with solid and water − Volume with water only

10. What is the volume of the solid?

Finally, you would place the object on an electronic mass scale and measure its mass.

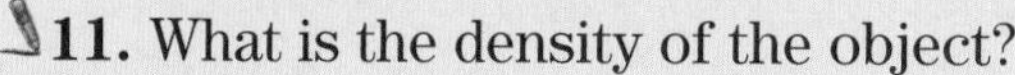

11. What is the density of the object?

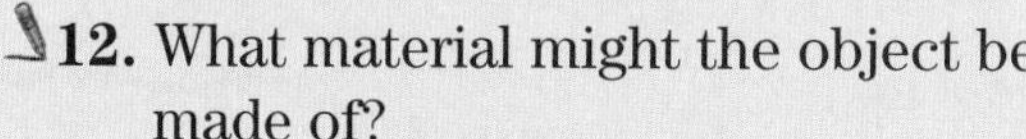

12. What material might the object be made of?

Your teacher will review the answers to these questions with the class.

What We Have Learned

Remember the key question for this activity:

How can you determine the density of an object?

Answer the key question on your record sheet.

Participate in the discussion in which your class reviews the answer to the key question.

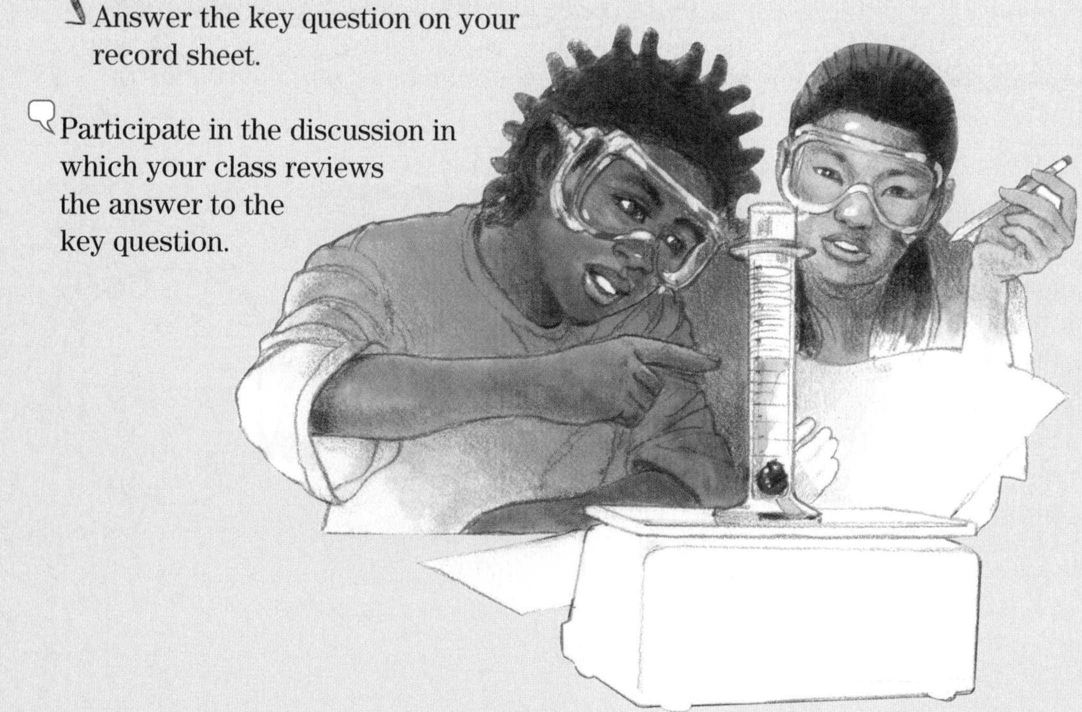

UNIT 2

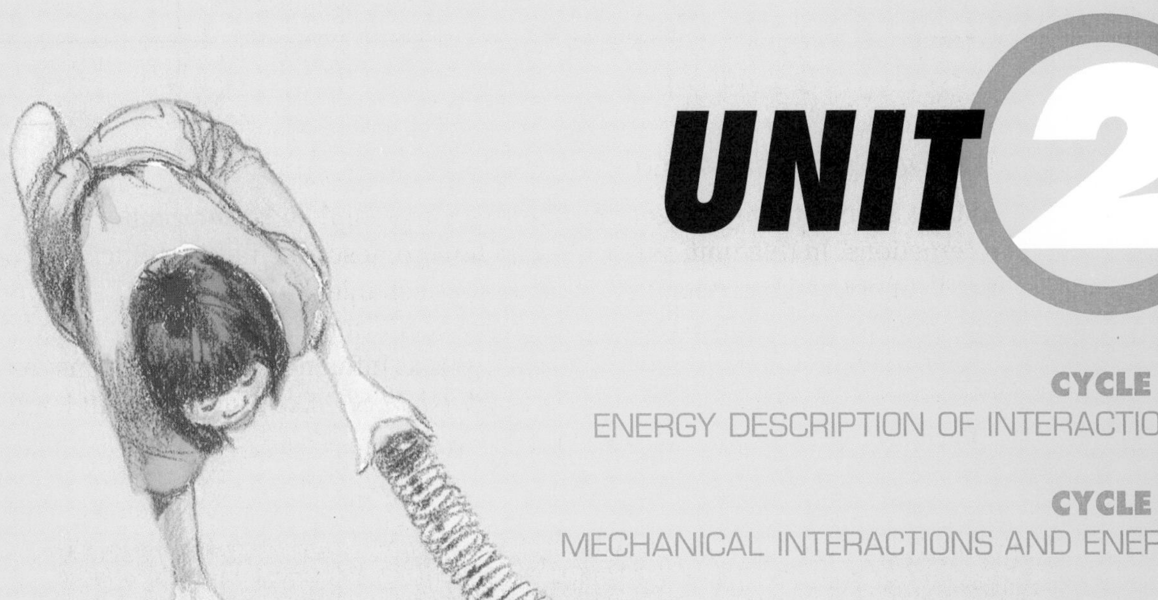

CYCLE 1:
ENERGY DESCRIPTION OF INTERACTIONS

CYCLE 2:
MECHANICAL INTERACTIONS AND ENERGY

INTERACTIONS AND ENERGY

What will you learn about in Unit 2?

In Unit 1, you studied magnetic, electric-charge, and electric-circuit interactions. In this unit, you will learn how to describe these interactions in terms of energy. You will also learn about mechanical interactions. Each type of interaction is unique. The changes that occur during the interactions can be large or small, depending on the variables that influence the interactions. As you work through the *InterActions in Physical Science* units, you will learn more about these and other interactions.

Water waves, sound waves, and earthquake waves are all examples of mechanical-wave interactions.

What types of activities are in Unit 2?

In Unit 2, you will begin each cycle with an *Our First Ideas* activity. You will develop and learn about new science ideas in *Developing Our Ideas* and *Learning About Other Ideas* activities. In this unit, you will also see two new types of activities. One is called *Idea Power* and one is called *Putting It All Together*. Science ideas are useful if they can help you understand interesting applications to everyday life. The *Idea Power* activities are designed to help

you apply the ideas you learned in the *Developing Our Ideas* activities to explain new and interesting situations involving everyday phenomena.

After working through the *Developing Our Ideas* activities, your class will have agreed on science ideas that are supported by experimental evidence. In the *Putting It All Together* activity (that first appears in Cycle 2 of this unit), you will compare your class ideas with a list of Scientists' Consensus Ideas that your teacher will give you. This will help you see that your ideas are similar to those that have been developed by scientists.

As you know, all *InterActions in Physical Science* units are divided into sections called "cycles of learning." In Unit 2, there are two cycles.

Cycle 1: Energy Description of Interactions

In Cycle 1, you will explore one type of mechanical interaction, the *mechanical-wave interaction.* You will read about how scientists use energy to describe interactions. You will use this knowledge to explore and describe several interactions in terms of energy.

Here is the key question for the first cycle:

How do scientists describe interactions in terms of energy?

Cycle 2: **Mechanical Interactions and Energy**

Actually, mechanical interactions are all around you. Kicking a ball, playing tennis, and flying a paper airplane all involve mechanical interactions. You will learn about these mechanical interactions in the second cycle. You will also learn about the energy changes involved in mechanical interactions.

Here are the key questions for the second cycle:

Motion energy: Where does it come from? Where does it go? Why does it change?

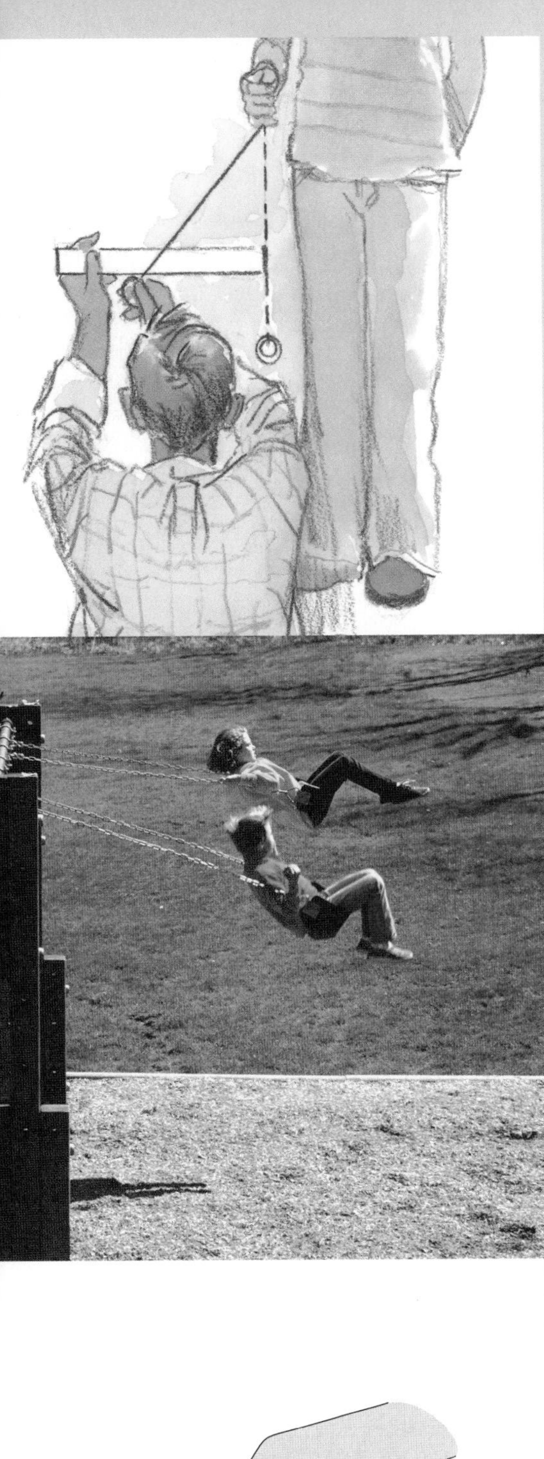

CYCLE 1:
ENERGY DESCRIPTION OF INTERACTIONS

Cycle Key Question

How do scientists describe interactions in terms of energy?

Activity 1: Interactions and Energy

Activity 2: Energy Description of Interactions

Activity 3: Mechanical Waves and Energy Transfer

Activity 4: Water, Sound, and Earthquake Waves

Activity 5: Interaction Chains, Energy Transfers, and the Fabulous Wake-Up System

Activity 6: Describing the Motion of an Object with Constant Speed

Activity 7: Speed of Waves

Activity 8: Objects and Waves with Changing Speeds

Activity 1
Interactions and Energy

Cycle Purpose

You have already studied several types of interactions.

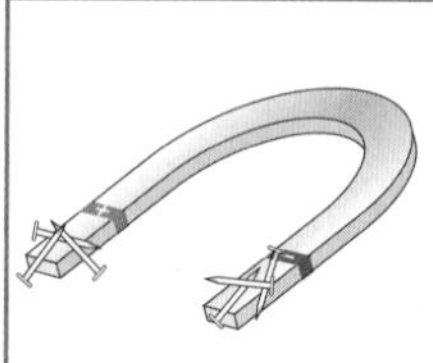

Magnetic interactions occur when a magnet is near a magnetic material or another magnet. Evidence is a change in motion of one or both objects.

Electric-Charge interactions occur when a charged object is near another charged or uncharged object. Evidence is a change in motion of one or both objects.

Electric-Circuit interactions occur when an electrical energy source is connected in a closed circuit. Evidence is an electric current in the circuit.

Mechanical interactions occur when one object pushes or pulls on another. Evidence is a change in motion or shape of one or both objects that are in contact.

Light interactions occur when a light source is near an object. Evidence is a change in the illumination of the object (including your eye).

One of the ways that scientists describe interactions is in terms of energy. Energy is the ability to change an object in some way; for example, by moving it or heating it. Each interaction can be described by considering which object provides energy, which object receives energy, and how the energy is transferred between the two objects. The energy description of an interaction is extremely useful and powerful. It offers scientists a way to analyze and explain phenomena they observe.

The purpose of this cycle is to answer the following question:

How do scientists describe interactions in terms of energy?

Record the key question for the cycle on your record sheet.

We Think

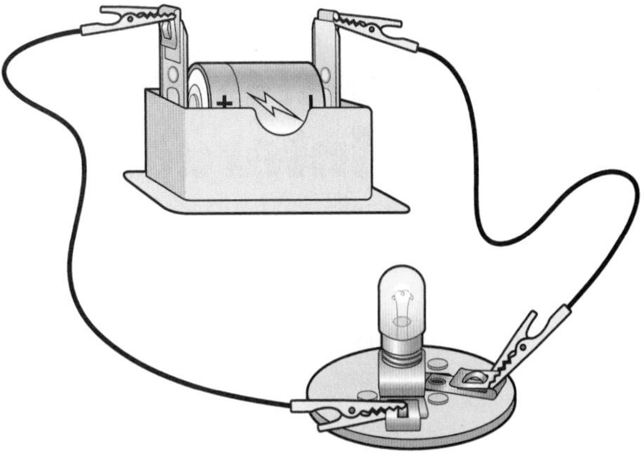

In Unit 1, you learned about the electric-circuit interaction. Consider an electric circuit consisting of just a cell, bulb, and connecting wires. The bulb glows.

1. How might you describe this interaction in terms of energy?

Imagine leaning over the edge of a swimming pool. You slam a soccer ball into the water and make a water wave. Some distance away, a toy boat floats in the still water.

A short time later, the water wave causes the toy boat to move up and down.

2. How might you describe this interaction in terms of energy?

Participate in the class discussion to share answers to the two questions.

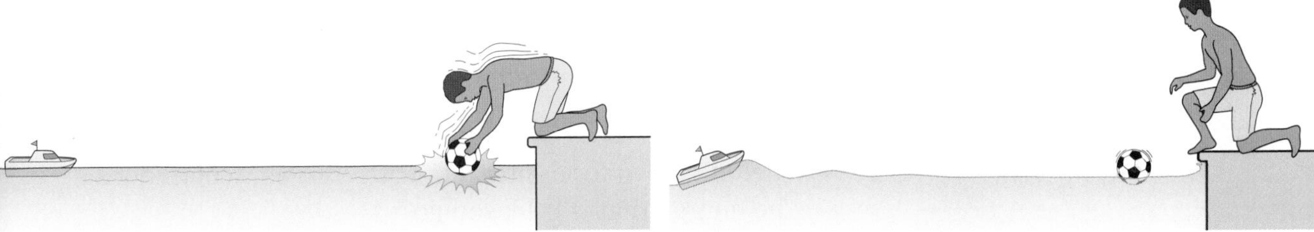

Our Class Ideas

In this activity, you considered two situations. One involved an electric circuit and one involved water waves. The class discussed some ways of describing these situations (interactions) in terms of energy. In the next activity, you will be introduced to the way that scientists describe interactions in terms of energy.

Participate in a class discussion to summarize ways the class has suggested to describe interactions in terms of energy.

Write the class ideas on your record sheet.

DEVELOPING OUR IDEAS

Activity 2
Energy Description of Interactions

Purpose

In this activity, you will learn how scientists describe interactions in terms of energy. You will then create some of your own interactions using different materials. You will also practice describing interactions in terms of energy.

The key question for this activity is:

How can you describe interactions in terms of energy?

Record the key question for the activity on your record sheet.

We Think

You need energy to sleep, sit and think, walk, run, play soccer, and so on. You get your energy from food. Digested food interacts with oxygen in your body to make new substances. At the same time, some energy is released.

Discuss the following question with your partner.

- When digested food interacts with oxygen, where does the energy that is released go?

Participate in a class discussion.

Where does the energy your body requires come from?

Explore Your Ideas

Part A: Energy Source, Energy Receiver, Energy Transfer, and Energy Diagrams

When two objects interact, they have an effect on each other. In terms of energy, the effect during an interaction is a *transfer of energy*. When there is an energy transfer, one object must be the supplier of the energy. This object is called the **energy source**. The other object gains the energy transferred to it. This object is called the **energy receiver**.

So the description of an interaction in *energy* terms is:

> *When two objects interact, the energy source object transfers energy to the energy-receiver object.*

Scientists sometimes use energy diagrams to describe an interaction. Consider the example of your body. The energy source for your body is the food you eat and oxygen. The rest of your body is the energy receiver. You can describe this interaction with the following energy diagram.

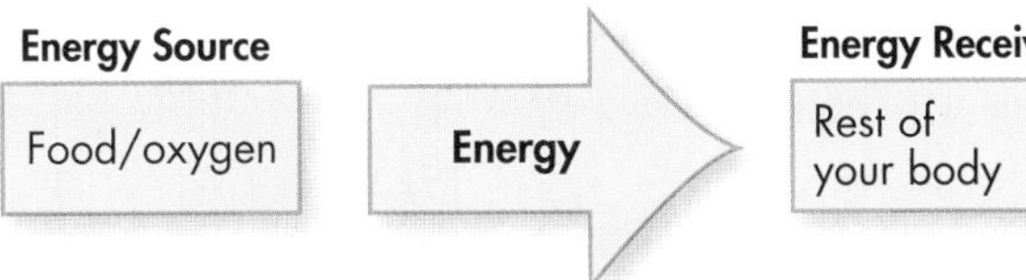

Example: Soccer Player Kicking the Ball

Consider what happens when a soccer player kicks a ball. Think about when the player is touching the ball. What is the evidence of this interaction?

To describe *any* interaction in terms of energy, ask yourself two questions.

- *Energy source:* Where did the energy for this interaction come from?
- *Energy receiver:* Where did the energy for this interaction go to?

Science Words

energy source: an object that is the supplier of energy

energy receiver: an object to which the energy is transferred

In the case of the soccer player and ball, the evidence for the interaction is the big change in motion of the ball. There is also a small change in the motion of the soccer player's foot. There is a change in motion while the interacting objects are touching each other, so this is a mechanical interaction. The energy for the interaction comes from the soccer player. This energy went to the ball, the energy receiver. The energy diagram for this interaction is shown here. The diagram also includes the type of interaction involved; in this case, it is a mechanical interaction. The evidence of the interaction is also included.

Mechanical Interaction

Energy Source: Soccer player → Energy → Energy Receiver: Ball

Evidence: change in motion (speeding up) of ball

STEP 1 Imagine aiming a flashlight directly into your eye. You would probably be dazzled by the light and have to squint. Suppose someone is observing your pupil (the black circle in the center of the eye). He or she might notice that when the light shines on your eye, your pupil gets smaller.

This interaction between the flashlight and the eye occurs whenever a person is looking at a source of light. Think about the interaction just after the flashlight is shone into the eye.

STEP 2 A cell is connected to a light bulb. Think about the few seconds just after the circuit between the cell and the bulb is closed.

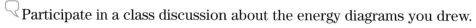

STEP 3 For the situations given in Steps 1 and 2, consider the following questions. Discuss the answers with your partner.

Record the answers to the following questions on your record sheet and *draw the energy diagram* for each interaction.

1. What is the *type of interaction*? (You should choose from *magnetic*, *electric charge*, *electric circuit*, *mechanical*, or *light*.)
2. What is the *evidence* for the interaction?
3. Which object is the *energy source*? (Where did the energy for this interaction come from?)
4. Which object is the *energy receiver*? (Where did the energy for this interaction go to?)

Participate in a class discussion about the energy diagrams you drew.

Do not leave the circuit connected for more than a few seconds. The wire can get hot.

Be careful around the moving blades of the motor and wear eye protection.

Part B: Exploring Electrical Energy Sources and Receivers

In the last unit, you explored the behavior of electric circuits. You used cells, bulbs, and switches. Now you will work with a team to build electric circuits using a cell as an energy source. You will use three different energy receivers. You will also practice drawing energy diagrams.

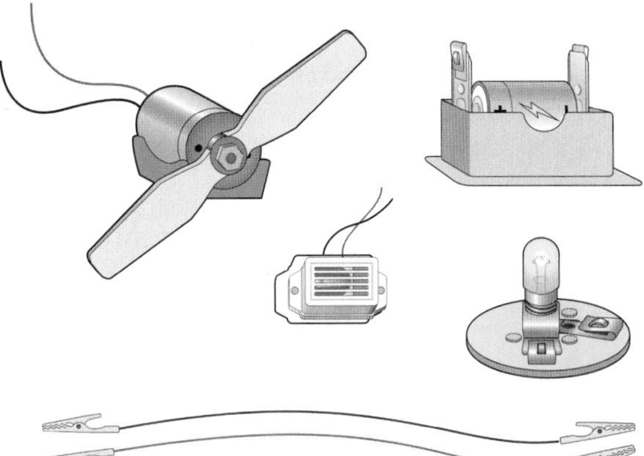

STEP 1 Select the battery cell and one of the three energy receivers.

STEP 2 Connect the source to the receiver so there is a transfer of energy.

5. Decide the type of interaction involved and write it down.
6. Draw a diagram of the circuit.
7. Draw an energy diagram.
8. Decide what evidence suggests there is an interaction and write that down.

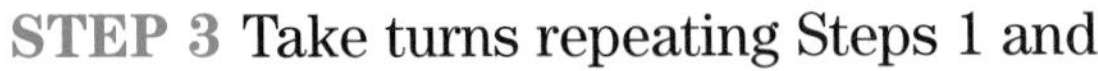

STEP 3 Take turns repeating Steps 1 and 2 until your team has completed the three examples of energy transfers.

Participate in the class discussion to review your energy diagrams.

Part C: Chains of Interactions

Your teacher may demonstrate or show you a movie about two other types of electrical energy sources: a generator and a solar battery. Both the generator and the solar battery involve *chains of interactions*. For example, with the hand-cranked generator connected to a light bulb, there is an interaction between the hand and the generator (causing the handle to be turned), and another interaction between the generator and the bulb (causing the bulb to glow).

In this case, the generator is both an energy receiver and an energy source. The generator is an energy receiver because it receives energy from the hand. The generator is also an energy source because the bulb receives energy from it.

9. On your record sheet, complete the energy diagram on the next page for the hand-cranked generator connected to the light bulb.

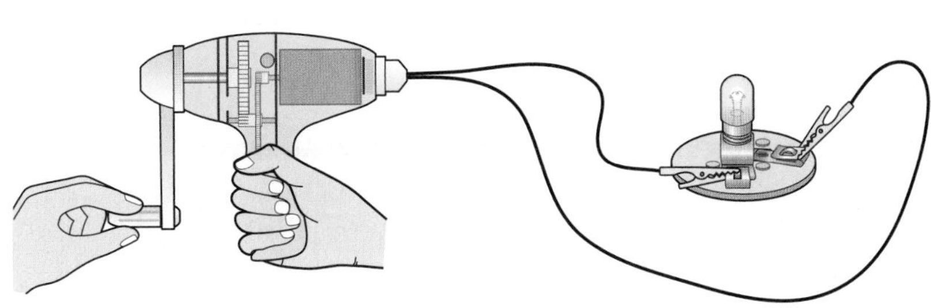

Your team will need:

- cell in a holder
- buzzer
- motor with fan blade attached to axle
- light bulb in socket
- hook-up wires

To Do

Team Member 1
Supply Master: gathers materials.

Team Member 2
Procedure Specialist: reads instructions and steps aloud.

Team Member 3
Team Manager: makes sure team stays on task and all team members are participating.

Team Member 4
Recycling Engineer: returns materials.

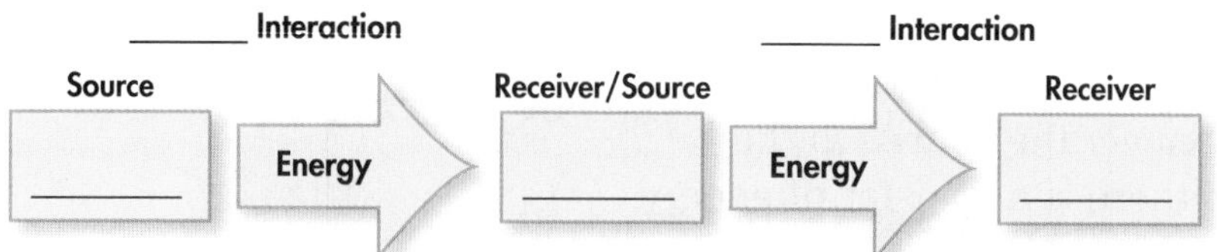

There is also a chain of interactions when a bulb shines light on a solar battery that is connected to a motor with a fan blade.

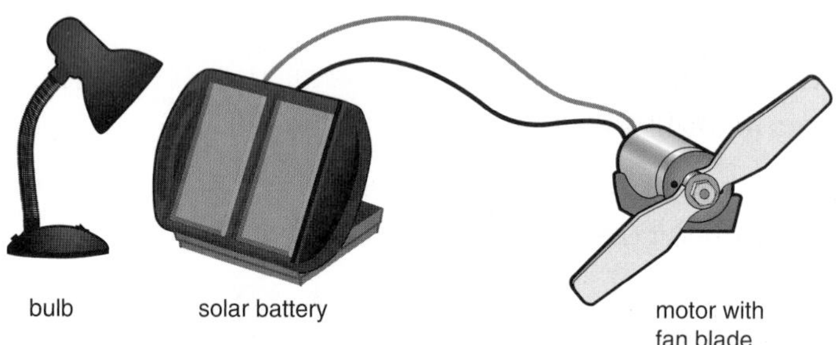

10. Complete the following energy diagram on your record sheet for the solar battery connected to a motor with a fan blade.

Participate in a class discussion to review your energy diagrams.

Make Sense of Your Ideas

When describing an interaction between two objects using energy ideas, what information do you need to know or what questions do you need to answer?

Our Consensus Ideas

The key question for this activity is:

How can you describe interactions in terms of energy?

1. Write your best answer to the key question.

Participate in a whole-class discussion to answer the key question.

2. Write the class consensus answer on your record sheet.

DEVELOPING OUR IDEAS

Activity 3
Mechanical Waves and Energy Transfer

Purpose

Waves are interesting phenomena because they provide a special way that energy can be transferred from one place to another. In this activity, a Slinky® spring will be used to study some types and properties of waves; in particular, *how* waves transfer energy from a source to a receiver. The key question for this activity is the following:

There are many examples of waves in your everyday experience.

What are some types and properties of waves?

Record the key question on your record sheet.

We Think

Suppose you and a friend stretch out a Slinky spring along the floor. If you hold some coils and suddenly swing your end back and forth, without letting go, your friend will *feel* the influence at the other end.

Discuss the following question with other members of your team.

- How does energy get transferred from you to your friend, since none of the coils you are holding onto travel all the way to your friend's end?

Participate in a discussion to answer the question.

Your class will need:

- Slinky®
- 2 pieces of tape

Do not release the spring while it is stretched.

Wear eye protection while holding the spring.

Explore Your Ideas

Your teacher will call upon some members of the class to help perform the following demonstrations with a Slinky spring. Be careful stretching the Slinky along the floor. It is easy to get the coils tangled and it may be very difficult to untangle them.

STEP 1 Carefully stretch out the Slinky along the floor. *Don't let go of the ends!*

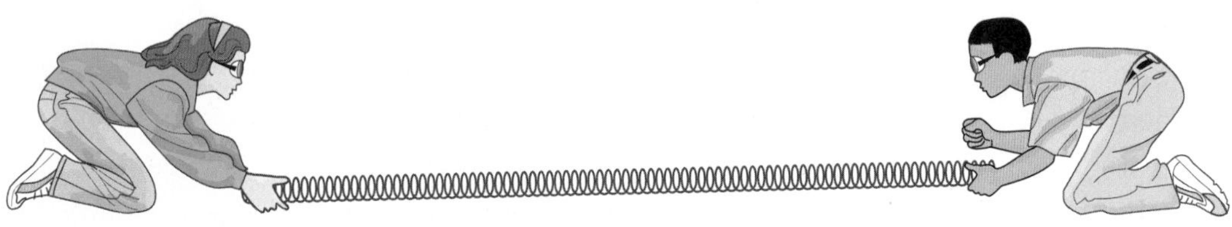

STEP 2 Place a piece of tape on the floor under each end of the stretched Slinky to mark the ends. This way the stretched Slinky can be kept the same length throughout the exploration.

You will first study a type of wave called a **transverse wave**. It is created by snapping the coils sideways.

STEP 3 Have the student at one end lay a hand on the floor, grab hold of a few coils, and quickly snap the wrist back and forth sideways *one time*. This generates a single pulse (or a *disturbance*) that moves along the Slinky coil to the other end.

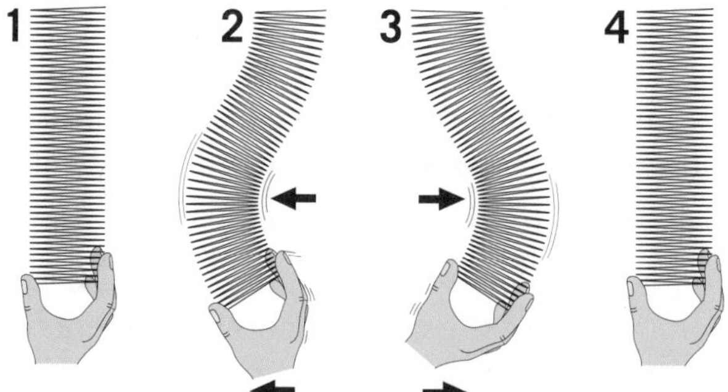

1. What does the person at the other end *feel* when the pulse reaches that end? (He or she should describe what is *felt*.)

2. Do you think energy is being transferred from the hand generating the pulse to the hand at the other end? Why do you think so?

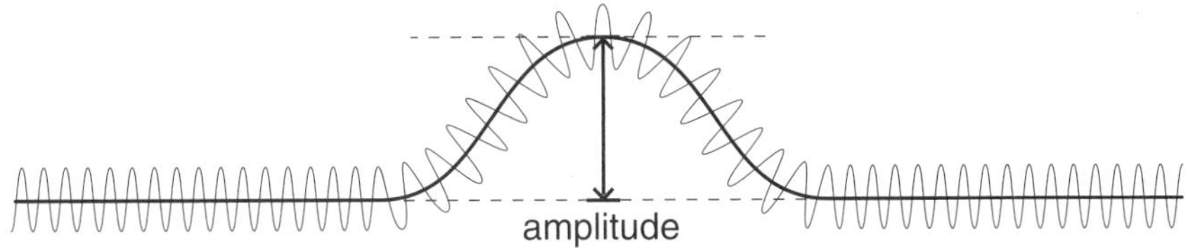

The **amplitude** is the maximum distance that each coil of the Slinky moves away from its original position.

The student at one end uses a certain amount of energy to generate the pulse. The student at the other end senses the pulse when it arrives. In this way, the two students can *feel* that energy is being transferred from the source to the receiver, along the Slinky coils. In the next step, you will investigate whether there is a relationship between the amplitude of the pulse and the energy that is transferred.

STEP 4 Start with a pulse that has small amplitude. After it stops moving, generate another pulse with larger amplitude. Generate one more pulse with even larger amplitude. The students holding the Slinky should comment on the amount of energy they *feel* is being transferred.

3. Does a transverse pulse with larger amplitude seem to transfer *more*, *less*, or *the same amount of energy* as a pulse with smaller amplitude? How do you know?

STEP 5 Your teacher will show you a movie of someone repeating the explorations you conducted. Because the movie can also show close-ups and slow motion, it may be easier to observe how the pulse actually moves along the Slinky. After watching the movie, answer the following questions.

4. As the pulse moves along the Slinky from one end to the other, do any of the individual coils actually travel along with the pulse from one end to the other?

5. What does each of the individual coils do as the pulse passes by?

A **wave** is a continuous succession of pulses. To generate a wave along the Slinky, you need to snap your wrist several times back and forth in quick succession. The diagram on the next page shows what the transverse wave might look like.

Science Words

wave: a continuous succession of pulses

transverse wave: a wave in which the motion of the material (medium) is perpendicular to the motion of the wave

amplitude: the height of a wave crest. It is related to a wave's energy.

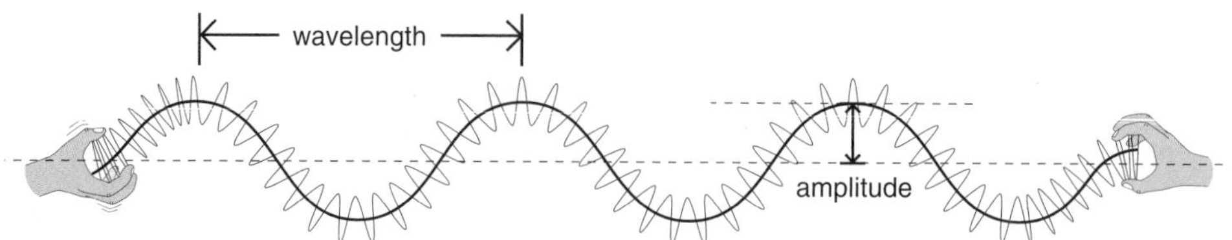

The distance along the wave between high points is called the **wavelength** of the transverse wave. Wavelength is measured in distance units, like centimeters or meters.

Science Words

wavelength: the distance between identical points along a wave

frequency: the number of waves produced per unit time

The number of times you snap your wrist back and forth each second is the **frequency** at which you generate the wave. Frequency is measured in units of hertz (Hz). If you move your hand back and forth two times each second, you will generate a wave with a frequency of 2 Hz. If you move your hand back and forth four times each second, the wave will have a frequency of 4 Hz. The frequency of the wave itself is the number of times each part of the Slinky moves back and forth each second. This is the same value as the number of times your hand moves back and forth each second.

STEP 6 Your teacher will show you a movie where the person generates transverse waves with the same amplitude but at different frequencies. Use your observations from the movie to answer the following questions.

6. Draw sketches of two transverse waves, one under the other. Each should have the same amplitude, but one is generated at a higher frequency and one at a lower frequency. Label which is the lower frequency and which is the higher frequency.

7. As the frequency of a transverse wave increases, what seems to happen to the wavelength? Does the wavelength *increase*, *decrease*, or *remain the same*?

STEP 7 Your teacher will show you a movie where the person generates transverse waves with the same frequency but at different amplitudes. Use your observations from the movie to answer the following questions.

8. Draw sketches of two transverse waves, one under the other. Assume each was generated at the same frequency, but one has larger amplitude and one has smaller amplitude. Label which has the larger amplitude and which has the smaller amplitude.

9. Based on your classmates' experiences with generating and receiving pulses, do you think a wave with larger amplitude transfers *more energy*, *less energy*, or the *same amount of energy* as a wave with smaller amplitude? Why do you think so?

A second type of wave is called a **compression wave**. A common name for a compression wave is a **longitudinal wave**.

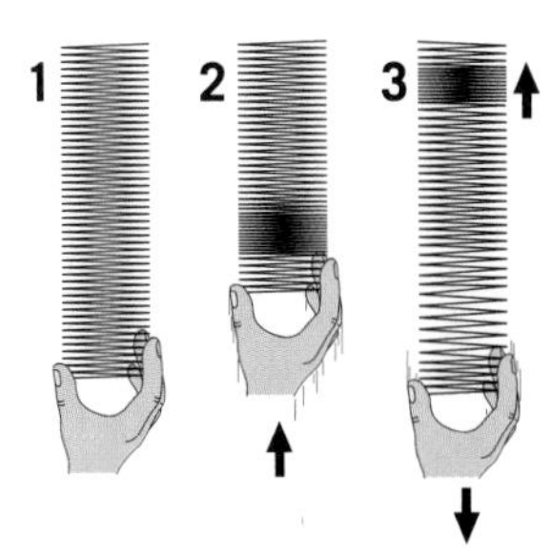

STEP 8 Drag out the Slinky along the floor, similar to what was done in Step 1. To generate a single compression pulse, one student holds a few coils on the floor, and then quickly moves the hand forward and backward *one time only*. (See the diagram.) The student's hand should not leave the floor. This process should first compress the coils and then spread them out. Try it!

STEP 9 The amplitude of the compression pulse is the maximum distance each coil moves forward from its original position as the pulse moves by. Try generating a compression pulse with small amplitude. After the pulse stops moving, generate one with larger amplitude. Repeat with even larger amplitude.

Science Words

compression (longitudinal) wave: a wave in which the motion of the material (medium) is parallel to the direction of the motion of the wave

10. Is energy transferred by the pulse from one end of the Slinky to the other? How do you know?

11. Does a compression pulse with larger amplitude seem to transfer *more*, *less*, or the *same amount of energy* as a pulse with smaller amplitude? How do you know?

STEP 10 Your teacher will show you a short movie of a person generating a compression pulse. After watching the movie, answer the following question.

12. What does each individual coil do as the compression pulse passes by?

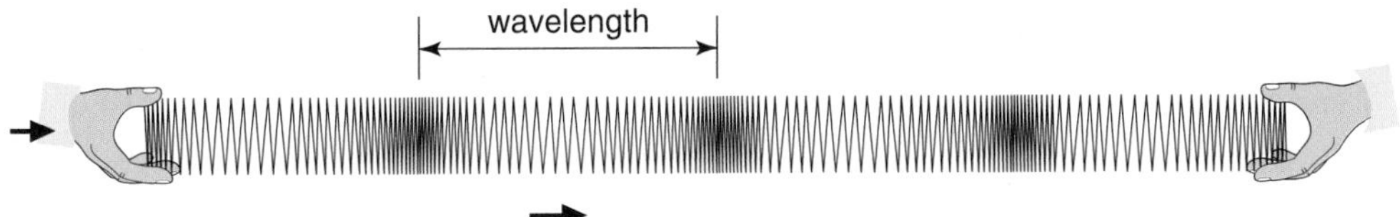

A compression wave is generated by quickly moving the hand forward and backward several times in a row. A compression wave might look like the one in the diagram above.

The distance along the wave between points where the coils are most compressed together (the darkest areas in the diagram) is called the *wavelength* of the compression wave. The *amplitude* of the compression wave is the maximum distance that each coil moves forward from its original position before the wave passes by. The amplitude of the wave is determined by how far the hand is moved forward and backward when the wave is generated. The *frequency* at which you generate the compression wave is the number of times the hand is moved forward and backward each second. The frequency of the wave itself is the number of times each part of the Slinky moves back and forth each second. This is the same value as the frequency at which you generate the wave.

STEP 11 Your teacher will show you a movie of a person generating a compression wave. Use your observations to answer the following.

13. Draw sketches of two compression waves, one under the other. Draw one that was generated at a higher frequency and one at a lower frequency. Label which is the higher frequency wave and which is the lower frequency wave.

14. As the frequency of a compression wave increases, what seems to happen to the wavelength? Does the wavelength *increase*, *decrease*, or *remain the same*?

STEP 12 Finally, your teacher will show you a movie where the person generates compression waves with the same frequency, first with small amplitude, and then with much larger amplitude. Use your observations from the movie to answer the following questions.

15. Do you think a compression wave with larger amplitude transfers *more energy*, *less energy*, or *the same amount of energy* as a wave with smaller amplitude? Why do you think so?

Make Sense of Your Ideas

Use your observations in this activity to answer the following questions.

1. What is the difference between transverse and compression waves?

2. When a wave travels along a Slinky, does each individual coil travel from one end of the Slinky to the other? If not, what does each coil do?

3. Which of the following statements about the relationship between frequency and wavelength is best supported by your evidence?

- **a)** As the frequency increases, the wavelength increases.
- **b)** As the frequency increases, the wavelength decreases.
- **c)** There is no relationship between frequency and wavelength.

4. What is your evidence that a wave transfers energy?

5. Which of the following statements about the relationship between the amplitude of a wave and the energy transferred seems best supported by your evidence?

- **a)** As the amplitude increases, the energy transferred increases.
- **b)** As the amplitude increases, the energy transferred decreases.
- **c)** There is no relationship between amplitude and energy transferred.

Participate in a discussion to answer the questions above.

Mechanical Waves and Energy Transfer

When a source is set into vibration, it creates a wave (a disturbance). The wave travels from the source to a receiver located some distance away. When the wave reaches the receiver, the receiver reacts in some way, usually changing its motion. Since the source influenced the receiver, there was an interaction between the two, but this is a special kind of interaction. In a normal mechanical interaction, the two objects (source and receiver) touch each other and at least one of them changes its motion.

In a wave, however, the source does not touch the receiver, nor do any of the Slinky coils actually travel from the source to the receiver. Instead, only energy travels from the source to the receiver in the form of a wave. This type of interaction is called a *mechanical-wave interaction*, and the general name for the wave produced is called a *mechanical wave*. In this activity, the mechanical wave traveled through the coiled Slinky spring. In Activity 4, you will study mechanical waves that travel through other substances, particularly water, air, and the surface of the Earth.

You have seen in Activity 2 that you can describe an interaction in terms of energy by drawing an energy diagram. For a mechanical-wave interaction along a Slinky where the source is a hand moving back and forth and the receiver is a hand at the other end, the energy diagram is as follows.

Mechanical-Wave Interaction (Slinky Coils)

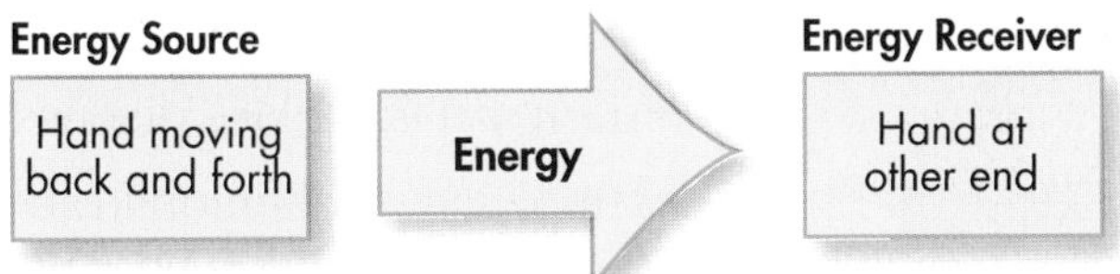

Our Consensus Ideas

The key question for this activity is:

What are some types and properties of waves?

1. Write your best answer to the key question by summarizing your answers to the Make Sense of Your Ideas questions.

Participate in a whole-class discussion to answer the key question.

2. Write the class consensus answer on your record sheet.

DEVELOPING OUR IDEAS

Activity 4
Water, Sound, and Earthquake Waves

Purpose

In the last activity, you used a coiled Slinky spring to study some general properties of waves. In this activity, you will extend that knowledge by investigating three important examples of waves.

The key question for this activity is:

What are some properties of water, sound, and earthquake waves?

Record the key question on your record sheet.

We Think

Imagine that a buzzer or bell is placed inside a large glass container. You can hear its sound loudly and clearly. Suppose a vacuum pump then removes all the air from the glass container (creating a vacuum). Discuss this question with your group.

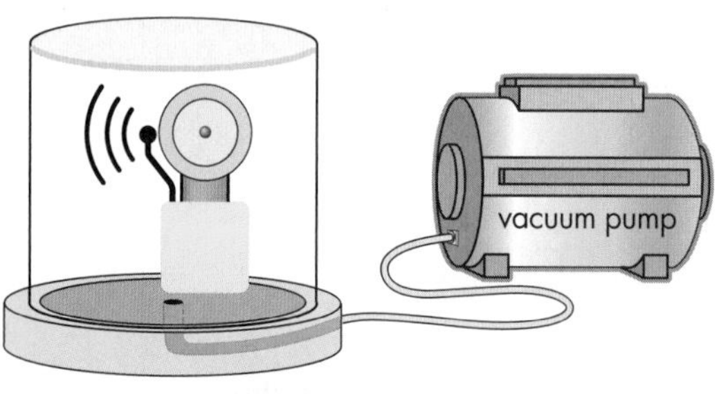

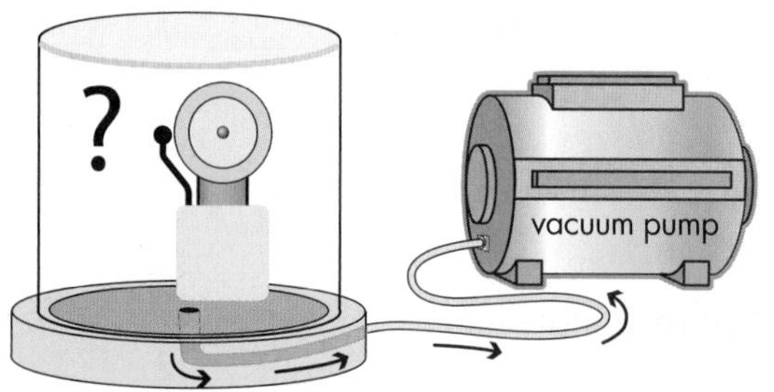

- Would you still be able to hear the sound from the buzzer or bell?

Participate in a class discussion.

You will observe a demonstration of this situation when you study sound later in the activity. First, however, you will explore some of the properties of water waves.

Explore Your Ideas

Part A: Explore Water Waves

STEP 1 Imagine you had a large container of water and you generated water waves by dipping your finger in and out at a certain frequency.

The diagrams on the right show what the surface of the water might look like, drawn both from the water level and from above. This is like a transverse wave that travels outward along the water surface in all directions away from your finger. (Actually, water waves are not exactly transverse waves. The particles of the water do not just move up and down, but instead move in circular-type paths.)

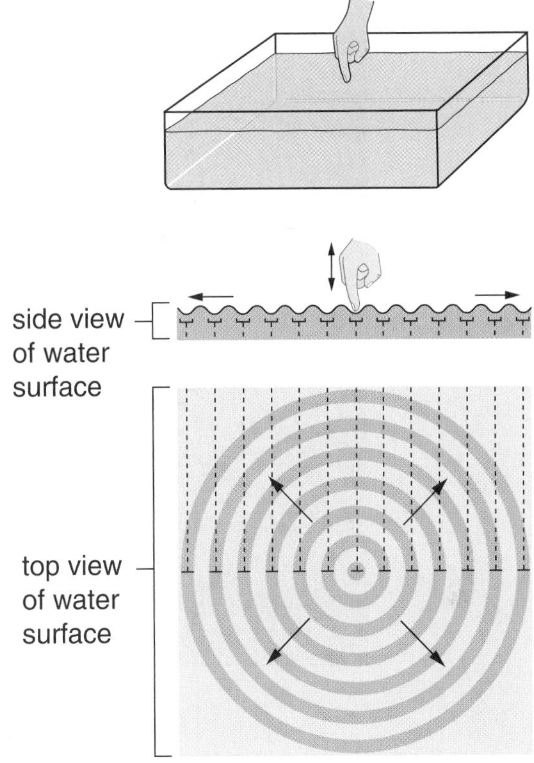

STEP 2 Your teacher will demonstrate a special simulator (or show you a movie of it) that shows how water waves travel outward along the surface of a water tank. A snapshot from the simulator is shown below. This presents a top view, looking down from above on the surface of the water. The dot at the center represents the source of the wave, which could be your finger moving in and out of the water. In the picture, the source is vibrating with a frequency of 1 Hz.

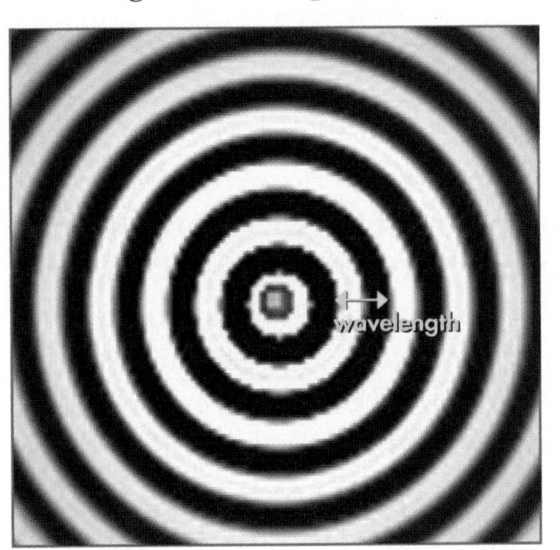

The distance between two white areas (or two dark areas) on the wave is the wavelength.

STEP 3 Imagine that you floated a piece of cork in the water tank far away from where your finger is moving in and out.

1. What do you think the motion of the cork would be like as the water wave passes it by?
2. Do you think energy is being transferred from your moving finger to the cork? Give your reasons.

STEP 4 With the simulator, you can easily change the frequency of the source. Look at the snapshots from the simulator when the frequency of the source has been set at four different values: 1 Hz, 2 Hz, 3 Hz, and 4 Hz.

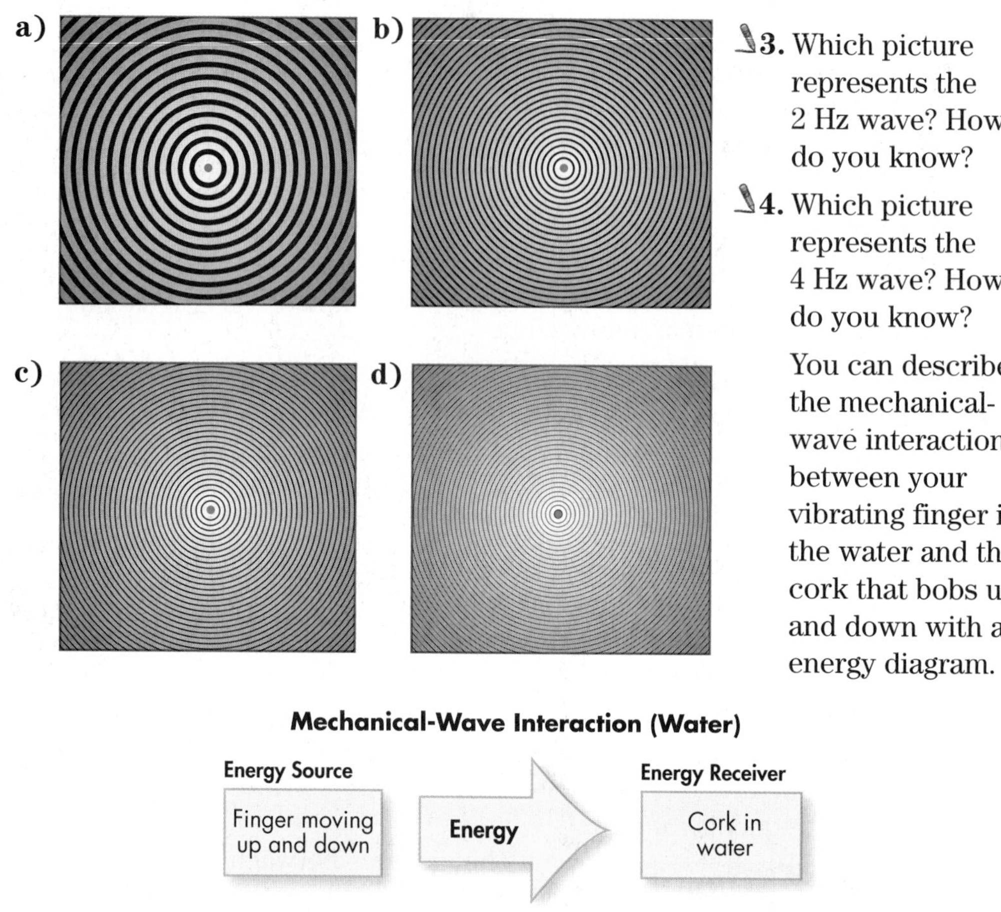

3. Which picture represents the 2 Hz wave? How do you know?

4. Which picture represents the 4 Hz wave? How do you know?

You can describe the mechanical-wave interaction between your vibrating finger in the water and the cork that bobs up and down with an energy diagram.

Part B: Explore Sound Waves

You learned that a buzzer produces sound by vibrating back and forth. You also learned that even though you cannot see the air, air has mass. When the buzzer vibrates, it generates a compression wave that travels through the air in all directions away from it. If you are nearby and the compression wave reaches your ear, your ear-brain system interprets the compression wave as sound. Sound travels from a source (like the buzzer) to a receiver (like your ear) as a compression wave.

When sound of a particular frequency reaches your ear, you interpret the sound as having a particular **pitch**. Humans can hear sounds in the range between about 20 Hz and 20,000 Hz (called the audio range). Many animals can hear sounds at higher frequencies than humans. Compression waves at much higher frequencies than either animals or humans can hear, for example in the range between 1 million Hz and 5 million Hz, are known as **ultrasound** and have many medical applications.

Science Words

pitch: the quality of a sound dependent mostly on the frequency of the sound wave

ultrasound: compression waves at much higher frequency than animals or humans can hear

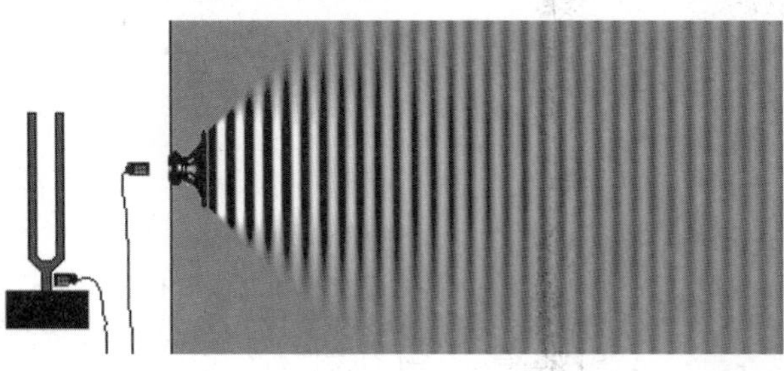

STEP 1 Your teacher will play a movie showing compression waves produced by vibrating speakers. The waves were generated with a computer simulator. Each speaker is connected through an amplifier to a tuning fork and vibrates at the same frequency as the tuning fork. The movie shows three different tuning forks vibrating at frequencies of 150 Hz, 300 Hz, and 600 Hz.

5. Which of the frequencies corresponds to the highest pitch sound?

6. Which of the frequencies corresponds to the lowest pitch sound?

7. What is the difference between the wave patterns produced by the 150 Hz, 300 Hz, and 600 Hz sounds?

You can also describe the compression (sound) wave that travels from the vibrating buzzer through the air to your ear with an energy diagram.

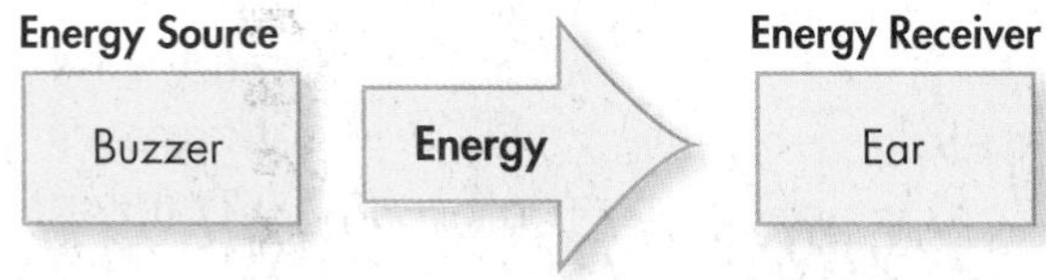

STEP 2 In the We Think section, you were asked whether you could hear the sound from a buzzer/bell if it were placed in a glass jar and all the air was removed (creating a vacuum). Your teacher will show you a movie of this situation.

8. What happens to the sound from the buzzer when air is removed from the jar?

A seismograph is a device used to measure the strength of earthquake (seismic) waves.

Earthquakes can cause severe damage to property, as you can see in the photos.

Science Words

fault: a fracture in rock, along which the rock masses have moved

earthquake: a sudden motion or shaking of the Earth

P wave: a seismic wave that involves motion in the direction in which it is traveling; it is the fastest of the seismic waves

S wave: a seismic wave that involves vibration perpendicular to the direction the wave is travelling; it arrives later than the P wave

Part C: Explore Earthquake Waves

The surface of Earth is covered by many plates. Where these plates meet is called a **fault**. Earthquakes are much more likely to occur on a fault line than anywhere else.

An **earthquake** happens when two plates of the Earth either move apart from each other, push towards each other, or slide against each other. This contact between plates releases an enormous amount of energy that travels away from the source of the earthquake in all directions. The energy travels outward from the source as earthquake (seismic) waves.

There are three different types of earthquake waves. One type, called the **P wave** (primary wave), is a compression wave and moves the fastest of the three types. As it moves through Earth from its point of origin, it compresses and expands material in the direction it is traveling.

A second wave, called the **S wave** (secondary wave), is a transverse type of wave. It is also called a shear wave because it shakes the ground up and down and sideways. It would be similar to taking the end of a long narrow block of Jell-O® and shaking it. This is the wave that is associated with the rolling part of an earthquake.

There are also surface waves, **L waves** (long waves), which are caused when the P and S waves reach the surface. They move the slowest and usually arrive at the end of an earthquake. L waves are very similar to water waves and they cause the most damage.

Earthquakes are recorded using an instrument called a **seismograph**. Earthquakes are classified in terms of their magnitude (the amount of energy that is released) using the Richter scale. The higher the number on the Richter scale, the more energy the earthquake released and the more damage may occur.

Earthquakes happen every day, but most are too small to notice. According to the United States Geological Survey (USGS), there are more than 3,000,000 earthquakes every year!

When an earthquake occurs on the ocean floor, it can cause a series of waves called a **tsunami**. (*Tsunami* comes from a Japanese word meaning "harbor wave.") Tsunamis travel very fast with small wave heights when they are out in deep water. However, when they reach shallow water, they slow down and their heights increase dramatically. When they reach land they can cause significant damage. Sometimes tsunamis can be as high as 30 m (about 100 ft.). They can travel very long distances, even across the entire ocean.

Science Words

L wave: a seismic wave that travels along the surface of the Earth; they are the last to arrive at a location

seismograph: an instrument that detects seismic waves

tsunami: a great sea wave produced by an earthquake (or volcanic eruption) on the ocean floor

Make Sense of Your Ideas

When a source in a material substance (like a Slinky, water, air, or the Earth) begins vibrating, the disturbance travels outward from the source in all directions as a wave. In this activity, you learned about transverse waves in water, compression waves in air (sound), and both transverse and compression waves that travel through Earth when there is an earthquake.

On the ocean, a ship cannot even tell a tsunami is passing by. As tsunamis come ashore, they build to great heights and are very destructive.

1. What is the relationship between the frequency of a wave and its wavelength?
2. What seems to be the relationship between the frequency of a sound wave and the pitch of the sound you hear?

3. Why didn't you hear the sound from the buzzer in the glass jar when the air was removed from the jar?

4. What do you think is the relationship between the amplitude of an earthquake wave and the reading of the earthquake on the Richter scale? Why?

Participate in a discussion to answer the Making Sense questions.

Our Consensus Ideas

The key question for this activity is:

What are some properties of water, sound, and earthquake waves?

You addressed this question when answering other questions throughout the activity.

Your teacher will hand out *Scientists' Consensus Ideas: The Mechanical-Wave Interaction*, which lists ideas that scientists use to think about mechanical waves. These ideas relate to what you have learned so far in this cycle.

Look over the ideas. Provide evidence from this and the previous activities that support each idea.

Participate in a discussion to review the evidence.

IDEA POWER

Activity 5
Interaction Chains, Energy Transfers, and the Fabulous Wake-Up System

Purpose

In this cycle, you learned how to describe interactions in terms of energy. Even though you have learned about six different types of interactions, the energy diagrams you drew showed only one type of energy-transfer arrow.

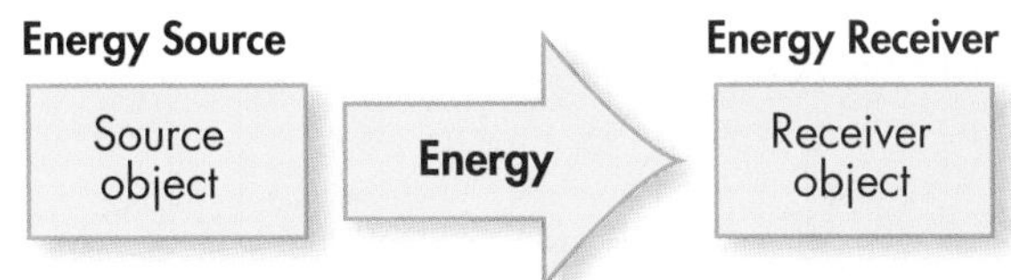

Scientists consider different types of energy transfers.

Table: Types of Interactions and the Corresponding Energy Transfers	
Type of Interaction	**Type of Energy Transfer**
Mechanical	Mechanical Energy
Mechanical Wave	Mechanical Energy
Magnetic	Mechanical Energy
Electric Charge	Mechanical Energy
Light	Light Energy
Electric Circuit	Electrical Energy

Whenever the interaction causes one or both of the objects to change its motion, the type of energy transfer involved is called a **mechanical energy** transfer. When there is a light interaction involved, the type of energy transfer is light energy. During the

Science Words

mechanical energy: the energy transfer involved in an interaction that causes one or both objects to change position electric-circuit interaction, the type of energy transferred is called electrical energy. The table on the previous page summarizes the types of interactions and energy transfers.

In this Idea Power activity, you will apply the ideas you have learned to construct an interesting and practical device and to describe it in terms of energy. You will also practice using the names for some of the types of energy transfers listed in the table.

Part A: The Fabulous Wake-Up Machine

Otis has a problem. He cannot wake up in the mornings. An alarm clock is just not enough. He dreams of a device that can wake him up by blowing air on him, shining lights into his eyes, and making a buzzing sound.

Your task is to design and build a prototype of a device that can do what Otis wants, at least on a small scale. You will then analyze the device in terms of interactions and energy.

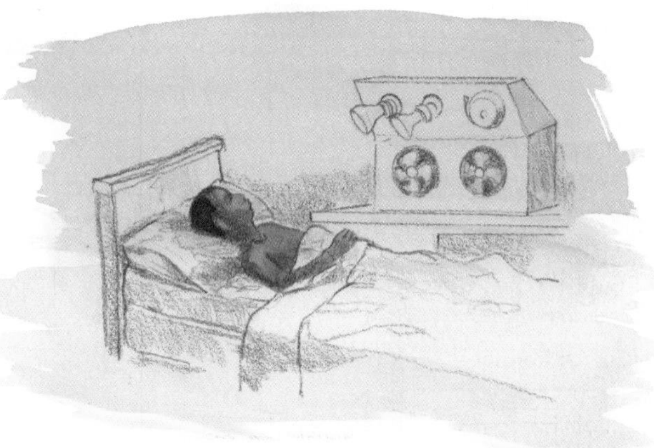

Your team will need:

- cell in holder
- switch
- buzzer
- bulb in holder
- motor with fan blade
- tape
- several hook-up wires

STEP 1 Think about how to design a circuit that consists of a battery cell (as an energy source) that will turn on the bulb, buzzer, and motor (with fan blade) all at the same time. Do *not* use the switch yet.

1. Draw a picture of the circuit that your team thinks will work. Do not include the switch.

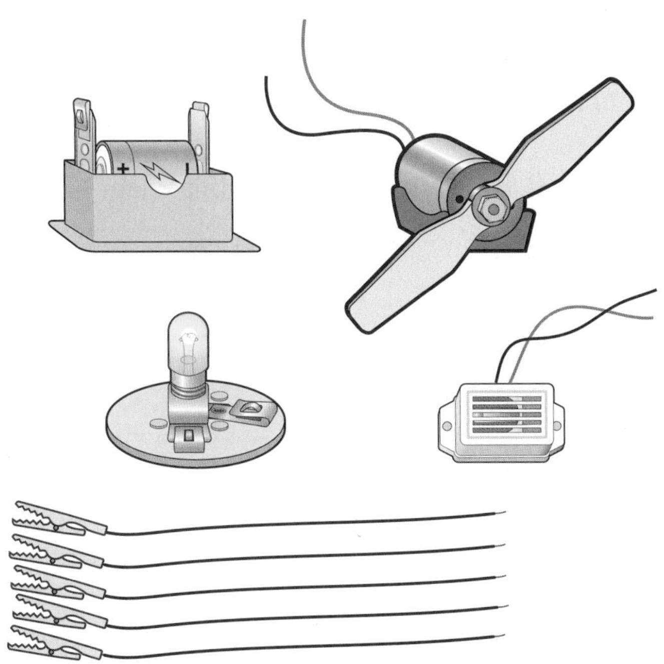

STEP 2 Try your circuit to see if it works. If you are having difficulty, look back at Unit 1 Cycle 2, Activity 4 where you learned about two different ways of hooking up a circuit.

Have your teacher check your ideas for safety before doing them.

Wear eye protection.

Do not leave the circuit closed for more than a few seconds. The wire can get hot.

2. Did your original circuit design work or did you have to change your design? If you changed it, draw the new design.

STEP 3 Now figure out how to add the switch to your circuit.

This is what should happen: When you lower the handle of the switch (switch closed), the circuit should be activated; the bulb should go on; the buzzer should begin buzzing; and the fan should begin blowing air. When you can make your circuit do that, it will be a prototype of a simple wake-up system. (In an actual device, a clock would activate the switch.)

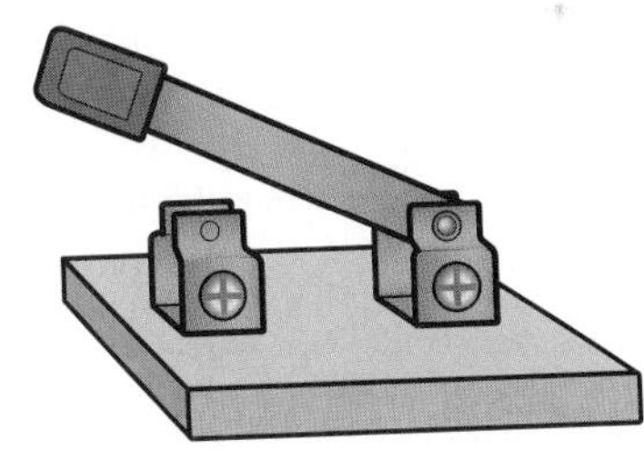

STEP 4 When you have successfully connected the switch and have everything working, ask your teacher to check it out.

3. Draw a picture of your team's final circuit, including the switch.

Part B: Analyze the Fabulous Wake-Up Machine

When you are asked to *analyze* a device or situation you need to identify the interacting objects and the types of interaction(s). You should draw an energy diagram to describe the interaction or chain of interactions. In drawing the energy diagram, you need to:

- label the types(s) of interaction(s) (use the table in this activity)
- label the source and receiver objects
- label the type of energy transfer in the arrow (use the table again)

There are three different systems involved in the wake-up machine.

The cell-bulb-eye system. The bulb is connected to the cell and you *see* the bulb glowing.

4. Fill in the energy diagram on your record sheet to describe this chain of interactions.

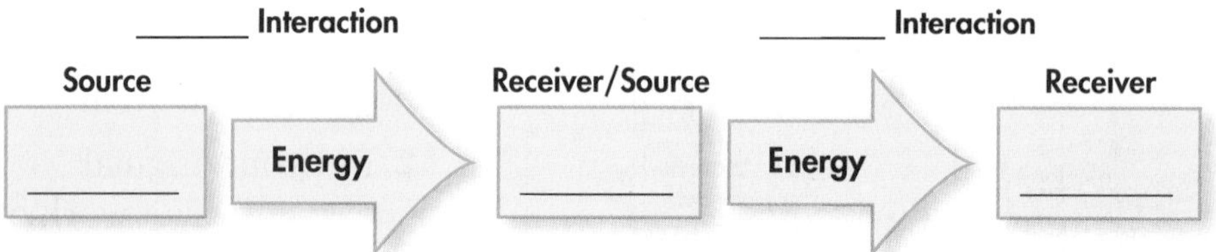

The cell-buzzer-ear system. The cell is connected to the buzzer and you *hear* the buzzer buzzing.

5. Fill in the energy diagram on your record sheet to describe this chain of interactions.

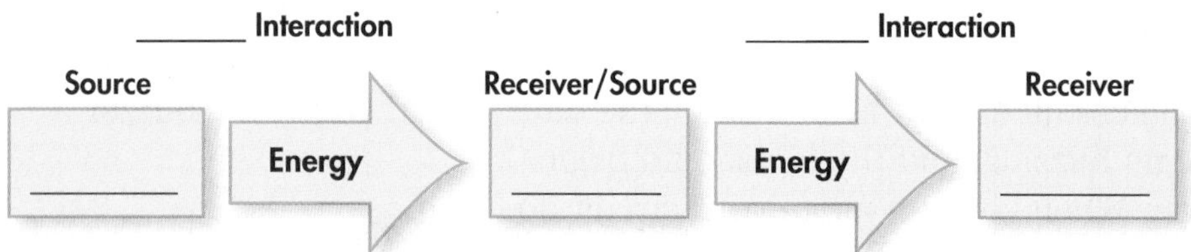

The cell-motor/fan blade-air system. The cell is also connected to the motor/fan. Treat the motor/fan as a single object. When the fan turns, it pushes air away from it.

6. Fill in the energy diagram on your record sheet to describe this chain of interactions.

Participate in a class discussion to go over the energy diagrams.

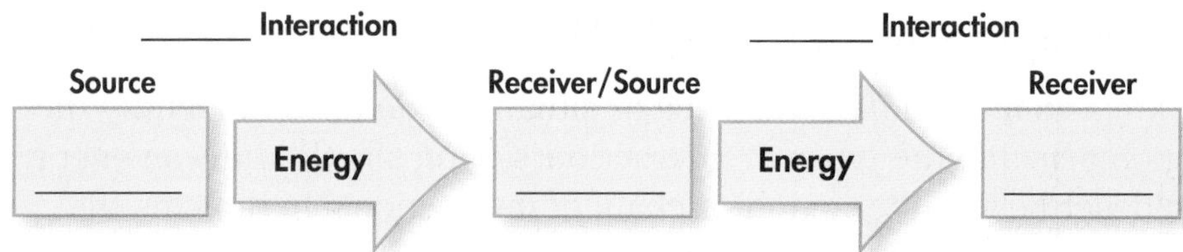

LEARNING ABOUT OTHER IDEAS

Activity 6
Describing the Motion of an Object with Constant Speed

Purpose

The world is filled with objects moving at different speeds: a truck passes your family's car on the freeway, a skateboarder races down the sidewalk toward you, a passenger jet flies high in the sky. All of these objects move at different speeds.

How can you tell how fast he is moving?

In the previous two activities, you studied wave motion. When a wave is generated, energy moves at a certain speed from the source to the receiver. Sometimes, it is important to know how fast the wave energy moves. For example, it might be important to know how fast sound travels through air or how fast a tsunami wave travels along the ocean water.

In this activity, you will examine different ways of finding the speed of an object with *constant* motion. In Activity 7, you will focus on the speed of waves that move at a constant speed. In Activity 8, you will examine ways of determining the speed of an object with *changing* motion.

The key question for this activity is:

How can you determine the speed of an object that has constant motion?

Record the key question for the activity on your record sheet.

Learning the Ideas

Measuring Position

Alex is riding his bicycle heading east along a straight highway. After an hour, he stops for a rest. He is curious about how far he has ridden, so he pulls out a map and studies it.

The map has a scale that begins at zero right under the City Hall of Alex's hometown. The scale shows +12 mi. under his current position. Alex's home, where he began his ride, is located along the highway at +3 mi.

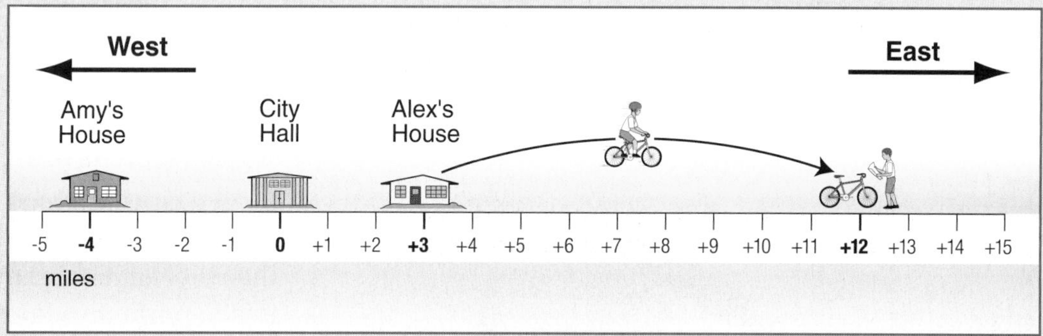

On the map, the positions of Alex and of his house are both defined as they relate to a *reference point*, which in this case is City Hall. An object's position can also have a negative value relative to a reference point. For example, Amy's house is 4 mi. west of City Hall, which on the map corresponds to a position of −4 mi. The minus sign indicates that Amy's house lies to the *west* of City Hall, just as the plus sign indicates a position *east* of City Hall.

To find the distance he has traveled, Alex needs to find the difference between his current position and his starting position. For the situation described above, the distance Alex has traveled is:

$$(+12 \text{ mi.}) - (+3 \text{ mi.}) = 9 \text{ mi.}$$

1. If Amy starts from her house and rides her bike to Alex's house, what would be the distance she traveled?

Your teacher will review the answer to this question with the class.

Calculating the Speed of an Object

Imagine you are in a car cruising on the highway. How would you know if you were traveling at a constant speed or at a changing speed (speeding up or slowing down)? One thing you could do is look at the speedometer on the car. The speedometer measures the speed of the car. If you were traveling at a **constant speed**, the speedometer needle would remain steady and not move around. That is because an object moving at a constant speed is neither speeding up nor slowing down.

Science Words

constant speed: neither speeding up nor slowing down

Your teacher will show you a movie or simulation of a car moving at a constant speed. Notice that the simulator has both a ruler and a clock on the screen. The ruler and clock allow you to collect data on the position of the car for different times. From this data, you can describe the motion of the car.

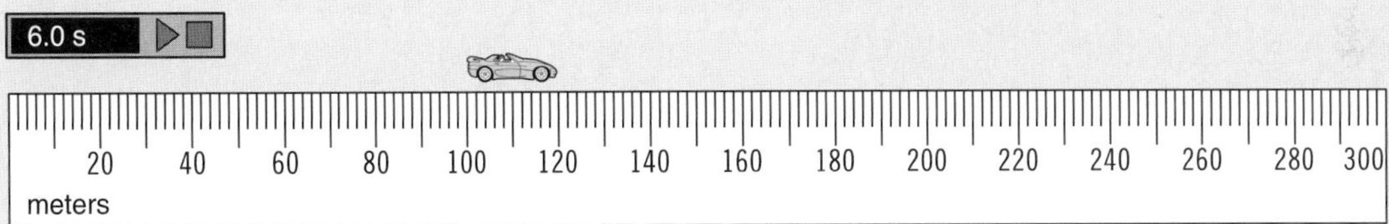

The data in the table below is similar to what you would record if you took data from the simulator.

Table: Position of Car versus Time

Time (s)	Position of Car (m)
0	0
2.0	40
4.0	80
6.0	120
8.0	160
10.0	200

Graph 1: Distance versus Time

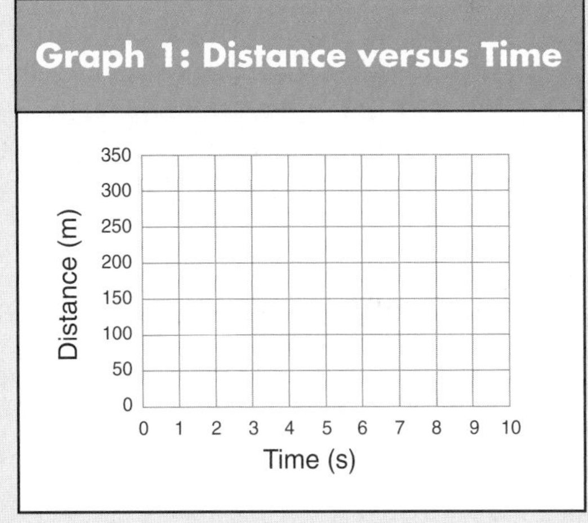

To determine the speed of the car from the data in the table, pick any two times and determine the time elapsed and the distance traveled. Then find the speed:

$$\text{Speed} = \frac{\text{Distance traveled}}{\text{Time elapsed}} = \frac{\text{Distance}}{\text{Time}}$$

For example, if you use the positions of the car at 2 s and 6 s, you can find the speed:

$$\text{Time elapsed} = 6\text{ s} - 2\text{ s} = 4\text{ s}$$

$$\text{Distance traveled} = 120\text{ m} - 40\text{ m} = 80\text{ m}$$

$$\text{Speed} = \frac{\text{Distance traveled}}{\text{Time elapsed}} = \frac{80\text{ m}}{4\text{ s}} = 20\text{ m/s}$$

2. What is the speed of the car between 4.0 s and 10.0 s? Show your work.

Because the speed of the car was *constant*, your answer should be the same as the example worked out above.

Now do the examples below, showing your work.

3. Ephraim walks to his job every day. His workplace is located 2 km (2000 m) from his home. He takes 20 min (or 1/3 h) to get to work. How fast is he walking? Give an answer in both meters per minute (m/min) and kilometers per hour (km/h).

4. In the distant future, Rachel flies a single-person, high-speed spacecraft between Earth and the Moon. Once she gets out of Earth's atmosphere, Rachel makes the 384,000 km trip in exactly 1000 s. How fast is Rachel moving?

Your teacher will review answers to these questions with the class.

Determining Speed from a Graph

Sometimes relationships between variables are difficult to find by simply examining a table or doing simple calculations. Another way of describing relationships in the data is by constructing a graph.

5. On Graph 1 on your record sheet, construct a graph of distance versus time using the data from the table (see the table and graph on page 139.) Draw a line through the data points. After you complete your graph, your teacher will show you a distance versus time graph drawn by the simulator. Compare it to your graph. Are they similar?

Your graph should look like a straight line. Whenever one quantity is plotted against another and the resulting graph is a straight line, there is a **linear relationship** between the two quantities. In this case, there is a linear relationship between distance and time.

Science Words

linear relationship: the relationship between two quantities that, when plotted against each other on a graph, produce a straight line

slope: the tilt or slant of a straight line on a graph; the rise divided by the run

6. Look at Graph 1: Distance versus Time that you sketched and, complete the following statement on your record sheet:

The distance that the sports car travels _________ *(increases, decreases, remains the same)* as time increases.

Your teacher will show you a graph of the speed of the car in the simulation versus time. Sketch the graph on Graph 2 on your record sheet.

7. Look at Graph 2: Speed of Car versus Time that you sketched and complete the following statement on your record sheet:

The speed of the car _________ *(increases, decreases, remains the same)* as time increases.

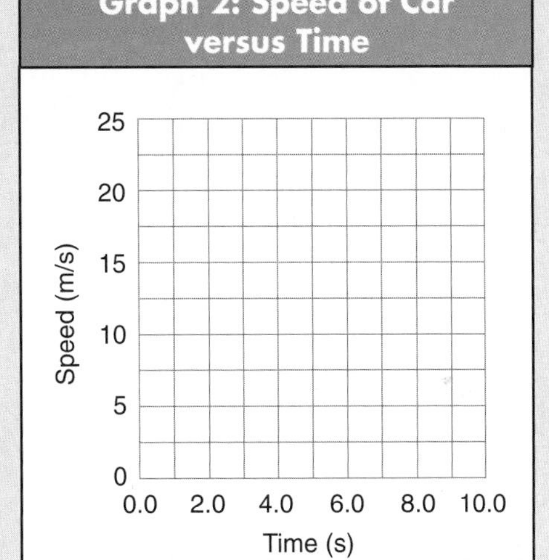

Your teacher will review answers to Questions 6 and 7 with the class.

Watch the simulation with two vehicles that are traveling at two different constant speeds. The simulator also displays the *distance* versus time graphs and the *speed* versus time graphs for each vehicle.

The tilt or slant of a straight line on a graph is called the **slope** of the line.

8. Compare and describe the slopes of the two plotted lines on the *distance* versus time graphs. (Graphs 3 and 4 on the next page.)

9. Compare and describe the slopes of the two plotted lines on the *speed* versus time graphs. (Graphs 5 and 6 on the next page.)

Your teacher will review answers to Questions 8 and 9 with the class.

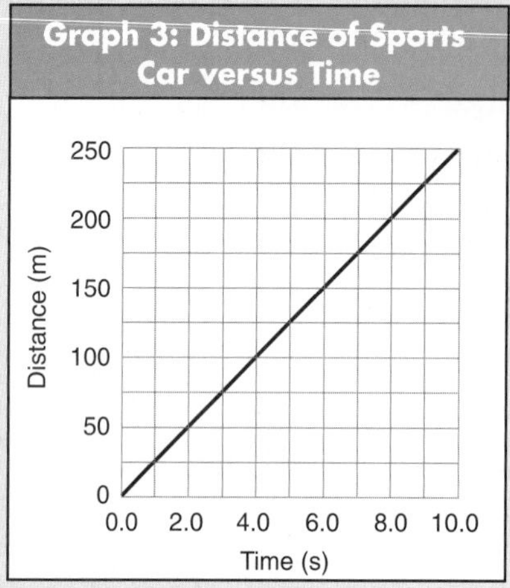

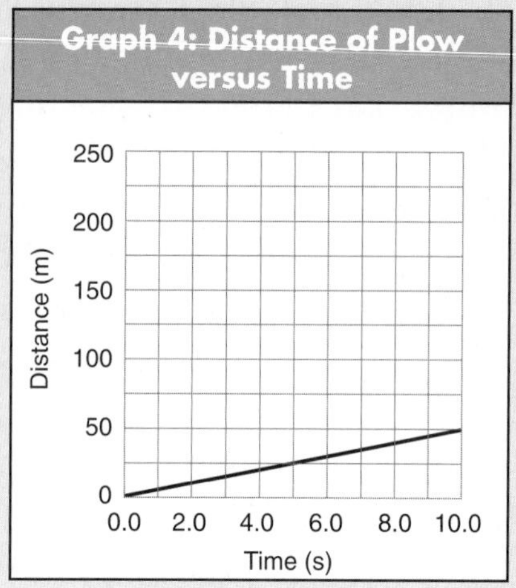

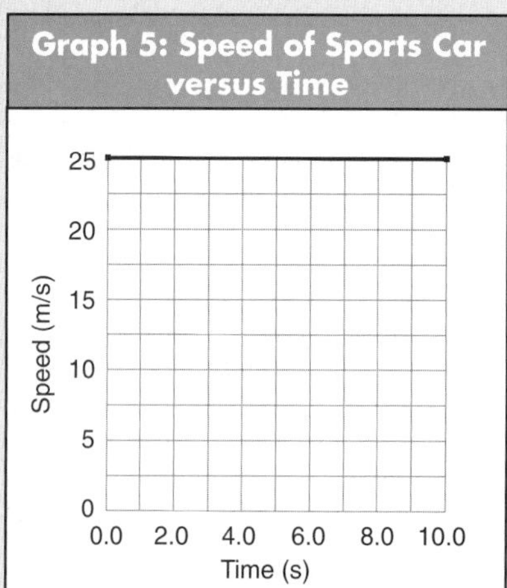

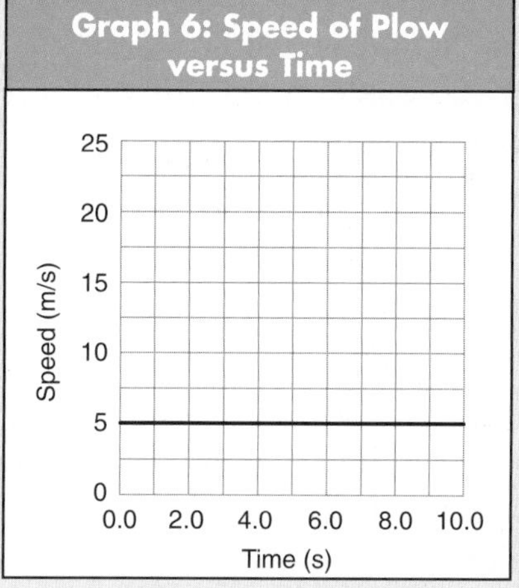

You can use the speed = distance/time equation to calculate the speed of an object. You can also calculate the speed directly from a distance versus time graph by calculating the slope of the line.

- Pick any two points on the graph.
- Subtract the distance-values of these two points. This is called the "rise."
- Subtract the time-values for the same two points. This is called the "run."
- Divide the rise by the run. The value you get is the slope of the line.

On Graph 7 below, for example, points (5, 20) and (10, 40) have been selected.

Graph 7: Finding the Rise and Run to Determine the Slope

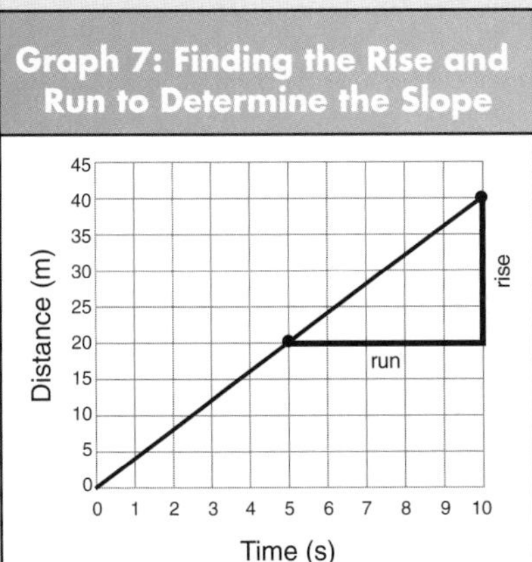

Subtract the two distance-values to find the rise:

$$\text{Rise} = 40 \text{ m} - 20 \text{ m} = 20 \text{ m}$$

Subtract the two time-values to find the run:

$$\text{Run} = 10 \text{ s} - 5 \text{ s} = 5 \text{ s}$$

Divide the rise by the run:

$$\frac{\text{Rise}}{\text{Run}} = \frac{20 \text{ m}}{5 \text{ s}} = \frac{4 \text{ m}}{\text{s}}$$

Therefore, the slope of the line for this graph is 4 m/s.

The rise value corresponds to distance traveled between two positions and the run value corresponds to time elapsed. So, the value of the slope (rise/run) is also equal to the speed of the object. Steeper lines mean larger slopes and, on distance versus time graphs, higher speeds.

Graph 8: Determining the Slope of Distance versus Time

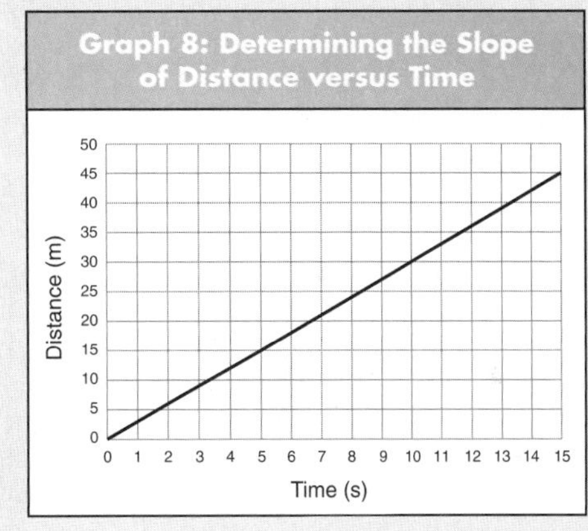

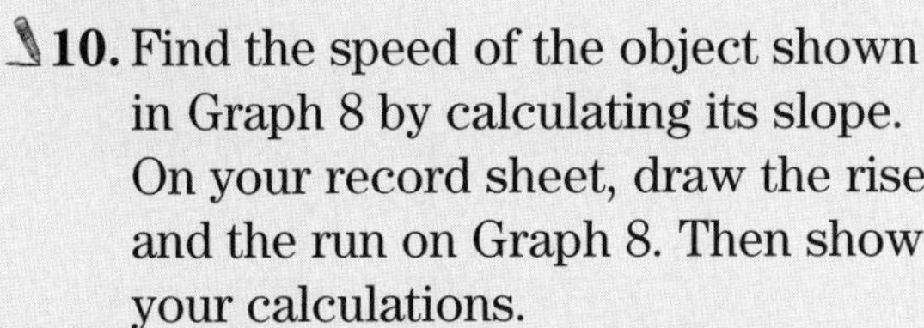

10. Find the speed of the object shown in Graph 8 by calculating its slope. On your record sheet, draw the rise and the run on Graph 8. Then show your calculations.

Your teacher will review answers to this question with the class.

What We Have Learned

Recall the key question for this activity:

How can you determine the speed of an object that has constant motion?

Participate in a whole-class discussion to review the answers to the key question.

Write the answer to the key question on your record sheet.

LEARNING ABOUT OTHER IDEAS

Activity 7
Speed of Waves

Most bats use echolocation to navigate in the dark and find food.

Purpose

In Activities 3 and 4, you learned that waves transfer energy and can be described in terms of their frequency, wavelength, and amplitude. Waves also travel at certain speeds and this fact is often used to determine distances (or speeds). Bats emit very high frequency compression waves (sound waves). They determine the distance to an object by measuring the delay between the waves they emit and the returning waves reflected from that object. (The process is called echolocation.) Cameras with automatic focusing work in a similar way to determine the distance to an object that needs to be put in focus. Radar guns used by police to detect the speed of an approaching car also work on this principle.

The speed of waves plays an important role in nature and in practical applications.

In this activity, you will first investigate what property or properties of a wave determine its speed. You will then apply what you learned in Activity 6 to calculate various distances based on the speed of waves.

The key question for this activity is:

What property or properties of a wave determine its speed?

Record the key question for the activity on your record sheet.

Learning the Ideas

Does the Speed of Waves Depend on Amplitude, Frequency, or Something Else?

To answer this question, your teacher will show you some movies. In the first movie segment, you will observe a person generating transverse waves along a Slinky, similar to the one you used in Activity 3. In a split screen, you will see the observer generating two waves at the same frequency but with different amplitudes. Pay attention to how the speeds of the waves in the two halves of the screen seem to compare.

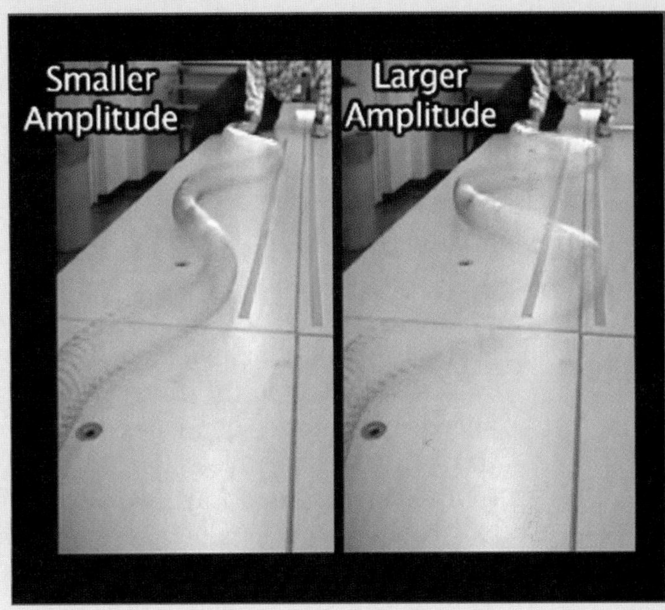

1. Does the larger amplitude transverse wave seem to move *much faster*, *much slower*, or *at about the same speed* as the smaller amplitude transverse wave?

The next movie segment shows another split screen where the person generates two compression Slinky waves of the same frequency but with different amplitudes.

2. Does the larger amplitude compression wave seem to move *much faster*, *much slower*, or *at about the same speed* as the smaller amplitude compression wave?

The next segment is a movie of a computer simulation, comparing sound waves generated with the same frequency but with different amplitudes. In the simulation, sound waves are generated by vibrating speakers placed in water.

3. Does the larger amplitude sound wave seem to move *much faster*, *much slower*, or *at about the same speed* as the smaller amplitude sound wave?
4. Based on the three examples shown, does the speed of either a *transverse* or *compression* wave seem to depend on its amplitude?

Your teacher will review the answers to these questions with the class.

You will observe another series of movies to determine whether the speed depends on the frequency of the wave. The first segment shows two transverse Slinky waves generated with different frequencies. The second segment shows two compression Slinky waves generated with different frequencies. The third segment shows a computer simulation of two sound waves generated with different frequencies but the same amplitude.

5. Based on the three movie segments, do higher frequency waves (either transverse or compression) seem to move *much faster*, *much slower*, or *at about the same speed* as lower frequency waves?

Your teacher will review the answers to this question with the class.

The last movie segment shows a computer simulation of sound waves in three different media (materials): air, water, and steel.

6. Does the speed of the sound waves seem to depend on the medium (material) through which they move?

7. Is the speed of sound in steel *greater than*, *less than*, or *the same as* it is in air?

Your teacher will review the answers to these questions with the class.

You drew your conclusions by looking at a computer simulation of sound waves. For all types of transverse and compression waves, however, the speed of the waves depends on the properties of the medium (material) through which it travels. For example, scientists have discovered that the internal structure of Earth consists of layers of different materials. They know this because earthquake waves originating inside Earth travel at different speeds through the different materials.

The speed of sound depends on the medium through which the sound travels. Generally, sound travels faster through solids than through liquids, and faster through liquids than through gases. The speed of sound in liquids and gases also depends somewhat on the temperature of the liquid or gas. The table on the next page lists the speed of sound in different media (plural of *medium*).

Calculating Speeds, Distances, and Times with Waves

Since waves travel through a medium at a constant speed, you can use the same relationship between speed, distance traveled, and time elapsed that you learned in Activity 6.

$$\text{Speed} = \frac{\text{Distance traveled}}{\text{Time elapsed}}$$

Table: Speed of Sound in Different Media*	
Medium	**Speed of Sound (m/s)**
Steel	5960
Pyrex Glass	5640
Gold	3240
Lucite Plastic	2680
Rubber (Gum)	1550
Sea (Salt) Water (25°C)	1531
Water (25°C)	1497
Kerosene (25°C)	1324
Helium (0°C)	965
Air (0°C)	331
Air (20°C)	343

**Data from the Handbook of Chemistry and Physics, 53rd Edition (1972-1973), published by The Chemical Rubber Company.*

Or, assuming distance means distance traveled, and time means time taken, it can be written as follows:

$$\text{Speed} = \frac{\text{Distance}}{\text{Time}}$$

If you know the speed and distance, you can rearrange the equation to solve for the unknown time.

$$\text{Time} = \frac{\text{Distance}}{\text{Speed}}$$

If you know the speed and time, you can rearrange the equation differently to solve for the distance.

$$\text{Distance} = \text{Speed} \times \text{Time}$$

Let's work through some problems involving the speed of waves.

Problem 1: The Slinky Wave

Suppose two students stretch out the Slinky so that it is 6 m long. One student generates a transverse pulse. It takes exactly 1.5 s to reach the student at the other end.

8. What is the speed of the transverse pulse? Show your work.

Problem 2: The Steel Beam

A steel beam is 20 m long. Someone strikes one end of the beam with a sledgehammer.

9. How long does it take for the sound of the hammer strike to travel from one end of the steel beam to the other? Show your work.

10. How long does it take for the sound of the hammer strike to travel the same distance through air (assume the temperature is 20°C)? Show your work.

Your teacher will review answers to these questions with the class.

What We Have Learned

Recall the key question for this activity:

What property or properties of a wave determine its speed?

Participate in a whole-class discussion to review the answer to the key question.

Write the answer to the key question on your record sheet.

LEARNING ABOUT OTHER IDEAS

Activity 8
Objects and Waves with Changing Speeds

Purpose

From your everyday experience, you know that moving objects do not usually have constant motion. If you are on your skateboard, you might build up speed as you whiz along a flat part of the sidewalk and slow down as you approach a crosswalk. Cars, bikes, and athletes are constantly speeding up and slowing down. In our everyday activities, each one of us speeds up and slows down.

Moving objects rarely travel at a constant speed.

In addition, waves travel at different speeds in different media. For example, as you saw in Activity 7, sound waves move much faster in steel than in air. In Activity 6, you learned several different ways to describe the motion of objects with constant speed. In this activity, you will learn different ways to describe the motion of objects and waves with changing speeds. Do they *speed up*, *slow down*, or *do both*?

The key question for this activity is:

How do you determine a speed for an object or a wave whose speed changes with time?

Record the key question for the activity on your record sheet.

Learning the Ideas

Calculating the Average Speed of an Object or Wave Not Moving at a Constant Speed

Science Words

average speed: the distance traveled divided by the time taken

You can calculate the **average speed** of an object or wave that speeds up or slows down in the same way that you calculate the speed of an object or wave with a constant speed. For an object or wave with changing speed:

$$\text{Average Speed} = \frac{\text{Distance traveled}}{\text{Time elapsed}} = \frac{\text{Distance}}{\text{Time}}$$

For example, the average speed of an object that travels 300 m in 10 s is 300 m/10 s = 30 m/s. This equation does not tell you what the speed of the object is at any given instant. It only tells you what the speed is averaged over the entire travel time. For example, a race car that speeds up from rest to 60 m/s in 10 s and travels 300 m has the same average speed as a car that has a constant motion of 30 m/s for the same 10 s.

Your teacher will show you a simulation of three cars: one that speeds up, one that moves at a constant speed, and one that slows down. Notice that the simulation includes a ruler and clock, like the simulations in Activity 6. Determine which car is speeding up, which one is moving at a constant speed, and which one is slowing down.

Watch the simulation a second time. Then answer the questions below.

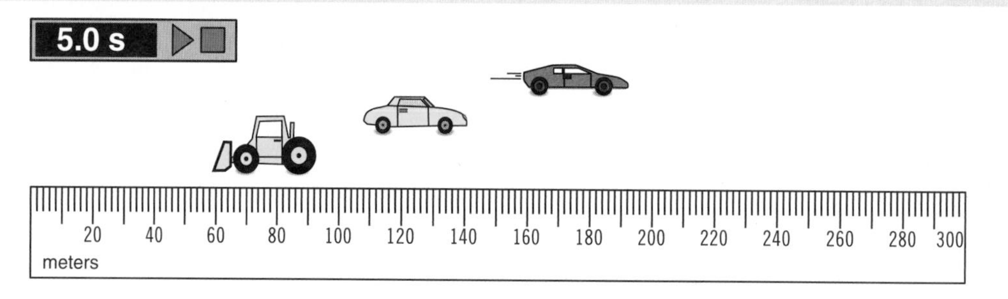

1. For the 10-s simulation run, how do the *average* speeds of the three cars compare? Explain your answer. *Hint*: You don't need to do any calculations to answer this question.
2. Pick one of the three cars and calculate its average speed.

Your teacher will review answers to these questions with the class.

Calculate the average speeds in the examples below. In each case, you only need to know the total distance traveled and the time elapsed to answer the question.

3. Alex's family travels 200 mi. by car to reach their favorite camping spot. Alex's mother is driving and repeatedly has to change the car's speed to adjust to the changing flow of traffic. The entire trip takes four hours. What is the average speed of the family's car? Answer in miles per hour.

4. Sofia is driving her car and needs to enter freeway traffic from an on-ramp. In five seconds and over a distance of 100 m, she speeds up from 13.0 m/s to 30.0 m/s. What is her average speed?

5. Richard is scuba diving in the Pacific Ocean 20 m below the surface. He hears the roars of two low-flying military jets passing overhead, 100 m above the surface of the ocean. The speed of sound in air is 332 m/s and in seawater, it is 1535 m/s. The sound waves from the jets take 0.30 s to reach Richard's ears. What is the average speed of the sound waves?

6. Refer to the diagram below. A seismograph detects P waves from an earthquake that originates in the upper mantle 73 km directly below the seismograph. Because the P waves move through different types of rock, their speed is different in each layer depicted in the picture. It is 8.0 km/s in the upper mantle, 7.0 km/s in the lower crust, and 6.0 km/s in the upper crust. The P waves take 10 s to reach the seismograph. What is the average speed of the P waves?

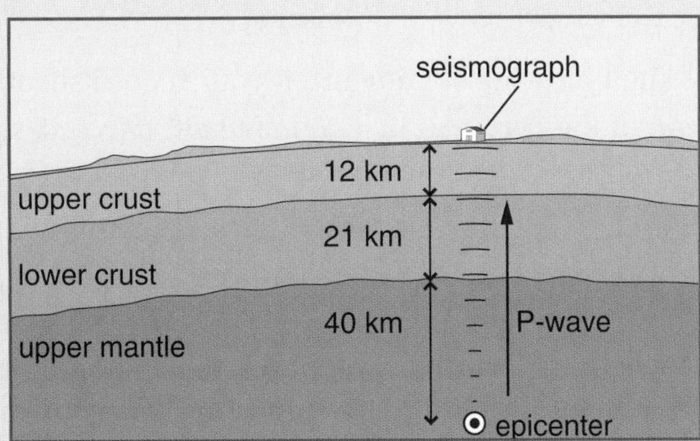

A P wave travels through 40 km of upper mantle, 21 km of lower crust, and 12 km of upper crust for a total distance of 73 km before reaching the seismograph.

Your teacher will review answers to Questions 3 - 6 with the class.

Distance versus Time and Speed versus Time Graphs

Watch a simulation of one car with changing speed. The simulator will display graphs of distance versus time and speed versus time. Watch the simulator a few times.

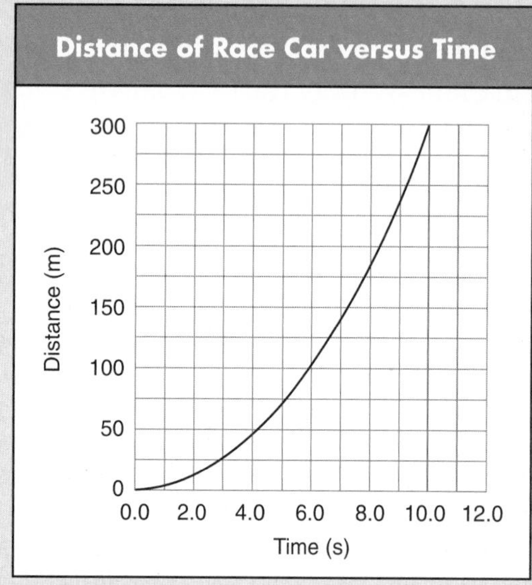

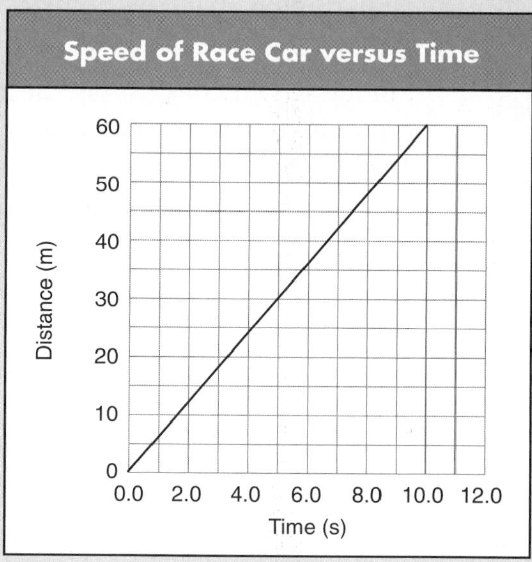

7. Complete the following statement on your record sheet:

The distance that the car travels __________ (*increases, decreases, remains the same*) as time increases.

8. Complete the following statement on your record sheet:

The speed of the car _________ *(increases, decreases, remains the same)* as time increases.

Your teacher will review the answers to these questions with the class.

Science Words

nonlinear relationship: the relationship between two quantities that, when plotted against each other on a graph, do not produce a straight line

When the speed of an object changes, as in this case, the graph of distance versus time is not a straight line as it was in Activity 6. Instead, it forms a curve. Whenever one quantity is plotted against another and the resulting graph is *not* a straight line, there is a **nonlinear relationship** between the two quantities. Note that while there is a nonlinear relationship between distance and time, the relationship between speed and time in this case is linear.

Average Speed when the Direction of Motion Changes

In the scenario described at the start of Activity 6, Alex had ridden his bicycle 9 mi. east of his home over the course of one hour. Thus, you can find his average speed:

$$\text{Average speed} = \frac{\text{Distance}}{\text{Time}} = \frac{9 \text{ mi.}}{1 \text{ h}} = 9 \text{ mi./h}$$

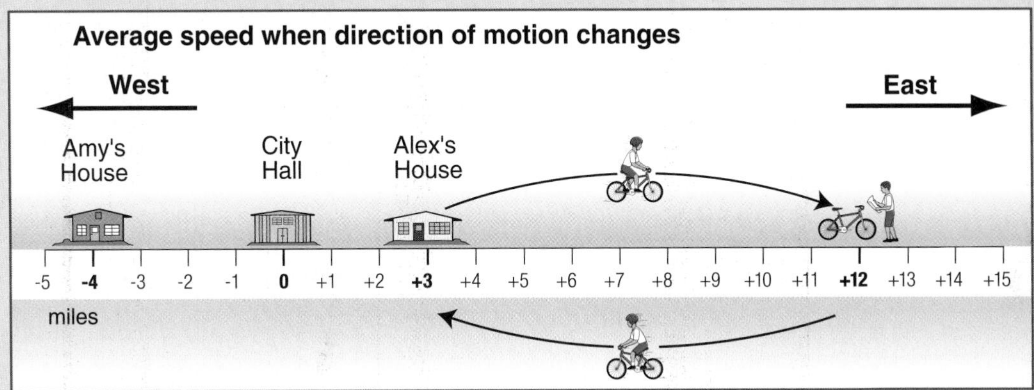

Suppose Alex then turned around and headed home. He takes another hour to ride home. What is his average speed for the entire two-hour trip?

$$\text{Average speed} = \frac{\text{Total distance traveled}}{\text{Time elapsed}} = \frac{\text{Total distance}}{\text{Time}}$$

To find the average speed for situations in which an object's direction changes, you need to use the *total* distance traveled:

You can find the total distance traveled by adding together the distances traveled between changes in direction. In Alex's case, he travels 9 mi. east, and then 9 mi. west, for a total distance of 18 mi. His total travel time is 2 hours. Therefore, we can find his average speed:

$$\text{Average speed} = \frac{\text{Total distance}}{\text{Time}} = \frac{(9 \text{ mi. east} + 9 \text{ mi. west})}{2 \text{ h}} = 9 \text{ mi./h}$$

9. Suppose that starting from home, Alex travels along the road 9 mi. east (to the right). He then turns around and, rather than going back home, he rides to City Hall. His total travel time is 3 h. What is his average speed?

Your teacher will review the answers to this question with the class.

Velocity and Acceleration

Science Words

velocity: how fast an object is moving in a given direction

acceleration: the change in velocity per unit time

The speed of an object will only tell you how fast it is moving, not the direction of its motion. To describe both how fast an object is moving and the direction of its motion, scientists use the term **velocity**. For example, consider the situation described before in which Alex is moving with the same speed traveling both east (away from home) and west (toward home). When he is moving east, his velocity is 9 mi./h to the east. When he is on his return trip, his velocity is 9 mi./h to the west.

In everyday life, the term **acceleration** means speeding up. But scientists use the term differently. They use acceleration to describe how quickly an object's velocity changes. Since velocity includes both speed and direction, an object has acceleration (or is accelerating) when any of the following occurs:

- its speed changes (either speeding up or slowing down)
- its direction changes (either turning or moving on a circular path)
- both its speed and direction change

What We Have Learned

Remember the key question for this activity:

How do you determine a speed for an object or a wave whose speed changes with time?

Participate in the class discussion of the answer to the key question.

Write the answer to the key question on your record sheet.

CYCLE 2:
MECHANICAL INTERACTIONS AND ENERGY

Cycle Key Question

Motion energy: Where does it come from? Where does it go? Why does it change?

Activity 1: Notions about Motion Energy

Activity 2: Mechanical Interactions and Motion Energy

Activity 3: Following the Energy Changes

Activity 4: Elastic Interactions

Activity 5: Mechanical Interactions and Energy Transfer

Activity 6: Analyze, Explain, and Evaluate

OUR FIRST IDEAS

Activity 1
Notions about Motion Energy

Cycle Purpose

In Cycle 1, you learned that energy can be transferred from one object to another. In this cycle, you will continue to learn about mechanical interactions and the energy changes involved in mechanical interactions. To get started, your teacher will show you a short sports and recreation video. Although you may not realize it, the video is packed with examples of mechanical interactions.

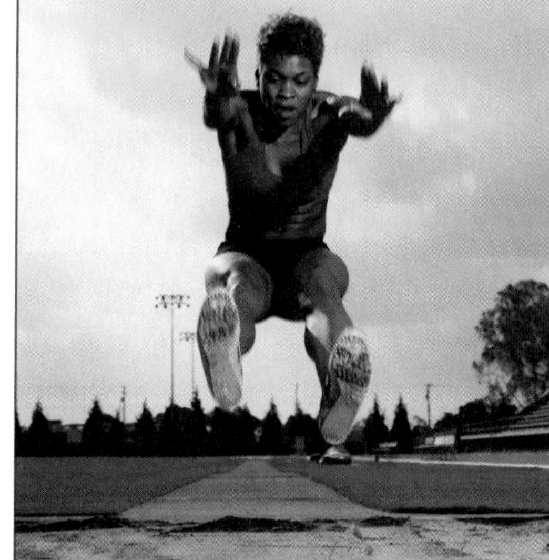

Activities in sports involve many mechanical interactions.

Otis wants to learn about sky diving and Isabel wants to learn about volleyball. I want to learn about swimming because I would like to be a better swimmer.
Xuan

Xuan wants to learn about swimming and Nguyen wants to lear about basketball. I want to learn about bike racing because I like racing bikes with my friends. Maybe if I know more about the interactions in bike racing, then I can win more often!
Carlos

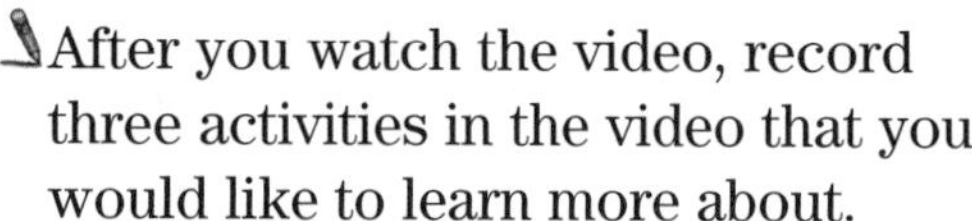

After you watch the video, record three activities in the video that you would like to learn more about.

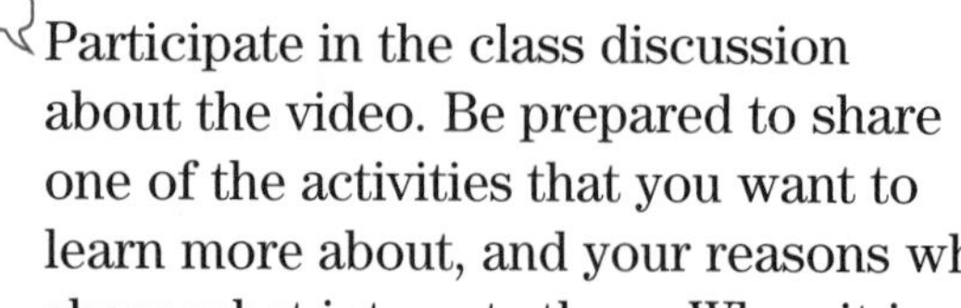

Participate in the class discussion about the video. Be prepared to share one of the activities that you want to learn more about, and your reasons why. Listen to your classmates when they share what interests them. When it is your turn to speak, you should be able to repeat the interests of two of your classmates.

Objects can have many different types of energy. One type is **motion energy**. Moving objects have motion energy. If an object is speeding up, it is increasing in motion energy. The faster the object moves, the more motion energy it has. If an object is slowing down, it is decreasing in motion energy. The slower the object moves, the less motion energy it has.

In this cycle, you will focus on these questions:

Science Words

motion energy: the energy an object has because of its motion

Motion energy: Where does it come from? Where does it go? Why does it change?

Record the key questions for the cycle on your record sheet.

In this activity, you will think about why motion energy changes and what happens to the motion energy.

We Think

Discuss these scenes with your team and answer the questions. Be prepared to share your answers with the class.

Scene 1

You are watching a world soccer championship game and you see the goalie make a spectacular save! The goalie then throws the ball with both of his arms, sending the ball flying through the air.

1. How did the ball get its motion energy? Where did it come from?

Later in the match, your favorite soccer player makes a goal. The ball is caught in the net.

2. What happened to the ball's motion energy? Where did it go?

Scene 2

A baseball player is trying to steal second base. He slides headfirst toward the base. He slows down as he approaches the base.

3. What makes the player slow down? What happened to his motion energy?

Scene 3

A parachute is let out behind a drag-racing car. The car slows down.

4. What makes the car slow down? What happened to the car's motion energy?

Participate in a class discussion of these four questions.

Under your written answer for each of the four questions, write down at least one other student's answer that is different from yours.

Our Class Ideas

The cycle questions are:

Motion energy: Where does it come from? Where does it go? Why does it change?

You started thinking about these questions today as you considered the soccer, baseball, and drag-racing car scenes. You will continue to think about them throughout this cycle. For now, answer the following questions:

1. If an object *increases* in motion energy, where do you think the additional energy comes from?

2. If an object *decreases* in motion energy, where do you think the energy goes?

3. In all of these scenes, the motion energy of an object *changes*. What do all of these scenes have in common that makes the motion energy of the objects change?

Participate in a class discussion about these questions.

DEVELOPING OUR IDEAS

Activity 2
Mechanical Interactions and Motion Energy

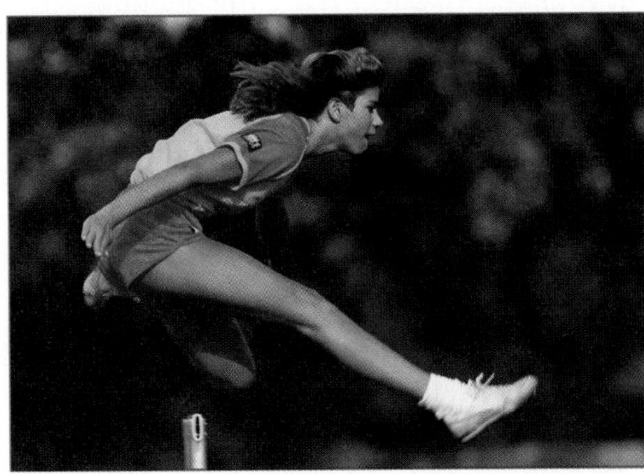

Mechanical interactions can cause changes in motion energy.

Purpose

In the last activity, you considered different examples of changing motion energy (throwing a soccer ball, a net catching the ball, a baseball player sliding into a base, and a drag-racing car slowing down). In this activity, you will explore mechanical interactions that cause changes in motion energy. The key question for this activity is:

How can you change the motion energy of an object?

Record the key question for the activity on your record sheet.

We Think

Discuss the activity key question with your team.

Write your response on your record sheet. Be prepared to share your response and your reasons in a class discussion.

Participate in a class discussion about the key question.

Science Words

mechanical interaction: an interaction in which objects touch each other while pushing and pulling each other over a distance

Explore Your Ideas

Scientists look at the world as interactions. When two objects interact, they influence each other somehow. In Cycle 1, you were introduced to the idea of interactions. In this cycle you will explore one particular type of interactions, called mechanical interactions. The defining characteristic of a **mechanical interaction** is that it occurs when objects touch each other while pushing or pulling on each other over a distance.

This is a very important idea that you will use over and over again.

Refer to *How To Identify Mechanical Interactions* in the Appendix. It describes the four different types of mechanical interactions. You will encounter these interactions throughout this course and throughout your everyday life.

STEP 1 The chart below shows the motion-energy scenes that you considered in the last activity.

1. Work with your partner. On your record sheet write the defining characteristic of each type of mechanical interaction.

Complete the chart on your record sheet.

Your team will need:

- variety of items supplied by your teacher

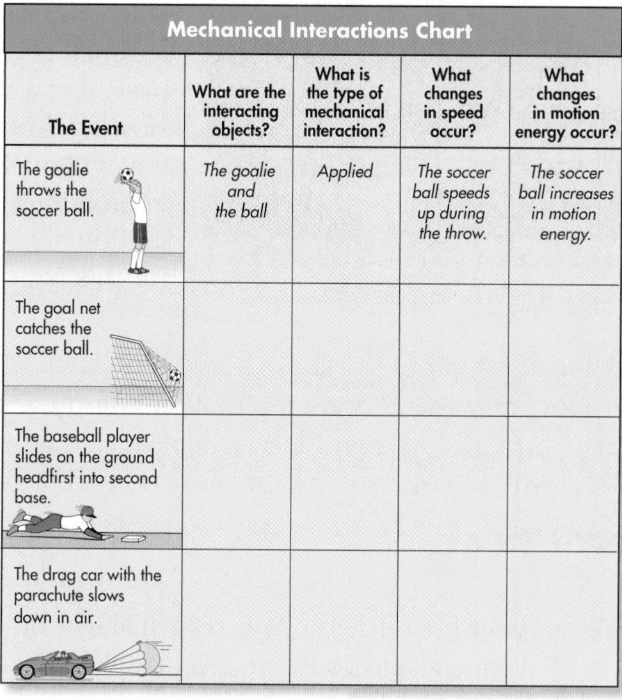

Mechanical Interactions Chart				
The Event	What are the interacting objects?	What is the type of mechanical interaction?	What changes in speed occur?	What changes in motion energy occur?
The goalie throws the soccer ball.	*The goalie and the ball*	*Applied*	*The soccer ball speeds up during the throw.*	*The soccer ball increases in motion energy.*
The goal net catches the soccer ball.				
The baseball player slides on the ground headfirst into second base.				
The drag car with the parachute slows down in air.				

Have your teacher check your plan before doing it.

Wear eye protection.

Participate in a class discussion and be prepared to share your chart.

STEP 2 Your team's mission is to create a setup using a variety of mechanical interactions. You must include at least one instance of *increasing* motion energy and one instance of *decreasing* motion energy. For example, you could roll a ball across the floor so that the ball speeds up. The ball then collides with a stretched rubber band and slows down. Don't use this example, but be creative! Try to create a setup with at least one of each of the four types of mechanical interactions.

2. Draw a sketch of your setup.

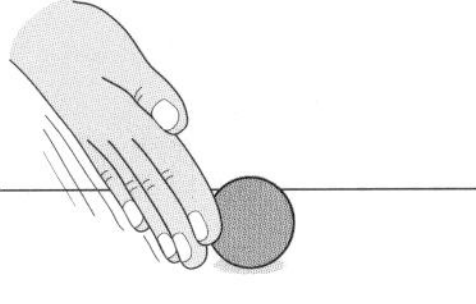

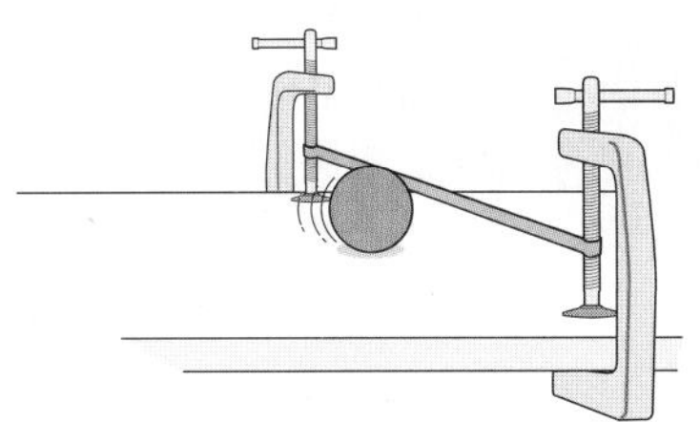

A chart similar to the one shown below is provided on your record sheet. Complete a row in the chart for each interaction. The first two rows are completed for the example below.

How to Identify Mechanical Interactions				
Describe the Event	What are the interacting objects?	What is the type of mechanical interaction?	What changes in speed occur?	What changes in motion energy occur?
A hand rolls the ball	The hand and ball	Applied	The ball speeds up	The ball increases in motion energy
The ball hits the rubber band	The ball and rubber band	Elastic	The ball slows down	The ball decreases in motion energy

Participate in a class discussion and be prepared to present a summary of your setup.

Make Sense of Your Ideas

1. In general, how can you increase the motion energy of an object?
2. In general, how can you decrease the motion energy of an object?

Our Consensus Ideas

Think about the key question for this activity and discuss it with your team. The key question is:

How can you change the motion energy of an object?

1. Write your answer on your record sheet.

Participate in the class discussion about the key question.

2. Write the class consensus ideas on your record sheet.

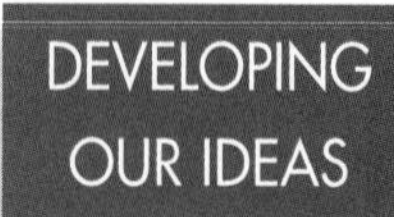
DEVELOPING OUR IDEAS

Activity 3
Following the Energy Changes

Purpose

In Cycle 1, you drew energy diagrams to describe energy transfers in some interactions.

In this activity, you will think about the source and receiver in mechanical interactions. The purpose of this activity is to answer the following key question:

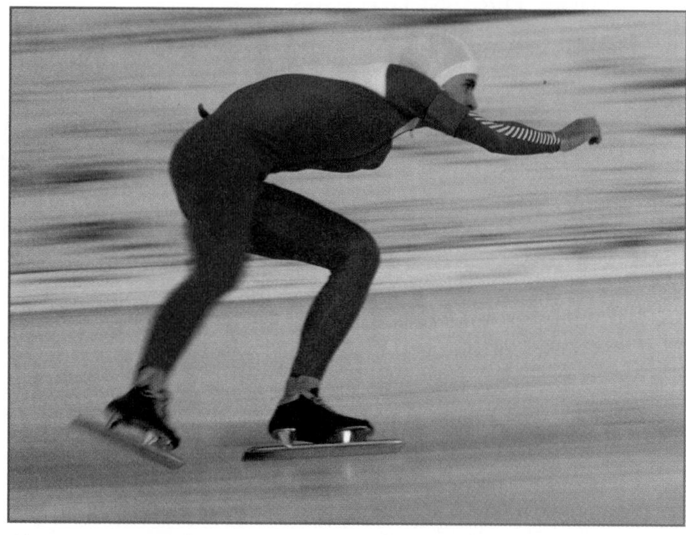

Most sports involve many energy changes as a result of mechanical interactions.

What happens to the energy in applied, friction, and drag interactions?

Record the key question for the activity on your record sheet.

We Think

Discuss the following questions with your team. Write your answers on your record sheet.

1. In a mechanical interaction (an applied, friction, drag, or elastic interaction), do you think the energy always changes in the source? If so, does it *increase* or *decrease*? Write your reasoning.

2. In a mechanical interaction, do you think the energy always changes in the receiver? If so, does it *increase* or *decrease*? Write your reasoning.

Participate in a class discussion about these questions.

Explore Your Ideas

You learned in Cycle 1 that the interacting objects (source and receiver) are written in the *rectangles* of an energy diagram. You also learned that the type of interaction is written above the energy diagram. In this cycle, you are learning about mechanical interactions. What is a mechanical interaction? If you don't remember, review *How To Identify Mechanical Interactions.*

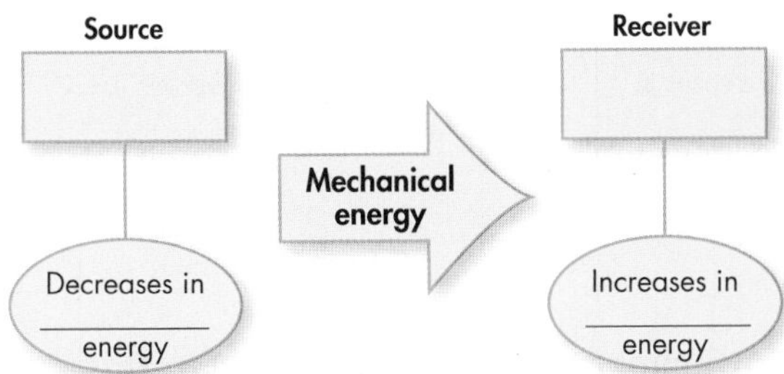

The type of energy that is transferred from the source to the receiver is written in the arrow. In a mechanical interaction, mechanical energy is transferred from the source to the receiver. In order for energy to be transferred from a source to a receiver, the source must *decrease* in some type of energy and the receiver must *increase* in some type of energy.

The energy change in the source and the energy change in the receiver are written in the ovals. In this cycle, you learned that objects can increase or decrease their motion energy. So one form of energy that can change for an object is *motion energy.*

Another type of energy is *stored chemical energy.* Your body stores chemical energy as a result of eating. This stored chemical energy is changed into other types of energy when you walk, run, etc. Batteries also store chemical energy.

Part A: Energy Diagrams for Applied Interactions

STEP 1 A boy is pulling a skateboarder faster and faster. The boy is interacting with the skateboarder on the skateboard. This is an example of an *applied* mechanical interaction. (In an applied interaction two non-stretchy objects push or pull on each other.)

What kind of energy *increases* in the skateboarder? Where does it come from?

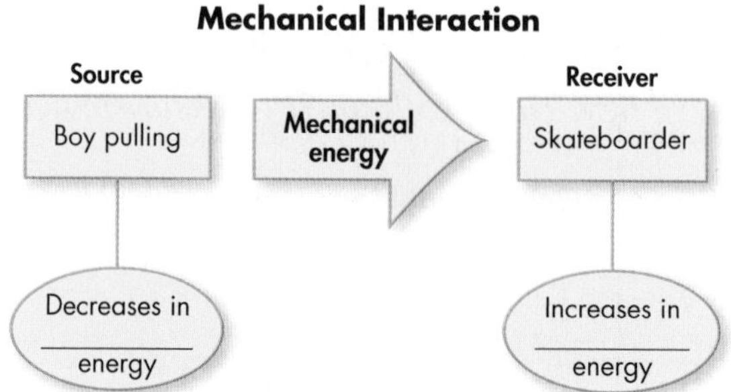

1. On your record sheet, complete the ovals in this energy diagram.

STEP 2 Working with your partner, discuss the following applied interactions and *draw energy diagrams* to describe them.

2. A person pushes a cart faster and faster.
3. A person pulls a sled faster and faster.
4. A bowling ball strikes a bowling pin.

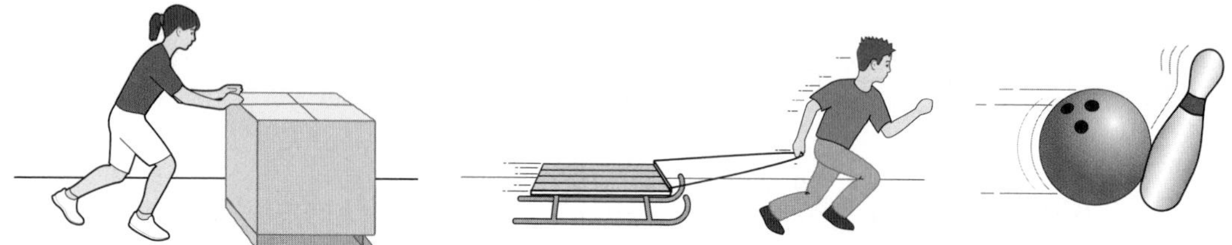

Part B: Energy Diagrams for Friction Interactions

STEP 1 To help you think about the energy in a *friction* interaction, your teacher will show you a video about friction. The video was made using a special infrared camera. Infrared cameras are sensitive to an object's warmth. In the video, you will see two friction interactions: hands rubbing together and a slat being rubbed along the top of a table.

5. In both cases, what was the evidence that a friction interaction had occurred?

STEP 2 Rub the palms of your hands together.

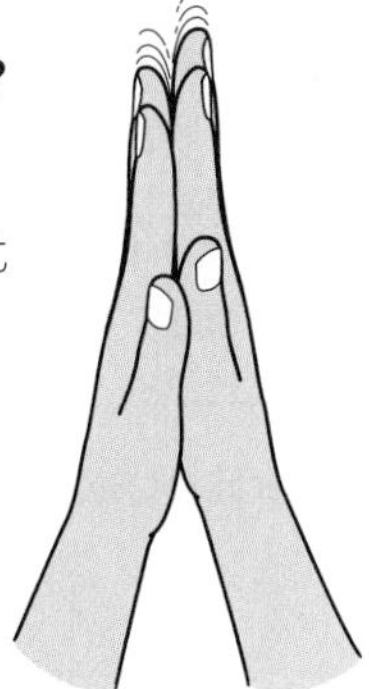

6. What is the evidence that a friction interaction is occurring?

When surfaces get warmer, that can be evidence for a friction interaction. When an object's temperature increases, the object is *increasing in thermal energy*. In all friction interactions two surfaces rub against each other and both surfaces become warmer. In other words, they increase in thermal energy.

What other evidence is there that a friction interaction has occurred?

STEP 3 Gently shove one of your books so that it moves across part of your desk. There is a friction interaction between the book and desktop.

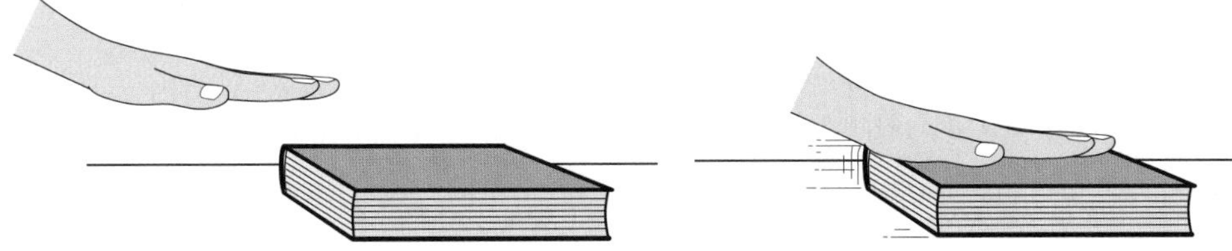

7. What do you observe?

After you shove the book, you observe the book slowing down to a stop. This is also evidence that a friction interaction has occurred. (If you had sensitive equipment, you could measure how much the book and desk warm up.)

8. On your record sheet, complete an energy diagram for the book sliding across the desk.

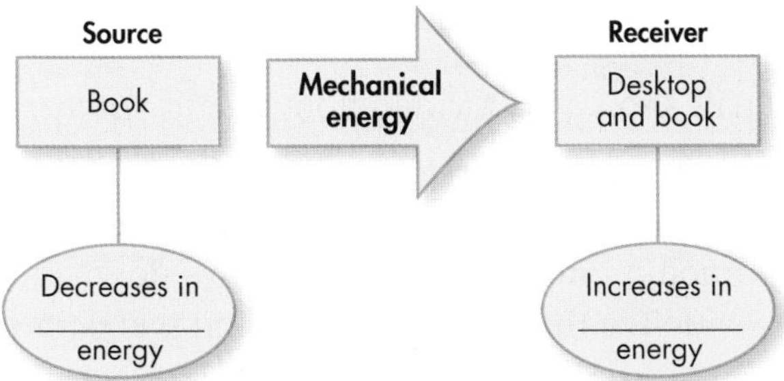

STEP 4 Working with your partner, discuss the following friction interactions and draw energy diagrams to describe them.

9. A snow skier is slowing down.

10. A skateboard is slowing down on smooth pavement. A friction interaction occurs between the skateboard and the rubbing parts in the wheels.

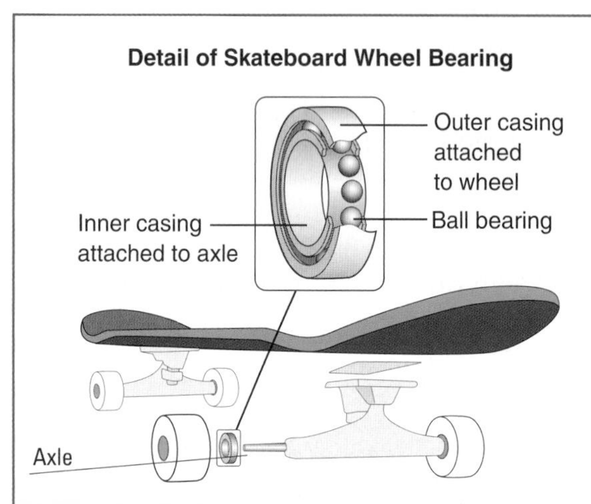

These rubbing wheel parts include the ball bearings and casings shown in the diagram. The inner casing is attached to the axle and does not spin. The wheel is attached to the outer casing, which rolls over the ball bearings and allows the wheel to spin.

As you draw your energy diagram, be aware that in this friction interaction, the skateboard is the energy source. The receiver is the rubbing wheel parts in the skateboard.

11. In general, what evidence is there that a friction interaction has occurred?

Part C: Energy Diagrams for Drag Interactions

Like friction, drag interactions are very common mechanical interactions in everyday life. *Drag* interactions occur when an object moves through a fluid like a gas or liquid. Evidence of a drag interaction is that the object slows down and the gas or liquid speeds up.

STEP 1 Your teacher will show you a video about *drag interactions* at a drag race.

In the video, a strong drag interaction occurs when the parachute is opened. The air and the car's parachute interact. The car slows down and the air speeds up. This is the evidence that a drag interaction has occurred.

12. Draw an energy diagram that describes the drag interaction while a racing car with a parachute is slowing down.

IDEA POWER

Activity 6
Analyze, Explain, and Evaluate

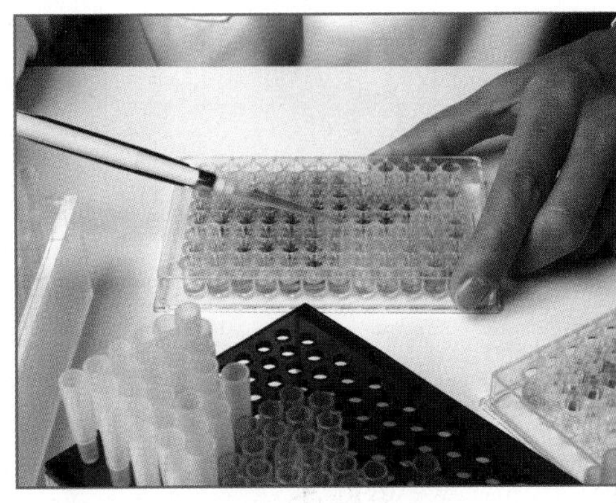

Scientists spend a lot of time trying to analyze and explain the world around them. Developers use what scientists learn to make more efficient cars, faster computers, better stereo speakers, glossier lipsticks, etc. In *InterActions in Physical Science*, you will learn how to analyze and explain some of the phenomena in the world around you.

Science Words

analysis: a procedure that helps you understand a situation

explanation: the use of science ideas and information from an analysis to answer questions about a situation

evaluation: a judgement of something

What is an **analysis?** An analysis is any procedure that helps you to understand a situation. Scientists always analyze a situation before answering questions about it. The type of analysis scientists perform depends on the question they are trying to answer. So far in this course, you too have analyzed situations by identifying interacting objects and how they are interacting with each other, and by drawing energy diagrams.

What is an **explanation?** An explanation uses information from the analysis and uses science ideas to answer questions about a situation. The science ideas that you will use are the ideas that you develop in this course. You can refer to the *Scientists' Consensus Ideas* forms to remind you.

What is an **evaluation?** An evaluation is a judgement of something. You performed evaluations in Unit 1 when you decided if experiments were fair tests or not. You also performed evaluations to determine whether experiment conclusions were valid. In *InterActions in Physical Science*, you will also evaluate whether analyses and explanations are good or poor.

In this activity, you will evaluate some analyses and explanations from a few other students. You will then perform some analyses and write your own explanations.

Here is a comic strip with lots of mechanical interactions. Look at the comic strip with your partner and identify as many mechanical interactions as you can.

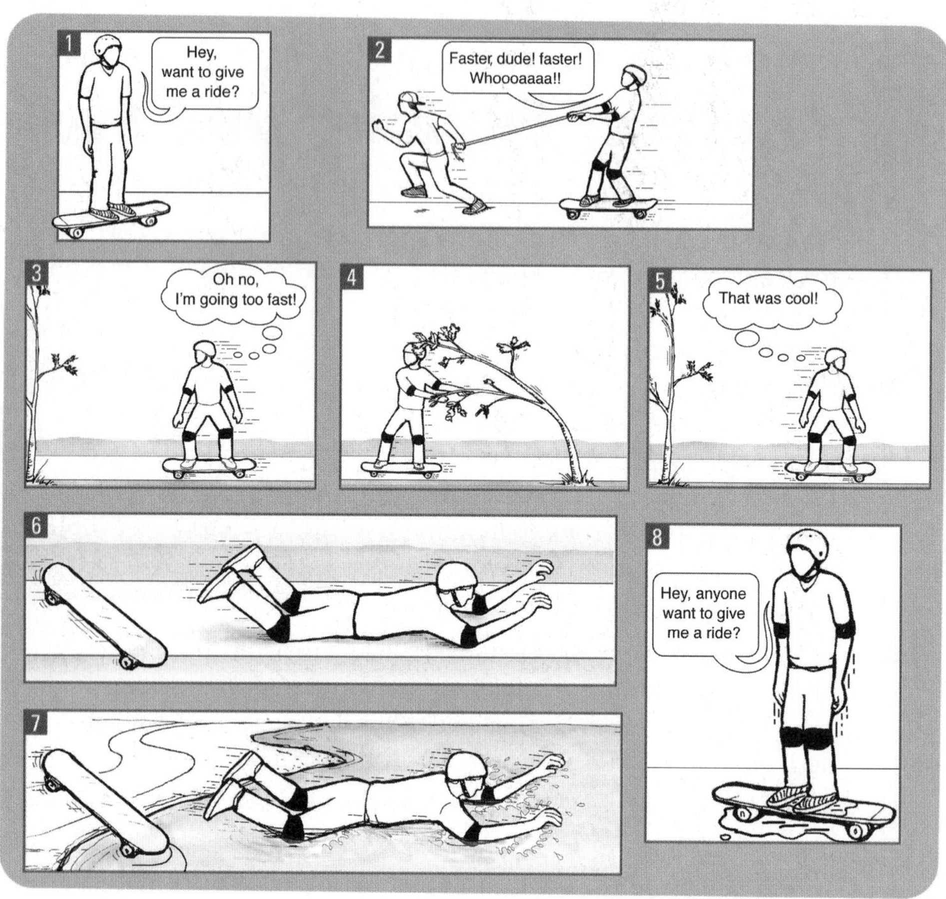

Otis wanted to analyze and explain why the skateboarder is speeding up in Frame 2 from the comic strip. Here is his work.

Task: Analyze and explain why the skateboarder is speeding up.

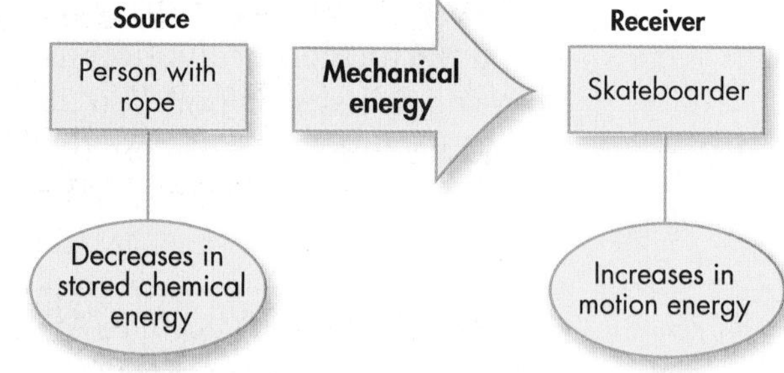

Analysis: There is an *applied interaction* between the skateboarder and the person with the rope.

Explanation: The skateboarder speeds up because the person with the rope transfers mechanical energy to the skateboarder during the applied interaction between them. The skateboarder's motion energy increases and the person's stored chemical energy decreases. Since the skateboarder's motion energy increases, he speeds up.

1. Look at #1 on *How To Evaluate an Analysis and Explanation.* Decide yes or no, and mark your answer on your record sheet. If you answered no, write the correct set of interacting objects and interaction type.
2. Look at #2 on *How To Evaluate an Analysis and Explanation.* Decide yes or no, and mark your answer on your record sheet. If you answered no, draw the correct energy diagram.
3. Look at #5 and #6 on *How To Evaluate an Analysis and Explanation* sheet. Decide yes or no, and mark your answers on your record sheet. If you answered no to either, rewrite the explanation to meet both criteria.

Here is an analysis and explanation written by Isabel of why the skateboarder slows down in Frame 4.

Task: Analyze and explain why the skateboarder slows down when he grabs the tree.

Analysis: There is an *elastic interaction* between the skateboarder and the tree.

Explanation: The skateboarder slows down because that's what always happens. The skateboarder's motion energy decreases and the tree's stored chemical energy increases.

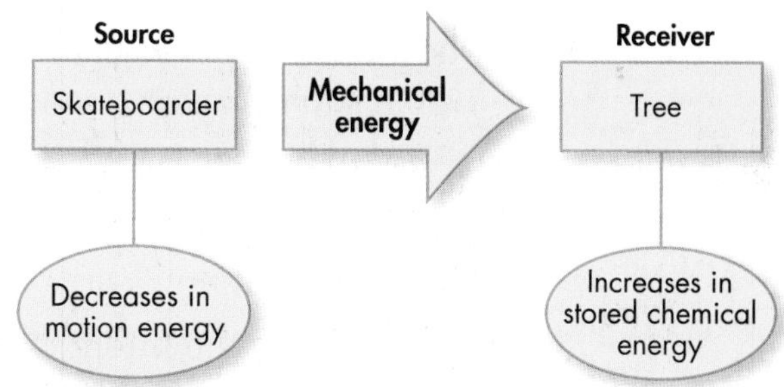

4. Look at #1 on *How To Evaluate an Analysis and Explanation.* Decide yes or no, and mark your answer on your record sheet. If you answered no, write the correct set of interacting objects and interaction type.
5. Look at #2 on *How To Evaluate an Analysis and Explanation.* Decide yes or no, and mark your answer on your record sheet. If you answered no, draw the correct energy diagram.
6. Look at #5 and #6 on *How To Evaluate an Analysis and Explanation.* Decide yes or no, and mark your answers on your record sheet. If you answered no to either, rewrite the explanation to meet both criteria.
7. *Now it's your turn!* Read the task, perform an analysis, and write an explanation.

Task: Analyze and explain why the skateboarder speeds up when he starts moving in the opposite direction in Frame 5.

Analysis:

- Identify the interacting objects and their type of interaction.
- Draw an energy diagram showing the energy transfers and changes.

Explanation:

- Use your analysis, and the science ideas you have developed or learned, to help you answer the question. Refer to earlier activities and to your *Scientists' Consensus Ideas* forms to review your science ideas.
- Write your explanation, using complete sentences. Your explanation should be able to pass an evaluation using *How To Evaluate an Analysis and Explanation.*

8. Choose *two* other events from the comic strip to analyze and explain. The events should involve mechanical interactions (applied, friction, drag or elastic). On your record sheet, complete the following for each event.

Task: Write the task.

Analysis:

- Identify the interacting objects and their type of interaction.
- Draw an energy diagram showing the energy transfers and changes.

Explanation:

- Use your analysis, and the science ideas you have developed or learned, to help you answer the question.
- Write your explanation using complete sentences.

Alert! In the Appendix, find *How To Write an Analysis and Explanation.* Notice that the steps are the same as you have followed in this activity. From now on, you will follow these steps when you write analyses and explanations. Acting like a scientist, you could also use *How To Write an Analysis and Explanation* to help you explain things in other science classes or in the world around you!

SECTION B

UNIT 3:
INTERACTIONS AND FORCES

UNIT 4:
INTERACTIONS AND CONSERVATION

NTERACTIONS, FORCES AND CONSERVATION

SECTION B
Interactions, Forces and Conservation

Interactions, Forces, and Conservation looks at interactions between large objects like an apple and a planet. These are objects that you can see with your eyes or with a telescope. You will learn how to use forces to describe these interactions, and you will learn how to use the conservation of mass and energy to describe these interactions. Forces and conservation of mass and energy are two of the most powerful ways that scientists use to understand interactions between large objects. As you learn about interactions, forces, and conservation, you will be practicing the skills you developed in Section A of *InterActions in Physical Science*.

Section B has two units.

UNIT 3
Interactions and Forces

UNIT 4
Interactions and Conservation

UNIT 3

CYCLE 1: MECHANICAL INTERACTIONS AND FORCES

CYCLE 2: GRAVITATIONAL INTERACTIONS

INTERACTIONS AND FORCES

What will you learn about in Unit 3?

In Unit 2, you learned how to describe mechanical interactions using energy ideas. In this unit you will learn about the mechanical and gravitational interactions that are all around you. In mechanical interactions, objects touch and push or pull each other over a distance. For example, throwing a baseball, snowboarding down a hill, sailing across a lake, and even riding a skateboard all involve mechanical interactions. In gravitational interactions, objects pull on each other even when they are not touching. Without gravitational interactions, life on Earth would be very different!

What types of activities are in Unit 3?

In Unit 3, you will once again begin each cycle with an *Our First Ideas* activity. You will also find *Developing Our Ideas* and *Learning About Other Ideas* activities. In the *Putting It All Together* activity, you will compare the ideas that your class developed with those of scientists. The *Idea Power* activities will give you an opportunity to apply the ideas you have developed to new situations.

In Unit 3, there are two cycles.

Cycle 1: Mechanical Interactions and Forces

In the last unit, you learned how to describe mechanical interactions using energy ideas. In Cycle 1 of this unit, you will learn how to describe mechanical interactions using *force ideas*. The ideas about forces were developed in the 1700s by Sir Isaac Newton.

Here is the key question for the first cycle:

How do forces affect motion?

Cycle 2: Gravitational Interactions

In Cycle 2, you will learn about gravitational interactions, like those that take place in skydiving. You will also explore why the Moon orbits the Earth, and why you don't fall out of a tilt-a-whirl at the fair!

Here is the key question for the second cycle:

How can you describe gravitational interactions?

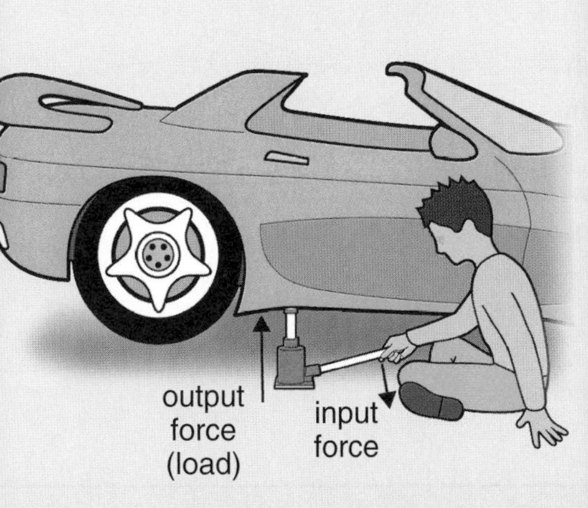

CYCLE 1:
MECHANICAL INTERACTIONS AND FORCES

Cycle Key Question

1. How do forces affect motion?

Activity 1: Forever Away?

Activity 2: Pushes, Pulls, and Motion

Activity 3: A Frictionless World?

Activity 4: Friction and Backward Forces

Activity 5: What Is Transferred?

Activity 6: Pushes, Pulls, and More Interactions

Activity 7: Multiple Forces and Motion

Activity 8: Forces and Direction of Motion

Activity 9: Mechanical Forces and Motion

Activity 10: Applying Force and Energy Ideas

Activity 11: Changing Force Strength and Mass

Activity 12: Simple Machines

OUR FIRST IDEAS

Activity 1
Forever Away?

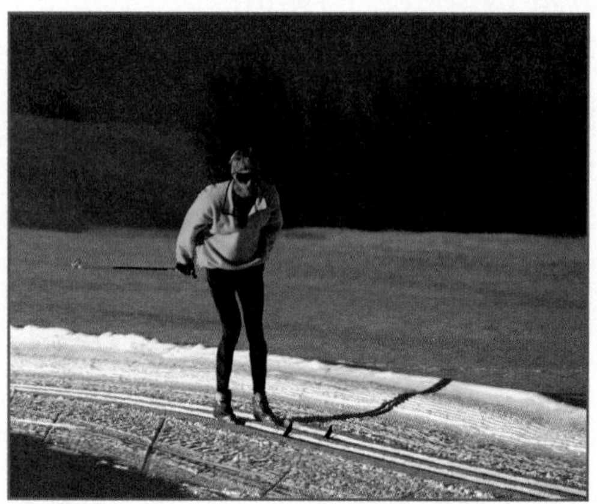

Cycle Purpose

In the last cycle, you learned how to use *energy ideas* to account for changes in motion during an interaction. For example, a skier slows down during the friction interaction between the skis and snow. The skier's motion energy decreases as the thermal energy of the skis and snow increases. As motion energy decreases, speed decreases.

Sometimes, it is more convenient to use *force ideas* to account for changes in motion during an interaction. Scientists define a **force** as a *push or a pull.* The purpose of this cycle is to learn how forces affect motion.

How do forces affect motion?

Science Words

force: a push or a pull

Record the key question for the cycle on your record sheet.

We Think

In this activity, you will discuss some ideas that you have about forces and motion. You will observe the motion of an air puck and think about what might cause that motion.

Your team will need:

- 18-cm (7″) balloon
- air puck
- balloon pump
- to launch the puck, at least 1 m (about 1 yd) of a flat, hard surface

Do not over inflate the balloon so it pops.

An air puck glides on a cushion of air. The base of the puck has a stem with a small hole drilled down the center. An air-filled balloon fits over the stem. The air from the balloon makes an air cushion for the puck to float on. The air does not push the puck—it just creates a frictionless surface for the puck to glide on.

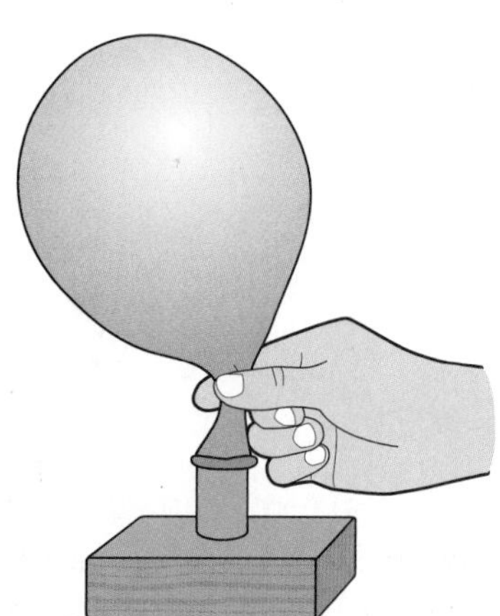

STEP 1 Let's see how the air puck moves on a cushion of air. Blow up the balloon with the balloon pump. Keeping the balloon pinched closed, slip it over the stem. *Keep the balloon pinched closed until you are ready to give the puck a push.*

STEP 2 Place the air puck on the floor or on a long clear table. Let go of the balloon and give the puck a gentle push. The air from the balloon provides an air cushion over which the puck can easily glide.

Observe the puck's motion. *Repeat this a few times so you can make very good observations.*

1. While gliding on a cushion of air, does the puck *speed up*, *slow down*, or *have constant motion*?

Read the following conversation among four students who are trying to account for the puck's motion as it glides on the cushion of air.

The puck looks like it moves with constant speed, but it is really slowing down a little. It has to slow down because the hand is no longer pushing it. To keep it moving, you would have to keep pushing it.
The puck keeps moving because the quick shove I gave it stays with it even after the puck is no longer touching my hand.
The puck keeps moving because nothing is stopping it.
The puck looks like it moves with constant speed, but it is really slowing down a little. Even if the air in the balloon never ran out, the puck would eventually stop because everything eventually slows down and stops.
Chantel
Than
Carlos
Xuan

2. What are your ideas? They may be similar to the ideas of these students, or different. In your own words, write why you think the puck had the motion that you observed.

Our Class Ideas

Participate in the class discussion about the activity.

DEVELOPING OUR IDEAS

Activity 2
Pushes, Pulls, and Motion

Purpose

Remember, in this cycle you are exploring how forces (pushes or pulls) affect motion. The purpose of this activity is to explore how a constant *forward* force affects the motion of an object. You will also learn about how scientists represent forces using force arrows. Here are the key questions for this activity.

1. How does a constant forward force affect motion?
2. How can an arrow be used to represent a force?

Record the key questions for the activity on your record sheet.

We Think

1. When air from a hair dryer blows on a ball, there is an applied interaction between the air and the ball. If you use a hair dryer to exert a constant force on a small ball, what kind of motion would the ball have? Does it speed up or move at a constant speed? Why?

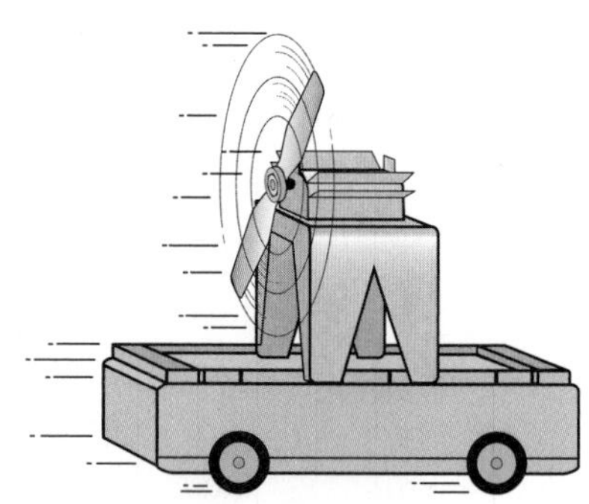

2. Suppose you have a fan connected to a car. (When the fan blades whirl around, they push the air backwards. In return, the air pushes steadily forward on the fan unit and car.) When the fan is on, the air provides a constant forward force on the car and fan unit. Would the car speed up or have a constant speed? Why?

Participate in a class discussion about these questions.

Explore Your Ideas

Part A: Constant Forward Force and Motion

1. How should your teacher conduct the ball and hair dryer demonstration so that the air from the hair dryer exerts a constant force on the ball?

Participate in a class discussion about the method that should be used.

STEP 1 Your teacher will do the ball and hair dryer demonstration.

2. What kind of motion did the ball have? Did it speed up or move at a constant speed?
3. Is this what you predicted in the We Think section?

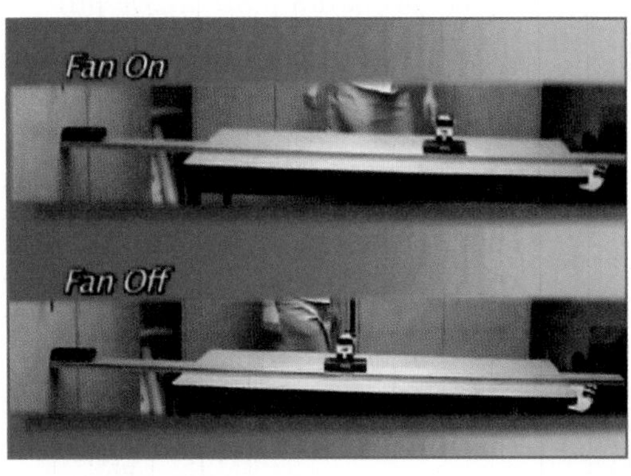

STEP 2 Now your teacher will show you a demonstration of a fan connected to a car. When the fan is turned on, there is a constant forward push on the car.

4. What kind of motion did the car have? Did it speed up or move at a constant speed?
5. Is this what you predicted in the We Think section?

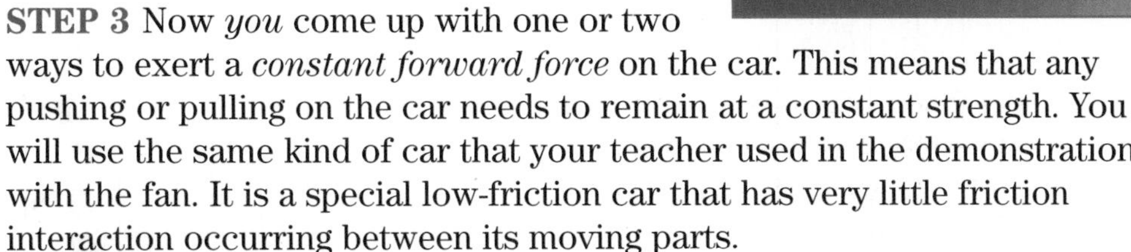

STEP 3 Now *you* come up with one or two ways to exert a *constant forward force* on the car. This means that any pushing or pulling on the car needs to remain at a constant strength. You will use the same kind of car that your teacher used in the demonstration with the fan. It is a special low-friction car that has very little friction interaction occurring between its moving parts.

Your team will need:

- low-friction car
- piece of string, about 30 cm long (optional)
- long area for the car to move

6. Describe how you will exert a *constant forward force* on the low-friction car.

STEP 4 Exert a constant forward force on the car and observe its motion.

7. What kind of motion did the car have? Did it speed up or move at a constant speed?

Wear eye protection.

Part B: Forces and Force Arrows

Now, you will learn how to represent forces (pushes and pulls) in a diagram. Scientists draw and label *force arrows* to represent pushes and pulls.

Force arrow labels are written in the following form:

Force exerted by ______________ on ______________.

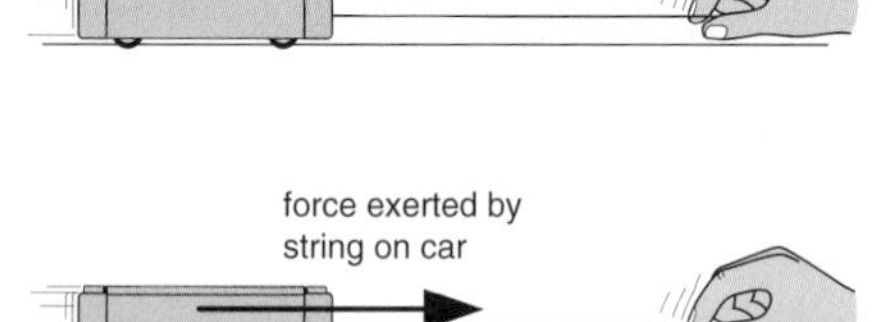

In the diagram at the left, the pull of the string has been represented with a force arrow. The force has been labeled as the *force exerted by string on car.*

The tail of the arrow is usually placed in the middle of the object and the *length* of the arrow represents the strength of the push or pull.

8. Which of the arrows on the right represents a stronger pull?

The *direction* of the arrow shows the direction of the push or pull.

9. Which of the force arrows represents a pull to the *left?*

Like scientists, you will draw and label force arrows to help you analyze many situations in this unit.

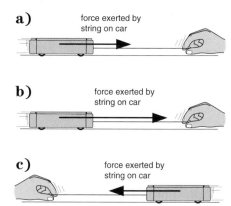

a) force exerted by string on car

b) force exerted by string on car

c) force exerted by string on car

Make Sense of Your Ideas

1. In each situation in this activity, what kind of motion did the low-friction car have when a constant forward force was exerted on it?

2. How can the direction and magnitude (strength) of a force be shown on a diagram?

Our Consensus Ideas

Think about the key questions for this activity and discuss them with your team:

1. How does a constant forward force affect motion?
2. How can an arrow be used to represent a force?

1. Write your answers on your record sheet.

Participate in a class discussion about the key questions.

2. Write the class consensus ideas on your record sheet.

DEVELOPING OUR IDEAS

Activity 3
A Frictionless World?

Purpose

You have been exploring how forces affect motion. So far, you have found that a constant forward force on an object during an interaction causes the object to speed up.

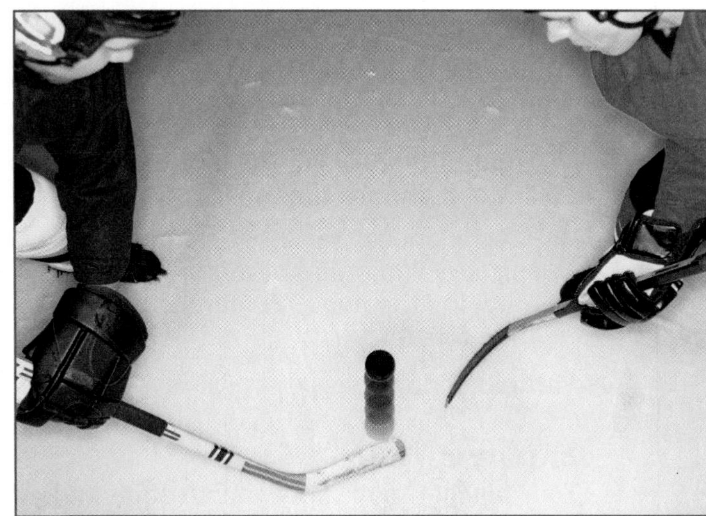

When a puck is moving along the ice, what affects its motion?

The key question for this activity is:

What kind of motion does an object have if there are *no interactions* affecting the object's motion?

Record the key question for the activity on your record sheet.

We Think

Suppose you slide a hockey puck over a series of different surfaces. First, you give it a shove so that it slides across a very rough surface. You then give it an identical shove to start it moving across a smoother surface. Next, you give it an identical shove to start it moving across an even smoother surface, etc. You keep repeating this procedure with each surface a little slicker than the previous one.

Discuss the following questions with your team. Then write your answers in complete sentences on your record sheet.

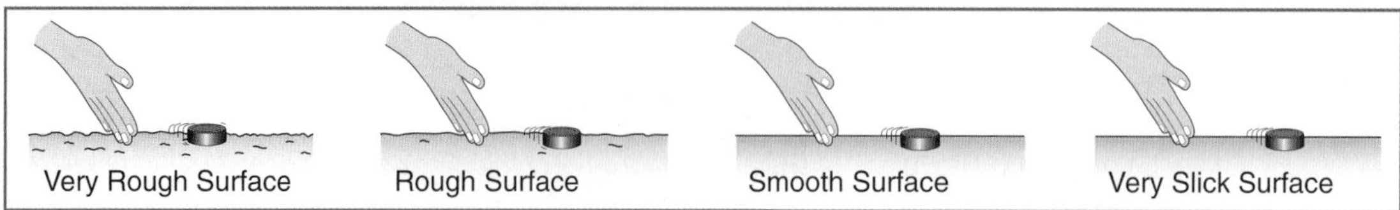

1. What differences in the motion would you observe as the surfaces become more and more slick?

2. Now suppose you shove the puck on a slick, completely frictionless surface. What kind of motion would you observe after you are no longer touching the puck? Assuming the surface is infinitely long, would the puck slide forever?

Participate in a class discussion about your answers.

Explore Your Ideas

Your teacher may show you an animated movie of students playing baseball on a frictionless field where there are no interactions affecting motion. Or, your teacher will show you a demonstration using battery-powered air pucks. Imagine the air pucks are students playing a game of baseball.

While the students are trying to play baseball, there are lots of examples of them sliding across the frictionless field. When they are not interacting with other people or objects, are they *speeding up*, *slowing down*, or *moving with a constant speed*?

Make Sense of Your Ideas

1. Suppose there are *no interactions affecting the hockey puck's motion.* What kind of motion will the hockey puck have? Will it *speed up*, *slow down*, or *have constant speed*?

2. In the absence of other interactions, do you need to continuously apply a force to an object to keep it moving? Write your reasoning.

Our Consensus Ideas

Think about the key question for this activity and discuss it with your team.

What kind of motion does an object have if there are *no interactions* affecting the object's motion?

1. Write your answer on your record sheet.

Participate in the class discussion about the key question.

2. Write the class consensus ideas on your record sheet.

DEVELOPING OUR IDEAS

Activity 4
Friction and Backward Forces

Purpose

In the previous activities, you found that *an object has constant speed when no interactions are affecting the object's motion*, and you discovered that *when a constant forward force is exerted on an object, the object speeds up.* How does a backward force affect motion? (A forward force is a force in the direction of the object's motion. A backward force is in the direction opposite to the object's motion.)

In the last cycle, you learned about friction interactions. In a friction interaction, moving objects slow down, but what exactly is friction? The key questions for this activity are:

1. How does a backward force affect motion?
2. What is friction?

Record the key questions for the activity on your record sheet.

We Think

With your team, discuss the following question:

How can you make an object slow down? What is your reasoning?

Participate in the class discussion about this question.

Explore Your Ideas

Your teacher will demonstrate the two situations pictured below. Carefully observe the demonstrations and answer these questions.

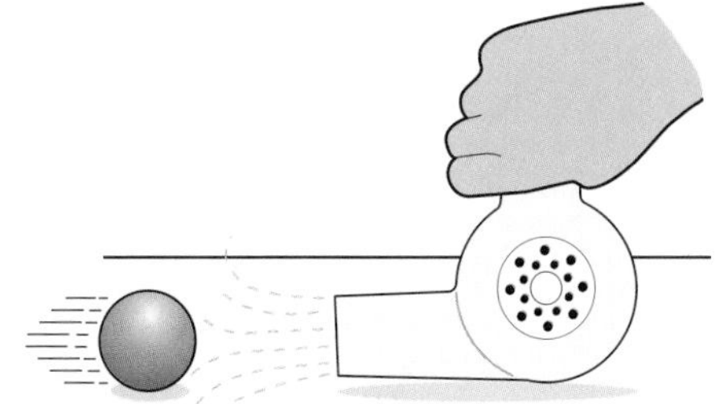

1. When the ball is rolling toward your teacher and your teacher blasts it with the hair dryer, what kind of motion does the ball have?

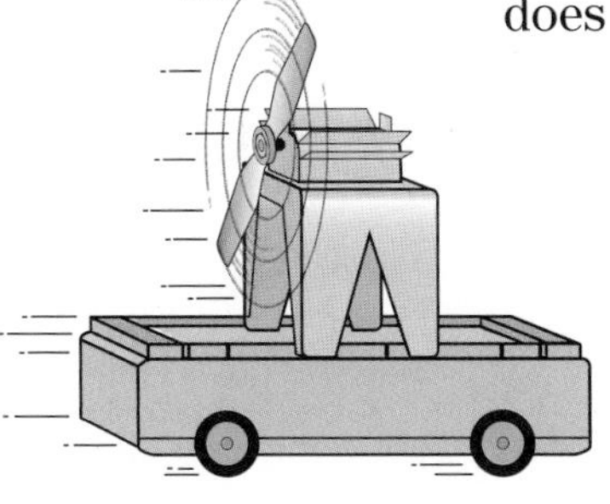

2. When the low-friction car is rolling forward and the fan is turned so that there is a backward force on the car, what kind of motion does the car have?

Here are some brief explorations for each team to do.

Your team will need:

- block of wood
- 15 small sticky notes
- piece of rough sandpaper

STEP 1 Give the block a push so that it slides across the desk. Observe the block's motion.

3. Describe the block's motion. Does it *speed up*, *slow down*, or *maintain a constant speed*?

STEP 2 Closely examine the sandpaper's surface. Look along the surface of the sandpaper.

4. Draw a picture of the sandpaper's surface as seen from the side.

STEP 3 Give the block a push so that it slides across the sandpaper. Observe the block's motion.

5. Describe the block's motion. Does it *speed up*, *slow down*, or *maintain a constant speed*?

STEP 4 Fold 15 sticky notes and place as shown, about 4 cm apart.

STEP 5 Give the block a shove, and let it slide over the sticky notes.

6. What kind of motion does the block have when it slides over the sticky notes? Does it *speed up*, *slow down*, or *maintain a constant speed*?

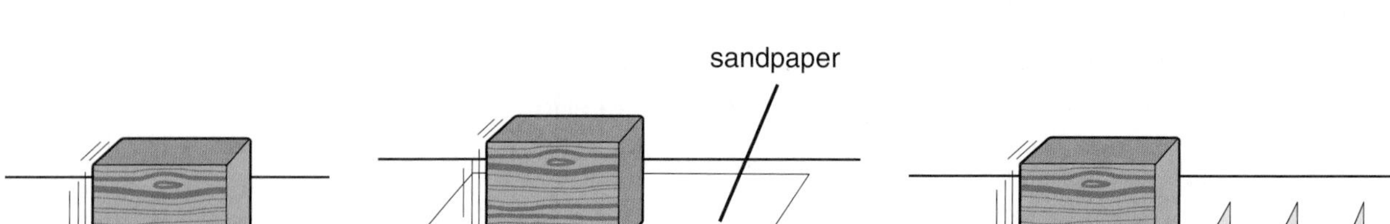

Make Sense of Your Ideas

1. What was similar about the two demonstrations your teacher showed you in Explore Your Ideas?

2. Based on your observations of these two demonstrations, how do backward forces affect motion?

3. Draw a force arrow to show the *force exerted by the air from the hair dryer on the ball.*

4. What was similar about the three situations your team explored in this activity?

5. Draw a force arrow to show the *force exerted by the sticky notes on the block.*

Like the sandpaper, the surfaces of the block and your desk are actually very bumpy. You just cannot see the bumps with your eye. You could see the bumps through a microscope. To the right is a highly magnified picture of what the surface of your desk and the surface of the block might look like. As the two bumpy surfaces slide across each other, the bumps push against each other. In other words, the bumps exert forces on each other.

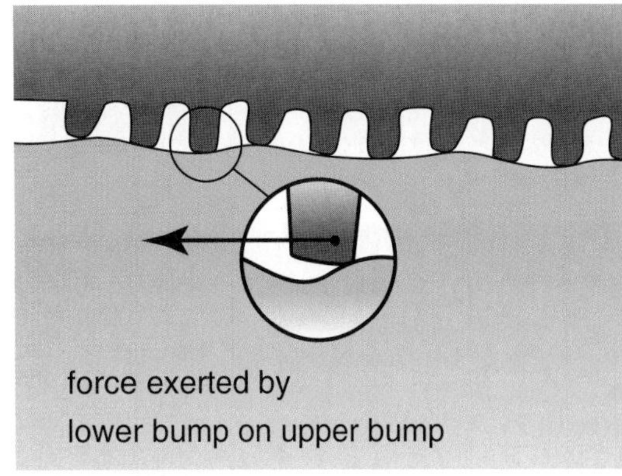

6. Think about the block sliding on the desk. As the bumps of the block and the desk slide across each other, how do you think that affects the motion of the block? Write your reasoning.

When the block slides along the desk, the bumps on the desktop push backward against the forward-sliding bumps on the block. This causes the block to slow down. In this course, we will call this the "bump model" for friction. This is a good model for friction that helps explain everyday situations. There are more complicated models that scientists use to describe friction.

7. What is similar about the interaction between the sticky notes and the block, and the interaction between the block and the desk?

8. Is it appropriate to think of forces happening during a friction interaction? Write your reasoning.

9. If you think of friction as a force, what is its direction?

Our Consensus Ideas

Think about the key questions for this activity and discuss them with your team.

1. How does a backward force affect motion?
2. What is friction?

1. Write your answers on your record sheet.

Participate in the class discussion about the Making Sense questions and the key questions for this activity.

2. Write the class consensus ideas on your record sheet.

DEVELOPING OUR IDEAS

Activity 5
What Is Transferred?

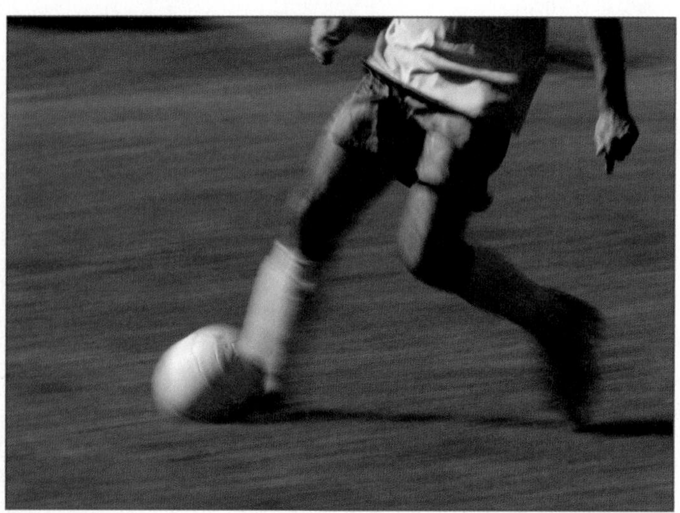
When the player kicks the ball, what will he transfer to the ball?

Purpose

In the previous activities, you learned that forward forces (pushes or pulls) cause objects to speed up and backward forces cause objects to slow down. When there is no interaction affecting an object's motion, the object's speed is constant.

Suppose you see a ball lying on the ground. You run up to it and give it a big kick. The ball then rolls rapidly across the ground. During the kick, what is being given, or transferred, to the ball?

In this activity, you will answer the following activity question:

What is given to an object during an interaction and stays with the object after the interaction is over?

Record the key question for the activity on your record sheet.

We Think

During the kick, what is transferred to the ball? Force? Energy? Neither force nor energy? Both force and energy? At the top of the next page read what other students think.

What are your ideas? Discuss the following question with your team members. Your ideas may be similar to those of the students or different. In your own words, write your team's answer to the following question on your record sheet.

What do you think is transferred to the ball during the kick and stays with the ball after the foot is no longer touching it? *Force*, *energy*, *neither*, or *both*? Write your reasoning.

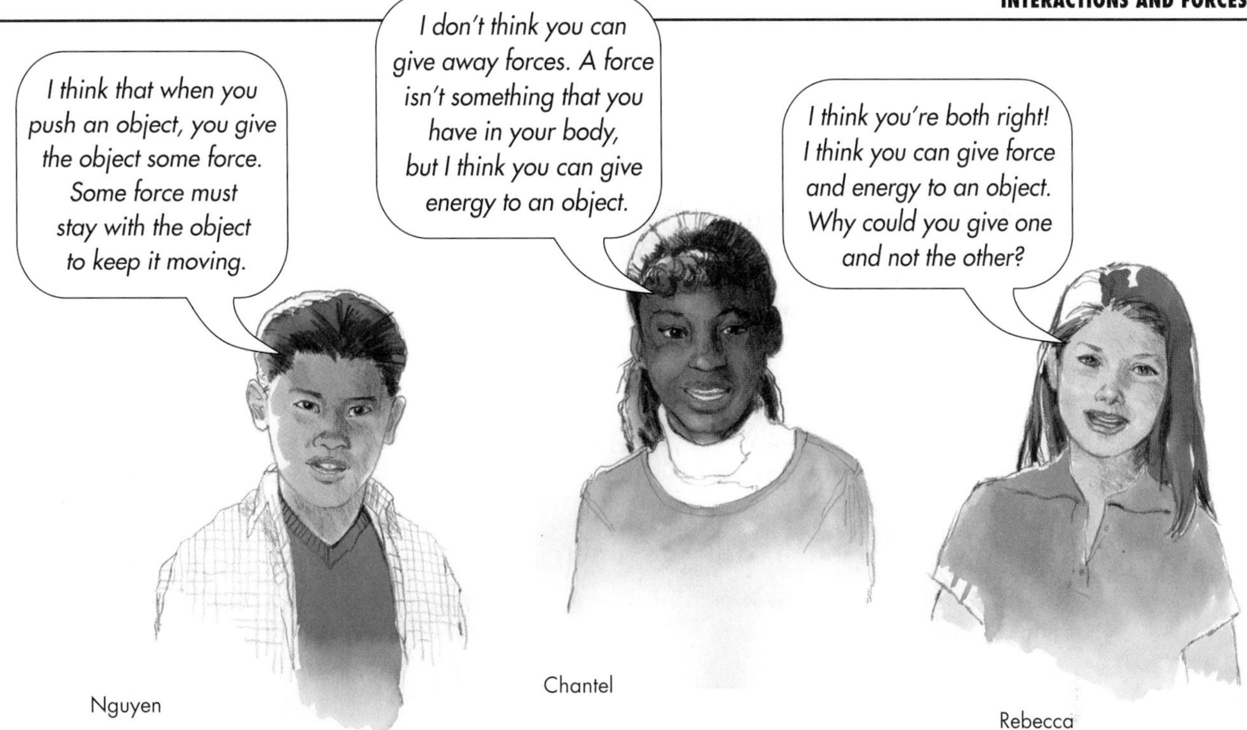

I think that when you push an object, you give the object some force. Some force must stay with the object to keep it moving.
I don't think you can give away forces. A force isn't something that you have in your body, but I think you can give energy to an object.
I think you're both right! I think you can give force and energy to an object. Why could you give one and not the other?
Nguyen
Chantel
Rebecca

Participate in a class discussion.

Explore Your Ideas

Part A: Motion of a Car during and after a Push

In this activity, you will give a *gentle, steady push* to a low-friction car, and observe the car's motion. You will then give a *quick shove* to the low-friction car. You will observe the car's motion during the time the car is in contact with your hand, and after the car leaves contact with your hand. You will use your observations to help you decide what, if anything, is transferred or given to the car from your hand.

Your team will need:

- low-friction car
- long area (at least 2 m) for the car to move

STEP 1 Exert a gentle, steady push (not too hard) on the low-friction car *so that it stays in contact with your hand as it rolls forward.*

Watch where you are going as you push the car. Be sure that the way is clear.

Try to keep the push the same strength as you move it along. *You will have to move your hand with the car to keep the push the same strength.* (Even if you have to move your hand faster, try to keep your hand's push the same strength as the car moves along.)

Each team member should try a push, so that the car runs several times. *Observe* the motion of the car as it stays in contact with your hand.

1. Describe the motion of the low-friction car when you exert a *steady push*. Does it speed up or have a constant speed?

STEP 2 Start with the low-friction car at rest (not moving). Exert a *quick push* so that your hand only stays in contact with the car for an instant. Let it roll for a while after it has left contact with your hand.

Each team member should try a push, so that the car runs several times. *Observe* the motion *during contact* with the hand, and *after leaving contact* with the hand.

2. How is the motion of the car different from its motion in Step 1?

Complete *rows one and two* of the table on your record sheet. (Friction and drag are so small that you can ignore them.)

Participate in a class discussion about your observations.

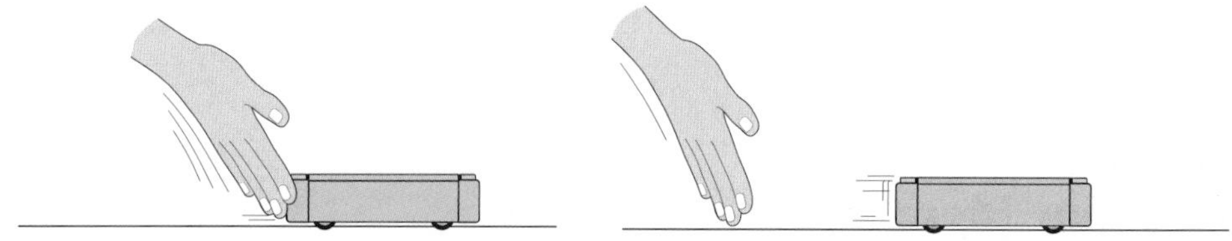

Table: Motion of Car during and after a Push	
	Did the car speed up or have a constant speed?
1. During the constant push (Step 1)	
2. After the quick push was over (Step 2)	
3. During the constant push from simulator (simulator running and space bar pressed)	
4. After the quick push from the simulator was over (simulator running and space bar pressed then released)	

Part B: Using a Simulator to Model Pushing a Car

Sometimes, it is difficult to make real-world observations, so scientists often rely on simulations to help them understand the world. You will simulate the explorations you did with the low-friction car. You may want to refer to *How To Use the Interactions and Motion Simulator*. Your teacher will show you how to open up the simulation file for this activity.

When you open the file, you see the gray force arrow before running the program. It is ready to push the car, but it is not pushing yet.

When you run the program, the gray force arrow disappears. The program waits for you to hold down the *space bar* to switch on the force.

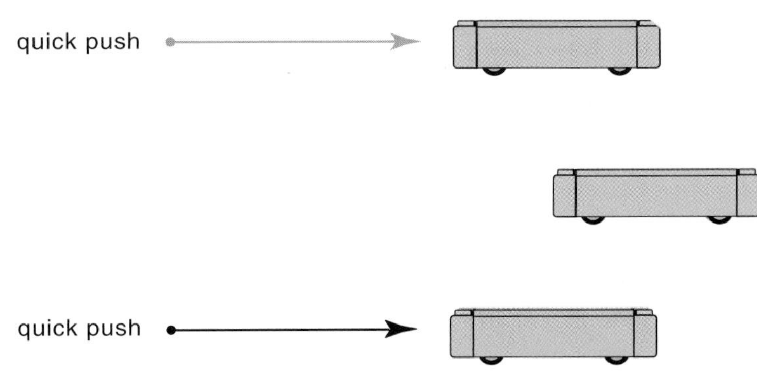

When you hold down the *space bar*, a force is applied to the car. A force arrow, now black, appears on the screen. When you release the *space bar*, the force is no longer applied to the car.

STEP 1 Hold the space bar down so the force is being exerted on the car. The red arrow in the simulator is a speed arrow. The red speed arrow represents the speed of the car. The longer the arrow is, the greater the speed.

3. Does the length of the red speed arrow *increase*, *decrease*, or *stay the same* while you are holding down the space bar? What does that indicate?

STEP 2 Let go of the space bar.

4. Does the length of the red speed arrow *increase*, *decrease*, or *stay the same* after you let go of the space bar? What does that indicate?

Complete rows 3 and 4 of the table on your record sheet.

5. Which of the two sketches below is best modeled by the simulator if you *press then release the space bar?* Explain why you chose that sketch.

6. Which of the two sketches below is best modeled by the simulator if you *press and hold down the space bar?* Explain why you chose that sketch.

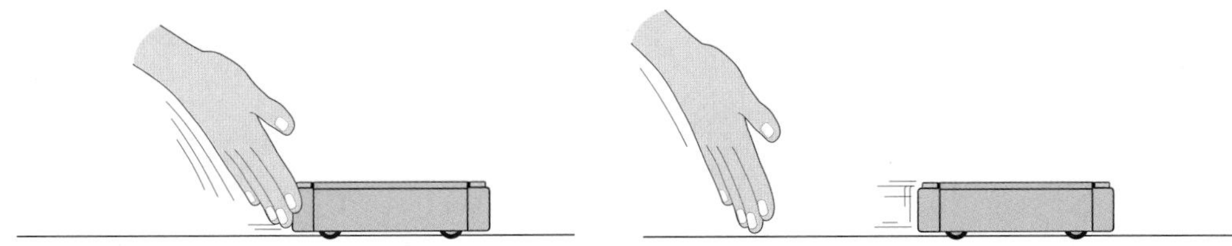

Make Sense of Your Ideas

1. Using the actual equipment and using the simulator, what happens to the car's motion when there is a constant push on the car?

2. On your record sheet, complete the energy diagram for the low-friction car while it is being pushed.

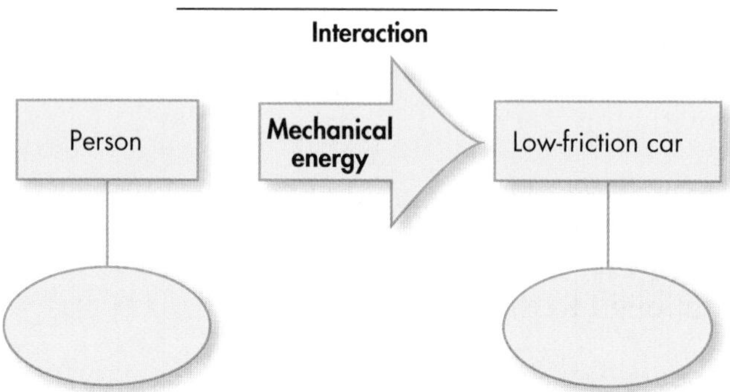

3. Based on the motion of the car, is there still a force from your hand being exerted on the car even after the car and your hand are no longer interacting?

4. Based on the motion of the car, does the car have motion energy even after the car and your hand are no longer interacting?

Participate in the class discussion about this activity.

Our Consensus Ideas

Think about the key question for this activity and discuss it with your team.

What is given to an object during an interaction and stays with the object after the interaction is over?

1. Write your answer on your record sheet.

Participate in class discussion about the key question.

2. Write the class consensus ideas on your record sheet.

DEVELOPING OUR IDEAS

Activity 6
Pushes, Pulls, and More Interactions

Purpose

In this cycle, you have examined how single forces and multiple forces affect the motion of an object. You have learned:

- When there is a forward force on an object, the object speeds up.
- When there is a backward force on an object, the object slows down.
- When there is no interaction affecting an object's motion, the object has constant speed.

Your team will need:

- 2 clamps
- rubber band
- low-friction car

Now that you have developed these force and motion ideas, you will practice applying these ideas. The key question for this activity is:

How can force ideas be used to account for changes in motion?

Record the key question for the activity on your record sheet.

Wear eye protection.

We Think

Suppose you set up the situation shown in the sketch, using the low-friction car and a rubber band stretched between two clamps.

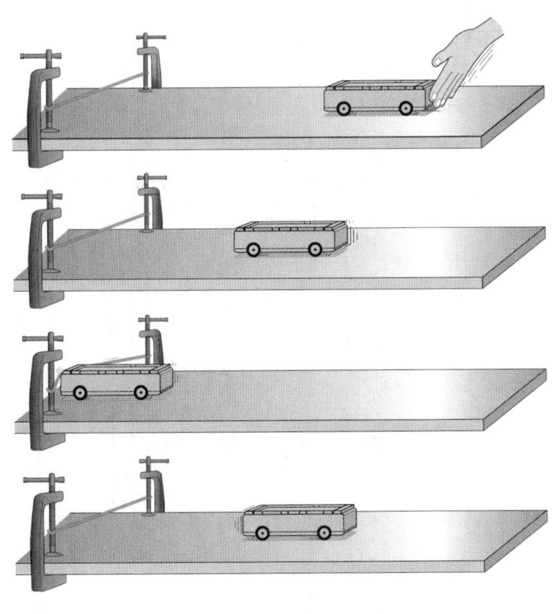

Discuss the following question with your team.

- In each sketch, will the car *speed up*, *slow down*, or *have a constant speed*? Why?

Participate in a class discussion.

Explore Your Ideas

STEP 1 Set up the equipment as shown in the diagram. Stretch the rubber band loosely between the two clamps. If the rubber band is stretched too tightly, it will be hard to make observations.

STEP 2 Give the low-friction car a push so that it rolls toward the rubber band and bounces back. Each team member should try a push so that your group runs the car several times.

Draw and label the force arrows for the interactions on the car. (Ignore friction and drag. They are very small in this situation.)	What interaction(s) affect the car's motion? (Ignore drag and friction.)	Is the car *speeding up, slowing down,* or does it *have a constant speed?*	Is there a *forward force, backward force,* or *no interaction* affecting the car's motion?
1. The car speeds up as the hand gives it a push.	*An applied interaction between the hand and the car*	*Speeding up*	*Forward force*
2. The car is moving with a constant speed toward the rubber band.			
3. The car slows down as it begins to interact with the rubber band.			
4. The car speeds up as the rubber band pushes it away.			
5. The car has a constant speed as it rolls away from the rubber band.			

STEP 3 Observe the car's motion and the different interactions that occur. Discuss with your partner how you would draw and label the force and speed arrows to complete your record sheet. *Ignore the effects of friction and drag because they are so small in this activity.*

Participate in the class discussion about your observations.

Complete the chart on your record sheet. The first row is done for you.

Make Sense of Your Ideas

1. What can you say about the force being exerted on the car

a) while the car was speeding up?

b) while the car was slowing down?

c) while the car had a constant speed?

2. After a car starts moving, is there always a force exerted on the car in the direction of motion to keep it moving? Write your reasoning.

3. After a car starts moving, can the car continue to roll along even if there are no forces being exerted on it? Write your reasoning.

4. Does a *push* (from either the hand or the rubber band) stay with the car? If not, is anything transferred to the car? Describe your reasoning.

Participate in the class discussion about these questions.

Our Consensus Ideas

Think about the key question for this activity and discuss it with your team.

How can force ideas be used to account for changes in motion?

1. Write your answer on your record sheet.

Participate in a class discussion about the key question.

2. Write the class consensus ideas on your record sheet.

DEVELOPING OUR IDEAS

Activity 7
Multiple Forces and Motion

Purpose

In this cycle, you have examined how a force affects the motion of an object. In many situations, an object is involved in *more than one* interaction at a time. For example, if someone is pedaling a bicycle, the bicycle experiences forces from many mechanical interactions at the same time. There is:

- an applied interaction from the rider's pedaling
- a friction interaction from the rubbing parts in the wheels
- a drag interaction with the air

In this activity, you will look at situations where there is more than one mechanical interaction causing *multiple* forces to act on an object at the same time. The key question for this activity is:

How do multiple forces affect motion?

Record the key question for the activity on your record sheet.

We Think

Discuss the following question with your team.

- Two fans are placed on a low-friction car so that they blow in opposite directions. What will happen to the motion of the car? Why?

Participate in the class discussion about this question.

Explore Your Ideas

Your teacher will show you a demonstration of two fans placed on a low-friction car. After watching the demonstration, answer the following questions.

1. What kind of motion does the car have when the fan pushing in the direction of motion is *stronger* than the fan pushing in the direction opposite the motion?
2. What kind of motion does the car have when the fan pushing in the direction of motion is *weaker* than the fan pushing in the direction opposite the motion?
3. What kind of motion does the car have when the fans push equally on the car and the car is initially at rest?
4. Suppose you shove the car while the fans are pushing equally. What kind of motion does the car have after the shove is over?

You will now simulate the fan-car demonstrations. You may want to use *How To Use the Interactions and Motion Simulator*. Your teacher will show you how to open up the simulation file for this activity.

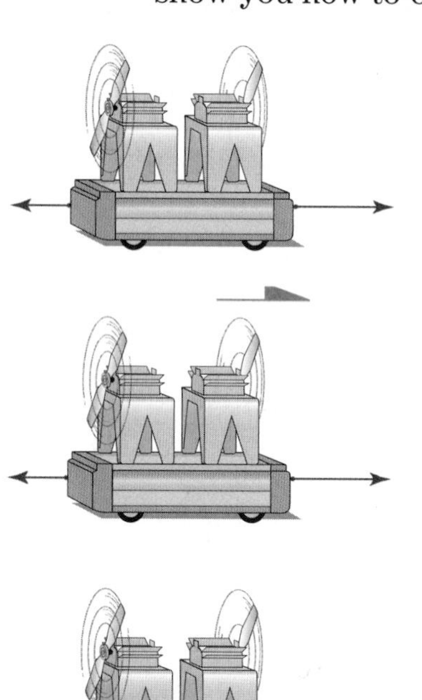

When you open the file, you will see a low-friction car and two force arrows.

Remember to rewind the simulator back to its original position before you start the next situation.

STEP 1 Run the simulator. Observe the red speed arrow that appears when the simulator starts running. The red speed arrow represents the speed of the car. The longer the arrow, the greater the speed.

5. What kind of motion does the car have? Does it *speed up, slow down,* or *move at a constant speed*?

STEP 2 With the mouse, click on the left force arrow and drag it out so that it is the longest arrow.

STEP 3 Run the simulator.

6. What kind of motion does the car have now? Does it *speed up, slow down,* or *move at a constant speed*?

STEP 4 With the mouse, carefully adjust the arrows so they *are exactly the same length.* (Your teacher will demonstrate if you are having difficulty.)

STEP 5 Run the simulator.

7. What kind of motion does the car have now? Does it *speed up, slow down,* or *move at a constant speed*?

STEP 6 Now use the thruster to exert a quick force on the car. Click on the button that looks like a rocket thruster (between the pull and speed buttons).

Click and draw out a force arrow like the gray one shown in the diagram. When you press the space bar and release it, you will exert a quick push on the car.

STEP 7 Run the simulator and give the car a quick push by pressing down and releasing the space bar.

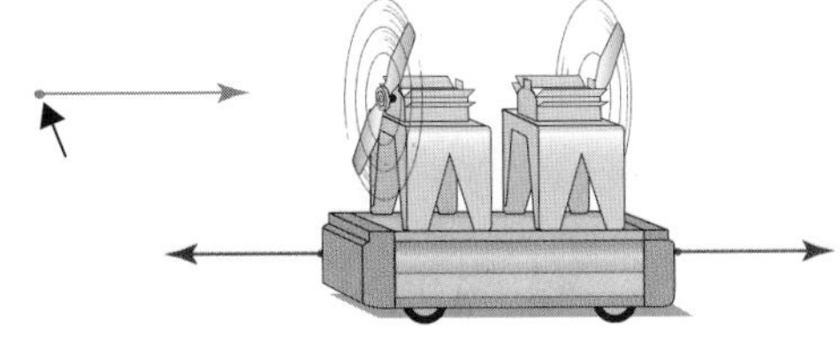

8. What kind of motion does the car have after the quick push? Is the car *speeding up*, *slowing down*, or *moving with a constant speed*?

STEP 8 With the mouse, click on the left force arrow and drag it out so that it is the longest arrow. Double-click somewhere in the white space (where the fan car is) and a window should pop up that says "Properties of Background." In this window, change the "Stop Time" from 10 s to 20 s. Click OK.

9. What do you think will happen now when you give the car a quick push with the thruster? What is your reasoning?

STEP 9 Run the simulator and give the car a quick push by pressing down the space bar and then releasing it after about 5 s.

10. Describe the motion of the car.

Make Sense of Your Ideas

A forward force is a force in the direction of motion. A backward force is a force that is in the direction *opposite* the motion.

When the push in the direction of motion is stronger, we say that the forces are *unbalanced in the direction of motion.* When the push opposite the direction of motion is stronger, we say that the forces are *unbalanced in the direction opposite the motion.* If fans push with equal force, we say that the forces are *balanced.*

1. If the car speeds up in the direction of motion, what conclusion can you make about the forces acting on it? Choose from the following responses. Write the correct one on your record sheet.

a) The forces are *balanced.*

b) The forces are *unbalanced* in the *direction of motion.*

c) The forces are *unbalanced* in the *direction opposite the motion.*

d) There are no interactions affecting the car's motion.

2. What is your evidence?

3. If the car slows down, what conclusion can you make about the forces acting on it? Choose from the following responses. Write the correct one on your record sheet.

a) The forces are *balanced.*

b) The forces are *unbalanced* in the *direction of motion.*

c) The forces are *unbalanced* in the *direction opposite the motion.*

d) There are no interactions affecting the car's motion.

4. What is your evidence?

5. If the car moves at a constant speed, what conclusion can you make about the forces acting on it?

a) The forces are *balanced.*

b) The forces are *unbalanced* in the *direction of motion.*

c) The forces are *unbalanced* in the *direction opposite the motion.*

d) There are no interactions affecting the car's motion.

6. What is your evidence?

7. If the car has zero speed, what conclusion can you make about the forces acting on it? Write the correct response(s) on your record sheet.

a) The forces are *balanced.*

b) The forces are *unbalanced* in the *direction of motion.*

c) The forces are *unbalanced* in the *direction opposite the motion.*

d) There are no interactions affecting the car's motion.

8. What is your evidence?

Our Consensus Ideas

Think about the key question for this activity and discuss it with your team.

How do multiple forces affect motion?

1. Write your answer on your record sheet.

Participate in the class discussion about the key question.

2. Write the class consensus ideas on your record sheet.

DEVELOPING OUR IDEAS

Activity 8
Forces and Direction of Motion

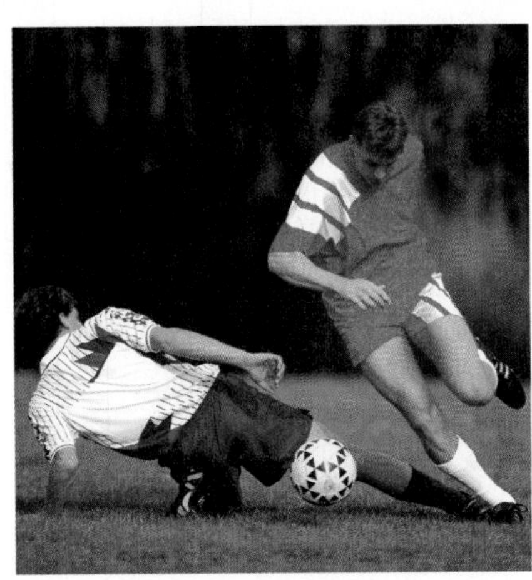

Purpose

So far in this cycle, you have developed some very important ideas about forces and motion. You have learned:

- An unbalanced force in the direction of motion causes an object to speed up.
- An unbalanced force in the direction opposite the motion causes an object to slow down.
- If there is no force on an object or the forces are balanced, the object either remains at rest or remains at a constant speed.

What happens to an object's motion if a force is exerted on it from a direction that is not along the object's path? The key question for this activity is:

What effect does a force have on a moving object when the force is not along the object's path?

Record the key question for the activity on your record sheet.

We Think

Discuss these questions with your team and write your answers.

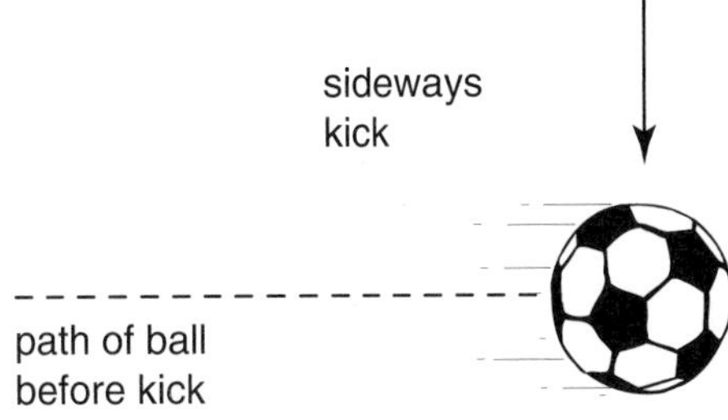

1. Suppose a soccer player gives a rolling ball a kick that is not along the ball's path. What, if anything, will happen to the ball's direction of motion?
2. Suppose a soccer player gives a rolling ball a sideways kick that is placed as shown in the diagram (perpendicular to the path of the ball). What, if anything, will happen to the ball's direction of motion?
3. Suppose the soccer player continues giving the ball a sideways (perpendicular to the path) kick. What, if anything, will happen to the ball's direction of motion?

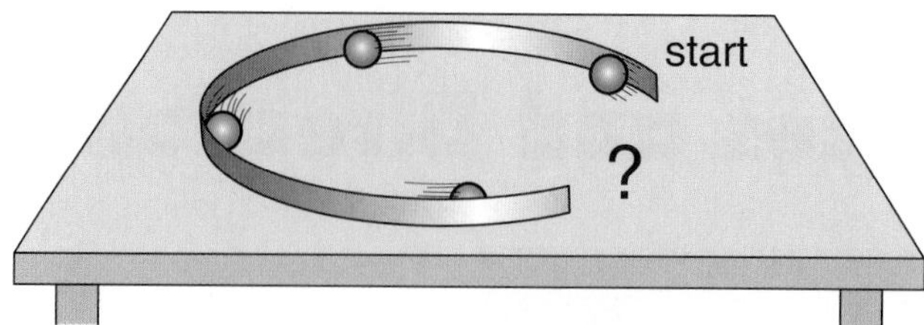

4. Suppose a ball rolls around a hoop as shown in the diagram. What will be the path of the ball when it reaches the end of the hoop?

Participate in the class discussion about this question.

Explore Your Ideas

You will explore what happens when a sideways force is exerted on an object while it is moving.

Your team will need:

- small ball
- area for the ball to roll, such as a tabletop

Wear eye protection.

STEP 1 Roll the ball.

STEP 2 Using your finger, or a pencil, give the ball a quick push. The push *should not* be along the direction of motion. Observe what happens to the ball's direction of motion.

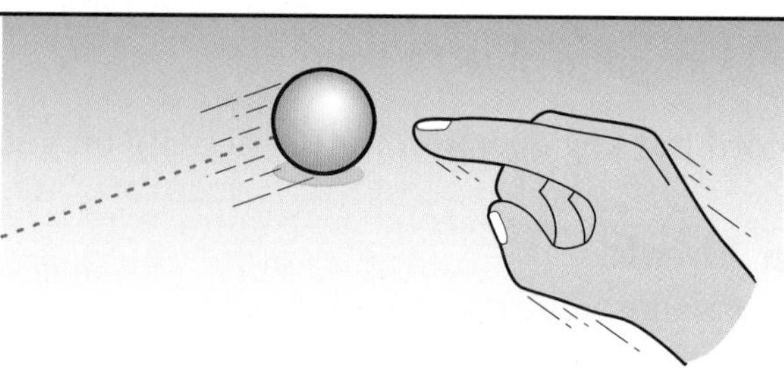

STEP 3 Repeat Step 2 until everyone on your team has had a turn.

1. What did you observe about the ball's direction of motion after the quick push? Draw a sketch to help you answer.

Participate in the class discussion about your observations.

STEP 4 Your teacher will show you two demonstrations:

- What happens when a quick sideways force (perpendicular to the path) is exerted on a moving object?
- What happens when a quick sideways force (perpendicular to the path) is repeatedly exerted on a moving object?

2. What did you observe about the path when a quick sideways force is exerted on a moving object? Draw a sketch to help you answer.

3. What did you observe about the path when a quick sideways force continues to be exerted on a moving object? Draw a sketch to help you answer.

STEP 5 Your teacher will show you a demonstration of a ball rolling around a hoop.

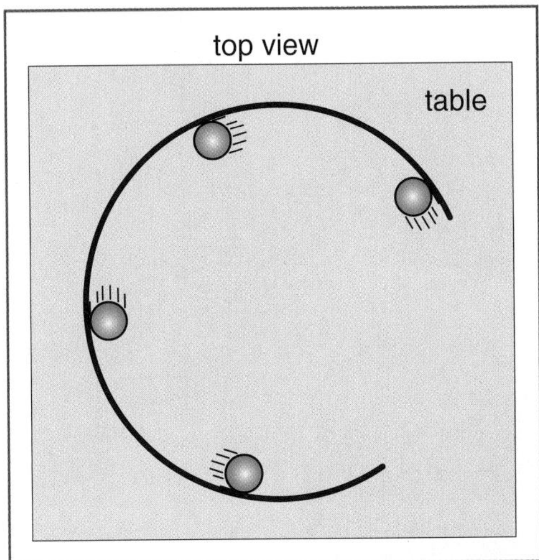

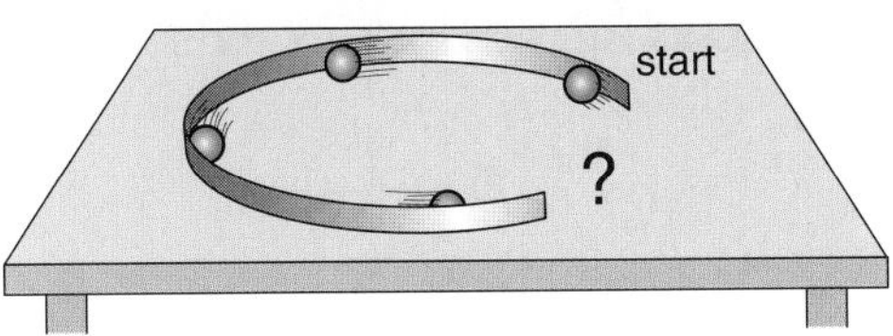

4. What was the path of the ball when it no longer rolled along the hoop?

Make Sense of Your Ideas

1. Draw and label force arrows showing the force exerted by the track on the ball. Do this for each position of the ball in the diagram.

2. Are any forces influencing the ball's path after it no longer touches the track?

3. How does a repeated or continuous sideways force cause an object to move in a circular path? Draw sketches to help you answer this.

4. When a force is not exerted on a moving object, what is the object's path?

Our Consensus Ideas

Think about the key question for this activity and discuss it with your team.

What effect does a force have on a moving object when the force is not along the object's path?

1. Write your answer on your record sheet.

Participate in the class discussion about the key question.

2. Write the class consensus ideas on your record sheet.

Activity 9
Mechanical Forces and Motion

Comparing Consensus Ideas

Recall the key question for this cycle:

How do forces affect motion?

Record your answer to the key question for the cycle on your record sheet.

Your teacher will review with you all of the ideas that the class developed during this cycle to help you answer the cycle question. Be prepared to contribute to this discussion.

Your teacher will distribute copies of *Scientists' Consensus Ideas: How Forces Affect Motion.* Read the scientists' ideas and compare them with the ideas your class developed during this cycle.

Write the evidence from activities in this cycle where it is requested on the *Scientists' Consensus Ideas* form.

IDEA POWER

Activity 10
Applying Force and Energy Ideas

Read the task, then perform an analysis and write an explanation for each of the following tasks. For guidance, use *How To Write an Analysis and Explanation.*

All of the analyses should *include the interacting objects and their interaction type*, and *labeled force arrows* showing the forces being exerted on the object named in the task.

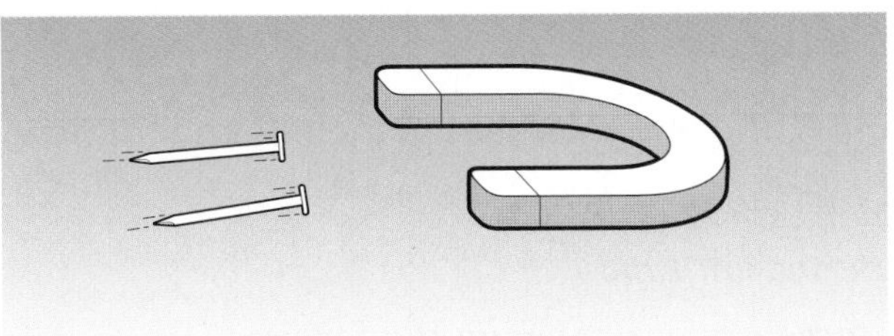

1. Some steel nails and a magnet are on a table. Analyze and explain why the nails speed up toward the magnet. Your analysis should include the *interacting objects and their interaction type*, and *labeled force arrows* showing the forces being exerted on the nails.

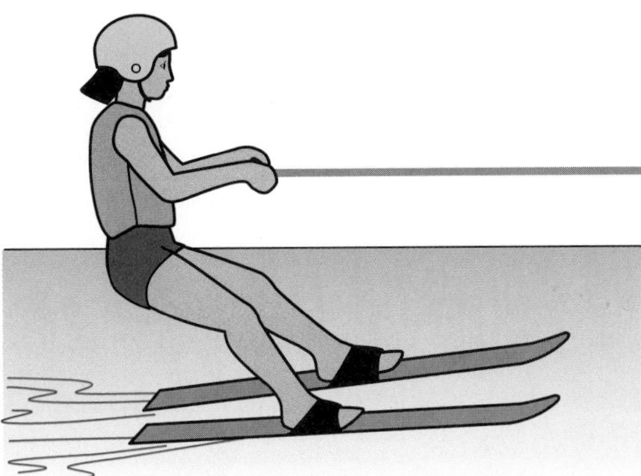

2. Analyze and explain why the water skier has a constant speed.

3. Analyze and explain why the ball slows down as it comes in contact with the net.

In Questions 4 and 5, the analyses should include the following:

- the *interacting objects*
- their *interaction type*
- *labeled force arrows* showing the forces being exerted on the Frisbee™
- an *energy diagram*

4. Analyze and explain why the Frisbee™ speeds up while the girl is throwing it.

5. Analyze and explain why the Frisbee™ slows down as it sails through the air.

LEARNING ABOUT OTHER IDEAS

Activity 11
Changing Force Strength and Mass

Purpose

In previous activities of this cycle, you explored how an unbalanced force acting on an object affects the motion of the object. You observed that when an unbalanced force acts in the direction of motion, the object speeds up. You also observed that when an unbalanced force acts opposite to the direction of motion, the object slows down. In your investigations, the strength of the force did not change and the mass of the object did not change. In everyday experiences, however, these may change.

Imagine you are pulling your friend in a wagon. Would his motion change if you pulled harder?

Now imagine that you pass by another friend who would also like to join in on the fun. She hops into the wagon and you pull both of your friends. Is there a difference in the motion of the wagon now that two friends are in it? That is, does the change in the mass of the wagon system affect its motion? The key questions for this activity are:

1. If a force acts on an object, how does its change in motion depend on the strength of the force?
2. How does its change in motion depend on the amount of mass?

Record the key questions on your record sheet.

Learning the Ideas

Speeding Up

In Activities 2 and 7 of this cycle, you explored the motion of low-friction cars being pushed along a track by spinning fans attached to them. You also learned that when an unbalanced force acts on an object in the direction of motion, it speeds up. In what follows, you will be considering low-friction cars speeding up.

Suppose you were to conduct an exploration with two identical low-friction cars, with attached fans. One of the fans is at a higher setting, causing a bigger push on the car. Would the motion of the two cars be the same or different?

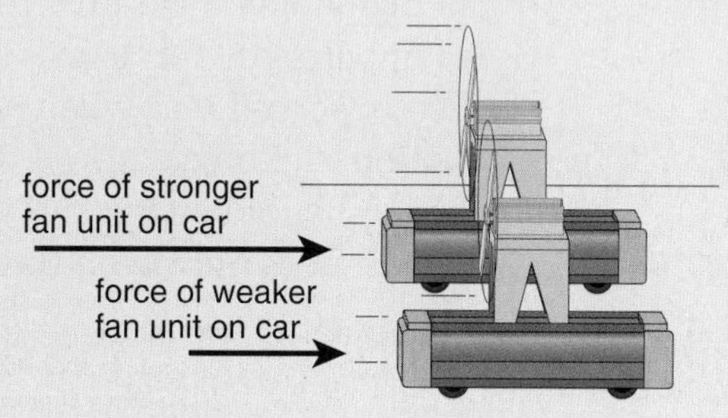

1. If the cars start out from rest, will one car speed up more quickly than the other car, or will both cars speed up in the same way? If one speeds up more quickly, which one will it be? Explain your answer.

Your teacher will show you a movie of two identical cars pushed by fans. One of the cars has a greater force on it because its fan is at a higher setting.

2. What did you observe? Which car sped up more quickly?

Now consider that the fans have the same setting, causing a push of the same strength. However, one car has more mass than the other car.

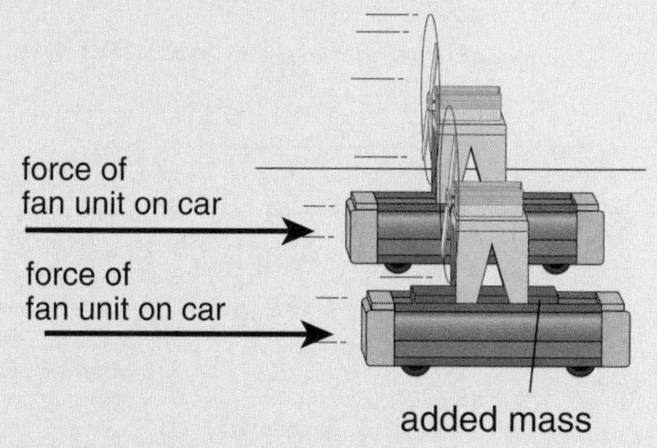

3. If the cars start out from rest, will one car speed up more quickly than the other car, or will both cars speed up in the same way? If one speeds up more quickly, which one will it be? Explain your answer.

Your teacher will show you a movie of two cars that have different masses being pushed by fans with the same strength.

4. What did you observe? Which car speeds up more quickly?

Your teacher will review answers to Questions 2 and 4 with the class. From the two examples you have just observed, you have seen that when a force is applied to an object in the direction of motion:

- Increasing the *strength* of the force increases how quickly the object speeds up.
- Increasing the *mass* of the object decreases how quickly the object speeds up.

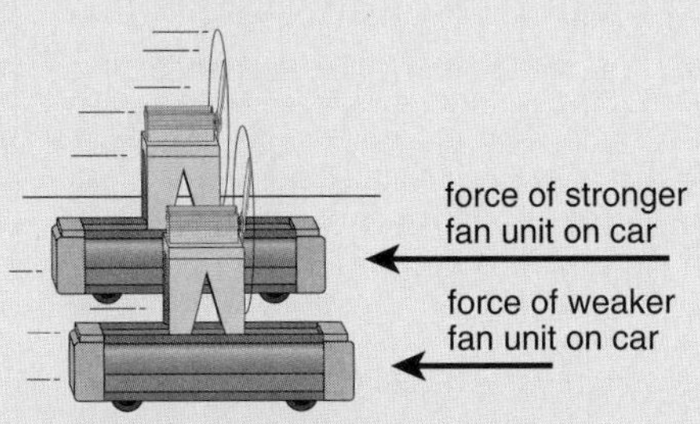

Slowing Down

Now consider what happens when forces slow down objects. That is, when there is an unbalanced force acting opposite to an object's motion. How does the strength of the force affect how quickly an object slows down? How does the object's mass affect its motion?

5. Imagine you have two identical low-friction cars, with fans attached. One fan exerts a greater force than the other fan. The cars start with about the same speed after a brief push in a direction opposite to the direction that the fans would push the cars. Which car would then slow down more quickly, or would they both slow down in the same way? Explain your answer.

Your teacher will show you a movie of this situation.

6. What did you observe? Which car slows down more quickly?

7. Now imagine that you repeat the exploration, but with both fans pushing with the same strength. You add more mass to one of the cars. The cars start with about the same speed after a brief push in a direction opposite to the direction that the fans would push the cars. Which car, the one with more mass or the one with less mass, would then slow down more quickly, or would they both slow down in the same way? Explain your answer.

Watch the movie and answer the question below.

8. What did you observe? Which car slows down more quickly?

Your teacher will review answers to Questions 6 and 8 with the class.

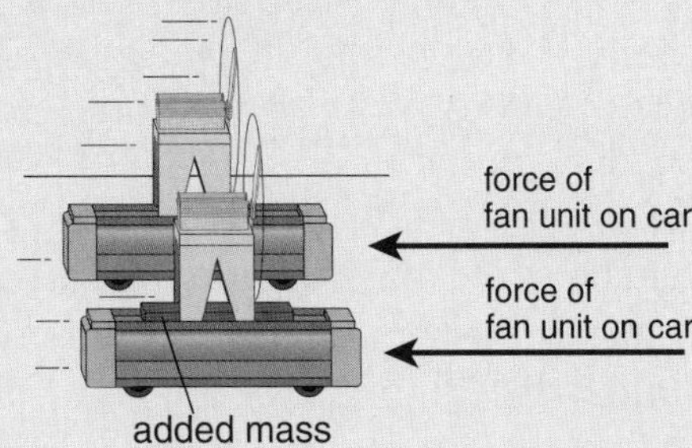

The two observations you have just made show that when a force is applied to an object in the direction opposite to its motion:

- Increasing the *strength* of the force increases how quickly the object slows down.
- Increasing the *mass* of the object decreases how quickly the object slows down.

Force, Mass, and Acceleration

The ideas that relate how the force strength and mass affect how quickly an object speeds up or slows down can be summarized with a mathematical sentence:

$$\text{how quickly object changes speed} = \frac{\text{strength of force applied to object}}{\text{mass of object}}$$

Science Words

acceleration: how quickly an object changes speed

The change in speed may refer to either slowing down or speeding up. Note that this equation only applies when the object moves in a straight line. Scientists also use the word **acceleration** to describe how quickly an object changes speed. (Recall that in Unit 2 Cycle 1, Activity 8, we defined acceleration as a change in an object's speed, its direction of motion, or both.)

Use the ideas introduced in this activity to answer the following questions.

9. Madeline drives a small, two-seat sports car, while Max drives a minivan. The minivan has three times the mass of the sports car. If Madeline and Max both start from rest, and forces of identical strength accelerate both cars, who will reach a speed of 65 mph first? What is your reasoning?

10. Identical twin brothers Tim and Harry are skateboarding along a sidewalk at the same speed. Suddenly, a wind comes up and pushes against the pair! While Tim is turned sideways into the wind, Harry faces forward and takes the full brunt of the wind. Because Harry presents a larger surface

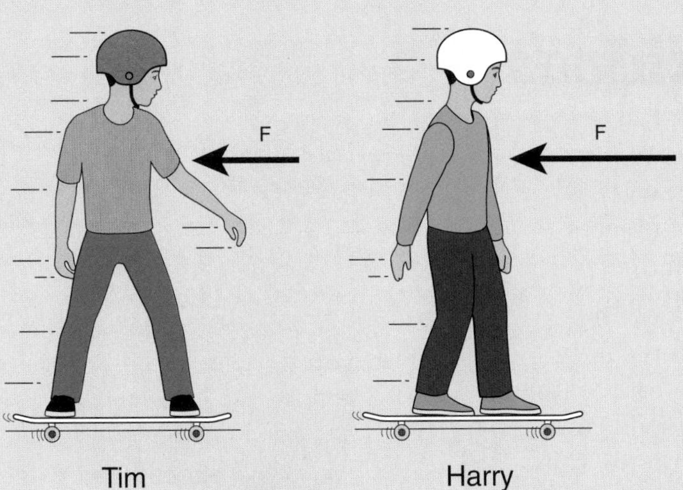

for the wind to push against, the wind exerts a greater force on him than it does on Tim.

Which skateboarder comes to a stop first? What is your reasoning? Assume Tim and Harry have the same mass and that neither twin tries to speed up or slow down on his own.

11. Sofia, a Little League pitcher, pitches baseballs and softballs with the same strength. In each case, she applies the same force to the ball. A baseball has a mass of 0.15 kg, while a softball has a mass of 0.20 kg. Which ball has a higher acceleration (that is, which ball speeds up more quickly) when Sofia pitches it? What is your reasoning? (Assume drag can be ignored.)

baseball
mass = 0.15 kg

softball
mass = 0.20 kg

Your teacher will review answers to these questions with the class.

What We Have Learned

Remember the key questions for this activity.

1. If a force acts on an object, how does its change in motion depend on the strength of the force?
2. How does its change in motion depend on the amount of mass?

Participate in the discussion in which your class reviews the answers to the key questions.

Answer the key questions on your record sheet.

LEARNING ABOUT OTHER IDEAS

Activity 12
Simple Machines

Purpose

In this cycle you have learned how forces affect the motion of objects. Forces are also involved in the operation of **simple machines**.

Science Words

simple machine: a simple device that affects the force required to perform a certain task

Did you know that you interact with simple machines every day? Each time you pull the tab to open a soda can, open a drawer, turn a doorknob, ride your bike, or go up a ramp, you are using simple machines! Simple machines make our lives easier by assisting with tasks that might otherwise require a lot of force or are difficult in another way. In this activity, you will learn about the six different types of simple machines and how they are useful.

The key questions for this activity are:

1. What are the types of simple machines?
2. How do simple machines help you accomplish tasks?

Record the key questions for the activity on your record sheet.

Learning the Ideas

Types of Simple Machines

Machines are used to change the force needed to move an object. Most machines change both the size and direction of the force applied to the machine. Machines are usually used to *decrease the amount of force a person would need to apply* to move an object, but they never decrease the total amount of energy needed. Simple machines make difficult tasks, such as moving heavy loads, easier. You cannot get something for nothing. In this activity, you will see what needs to be increased to decrease the amount of force you use.

There are six types of simple machines, as shown in the diagrams.

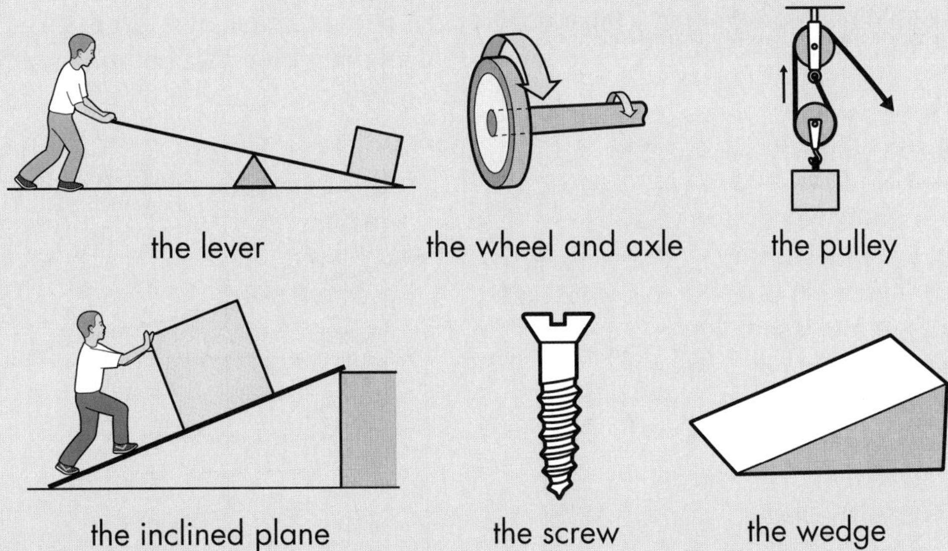

Scientists call the force that a person exerts on a machine the **input force** (sometimes called the **effort**). The force that the machine exerts on the object is called the **output force** (sometimes called the **load**). Usually, the output force is greater than the input force, allowing you to do things like lift a car! Of course, not many people can lift a car without a jack. The output force is usually equal to the force you would have to apply without the assistance of a machine.

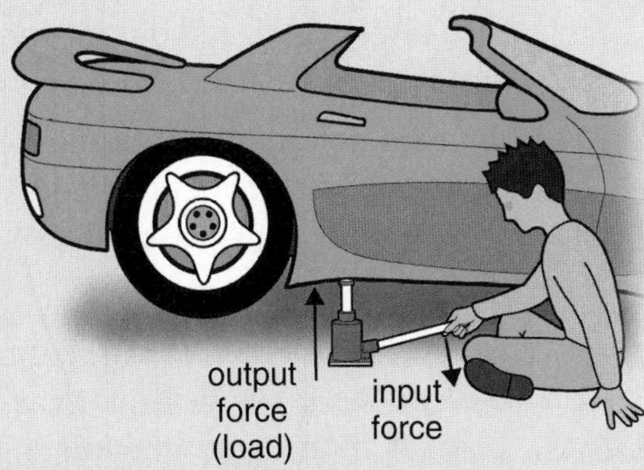

Science Words

input force (or effort): the force exerted on a machine

output force (or load): the force a machine exerts on an object

If you look closely at different objects around you, you can find examples of these simple machines everywhere. In the remainder of this activity, you will study the lever, the wheel and axle, and the pulley, and see how they are useful. You will also read about the inclined plane, screw, and wedge.

The Lever

The lever, one of the first simple machines, is made up of a bar that rotates or turns around a support point (called a fulcrum). The main use of a lever is to move an object. Levers allow you to move objects that you could not move on your own.

Look at the diagram of a lever lifting a box. The lever allows a person to exert an input force (on the longer side of the lever) that is less than the output force required to lift the heavy box. Because the input force is less than the output force, the lever makes the task of lifting the load easier for the person to accomplish. The distance from the fulcrum (support point) to the source of the input force is called the lever arm.

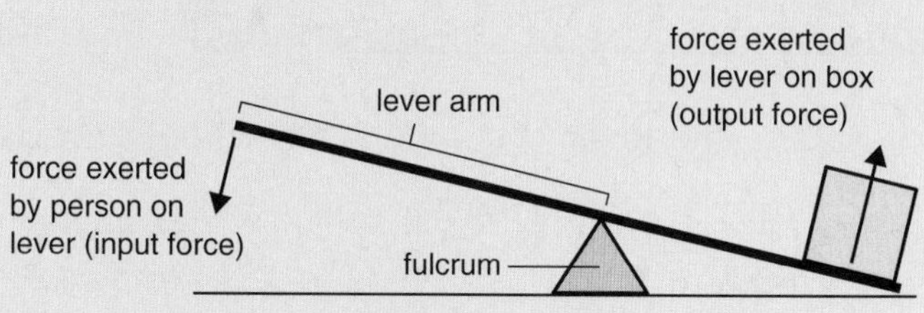

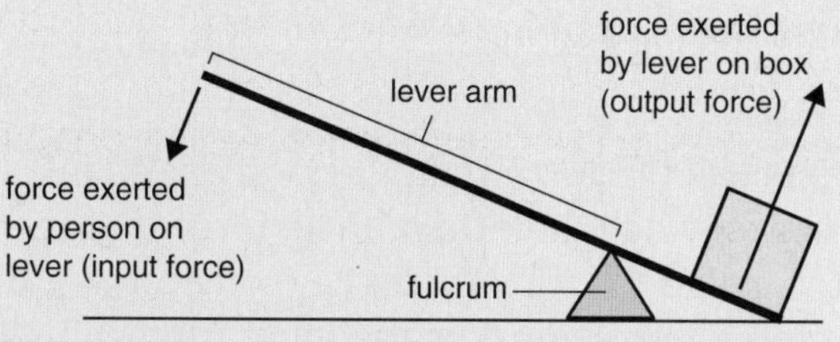

If the fulcrum were moved closer to the end of the lever, increasing the lever arm, the output force would increase. See the diagram. Note also that the input force is in the opposite direction to the output force, so a lever can also result in a change in the direction of force.

If the force to move the object is not an issue but moving the object over a large distance is, one could use a short lever arm. The input force would be greater than the output force, but the object would move over a much greater distance, as shown in the diagram.

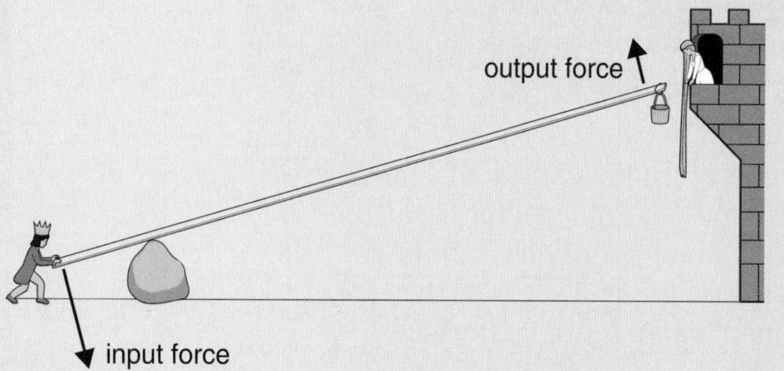

You are probably familiar with the lever from many different situations. One example is a claw hammer. Have you ever tried to pull a nail straight out of a piece of wood? It is very difficult because it requires a large force. A claw hammer helps by allowing you to use a small input force to pull out the nail. The handle of the hammer is much longer than the head. By using the handle as the lever arm and moving it a larger distance, the head of the hammer exerts a larger output force over a shorter distance.

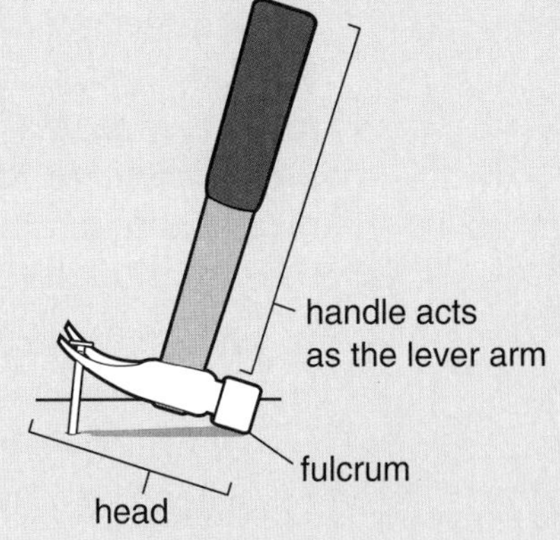

Some common uses of levers are in scissors, bolt-cutters, pull-tabs on soda cans, toasters, and wheelbarrows.

1. Identify an example of a lever. Draw a picture of the object and label the fulcrum. Draw force arrows for the input force and the output force and label them.

The Wheel and Axle

The wheel and axle is probably one of the most important inventions in human history. An axle is a cylinder or rod that goes through the center of the wheel. The wheel is attached to the axle. Depending on the use of this simple machine, the axle may or may not turn as the wheel turns. The wheel-and-axle system allows you to move heavy things from one place to another easily and quickly. Wheels assist in moving an object because they reduce the amount of friction between the object you want to move and the surface you are moving it over. Wheel-and-axle systems are all around you. You see wheels and axles on cars, bikes, skateboards, and rollerblades. However, you may not be aware that doorknobs, most drawers, screwdrivers, and the knobs on your radio are also wheel-and-axle systems.

The amount of force required to move the object is reduced further if the radius of the wheel is increased. A doorknob is an example of this. The larger the diameter of the doorknob (the "wheel") is compared to the diameter of the shaft (the "axle"), the smaller the input force is compared to the output force. When you turn a doorknob, it turns an axle attached to it. If you took apart the doorknob and tried to turn the axle on your own, you would find it extremely difficult!

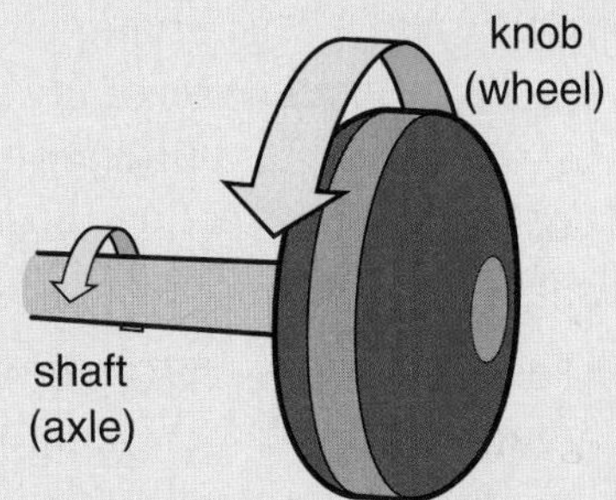

The fact that the doorknob's diameter is several times greater than the axle's diameter makes it easier to turn the axle and open the door. As you increase the diameter of the doorknob, the input force required to open the door decreases, although the distance you need to turn the doorknob increases.

Gears are a good example of how a wheel-and-axle system can change the direction of a force. A gear is just a wheel that has "teeth" on it. When two gears are combined, they must rotate in opposite directions. Therefore, the direction of the force is changed.

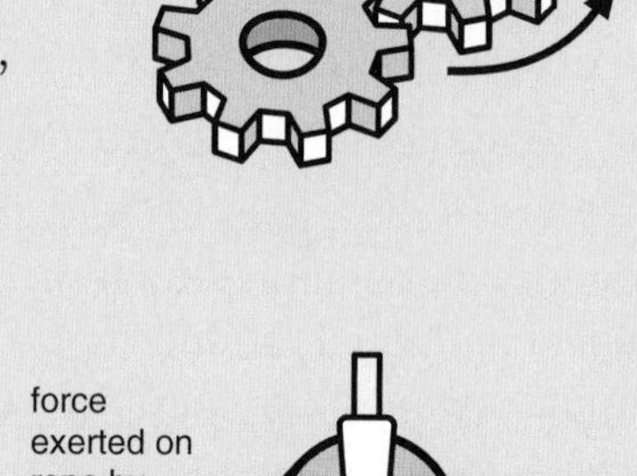

2. Identify an example of a wheel-and-axle system and draw a picture of the object. Label the "wheel" and "axle." Indicate the directions of the input and output forces.

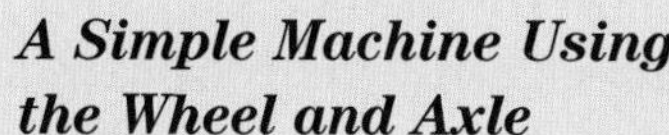

A Simple Machine Using the Wheel and Axle

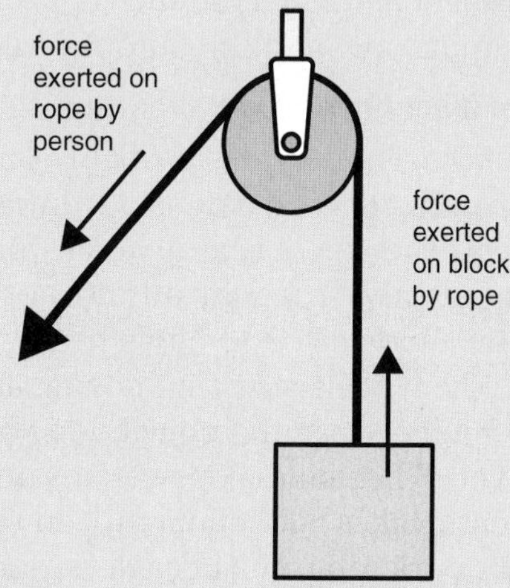

THE PULLEY

Another type of simple machine is the pulley. Pulleys make use of the wheel-and-axle system. When a single pulley is used, as shown in the diagram, it only changes the direction of the force that you exert, not its strength. The force you exert on the end of the rope (the input force) is equal to the force that the rope exerts on the block (the output force). This change in direction is useful, for example, if you are lifting a bucket of water out of a well. It is easier to pull down on a rope than it is to pull up on the rope.

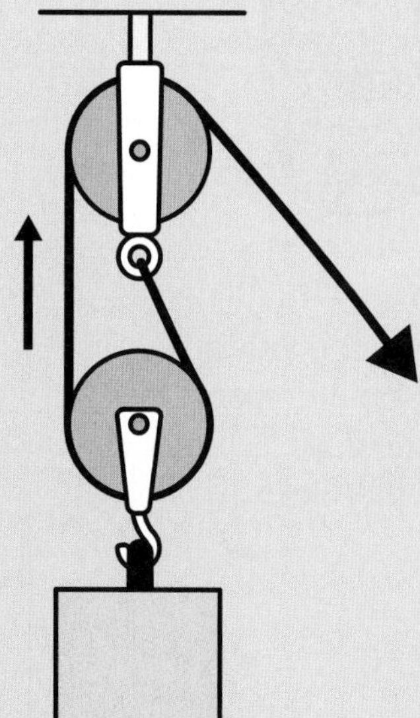

If you want to reduce the strength of force required to lift an object, you need to use a system of pulleys known as a compound pulley.

Compound pulleys, like the one shown on the left, are used both to change the direction of the force and to reduce the force required to move an object. The multiple pulleys effectively spread the force required to move the object over the entire system. The more pulleys in the system, the smaller the input force required. As with levers and wheels and axles, the input force has to be applied over a greater distance than the object is moved. A compound pulley is often called a block and tackle.

Some common uses of pulleys are in cranes, elevators, and fishing poles (the reel and line).

Exploration 2: Is gravity caused by Earth's magnetic field?

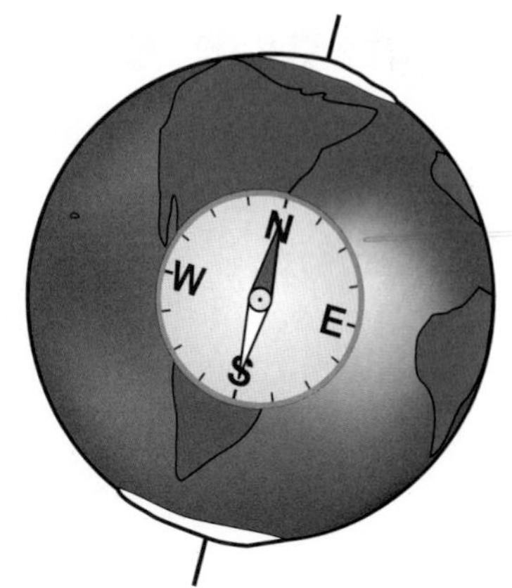

STEP 1 With your team, design a quick exploration to test whether gravity is caused by Earth's magnetic field.

STEP 2 Conduct your exploration. (You should complete this exploration in about 10 min or less!)

4. Write a paragraph to describe your exploration and your results.

5. Do you think gravity is caused by Earth's magnetic field? Why or why not?

Exploration 3: Is gravity caused by Earth's rotation?

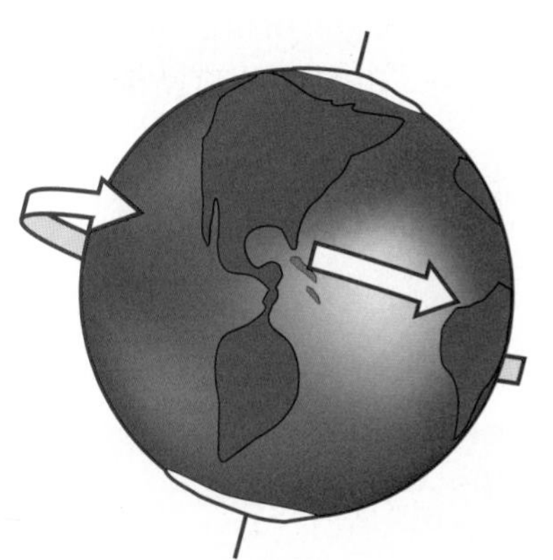

STEP 1 With your team, design a quick exploration to test whether gravity is caused by the Earth's rotation.

STEP 2 Conduct your exploration. (You should complete this exploration in about 10 min or less!)

6. Write a paragraph to describe your exploration and your results.

7. Do you think gravity is caused by Earth's rotation? Why or why not?

Have your teacher check your explorations before you begin.

Make Sense of Your Ideas

Participate in the class discussion of the explorations that you and your class conducted.

Have you discovered yet what causes gravity? If not, what have you eliminated as causes of gravity?

Our Consensus Ideas

The key question for the activity was:

What are possible causes for gravity?

Are you and your class able to answer the key question? Explain why or why not.

DEVELOPING OUR IDEAS

Activity 3
More on Gravitational Interactions

Purpose

The goal of this cycle is to describe gravitational interactions.

The purpose of this activity is to consider the following key question:

What is interacting in a gravitational interaction?

Record the key question for the activity on your record sheet.

We Think

Black holes in space are very interesting phenomena. At the center of a black hole is a very, very massive object. Anything near a black hole is pulled into it and is never seen or heard of again! This includes planets, stars, and any other nearby space objects.

Discuss the following question with your team.

How do black holes attract space matter?

Participate in a class discussion.

Even light cannot escape the gravitational pull of a black hole.

Explore Newton's Ideas

In Cycle 1, you encountered the famous scientist Isaac Newton (1642–1727). Newton thought about many areas in physics, including what causes gravity. Newton suggested an explanation for gravity that has been used by scientists for over 300 years.

According to Newton's explanation, gravitational interactions not only happen between the Earth and other objects, but gravitational interactions occur between *every pair of objects* in the universe. This is a very strange idea since you do not see objects around you sliding toward each other!

In Newton's theory on gravitational interactions, there are two variables that influence the strength of a gravitational interaction. These variables are discussed on the next page.

- One variable is the *mass of the objects*. Mass measures the amount of material an object has. If the mass of either of the interacting objects increases, then there is a stronger attraction between the objects.

- The other variable is the *distance between the objects*. Gravitational interactions are stronger for objects that are closer together. Scientists have verified that the gravitational interaction between an object and the Earth is a *very tiny* bit stronger on the ground floor of a skyscraper than at the top.

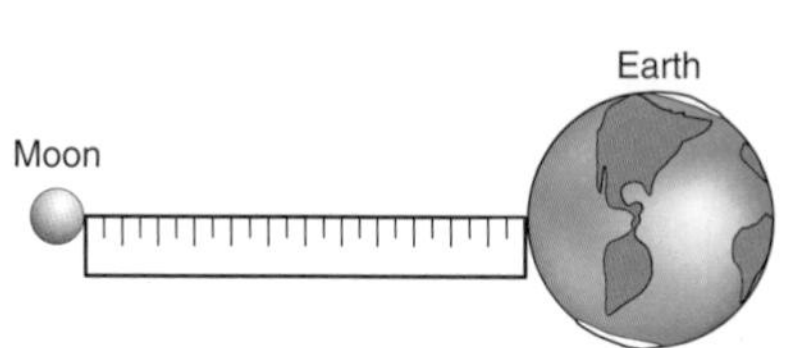

1. Which do you think would be stronger, the gravitational interaction between an apple and the Earth, or the gravitational interaction between an apple and the Moon? Write your reasons.

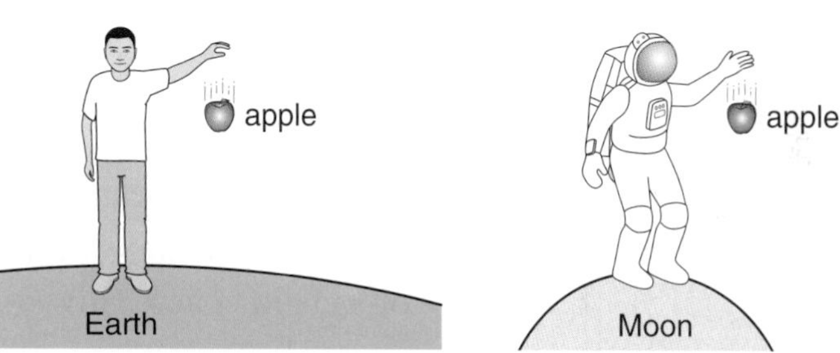

2. Look at the picture below. Which do you think would be stronger, the gravitational interaction between the Sun and Mercury, or the gravitational interaction between the Sun and Pluto? Write your reasons.

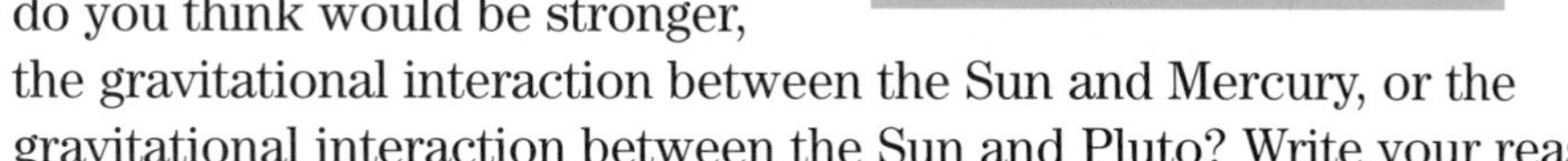

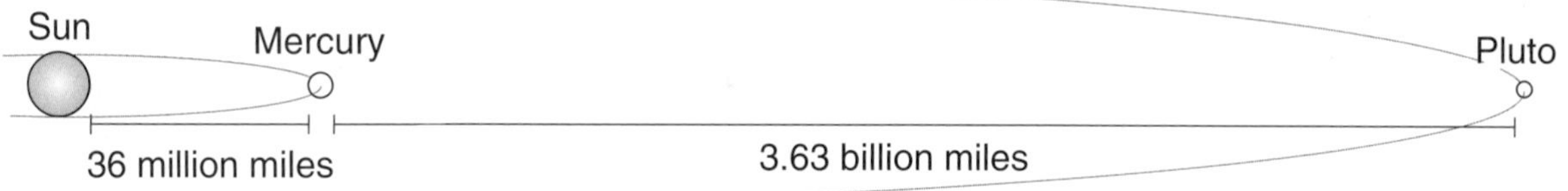

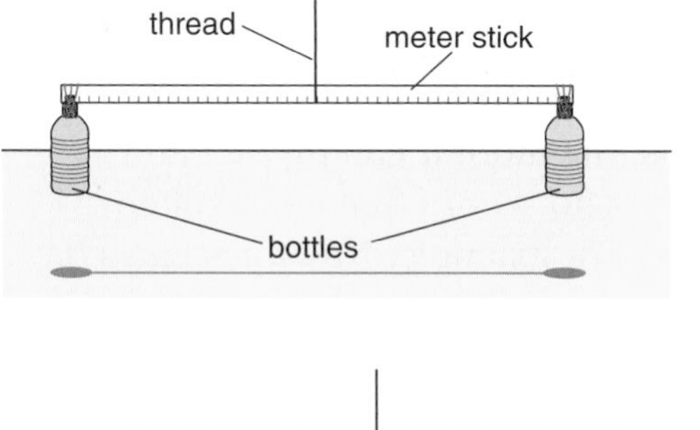

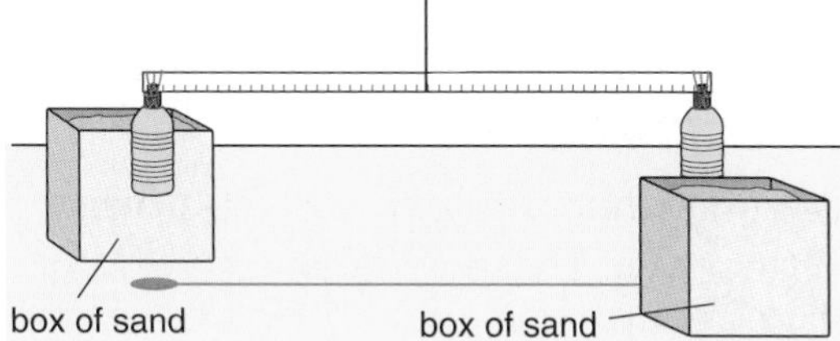

In 1798, Henry Cavendish did an experiment that verified Newton's explanation of gravitational interactions.

To set up the Cavendish experiment, a meter stick is suspended by a fine thread. Bottles are attached to each end of the meter stick.

To test Newton's idea about what causes gravity, heavy boxes of sand are brought close to the bottles, but the bottles and boxes do not touch each other.

You will watch a video of an experiment that is similar to Henry Cavendish's experiment. The film footage is old, but just think of it as a classic black and white movie. For a scientist, Cavendish's experiment *really* is a classic!

3. After viewing the video, write a paragraph to describe the experiment and its results.

Make Sense of Your Ideas

1. Why do you observe gravitational interactions between the water bottles and the boxes of sand in the Cavendish experiment, but not between two apples on a table?

2. A *similarity* between gravitational and magnetic interactions is that objects do not need to touch each other. What are some *differences* between these interactions? Use complete sentences to describe at least three differences.

3. How does the strength of gravitational interactions compare to the strength of magnetic interactions?

4. Scientists believe that black holes are caused by *very, very, very massive* objects in space. How do black holes attract space matter?

Participate in the class discussion of this activity.

Our Consensus Ideas

Think about the key question for this activity and discuss it with your team.

What is interacting in a gravitational interaction?

1. Write your answer on your record sheet.

Participate in the class discussion about the key question.

2. Write the class consensus ideas on your record sheet.

In scientific terms, a theory implies that an idea has been strongly supported by evidence. The more successful a theory is, the more it can predict and explain the physical world. At present, there is no theory that can explain all the forces in the universe in all situations. For example, Newton's Laws on **gravity** work well for massive things in the universe like apples and slow-moving planets. However, they cannot explain gravity interacting on tiny or fast-moving objects. They also cannot explain the gravitational forces that move galaxies. Scientists are always working on improving or finding new theories.

Science Words

gravity: the force of attraction between two bodies due to their masses

LEARNING MORE IDEAS

Activity 4
Weight

Purpose

A concept that you encounter in your everyday life is weight, but you may not have thought about it very much. You use the word weight in many different ways. For example, the doctor measures your weight; you lift weights at the gym; and in the lab, you might use a weight to help you determine how heavy an object is. In this activity, you will learn what *scientists* mean by the word *weight*. The key questions for this activity are:

1. **How do scientists define weight?**
2. **What is the direction of the force associated with the weight of an object?**

Record the key questions for the activity on your record sheet.

What does weight mean to scientists?

We Think

Discuss the following question with your team.

- What is weight?

Write your thoughts on your record sheet.

Participate in the class discussion about this question.

Explore Your Ideas

What is Weight?

Weight is related to gravitational interactions. You learned in the last activity that gravitational interactions happen between any two objects. The mass of one object is attracted to the mass of another object, and vice versa. Using forces to describe gravity, scientists say that the gravitational interaction produces a force between objects. The strength of this force is determined by the mass of the two objects interacting (for example, an apple and Earth) and how far apart they are from each other.

Science Words

weight: the force exerted by a planet on an object

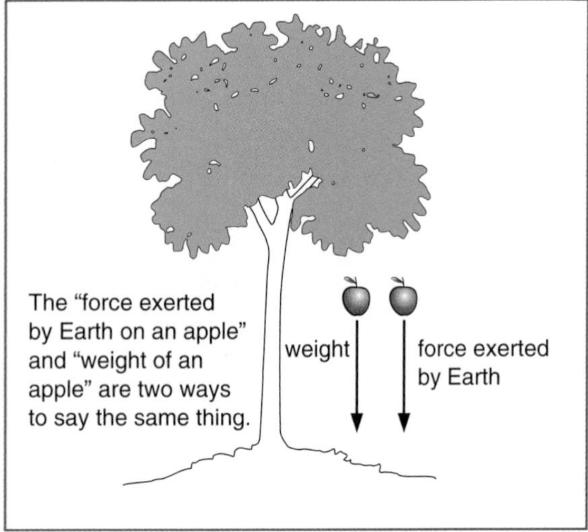

The "force exerted by Earth on an apple" and "weight of an apple" are two ways to say the same thing.
weight
force exerted by Earth

In everyday experiences, the gravitational interaction is noticeable only when at least one of the masses is very large. For example, you notice the attraction between an apple and the Earth because the Earth is so massive. Scientists call the force exerted by a planet on an object its *weight.*

When you hold an apple, you feel the strength of the Earth pulling on it. When you hold an apple in one hand and a melon in the other, you can compare the strength of the Earth pulling on each one. You can tell that the weight of the melon is greater than the weight of the apple.

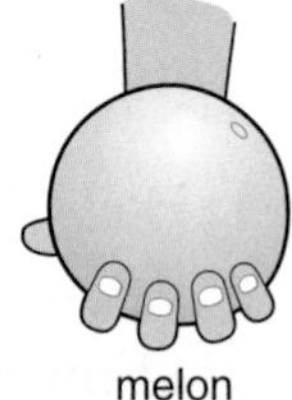

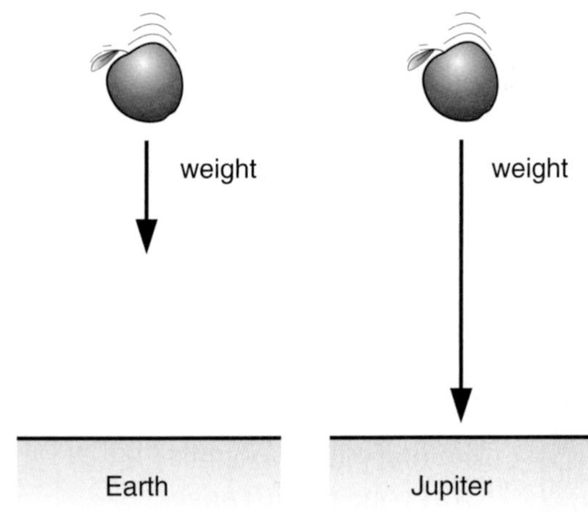

Compare the weight of an apple on two different planets. The mass of the apple (the amount of material in it) will not change from place to place. Jupiter has a stronger gravitational pull than Earth (about 2 ½ times as strong), so the gravitational interaction between the apple and Jupiter will be much greater than the gravitational interaction between the apple and Earth. The weight of the apple will be greater on Jupiter than on Earth, but the apple's mass will be the same on both planets.

1. Mercury has a weaker gravitational pull than Earth (about 4⁄10 as much). Will an apple weigh more on Earth or on Mercury? Why?

You just read that the gravitational interaction produces a force between objects, and that weight is the force exerted by a planet on an object. What is the *direction* of the force exerted by the planet on an object? To help you think about this, answer the following questions.

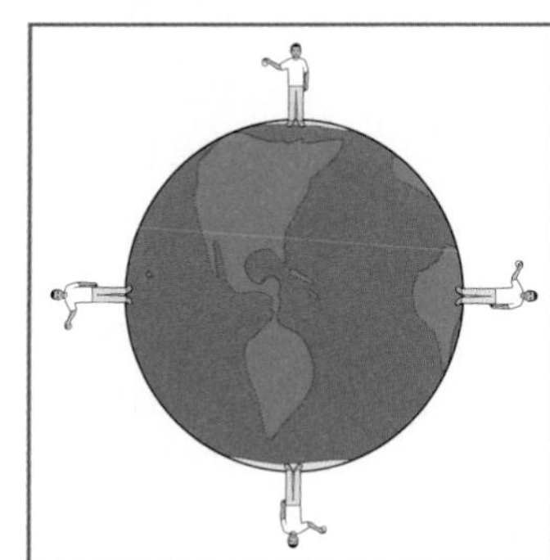

2. In the picture at right, four people holding balls are standing on different parts of the Earth. Suppose each person lets go of the ball. On a picture on your record sheet draw the path that shows how each ball would fall. Write your reasons for your drawing.

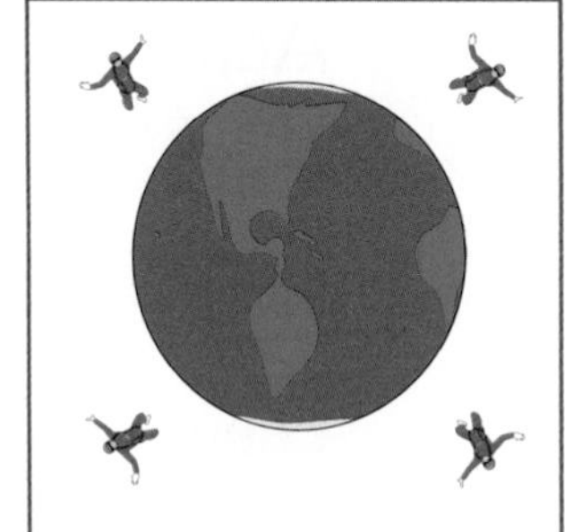

3. Four sky divers travel 32 km (20 mi.) above the Earth in hot-air balloons. The balloons hover over landing sites while the sky divers jump from the balloon. On a picture on your record sheet draw the path that shows how each sky diver would fall. Write your reasons for your drawing.

4. An explorer has two half-filled water bottles at the North Pole. One bottle is capped and the other is not. Suppose the explorer travels to the South Pole and sets her bottles down. Show on a sketch on your record sheet what will happen to the water in the bottles. Write your reasons.

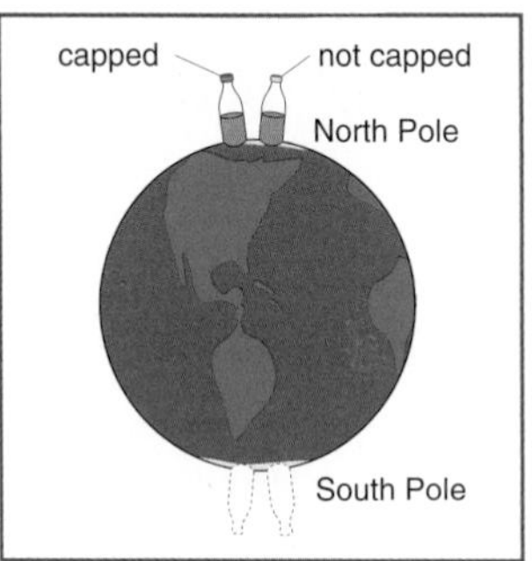

Make Sense of Your Ideas

In each of the previous three situations, the force exerted by the Earth on each object is much larger than any other force on each object. Drag, wind, etc., are much smaller forces.

1. Look back at your answers to Questions 2 through 4. In general, what is the direction of the force exerted by the Earth on nearby objects?

The weight of an object depends on the masses of the object and the planet. The weight also depends on the distance between the object and the planet. If the distance is "small" enough, the weight does not change much.

For example, your weight at the doctor's office and your weight in a flying airplane is almost the same. This difference in distance from the center of the Earth does not make much difference in weight.

2. Why do astronauts traveling from the Earth to the Moon feel weightless?

Participate in a class discussion about your answers to these questions.

Our Consensus Ideas

Think about the key questions below and discuss them with your team.

1. How do scientists define weight?

2. What is the direction of the force associated with the weight of an object?

1. Write your answers on your record sheet.

Participate in the class discussion about the key questions.

2. Write the class consensus ideas on your record sheet.

Activity 5
Putting Together Gravitational Interaction Ideas

Comparing Consensus Ideas

Recall the key question for this cycle:

How can you describe gravitational interactions?

In this cycle's activities, you learned about gravitational interactions. You also used gravitational interaction ideas to help you understand many everyday events. For example, you now understand why objects fall on Earth.

Your teacher will review with you all of the ideas that the class developed during this cycle to help you answer the cycle question. Be prepared to contribute to this discussion.

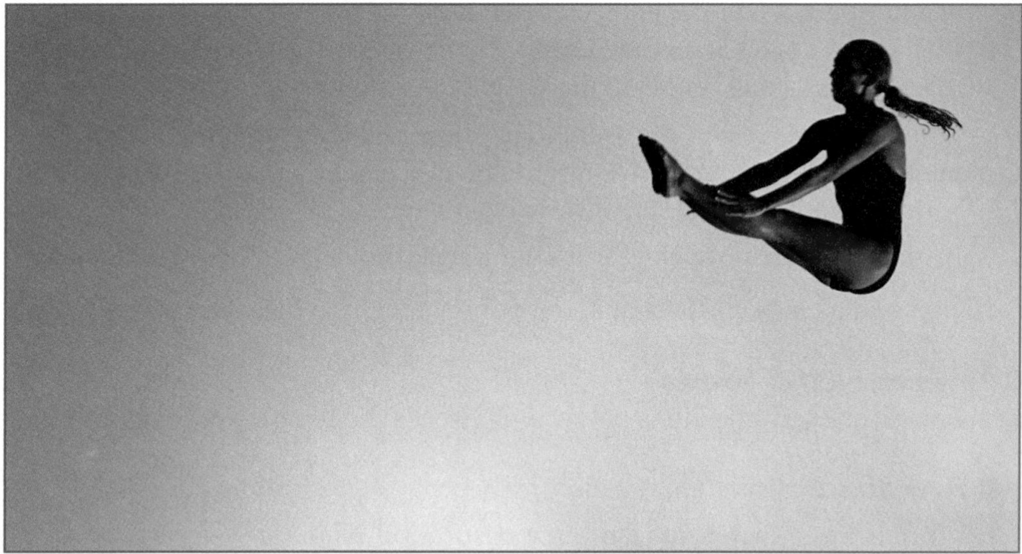

Your teacher will now distribute copies of *Scientists' Consensus Ideas: The Gravitational Interaction.* Read the scientists' ideas and compare them with the ideas your class developed during this cycle.

Write the evidence from the activities in this cycle where it is requested on the *Scientists' Consensus Ideas* form.

Your teacher will lead a whole-class discussion.

IDEA POWER

Activity 6
Orbital Motion

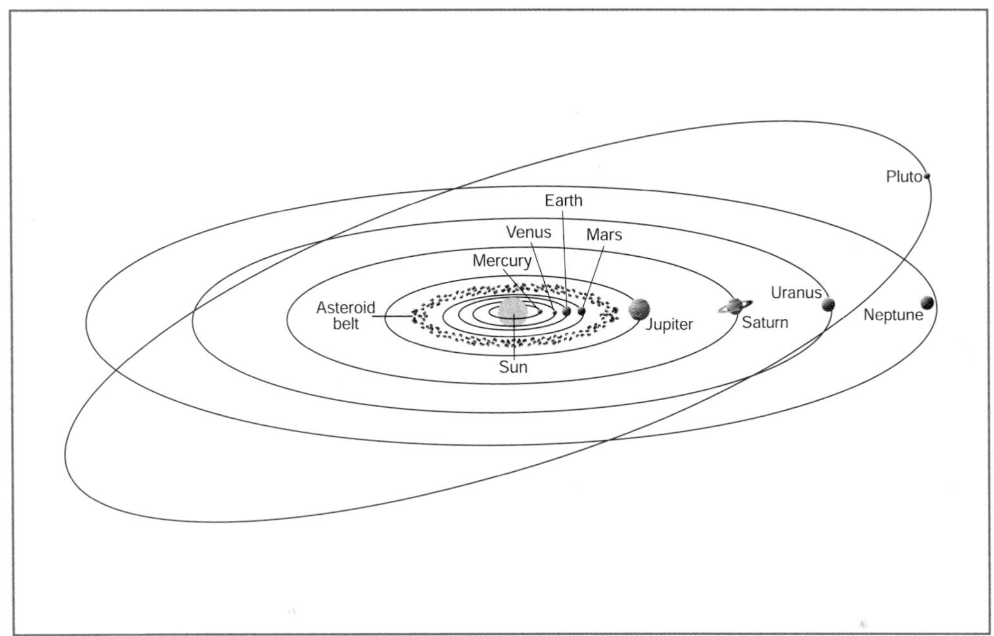

Why do objects follow an orbital path?

Orbital Motion

Planets orbit the Sun, moons orbit planets, and communication satellites orbit Earth. Orbital motion is all around you! The path that one object takes as it moves around another object is called an **orbit**. The purpose of this activity is to apply your ideas about forces and about gravitational interactions to help understand orbital motion.

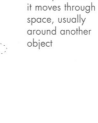

Science Words

orbit: the path that an object takes as it moves through space, usually around another object

You studied circular motion in Unit 3, Cycle 1. A hopper ball whirled around on a string has circular motion. The ball makes a circle around the leg. The force exerted by the string on the ball keeps the ball in a circle.

Like the hopper ball's path, an orbit can be circular. Planets, moons, and other heavenly bodies frequently have orbits that are almost circular. Whether the orbit is a perfect circle is not important for you to think about now.

1. Think about the Earth orbiting the Sun. What type of interaction occurs between the Earth and Sun?
2. On a sketch on your record sheet, draw a force arrow *for each position of the Earth* to represent the force exerted *on the Earth* by the Sun.
3. What are some similarities about the hopper ball twirling around the leg and the Earth orbiting the Sun?
4. Use interaction ideas and force ideas to write an explanation of what keeps the Earth in orbit around the Sun.

Participate in the class discussion about orbital motion.

Analyze, Explain, and Evaluate

In this part of the activity, you will explain some other interesting situations involving orbital motion. The procedure for doing this can be found in *How To Write an Analysis and Explanation.*

All of the analyses should include the *interacting objects and their interaction type*, and *labeled force arrows* showing the forces being exerted on the orbiting object. You will not draw energy diagrams, however. Be sure that what you write would get a good evaluation using *How To Evaluate an Analysis and Explanation.*

5. Saturn has fascinated astronomers for centuries because beautiful rings encircle it. At some stage of Saturn's formation, pieces of ice and debris moving in its vicinity were attracted to the planet, forming the rings. These pieces range in size from smaller than an ant to as large as a boulder. Why do the particles circle Saturn?
6. Comets have very complicated paths through the universe. They speed up, slow down, and even make turns on their own galactic

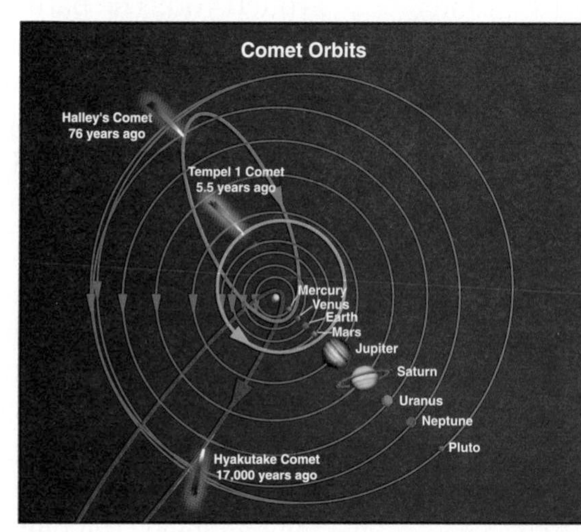

highways. A comet's path curves in a partial circle when it travels near something massive like our Sun. Why does a comet's path curve around a large object like our Sun?

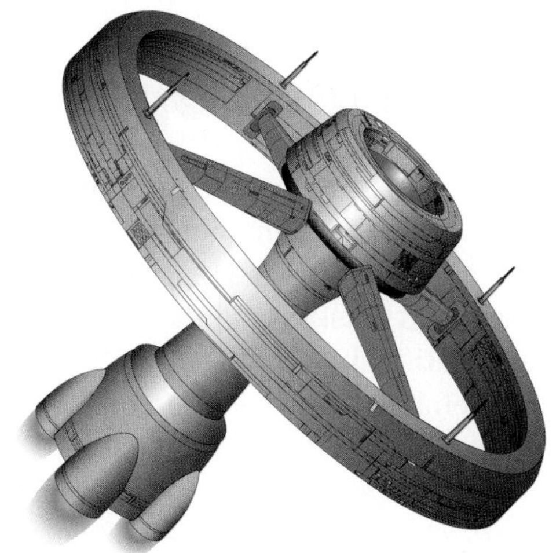

Simulating Gravity

Space stations are often found in science fiction books and movies. However, a space station would not be massive enough to have a very strong gravitational interaction with its inhabitants. To simulate gravity, the space stations often rotate like a bicycle wheel. How does the rotation simulate gravity?

As you stand on Earth, you feel pressure on the bottom of your feet from the force exerted by the Earth on you. This is one way that you are aware of gravity. A rotating space station simulates this feeling by pushing you in a circular path as it rotates. As the space station pushes you along, you feel pressure on the bottom of your feet just as you would standing on Earth.

What about objects on the space station falling to the floor? If you let go of an apple on Earth, it drops to the ground. This is another way that you are aware of gravity. If you let go of an apple in a space station, it doesn't drop. The floor moves up to meet the apple as the space station rotates. However, since the space station is pushing you around in a circle at the same time, it appears to you that the apple has dropped!

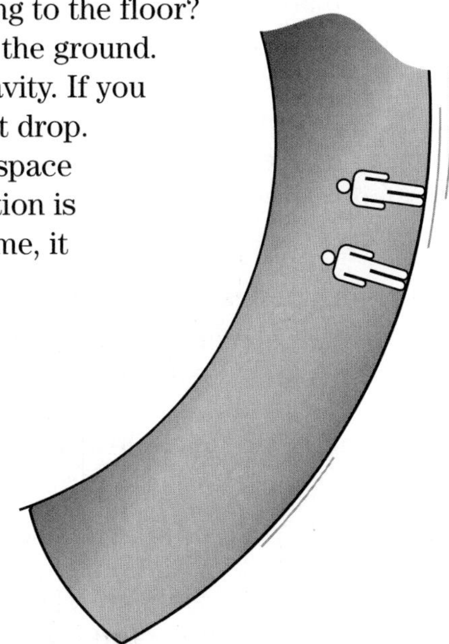

How fast the space station needs to rotate depends on its diameter and the amount of gravity it is trying to simulate. Although inhabitants in the space station will "feel gravity," they also may get woozy watching the stars swing by. Perhaps space stations should have few windows. Think about this the next time you watch your favorite science fiction show!

IDEA POWER

Activity 7
Terminal Speed

In this unit, you have explored four types of mechanical interactions (applied, friction, drag, and elastic) and the gravitational interaction. Remember, *in a mechanical interaction, two objects push or pull each other over a distance, and in a gravitational interaction, the masses of two objects attract each other*. In mechanical interactions, objects must touch each other, but objects in gravitational interactions do not need to touch each other.

As objects fall through the air, their motion is affected primarily by two interactions: drag and gravitational. This is especially important to sky divers. They are *very* concerned about the kind of motion that they have while falling through the air! Sky divers rely on the gravitational interaction between their bodies and Earth to return them from their aircraft to the ground. They rely on the drag interaction between their parachutes and the air to help them land on the ground safely! The purpose of this activity is to further explore the motion that objects have as they fall through the air.

Your team will need:

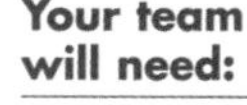

- 2 coffee filters

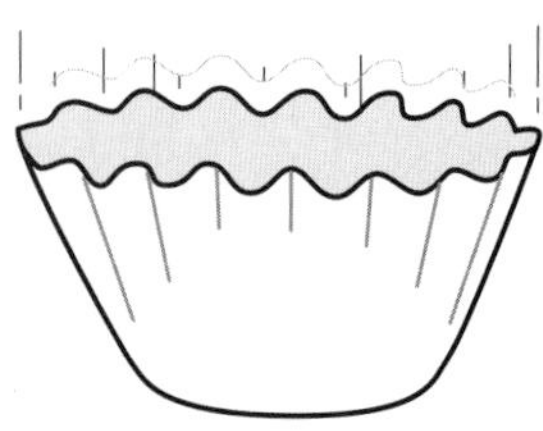

Explore the Ideas

Let's explore what happens as sky divers (or any objects) fall through the atmosphere.

STEP 1 Hold one of the coffee filters up high, positioned like a bowl of soup. Drop the filter and watch it fall.

1. What interactions affect the coffee filter as it falls?

STEP 2 Crumple the other coffee filter. Hold it up high. Drop the crumpled filter and watch it fall.

2. What interactions affect the crumpled coffee filter as it falls?

STEP 3 Hold both filters at the same height. Drop them at the same time and watch them fall.

3. What did you observe?

4. Analyze and explain why one coffee filter hit the ground before the other. Use *How To Write an Analysis and Explanation* as an outline. The analysis should include the *interacting objects and their interaction type*, and *labeled force arrow(s)* showing the force(s) being exerted on the coffee filter. Do not draw energy diagrams. You will learn about analyzing gravitational interactions using energy ideas later in this course or in other science courses. Be sure that what you write would get a good evaluation using *How To Evaluate an Analysis and Explanation.*

5. When either of the coffee filters is dropped, it speeds up as it falls. If you could drop a filter from a very high place (such as the top of a tall building or an airplane), do you think the filter would continue to speed up as it falls? Or do you think that it will reach some final speed?

6. Analyze and explain your answer to Question 5. (Do you think the filter would continue to speed up as it falls?) Use *How To Write an Analysis and Explanation* as an outline. The analysis should include the *interacting objects and their interaction type*, and *labeled force arrow(s)* showing the force(s) being exerted on the coffee filter. Again, do not draw energy diagrams. Be sure that what you write would get a good evaluation using *How To Evaluate an Analysis and Explanation.*

I think that the coffee filter will keep speeding up as it falls. There is no speed limit for the coffee filter. It just depends on how high it is when it is dropped.
Nguyen

No way! That coffee filter is going to reach some final speed, then finish falling to Earth with that final speed.
Isabel

Sky divers are familiar with the phrase terminal speed. Usually terminal speed is used to refer to the fastest speed that can be achieved by falling through the atmosphere. As a sky diver falls towards Earth, there are two forces exerted on the sky diver:

- the downward force from the gravitational interaction between the sky diver and the Earth,
- the upward force from the drag interaction between the sky diver and the air.

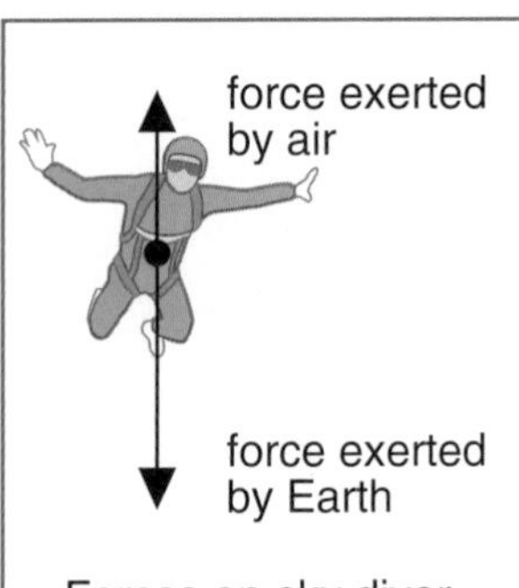

Forces on sky diver are unbalanced. Sky diver speeds up while falling.

As long as these two forces are unbalanced, the sky diver continues to speed up. However, as the sky diver speeds up, the force from the drag interaction increases. When the two forces on the sky diver become balanced, the sky diver will not speed up any more. The sky diver has reached terminal speed.

Some sky divers take on the daring challenge of reaching a maximum terminal speed during their dives before opening their parachutes. To reach the highest possible terminal speed, sky divers try to minimize the drag interaction. The three variables that affect the drag interaction between the sky diver and the air are:

- the thickness of the air
- how the sky diver is positioned while falling through the air
- the speed of the sky diver (the greater the sky diver's speed, the greater the drag interaction)

Forces on sky diver are balanced. Sky diver has constant speed while falling.

Typically sky divers try to minimize the drag interaction by wearing slick clothing and positioning their bodies like a downward arrow. They are able to obtain terminal speeds of about 125 mph (that's fast)! The world record terminal speed was set in 1960 by Captain Joseph W. Kittinger, Jr of the United States Air Force. He rose to an altitude of about 20 mi. above Earth in a balloon and reached a terminal speed of 614 mph!

7. Analyze and explain how a sky diver reaches terminal speed. Use *How To Write an Analysis and Explanation* as an outline. The analysis should include the *interacting objects and their interaction type*, and *labeled force arrow(s)* showing the force(s) being exerted on the sky diver. Do not draw energy diagrams. Be sure that what you write would get a good evaluation using *How To Evaluate an Analysis and Explanation.*

LEARNING ABOUT OTHER IDEAS

Activity 8
Unbalanced and Balanced Forces

Purpose

In the activities in this unit, you have examined different types of forces, such as elastic, friction, and gravitational forces. You have also studied what happens in the case of multiple forces acting on an object, and examined situations in which forces are balanced and unbalanced.

For the sled to speed up,the multiple forces acting on it must be unbalanced.

Suppose the multiple forces acting on an object are unbalanced. If you know the strength and direction of each of the forces, how do you combine them to determine their total effect on the object? If the multiple forces are balanced, and you know the strengths and directions of all but one of the forces, how can you determine the strength and direction of the remaining force? The purpose of this activity is to help you answer these key questions:

1. How do you combine multiple forces to determine their total effect on an object?

2. How do you determine the force necessary to balance other forces?

Record the key questions for the activity on your record sheet.

Learning the Ideas

Standard Unit of Force Strength

To solve problems in which you need to find the strength of a force, you have to know the standard units used to measure force strengths. Scientists measure force strengths in **newtons**, a standard unit named after Isaac Newton. The symbol for a newton is N.

Science Words

newton: a unit of force

A force of 1 N (one newton) is about the weight of a tangerine (on Earth). A middle-school student about one-and-a-half meters tall typically weighs around 400 N.

Unbalanced Forces

Suppose a girl pulls a shopping cart across the floor with strength 6 N. As the shopping cart moves there is a friction force of strength 5.6 N that opposes the motion.

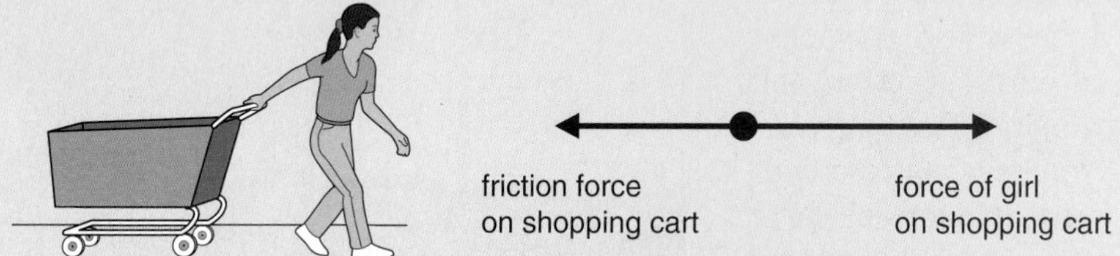

Since the force directed to the right (force of the girl on shopping cart) is greater in strength than the force directed to the left (friction force), the forces on the shopping cart are *unbalanced.* You can combine these forces in order to find the strength and direction of the single unbalanced force that would have the same effect on the shopping cart as the two opposing forces. To determine the unbalanced force, follow this procedure:

- Choose a positive direction and a negative direction.
- Assign a plus sign to all the forces pointing in the positive direction.
- Assign a minus sign to all the forces pointing in the negative direction.
- Add all the forces, taking into account the plus and minus signs. The *size* of the answer determines the strength of the unbalanced force, and the *sign* of the answer (+ or –) determines the direction of the unbalanced force.

Follow this procedure to determine the unbalanced force on the shopping cart.

- Let the positive direction be to the right. Therefore, the negative direction is to the left.
- The strength of the pulling force of the girl on the shopping cart is + 6 N. The + sign means this force is directed to the right.
- The friction force on the wagon is – 5.6 N. The minus sign means this force is directed to the left.
- Add all the forces.

Unbalanced force on shopping cart = + 6 N + (– 5.6 N) = + 0.4 N

The answer of + 0.4 N means the size of the unbalanced force is 0.4 N and its direction is to the right (because of the + sign). This unbalanced force will cause the shopping cart to speed up to the right in the exact same way as the two separate forces acting together will cause the shopping cart to speed up to the right.

Here are two more examples.

Example 1: Imagine that Jocelyn is pushing a box across a floor to the right. Her little brother Kenny wants to play and he tries to push the box toward the left. Jocelyn pushes with force strength 26 N and Kenny pushes back with force strength 11 N. A friction force of 6 N also opposes the motion of the box. What is the strength and direction of the unbalanced force on the box?

Let the positive direction be to the right, and the negative direction to the left.

Answer the following on your record sheet.

1. What is the pushing force on the box by Jocelyn (include sign)?
2. What is the pushing force on the box by Kenny (include sign)? What is the friction force on the box (include sign)?
3. Add all the forces and determine the strength and direction of the unbalanced force on the box. Show your work.

Example 2: Tomás, a sky diver, who weighs 852 N, experiences a drag force (also called air resistance) of 302 N shortly before opening his parachute. What is the direction and strength of the unbalanced force on the sky diver? Remember that the weight force equals the gravitational force on the sky diver.

Let the positive direction be up, and the negative direction be down.

4. What is the drag force (air resistance) on the sky diver (include sign)?
5. What is the gravitational force on the sky diver (include sign)?
6. Add all the forces and determine the strength and direction of the unbalanced force on the sky diver. Show your work.

Your teacher will review the answers to these questions with the class.

Solve the following problem on your own.

7. Gillian exerts an upward force of 300 N on a box that weighs 256 N. What is the strength and direction of the unbalanced force on the box?

Balanced Forces

As you learned in Activity 7 of this unit, if an object is at rest and the multiple forces acting on it are *balanced*, then the object will remain at rest. When dealing with problems involving balanced forces, the following mathematical relationship is useful.
For balanced forces only:

Sum of the forces in one direction = Sum of forces in opposite direction

Consider Example 3 where the forces are balanced.

Example 3: A sphere hangs motionless from a rubber band. The sphere has a weight of 5 N. The rubber band is stretched, which means that the rubber band exerts an upward elastic force on the sphere. A picture of the situation, along with a force diagram, is shown. What is the strength of the force that the rubber band exerts on the sphere?

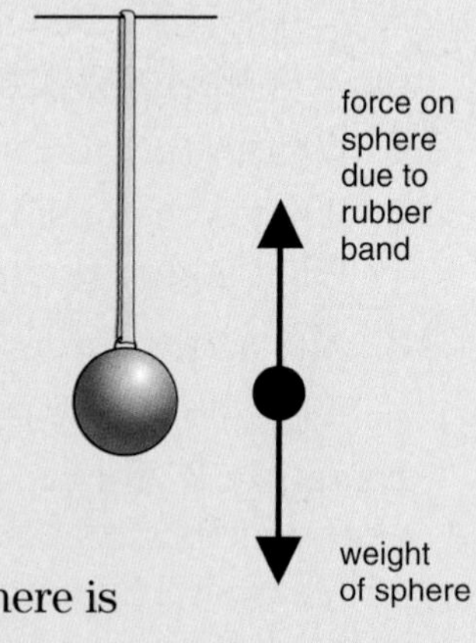

In this problem, there is only one upward force and one downward force on the sphere. The upward force is the elastic force of the rubber band on the sphere. The downward force is the weight of the sphere. Since the sphere is motionless, the forces must be balanced. Therefore:

Upward force on sphere = Downward force on sphere

Force on sphere due to rubber band = Weight of sphere

8. The weight of the sphere is 5 N. What is the strength of the elastic force on the sphere due to the rubber band?

Solve the following problems on your own.

9. A heavy book sits on top of a piece of foam rubber. The book is subject to a downward force of 12 N due to its weight, and to a force exerted by the foam rubber. Because the foam rubber is compressed, the force it exerts is known as a compression force, which is a special type of elastic force. What is the strength and direction of the compression force that the foam rubber exerts on the book?

10. Alex is walking back from fishing in a river. The two fish he caught hang motionless from a string he carries. The larger fish weighs 20 N, and the smaller fish weighs 7 N. What is the strength and direction of the force that balances the weights of the two fish? (This is similar to the situation in Question 8, but in this case, the string stretches only a small bit, and it is harder to notice.)

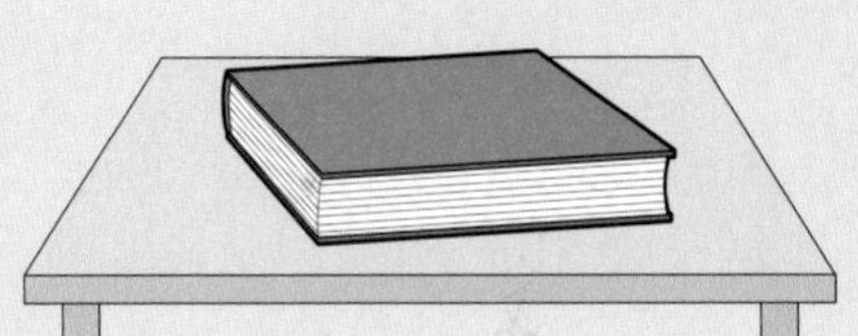

11. A heavy book sits on top of a table. The book weights 12 N. What is the strength and direction of the force balancing the weight of the book? (This is somewhat similar to the situation in Question 9, but in this case the table is compressed only a very tiny bit. The compression is much too small to see except with an extremely powerful microscope.)

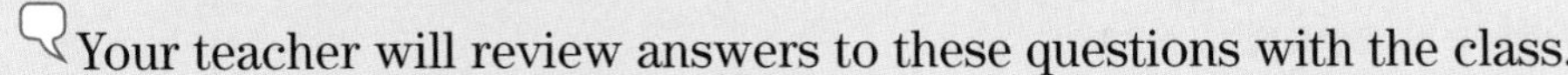

Your teacher will review answers to these questions with the class.

What We Have Learned

Recall the key questions for this activity:

1. How do you combine multiple forces to determine their total effect on an object?

2. How do you determine the force necessary to balance other forces?

Participate in the class discussion about your answers to the questions above.

Write the answers to the key questions on your record sheet.

LEARNING ABOUT OTHER IDEAS

Activity 9
Buoyancy

What keeps this boat afloat?

Purpose

You have observed that when placed in water, some objects float and others sink. You may have also observed that objects seem lighter when you hold them under water. The explanation for these observations has to do with a type of force called the *buoyant force*. The purpose of this activity is to learn about buoyancy and then use that idea to account for why things float or sink. The key question for this activity is:

Why do things float or sink?

Record the key question for the activity on your record sheet.

Learning the Ideas

The Buoyant Force

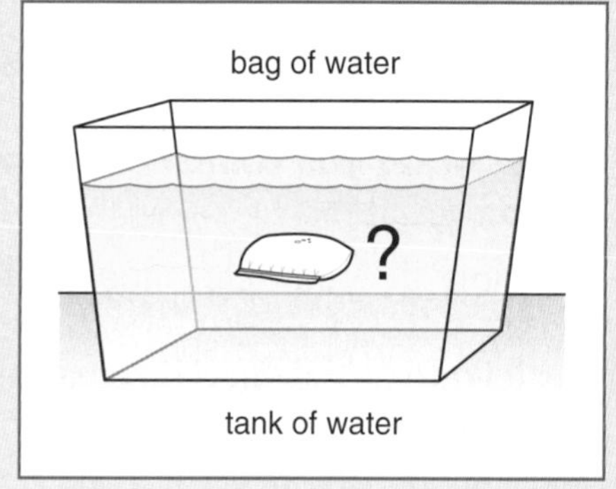

Suppose you fill a small plastic bag full of water and immerse it in a large tank of water. What will happen to the bag of water? Will it rise to the top and float? Will it sink to the bottom? Your teacher will show you a movie or do a demonstration.

1. What happens to the bag of water?

You have learned that if an object remains still, the forces acting on it must be *balanced*.

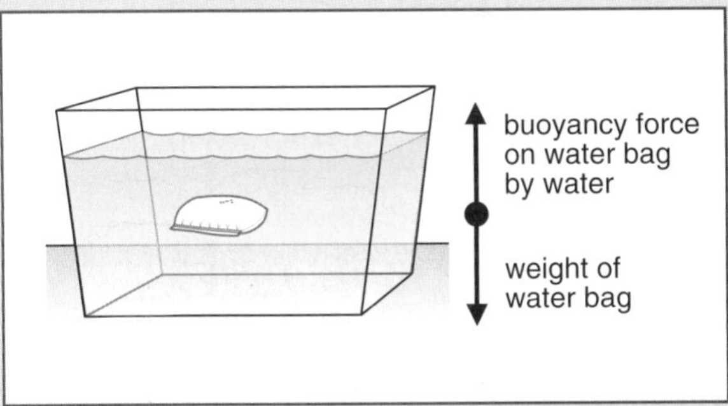

One of the forces acting on the bag of water is the gravitational force of Earth pulling down on the water bag. That force is called the weight of the water bag. For the water bag to remain still, there must be an upward force acting on the bag to balance its weight.

Science Words

buoyant force: for an object that is placed in a liquid, the force exerted on the object by the liquid

This situation is similar to situations you studied in Activity 8. For example, when a book sits motionless on a table, the table exerts an upward force on the book that balances the weight of the book. In this case, the upward force is due to the surrounding water and is called the **buoyant force**. Whenever any object is placed in water (or any liquid), there is always an upward-pointing buoyant force.

Weighing Objects

Scales are used to weigh objects. The diagram shows a 200-g object hanging from a spring scale. The scale reads 2 N. The force diagram for this situation is also shown. The upward force on the object due to the spring scale just balances the weight of the object. The scale thus *reads* the weight of the object; in this case, 2 N.

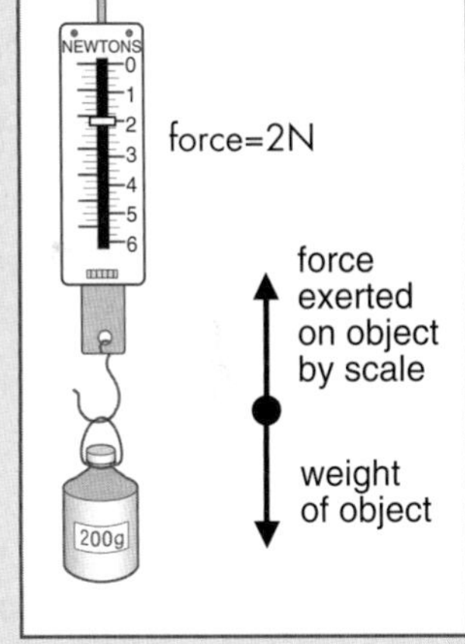

If you were to hold the same object in your hand, the upward force of your hand on the object would just balance the weight of the object. The balanced forces in this situation are shown in the diagram. Because of the upward force you exert on the object, you *feel* its weight of 2 N.

Suppose you have a spring scale holding up an object (which may be different from the 200-g object mentioned above). Imagine that the object is totally immersed in a tank of water. See the diagram on the next page.

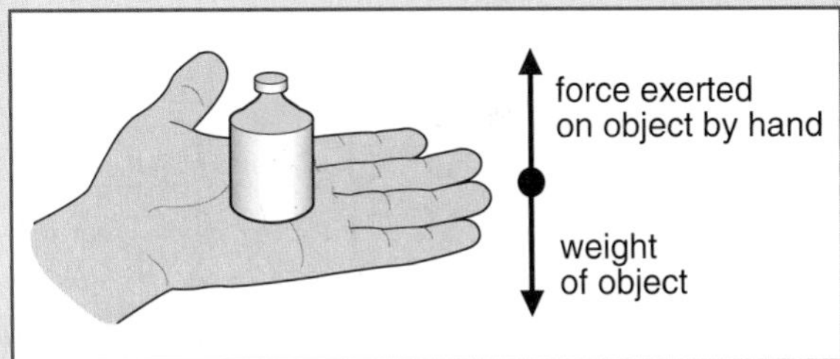

2. As the object is immersed in the water, do you think the scale reading will *increase*, *decrease*, or *remain the same?* Explain why you think so.

Your teacher will show you a movie or do a demonstration.

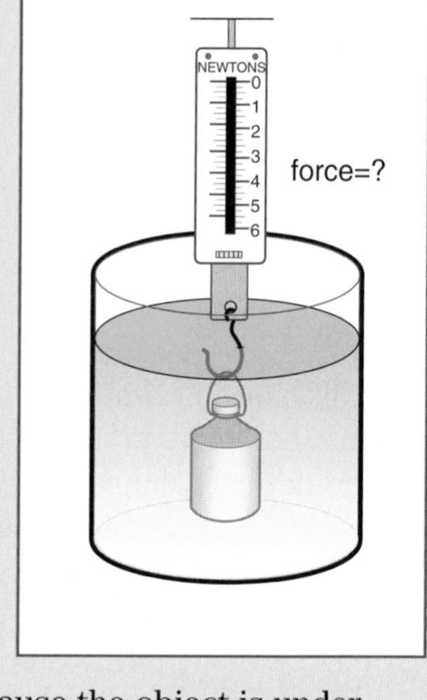

3. Record the following data on your record sheet:

   a) scale reading of object in air

   b) scale reading of object after it is fully submerged

4. What happens to the scale reading when the object is submerged in the water?

Because the submerged object is motionless, all the forces acting on it must be balanced. Consider what those forces are. First, there is the downward-pointing force on the object by the Earth (the object's weight). Second, there is the upward-pointing force on the object by the scale (since it is still pulling up on the object). Third, because the object is under water, there is also the upward-pointing buoyant force. These forces are included in the diagram. In this case, the spring force and the buoyant force are *both* acting to balance the weight force. Therefore, the spring scale does not pull upward as much as it did when the object was hanging in the air, so it *reads* a smaller value.

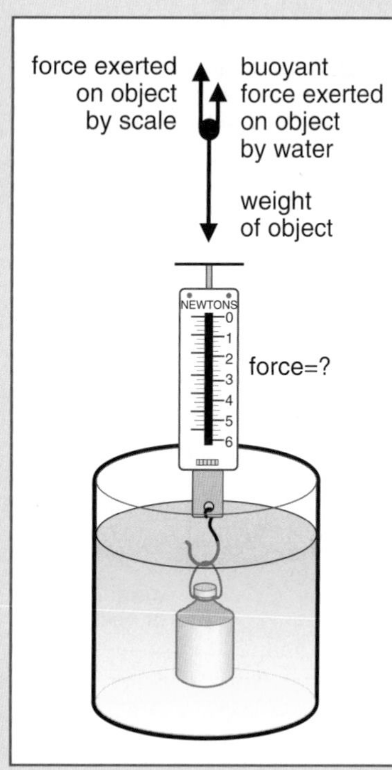

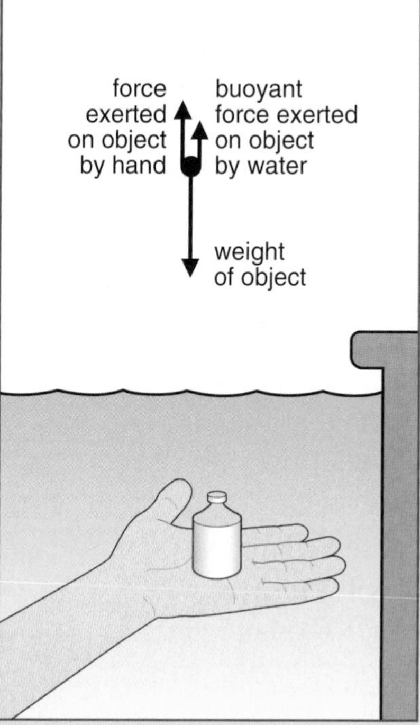

If you hold the object under water, the upward force you exert on the object is less than it would be if the object was in air. This is because the buoyant force due to the surrounding water is also pointing upward and is therefore helping you balance out the weight of the object. Therefore, the object *feels* lighter under water than it does in the air. The situation is similar when you stand on the bottom of a pool. Because of the upward buoyant force on you due to the water, you *feel* lighter than you do when standing on the ground outside the pool.

Strength of the Buoyant Force

What determines the strength of the buoyant force? Your teacher will show you a movie or do a demonstration that will help answer this question. In the movie, an object hanging from a spring scale is immersed totally in water. When the object enters the water, some of the water is displaced (pushed aside). The displaced water is collected in a separate container.

5. On your record sheet, record the following data:

a) scale reading of object in air (its weight)

b) scale reading of object after it is fully submerged

c) weight of empty container

d) weight of container plus displaced water

The strength of the upward-pointing buoyant force is the difference between the scale reading of the object in air and the scale reading of the object fully submerged.

6. Calculate the strength of the buoyant force.

The weight of the displaced water is the difference between the weight of the container plus displaced water and just the weight of the empty container.

7. Calculate the weight of the displaced water.

8. How does the strength of the buoyant force compare to the weight of the displaced water?

Your teacher will review the answers to these questions with the class.

Suppose you had performed the same experiment done in the movie many times with many different objects. You would have gathered evidence to support the following relationship between the strength of the buoyant force and the weight of the displaced water.

Strength of the buoyant force = Weight of the displaced water

Sinking and Floating

Suppose you placed an object completely under water and then let it go. Would it stay where it is, would it rise to the top and float, or would it sink to the bottom?

After you let go of the object, there are only two forces acting on the object. There is the upward buoyant force due to the surrounding water. There is also the downward force of the weight of the object. Whether the object sinks, floats, or remains submerged depends on how the strength of the upward buoyant force compares to the strength of the downward weight force.

9. Suppose the upward buoyant force was *greater* than the downward weight force. What would the object do?

10. Suppose the upward buoyant force was *less* than the downward weight force. What would the object do?

11. Suppose the upward buoyant force was *equal* to the downward weight force. What would the object do?

Your teacher will review the answers to these questions with the class.

Recall that the strength of the buoyant force equals the weight of the water displaced. To decide whether the object sinks or floats, you compare the *weight* of the object to the *weight* of the displaced water. Recall from Unit 1 Cycle 3, Activity 7 that when you place the object *completely* under water, it displaces a volume of water equal to its own volume. Both the object and displaced water have the same volume. So, comparing the weight of the object to the weight of the water displaced is the same as comparing the density of the object to the density of the surrounding water.

- If the density of the object is greater than the density of water, the object will sink.
- If the density of the object is less than the density of water, the object will rise up and float.
- If the density of the object is equal to the density of water, the object will remain where it is.

To test these ideas, your teacher will show you a movie. The person places three small plastic bags filled with different liquids in a tank of water. The water in the tank has a density of 1.0 g/mL. The names and densities of the three liquids are listed in the table on the next page.

LEARNING ABOUT OTHER IDEAS

Activity 10
Potential Energy

Purpose

This activity will focus on a new kind of energy, potential energy. You will consider the following key questions:

1. **What is potential energy?**
2. **How does it relate to motion energy?**

Record the key questions for the activity on your record sheet.

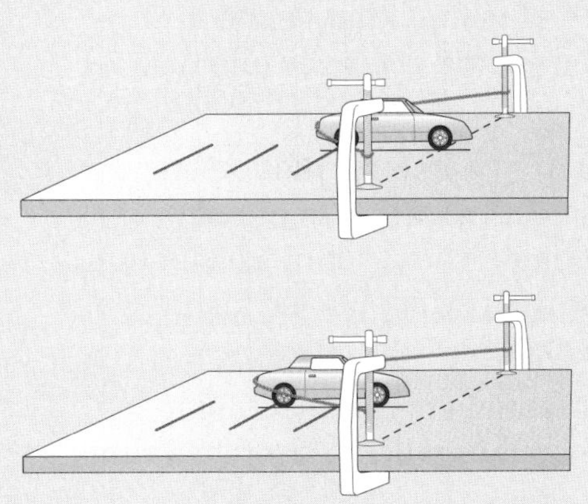

Learning the Ideas

Think back to Unit 2 Cycle 2, Activity 4 when you studied elastic interactions. You did an exploration with a rubber band that you stretched to different lengths. You used the rubber band to push a toy car across the floor or a tabletop. In that activity, you learned that a rubber band, along with other elastic objects, could store a certain type of energy called *stored elastic energy*. The rubber band increased in stored elastic energy when you stretched it.

You also observed in your exploration that as the rubber band shortened, the car sped up. This interaction can be described in an energy diagram.

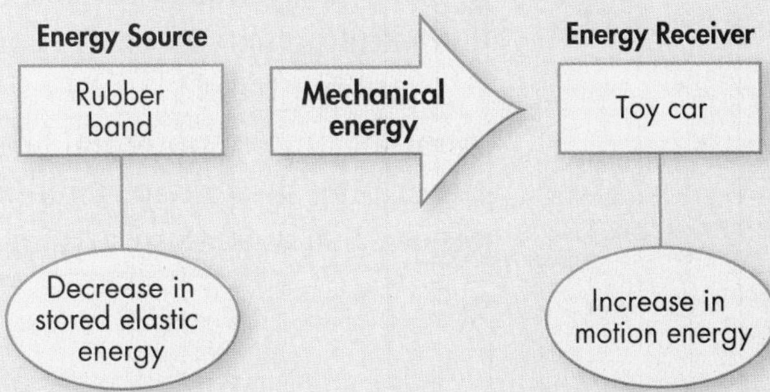

The energy diagram indicates that during the elastic mechanical interaction between the rubber band and the toy car, the stored elastic energy in the rubber band was changed into motion energy of the toy car. In other words, during this interaction, the stored elastic energy in the rubber band decreased and the motion energy of the car increased.

Science Words

potential energy: the energy of an object that is dependent on its position

kinetic energy: the energy an object possesses because of its motion

In the same exploration, you also found that the more the rubber band was stretched before releasing the car, the faster the car moved when leaving the rubber band. As you stretched the rubber band more, its stored elastic energy increased.

The amount of stored elastic energy in the rubber band depends on the distance that you pull it. Therefore, you can say that the amount of stored elastic energy depends upon the rubber band's *position* (how far you stretch it). Scientists have a name for any type of energy that depends on position. They call it **potential energy**. Stored elastic energy is one type of potential energy.

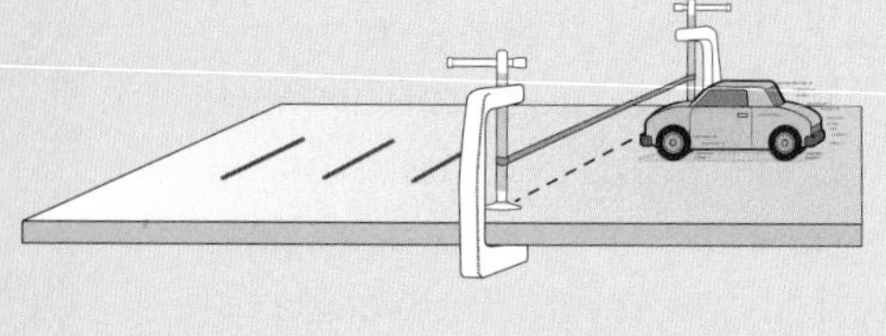

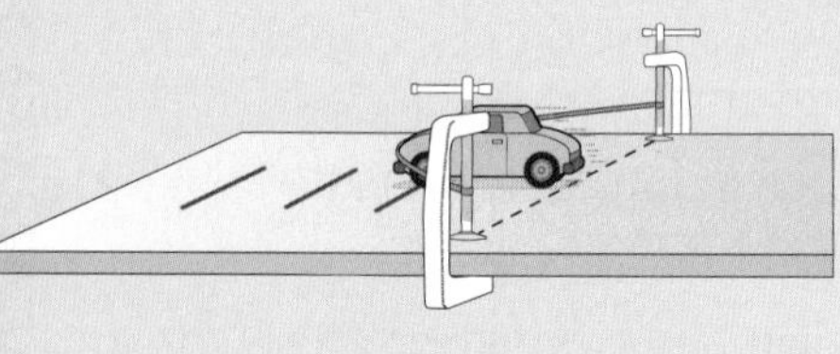

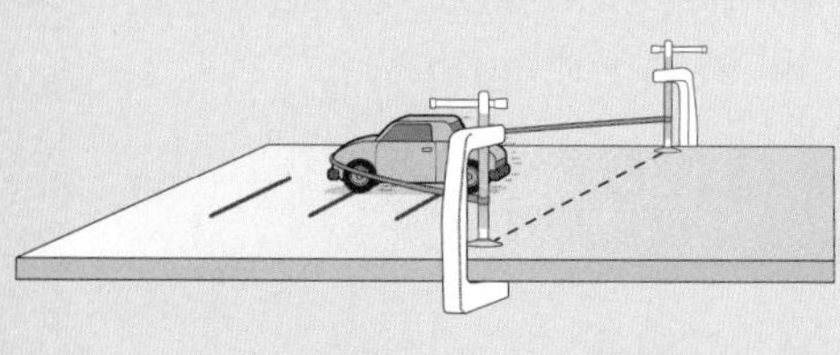

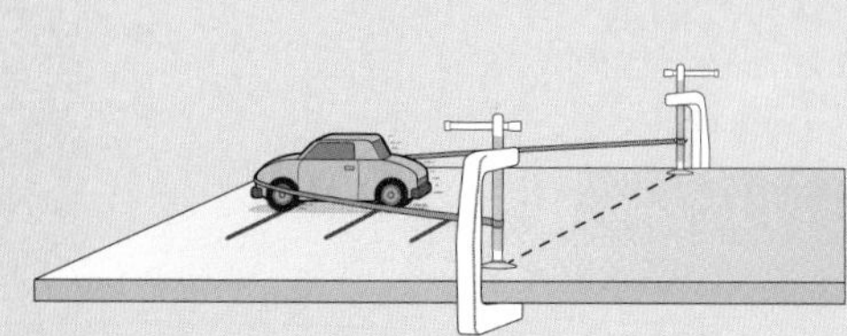

Scientists also have a different name for motion energy. They say that a moving object has **kinetic energy** ("kinetic" comes from a Greek word that means motion). In the remainder of this activity, you will use kinetic energy instead of motion energy.

You can use the terms potential energy and kinetic energy to describe what happens when a stretched rubber band launches a car, as follows:

As the elastic potential energy in the rubber band decreases, the kinetic energy of the car increases.

Now consider the reverse case, when a moving car runs into the rubber band.

1. When the moving car is being slowed down by the rubber band, what is the relationship between the kinetic energy of the car and the elastic energy of the rubber band?

Another type of potential energy comes about when describing the gravitational interaction between two objects, such as the Earth and a ball.

This type of energy is called the **gravitational potential energy** of the ball-and-Earth system. The higher the ball is above the ground (Earth's surface), the more gravitational potential energy there is.

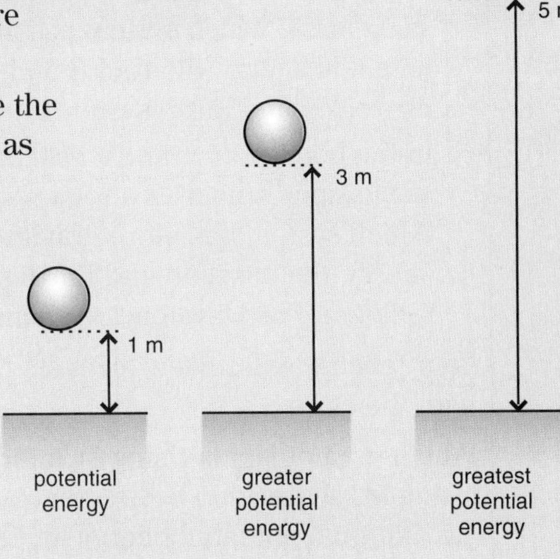

If you hold the ball a certain distance above the ground and then drop it, the ball speeds up as it falls. The higher the ball starts out above the ground, the greater its speed when it hits the ground. You can describe what happens as the ball falls to the ground in terms of energy:

As the gravitational potential energy of the ball-and-Earth system decreases, the kinetic energy of the ball increases.

Answer the questions below.

2. Consider a situation in which a girl tosses a baseball straight up into the air. Think about the time *after* the ball leaves her hand and *before* it begins coming back down. What is the relationship between the kinetic energy of the baseball and the gravitational potential energy of the baseball-and-Earth system?

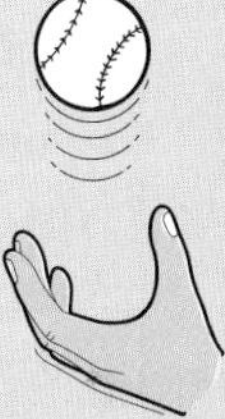

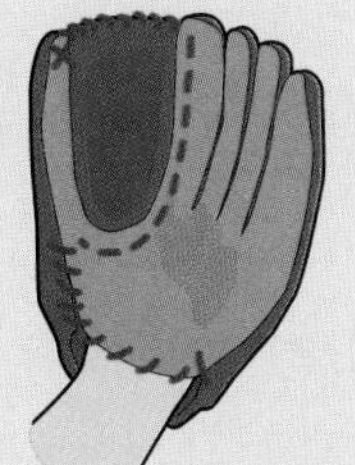

3. Think about the time *after* the baseball starts coming back down toward the ground. What is the relationship between the kinetic energy of the baseball and the gravitational potential energy of the baseball-and-Earth system?

Your teacher will review answers to these questions with the class.

Not only does the amount of gravitational potential energy depend on the position of an object above the ground, it also depends on the mass of the object. When several objects are located at the same distance above the ground, the objects with larger mass have more potential energy than objects with smaller mass.

Science Words

gravitational potential energy: the energy of a system with two objects interacting through gravity. The energy depends on the distance between the objects and their masses.

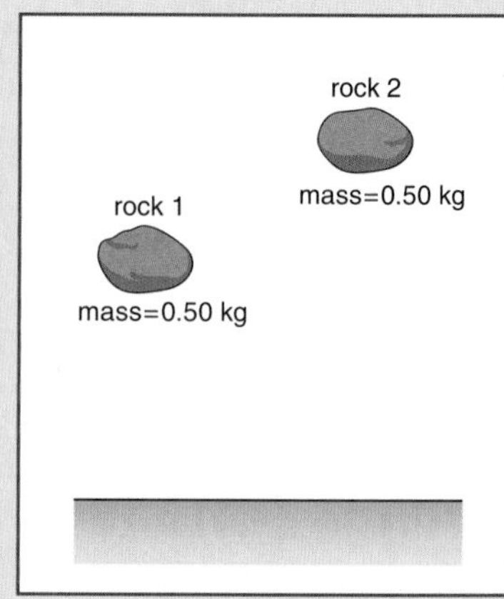

4. In the top diagram, rock 1 is located closer to the ground than rock 2. The two rocks have the same mass. Which system has more gravitational potential energy, the rock 1-and-Earth system or the rock 2-and-Earth system? Explain your answer.
5. In the bottom diagram, a baseball is positioned at the same height as a rock with a greater mass. Which system has more gravitational potential energy, the baseball-and-Earth system or the rock-and-Earth system? Explain your answer.

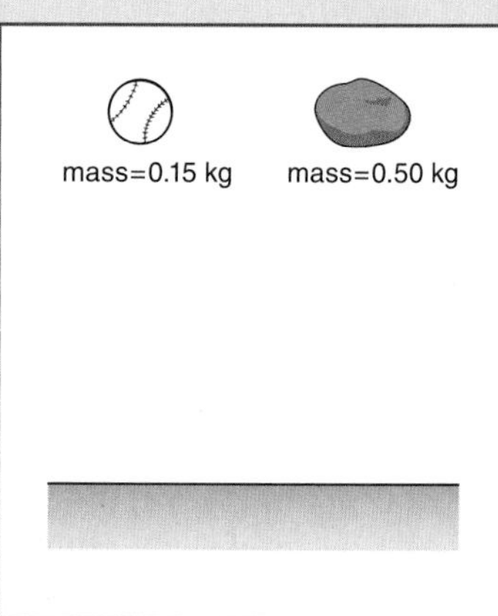

Your teacher will review answers to these questions with the class.

Whether an object is rising or falling, there is a simple relationship between changes in the gravitational potential energy and kinetic energy. As one increases, the other decreases.

There are many examples of gravitational potential energy in everyday life. For example, at the very top of a waterfall, the water has gravitational potential energy (it is high above the surface of the Earth). As the water falls down, its potential energy decreases and its kinetic energy increases.

A dam makes use of this potential energy. Dams block off a river, creating a lake. The water coming through the dam can then be controlled. The water falls onto a turbine (which is like a wheel with paddles on it) that is connected to a generator. The generator then turns that mechanical energy of the falling water into electrical energy.

What We Have Learned

Recall the key questions for this activity:

1. What is potential energy?

2. How does it relate to motion energy?

Participate in the class discussion about your answers to the questions.

Write the answers to the key questions on your record sheet.

LEARNING ABOUT OTHER IDEAS

Activity 11
The Solar System

Purpose

The interaction you have studied in this cycle, the gravitational interaction, plays a central role in interactions between objects in space. It is essential in both the formation and maintenance of stars, planets, and solar systems. In Activity 13, you will study the role of gravity in space. This activity introduces our Solar System.

Gravitational interactions play a key role in the Solar System.

The Solar System is home to not just the planets but over a hundred moons, millions of asteroids, and billions of comets. In recent years, scientists have even found another group of icy bodies that lie out beyond the orbit of Neptune. A solar system is all the stuff that orbits around a star, called its sun.

Astronomers have discovered in recent years that our Solar System is just one of many solar systems in our galaxy. The key questions for this activity are:

1. What kinds of astronomical bodies are found in our Solar System?
2. What are the two major types of planets?

Record the key questions for the activity on your record sheet.

Learning the Ideas

Our knowledge of the Solar System comes mostly from observations using Earth-based and space-based telescopes, and from dozens of unmanned spacecraft that have traveled throughout the Solar System since 1962. All of the photographs in this activity were taken by spacecraft or space-based telescopes.

The Sun

The Sun's Central Role

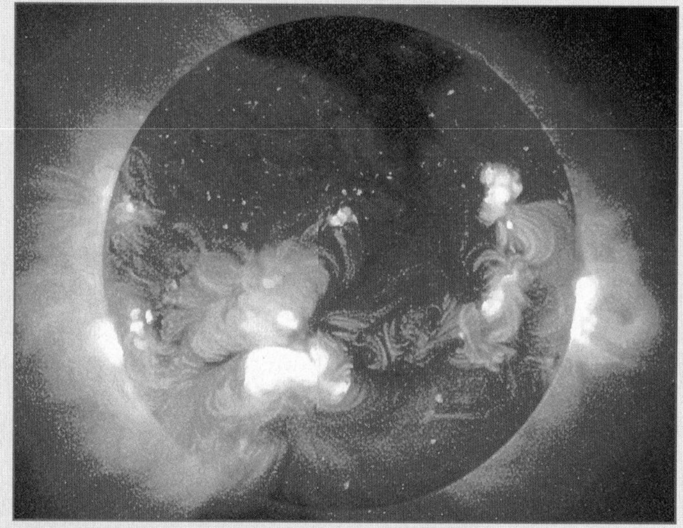

The Sun is the only body in the Solar System that produces its own light.

The Solar System would not exist without the Sun. Its gravitational attraction holds everything in the Solar System together. The Sun is only a small star compared to the brighter stars you see in the night sky. However, it is by far the largest body in our Solar System. It has a diameter about 110 times Earth's diameter, and a mass that is about 333,000 times greater than Earth's mass. In fact, the Sun makes up about 99.8% of the total mass of the Solar System. The Sun consists entirely of hot gas. Most of the Sun is hydrogen and helium, although small quantities of other substances, like oxygen and iron, are also present.

In the center of the Sun, nuclear-fusion reactions produce the energy that is responsible for keeping the Sun shining and making it a source of warmth. Nuclear-fusion reactions occur when the small centers of hydrogen atoms, called nuclei (the plural of nucleus), combine to form helium nuclei. These nuclear reactions keep the center of the Sun extremely hot, about 15,000,000°C, compared to about 5000°C on the Sun's surface.

Lighting up the Solar System

The Sun is the only body in the Solar System that produces its own light. Although people have observed the Moon and planets in the night sky for thousands of years, they have seen them only because of sunshine. No planet, moon, asteroid, or comet produces its own light. All objects in the Solar System, except the Sun, appear to shine because they *reflect* light from the Sun. This fact is demonstrated by eclipses of the Moon. Eclipses occur when Earth's shadow hides part of the Moon or the entire Moon. You will explore reflection of light in more depth in Unit 4 Cycle 2, Activity 9.

The brightness of an object in the sky depends on three factors:

- the size of the object
- the distance of the object from Earth
- the percentage of sunlight the object reflects

After the Sun itself, the Moon is the brightest object in the sky because it is the closest to Earth, although it is smaller than every planet except Pluto. Although Mars is much closer to Earth than Jupiter, it usually appears dimmer than Jupiter because Jupiter is much larger than Mars.

The percentage of sunlight that an object reflects is a characteristic property of the object. For example, Mars reflects about 15% of sunlight and Venus reflects about 65%. Although Mars is sometimes much closer to Earth than Venus, it is usually less bright because the percentage of sunlight it reflects is much smaller.

1. Mars is occasionally brighter than Jupiter in the night sky. But Jupiter is not only much larger than Mars, it also reflects 52% of sunlight compared to 15% for Mars. Why is Mars sometimes brighter than Jupiter?

Your teacher will review answers to this question with the class.

The planets that make up the Solar System (left to right) include Mercury, Venus, Earth, Mars, Jupiter, Saturn, Uranus, Neptune, and Pluto.

Overview of the Solar System

Planets in the Solar System

The illustration on the previous page shows that in addition to the Sun, the Solar System contains the nine planets. These planets orbit the Sun at distances between sixty million kilometers (Mercury) and six billion kilometers (Pluto). They range in size from tiny Pluto (only 0.2% of Earth's mass) to Jupiter (over 300 times Earth's mass). The picture on the previous page also shows the sizes of the planets relative to each other and the Sun (left edge).

Planets fall into two main groups: terrestrial planets and gas giants.

Terrestrial planets include Mercury, Venus, Earth, and Mars. Terrestrial means earthlike. These planets are small and dense, and composed primarily of rock and metal. Compared to the larger planets, they have only shallow atmospheres or none at all (Mercury). They have only a few moons and no rings.

The terrestrial planets (left to right) include Mercury, Venus, Earth, and Mars.

Science Words

terrestrial planet: a small, dense planet similar to Earth that consists mainly of rocky and metallic material; terrestrial planets include the inner Solar System planets Mercury, Venus, Earth, and Mars

gas-giant planet: a large planet with a deep atmosphere that is mostly hydrogen and helium, and a core of icy and rocky material; gas-giant planets include the outer Solar System planets Jupiter, Saturn, Uranus, and Neptune

Gas-giant planets are many times more massive than Earth. They have deep atmospheres that account for much of the planets' masses. They are orbited by many moons and have rings.

The gas-giant planets (left to right) include Jupiter, Saturn, Uranus, and Neptune.

- The *large gas giants*, Jupiter and Saturn, are mostly hydrogen and helium like the Sun.
- The *small gas giants*, Uranus and Neptune, have hydrogen/helium atmospheres and interiors made of ice and rock that account for most of their masses.

The most distant planet, Pluto, is a small world of rock and ice that is probably the largest known member of a group of icy bodies known as Kuiper (pronounced KYEP-ER) Belt Objects (described on the next page).

Other Bodies in the Solar System

The Solar System also contains a huge number of bodies that are smaller than planets for the most part. They include: moons or satellites, asteroids, Kuiper Belt objects, and comets.

Science Words

moon (satellite): a body that orbits a planet

More than 150 **moons** or **satellites** orbit the planets and range in size from rocks a couple of kilometers long to moons as large as Mercury. Millions of rocky *asteroids* mostly orbit the Sun in the *Asteroid Belt*, a ring-shaped region of space between Mars and Jupiter. Hundreds of thousands of Pluto-like, icy/rocky bodies called *Kuiper Belt Objects* (KBOs), mostly orbit the Sun in another ring-shaped region of space (the Kuiper Belt) beyond Neptune. Billions to trillions of *comets*, small icy/rocky bodies also orbit the Sun. They have highly elliptical orbits and put on spectacular light shows when they come near the Sun.

Bodies in Orbit in the Solar System

If you could look down on the Solar System from high above Earth's North Pole, you would observe that all planets orbit the Sun counterclockwise, along with most comets, asteroids, and KBOs. Most moons orbit their planets in the same way, and nearly all bodies in the Solar System rotate counterclockwise as well (Venus and Uranus are two important exceptions).

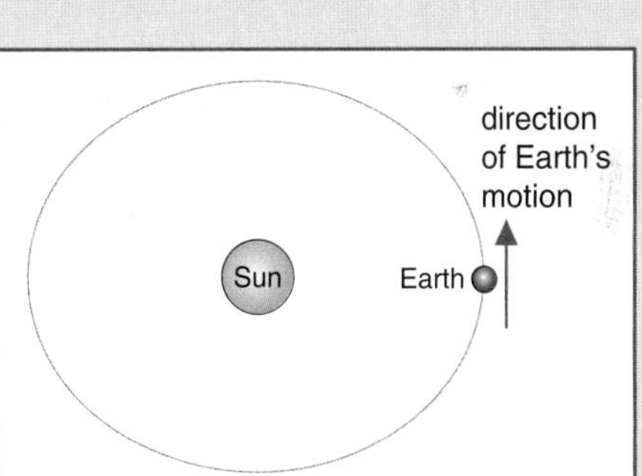

The table on the next page summarizes some characteristics of the nine planets, the Sun, and the Moon.

2. Many planets have one or more characteristics that make them "earthlike" in some way. Identify the planets that are most earthlike in terms of the characteristics listed below. Name only one planet for each characteristic.

a) mass

b) diameter

c) density

d) day length (rotation period)

e) percent of sunlight reflected

f) number of moons

Table: The Nine Planets, Sun and Moon								
Planet/ Body	Type of Planet or Body	Mass (Earth=1)	Diameter (km)	Density (g/cm^3)	Rotation (days)	Percent of Sunlight Reflected	Moons	Rings
Sun	Yellow Star	333,000	1,394,000	1.41	24.6	—	—	—
Mercury	Terrestrial	0.055	4880	5.43	58.6	11%	0	No
Venus	Terrestrial	0.82	12,104	5.24	243.0	65%	0	No
Earth	Terrestrial	1.00	12,756	5.52	0.99	30%	1	No
Moon	Terrestrial (Moon)	0.012	3476	3.34	27.32	12%	—	No
Mars	Terrestrial	0.11	6794	3.93	1.03	15%	2	No
Jupiter	Large gas giant	318.2	142,984	1.33	0.41	52%	63+	Yes
Saturn	Large gas giant	95.1	120,536	0.69	0.45	47%	47+	Yes
Uranus	Small gas giant	14.5	51,118	1.32	0.72	51%	27+	Yes
Neptune	Small gas giant	17.2	49,532	1.64	0.67	41%	13+	Yes
Pluto	KBO	0.0021	2274	2.06	6.39	55%	1	No

3. Which planet could float in water? Explain your answer.

4. Which planets have the shortest and longest days?

Your teacher will review answers to these questions with the class.

Description of the Solar System

Visiting Our Neighbors

The Solar System can be divided into three large regions that vary in terms of population and composition.

Except for the Sun, the *inner Solar System* consists mainly of bodies made of rock and metal, like the terrestrial planets, their moons, and the asteroids. All of the terrestrial planets, their moons, and a few asteroids have been visited by spacecraft.

The *outer Solar System* is dominated by the four gas giants, worlds with deep atmospheres of hydrogen and helium gases. Icy rings and (mostly) icy moons circle these giant planets. All of the gas-giant planets and many of their moons have also been visited by spacecraft.

The *distant Solar System* beyond Neptune consists of small icy bodies like Pluto, its moon Charon (pronounced KAREN), other Kuiper Belt Objects, and billions of comets. No spacecraft have yet visited the distant Solar System.

The Inner Solar System: Bodies of Rock and Metal

The closest planet to the Sun and second smallest, Mercury has a heavily cratered surface that resembles the Moon's. On the inside, Mercury more closely resembles Earth, with a molten iron core that's probably responsible for Mercury's weak magnetic field. Because Mercury's rotational period is two-thirds of its three-month year, its temperature varies from –180°C at night to over 400°C during its daytime. Mercury has no atmosphere.

Venus is nearly the same size and mass as Earth, and has a similar density and composition. Unlike Earth, Venus is completely covered by clouds and its atmosphere is so thick that if you stood on Venus's surface, the air would crush you! This thick atmosphere developed because the Sun heats Venus more than Earth. This extra heating is enough to trigger runaway global warming due to the greenhouse effect, raising Venus's surface temperatures to about 470°C. This temperature is hot enough to melt lead and is warmer than Mercury's surface. Venus rotates more slowly than any planet in the Solar System. Its rotational period is actually longer than its year.

Venus

The largest terrestrial planet, Earth, has a molten iron core and a stronger magnetic field than Mercury. Earth's atmosphere is thick enough to warm the planet through the greenhouse effect, raising its surface temperature by about 35°C above what it would be without an atmosphere. Without the warming provided by the greenhouse effect, water would be locked up in ice and life would not exist. Earth's short day (24 h) also helps make life possible. If Earth rotated as slowly as Mercury or Venus, the variation in temperature would be too great for complex life forms like humans to exist.

Earth and the Moon

The Moon is the only body, other than Earth, to be visited by humans. It has a heavily cratered surface and no atmosphere, like Mercury. Unlike Mercury, the Moon is much less dense than Earth and appears to have a different composition. This has led to a hypothesis that the Moon formed when a Mars-sized planet collided with Earth shortly after the Solar System formed. The Moon then formed from the debris.

Mars

Mars is the planet most frequently visited by Earth spacecraft. Mars is a curious planet that combines features of the Moon and Earth. Like the Moon, Mars is heavily cratered. Mars also has a thin atmosphere and has surface features that are more earthlike. These features include dry channels similar to Earth river systems, canyons that dwarf the Grand Canyon, the Solar System's largest (and currently inactive) volcanoes, and polar ice caps.

With its relatively mild temperatures (between −100°C and 27°C) and a day nearly identical to Earth's, people have long wondered whether Mars supports life. Evidence (the dry channels) that Mars once had flowing water has encouraged that speculation. However, no clear evidence has yet been found for life on Mars by spacecraft that have landed on the planet.

Asteroids mostly orbit the Sun in the Asteroid Belt between Mars and Jupiter. Others orbit outside of the Belt, including some that pass close to Earth. Asteroids are mostly small collections of rock and metal only a few dozen kilometers long or smaller. Out of hundreds of thousands of asteroids found to date, only 250 are more than 100 km across. The total mass of asteroids is only about 20% of Pluto's mass.

The Outer Solar System: Gas Giants and Icy Moons

Jupiter is by far the most massive of the planets. It has a mass over 300 times Earth's mass and more than twice the mass of all the other planets combined. Like the Sun, Jupiter is primarily composed of hydrogen and helium, the two gases that make up nearly its entire, huge atmosphere. The only "surface" of Jupiter that can be seen is colored bands of clouds that speed around the planet, with adjacent bands moving in opposite directions. Jupiter has the most turbulent weather of any planet, with storms like the Great Red Spot that can last for centuries. Deep inside Jupiter, hydrogen gas is so compressed that it forms a metallic liquid. This "metallic hydrogen" acts like a huge magnet and gives Jupiter by far the strongest magnetic field among the planets.

Jupiter

First discovered by Galileo in 1610, Jupiter's four largest "Galilean" moons, Io, Europa, Ganymede, and Callisto are separate little worlds. They range in size from slightly smaller than the Moon (Europa) to the size of Mercury (Ganymede). Like the terrestrial planets, Io and Europa are mostly rock and metal. Ganymede and Callisto, like objects in the distant Solar System, are mostly ice and rock. Callisto is the most heavily cratered object in the Solar System, but Europa is nearly as smooth as a marble, and may have oceans underneath its icy surface. Io is the most volcanic world in the Solar System.

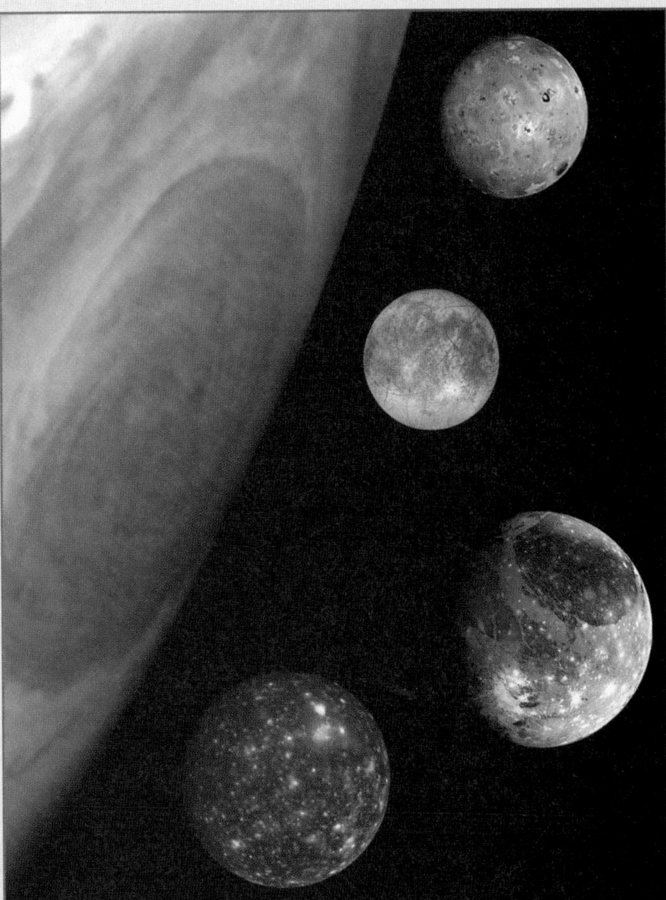

Jupiter's four largest "Galilean" moons (top to bottom) are Io, Europa, Ganymede, and Callisto.

Saturn has about one-third of Jupiter's mass and in many ways is just a smaller version of Jupiter. It has a composition similar to Jupiter, colored bands of clouds sweeping around the planet, and a strong magnetic field. Saturn's icy largest moon, Titan, is about the same size as Mercury and has an atmosphere thicker than Earth's, and has rivers and lakes of liquid methane.

Although all four gas-giant planets have rings, Saturn's rings are the only ones visible from Earth through a small telescope. Though they are over 250,000 km in width (about 60% of the distance between Earth and the Moon), Saturn's rings are less than 1 km thick—about a 15-minute walk.

Saturn

Though they look solid, the rings are actually composed of many small pieces each orbiting Saturn independently. These pieces range in size from 1 cm across to several meters across, with a few larger pieces. The pieces appear to be chunks of ice or rocks coated with ice. The mass of the rings is a tiny fraction of the mass of Pluto.

The two small gas giants, Uranus (pronounced YOU-RAIN-US) and Neptune, are nearly the same size. Neptune has a slightly larger mass and Uranus has a slightly larger volume. Both planets have strong magnetic fields. Like Jupiter and Saturn, Uranus and Neptune have atmospheres that are mostly hydrogen and helium, and have bands of clouds. The interiors of the small gas giants are composed of ice and rock. This "solid" interior accounts for about 85% of the two planets' masses.

Uranus is the only planet that rotates nearly flipped on its side. Its north and south poles regularly face directly into the Sun during its summer and winter.

Neptune has a moon called Triton that is larger than Pluto, has about the same ice/rock composition as Pluto, and orbits the planet clockwise, rather than counterclockwise like all nine planets and most moons. Some scientists think that Triton might once have orbited the Sun like Pluto, but was captured by Neptune's gravity and became a moon.

Uranus shown in false color with its rings and moons.

The Distant Solar System: Bodies of Ice and Rock

The smallest planet, Pluto, is a world of ice and rock with only 20% of the Moon's mass. Observations of Pluto have shown that it has a very thin atmosphere and polar caps. Pluto's icy moon, Charon, is half Pluto's diameter and 15% of its mass, making it the largest moon in the Solar System compared to the planet it orbits.

Pluto appears to be simply the largest Kuiper Belt Object (KBO) that humans have so far discovered. The Kuiper Belt (discovered in 1992) consists of objects of ice and rock similar to Pluto and Charon. Some KBOs even have Pluto-like, highly elliptical orbits. Most KBOs have roughly circular orbits located beyond Neptune's orbit. KBOs are more than simply icy asteroids. The largest KBOs are much larger than any asteroids, and a few are close to Pluto in size. The total mass of KBOs is at least a few dozen times the total mass of the asteroids.

The Kuiper Belt is also important as a probable source of comets. Comets flare up as they approach the Sun because light from the Sun boils off icy material on the surface of the comet and forms the bright halo and tail that we see. Comets come in two groups. *Short-period comets* are the comets, such as Halley's Comet (orbital period = 76 years), that humans have seen repeatedly throughout their history. These comets may be remnants of KBOs broken off in collisions.

The comet Kohoutek

Long-period comets such as Kohoutek (pictured), which "visited" Earth in 1973-74, have orbital periods lasting thousands of years. Scientists believe these comets may come from the *Oort Cloud*, a hypothetical sphere of comets located about 50,000 times farther away from the Sun than Earth.

What We Have Learned

Think about the key questions for this activity:

1. What kinds of astronomical bodies are found in our Solar System?

2. What are the two major types of planets?

Participate in the class discussion about your answers.

Write the answers to the key questions on your record sheet.

LEARNING ABOUT OTHER IDEAS

Activity 12
Distances in Space

Purpose

By human standards, distances between objects in space are enormous. For example, the distance from Earth to the Moon is approximately 384,000 km. If you tried driving that distance in a car moving at 100 km/h (62 mph), it would take you 160 days to drive that distance. That distance is small compared to the distance between Earth and the Sun, which averages 149.6 million kilometers. In the same car, you would need 171 years, or about two human lifetimes, to cross that distance!

The distance between Earth and the Moon seems so great. However, it is extremely small compared to other distances in space.

Even that distance is dwarfed by the distances between stars. The nearest star to Earth other than the Sun, Alpha Centauri, is about 265,000 times farther away. In the same car, a round trip to Alpha Centauri would take over 100 million years. If you had left during the time dinosaurs ruled the world, 40 million years before they became extinct, you would just now be coming back home!

Because the distances between objects in space are so large, scientists do not use units of distance measurement like yards, miles, meters, or kilometers. Instead, they define new units for measuring distance. They use one unit for measuring distances within solar systems, and another for measuring distances between stars and galaxies.

In this activity, you will answer the following key question:

What are the units used to measure distances within the Solar System and between stars?

Record the key question for the activity on your record sheet.

Learning the Ideas

Distances in the Solar System

To measure the distances between objects in the Solar System, scientists use a unit of measurement called the **astronomical unit**, or AU. An astronomical unit is equal to the average distance between Earth and the Sun.

Science Words

astronomical unit: a unit of measurement equal to the average distance between the Sun and Earth

The table below gives the distances of the nine planets, the Asteroid Belt, and the Kuiper Belt from the Sun in terms of AUs, along with the time that it takes to orbit the Sun in Earth years. Because the Asteroid Belt and Kuiper Belt are

Table: Solar System Orbital Data

Planet	Distance from Sun (AU)	Orbital Period (Years)
Mercury	0.39	0.24
Venus	0.72	0.62
Earth	1.00	1.00
Mars	1.52	1.88
Asteroid Belt	*2 to 4*	*3 to 8*
Jupiter	5.20	11.86
Saturn	9.55	29.46
Uranus	19.19	84.01
Neptune	30.11	164.79
Pluto	39.53	247.91
Kuiper Belt	*40 to 50*	*250 to 350*

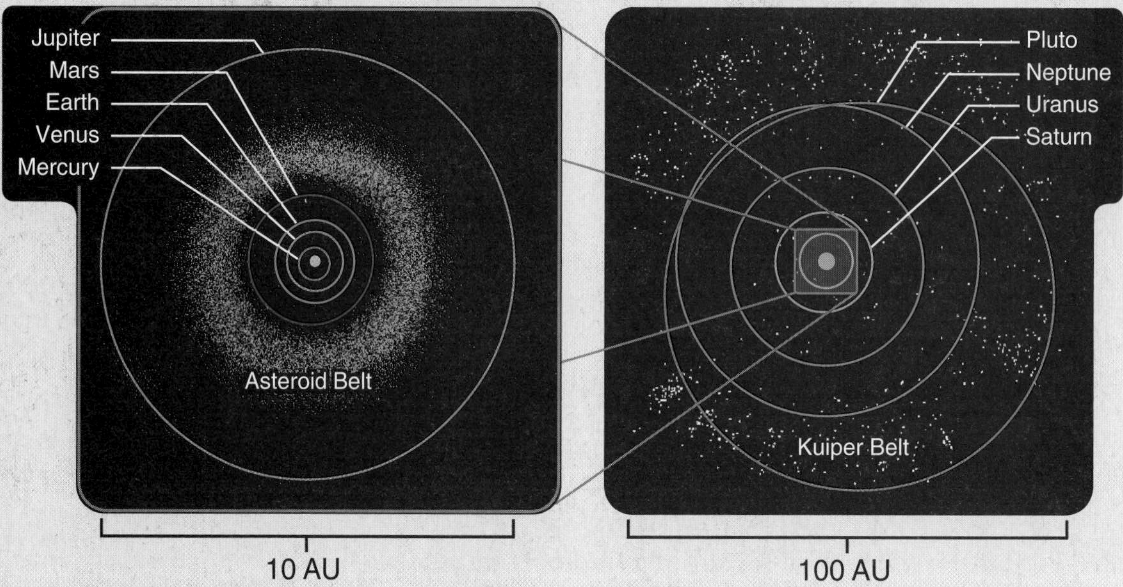

composed of thousands of objects, their distances from the Sun and their orbital periods are given as ranges rather than as specific values. The pictures above show the orbits of the nine planets, the Asteroid Belt and the Kuiper Belt.

For the nine planets, the distances shown in the table are *averages*. Because the planets have *elliptical* orbits, their distance from the Sun varies depending on where they are in their orbits. Pluto's orbit is the most elliptical. It comes as close to the Sun as 29.6 AU—inside Neptune's

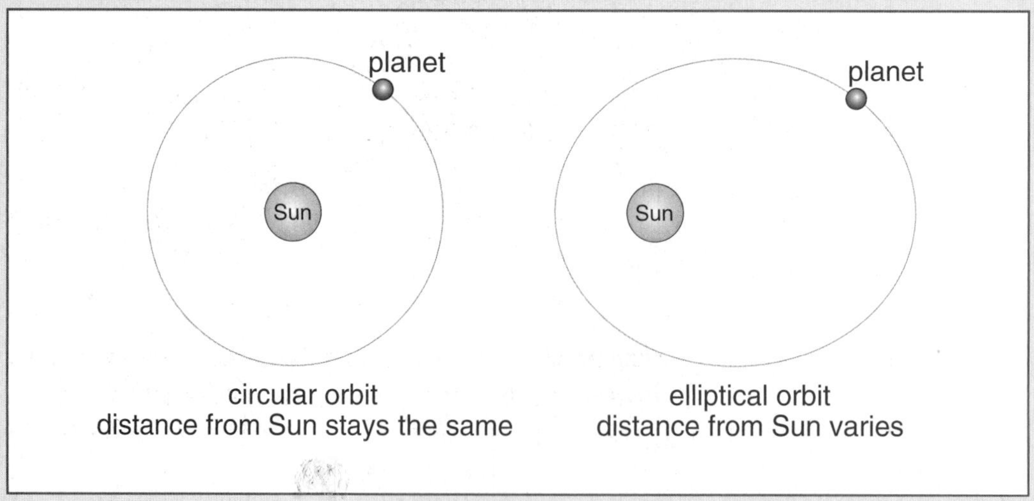

orbit—and as far from the Sun as 49.3 AU. Many asteroids and Kuiper Belt Objects orbit the Sun outside of their "belts." All known comet orbits are highly elliptical, with some going out as far as 50,000 AU from the Sun.

1. If you were driving in a car at 100 km/h, about how long would it take you to drive from the Sun to Pluto? To make the calculations easier, *round off* Pluto's distance in AUs to the nearest whole number. Remember that it would take about 170 years to go from Earth to the Sun.

Your teacher will review answers to this question with the class.

Distances between Stars and Galaxies

Because distances between stars are large in Solar System terms, scientists use a different unit of length to measure those distances. This unit of length is the **light-year** which despite its name is *not* a unit of time. A light-year is the distance a beam of light travels in space in one Earth year. One light-year is equal to 9.46 trillion kilometers (9,460,000,000,000 km) or 63,240 AU.

Science Words

light-year: a unit of measurement equivalent to the distance light travels in one year

2. If you were driving in a car at 100 km/h, about how long would it take you to drive one light-year? To make the calculation easier, *round off* one light-year to either 10 trillion kilometers or 60,000 AU. Remember that it would take about 170 years to go from Earth to the Sun.

Your teacher will review answers to this question with the class.

Here are a few distances in light-years to give you some idea just how far away the stars you can see at night really are.

- The Oort Cloud (the source of long-period comets mentioned in Activity 11) is located about one light-year from the Sun.
- The closest star (other than the Sun) to Earth, Alpha Centauri, is about 4.3 light-years away.
- Two bright stars in the constellation Orion, Betelgeuse (pronounced BEETLE-JOOZ) and Rigel (pronouced RYE-JEL), are 427 and 733 light-years away, respectively.
- The Sun is about 20,000 light-years from the center of the Milky Way galaxy.
- The Milky Way galaxy is about 100,000 light-years across.
- The closest large galaxy to the Milky Way, the Andromeda Galaxy, is about 2,000,000 light-years away.

Your teacher will now show you a series of slides showing the objects in Earth's "neighborhood." Each slide will show a region that is about ten times farther across than the slide before.

3. How much does the *volume* of space increase between each slide?

Your teacher will review answers to this question with the class.

What We Have Learned

Think about the key question for this activity:

What are the units used to measure distances within the Solar System and between stars?

Participate in the class discussion about your answers.

Write the answer to the key question on your record sheet.

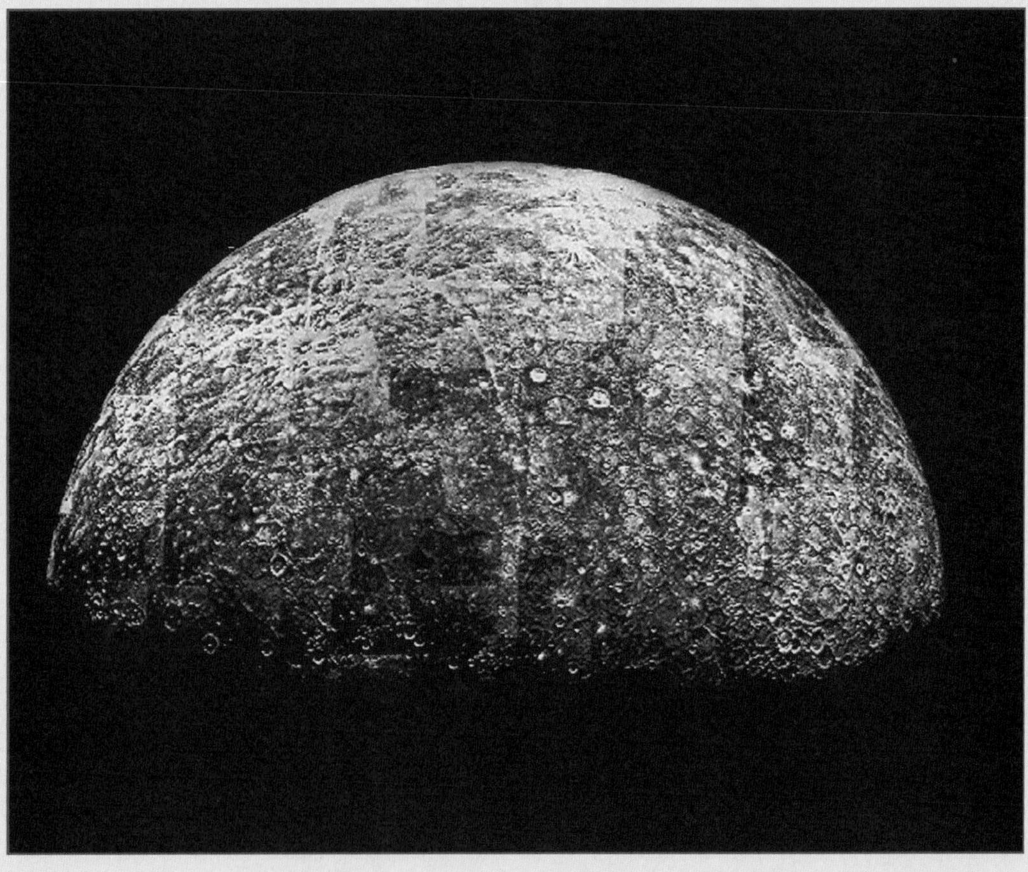

LEARNING ABOUT OTHER IDEAS

Activity 13
Gravity, Stars, and Planets

Purpose

You have learned that when you jump up, gravity is responsible for bringing you back down to Earth. Gravity is what keeps objects close to the surface of planets and moons. You have also learned that the gravitational interaction keeps planets in orbit about the Sun, and moons in orbit about planets. The importance of the gravitational interaction extends even further. It plays the same role for other stars and solar systems throughout the Universe. Gravity helps determine the shapes of stars and planets and maintains those shapes. Gravity also played the central role in the formation of the Sun and Solar System. In this activity, you will look for answers to the key questions:

1. How does gravity help form stars, planets, and solar systems?
2. How does gravity determine and maintain their shapes?

Record the key questions for the activity on your record sheet.

Learning the Ideas

Formation of Stars and Solar Systems

Stars, planets, and solar systems form from huge clouds. These clouds are scattered through the galaxy in the huge, seemingly empty spaces between the stars. The clouds consist of gases (mostly hydrogen and helium) and small specks of dust with extremely tiny masses. These gases and dust specks can be pushed together by some event. The event may be the explosive death of a giant star. The gas and dust specks can then be pulled together enough by gravity to form stars and solar systems.

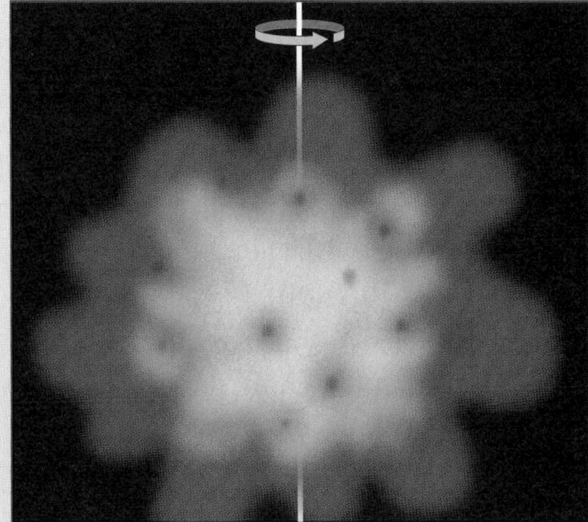

Solar systems form from huge clouds that are spinning slowly.

1. What causes tiny objects, like dust specks and clumps of gas, to be pulled closer together?

Before the cloud starts pulling together, it is spinning very slowly. As the cloud grows smaller, it spins faster and faster. This is what happens when a spinning figure skater pulls her or his arms in. The cloud also flattens out into a disc. Most of the mass of the cloud collects in the center of the spinning disc.

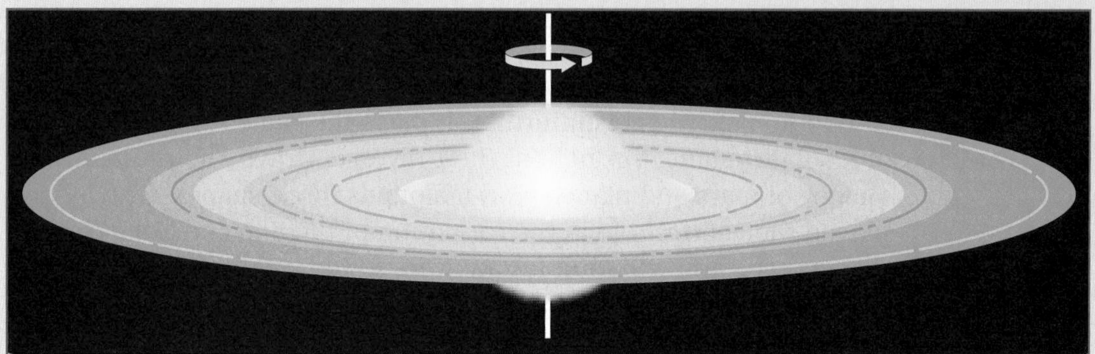

The gravitational potential energy of the gas and dust specks in the cloud becomes motion (kinetic) energy. In the dense center of the disc, the gas and dust specks collide with each other and transform that motion energy into thermal energy. When the temperature and density of the gas is high enough, nuclear-fusion reactions begin as in the center of the Sun. The center of the disc becomes a star. The nuclear reactions produce radiation that is the source for all the light energy and heat energy that the star produces.

In the part of the disc just beyond the star, the gas is much cooler. Dust specks combine through gravity to form small pieces of metal and rock. These small pieces combine in turn to become larger pieces. The larger pieces combine again. These chunks of metal and rock eventually attract all the rock and metal in their orbits (the "rings" of material) until they become terrestrial planets. The gravity of some of these planets (such as Earth) is strong enough to attract some gases, such as oxygen, nitrogen, and carbon dioxide, and give the planets atmospheres.

Farther out in the disc, the rocky centers of giant planets form in the same way. However, this area cools down enough so that some gases, such as water vapor, freeze and become ice. These ices surround the rocky centers and cause a large increase in the masses of the forming planets. These massive ice-and-rock bodies then attract gases. In this way, they can form atmospheres much thicker than those of terrestrial planets. The largest of these ice-and-rock bodies, like the centers of Jupiter and Saturn, have enough mass to attract thick atmospheres of helium and hydrogen. The atmospheres of these planets consist of so much of these two gases that the gases eventually make up most of the planet's mass. Smaller giants, such as Uranus and Neptune, are mostly rock and ice despite their thick atmospheres.

2. Why is Earth not mostly made of gases as Jupiter is?

The rocky and metallic asteroids are a bit like tiny terrestrial planets. They may have formed in a similar way. Also, the icy, rocky objects of the Kuiper Belt (including Pluto) are like small versions of the centers of the giant planets that are simply too small to attract many gases.

Once a solar system forms, gravity works to keep it together. Gravity keeps planets (along with asteroids, comets, and Kuiper Belt Objects) in orbits around the central star. It also keeps moons in orbits around planets. Gravitational interactions between objects in a solar system can change its appearance. However, it would probably take a collision with another star (an extremely rare event) or something similar to tear the solar system apart.

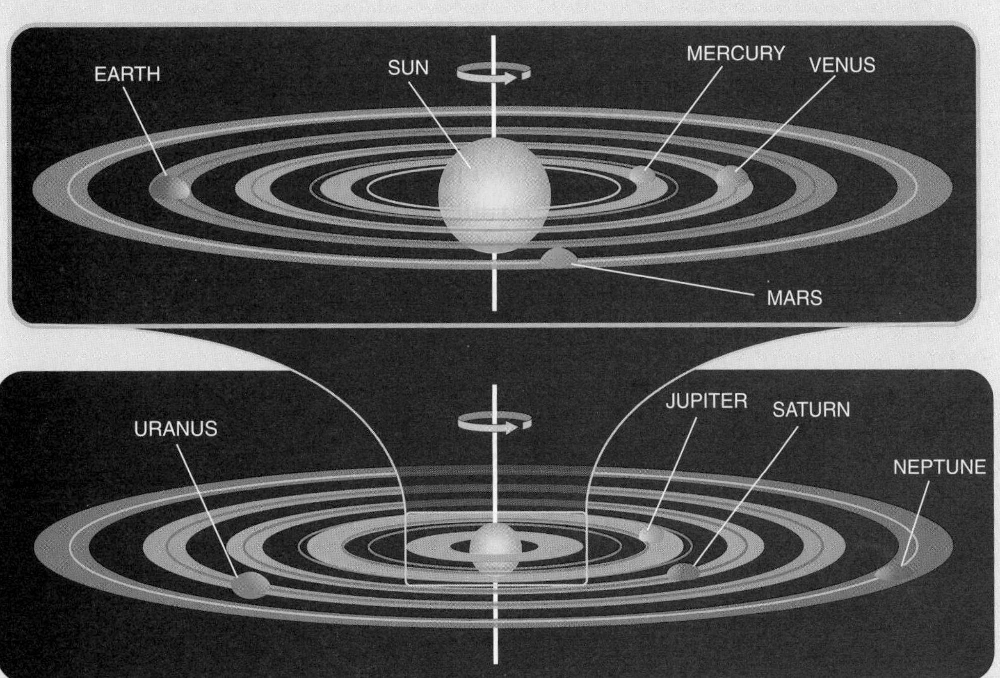

3. Based on what you've just read, list two ways in which gravity contributed to the formation of the Solar System.

Your teacher will review answers to these questions with the class.

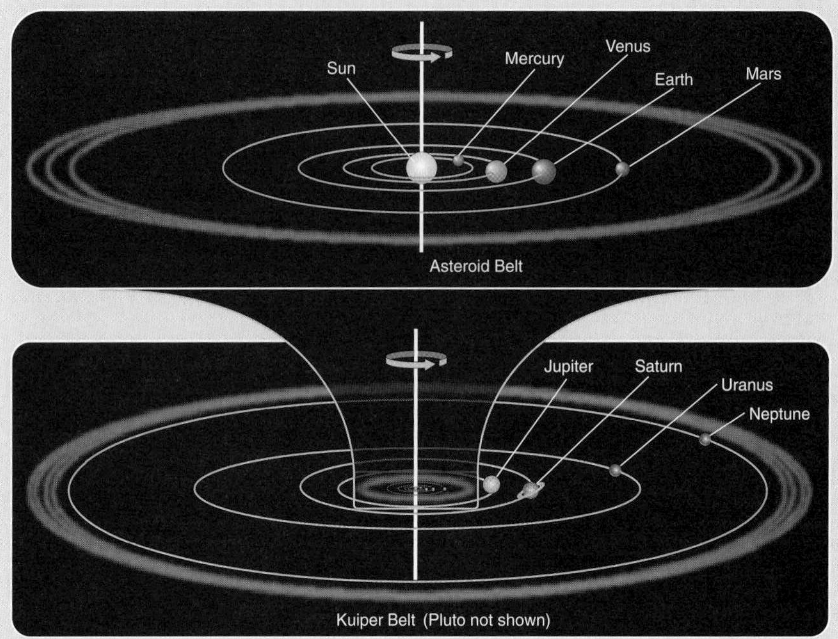

How Gravity Determines Shapes of Planets and Stars

In addition to its role in forming stars and solar systems, gravity is important in determining the shapes of planets, moons, and stars. Most bodies with a diameter or width of about 400 km are roughly spherical in shape. In the Solar System, these nearly spherical bodies include the Sun, the nine planets, the nineteen largest moons, a couple of large asteroids, and probably several large Kuiper Belt Objects. The gravitational interaction both causes the spherical shape and maintains it.

For example, on Earth the total gravitational forces are directed toward the center of the planet. This draws the pieces of Earth inward. However, an object cannot fall through the solid part of the Earth. This solid part exerts an upward force to balance the gravitational force. In this way, the forces on different parts of the planet are balanced.

If gravity is strong enough to overcome the forces holding pieces of an object together, the object will gradually be pulled into a ball-like shape until gravity is balanced by the upward forces. If gravity is not strong enough, the object will not be pulled into a ball-like shape.

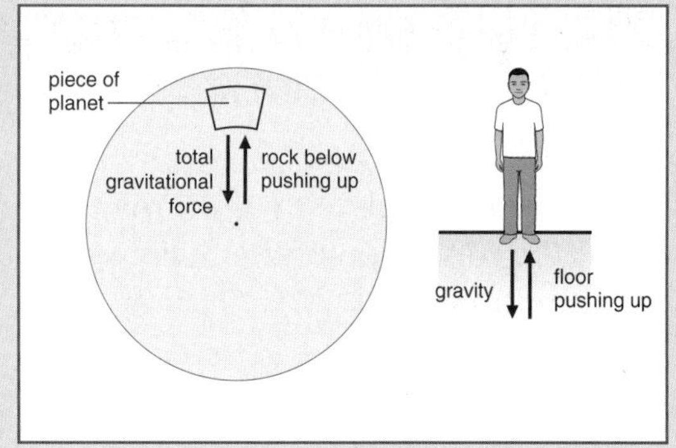

The picture shows three examples of small bodies that are not shaped like balls.

4. Why is the Moon round and Phobos shaped like a potato?

Your teacher will review answers to this question with the class.

The asteroid Gaspra (left), with the two moons of Mars: Deimos (top) and Phobos (right).

Stars like the Sun keep their spherical shape for a different reason than planets do. Nuclear reactions keep the center of the star extremely hot (several million degrees Celsius). This hot gas, combined with the nuclear radiation, produces a force that tries to make the star expand. This force perfectly balances gravity, so once nuclear reactions start, the star stops contracting (getting smaller and denser). As long as nuclear reactions continue, the star stays about the same size.

Additional Effects of Gravity on Planets and Moons

In addition to keeping bodies in the Solar System in orbit and giving larger bodies a round shape, the gravitational interaction is responsible for other phenomena. Because of gravity, only one side of the Moon is visible and there is volcanic activity on Jupiter's moon Io. Gravity is also responsible for wandering planets, both in our Solar System and in solar systems around other stars.

Why the Moon Has Only One Face

When you look up at the Moon in the night sky (or the day sky), you probably have noticed that the face of the Moon never seems to change. Most of the time only part of the Moon's face is visible.

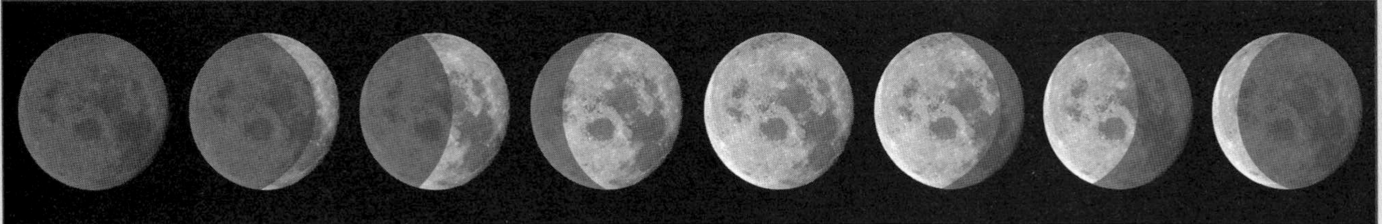

The reason that you never see the "far side of the Moon" is that the Moon has become locked into position so that one side of the Moon always faces Earth, and one side always faces away. Its rotation period is exactly equal to the period of its orbit around Earth. Scientists call this type of rotation synchronous.

Photograph of the far side of the Moon.

The Moon hasn't always been locked into synchronous rotation. In the past, the Moon rotated much faster than it does now. At any given time, because it was closer to Earth, the side of the Moon facing Earth felt more of a gravitational tug from Earth than the side facing away. This difference in the force of the gravity on the two sides of the Moon created a slight bulge in the Moon's surface in the direction of Earth. The Moon still rotated under this bulge, but friction between the bulge and the Moon gradually slowed down the Moon's rotation. Over hundreds of millions of years, this friction interaction reduced the Moon's rotation until it matched the rotation of the bulge. Now one side of the Moon was always lined up with the bulge facing Earth.

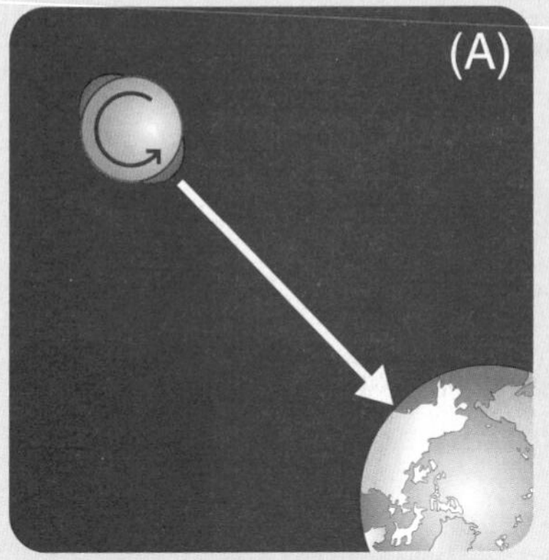

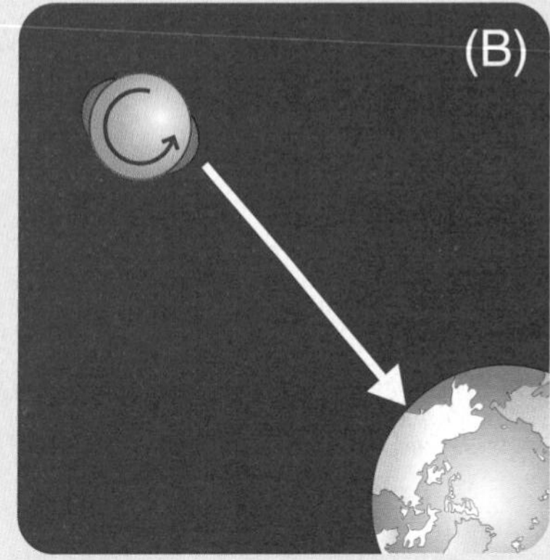

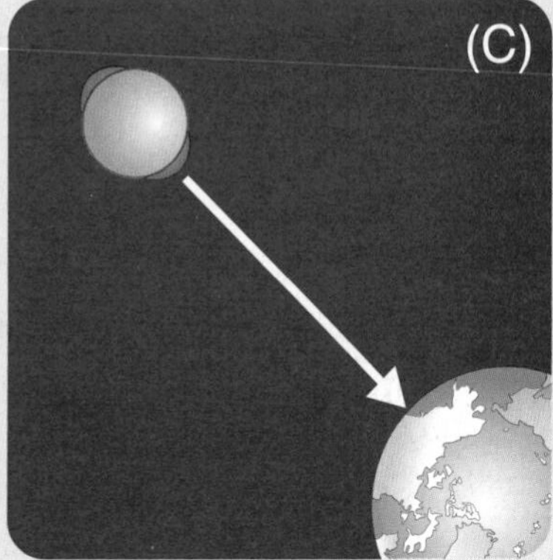

(A) Bulge lines up with Earth due to Earth's gravity. **(B)** Moon tries to rotate bulge away, but Earth pulls it back. This causes friction between the bulge and the Moon. **(C)** Over time, friction slows down the Moon's rotation, until the bulge always faces Earth.

In exactly the same way, Earth's gravitational interaction with the Moon is gradually slowing down Earth's rotation. On Earth, we call the bulges produced by the Moon's gravity, tides. In the oceans, tides are particularly large, ranging up to several meters in height. High tide comes when the Moon is at the height of its daily trips across the sky. Low tide comes when the Moon is on the other side of Earth.

Like the Moon, the friction between Earth's surface and the tides are gradually slowing down Earth's rotation. Fortunately, the process is extremely slow. It takes 67,000 years for Earth's day to grow longer by one second. Billions of years will pass before the same side of Earth always faces the Moon.

In the case of Pluto and its moon Charon, this has already happened. Both Pluto and Charon rotate in 6.4 days, which is exactly equal to Charon's orbital period. As a result, the same side of Pluto always faces Charon, and the same side of Charon always faces Pluto. Pluto and Charon reached this state before Earth and the Moon because Pluto and Charon are much closer together. That means that Charon's gravitational pull on Pluto has more effect on Pluto than the Moon's gravitational pull has on Earth.

Volcanic Moons

In orbit around Jupiter is a moon slightly larger than our Moon, called Io. Like the Moon, Io always keeps the same face toward Jupiter. Unlike the Moon, Io also interacts with two other large moons of Jupiter, Europa and Ganymede. These two moons orbit Jupiter at greater distances. They tug and pull on Io with their gravity, causing tides just as the Moon causes tides on Earth.

Close-up of a volcanic eruption on Io.

The gravitational pulls of Europa and Ganymede on one side, and massive Jupiter on the other side, cause a great deal of frictional heating in Io's interior. Io's internal temperature increases until it becomes hot enough to melt rock. Io then spews forth the molten rock in volcanic eruptions. Io is the most volcanic body in the Solar System, much more so than Earth.

Because of the gravitational interactions between the three moons, there is also a simple relationship between their orbital periods. Europa takes exactly twice as long to orbit Jupiter as Io does. Ganymede takes exactly four times as long.

Little Plutos and Wandering Planets

In the outer Solar System, Pluto circles the Sun in an orbit that is so elliptical that Pluto sometimes comes closer to the Sun than Neptune. Pluto's year is also exactly 50% longer than Neptune's. Like the relationship between the orbital periods of Jupiter's large moons, this relationship suggests that Neptune determined the period (as well as the shape) of Pluto's orbit through its gravitational interaction with Pluto.

In recent years, scientists have discovered dozens of icy worlds (part of the Kuiper Belt) similar to Pluto. They have the same kind of orbit and practically the same orbital period as Pluto. The similarity between their orbits and Pluto's orbit, as shown in the diagram on the next page, is so close that scientists call these icy worlds "Plutinos," which means "little Plutos." They believe that Neptune determined the shape of the orbit of every Plutino.

The interactions between Neptune and Kuiper Belt Objects (KBOs) are not completely one-sided. The KBOs also pull Neptune toward them. This may have changed Neptune's orbit. Some scientists think that Neptune may have once been much closer to the Sun than it is now, but that its gravitational interactions with KBOs and comets pulled it farther away from the Sun. If that is true, Neptune has a history of being a wandering planet.

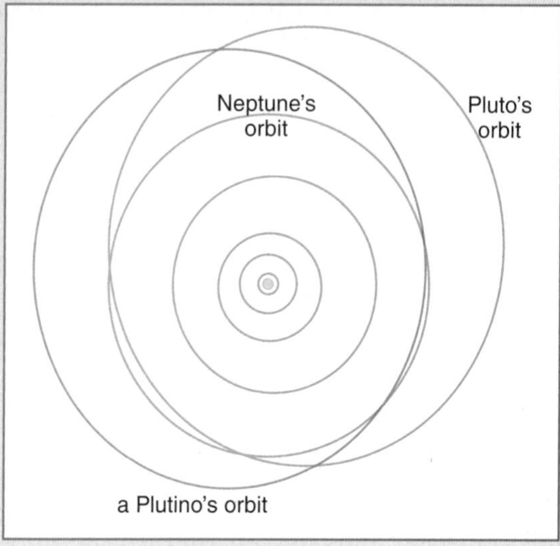

Planets can wander inward, too. For example, the interaction of Jupiter with comets may be causing Jupiter to slowly move closer to the Sun. That may help explain another observation that has puzzled scientists. In recent years, scientists have discovered about 120 giant planets (around Jupiter's size) orbiting other stars. Many of these planets orbit their stars at distances much closer than Jupiter. Some orbit closer than Mercury; one planet larger than Jupiter takes only three Earth days to orbit its star!

Scientists have not been able to explain how solar systems could form with giant planets so close to a star. Most theories predict that, as in our Solar System, giant planets will only form several Astronomical Units (AUs) from their stars. If these giant planets didn't form close to their stars, then they must have wandered closer through gravitational interactions. Wandering planets may be common in the Universe.

What We Have Learned

Think about the key questions for this activity:

1. How does gravity help form stars, planets, and solar systems?

2. How does gravity determine and maintain their shapes?

Record your answers to these questions on your record sheet.

Participate in the class discussion about your answers.

CYCLE 1:

Mass Conservation

CYCLE 2:

Energy Conservation

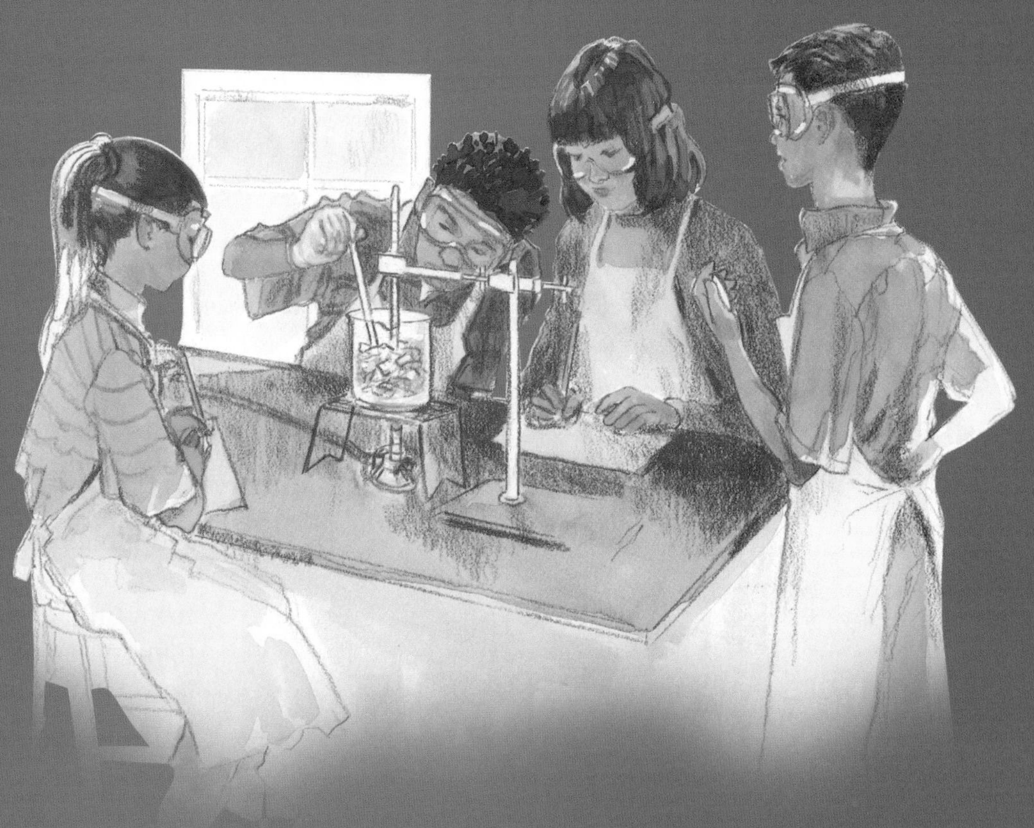

NTERACTIONS AND CONSERVATION

INTRODUCTION

What will you learn about in Unit 4?

In Units 1 through 3, you learned about different kinds of interactions: magnetic interactions, electric-charge interactions, electric-circuit interactions, electromagnetic interactions, light interactions, gravitational interactions, and different kinds of mechanical interactions such as friction and elastic interactions. Your investigations have focused on *changes* as the evidence of interactions.

Hardly a minute goes by when you don't see some sort of change. The traffic outside changes as cars and trucks speed up, slow down, and turn. Ice melts in your glass of soda. The leaves on trees move and rustle.

Despite all this change, there are some things that stay the same. For example, once planted, a tree's location does not change. The identities of a child's birth parents do not change. Although the foods that you eat each day change, the fact that you get hungry every day stays the same. Although your favorite band may change from time to time, the fact that you enjoy listening to music stays the same.

In Unit 4, you will take a closer look at what kind of changes occur for many different types of interactions and what stays the same.

What types of activities are in Unit 4?

There are no new types of activities in Unit 4. You will once again begin each cycle with an *Our First Ideas* activity. You will also find *Developing Our Ideas* and *Learning About Other Ideas* activities. In the *Putting It All Together* activity, you will compare the ideas that your class developed with those of scientists. The *Idea Power* activities will give you an opportunity to apply the ideas you have developed to new situations.

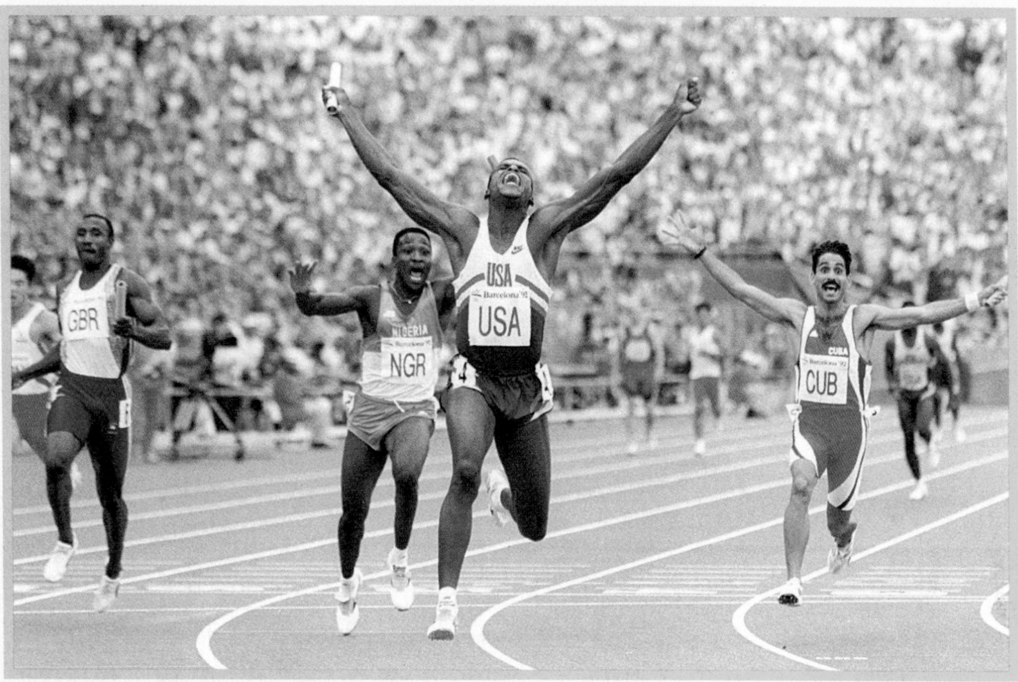

In Unit 4, there are two cycles.

Cycle 1: Mass Conservation

In this cycle, you will investigate several different types of interactions to see if the amount of stuff (material) changes or stays the same.

Here is the key question for the first cycle:

Can interactions cause the amount of stuff (material) to change?

Cycle 2: Energy Conservation

In this cycle, you will investigate two new kinds of interactions involving warmer and cooler objects. You will also see what happens to energy when solids melt to become liquids and liquids boil to become gases. You will then examine whether energy can be created or destroyed in interactions. In other words, whether the amount of energy changes or stays the same.

Here are the key questions for the second cycle:

1. Can energy be created from nothing?
2. Can energy be destroyed into nothing?

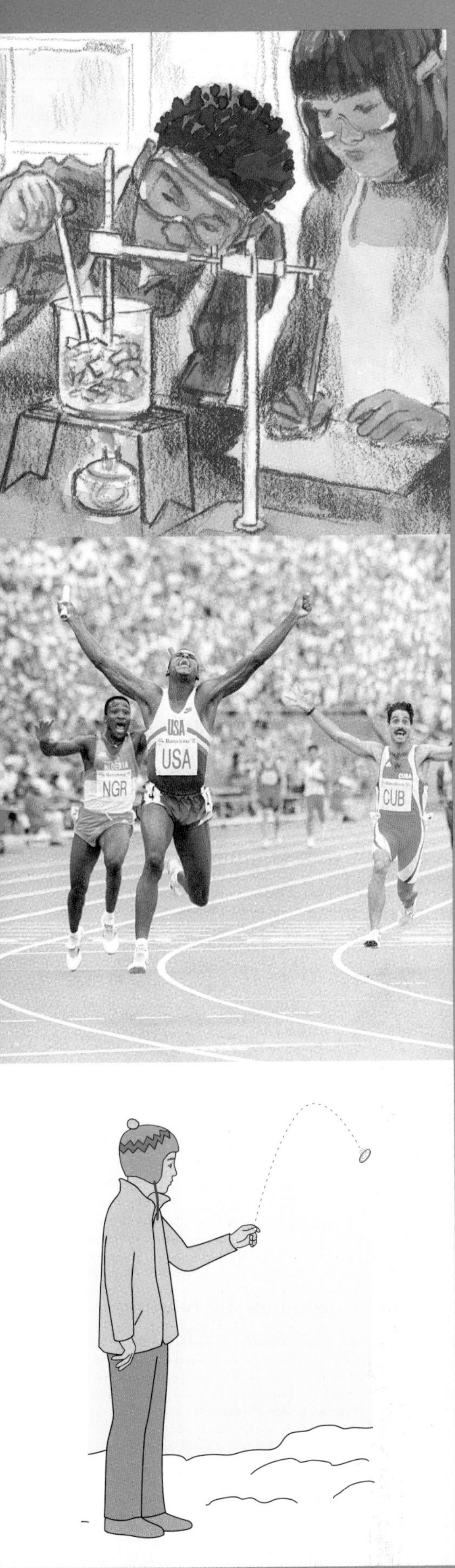

CYCLE 1: MASS CONSERVATION

Cycle Key Question

Can interactions cause the amount of stuff (material) to change?

OUR FIRST IDEAS

Activity 1
What Happens to the Amount of Stuff?

Cycle Purpose

The purpose of this activity is to find out your ideas about what happens to the amount of stuff as the result of an interaction. In Unit 1, you learned that the amount of stuff can be described by mass or volume. How much room or space an object occupies is called the object's volume. Mass describes how much material an object is made of. The more mass an object has, the heavier it is. The key question for this cycle is:

Can interactions cause the amount of stuff (material) to change?

Record the key question for the cycle on your record sheet.

We Think

Discuss the following situations with your team and write your responses on your record sheet.

Dissolving sugar in water: You find the mass of all your equipment and ingredients. You then dissolve 15 g of sugar in 250 mL of water, stirring until the sugar is completely dissolved (as in the diagram).

1. After the interaction between the sugar and the water, has the reading on the mass scale *increased*, *decreased*, or *stayed the same*? What is your reasoning?

Mixing chemicals: Your friend is conducting a chemistry exploration. She starts with two clear solutions in two beakers. One beaker contains 20 mL of sodium sulfate and the other contains 20 mL of calcium chloride.

Both beakers are sitting on a mass scale, and both have pieces of cardboard over the top. Your friend pours the contents of one beaker into the second beaker and replaces the cardboard. A whitish solid forms that settles to the bottom of the beaker.

2. After the interaction between the two chemicals, has the reading on the mass scale *increased, decreased,* or *stayed the same*? What is your reasoning?

Mixing alcohol and water: You have a beaker of 30 mL of water, and a beaker of 30 mL of alcohol. You pour the beaker of alcohol into the beaker of water.

3. What is the *volume* of the mixture?

Be prepared to share your team's ideas for all of these interactions in a class discussion.

Our Class Ideas

The cycle key question is:

Can interactions cause the amount of stuff (material) to change?

What are the class ideas about the cycle key question? Summarize them on your record sheet.

DEVELOPING OUR IDEAS

Activity 2
Keeping Track of Stuff in a Closed System

Purpose

The key question for this cycle is: "Can interactions cause the amount of stuff (material) to change?" In Unit 1, you learned two ways to describe the amount of stuff: by measuring mass or volume. In this activity, you will explore what happens to the amount of *mass* as a result of an interaction. The key question for this activity is:

In a closed system, can interactions cause the amount of mass to change?

Record the key question for the activity on your record sheet.

What is a closed system? Think about a relay team that has four runners. The coach requires the runners to agree to participate on the team for one season. During that season this team is a system that is closed to other runners. No other runners are allowed to join the team and none of the runners leave the team.

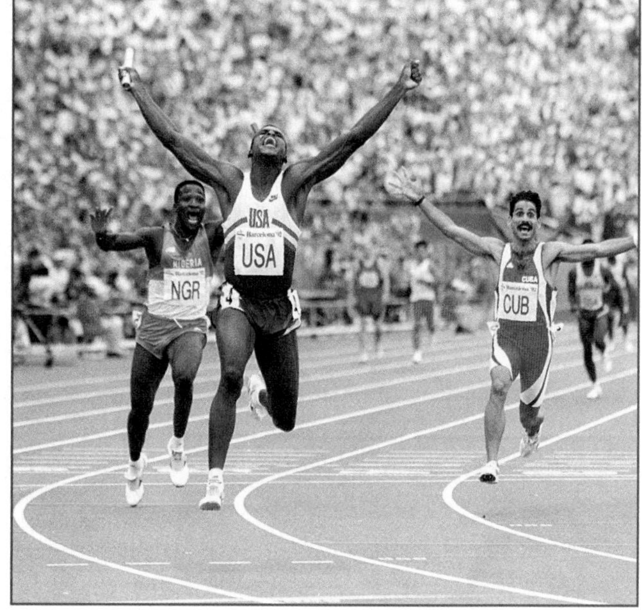

Now think about a different relay team with a different coach. During the season this coach lets members on her team come and go. Sometimes the team has many members and sometimes only a few, and the members are not always the same people. This team is a system that is open to other runners.

Scientists also use the notion of open systems and closed systems. In this activity you will explore closed mass systems. That is, you will explore systems that do not have any mass entering or leaving.

We Think

Review your answers in the last activity for dissolving sugar in water, and mixing chemicals. Both of these interactions occurred in *closed mass systems*. No mass enters or leaves the system once the equipment is set up.

Discuss the following students' answers with your team.

When the sugar is dissolved in water, the mass decreases.
The mixture of two clear solutions will become more massive when the white solid forms.
Jason
Chantel

Follow standard safety rules and school safety rules for laboratory activities.

Your teacher may review with you *How To Follow Safety Rules During An Experiment.*

Do you agree or disagree with these students? Explain your reasons.

Be prepared to share your team's ideas in a class discussion.

Explore Your Ideas

Exploration 1: What happens to mass when chemicals are mixed in a closed system?

Your team will need:

- plastic cup with 250 mL water
- plastic cup with 15 g sugar
- stirrer
- plastic cup with 20 mL of sodium sulfate solution
- plastic cup with 20 mL of calcium chloride solution
- 3 x 5 index cards
- mass scale

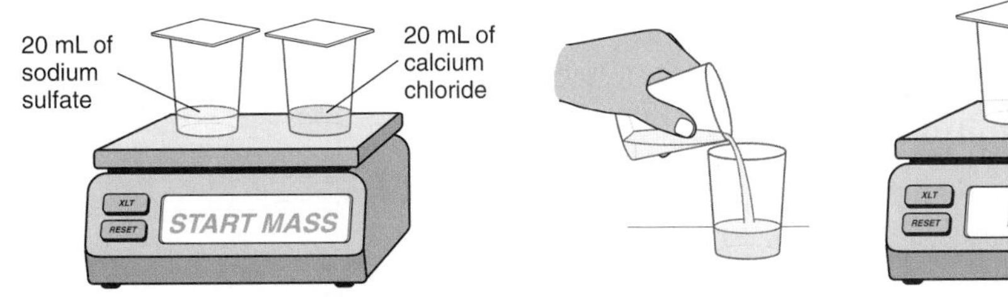

STEP 1 Set the plastic cups with 20 mL of sodium sulfate solution and 20 mL of calcium chloride solution on the mass scale. Put the cards on top of each cup.

Follow instructions carefully when using chemicals. Report spills.

Do not consume the sugar. It is not safe to eat.

Record this *start mass* in the Mass Measurements table on your record sheet.

Table: Mass Measurements

	Dissolving Sugar in Water	**Mixing Chemicals**
End mass	(After mixing) ________ g	(After mixing) ________ g
Start mass	(Before mixing) ________ g	(Before mixing) ________ g
Change in mass = End mass – Start mass	________ g	________ g
Class average Change in mass	________ g	________ g
Uncertainty in mass	________ g	________ g
Taking into account the uncertainty in mass, does the mass *increase, decrease,* or *stay the same?*		

STEP 2 Pour one solution carefully into the other. Replace the cards.

STEP 3 Remove the cups and cards carefully from the scale. Begin Exploration 2. You will complete Exploration 1 after doing Exploration 2.

Exploration 2: What happens to the mass when sugar is dissolved in water?

STEP 1 Set the plastic cups with 250 mL water and 15 g sugar and the stirrer on the mass scale. Cover the cups with the cards.

Record this *start mass* in the table on your record sheet.

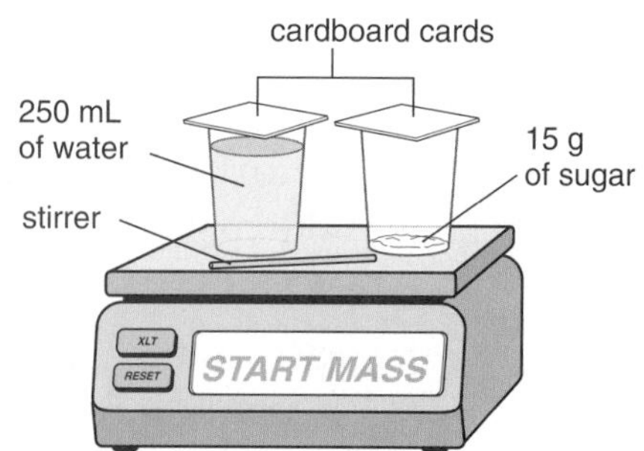

STEP 2 Pour the sugar carefully into the water and stir until all of the sugar is dissolved. Cover the cups with the cards and put the stirrer on the mass scale.

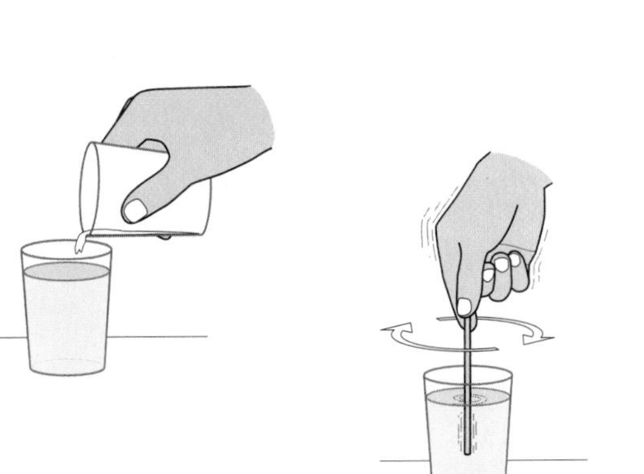

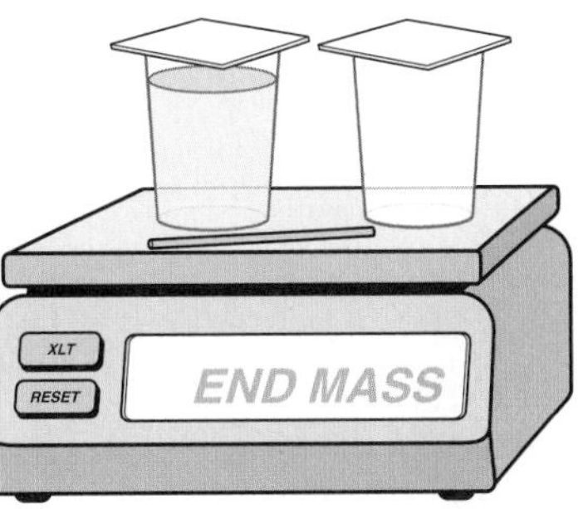

Wash your hands well after the activity.

Record this *end mass* in your table.

STEP 3 Return to Exploration 1. Replace the cups and cards on the mass scale. Record the *end mass* in your table.

STEP 4 Post your team results in a class data table.

When everyone has finished posting results, copy the class averages into your table.

Record in the table the *uncertainty in mass* that you calculated in Unit 1 for the equipment in your classroom.

Participate in the class discussion about your data.

Make Sense of Your Ideas

1. What does the data in this activity indicate about mass in a closed system? Write your reasoning and include evidence from the activity.
2. How does the uncertainty in the mass measurements affect your conclusion about the explorations?
3. How do the results in this activity compare with your predictions?
4. Read the conversation on the next page. How is it that scientists ever come to accept an idea?

I think we can conclude from the explorations that the total mass always stays the same in mixing interactions.

Rebecca

I disagree. We haven't tried all of the different kinds of substances that we could mix together. So we can't conclude that total mass always stays the same.

Carlos

But Carlos, we can never do all of the mixing explorations possible!

Otis

Maybe we should compromise. We could conclude that the total mass sometimes stays the same in mixing interactions.

Xuan

Well I know something I can conclude for sure. I'm confused! How is it that scientists ever come to accept an idea?

Carlos

Our Consensus Ideas

The key question for this activity is:

In a closed system, can interactions cause the amount of mass to change?

1. Write your best answer on your record sheet.

Participate in the class discussion about your answer.

2. Record the class consensus idea(s) on your record sheet.

DEVELOPING OUR IDEAS

Activity 3
Mass and Open Systems

Purpose

In the last activity, you found that in a *closed mass system*, the mass in the system stays the same no matter what changes happen in the system. In this activity, you will think about mass and *open systems*. An open system is one in which stuff can enter or leave the system. The key question for this activity is:

What can happen to the amount of mass in an open system during an interaction?

Record the key question for the activity on your record sheet.

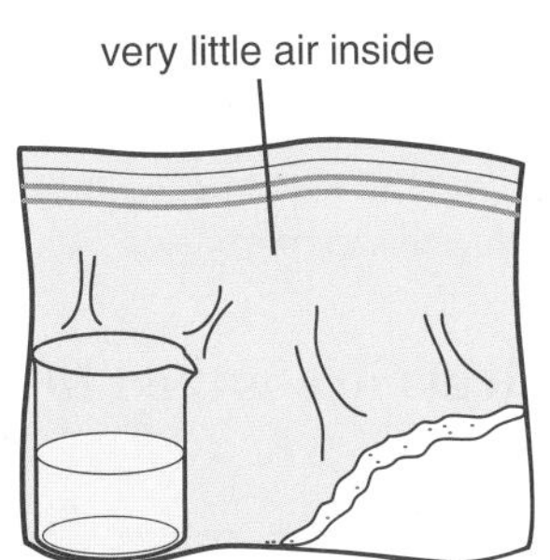

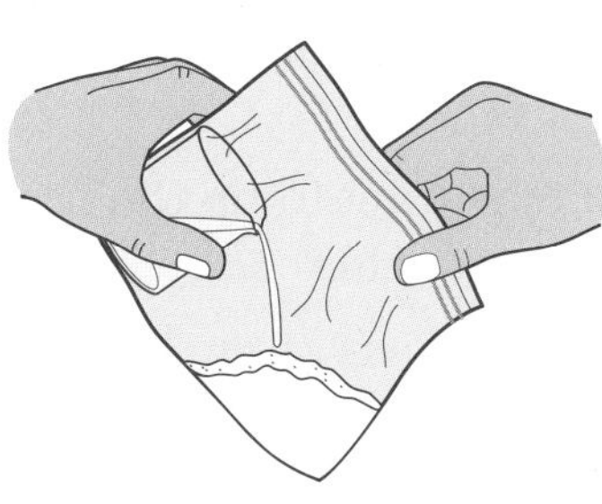

We Think

In a chemistry exploration, a student put baking soda in one side of a re-sealable plastic bag and a small jar of vinegar in the other side. She then carefully pushed as much gas as she could out of the bag (without spilling the vinegar) and sealed the bag.

When she tipped the liquid into the solid (without opening the bag), the student observed bubbling. The bag inflated as it filled with the gas from the bubbles.

After the bag inflated, the student measured the mass of the bag and its contents.

She released the gas from the bag, and measured the mass of the bag and its contents again.

Discuss the following question with your team.

What do you think will happen to the mass of the plastic bag and its contents after the gas is released? Why?

Participate in a class discussion about your answers.

Explore Your Ideas

Your team will do the chemistry exploration and find out what happens to the mass. You will record your results in a Mass Measurements table on your record sheet.

Follow instructions carefully when using chemicals. Report spills.

Wear eye protection.

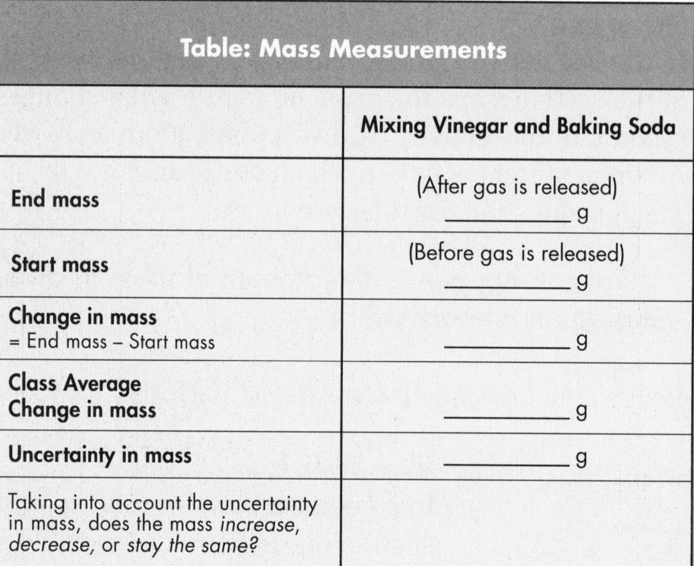

Table: Mass Measurements	
	Mixing Vinegar and Baking Soda
End mass	(After gas is released) ________ g
Start mass	(Before gas is released) ________ g
Change in mass = End mass – Start mass	________ g
Class Average Change in mass	________ g
Uncertainty in mass	________ g
Taking into account the uncertainty in mass, does the mass *increase*, *decrease*, or *stay the same*?	

Exploration: What happens to mass when chemicals are mixed in an open system?

Your team will need:

- re-sealable large plastic bag (2-gallon size)
- plastic cup or beaker
- 45 g baking soda
- 400 mL vinegar
- mass scale

STEP 1 Prepare your plastic bag with the baking soda in one corner and the cup with vinegar in the other corner. Push the air out of the bag and seal the bag. *Make sure the seal is tight!*

STEP 2 Tip the vinegar onto the baking soda. You may want to tip the bag back and forth to mix the vinegar and baking soda thoroughly.

STEP 3 When there are no more bubbles, measure the mass of the bag. *Make sure the bag is completely on the balance pan and does not touch anything else.*

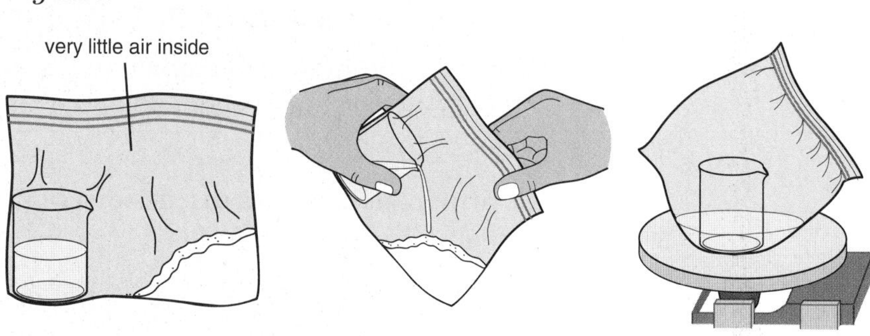

Record this *start mass* in the table on your record sheet.

STEP 4 Open the bag. Release the gas, pushing as needed to get all gas out of the bag. Reseal the bag so that air does not get back in.

Wash your hands well after the activity.

STEP 5 Measure the mass of the bag again.

Record this *end mass* in your table.

STEP 6 Calculate the mass change.

Record the mass change in your data table.

STEP 7 Post your team results in a class table.

When everyone has finished posting results, copy the class averages into your table.

Record in the table the *uncertainty in mass* that you calculated in Unit 1 for the equipment in your classroom.

Participate in a class discussion about your answers.

Make Sense of Your Ideas

1. Is the system used in the chemistry exploration a closed or an open system? Why?
2. What happened to the mass in the exploration? Include your evidence from the class data and take the uncertainty into account.
3. What would have happened to the mass during the interaction if the bag had been left open?
4. What do you think happens to the mass of an open system if you add mass to the system?

Our Consensus Ideas

Think about the key question for this activity and discuss it with your team.

What can happen to the amount of mass in an open system during an interaction?

1. Write your best answer on your record sheet. Include your reasoning.

Participate in the class discussion about your answer.

2. Record the class consensus idea(s) on your record sheet.

DEVELOPING OUR IDEAS

Activity 4
Keeping Track of Volume in a Closed System

Purpose

In previous activities, you learned that mass stays the same in a closed system and that mass may change in an open system. In this activity, you will explore what happens to the *volume* in a closed system. The key question for this activity is:

In a closed system, can interactions cause the volume to change?

Record the key question for the activity on your record sheet.

We Think

You have a beaker with 30 mL of water and a beaker with 30 mL of alcohol. You pour the beaker of alcohol into the beaker of water. What is the *volume* of the mixture?

Follow instructions carefully when using chemicals. Report spills.

Wear eye protection.

You answered this in the first activity in this cycle. Review your answer.

Explore Your Ideas

You will collect data on what happens to volume when *alcohol and water* are mixed, and what happens to volume when *vinegar and water* are mixed. When you mix the liquids, you will use narrow clear plastic tubing. The narrow tubing makes it easier to see if there are changes in the volume of the liquids before and after mixing by looking at changes in the *height* of the liquids in the tubing.

Exploration 1: What happens to volume when alcohol and water are mixed in a closed system?

STEP 1 Label one end of the clear plastic tubing *B*, for bottom. Put the stopper securely in the bottom end of the plastic tubing.

Your team will need:

- long piece of clear plastic tubing, about 45 cm long
- 2 stoppers (optional: rub with a little liquid soap solution)
- ruler with metric markings
- 30 mL colored water
- 30 mL colored alcohol
- funnel
- grease pencil

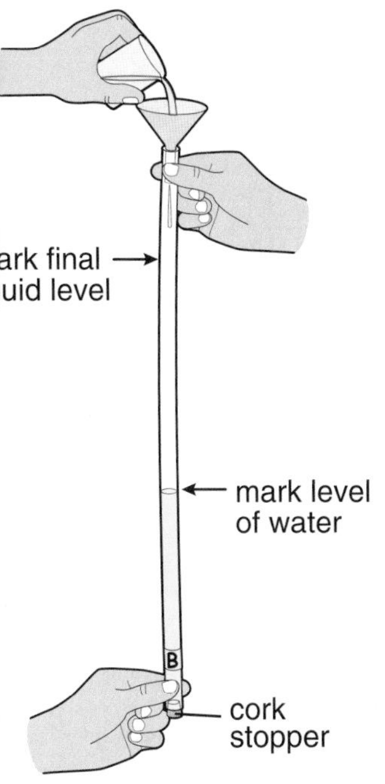

STEP 2 With one team member holding the bottom of the tubing, *slowly* pour in the colored water until the tube is filled halfway. Use the grease pencil to draw a line at the level of the water.

STEP 3 *Gently and slowly* pour alcohol into the tubing until the alcohol is about 10 cm (4 inches) below the top of the tubing.

STEP 4 If there are any air bubbles, let them rise to the surface. Use a grease pencil to *carefully* mark the top level of the alcohol.

STEP 5 Put a stopper in the open end of the tubing so that no liquid will leak out. Use a towel to wipe off all water or alcohol on the outside of the tubing.

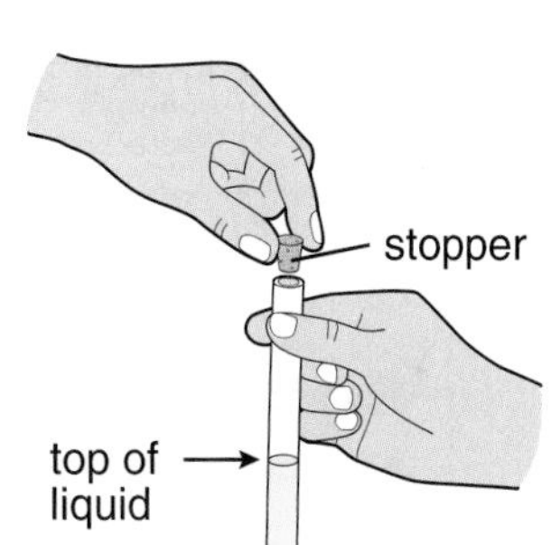

STEP 6 Gently tilt the tubing back and forth about five times until the alcohol and water are thoroughly mixed.

Caution: The alcohol is very flammable.

STEP 7 Tap the tubing to shake all of the air bubbles loose and wait for them to rise to the top of the mixed liquid. Use a grease pencil to mark the new level of the mixed liquid. With another team member holding the tubing vertically and stretched out, measure the change in the height of the mixed liquids (the distance between the two marks that you made).

Record the *change in height* in Table 1 on your record sheet.

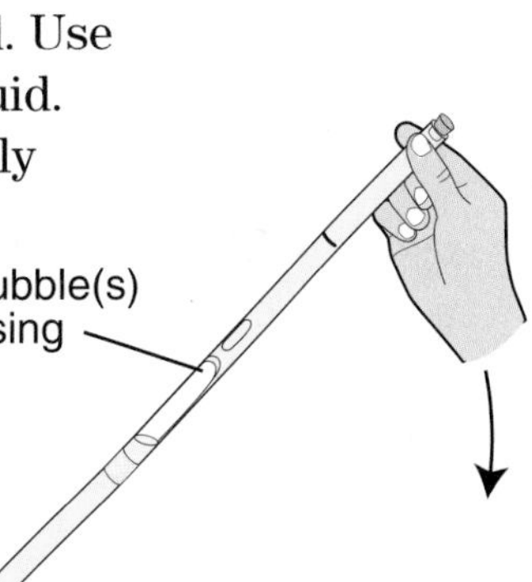

STEP 8 Post your team results in a class data table.

When everyone has finished posting results, copy the class averages into Table 1.

Record in Table 1 on your record sheet the *uncertainty in the class average change in height.*

Wash your hands well after the activity.

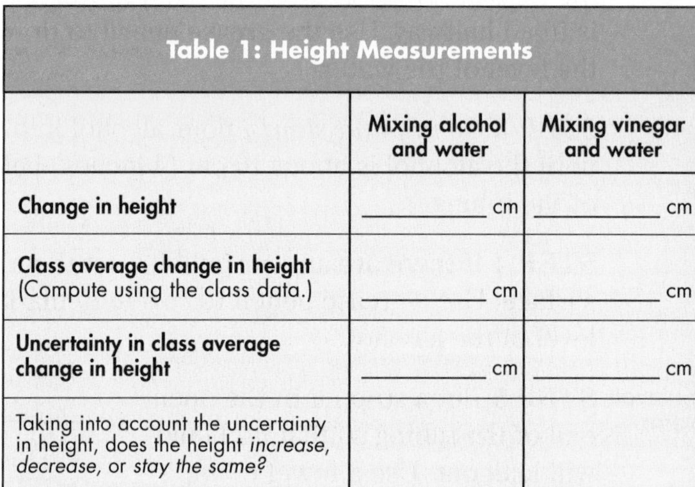

Table 1: Height Measurements		
	Mixing alcohol and water	**Mixing vinegar and water**
Change in height	________ cm	________ cm
Class average change in height (Compute using the class data.)	________ cm	________ cm
Uncertainty in class average change in height	________ cm	________ cm
Taking into account the uncertainty in height, does the height *increase, decrease,* or *stay the same?*		

Your team will need:

- long piece of clear plastic tubing, about 45 cm long
- 2 stoppers (optional: rub with a little liquid soap solution)
- ruler with metric markings
- 30 mL colored water
- 30 mL vinegar
- funnel
- grease pencil

Exploration 2: What happens to volume when vinegar and water are mixed in a closed system?

STEP 1 If you are using the same tubing and other items from the previous exploration, rinse and dry them before using them for this exploration. Wipe off grease pencil marks.

STEP 2 Repeat Steps 1 through 8 in the previous exploration, *substituting 30 mL of vinegar for the alcohol.*

Your teacher will show you a demonstration of these explorations and will measure the mass before and after mixing.

Record the results in Table 2 on your record sheet.

Table 2: Mass Measurements		
	Mixing alcohol and water	**Mixing vinegar and water**
End mass	(After mixing) ________ g	(After mixing) ________ g
Start mass	(Before mixing) ________ g	(Before mixing) ________ g
Change in mass = End mass – Start mass	________ g	________ g
Uncertainty in mass Determined earlier in this course	________ g	________ g
Taking into account the uncertainty in mass, does the mass *increase, decrease* or *stay the same?*		

Participate in a class discussion about your data and its uncertainty.

Make Sense of Your Ideas

What do the results in this activity indicate about volume in a closed system? Write your reasoning and include evidence from the activity.

Our Consensus Ideas

The key question for this activity is:

In a closed system, can interactions cause the volume to change?

1. Write your best answer on your record sheet.

Participate in the class discussion about your answer.

2. Record the class consensus idea(s) on your record sheet.

PUTTING IT ALL TOGETHER

Activity 5
Interactions and Mass

Comparing Consensus Ideas

Recall the key question for this cycle:

Can interactions cause the amount of stuff (material) to change?

Your teacher will review with you all of the ideas that the class developed during this cycle to help you answer the cycle question. Be prepared to contribute to this discussion.

Your teacher will distribute copies of *Scientists' Consensus Ideas: Mass Conservation.* Read the scientists' ideas and compare them with the ideas your class developed during this cycle.

1. Write the evidence from activities in this cycle where it is requested on the *Scientists' Consensus Ideas* form.

2. Why must scientists do many experiments to prove that mass is always conserved, but just one experiment is enough to disprove that volume is always conserved?

Your teacher will lead a whole-class discussion.

IDEA POWER

Activity 6
Mass Conservation Problems

Analyze, Explain, and Evaluate

In this activity, you will use the ideas that you learned in this cycle to analyze and explain some more interesting situations. The procedure for doing this is found in *How To Write an Analysis and Explanation.*

All of the analyses for these situations should include *the system that you are analyzing* and *whether it is an open or closed system*. You will not draw energy or force diagrams.

Be sure that what you write would get a good evaluation using *How To Evaluate an Analysis and Explanation.*

On Matthew's birthday, his grandfather gave him a silver dollar. Matthew measured a mass of 30.0 g for the silver dollar.

Later, Matthew stood in his snow-covered front yard flipping the silver dollar. On one flip, he missed the coin and it sank into the snow. Matthew tried to find the coin, but he could not.

Matthew eventually did find the silver dollar after the snow melted. The coin was still wet from the melted snow. Matthew measured the mass of the coin again.

1. Is the mass of the wet coin *greater than*, *less than*, or *equal to* the start mass of 30.0 g? Analyze and explain.

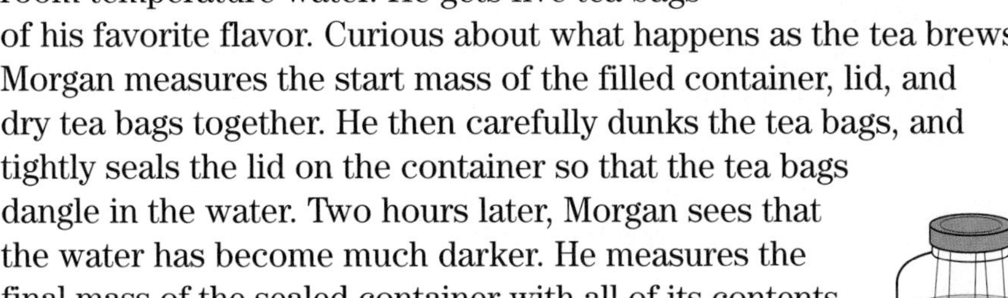

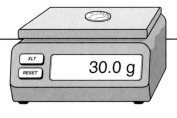

On a hot summer day, Morgan wants to brew some tea. He fills a large clear container with room-temperature water. He gets five tea bags of his favorite flavor. Curious about what happens as the tea brews, Morgan measures the start mass of the filled container, lid, and dry tea bags together. He then carefully dunks the tea bags, and tightly seals the lid on the container so that the tea bags dangle in the water. Two hours later, Morgan sees that the water has become much darker. He measures the final mass of the sealed container with all of its contents.

2. Is the final measured mass of the tea, tea bags, and container *greater than*, *less than*, or *equal to* the start mass? Analyze and explain.

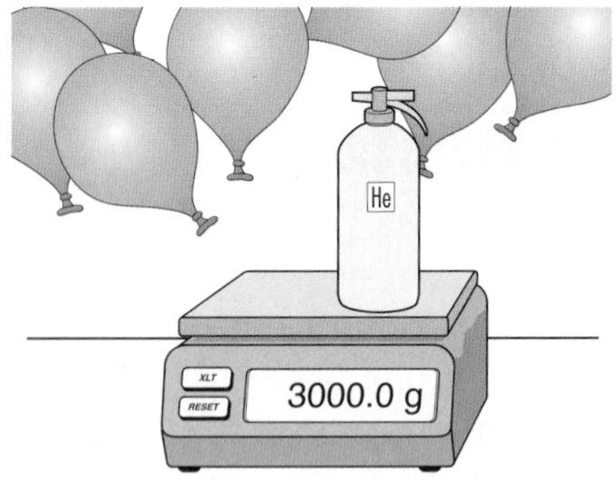

Kaitlin does an exploration to find out if decompressing a gas changes its mass. Kaitlin measures the mass of a tank containing compressed helium gas. The system (tank plus helium) has a mass of 3250 g. Kaitlin records this mass as the system start mass.

Kaitlin pumps 20 balloons full of helium gas. She then measures the mass of the tank and records it as 3000 g.

Kaitlin explains the change in mass by concluding that decompressing a gas reduces the mass of the gas.

3. Evaluate Kaitlin's analysis and explanation using *How To Evaluate an Analysis and Explanation.* If Kaitlin's analysis and explanation are not good, write a good one.

Linus weighs a glass of water and two fizzy tablets. They have a mass of 305 g.

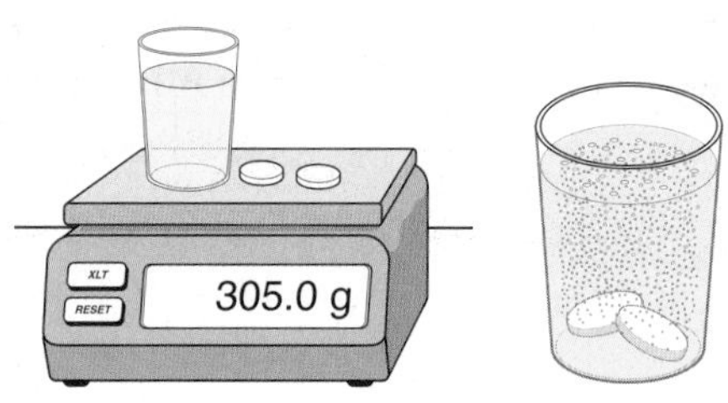

He drops the tablets into the glass and watches as they dissolve in the water. Bubbles of gas form, rise to the top of the water, and burst, releasing the gas they hold into the air.

After the tablets have dissolved and all the bubbles have popped, the glass and all it contains has a mass of 300 g.

4. Why does the solution have less mass after all of the bubbles have popped? Analyze and explain.

Popcorn kernels cook from the inside out. The kernels contain a small amount of water stored in the softer starch inside the harder outer casing. A microwave oven heats the water inside the kernels, causing steam to build up inside each kernel. The steam expands and builds up pressure inside until the casing explodes open in a loud "pop." The steam quickly escapes, pushing the inside of the kernel to the outside.

5. Is the mass of the popcorn after it has popped *greater than*, *less than*, or *equal to* the mass before it has popped? Analyze and explain.

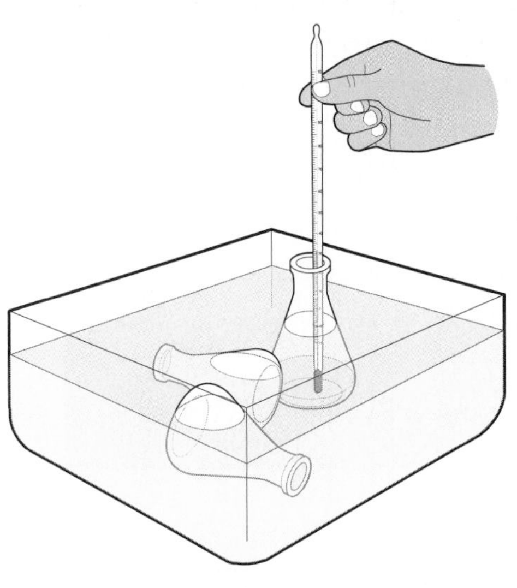

CYCLE 2:
ENERGY CONSERVATION

Cycle Key Questions

1. Can energy be created from nothing?
2. Can energy be destroyed into nothing?

OUR FIRST IDEAS

Activity 1
Energy and Interactions

Purpose

Think about a metal cube in a beaker of boiling water. While the cube is in the water, the cube will get hot. After the cube is removed from the boiling water, the cube will cool down. In other words, while the metal cube is interacting with the boiling water, the temperature and thermal energy of the cube increase. Where does the additional thermal energy come from? When the cube cools down, where does the thermal energy go? Or is the cube's thermal energy *created from nothing*, and *destroyed into nothing* as the cube first warms up and then cools off?

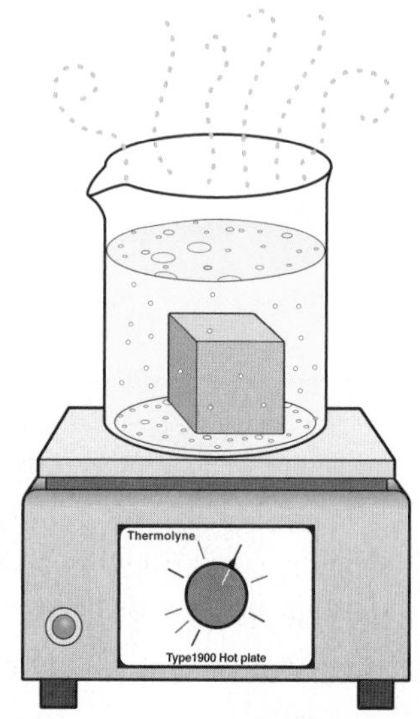

When trying to understand what happens to the energy of objects during interactions, you need to consider whether energy can be created or destroyed. The key questions for this cycle are:

1. Can energy be created from nothing?
2. Can energy be destroyed into nothing?

Record the cycle key questions on your record sheet.

We Think

Before discussing these questions with your team, write your answers on your record sheet.

1. Do you think energy can sometimes be destroyed into nothing, no longer existing as anything? If yes, give at least one example. If no, what are your reasons? Use your everyday experiences and/or the explorations you have done to support your ideas.
2. Do you think energy can sometimes be created from nothing? If yes, give at least one example. If no, what are your reasons? Use your everyday experiences and/or the explorations you have done to support your ideas.

Discuss your responses to these questions with your team. Come up with a team response to each question. Prepare a presentation board to explain your answers to the class. Take turns writing your team's response to each question. Be sure to include your examples and/or reasons. *Make sure each team member is prepared to explain your responses to the class!*

Our Class Ideas

Participate in the class discussion to summarize the class ideas about energy being created or destroyed.

For each question on your record sheet, write at least *one* idea that is different from what you have already written.

DEVELOPING OUR IDEAS

Activity 2
Energy and Heat-Conduction Interactions

Purpose

In this course, you have learned many types of interactions. In three of them the temperature and thermal energy of at least one of the interacting objects changes. *Remember, as an object's temperature increases, its thermal energy also increases.* Think about the following examples.

- In a *light* interaction, when light shines on an object, the object usually becomes warmer and its thermal energy increases.
- In a *friction* interaction, when two objects rub against each other, both objects get warmer and their thermal energy increases.
- In an *electric-circuit* interaction, when an electric circuit is connected, the wires, bulbs, and other items in the circuit get warmer. This means that in each of these interactions there is an increase in thermal energy for at least one of the interacting objects.

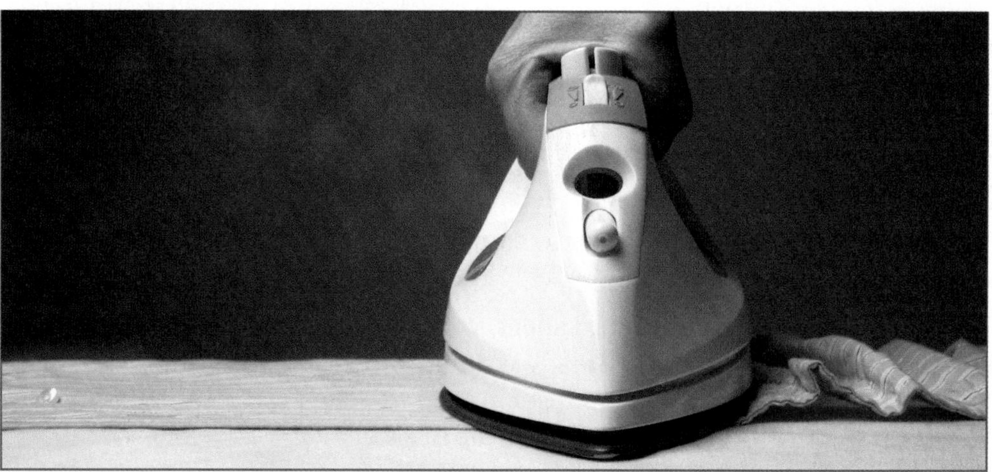

Think of situations, from your everyday experiences, in which the thermal energies of objects change. What happens when warm and cold objects touch each other?

The key question for this activity is:

Does an interaction occur when warm and cold objects touch each other?

Record the activity key question on your record sheet.

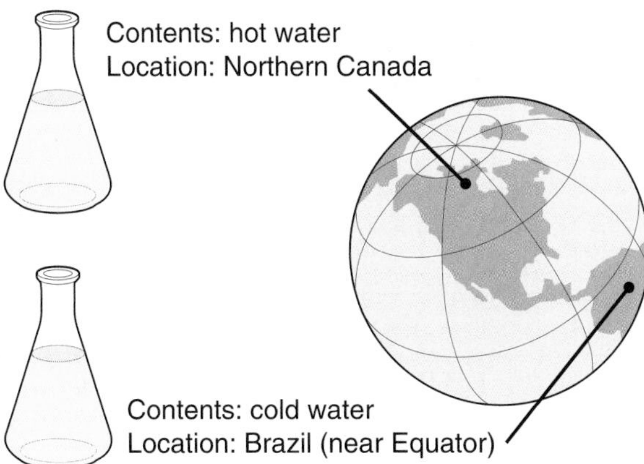

We Think

A student in northern Canada put a flask of *hot* water on a counter for several hours. Another student, near the Equator in Brazil, put a flask of *cold* water on a counter for several hours. Assume that the temperature in Canada is cold and the temperature in Brazil is hot. What will happen to the temperature of the water in the flasks?

Read Rebecca's answer to the We Think question. Discuss it with your team.

Do you agree or disagree with Rebecca? What do you think will happen? Write your response and reasoning on your record sheet.

Participate in the class discussion about your responses and reasons.

The hot water in northern Canada will cool down until it is a little cooler than the area where it is sitting. The cold water near the Equator in Brazil will warm up until it is almost as warm as where it is sitting. I think this because water always feels a little colder than its surroundings.

Rebecca

Explore Your Ideas

You will do an exploration to help you answer the activity key question, and the We Think question. In the exploration, you will use hot water in a flask and cold water in a beaker. You will put the flask in the beaker, and see what happens.

Exploration: What happens to the temperature of the water when a flask of hot water is set inside a beaker of cold water?

Your team will need:

- 250 mL beaker
- foam-rubber sleeve for the beaker
- 50 mL Erlenmeyer flask
- 2 thermometers
- clock or stopwatch
- access to a cold-water bath or cold tap water
- access to a hot-water bath

Follow standard safety rules and school safety rules for laboratory activities.

To Do

Team Member 1
Supply Master: gathers materials. Be the *Timekeeper* for the exploration.

Team Member 2
Procedure Specialist: reads instructions and steps aloud. Be *Temperature Taker A* (keep track of the temperature of the *hot water* for 10 minutes).

Team Member 3
Recycling Engineer: returns materials and supervises cleanup. Be *Temperature Taker B* (keep track of the temperature of the *cold water* for 10 minutes).

Team Member 4
Team Manager: makes sure everyone stays on task and all team members are participating. Be *Recorder* of the temperature data for the team.

Table: Time and Temperature of the Hot and Cold Water

Time (min)	Hot Water Temperature (°C)	Cold Water Temperature (°C)
0		
1		
2		
3		
4		

STEP 1 Read your jobs in the *To Do* list shown.

To complete this exploration successfully, you must work together and coordinate your tasks smoothly. To practice the smooth coordination of an experiment like this one, scientists first do dry runs of an experiment. *Before you do the exploration, you will first do a dry run.* Read the directions and complete Steps 2 through 6 *using tap water only.*

When you and your team members are confident of the steps, arrange yourselves at the table so everyone can perform their jobs quickly and smoothly. Start the exploration using both hot and cold water.

STEP 2 *Temperature Taker A:* Place 50 mL of hot water *from the hot water bath* into the small flask.

While the flask is still in the hot water bath, measure the temperature of the water inside the flask. Read the thermometer at eye level. *Do not let the thermometer bulb touch the sides or bottom of the flask!* If the bulb touches the glass of the flask, the thermometer will record the temperature of the glass, not the water.

Recorder: Record the temperature for the hot water in the first row (Time = 0) in the table on your record sheet.

STEP 3 *Temperature Taker B:* Put 50 mL of cold water into the beaker. If your beaker has two scales, measure 50 mL by using the scale that goes up from the bottom.

Place the beaker in the foam-rubber sleeve and measure the temperature of the water in the beaker. *Don't let the thermometer bulb touch the sides or bottom of the beaker.*

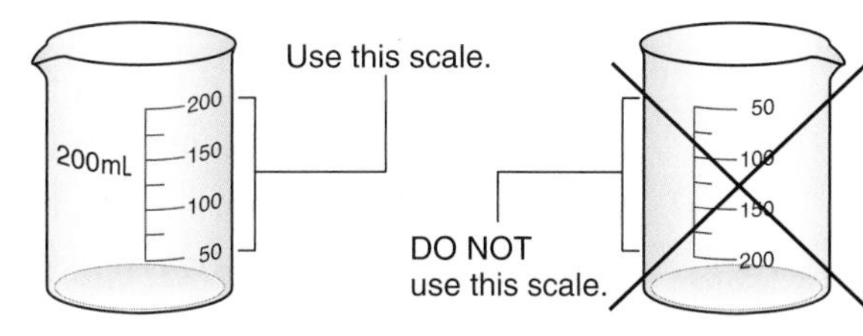

Recorder: Record this temperature for the cold water in the first row (Time = 0) in your table.

STEP 4 *Temperature Taker A:* Place the flask in the beaker. Try not to let the flask touch the sides of the beaker.

STEP 5 *Timekeeper:* Immediately start the stopwatch or note the time on a clock. Every minute, tell the *Temperature Takers* that it is time to read the thermometers.

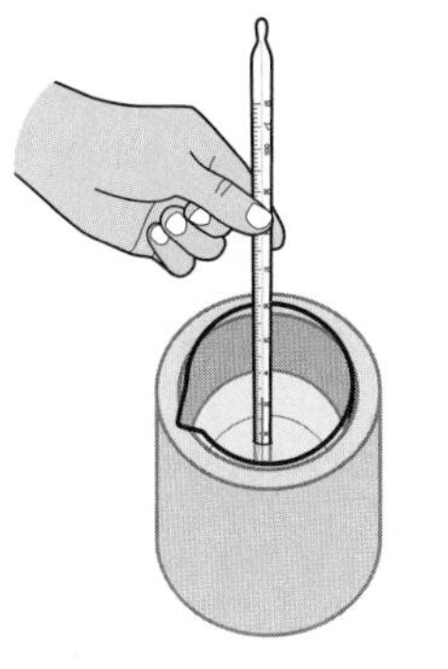

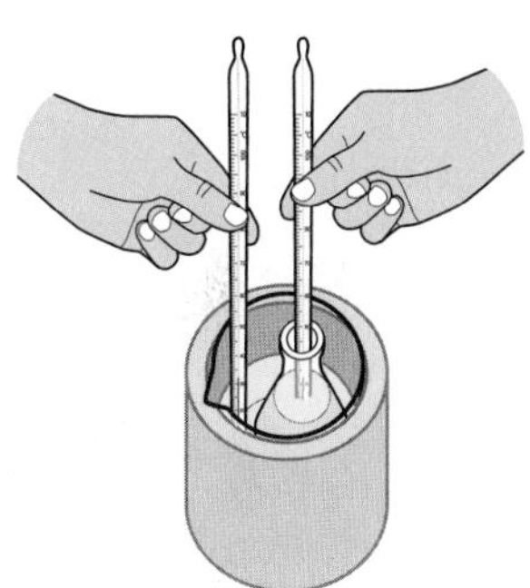

STEP 6 *Temperature Takers:* Hold your thermometers in the water for the entire 10 min. *Do not let the thermometer bulb touch the sides or bottom of the flask or beaker!* Keep your eyes on the thermometer so that you can call out the temperature readings to the *Recorder* as soon as the *Timekeeper* says "Time!"

Recorder: Write the temperatures in the data table every minute for 10 min.

Copy the *Recorder's* temperature data into your table.

Plot your hot-water data on the time graph on your record sheet. Draw a line through your data points that best fits the data. Label the line connecting your data points *hot water.* Think about what happened to the temperature of the hot water during the time you collected data.

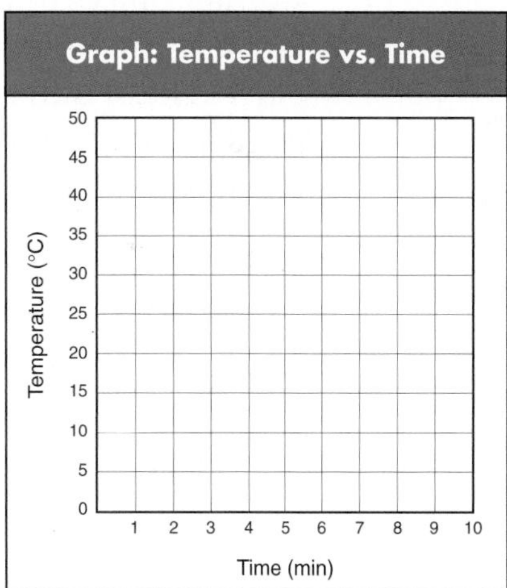

Plot your cold-water data on the same graph. Draw a line through your data points that best fits the data. Label the line connecting your data points *cold water.* Think about what happened to the temperature of the cold water during the time you collected data.

Make Sense of Your Ideas

1. Answer the exploration question: What happens to the temperature of the water when a flask of hot water is set inside a beaker of cold water? Use evidence from your exploration to support your answer.
2. Does the evidence from the exploration support Rebecca's answer to the We Think question? Why or why not?
3. Does the evidence from the exploration support your answer to the We Think question? Why or why not?
4. If two objects that are in contact with each other have exactly the same temperature, do you think they interact? Write your reasoning.
5. Read the We Think question and your answer again. Now that you have performed this exploration, is your answer the same? If not, write your new answer and explain why you changed it.

Our Consensus Ideas

Think about the key question for this activity and discuss it with your team.

Does an interaction occur when warm and cold objects touch each other?

1. Write your answer on your record sheet.

Participate in the class discussion about your answer.

2. Write the class consensus idea on your record sheet.

Scientists call the interaction you examined in this activity a *heat-conduction interaction*. Heat-conduction interactions take place when two objects with different temperatures are touching. Heat energy is transferred from the warmer object to the colder object. In Unit 6, you will learn why heat energy is transferred. Heat-energy transfers take place only when the interacting objects have different temperatures. The defining characteristics of a heat-conduction interaction are that the interacting objects must be in contact, and one must be warmer than the other.

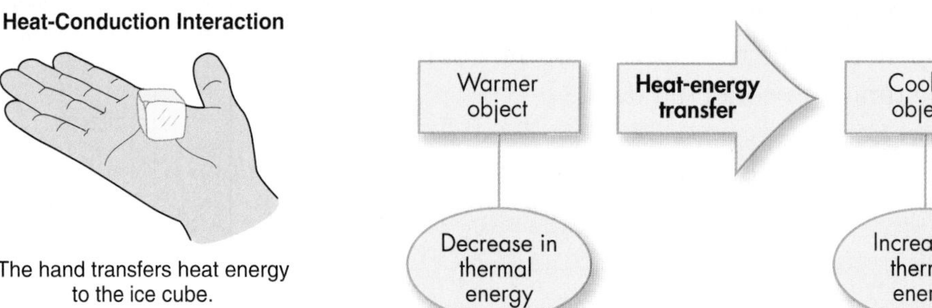

The hand transfers heat energy to the ice cube.

DEVELOPING OUR IDEAS

Activity 3
Energy and Infrared-Radiation Interactions

Purpose

In Activity 2, you placed a flask of hot water in a beaker of cold water and found that after several minutes, the flask water and beaker water became the same temperature. That exploration was an example of a *heat-conduction interaction*, which occurs when two objects with different temperatures are touching. In heat-conduction interactions, there is a transfer of *heat energy* from the warmer to the cooler object.

The purpose of this activity is to answer the question:

What happens when a warmer object is near (but not touching) a cooler object?

Record the key question for the activity on your record sheet.

We Think

Imagine holding a cup of hot chocolate in your hand. During this heat-conduction interaction, the hot chocolate transfers heat energy to your hand. The thermal energy of the hot chocolate decreases, while the thermal energy of your hand increases.

Heat-Conduction Interaction

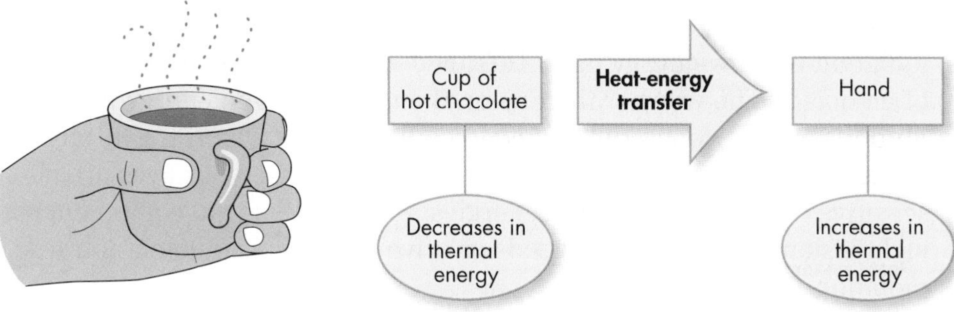

Now imagine that you hold your hand close to a cup of hot chocolate, but you do not touch it. You feel your hand becoming warmer.

Are the cup of hot chocolate and your hand interacting now? Write your reasoning.

Discuss this question with your team. Write your answers on your record sheet. Be prepared to share your team's ideas with the class.

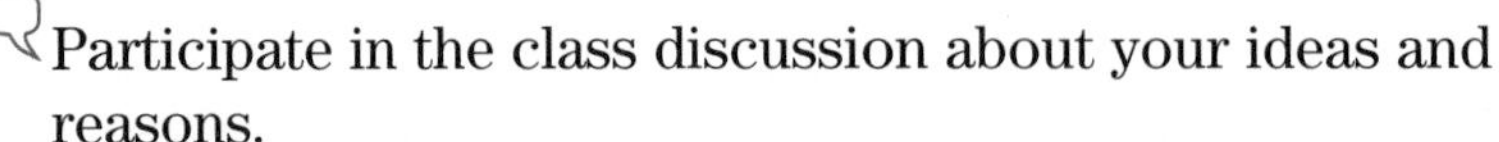

Participate in the class discussion about your ideas and reasons.

Explore Your Ideas

You have probably already learned a little about light in other science courses. Scientists often refer to light as *radiation*. Light or radiation has a large range of energies. Your eyes detect light in just a small segment of these energies and each color is a slightly different energy from the other colors.

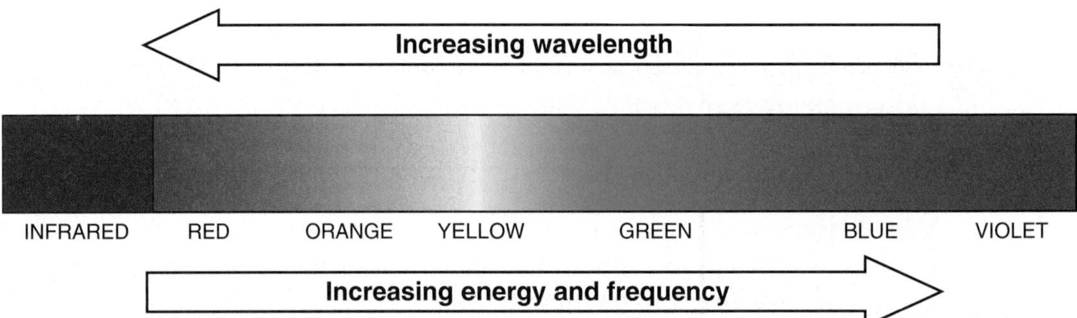

Science Words

ultraviolet radiation: radiation with energies greater than visible light

infrared radiation: radiation with energies lower than visible light

Are there any radiation energies that you cannot see? Yes, there are radiation energies higher than visible light. **Ultraviolet radiation** is high-energy light that can cause your skin to burn, for example, when you get a sunburn. There are also radiation energies lower than visible light. For example, infrared radiation is a low-energy light. **Infrared radiation** is emitted by the surface of an object. The warmer the object, the more infrared energy it emits. Some objects, such as your skin, can become warmer by absorbing infrared radiation. You cannot see infrared-radiation because your eyes are not sensitive to infrared-radiation energies. However, special equipment such as night-vision goggles and infrared-sensitive cameras can detect infrared radiation.

Recall from your explorations with mechanical waves and energy transfer that as the frequency of a wave increases, the wavelength decreases. The same is true for light waves. Light waves in the infrared and red areas of the electromagnetic spectrum have lower energy and frequency but greater wavelengths. Light waves in the violet and ultraviolet areas have higher energy and frequency but shorter wavelengths.

All radiation travels across space from its source to a receiver. The interaction between the source of the infrared radiation and its receiver is called an *infrared-radiation interaction.* Just like heat-conduction interactions, infrared interactions happen between objects of different temperatures. The defining characteristic of an infrared-radiation interaction is that one of the interacting objects must be warmer than the other.

Whenever an interaction happens between objects of different temperatures, the type of energy transferred between the objects is *heat energy.*

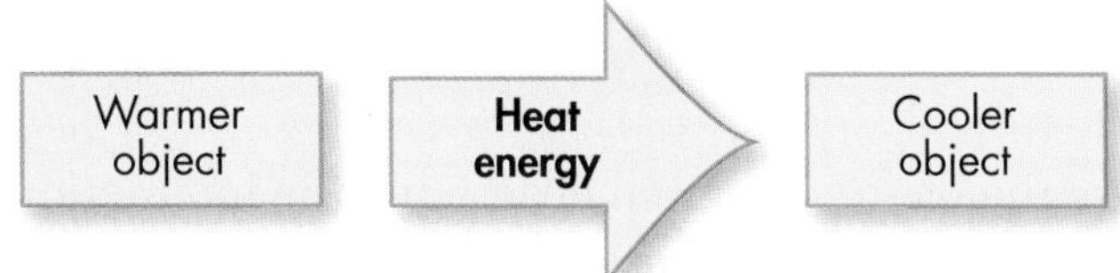

Special cameras can detect infrared radiation. These cameras usually produce images that are shades of a single color, often gray or green. *The greater the amount of infrared radiation emitted by the source, the brighter the image in the camera.*

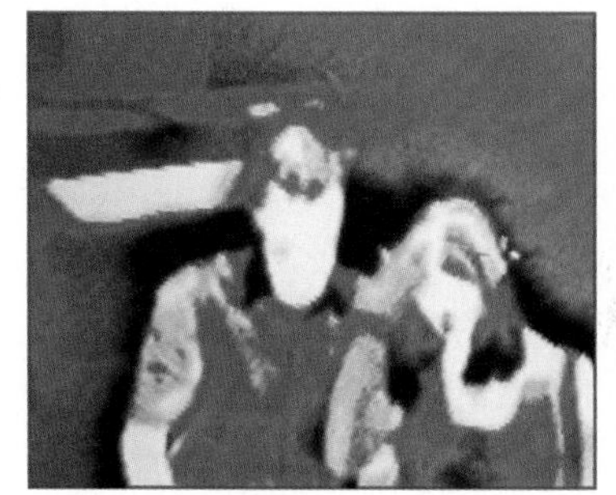

Your teacher will show you a video of several examples from everyday life taken with both an infrared video camera and a regular video camera. You will be able to see how objects appear both in visible and infrared radiation.

As each section of the video is shown, the class will stop and discuss the questions about that section. Be prepared to participate in the discussion and write down the answers to the questions.

1. After the hands are rubbed together, the skin of the hands glows brightly when viewed through an infrared camera. Complete an energy diagram on your record sheet to describe what happens between the glowing hands and the infrared camera.

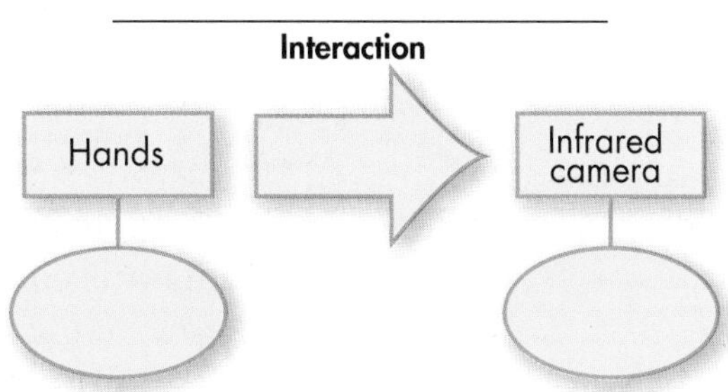

2. Complete the energy diagrams describing the infrared video of the hot water being poured into the glass.

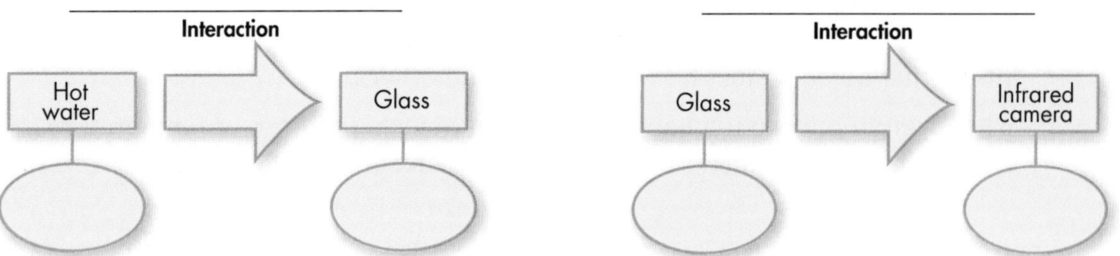

3. Complete the energy diagram describing the interaction between the *cold bottle* and the *warm hands*.

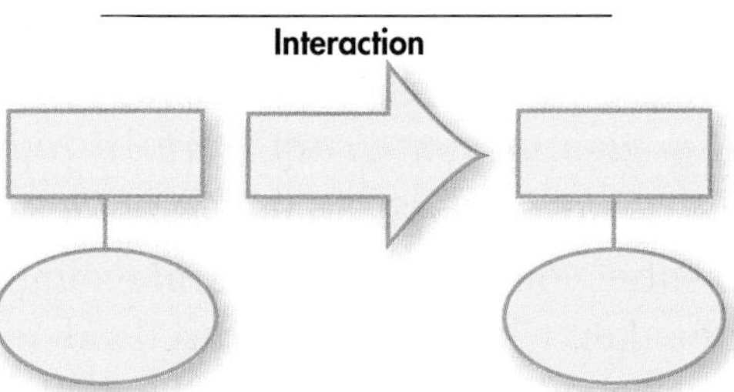

4. Now think carefully about the hand after the bottle has been removed and complete the energy diagram on your record sheet.

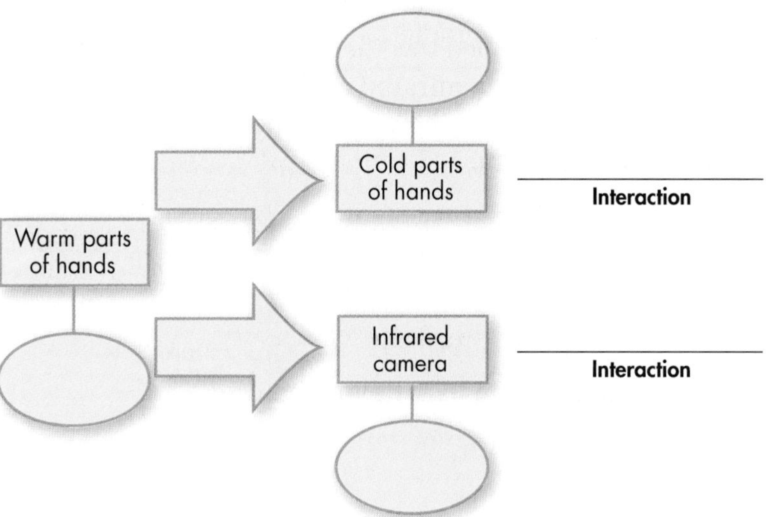

5. In the infrared video, after the toaster is turned on, the toaster coils glow extremely bright. What interactions and energy transfers cause the glowing coils to be recorded by both the regular camera and the infrared camera? (Light must enter a regular camera in order for it to record an image.) To answer this question, complete the energy diagram on your record sheet.

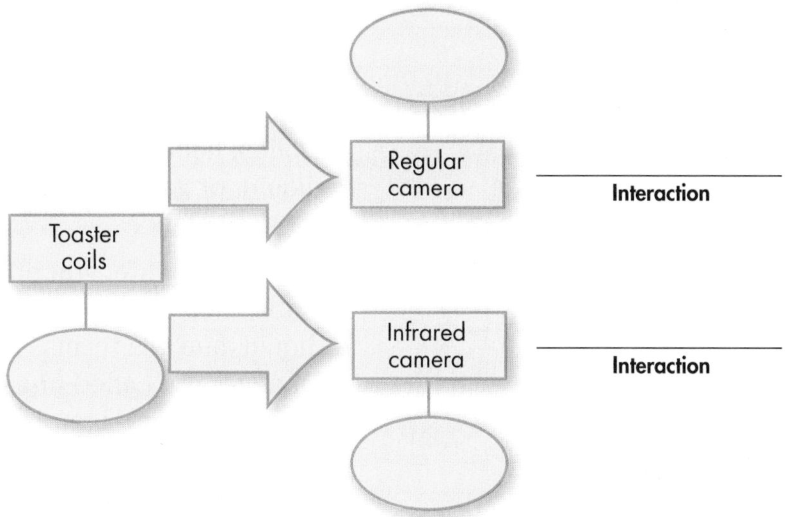

Make Sense of Your Ideas

Discuss each of the questions below with team members. Then write your answers.

1. In what way(s) is the infrared-radiation interaction similar to the heat-conduction interaction?
2. In what way(s) is the infrared-radiation interaction different from the heat-conduction interaction?

Our Consensus Ideas

Think about the key question for this activity and discuss it with your team.

What happens when a warmer object is near (but not touching) a cooler object?

1. Write your answer on your record sheet.

Participate in the class discussion about your answer.

2. Write the class consensus idea on your record sheet.

DEVELOPING OUR IDEAS

Activity 4
Thermal Energy and Phase Change

Purpose

At the range of temperatures on Earth, most substances exist as solids, liquids, or gases. Solid, liquid, or gas describes the *phase* of a substance.

In everyday life, you often see a substance change between solid, liquid, and gas forms. These changes are called **phase changes**. For example, the wax in a candle changes from solid to liquid. Water changes from liquid to gas. Water drops form on a window as it changes from gas to liquid. In your freezer, water changes from liquid to solid.

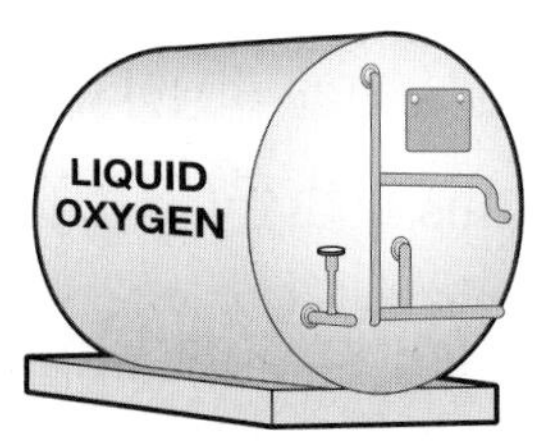

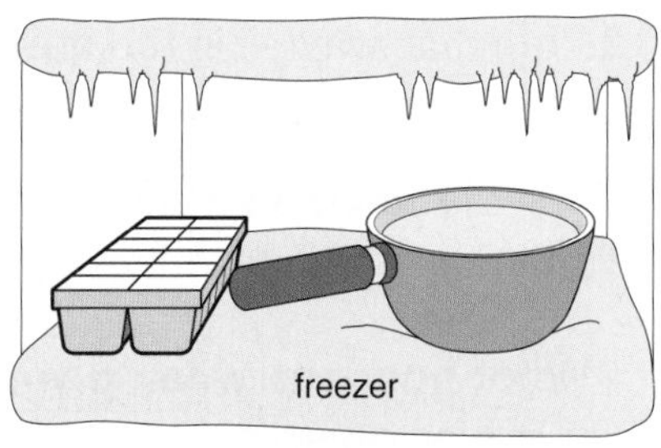

Science Words

phase change: the conversion of a material from one phase to another phase. Example: solid to liquid, liquid to gas.

All substances can be a solid, a liquid, or a gas if they are cooled enough or heated enough! If you cool air to −183°C, the oxygen in the air turns into liquid oxygen. If you heat a solid piece of aluminum to 2467°C, it melts and turns into a liquid.

In this activity, you will explore what happens to the energy transferred to a substance during a phase change. You will collect data from water as it changes phases from ice to liquid to vapor (gas). The key question for this activity is:

What happens to energy that is input during a phase change?

Record the key question for the activity on your record sheet.

We Think

Discuss these questions with your team.

1. Imagine some ice melting in a glass. Do you think that the temperature of the ice and water *increases*, *decreases*, or *stays the same* while the ice is melting? Write your reasoning.
2. Imagine a pot of water boiling on a stove burner. Do you think that the temperature of the water *increases*, *decreases*, or *stays the same* while the water is boiling? Write your reasoning.

Participate in a class discussion about your answers.

Explore Your Ideas

Exploration: What happens to the input energy as solid water changes to liquid water then to water vapor?

Part A: Heating Water Using a Hot Plate or Burner

To answer the question, you will heat solid water (crushed ice). You will measure the temperature of the water once each minute until the ice has melted and continue heating it until the water has boiled for 3 to 4 min. The responding (dependent) variable is the temperature of the water. The manipulated (independent) variable is the time.

STEP 1 To complete this exploration successfully, you must work together in a smooth, coordinated fashion.

Assign roles. One team member must be the *Timekeeper*, one must be the *Temperature Taker*, one must be the *Recorder*, and one must be the *Observer*.

Observer: Read all the directions out loud while the other team members read along. *Do not start the exploration until each team member knows what to do.*

Wear eye protection.

Report any spills immediately.

Moisten burned matches before throwing them away.

Table: Phase Changes of Water		
Time (minutes) *(Heat-Energy Transfer)*	Temperature (°C)	Observations
0		
1		
2		
3		
4		
Etc.		

Your team will need:

- 100 mL beaker
- *finely* crushed ice (about 50 mL)
- hot plate or alcohol burner, stand, and starter
- ringstand with clamp
- thermometer
- glass stirring rod
- stopwatch or clock with a second hand
- 4 colored pencils: green, blue, yellow, and red

STEP 2 Set up the equipment as shown in the diagram on the next page. Make sure the thermometer bulb stays completely surrounded by ice. Do not let the thermometer bulb touch any part of the beaker! Be sure you can read the thermometer.

Measure the temperature of the crushed ice. Record this temperature in the first (Time = 0) row in your table.

STEP 3 Light the alcohol burner. The distance from the flame to the stand should be about 2 or 3 cm. (If you are using a hot plate, choose a medium heat and do not change the heat setting during the exploration.)

STEP 4 *Timekeeper: Immediately* start the stopwatch or note the time on a clock. Every 60 s, say "Time!" to indicate to the *Temperature Taker* that it is time to read the temperature.

STEP 5 *Temperature Taker:* Gently stir the ice with the glass stirrer and watch the thermometer. Call out the temperature readings as soon as the *Timekeeper* says "Time!"

Recorder: Record the temperature in the table.

STEP 6 *Observer:* Watch what is happening in the beaker for the entire time.

Observer: Record the times when the ice starts to melt, when all the ice is melted, and when the water is vigorously boiling. If the ice starts melting before you light the alcohol burner or turn on the hot plate, write that the ice was melting at Time = 0.

Continue until the water has boiled vigorously for three to four minutes.

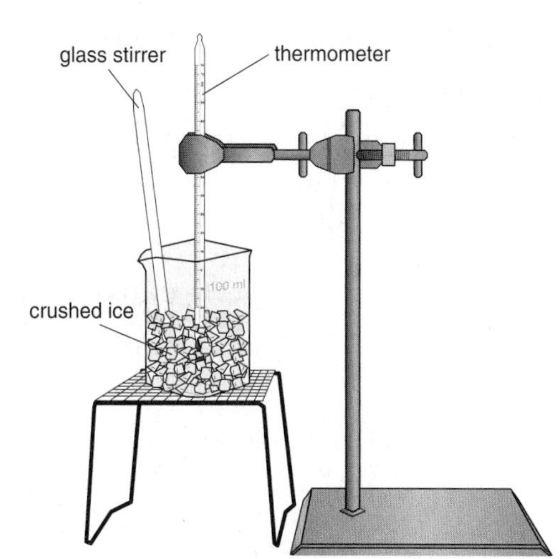

STEP 7 After the water has been boiling for 3 to 4 min, extinguish the flame on the alcohol burner or turn off the hot plate. *Be very careful with the hot glassware and hot or boiling water.*

Team members: Copy the data recorded by *both* the *Recorder* and the *Observer* on your table.

Plot your data on a graph of Temperature versus Time.

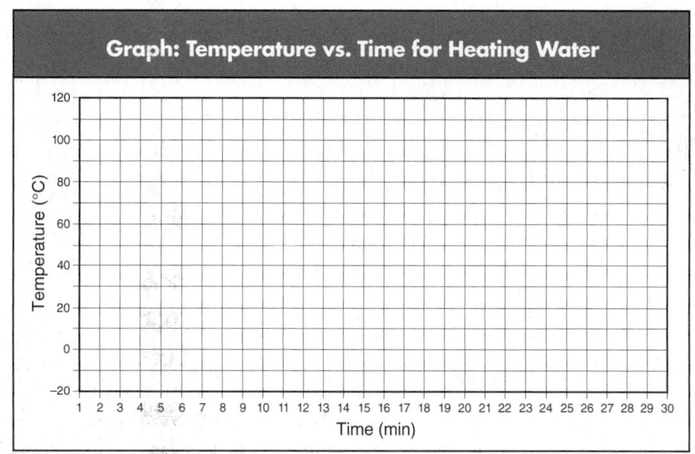

Follow these directions for drawing over your graphed data using colored pencils:

- Circle the *ice melting* part and label it *ice melting.*
- Circle the *water warming* part and label it *water warming.*
- Circle the *boiling* part and label it *water boiling.*

1. Compare the different parts of the graph. What is similar and what is different?

Part B: Observing Heating Water Using a Simulator

Sometimes, it is difficult to record very accurate data, either because of avoidable errors or unavoidable uncertainties. Your teacher will show you a simulation of ice melting and water boiling. Read the questions below and watch the simulation to find the answers.

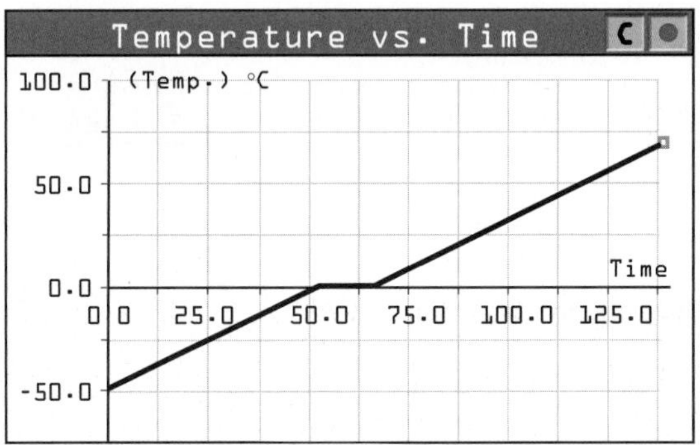

2. What happens to the graph at about Time = 50 s?

3. What happens to the graph at about Time = 65 s?

4. Compare your graph to the simulator graph. What is similar and what is different? If there are differences, write why you think the differences exist.

5. What happened to the *temperature* and *thermal energy* while the water was changing phases? Support your conclusion with evidence.

6. What happened to the *temperature* and *thermal energy* while water was not changing phases? Support your conclusion with evidence.

Make Sense of Your Ideas

During the entire time that the data was collected, heat energy was being transferred from the energy source (hot plate or alcohol burner) to the contents of the beaker.

1. Can you make sense of what happened to the energy transferred to the beaker during the time that the data was collected? Does anything seem strange to you?

So far, you have learned about many kinds of energy that an object can have: motion energy, thermal energy, stored elastic energy, stored chemical energy, etc. In this activity, you encountered another type of energy that an object can have: *stored phase energy*.

Science Words

stored phase energy: the energy associated with the phase of a material

Stored phase energy is the energy associated with the phase (gas, liquid, or solid) of a substance. Gases have more stored phase energy than liquids and liquids have more stored phase energy than solids. Like other forms of stored energy, stored phase energy increases when you transfer energy into a system. For example, consider stored chemical energy. Suppose you eat some fruit. As your body digests the fruit, your body increases in stored chemical energy. Similarly, putting heat energy into a substance so that it changes from a solid to a liquid increases the stored phase energy of the substance.

stored chemical energy increases

Consider what happened in your exploration. As solid water (crushed ice) was heated to 0°C, the heat energy transferred to the solid water increased its thermal energy and temperature. When the solid water reached 0°C, it began to melt. The energy transferred to the water system (water + ice) while the ice melted increased the system's stored phase energy. While the system increases in stored phase energy (that is, while the ice is melting), its thermal energy and temperature does not change.

stored phase energy increases

In the same way, when liquid water changes to water vapor at 100°C, the input energy increases the stored phase energy in the water system. While the system is increasing in stored phase energy (that is, while the liquid water is boiling), its thermal energy and temperature stay the same.

Here is an energy diagram that describes the solid water as it warms to its melting point at 0°C.

Heat-Conduction Interaction

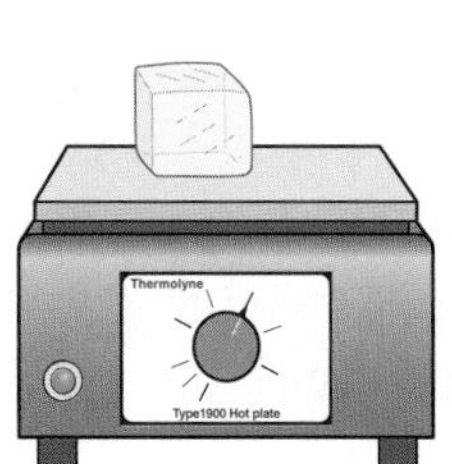

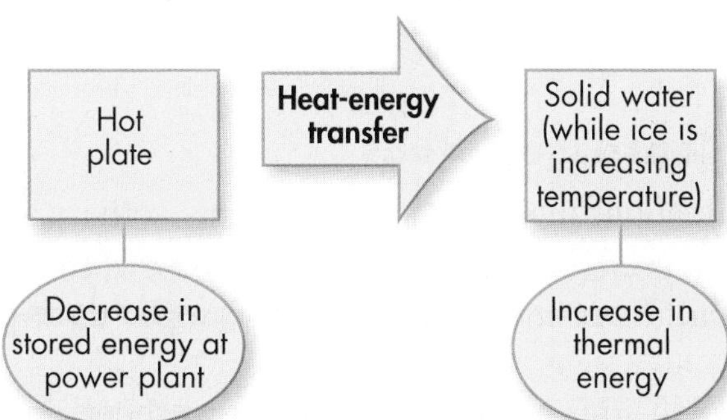

2. Complete the energy diagram that describes ice as it is melting.

Interaction

Solid water and liquid water (while ice is melting)

3. Complete the energy diagram that describes liquid water increasing in temperature.

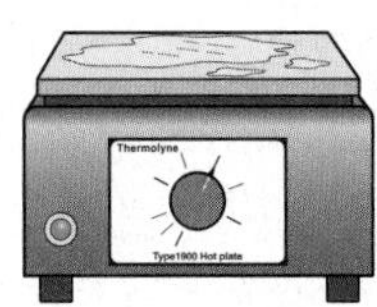

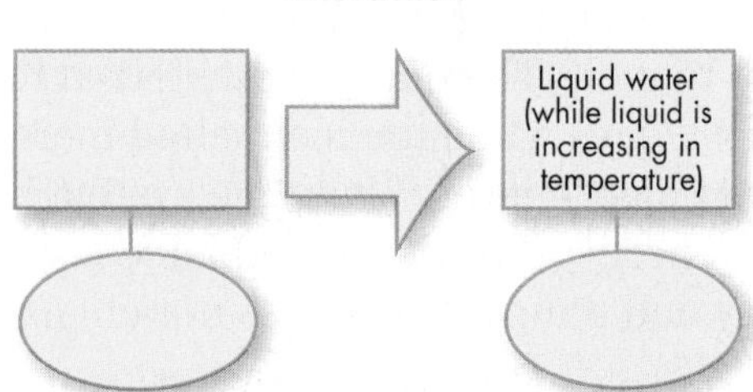

4. Complete the energy diagram that describes liquid water boiling.

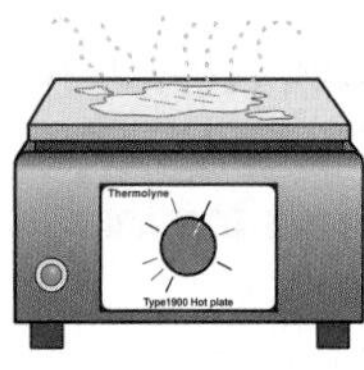

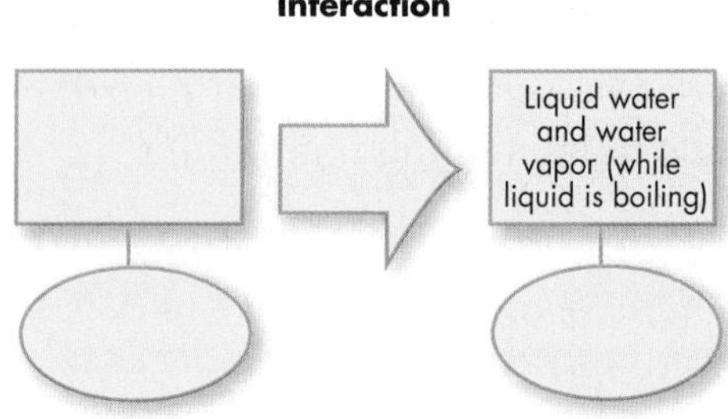

Our Consensus Ideas

Think about the key question for this activity and discuss it with your team.

What happens to energy that is input during a phase change?

1. Write your answer on your record sheet.

Participate in the class discussion about your answer.

2. Write the class consensus idea on your record sheet.

DEVELOPING OUR IDEAS

Activity 5
Conservation of Energy

Purpose

Energy is one of the most central and fundamental concepts in all of science. Therefore, it is critical to have an understanding of the behavior of energy. In this activity, you will consider two very important ideas about energy. These ideas also make up the cycle key questions.

1. Can energy be created from nothing?
2. Can energy be destroyed into nothing?

We Think

Now that you have worked through this cycle, consider again the cycle key questions. Discuss them with your group.

Write your answer to the key questions on your record sheet. Write at least four examples as evidence to support your answers.

Participate in the class discussion about your answers.

Explore Your Ideas

Law of Conservation of Energy

Can energy be created from nothing? Can energy be destroyed into nothing? These questions are difficult to answer by doing explorations in a classroom. However, they are very important questions that physicists have pondered.

By doing very careful experiments, physicists have concluded that energy is *never* created or destroyed, but only changed from one form to another and/or transferred from one object to another. This is a fundamental idea in physics that scientists use to help them understand and make new discoveries about the world around them. Physicists call this the *Law of Conservation of Energy.*

You have already encountered the idea that energy can be changed from one form to another. For example, think about what happens to the energy in an elastic mechanical interaction when a slingshot launches a ball. Stored elastic energy in the rubber band is transferred to the ball and the ball increases in motion energy. The Law of Conservation of Energy tells you that no energy is created or destroyed in the interaction between the rubber band and the ball. This means that the *exact amount* of energy decrease in the rubber band shows up as increases in energy somewhere else.

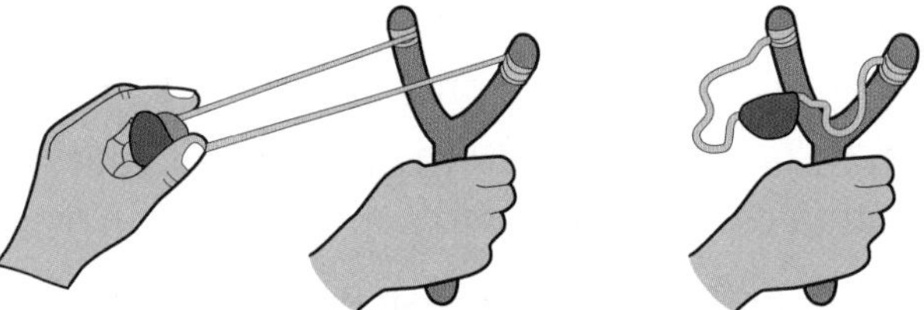

1. Suppose that before the launch 100 units of energy were transferred to the slingshot as it was stretched for the launch. After the launch, the slingshot has 0 units of stored elastic energy. According to the Law of Conservation of Energy, how much energy would you expect to be transferred away from the slingshot?

2. Suppose energy has been transferred to a book so that it has 50 units of motion energy. As it slides across a table, the book slows to a stop due to the friction interaction with the table. The friction interaction causes the book and table to gain 45 units of thermal energy. According to the Law of Conservation of Energy, how many units of motion energy were transformed into other types of energy?

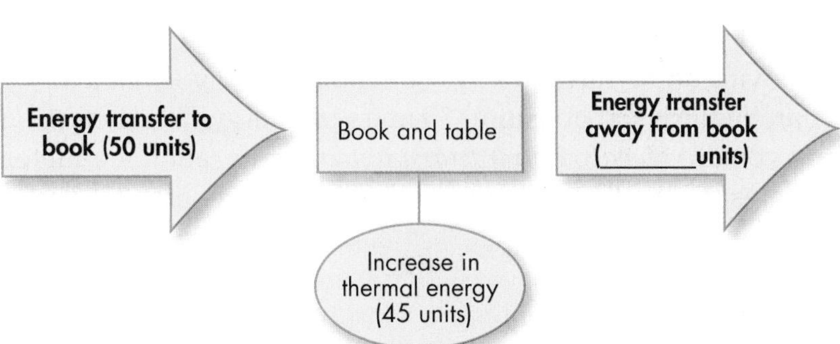

3. Consider a car that has 75,000 units of stored chemical energy in its fuel. Otis drives the car until it has 50,000 units of stored chemical energy in its fuel. According to the Law of Conservation of Energy, how much energy has been transferred away from the car?

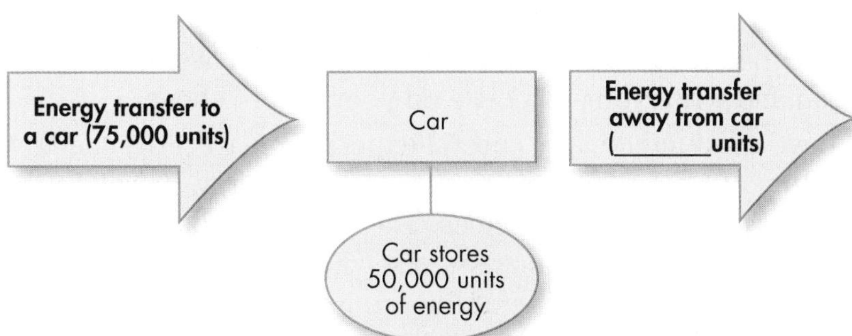

4. If you are told the amount of energy that is transferred to an object and the amount of energy that stays with the object, how can you determine the amount of energy transferred away from the object?

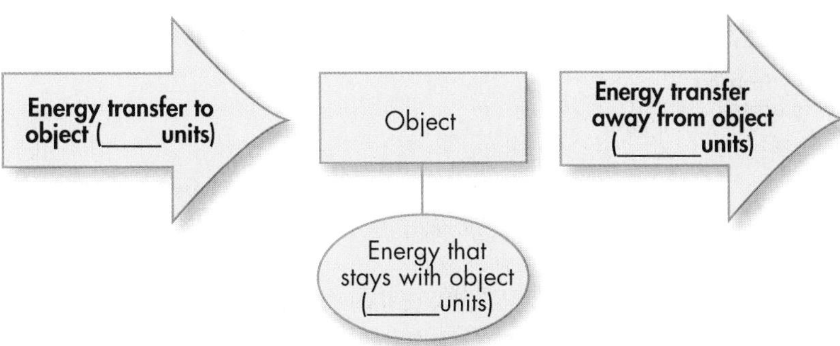

Participate in a class discussion about this section.

Unavoidable Heat-Energy Transfer

Think about moving some furniture in your room. First, you eat a snack because you are hungry and consume 75 units of energy. You then move your furniture until you are *sweaty* and *hungry again*. Your "energy-o-meter" on your belt indicates that you transferred only 70 units of energy to the furniture.

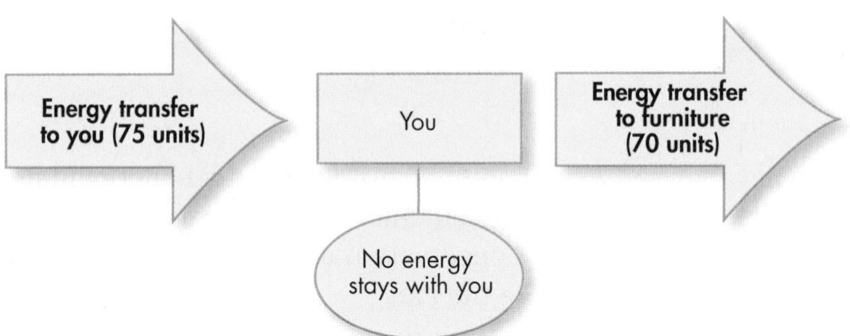

5. What happened to the other 5 units of energy? Discuss this question with your team and write your response on your record sheet.

Scientists have conducted many careful experiments to verify the Law of Conservation of Energy. For every decrease in energy, there is a corresponding increase in other places. The amount of decrease in the source always equals the sum of the increases elsewhere. Interestingly, energy in the form of heat is almost always one of the products in an interaction. In other words, *there is almost always at least a small amount of heat energy transferred away from the interacting objects.*

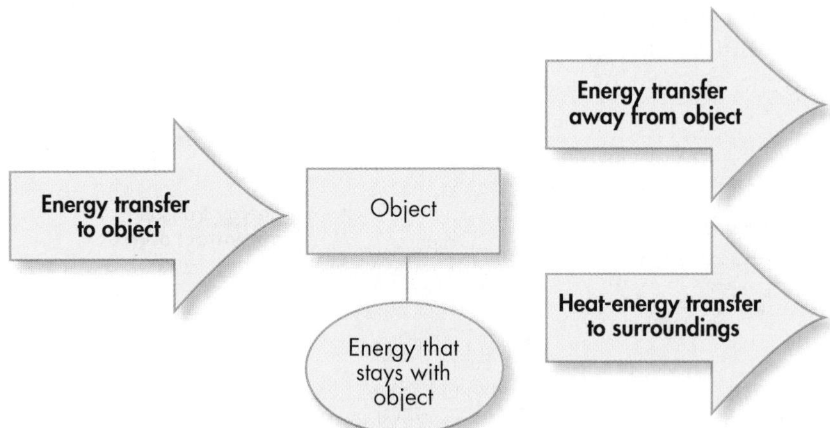

Think again about moving furniture in your room. In the process of moving the furniture, you get hot and sweaty. So, some of the energy from your snack is transferred away in a heat-energy transfer.

6. What kinds of interactions could occur between you and the surroundings to transfer heat energy away from your body?

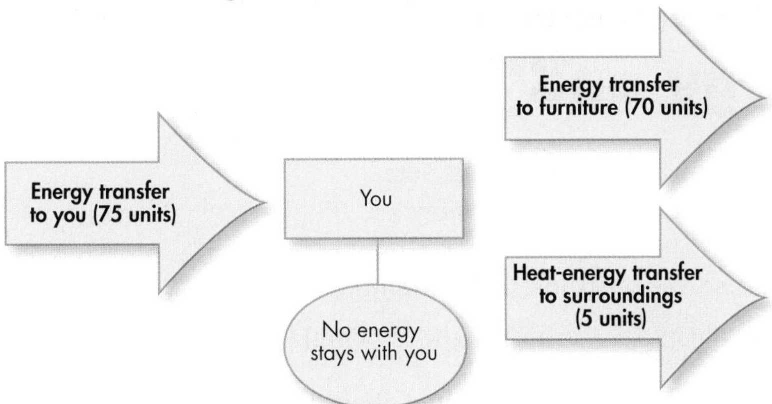

Make Sense of Your Ideas

Look at the earlier examples in this activity. In each example, think about the heat-energy transfer to the surroundings. Heat transfers are often very tiny. You might need to think of an exaggerated situation to help you recognize the heat transfer in the example.

Complete the table on your record sheet. The first row is done for you.

Table: Heat-Energy Transfers

Example	Similar Exaggerated Situation	How could heat be transferred away from the interaction in the example?
Slingshot launches the ball	What happens to the temperature of a wire as you bend it back and forth, or to the temperature of a rubber band as you stretch and release it? *The wire or rubber band gets warmer.*	*The slingshot's rubber band gets warmer as it returns to its natural shape. Heat can be transferred away from the rubber band in a heat-conduction or infrared-radiation interaction.*
Book slides on table	What happens to the temperature of your hands as you slide them back and forth many times?	
You eat a snack then throw a ball	What happens to the temperature of a pitcher's arm after she throws a ball many times?	
Car drives	What do you observe about the temperature of a recently driven car?	

Now complete a similar table using your examples from the We Think section of this activity.

Participate in a class discussion of this section.

You will answer the key questions in the next activity.

PUTTING IT ALL TOGETHER

Activity 6
Wrapping up Energy and Interactions

Comparing Consensus Ideas

Recall the key questions for this cycle:

1. Can energy be created from nothing?
2. Can energy be destroyed into nothing?

In this cycle's activities, you learned about interactions and energy. In the previous activity, you explored scientists' ideas about the Law of Conservation of Energy.

Your teacher will review with you all of the ideas that the class developed during this cycle to help you answer the cycle questions. Be prepared to contribute to this discussion.

Your teacher will distribute copies of *Scientists' Consensus Ideas: Interactions and Energy.* Read the scientists' ideas and compare them with the ideas your class developed during this cycle.

Write the evidence from the activities in this cycle where it is requested on the *Scientists' Consensus Ideas* form.

Your teacher will lead a whole-class discussion.

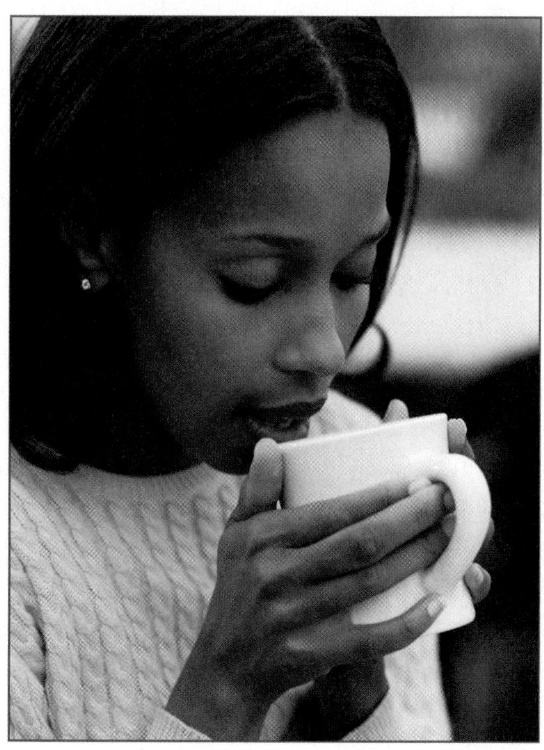

The Conservation of Mass-Energy Law

Understanding what things are conserved is a big help to scientists as they try to describe how the universe works. In this unit, you have seen evidence for mass conservation and read about energy conservation. Before modern physics, scientists believed that there were two separate conservation laws: the *conservation of mass* and the *conservation of energy*.

The *Law of Conservation of Mass* stated by Antoine Lavoisier in 1785 is:

Mass is neither created nor destroyed.

In 1842, Julius Robert Mayer stated the *Law of Conservation of Energy*:

Energy is neither created nor destroyed.

However, in the early 1900s, Albert Einstein proposed the famous formula:

$$E = mc^2$$

The E stands for energy, m for mass, and c for the speed of light.

This formula indicates that mass and energy are two different views of the same thing. The mass of a body is actually a measure of its energy content. Scientists now believe the *Law of Conservation of Mass-Energy*:

Mass can be created or destroyed, but when this happens, an equal amount of energy vanishes or comes into being and vice versa.

For example, as the Sun bathes you in light and heat energy, some of its mass is destroyed. Don't worry though about the Sun burning away anytime soon. It is just a tiny bit of mass that represents a huge amount of energy. For example a *trillionth of a gram* of mass represents roughly the amount of energy needed for one large fireworks explosion. Therefore, the Sun should be able to burn brightly for a few more billion years! Because of the tiny amounts of mass involved, only recently have scientists been able to measure the mass change due to conversion to energy. Many experiments have now confirmed the *Law of Conservation of Mass-Energy*.

The old conservation laws are generally accurate enough for everyday life, and they still help you analyze and explain many of the events around you. In Unit 7, you will find that the *Law of Conservation of Mass* is still a very useful idea in chemistry.

IDEA POWER

Activity 7
Analyzing and Explaining Energy Interactions

Analyze, Explain, and Evaluate

Use *How To Write an Analysis and Explanation* to help you in the following problems. In your analyses, you should draw energy diagrams. Check that your analyses and explanations are *good* using the criteria in *How To Evaluate an Analysis and Explanation.*

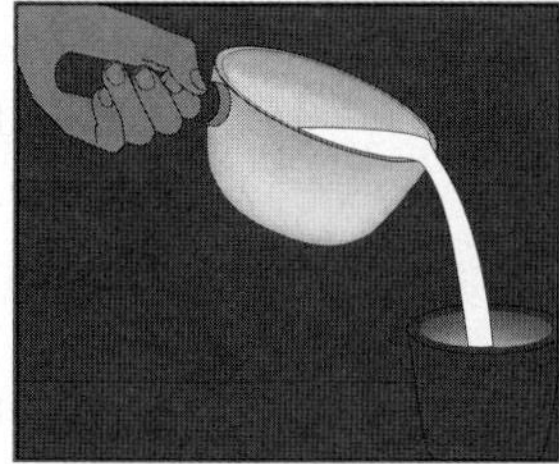

1. "Hey, Shannon!" Gareth said. "Look at these night-vision goggles! When you wear them, you can detect infrared radiation that your eyes can't detect by themselves. Put them on and watch me pour this hot chocolate." Using the goggles, Shannon could see a stream of hot chocolate glowing brightly as Gareth poured it.

Question: Why is the hot chocolate glowing?

2. Shannon continued to watch Gareth pour the hot chocolate in the cup. Suddenly, she called, "Gareth, you should see this. At first, all I could see was the stream of hot chocolate, but now I can see the cup too. It's glowing!"

Question: Why does the cup begin to glow? How does Shannon see it?

3. After he poured the hot chocolate into the cup, Gareth picked up the cup. Through the night goggles, Shannon saw a bright image on the table on the spot where the cup sat, although the rest of the table appeared dark.

Question: How did the spot on the table get warm and why does the spot on the table appear bright in the night goggles?

4. Shannon and Gareth decided to make another pot of hot chocolate. When the hot chocolate was ready, they poured it into two cups. They then let the drink cool for a few moments before they drank it.

Question: As the hot chocolate warmed on the stove, was energy being created? As the hot chocolate cooled, was energy being destroyed?

IDEA POWER

Activity 8
Efficiency

Machines play an important role in modern life. Homes often have many electrical devices, large and small, to make life easier. In a kitchen, you might find a stove, a refrigerator, a dishwasher, and smaller electric devices such as a toaster, waffle iron, coffee maker, can opener, blender, or mixer. Homes also contain lamps, TVs, radios, electric clocks, hair dryers, fans, electric drills. Don't forget the other machines such as cars, computers, photocopy machines and more that affect your daily life.

Engineers try to design these machines to be *energy efficient.* In this activity, you will apply what you learned in this cycle to help you understand the energy efficiency of machines.

Energy Efficiency

Think about a car. Often, about 70% of the energy in the gasoline is converted to wasted heat energy. *The larger the amount of wasted energy, the less efficient a machine is.* The main job of the car's cooling system is to keep the engine from overheating by transferring the wasted heat energy from the car to the surroundings. Car engines run best at about 93°C (200°F), partly so the oil used to lubricate the engine stays thinner and the engine parts move freely, minimizing friction interactions between the moving parts. So the car's cooling system maintains a balancing act between keeping the engine from melting but also not over cooling the engine.

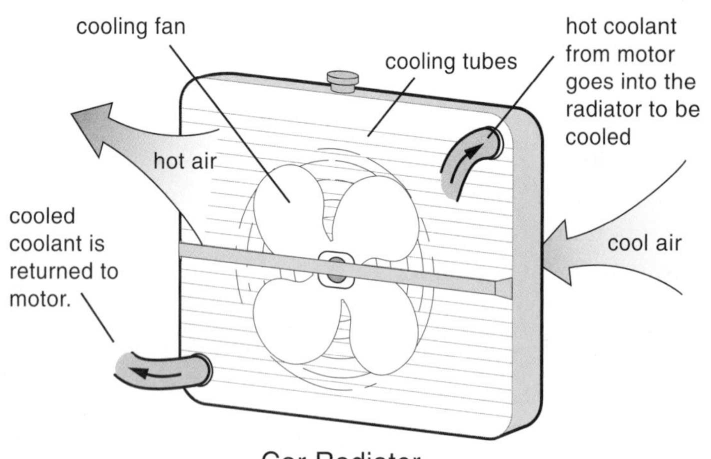

Car Radiator

To keep the engine area at an appropriate temperature, cool coolant fluid travels in tubes through the engine area. Heat energy from the engine area is transferred to the coolant fluid, making the fluid's thermal energy increase. The warmed fluid then travels through thin cooling tubes in the radiator. A radiator fan blows cool air over the tubes. Heat energy from the fluid is transferred to the air, and the cooled coolant travels back through the engine area again to start the process all over!

The table shows the efficiencies for some common items.

Table: Approximate Efficiencies for Some Energy Transformations

System	Energy Input	Energy Output	Efficiency (%)
Incandescent bulb	Electrical	Light and heat	5
Steam locomotive	Chemical	Mechanical and heat	8
Fluorescent lamp	Electrical	Light and heat	20
Solar cell	Stores light energy	Electrical and heat	25
Automobile engine	Chemical	Mechanical and heat	25
Coal-fired power plant	Chemical	Electrical and heat	30
Nuclear power plant	Nuclear	Electrical and heat	30
Steam turbine	Heat	Mechanical and heat	47
Battery	Stores chemical energy	Electrical and heat	60
Small electric motor	Electrical	Mechanical and heat	63
Dry cell battery	Stores chemical energy	Electrical and heat	90
Large electric motor	Electrical	Mechanical and heat	92
Electric generator	Mechanical	Electrical and heat	99

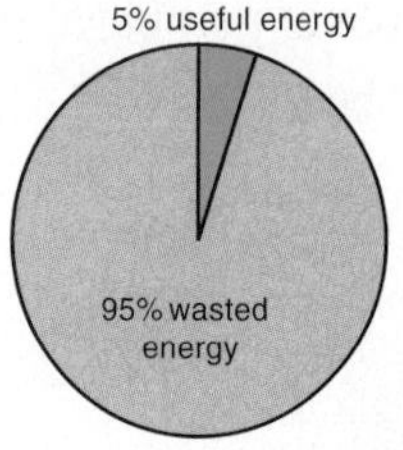

An incandescent light bulb has only a 5% efficiency. Most of the electrical energy input is wasted!

Here is an energy diagram describing the energy inputs, outputs, and changes for the incandescent light bulb:

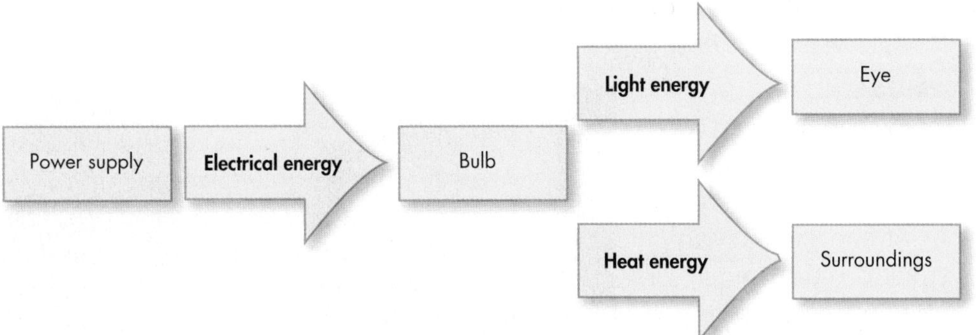

In this activity, you will examine two electrical devices that have been opened so you can see how they work. For each device, you will complete an energy diagram and comment on features of the design that you think help maximize the useful energy output.

Your team will need:

- 2 dismantled electrical devices

Part A: Device 1

1. What is the name of your device?

2. What is the purpose of the device?

3. Complete an energy diagram for your device.

Look at the inside of the device. Identify some of the parts. Sometimes, there are features that help a device work more efficiently, such as a fan to keep a motor cool.

4. What do you think are some features of the device that maximize its efficiency? How do the features maximize efficiency? If you cannot identify some features that maximize the efficiency, suggest one or two features that could be added to maximize the efficiency.

5. Do you think your device is very efficient? Give your reasons.

Part B: Device 2

6–10. Repeat Questions 1 – 5 for a second electrical device.

Participate in the class discussion about your devices.

11. Write down which item in the class you think is the most efficient and which is the least efficient. Include your reasons.

LEARNING ABOUT OTHER IDEAS

Activity 9
Reflection of Light

Purpose

In this cycle and in previous cycles you have learned about different types of energy that can be transferred from sources to receivers during interactions. For example, back in Unit 2 you learned that during a light interaction, light energy is transferred from a light source to an object receiver. In the next three activities you will learn different ways that light can interact with objects. You will start by looking at what happens when light strikes a surface. You are probably familiar with the fact that a mirror reflects light. But how does the light reflect? Also, what happens when light strikes a piece of paper? Does light also reflect from the paper, or does the light just illuminate the paper, making it brighter? The key question for this activity is:

How does light interact with surfaces?

Record the key question for the activity on your record sheet.

Learning the Ideas

How Light Travels

Before investigating how light interacts with surfaces you first want to describe how light travels. When a light bulb is turned on in front of a screen, the screen is illuminated. That means that light travels outwards from the bulb in all directions. The light that reaches different parts of the screen illuminates it.

Three students drew diagrams showing how light from the bulb illuminates the screen.

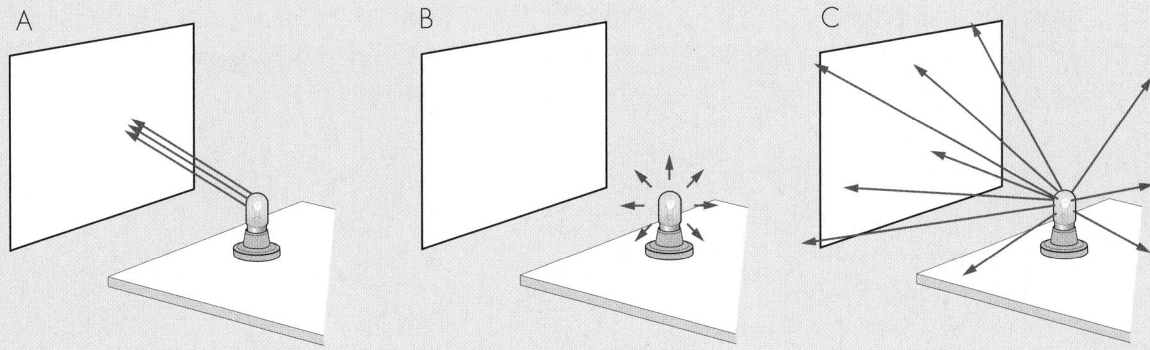

1. Which one of the above pictures makes the most sense to you? Why don't the others make as much sense? Explain briefly.

Your teacher will have a brief discussion about your choices.

Have you had to squint when looking directly at a bright light? This is evidence that for you to see the bulb, light traveling from the bulb had to enter your eye. The light ray diagram to the right represents many people looking at a light bulb. To suggest that each person sees the bulb, you can draw light rays from the bulb going to their eyes.

Scientists represent light rays by drawing straight lines with arrows. The direction of the arrow represents the direction the light travels. In the picture, the light travels outward from the bulb to the eyes of the four students. It would not be appropriate to draw rays the other way, from the eyes to the bulb. Although such a backward arrow might suggest where each person is looking, it does not show the direction that the light travels. Light rays are always drawn to show the direction the light travels. Scientists also draw straight lines because they assume light travels in straight lines. One piece of evidence that supports this idea is the formation of sharp shadows.

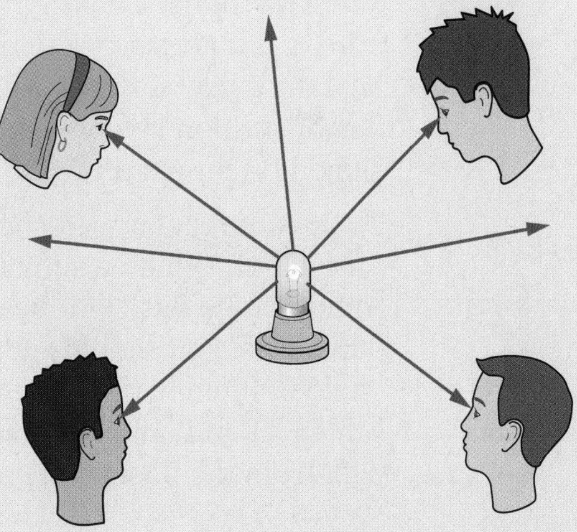

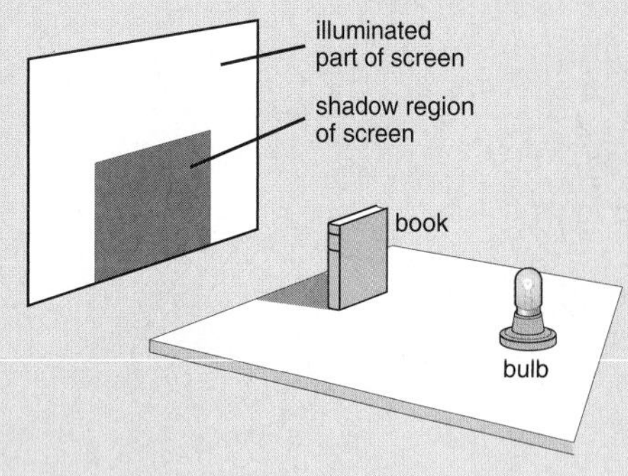

Look at the picture of a bulb, book, and screen. When the book is placed between the bulb and screen, a shadow of the book appears on the screen.

Three students sketched how they thought the shadow is formed.

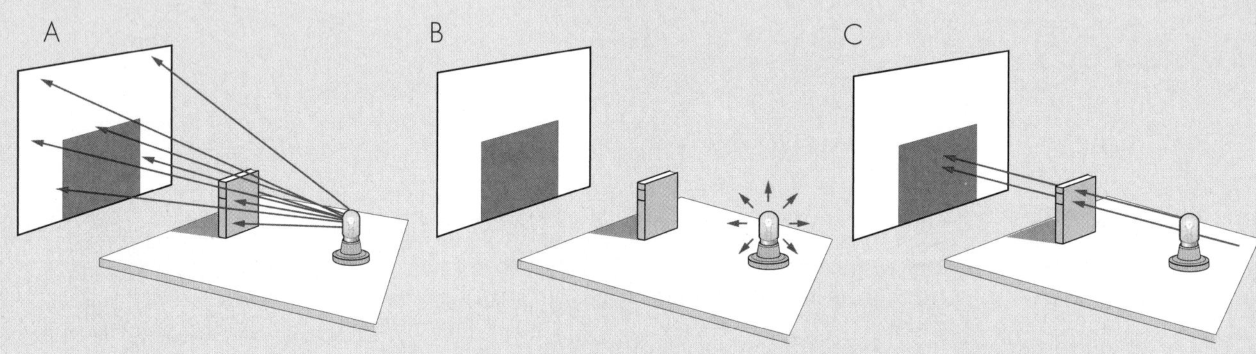

2. Which one of the above pictures makes the most sense to you? Why don't the others make as much sense? Explain briefly.

Your teacher will have a brief discussion about your choices.

How Light Interacts with a Shiny Surface

You have all been dazzled by light reflecting from a shiny surface like a mirror. It makes sense then, that light reflects when it strikes a mirror. But how does light actually reflect from a mirror? Your teacher will show a demonstration involving a flashlight, mirror, and a screen. The room lights should be turned off for this demonstration.

First, your teacher will shine a flashlight beam on a screen. Although the flashlight beam spreads out a little, you should see a bright circle of illumination. The rest of the screen should not be illuminated by light from the flashlight.

Next, your teacher will aim the beam at a mirror and reflect the light onto the screen. Observe the pattern of illumination on the screen. Then your teacher will tilt the mirror. Observe how the illumination changes on the screen, if at all.

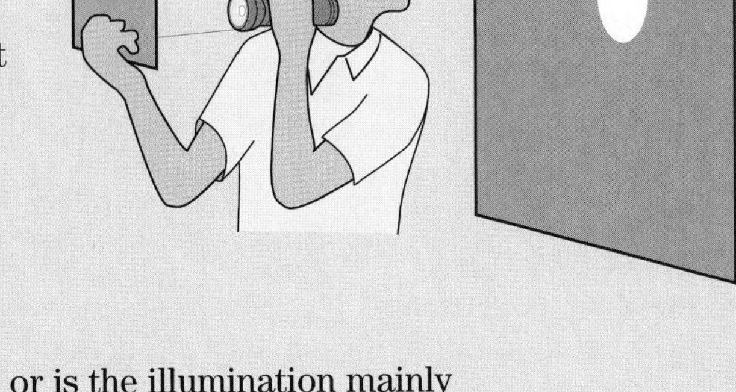

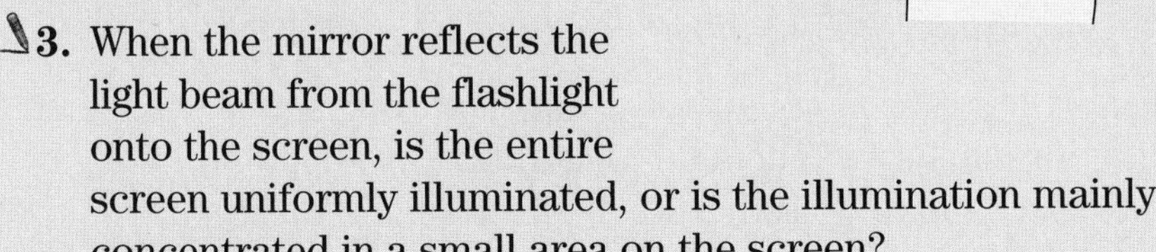

3. When the mirror reflects the light beam from the flashlight onto the screen, is the entire screen uniformly illuminated, or is the illumination mainly concentrated in a small area on the screen?

4. When the mirror was tilted, what happened to the illumination on the screen? Does this observation suggest that when light reflects from a mirror, the reflected light goes in all directions away from the mirror or mainly in a particular direction away from the mirror?

Your teacher will have a brief discussion about the answers to these questions.

In order to know how light reflects from a mirror, you need to know if there is a relationship between the direction that the light beam strikes the mirror and the direction that the light beam reflects from the mirror. To gather data, your teacher will use a computer simulator. Below are four snapshots from the simulator. Each picture shows a projector shining a light beam so it strikes a mirror at a different angle (the mirror surface is black in these pictures).

5. What seems to be the relationship between the angle at which the light strikes the mirror and the angle at which the light reflects from the mirror?

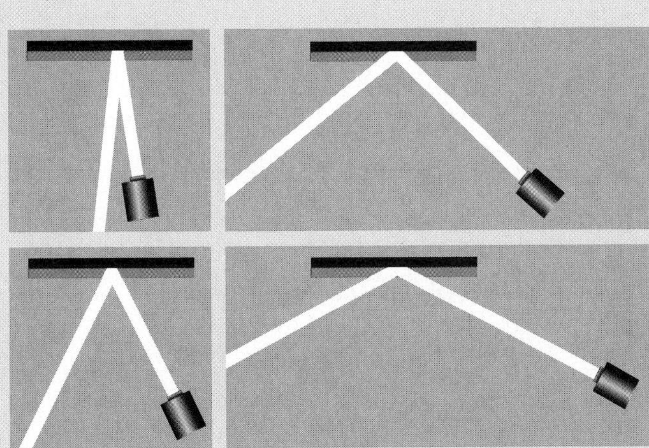

Look at the picture of a boy looking into a mirror. He sees the image of the light bulb.

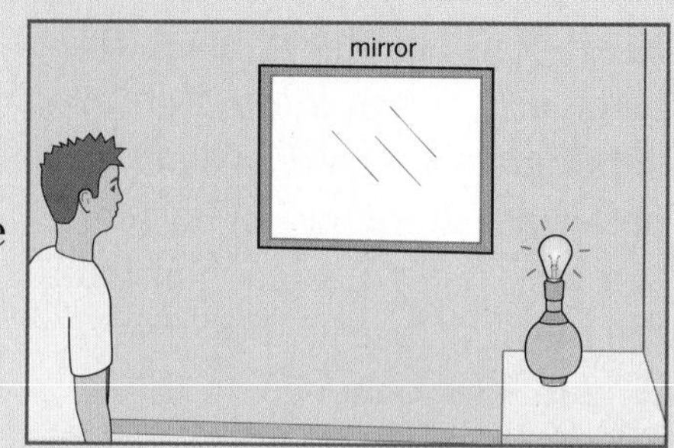

6. On your record sheet, draw light rays to show how he sees the mirror image of the bulb.

Your teacher will have a brief discussion to go over answers to these questions.

How Light Interacts with a Non-Shiny White Surface

What happens when light strikes a non-shiny surface, like a sheet of white paper? To find out, your teacher will do a demonstration by shining the flashlight on a white piece of paper. If the paper reflects light, then you would expect to see some of that reflected light on the screen. If it does not reflect light, then the screen should not be illuminated. The room needs to be darkened for this demonstration.

7. Does light seem to reflect from the piece of white paper? If so, does it reflect in one particular direction, or does it seem to reflect in all directions away from the paper?

Imagine looking at a white surface on a bright, sunny day. You would probably squint because the surface is so bright. That is evidence that when looking at a white surface that is not a mirror, light reflects from the surface and enters your eye.

The picture shows a boy looking at a piece of white paper. A light source is on the ceiling of the room.

8. On your record sheet, draw a light ray diagram to show how the boy can see the paper.

Your teacher will have a brief discussion about the answers to these questions.

How Light Interacts with a Very Dark Surface

So far you have seen how light reflects from a mirror and from a white non-shiny surface, like a sheet of paper. What happens if the non-shiny surface is very dark in color? Your teacher will show you a demonstration.

9. Does light seem to reflect from the piece of black paper? If so, does it reflect in one particular direction, or does it seem to reflect in all directions away from the paper?
10. Why do you think the light behaves differently when striking a dark-colored piece of paper versus striking a white piece of paper?

Have you noticed any difference wearing light-colored clothing or dark-colored clothing on a hot, sunny day? Most people feel the light-colored clothing is more comfortable. The dark-colored clothing usually make them feel much warmer, sometimes uncomfortably warm. Why does this happen? The dark-colored clothing absorbs much of the light, whereas the light-colored clothing reflects it. Light is a form of energy that can be transferred from a source to a receiver. If the receiver is dark colored, like a black piece of paper, or a black T-shirt, the light energy is absorbed. The thermal energy of the receiver increases. This causes the receiver, like the black paper or T-shirt, to become warmer.

What We Have Learned

Recall the key question for this activity:

How does light interact with surfaces?

Your teacher will lead a class discussion to answer the key question.

Write the answer to the key question on your record sheet.

LEARNING ABOUT OTHER IDEAS

Activity 10
Refraction of Light

Purpose

In the last activity you learned that light travels in straight lines and that it can reflect from different surfaces. Light can also travel through materials. Such materials are called **transparent**. Does light continue to travel in straight lines when it passes through a transparent material (like water or glass)? Have you ever observed that a pencil appears bent when it is placed in a cup of water? Have you noticed that you can focus a camera on both near and distant objects? You can also focus your own eyes to see both near and distant objects clearly. The reason the pencil appears bent, and the reason why you can focus both a camera and your eyes, has to do with the way light behaves when traveling through transparent materials. The key question for this activity is:

Science Words

transparent: a material through which light can travel

What happens when light passes through transparent objects?

Record the activity key question on your record sheet.

Learning the Ideas

Light Traveling between Air and a Clear Material

What happens when light strikes a clear, transparent material like glass? To investigate this, your teacher will use a computer simulator.

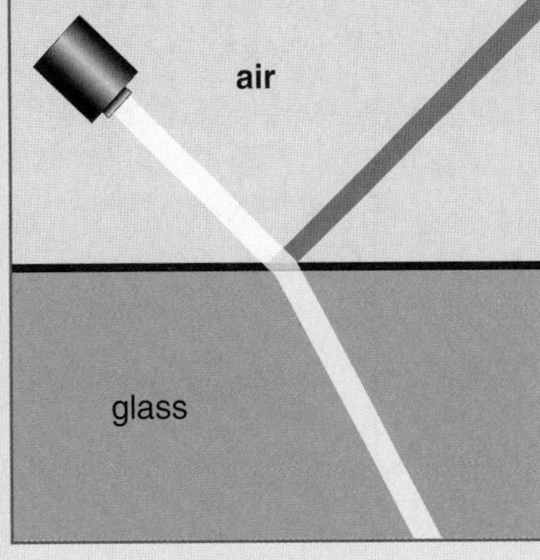

In the simulator, a beam projector is aimed at a solid piece of transparent glass. Notice, when the light strikes the glass surface, some of the light is reflected from the shiny surface (the reflected light is shown in gray) and some of the light passes into the glass.

1. Consider the light that travels from the air into the glass. Does that light travel in a straight line, or does it change direction?

What happens when light travels from the transparent glass out into the air? Again, your teacher will use the computer simulator to demonstrate what happens.

2. Does the light travel in a straight line when going from glass into the air, or does it change direction?

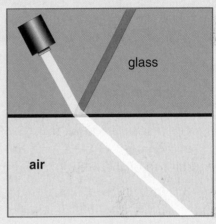

The change in direction of light when traveling between air and a transparent material (like glass or water) is called **refraction**. We say that the light refracts when going from the air into the glass or from the glass into the air.

Converging Lenses

Because light refracts when traveling between glass and air, a piece of glass can be shaped to make the light change direction in certain ways. A good example of this is the **converging lens**, which is shaped so its middle is thicker than its edges. With this type of lens you can form images of objects.

Your teacher will show you a demonstration of a light bulb, converging lens, and screen. You should be able to observe a sharp upside-down image of the bulb on the screen.

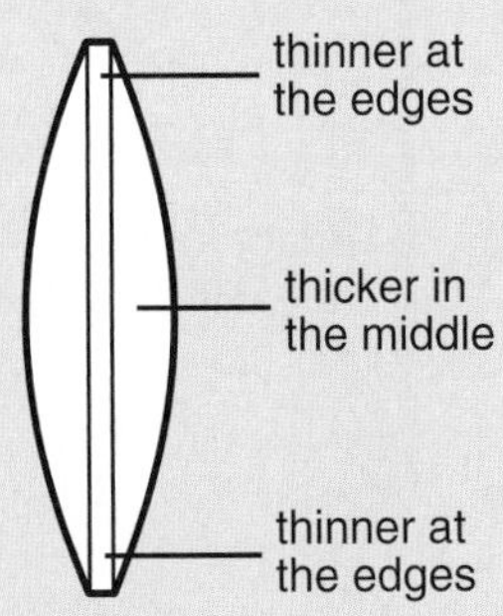

How does the lens form the image that can be observed on the screen? Your teacher will use the computer simulator to help you understand how this happens.

Science Words

refraction: the change in direction (bending) of light when it passes at an angle from one material to a different material

converging lens: a lens shaped so that its middle is thicker than its edges. This type of lens allows you to form images of objects.

First consider how the image of the top of the light bulb is formed at the bottom of the screen. The diagram below comes from the simulator. Light from the top of the bulb travels outwards in all directions. The light refracts when passing through the lens. (The light actually refracts twice, once when going from air into the glass, and again when going from the glass back into the air. However, the simulator simplifies the diagram by showing refraction as if it only occurred in the middle of the lens.) Because of refraction, all the light that spread out from the top of the bulb and passed through the lens is converged (brought together) by the lens at a point on the screen. That point is the image of the top of the light bulb. As the diagram shows, light leaving the top of the bulb forms an image on the lower part of the screen.

3. On your record sheet add at least two more light rays to this diagram to show how light that spreads out from the top of the bulb is refracted by the lens and converges to the image point on the lower part of the screen.

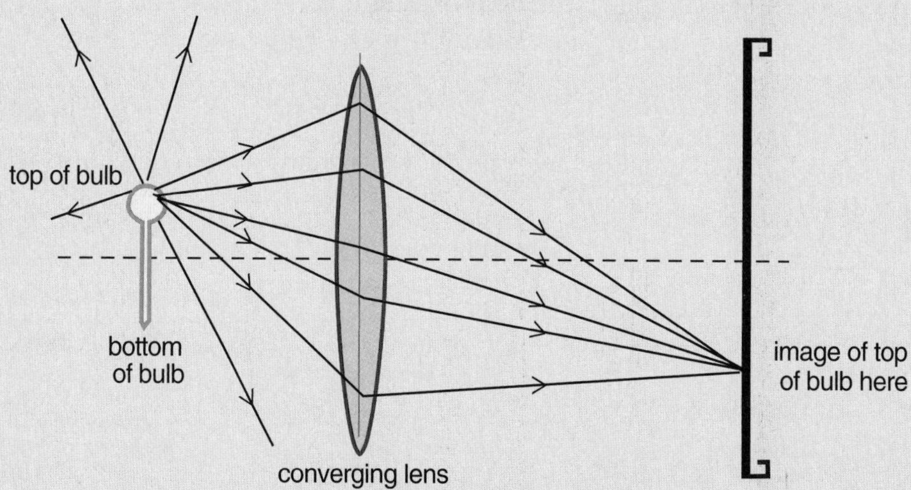

What about light spreading out from the bottom part of the bulb? The next diagram shows one light ray leaving the bottom of the bulb, passing through the lens, being refracted, and then going to the screen. The place where that light ray hits the screen is where the image of the bottom of the bulb is located.

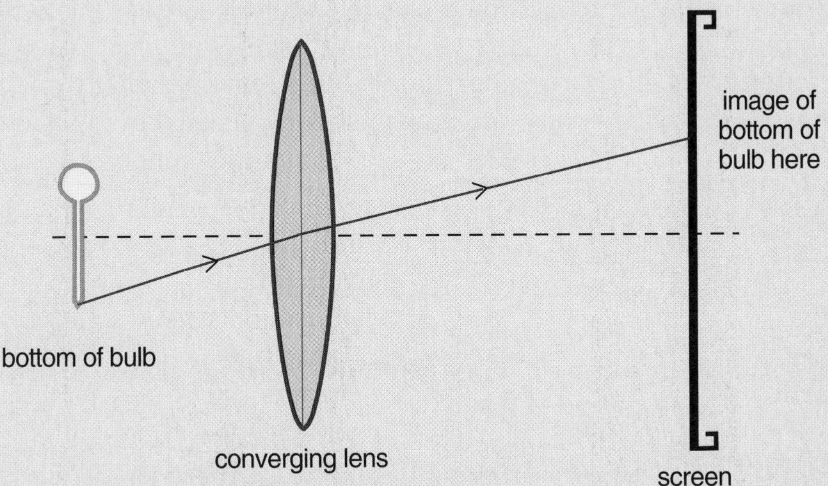

4. On your record sheet add at least four more light rays to this diagram to show how light that spreads out from the bottom of the bulb is refracted by the lens and converges to the image point on the upper part of the screen.

Your teacher will have a brief discussion about your diagrams.

In the above diagrams only light spreading out from the top and bottom of the bulb was shown. In reality, light that spreads out from every other point on the bulb is converged by the lens to another point on the screen. In this way, the entire upside-down image of the bulb is formed.

The Camera and the Eye

Returning to the bulb, lens, and screen demonstration, your teacher will show you that the bulb needs to be a certain distance from the lens in order for the lens to converge the light to a sharp image on the screen. If the bulb is moved either closer to the lens, or farther from it, the image on the screen becomes blurry.

When the image becomes blurry, there are two ways of fixing the situation so the image on the screen becomes sharp again. One way is to move the lens. The other way is to change the thickness of the lens. A camera uses the former method when "focusing" on an object. Your eye uses the latter method when "focusing" on an object.

Look at the simple diagram of a film camera. When you take a picture, a shutter opens for a very short period of time, allowing light from the object to pass through an aperture opening. This light is converged by the camera lens to form a sharp image on the film. (If the camera is digital, the image is formed on a piece of magnetic media.)

If the object is farther away from the camera, or closer to it, the lens will not form a sharp image on the screen. In that case, you "focus" the camera by moving the lens either closer to the film or farther from it. Your instructor will demonstrate how this works, using both the actual setup and the computer simulator.

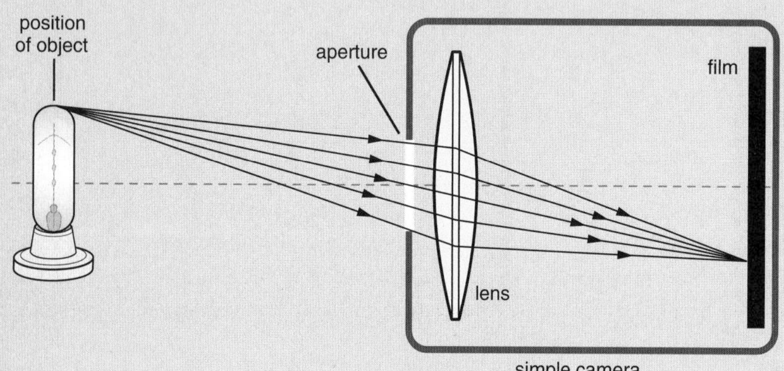

Now look at the simple diagram of the human eye below. Only a few important parts are included. The inside of the eyeball is spherical in shape and consists of transparent, jelly-type material. Imagine you are "looking" at an object, like a bulb. Light spreading out from the top of the bulb enters the outer transparent surface of the eye, called the cornea. The cornea refracts the light. The light passes through a small circular opening, called the pupil. It then goes through the lens, which refracts the light farther and converges it to a point on the retina. The retina consists of special cells, called rods and cones. When light strikes these special cells, they absorb the light and send electrical signals on to the part of the brain responsible for vision. You then "see" the bulb.

If the object you are looking at moves closer to the eye, or farther from it, the lens of the eye automatically changes its shape so the light converges the proper amount to continue to form a sharp image on the retina. The images shown are taken from the computer simulator. They show that for an object at different distances from a lens, the shape of the lens needs to change to converge light to the same screen. The lens needs to become thinner in the middle

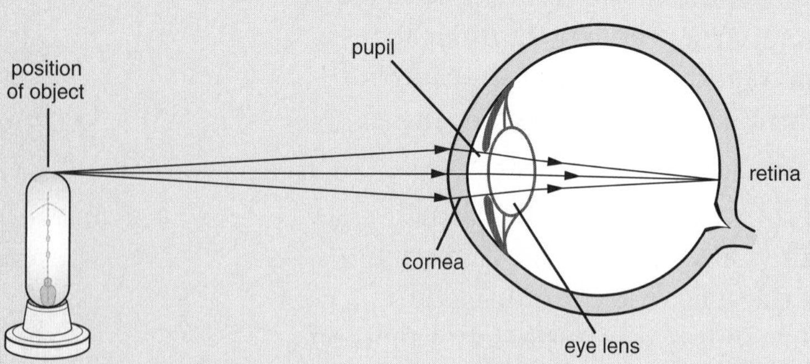

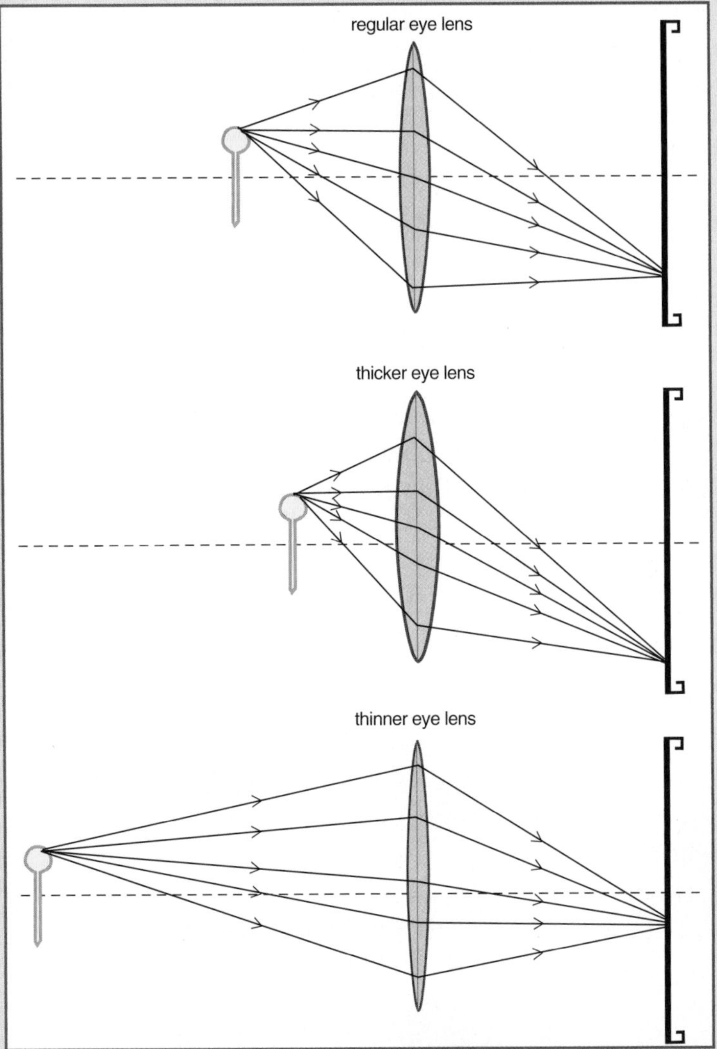

when looking at distant objects, and thicker in the middle when looking at nearby objects.

For some people, the lens of their eye cannot become thin enough to view distant objects. They are nearsighted (they can only see nearby) and need corrective lenses to help them see far. For others, their lenses cannot become thick enough to see nearby objects. They are farsighted (they can see far) and need corrective lenses to help them see close-by.

What We Have Learned

Recall the key question for this activity:

What happens when light passes through transparent objects?

Your teacher will lead a class discussion to answer the key question.

LEARNING ABOUT OTHER IDEAS

Activity 11
Color

Purpose

In the previous two activities you learned about reflection and refraction of light. You discussed reflection from mirrors, and from white and black paper. You also learned what happens when light passes through colorless, transparent materials, like glass or water. Most objects around you, however, are colored. In this activity you will learn what happens when light passes through a transparent colored piece of glass or plastic, and what happens when light strikes an **opaque** colored object. (An opaque object is one that light cannot pass through.) The key question for this activity is:

Science Words

opaque: a material through which light cannot travel

How does light interact with colored objects?

Record the activity key question on your record sheet.

Learning the Ideas

Spectrum of White Light

In Activity 3 of this cycle you learned that white light consists of a spectrum of many colors. Your teacher will show you a demonstration. He will project a narrow strip of white light onto the screen. Then he will place a special piece of plastic, called a diffraction grating, in front of the light source. The diffraction grating breaks up the white light into its component colors. What you observe on the screen is called the color spectrum of white light. Notice the different colors that are present. To make things simple, you can label the major color bands: **R**ed, **O**range, **Y**ellow, **G**reen, **B**lue, and **V**iolet. These bands are not of equal width. For example, the red band is about the same width as the green band, but each is much wider than the yellow band. The diagram shows an approximate sketch of the different colored bands. The colors of the spectrum are sometimes simply referred to as **ROYGBV**.

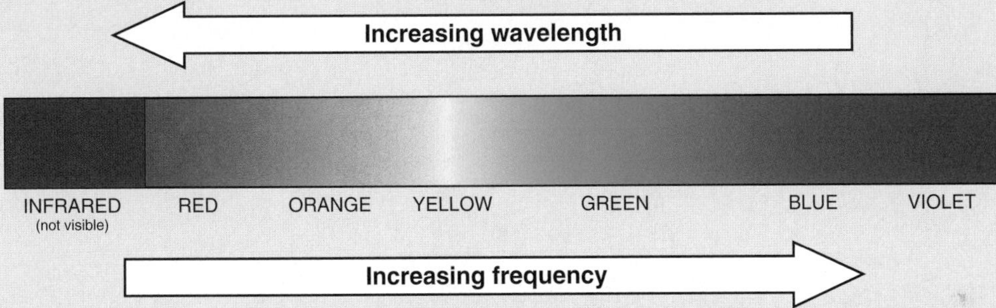

Scientists find it useful to think of light as a form of wave motion, that light travels as waves. You studied wave motion in Unit 2 Cycle 1 and learned that one of the characteristics of a wave is its wavelength. The different colors of light correspond to different wavelengths, with red being the longest and violet being the shortest. Note that as the wavelength of a light wave increases, its frequency decreases.

Transparent Colored Objects

Your teacher will hold a piece of red plastic in front of a flashlight, and will project a red circle of illumination on the screen.

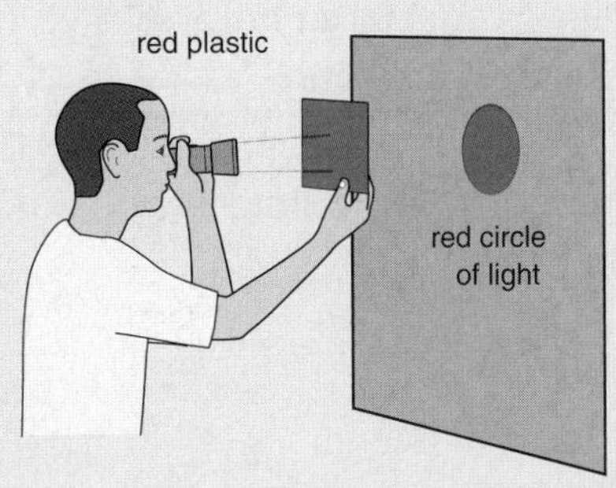

Consider the following conversation among three students, who are trying to understand what the red plastic is doing to the light.

1. Which of these three students' ideas makes the most sense to you at this time? Why do you think so?

The plastic is adding its color to the white light, making it red.
No, the red plastic is taking away some color from the white light, leaving it red.
I think that you are both right. The plastic is adding color to the light and also taking something away.
Carlos
Rebecca
Xuan

To help you decide which response makes the most sense, your teacher will do a demonstration. He will again use the diffraction grating to project a white-light spectrum on the screen. Then he will place the red plastic over the light source.

2. Is the entire spectrum colored red, or are some components of the spectrum now missing?
3. Does this observation suggest that the colored plastic adds color to the spectrum, or does it suggest the colored plastic removes some colors from the spectrum?

Look at the picture of a student looking at a white-light source. When a piece of red plastic is placed in front of the source, only red light travels to the student's eye. He will say that the source "appears red."

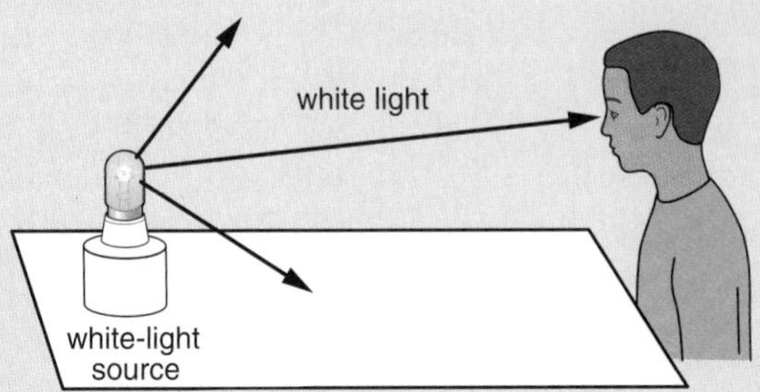

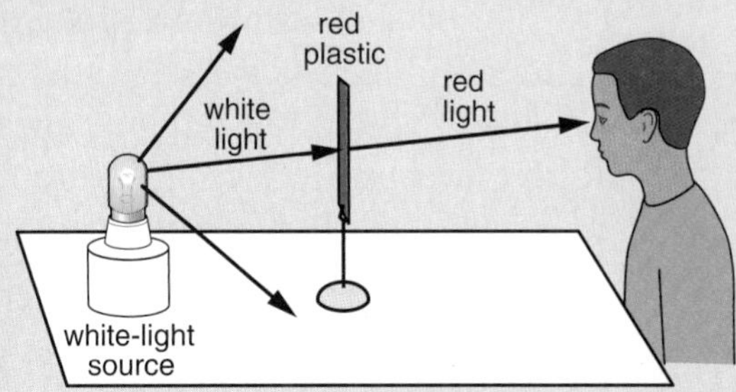

4. Suppose the student places green plastic in front of the bulb. Draw a diagram, similar to the one above to the right, showing how the student sees the bulb as green.

Previously you learned that light is a form of energy. In the Reflection of Light Activity, you learned that when light strikes a black surface, very little light is reflected. Instead, most of it is absorbed. This increases the thermal energy of the black object. The black object becomes warmer. A similar thing happens when white light passes through colored plastic. In that case, however, some of the energy is absorbed, and the rest is transmitted through the plastic (and reaches your eye). The part of the light energy that is absorbed causes the thermal energy of the plastic to increase. This makes the plastic warmer.

Opaque Colored Objects

In the Reflection of Light Activity, your teacher showed you a demonstration of how white light reflects from white paper. Suppose your teacher were to use a red piece of paper instead. What do you think will happen?

After the room is darkened, your teacher will show you the demonstration. Pay attention to the illumination on the screen caused by light reflecting from the red paper.

5. Does the illumination on the screen appear to be colored? If so, what color?

When white light strikes an opaque red object, only red light is reflected. Just as in the case with the red plastic, the other colors in the light are absorbed. The red object increases in thermal energy and becomes warmer. Sometimes the increase in thermal energy is very small and you don't even notice that the object warms up. Other colored objects behave in a similar way when light strikes them. For example, when white light strikes an opaque blue object, blue light is reflected and the other colors are absorbed.

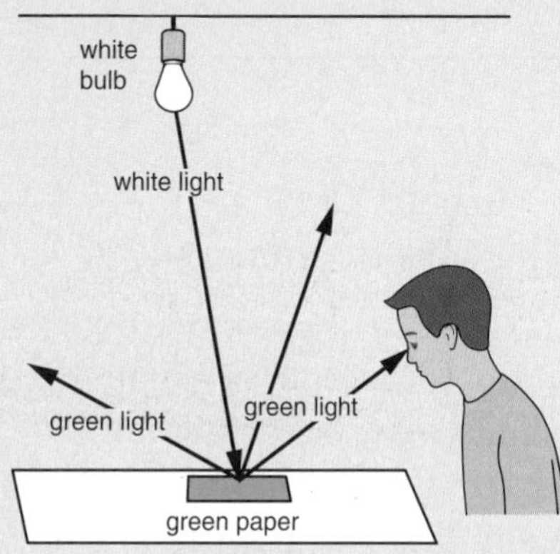

Look at the diagram showing how a boy sees a green piece of paper. White light from the ceiling lamp travels down to the green paper. When it strikes the paper, the green part is reflected in all directions. The other components are absorbed in the paper. The green light goes into his eye and he sees the paper as green.

Now consider the following problem. Humberto is sitting on his bed looking at a red apple. The lamp on his night table, which is the only light source in the room, is directly behind him. It gives off white light. The ceiling and walls are painted a non-shiny white color.

6. On a diagram on your record sheet, draw light rays to show how Humberto can see the red apple.

What We Have Learned

Recall the key question for this activity:

How does light interact with colored objects?

Your teacher will lead a class discussion to answer the key question.

Write the answer to the key question on your record sheet.

LEARNING ABOUT OTHER IDEAS

Activity 12
Energy Resources

Purpose

In this cycle, you learned that energy is always conserved in interactions between objects. You also learned that heat energy is always transferred away from the interacting objects. You then found out that the energy efficiency of a device usually is higher when this heat-energy transfer is kept small.

Of all fossil fuels, the largest reserves are contained in coal deposits.

Devices that are energy efficient also help to conserve *energy resources*. Energy resources include oil, natural gas, uranium and the Sun itself. They provide electrical energy, heat energy, and other types of energy for human activities. Only limited supplies of some energy resources, such as coal and oil, exist. These energy resources are **nonrenewable**, which means that they cannot be replaced. Other energy resources are either nearly unlimited (for example, sunlight), or can be replaced (if you use trees as an energy resource, you can grow more). These energy resources are **renewable**.

Science Words

nonrenewable energy resource: an energy resource that cannot be replaced

renewable resource: an energy resource that can be replaced in a short period of time

In this activity, you will look at both nonrenewable and renewable energy resources. You will explore how some energy resources provide electrical energy, and how energy resources can be managed and conserved. The key questions for this activity are:

Which energy resources used in our society are renewable, and which are nonrenewable?

Record the activity key questions on your record sheet.

Learning the Ideas

Energy Resources and the Sun

Most of the resources that provide energy for your daily life originally came from the Sun. This includes the food that fuels your body and the gasoline that fuels cars. The Sun provides energy for living things to live and grow. Most of our energy resources were living organisms at one time, or came from living things. In the case of fossil fuels, the animals and plants that are the source of oil, coal, and natural gas lived tens or hundreds of millions of years ago. Energy from the Sun can also be transformed directly to electrical energy or into heat energy for warming buildings, water, etc.

Renewable Resources

Renewable energy resources are those that can be replenished in a short amount of time. These include solar energy (energy directly from the Sun), energy from the motions of air (wind) and water, and geothermal energy from the hot interior of Earth. All of these are steady sources of energy. Biomass energy, which is energy derived from organic sources (primarily plants), is also renewable because you can always grow more plants.

Solar Energy

Humans directly utilize energy from the Sun in two ways:

- through light and infrared-radiation interactions with the Sun, energy is transferred to buildings or water to warm them.
- electricity is generated by converting sunlight into electrical energy with solar cells or by using sunlight to boil water and drive steam turbines.

When sunlight is used to warm up water or air, mirrors are often used to focus sunlight into a small absorbing area. The air or water rises to a high temperature. If used for heating, the hot air or water circulates through a building and transfers heat energy to the building's interior, keeping it warm. The method used for generating electricity from water boiled by focused sunlight is essentially the same as used by geothermal systems (see right).

Solar cells absorb light energy from the Sun, store it as chemical energy, and then output that energy as electrical energy when connected to electrical devices in a closed circuit.

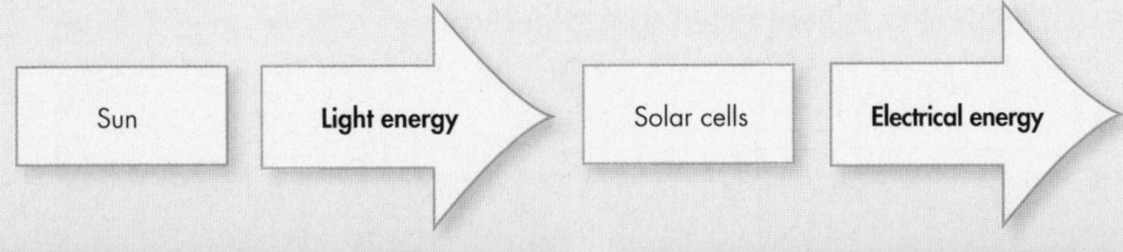

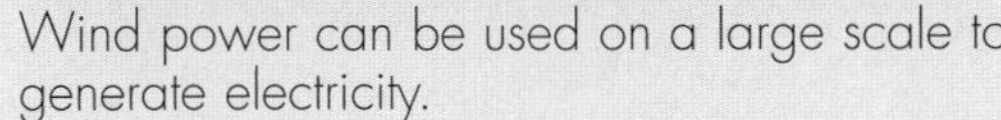

Wind power can be used on a large scale to generate electricity.

Water is the leading renewable energy source used to generate electric power.

Energy from Wind and Water

Humans have been utilizing wind and water (hydro) power for thousands of years. For example, both windmills and watermills have been used for purposes such as grinding grain. Today, humans use the motion energy of wind and water to generate electricity. "Wind farms" consisting of dozens to hundreds of windmills are becoming increasingly common in high-wind areas throughout the United States and other countries. Hydroelectric power plants are located next to swiftly running rivers, or located in dams where they take advantage of the motion energy of falling water.

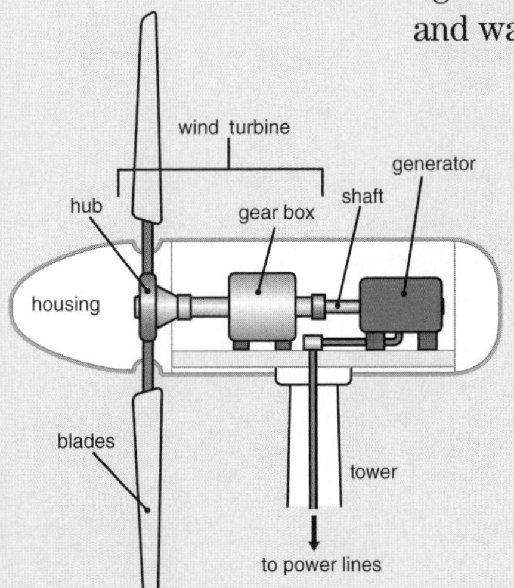

Both windmills and hydroelectric plants work on similar principles. The moving fluid (air or water) interacts with a windmill or turbine that is connected to a generator. A turbine is a circular wheel or hub with fan-like blades attached, like a windmill. When a fluid flows past the blades, it transfers mechanical energy to them and causes the blades and hub to move in a circle. The hub may be connected to a gear assembly that transfers the motion energy of the spinning blades to a shaft that is connected to a generator.

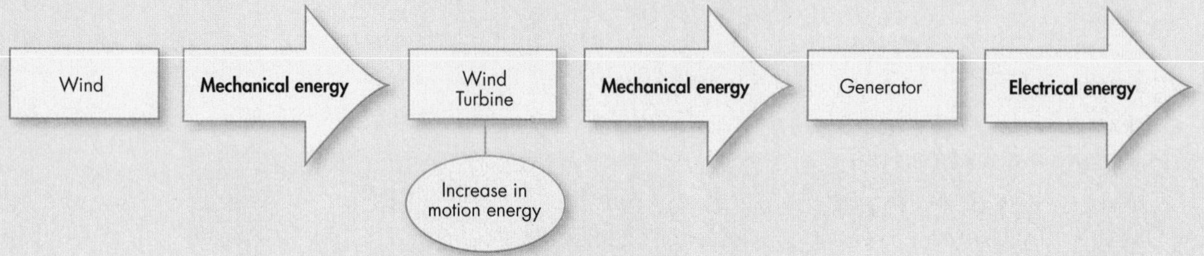

The blades, hub, and gears together constitute the turbine. Through the connecting shaft, the turbine transfers mechanical energy to the generator. The generator then transforms the mechanical energy input into electrical energy output. One example of a generator is the hand-held generator that you saw in Unit 2 Cycle 1, Activity 2. A turbine-driven generator works on the same principle, with the moving fluid replacing the hand and the turbine replacing the crankshaft and the gear assembly inside the generator housing.

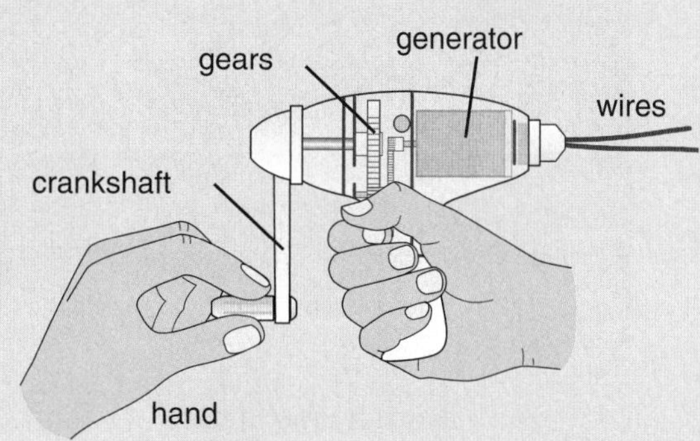

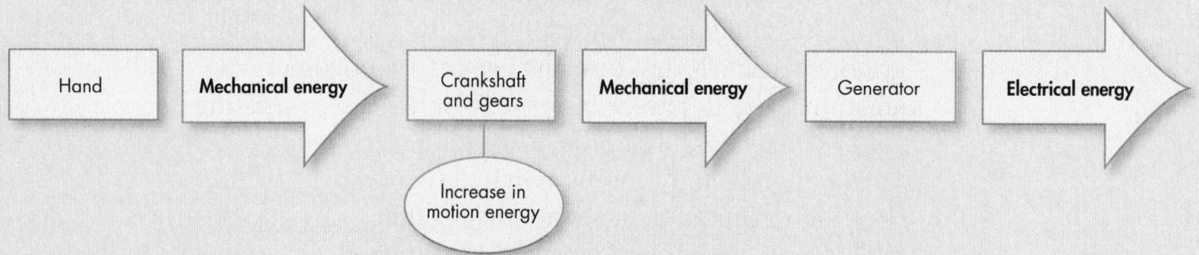

We can simplify the energy diagram for wind or water generation of electrical energy by simply combining the turbine and the generator. We will use this shortcut in the sections on the next page as well.

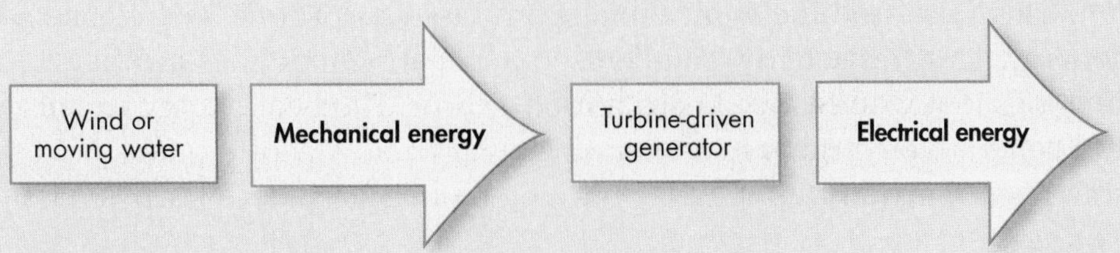

Geothermal Energy

Geothermal energy can be used to generate electricity.

Geothermal power plants tap the thermal energy of Earth itself. The same energy source that powers volcanic eruptions provides geothermal energy. Heat-conduction interactions from the hot interior of Earth sometimes transfer heat energy into underground bodies of water, warming the water to high temperatures. This warming causes some of the water to vaporize or become gas. The gas tries to expand but has nowhere to go. Like steam escaping a teakettle, the hot water vapor and water can erupt from their underground confinement at high speed, with spectacular results. This is the explanation for hot-water geysers like Old Faithful in Yellowstone National Park. It erupts regularly once every hour or two, with an average of 76 min between eruptions.

Geothermal power plants generate electricity by tapping the motion energy of the expanding water vapor. They use pipes to pump the hot water and water vapor from underground, and allow the water vapor to expand explosively on the surface. The water vapor transfers mechanical energy to a turbine connected to a generator, which then produces electrical energy.

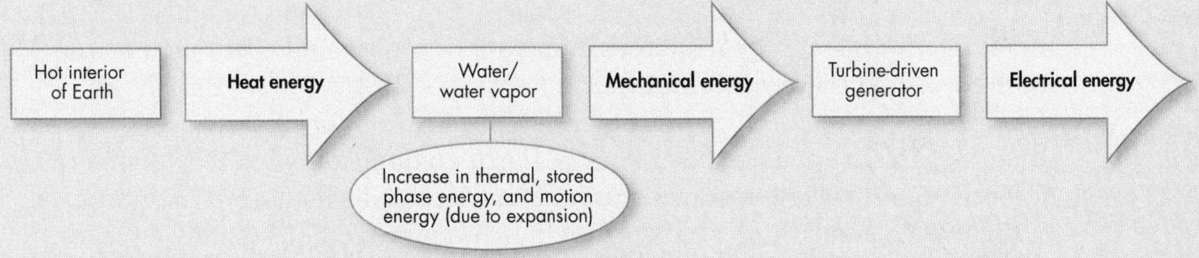

Biomass Energy

Biomass is organic material that has stored energy from the Sun in the form of chemical energy. Biomass includes wood, straw, manure, sugar cane, other byproducts of agriculture, and organic garbage, such as raked leaves or weeds. It also includes fuels such as ethanol, a type of alcohol distilled from corn that is often mixed with gasoline to help power motor vehicles.

When burned, the stored chemical energy in biomass products is released as heat energy. This heat energy can be used for heating buildings, just as burning logs in a wood stove keeps a room warm. The heat energy can also be used to generate electricity in a manner similar to how geothermal sources generate electricity. In this case the burning biomass replaces Earth's hot interior as the source of heat energy.

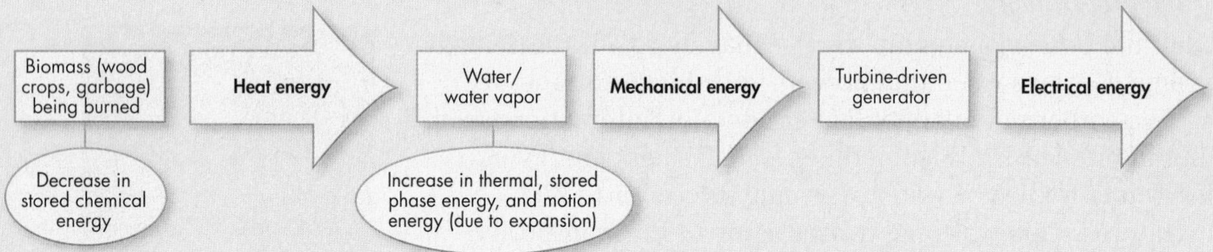

Before you answer the questions below, think about the area in which you live (say, your home and the area within about 100 km [60 mi.] of it). Consider how sunny it normally is. Is there a fast-moving river, a high waterfall, or a dam in your area? Is there an area that is really windy?

1. Given the choices of sunlight, wind, and water, which renewable energy resource do you think would provide the *greatest* amount of electrical energy for your area?
2. Given the same three choices, which renewable energy resource do you think would provide the *smallest* amount of electrical energy for your area?

Your teacher will review answers to these questions with the class.

Nonrenewable Resources

All nonrenewable energy resources come out of the ground. They include the fossil fuels—oil (liquid), natural gas and coal (solid)—and uranium. These resources contain stored chemical energy (fossil fuels) or nuclear energy (uranium) that can be used by humans. Earth contains only limited supplies of these nonrenewable resources. If humans depleted every oil reserve on Earth and consumed the oil, there would be no more oil for millions of years.

Fossil Fuels

Coal, oil and natural gas are called fossil fuels because they formed from the remains of plants and animals that lived in swamps and oceans tens of millions to hundreds of millions of years ago. These remains were buried under layers of mud, sand, and silt. Over millions of years, the remains were buried deeper and deeper. The intense heat and pressure on the fossil remains turned them into coal, oil, and natural gases such as methane. Humans retrieve fossil fuels by drilling wells (oil and natural gas), digging mines deep underground (coal), or mining along Earth's surface (also coal).

Like biomass sources of energy, fossil fuels contain energy that originally came from the Sun and was stored in the form of chemical energy when the "fossil" plants and animals were still alive. Burning fossil fuels releases this stored chemical energy as heat energy. This heat energy can be used directly to heat buildings, or to generate electricity in a manner similar to how biomass is used to generate electricity. Although oil and natural gas are sometimes burned to generate electricity, coal is the fossil fuel burned most often to generate electricity. In fact, coal is used to generate more than one-half of all the electricity used in the United States. Coal plants have major health and safety issues related to the chemical wastes produced by burning coal.

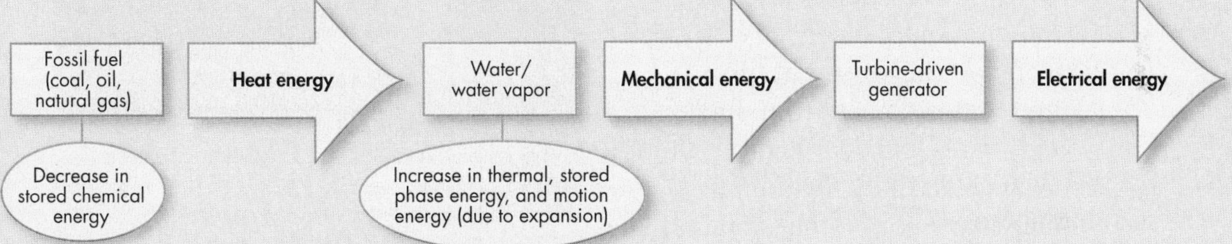

Oil and natural gas, along with biomass products like ethanol, are also used to produce fuels for such purposes as cooking (propane, developed from natural gas) and powering motor vehicles (gasoline, developed from oil). In these cases, the energy released from burning the fuel is used directly. In automobiles, gasoline burns under high compression in a mixture with air. Gasoline undergoes a chemical change when it combusts and releases its stored chemical energy as heat energy. (You will study interactions of this type in Unit 7.) The air-gas mixture absorbs the heat energy and its temperature increases sharply. When it expands, the air-gas mixture transfers mechanical energy to the car's crankshaft. The crankshaft is connected to the car's wheel and axle system.

Nuclear Energy (Uranium)

Nuclear power plants extract heat energy from uranium in a process called nuclear fission. Nuclear reactions in power plants pose a threat due to exposure to nuclear radiation. Thus, nuclear power plants are designed with numerous safeguards, and government regulations control the disposal of materials contaminated with nuclear waste.

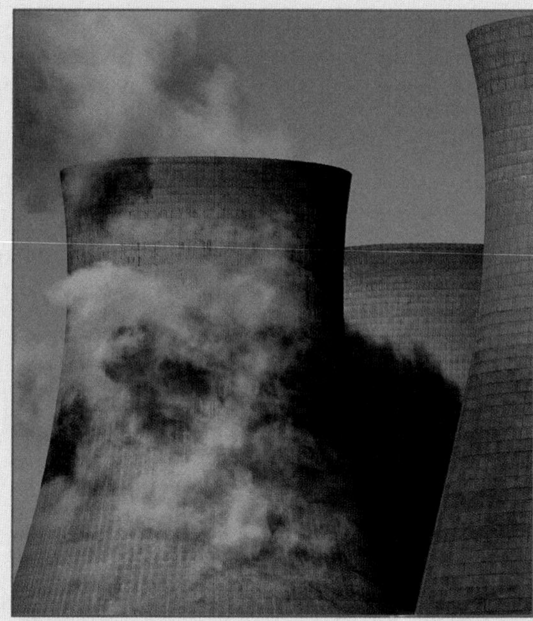

As the population of the world increases there is likely to be a demand for more energy. Nuclear power is one source of energy.

Nuclear power is used solely for the generation of electricity. About 20% of electricity used in the United States comes from nuclear power plants. Nuclear power plants generate electricity in a manner similar to coal or biomass plants. The only difference is the source of heat energy. This comes from the energy required to hold the particles together inside the uranium.

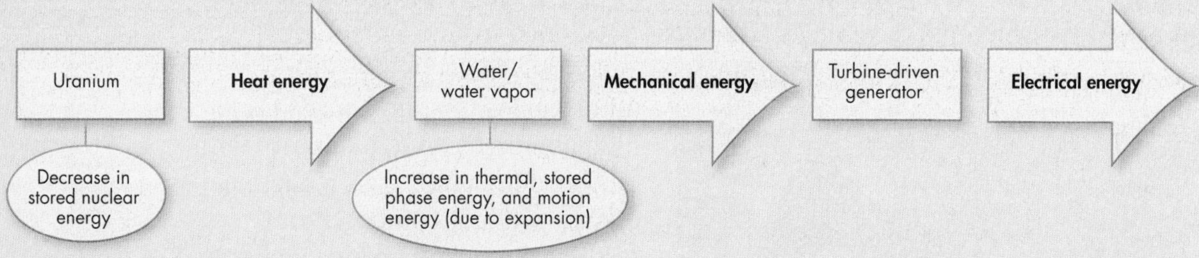

3. What are the **similarities** between how electricity is generated from coal, and how electricity is generated from uranium?

4. What are the **differences** between how electricity is generated from coal and uranium, and how electricity is generated from moving water (such as a river)?

Your teacher will review answers to these questions with the class.

Management of Energy Resources

Each kind of energy resource has advantages and disadvantages associated with its use. Generally, nonrenewable sources have the advantage of producing energy at relatively low cost. Electric power plants using nonrenewable sources can also be built in locations far from where the nonrenewable sources are drilled or mined. The major disadvantages of nonrenewable sources are the limited supplies of fossil fuels and uranium. There are also potentially more severe environmental costs, both when drilling or mining for nonrenewable sources, and when using them. Nuclear power produces radioactive waste that has to be carefully stored to prevent environmental damage. However, the use of nuclear energy does not contribute much to global warming. Using fossil fuels, particularly oil and coal, leads to both water and air pollution.

Government regulations have significantly reduced the amount of pollution resulting from the use of fossil fuels in the past 30 years. The burning and consumption of fossil fuels produces carbon dioxide, a gas that contributes to global warming. Many scientists consider human consumption of fossil fuels to be the leading cause of global warming in the past century.

Renewable resources, on the other hand, have a smaller environmental impact. The production of solar, geothermal, wind, and water energy cause little or no water or air pollution. Biomass resources normally burn cleaner than coal or oil. In particular, the use of renewable energy resources contributes little to global warming. Some environmental problems do come from using renewable resources. For example, if a dam is built on a river, land is flooded. Like other industrial activities, the production of technology (such as solar cells) used to generate electricity from renewable sources can create industrial waste and contribute to global warming.

Renewable resources have the disadvantage that, except for biomass resources, their energy can only be harnessed at specific locations. For example, geothermal power can only be harnessed close to "hot spots." Cost is also a factor. Currently, in the United States, obtaining energy from renewable sources is still more expensive than obtaining energy from nonrenewable sources.

In the long term, people will need to rely on renewable energy resources. Their supply is essentially unlimited. Also, they do not affect the environment as much as nonrenewable resources. How fast human society is converting to using renewable resources is not certain. For now, humans can reduce the environmental impact of using fossil fuels and nuclear power by:

- Using electrical devices that are more energy efficient (see Activity 8).
- Recycling as much as possible. In most cases, the energy used to produce products like aluminum cans and plastic bottles from recycled materials is smaller than the energy used to produce them from raw materials.
- Using renewable energy resources when possible.

5. Identify other ways in which humans could reduce the environmental impact of using energy resources.

Your teacher will review the answer to this question with the class.

What We Have Learned

Remember the key questions for this activity:

Which energy resources used in our society are renewable, and which are nonrenewable?

Participate in the discussion in which your class reviews the answer to the key questions.

Answer the key questions on your record sheet.

LEARNING ABOUT OTHER IDEAS

Activity 13
Stars and Galaxies

Purpose

In this cycle, you have studied light and infrared radiation, two types of radiation that can move through space and transfer energy (light energy or heat energy) from one object to another object. On Earth, the main source of light and infrared radiation is the Sun. Every second of the day, the Sun transfers both light and heat energy to Earth through the vacuum of space. The Sun itself is just one of trillions of stars in the universe. All of these stars are sources of light and heat energy. Exactly how much energy a star produces depends on a couple of different factors, one of which is the size of the star. Compared to Earth, the Sun is a huge object. It has a diameter more than 100 times Earth's diameter. It has a mass more than 300,000 times Earth's mass. When compared to other stars, however, the Sun is a bit on the small side. For example, two bright stars in the constellation of Orion, the blue star Rigel and the red star Betelgeuse, both have masses nearly 20 times the mass of the Sun. Rigel's diameter is 60 times the diameter of the Sun. Betelgeuse, a star almost ready to explode in a supernova, has a diameter over 600 times larger than the Sun!

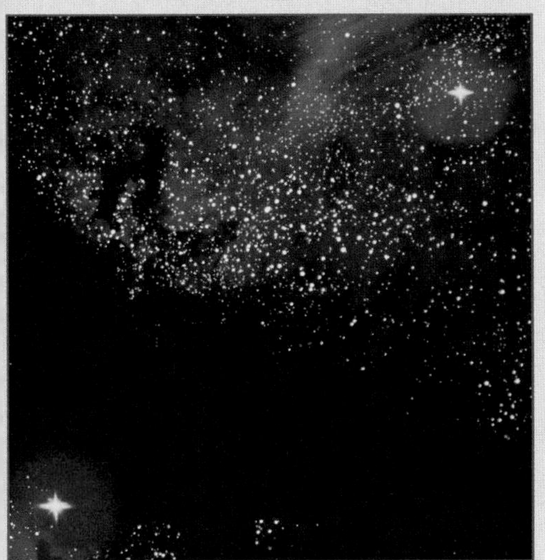

The Sun, Rigel, and Betelgeuse are just three stars out of about 100 billion that form the galaxy (a giant cluster of stars) that we call the Milky Way. Like stars, galaxies also come in a variety of shapes and sizes. The Milky Way is a large galaxy with a spiral shape. Although some galaxies are larger than the Milky Way, over 90% are smaller, and many have non-spiral shapes.

This activity discusses stars and different types of galaxies. The key questions for the activity are:

1. What determines the luminosity (brightness) of a star?
2. What are the different types of galaxies?

Record the key questions for the activity on your record sheet.

Learning the Ideas

Stars

Like the Sun, all stars are composed mostly of hydrogen and helium, plus smaller amounts of other chemicals. Stars are "powered" by nuclear fusion reactions in their hot cores. The nuclear reactions produce the energy that the stars emit as light, infrared radiation, and other types of radiation. Stars vary in terms of size, mass, temperature, and brightness.

Astronomers classify stars based upon two factors: their *color* and their *luminosity*. The color of a star is an indication of its temperature. The relationship between temperature or thermal energy of a star and its color is similar to the relationship between the energy of a light wave and its color (Activity 3). Just as blue light waves are higher energy light waves than red light waves, blue stars have much higher temperatures than red stars. Stars emit light in all wavelengths, however. A blue star looks blue because it emits more blue light than red light, and some stars appear white (they are the second hottest stars, after blue stars).

Science Words

luminosity: (of a star) a measure of the star's brightness that does not depend on the distance between the star and an observer

The **luminosity** of a star is a measure of the star's brightness that does not depend on the distance between the star and an observer. For example, if you lived on a planet orbiting Alpha Centauri at the same distance that Earth is from the Sun, Alpha Centauri would appear about 60% brighter than the Sun, because Alpha Centauri's luminosity is about 60% greater than the Sun's. That's another way of saying that Alpha Centauri emits 60% more light than the Sun.

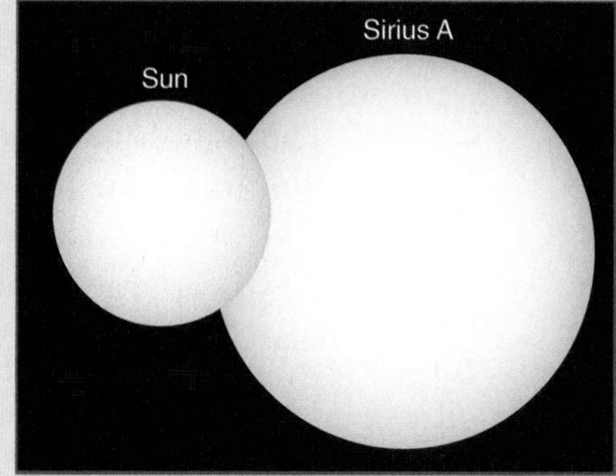

The same principle holds if you are several light-years away from a star. For example, the brightest star in the night sky is Sirius, which you can see in the constellation Canis Major. Sirius is about 8.6 light-years from Earth. If you were 8.6 light-years away from both Sirius and the Sun, Sirius would look the same as it does from Earth, but the Sun would now be much dimmer than Sirius. That's because the luminosity of Sirius is about 23 times that of the Sun. Sirius only looks dimmer from Earth because it is 544,000 times farther away than the Sun.

Sirius has a greater luminosity than the Sun for two reasons. First, Sirius is simply larger than the Sun. If two stars have the same color but different sizes, the larger star will have a greater luminosity simply because it shines from a larger surface.

Second, Sirius is a blue-white star, which means that it is hotter than the Sun. If two stars are the same size and have different temperatures, the hotter star always has a greater luminosity (it is brighter). This is much like an electric stove burner. It is brighter when it is set on "high" compared to when it is set on "medium."

The size and color of stars depend primarily on their masses. (That is, if they are neither close to birth nor close to death.) Stars with masses much higher than the Sun's mass have larger volumes and are much hotter than the Sun, and consequently are much brighter. Stars with masses lower than the Sun's mass have smaller volumes and are cooler than the Sun, and therefore are dimmer than the Sun. The table below describes how stars of different colors differ in terms of their masses, temperatures, diameters, and luminosities. The Sun is a slightly smaller than average yellow star.

Table: Stars of Different Colors

Color	Blue	Blue-White	White	Yellow-White	Yellow	Orange	Red
Temperatures	Over 25,000°	11,000°-25,000°	7500°-11,000°	6000°-11,000°	5000°-11,000°	3500°-11,000°	under 3500°
Average Diameter (Sun = 1)	15	7	2.5	1.3	1.1	0.9	0.4
Average Mass (Sun = 1)	60	18	3.2	1.7	1.1	0.8	0.3
Average Luminosity (Sun = 1)	1,400,000	20,000	80	6	1.2	0.4	0.04

A young star that is still forming tends to be brighter than the "mature" star it will become one day. Near the end of a star's life, when it has burned up most of its nuclear fuel, the outer parts of the star balloon outward. The star becomes both much larger and cooler. For example, several billion years from now the Sun will become a red giant like Betelgeuse (though not nearly as big!) and be much brighter than it is now. Large, bright stars like red giants or giant blue and white stars can be seen from hundreds or thousands of light-years away. Although they are far more numerous, stars the size of the Sun or smaller become invisible to your eyes after a few dozen light-years. For this reason, nearly all the stars you see in the sky are giant red, blue, or white stars much larger than the Sun.

When a star like the Sun becomes a red giant, the red star you see is actually just the low-density outer layer of the star. This low-density layer covers a

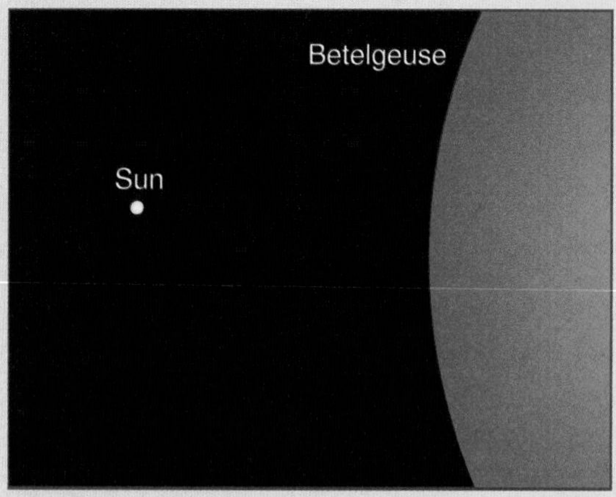

very dense core only about the size of Earth. Over time, the low-density layer drifts away, leaving a small, hot white star at the center. This leftover or remnant star is known as a white dwarf, and it is extremely dense. One teaspoonful of matter from a white dwarf has as much mass as a five-ton truck.

A star with a mass several times the mass of the Sun has a different fate. When all of its nuclear fuel runs out, the star explodes in a brilliant flash of light and energy called a supernova. A supernova is such a powerful blast that the light from the explosion can briefly outshine an entire galaxy. The force of a supernova explosion can push together large clouds of dust and gas so that they form new solar systems, as discussed in Unit 3 Cycle 2, Activity 13.

Left behind by the supernova is the core of the star, which has been compressed by the star's extremely strong gravitational force into an asteroid-sized body. The leftover core may be a neutron star, which is only about 10 km across but more massive than the Sun. Or, the leftover core may be a black hole, which you learned about briefly in Unit 3 Cycle 2, Activity 3. The gravity of a black hole is so strong that not even light can escape its pull.

1. The picture of the Sun and Betelgeuse shows their relative sizes. Which star has a greater luminosity? Explain your answer.
2. Suppose that Betelgeuse and the Sun were the same size. Which star would then have the greater luminosity? Explain your answer.

Participate in the class discussion about your answers.

Galaxies

On a clear, moonless night in areas far away from city lights, you can look up in the sky and see the Milky Way, a large "milky" band of light across the night sky that is caused by billions of stars that are part of our galaxy, which is also called the Milky Way. You view the Milky Way from its edge, looking inward toward the center. If you could see all the stars of the Milky Way, you would see something that looks like a disc on its edge with a bulge in the middle. However, the presence of interstellar dust clouds in the Milky Way blocks the view of many stars in the disc. If you were to look at the Milky Way using instruments that can detect types of light that your eyes cannot see, such as infrared radiation, you could see the disc of the Milky Way clearly. You

The Milky Way Galaxy.

could also see the large bulge that forms the center of the galaxy.

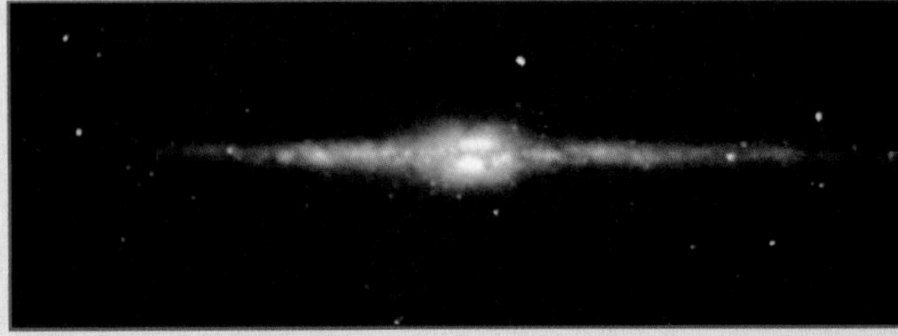
The Milky Way Galaxy.

NGC 7331 Galaxy.

The Milky Way Galaxy contains about 100 billion stars, including the Sun. Nearly all stars in the universe are collected into galaxies, ranging in size from tiny galaxies containing only hundreds of millions of stars to galaxies larger than the Milky Way. *Large galaxies*, which contain tens of billions to hundreds of billions of stars, make up about 10% of all galaxies in the universe. The Milky Way is an example of a typical large galaxy, and has roughly the same chemical makeup and types of energy as other galaxies. The remaining 90% of galaxies are *dwarf galaxies*, which contain several hundred million to a few billion stars. Examples include the Large Magellanic Cloud and Small Magellanic Cloud. They are two small galaxies that orbit the Milky Way.

In addition to categorizing galaxies by size, astronomers classify galaxies by shape. The Milky Way is an example of a *spiral galaxy*, because, from a distance, the Milky Way forms a spiral shape. In fact, the Milky Way probably looks similar to NGC 7331. This galaxy has a mass, number of stars, and structure that make it almost a twin of the Milky Way. Another example of a spiral galaxy is the Andromeda Galaxy. It is a spiral galaxy about two million light-years away. It is the closest large galaxy to the Milky Way.

Elliptical galaxy (NGC 4881).

Other categories of galaxies include *elliptical* galaxies such as NGC 4881 and *irregular* galaxies. Elliptical galaxies are elliptical in shape and do not have a spiral structure.

Examples include the two dwarf galaxies orbiting the Andromeda Galaxy and the nearly circular galaxy shown in the photograph. Irregular galaxies such as the Large Magellanic Cloud don't have a regular shape.

Large Magellanic Cloud.

Andromeda Galaxy.

Virgo Cluster.

Just as stars have a tendency to collect in large clusters called galaxies, galaxies themselves collect in clusters. The Milky Way is part of a cluster of about forty known galaxies, plus perhaps a few dozen dwarf galaxies that have not yet been discovered, called the Local Group. All except three of the galaxies in the Local Group are small dwarf galaxies. The only large galaxies are the Milky Way, the Andromeda Galaxy, and another galaxy called the Triangulum Galaxy. With forty or so galaxies, the Local Group is a relatively small galactic cluster. By comparison, the Virgo Cluster, located about 52 million light-years away, contains about 160 large galaxies and nearly two thousand small galaxies.

Although vast amounts of mostly empty space separate galaxies, the universe still contains a huge number of galaxies. Scientists believe that there are at least several hundred billion galaxies, most of them (90%) dwarf galaxies like the Magellanic Clouds, and only 10% large galaxies like the Milky Way. You can find them anywhere in the sky.

What We Have Learned

Think about the key questions for this activity:

1. What determines the luminosity (brightness) of a star?

2. What are the different types of galaxies?

Participate in the class discussion about your answers.

SECTION C

UNIT 5:
MATERIALS AND THEIR INTERACTIONS

UNIT 6:
PHYSICAL INTERACTIONS AND PHASES

UNIT 7:
CHEMICAL INTERACTIONS

NTERACTIONS OF MATERIALS

SECTION C

Interactions of Materials

In the section you just completed, you studied interactions between large objects like a person and a ball, and you used force and conservation ideas to describe the interactions. In *Interactions of Materials* you will learn how to describe the different materials that make up objects. You will also learn the scientists' theory of what different materials and their interactions are like on a scale too small to see with your eyes.

The activities in the units in this section are different. The explorations that provide evidence for some of the ideas in these units are very complex. Some also require very expensive equipment. You cannot really do these explorations in your classroom laboratory. Therefore, you will be doing a lot of explorations to help you make sense of these ideas, but you will also do more reading and answering questions about scientists' ideas. Since many of the activities are written to help you make sense of scientists' thinking, you will also find fewer We Think sections.

Section C has three units.

UNIT 5
Materials and Their Interactions

UNIT 6
Physical Interactions and Phases

UNIT 7
Chemical Interactions

MATERIALS AND THEIR INTERACTIONS

INTRODUCTION

What will you learn about in Unit 5?

In the previous units, you investigated many questions about different types of interactions. You learned about magnetic, electric-charge, electric-circuit, mechanical, gravitational, heat-conduction, and infrared-radiation interactions. You also learned two of the ways that scientists describe interactions. Scientists describe interactions using Newton's Laws of Motion and using the Conservation of Energy Law.

You learned in Unit 4 that stuff has both mass and volume. In this unit, you will explore some of the basic questions scientists ask about stuff and the different kinds of interactions of stuff. There are many names for stuff. For example, stuff can be called materials, substances, and chemicals. Scientists sometimes call stuff "matter." Scientists who investigate matter and the interactions of matter are called chemists.

You are surrounded by classification schemes everywhere you go. Classification schemes are used to arrange things in an organized manner. Books in a library are organized so they are easier to find.

Materials and Their Interactions

The key questions for this cycle of learning are:

1. What are physical and chemical interactions?
2. How do chemists describe and classify materials (matter)?

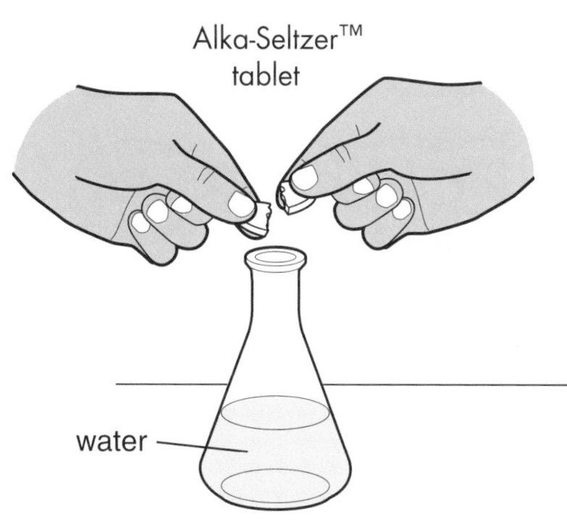

UNIT 5: MATERIALS AND THEIR INTERACTIONS

Key Questions

1. What are physical and chemical interactions?
2. How do chemists describe and classify materials (matter)?

Activity 1: Describing and Classifying Materials

Activity 2: What Is It?

Activity 3: How to Use a Hand-held Microscope

Activity 4: Chemical Interactions

Activity 5: Physical Interactions

Activity 6: Mixtures

Activity 7: Solutions and Chemicals

Activity 8: Classifying Chemicals

Activity 9: Metals, Nonmetals, and the Periodic Table

Activity 10: Describing and Classifying Materials

Activity 11: Analyze and Explain

Activity 12: Acid, Base, or Neutral?

Activity 13: Groups and the Periodic Table of the Elements

DEVELOPING OUR IDEAS

Activity 1
Describing and Classifying Materials

Purpose

Have you ever collected things like rocks, coins, stamps, or cards? If so, you probably know how important it is to describe and group, or classify, the objects in your collection.

The produce in a grocery store is arranged in an organized classification scheme.

Chemists, however, are not interested in describing and classifying *objects*. They ask questions about the stuff or materials that make up the objects. The key question for this activity is:

How can you describe and classify materials?

Record the key question for the activity on your record sheet.

We Think

Suppose you were given the eight objects shown in the diagram.

Discuss the following with your team.

How would you classify these objects?

Be prepared to share your team's classification in a class discussion.

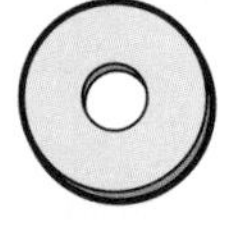
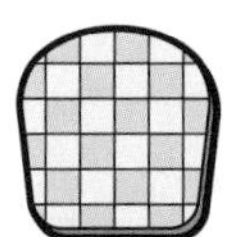

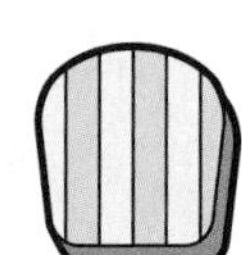
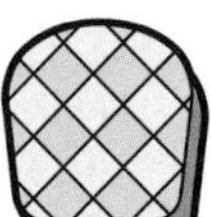

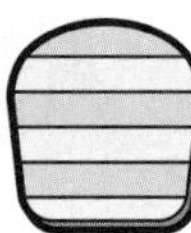

Explore Your Ideas

You will do two explorations in this activity. In the first exploration, you will examine 12 different materials to determine some properties and group names. In the second exploration, you will use some of these properties to sort the materials into groups.

Wear eye protection.

Clean up spills immediately.

Exploration 1: Properties and Groups

You will record your properties and groups on your record sheet in a table like the one shown on the next page.

STEP 1 Look at some of the properties of the 12 materials.

Do not open the containers with the gases: air, carbon dioxide, and helium.

Supply Master: Open the containers of aluminum, brass, charcoal, and Styrofoam™. Place a small amount (less than a teaspoon) of the muddy water, pure water, salt water, sugar, and Tabasco® sauce on a sheet of wax paper so you and your team can examine their properties more closely.

Your team will need:

- wax paper
- plastic tray
- plastic spoon
- set of 12 labeled materials
  - *air*
  - *aluminum*
  - *brass*
  - *carbon dioxide*
  - *charcoal*
  - *helium*
  - *muddy river water*
  - *pure water*
  - *salt water*
  - *Styrofoam™*
  - *sugar*
  - *Tabasco® sauce*

STEP 2 Take turns identifying at least *four* properties that you could use to classify the materials into groups.

Write the four property names, group names, and number of groups in Table 1 on your record sheet.

Table 1: Names of Properties and Groups

Name of Property	Names of Groups	No. of Groups
Example A: See-through?	transparent, opaque	2
Example B: Moisture	wet, dry	2
1.		
2.		
3.		
4.		

Exploration 2: Using the Properties to Classify the Materials

Your team will need:

- scissors
- string
- set of 12 labeled materials (in containers)

You will sort the 12 materials into groups. Take turns using a property in *Table 1: Names of Properties and Groups* as a defining property.

STEP 1 *Team Manager:* Use Property 1 on your table as the defining property to classify the 12 materials into groups. Put a string around each group.

1. Draw a diagram of this classification scheme following this procedure:

- Draw a circle or oval for each group.
- Write the group name at the top of each circle.
- List the materials that belong in each group.
- Label your diagram with the name of the defining property.

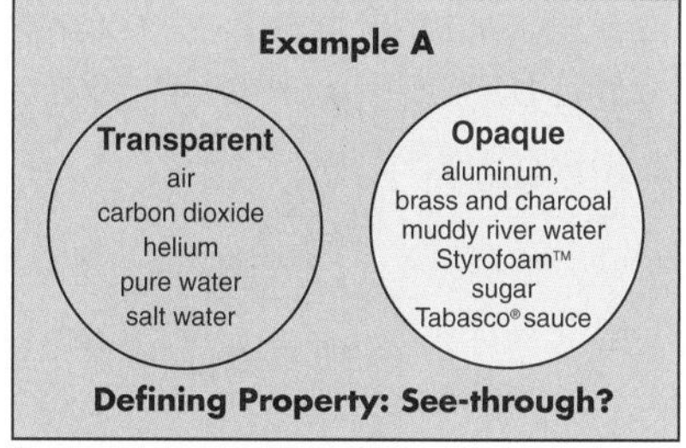

Example A
Transparent
air
carbon dioxide
helium
pure water
salt water
Opaque
aluminum,
brass and charcoal
muddy river water
Styrofoam™
sugar
Tabasco® sauce
Defining Property: See-through?

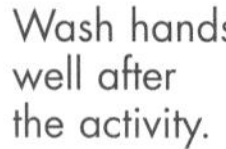

Wash hands well after the activity.

STEP 2 Repeat Step 1. Use Properties 2, 3, and 4 on your table to classify the materials. Put a string around each group.

2. Draw a diagram of each classification scheme. Follow the same procedure as before.

Make Sense of Your Ideas

Participate in a class discussion about the different defining properties teams used to classify the 12 materials.

Copy the properties your class used into the left column of *Table 2: Properties and Functions* on your record sheet.

Table 2: Properties and Functions	
Property How materials interact with your senses.	**Functions** How you use the materials or the purpose of the material.
Interaction with eyes (seeing) see-through? (transparent, opaque) ***Interaction with skin (feeling)*** moisture (wet, dry)	***The materials are used to:*** eat (edible or non-edible) burn (fuel, not for fuel) make things (jewelry, appliances, etc.)

Scientists describe and classify materials by their properties. In everyday life, however, people sometimes describe and classify materials by their function (their purpose, or how they are used). For example, you may use materials to eat so you could describe them as *edible* or *not edible*. You may use them to burn and describe them as *burn for fuel* or *do not burn for fuel*. You may use them to make things (*jewelry*, *appliances*, *buildings etc.*).

Participate in a class discussion about some additional functions that could be used to classify materials.

Write the additional functions in your table.

Our Consensus Ideas

The key question for this activity is:

How can you describe and classify materials?

1. Write your answer on your record sheet.

Participate in the class discussion about the key question.

2. Write the class consensus ideas on your record sheet.

DEVELOPING OUR IDEAS

Activity 2
What Is It?

Purpose

In Activity 1, you learned that chemists describe and classify materials by their properties. A property of a material is a description of how the material interacts with other things, such as your senses and instruments.

Suppose you lost the labels on three containers holding air, carbon dioxide, and helium. How could you tell which gas was which? Or, do *all* gases have the same properties? Describing materials by how they interact with your senses (or instruments that extend your senses) is not always enough to decide if materials are the same or different. In this activity, you will investigate air and three other gases made in different ways. The key questions for this activity are:

1. Are there different kinds of gases or are all gases some form of air?

2. How do scientists identify unknown materials when they have the same sense properties?

Record the key questions for the activity on your record sheet.

I don't think there are really different kinds of gases. I think all gases are some form of air.
I disagree. I think there are different kinds of gases, just as there are different kinds of liquids and solids.
Chantel
Carlos

I Think

Imagine that you heard the conversation between Chantel and Carlos.

Answer the following question and be prepared to share your answer with the class.

Do you agree with Chantel or with Carlos, or do you have a different idea about whether there are different kinds of gases? Explain your reasoning.

Participate in a class discussion about the answer to this question.

Explore Your Ideas

Your class will investigate four gases:

- air
- the gas made when hydrogen peroxide interacts with manganese dioxide (Gas A)
- the gas made when vinegar interacts with baking soda (Gas B)
- the gas made when Alka-Seltzer™ interacts with water (Gas C)

Wear eye protection.

Clean up spills immediately.

Exploration: Are the three unknown gases really the same gas (air) or are they different gases?

To answer this question, you will observe two different properties of the gases. The first property is a description of how the gas interacts with a substance called bromothymol blue (BTB). BTB is called an *indicator* because it can change color when it interacts with a gas. The second property is a description of what happens when a gas interacts with a flame. You will record your observations in a table on your record sheet.

Wash hands well after the activity.

Your team will be assigned one gas (Gas A, B, or C) to make and test with BTB. You will follow one of two options. You will do the flame test for your assigned gas or you will watch a demonstration or a video of the flame tests for Gases A, B, and C.

Participate in a class discussion about additional safety rules for this activity.

Testing Air

Your teacher will demonstrate testing air with the BTB indicator.

Observe what happens (if anything) when air bubbles through the BTB indicator. Record your observations in Table 1 on your record sheet.

Your teacher will demonstrate testing air with a flame.

Record your observations of what happens when air interacts with a flame in Table 1 on your record sheet.

Your teacher will tell you which gas you will produce and test with BTB, and which option you will follow for the flame test.

Table 1: Properties of Air and Three Unknown Gases

Property	Air	Gas A ______	Gas B ______	Gas C ______
1. Interaction with your senses	transparent, colorless, odorless	transparent, colorless, odorless	transparent, colorless, odorless	transparent, colorless, odorless
2. Interaction that makes the gas	None	Hydrogen peroxide + manganese dioxide	Vinegar + baking soda	AlkaSeltzer™ + water
3. Interaction with BTB				
4. Interaction with a flame				

All teams will need:

- 250 mL flask
- 150 mL beaker (for BTB)
- plastic baggie taped to plastic tubing
- connector to one-hole stopper
- 30 mL BTB

optional (for flame test)

- splint
- 2 matches
- test tube and stopper

Gas A:

- 250 mL beaker
- mass scale
- teaspoon
- 100 mL hydrogen peroxide
- 2 g manganese dioxide

Gas B:

- 100 mL beaker (for vinegar)
- teaspoon
- 60 mL vinegar
- 12 g (about 2 tsp.) baking soda

Gas C:

- 2 Alka-Seltzer™ tablets

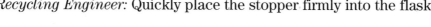

Producing Gases A, B, and C

Read Steps 1 through 3 *before* you start. The procedure for Step 2 will vary depending on your assigned gas, A, B, or C. The other procedure steps are the same for all the gases.

STEP 1 *Supply Master:* Attach the free end of the plastic tubing to the glass tube through the top of the rubber stopper. Push the air out of the baggie and seal the baggie. *Be sure the seal is tight so gas will not leak out of the baggie!*

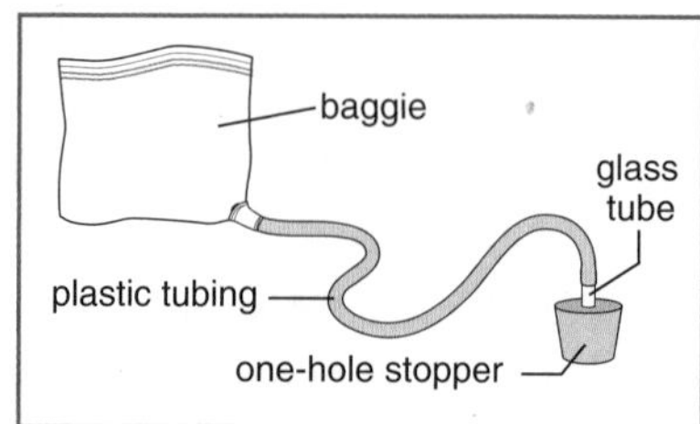

Recycling Engineer: Measure and pour 30 mL of BTB into the 150-mL beaker.

Producing Gas A

STEP 2 *Procedure Specialist:* Use the mass scale to measure about 2 g of manganese dioxide. Pour the manganese dioxide into the flask.

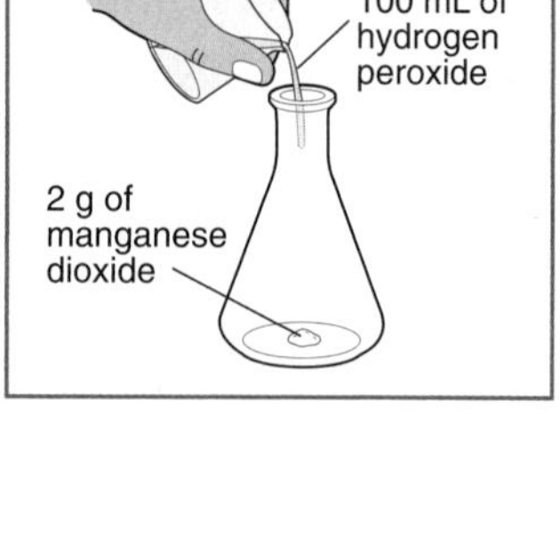

Team Manager: Measure 100 mL of hydrogen peroxide with the 250 mL beaker. Pour it into the flask.

Recycling Engineer: Quickly place the stopper firmly into the flask.

Producing Gas B

STEP 2 *Procedure Specialist:* Place 12 g (2 level teaspoons) of baking soda into the flask.

Team Manager: Measure 60 mL of vinegar with the 100 mL beaker. Pour it into the flask.

Recycling Engineer: Quickly place the stopper firmly into the flask.

Producing Gas C

STEP 2 *Procedure Specialist:* Pour water into the flask to about the 175 mL level.

Recycling Engineer: Open the packet of 2 Alka-Seltzer™ tablets and break each tablet into 4 pieces.

Team Manager: Drop the pieces of the Alka-Seltzer™ tablets into the flask with the water.

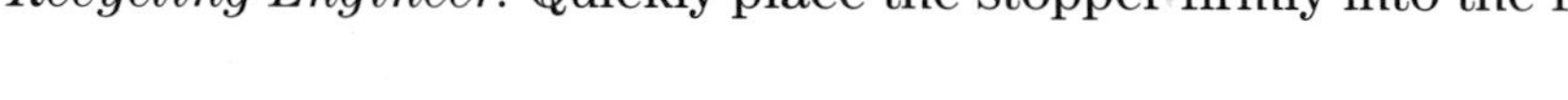

Recycling Engineer: Quickly place the stopper firmly into the flask.

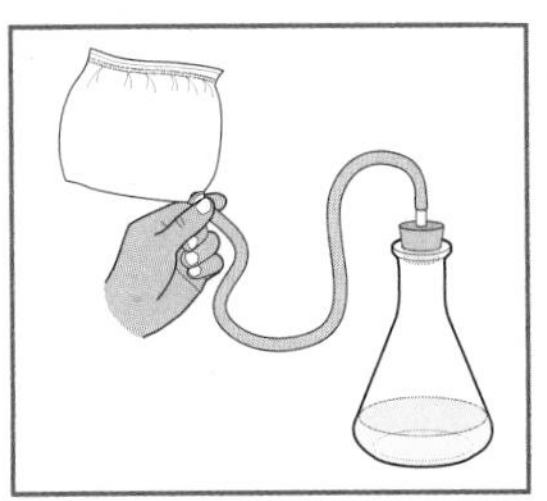

The gas produced in the flask collects in the baggie. Collect the gas until the baggie is almost full. *Do not let the baggie get too full or it will burst open and you will lose all the gas.*

Testing Gases A, B, and C

STEP 3 *Team Manager:* Hold down the flask.

Recycling Engineer: While the *Team Manager* is holding the flask, fold or pinch the tube closed. Keep it closed until the end of Step 5.

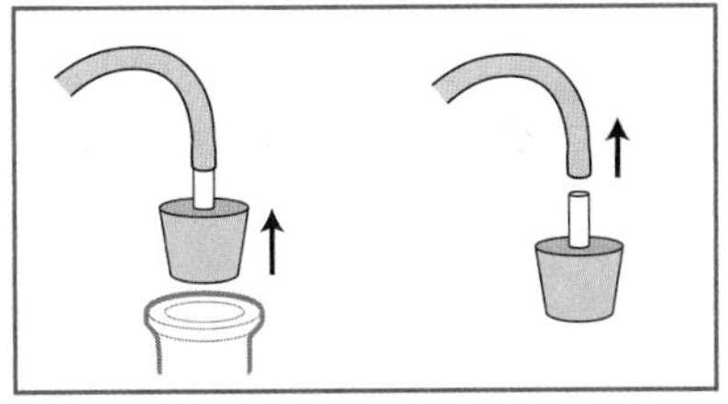

STEP 4 *Procedure Specialist:* Gently remove the one-hole stopper from the flask. Then remove the plastic tubing from the one-hole stopper.

1. What *two* observations provide evidence that the original substances interact when they are mixed together?

Read Steps 5 and 6 *before* you continue.

Avoid breathing the gas directly.

STEP 5 *Supply Master:* Place the end of the tubing into the beaker of BTB. Be sure to hold the tubing so the end of the tubing is *below* the surface of the BTB indicator.

Recycling Engineer: Once the tubing is in place, release the pinch in the tubing.

STEP 6 *Supply Master:* Give the baggie a gentle squeeze so the gas bubbles through the BTB indicator. *Do not squeeze all of the gas out of the baggie* because you may need some for the flame test.

Team Manager: Pinch the tubing closed to keep the rest of the gas from escaping. Remove the tubing from the beaker of the BTB indicator.

If you are going to do the flame test, do *not* let go of the tubing because you will need the rest of the gas for the flame test. Proceed to the flame test.

If you are not doing the flame test, squeeze the baggie gently so the gas escapes into the air. Clean up as instructed by your teacher.

BTB indicator

Observe what happens (if anything) when the gas bubbles through the BTB indicator. Record your observations on Table 1.

Wear eye protection.

Be very careful with the flames.

Flame Test of Your Gas

Your teacher may wish to show you a video or do a demonstration.

If you are testing your own gas, carefully read Steps 7 through 12 before you start.

STEP 7 *Procedure Specialist:* Hold the test tube on its side. Hold the stopper in your other hand.

STEP 8 *Team Manager:* Place the tubing into the test tube *all the way to the bottom of the test tube.* Then release your pinch in the tubing.

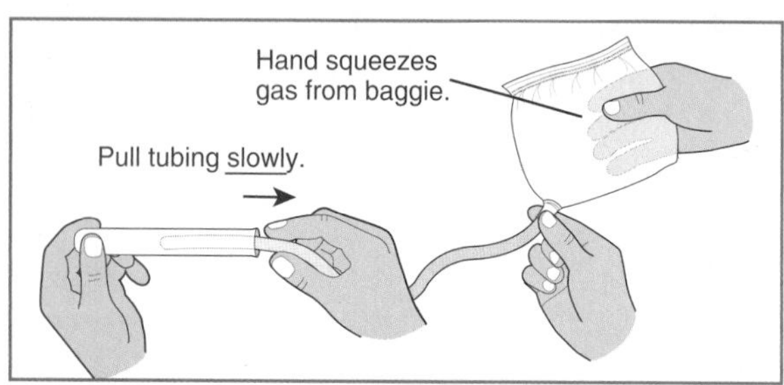
Hand squeezes gas from baggie.
Pull tubing slowly.

STEP 9 This step requires good cooperation!

Team Manager: Squeeze the baggie gently.

Procedure Specialist: While the *Team Manager* is squeezing the bag, slowly move the plastic tubing from the test tube. *All the gas should be squeezed from the baggie when the end of the tubing is at the opening of the test tube.*

STEP 10 *Procedure Specialist:* Quickly place a stopper in the test tube to keep the gas from escaping.

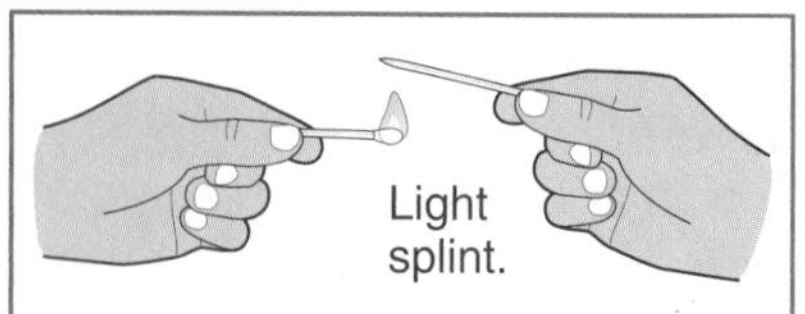
Light splint.

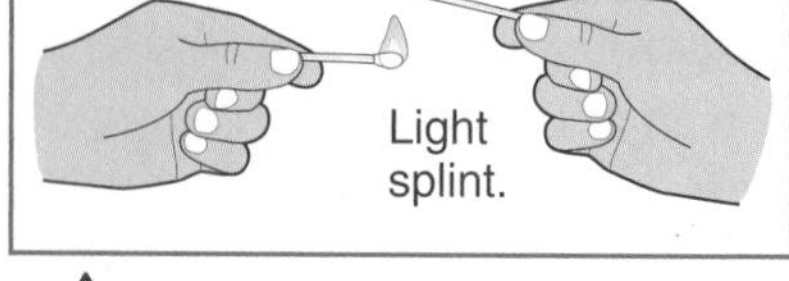

STEP 11 *Recycling Engineer:* Hold one end of the wooden splint.

Team Manager: Light the match and use it to light the splint.

Tie back hair and loose sleeves before lighting the flame.

STEP 12 *Procedure Specialist:* Hold the test tube with the end pointing a little downward, as shown in the diagram. Then remove the stopper from the test tube.

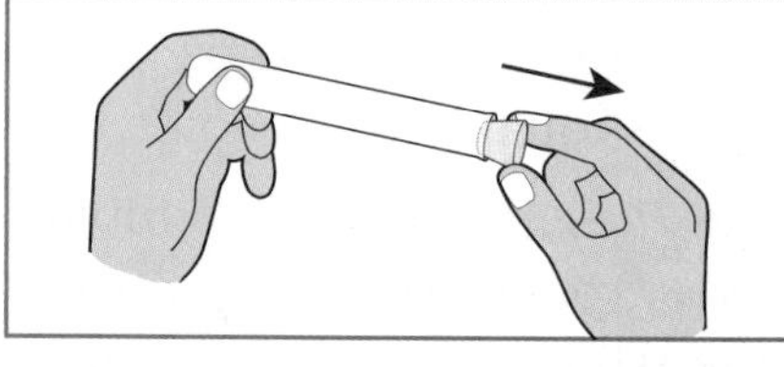

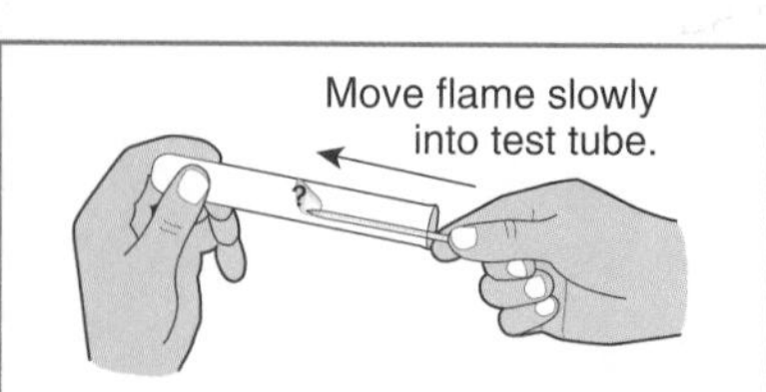
Move flame slowly into test tube.

STEP 13 *Recycling Engineer:* Move the flame slowly into the test tube, as shown in the diagram. Observe what happens to the flame.

Record in Table 1 your observations of what happens when the gas interacts with a flame.

STEP 14 Follow your teacher's directions for cleaning up.

Participate in a class discussion to summarize the team observations for each of the three gases.

Record in Table 1 the observations for the gases tested by other teams.

Make Sense of Testing Gases

Discuss the following questions with your team and record your answers.

1. Which of the three unknown gases do you think are different gases? What is your evidence?

2. Do you think any of the three unknown gases are the same gas? What is your evidence?

It is not easy to know for certain whether unknown materials are the same material *when your decision is based on only four properties.* For example, Gas B and Gas C *could* be the same gas because:

- they have the same interaction with the BTB indicator (change of color).
- they have the same interaction with a flame.

Gas B and Gas C have one property that is *different,* however. The interactions that made the gases are different.

Over the years, scientists learned that you need a *set* of properties to be certain that two unknown gases are really the same. These properties should be useful for identifying different materials. In Unit 1 Cycle 3, you learned that the most useful properties for identifying materials are called *characteristic properties.* Most characteristic properties of materials are measurements (numbers) that are different for different kinds of materials.

One characteristic property of materials is their density. *Density is the mass of a standard-unit volume* of a substance. The temperature at which a substance changes phase is also a useful characteristic property to help you identify unknown materials. Look at the *Table of Melting and Boiling Points* in the Appendix. This table shows the constant melting points (abbreviation M.P.) and boiling points (abbreviation B.P.) of some gases.

Make Sense of Scientists' Ideas

Suppose that someone measured these two additional characteristic properties, density and boiling/condensing temperature, of the three gases you tested. The measurements are shown in Table 2. Now you have six properties to compare instead of only four properties.

Discuss and answer the following question.

3. How certain are you *now* that Gas B and Gas C are the same gas? Explain your reasoning.

Table 2: Properties of Three Unknown Gases			
Property	**Gas A**	**Gas B**	**Gas C**
1. Interaction with your senses	transparent, colorless, odorless	transparent, colorless, odorless	transparent, colorless, odorless
2. Interaction that makes the gas	Hydrogen peroxide + manganese dioxide	Vinegar + baking soda	AlkaSeltzer™ + water
3. Interaction with BTB	No change	BTB turns yellow	BTB turns yellow
4. Interaction with a flame	Flame becomes larger and brighter, then goes out.	Flame goes out *right away.*	Flame goes out *right away.*
5. Interaction of equal volumes with a mass scale (density)	1 L has a mass of 1.33 g	1 L has a mass of 1.83 g	1 L has a mass of 1.83 g
6. Interaction with thermometer during boiling (B.P. is boiling point)	B.P. Temp = −183 °C	B.P. Temp = −78 °C	B.P. Temp = −78 °C

Compare the boiling temperatures of each unknown gas with the boiling temperatures of different gases listed in the *Table of Melting and Boiling Points.* Also, compare the densities (mass of a one-liter volume) of each unknown gas with the densities listed in your *Table of Densities.* Discuss the comparisons with your team and answer the following questions.

4. What is Gas A? Support your answer with evidence from the two tables.

5. What are Gas B and Gas C? Support your answer with evidence from the two tables.

Participate in the class discussion about team answers to these questions.

Write the name of the Gas A, Gas B, and Gas C in the first row of Table 1.

Our Consensus Ideas

The first key question for this activity is:

1. Are there different kinds of gases or are all gases some form of air?

Participate in the class discussion to determine the class consensus ideas for this key question.

1. Answer the first key question on your record sheet.

The second key question for this activity is:

2. How do scientists identify unknown materials when they have the same sense properties?

Discuss this question with your team and summarize your team's answer. Be sure that your summary includes the answers to the following questions.

2. Is one property of a material always enough to identify the material? Why or why not?

3. What is the most useful type of property for identifying unknown substances? Why?

Participate in a class discussion to determine the class consensus ideas for this key question.

4. Record the class consensus ideas on your record sheet.

DEVELOPING OUR IDEAS

Activity 3
How to Use a Hand-held Microscope

Purpose

You will use a hand-held microscope several times in this unit. A microscope magnifies objects. It makes objects look bigger than they really are. The magnification tells you *how many times bigger* an object looks through the microscope. A microscope that has a magnification of 6x (6 times) means an object would look 6 times bigger than it really is. The field of view is the circle you would see through the microscope.

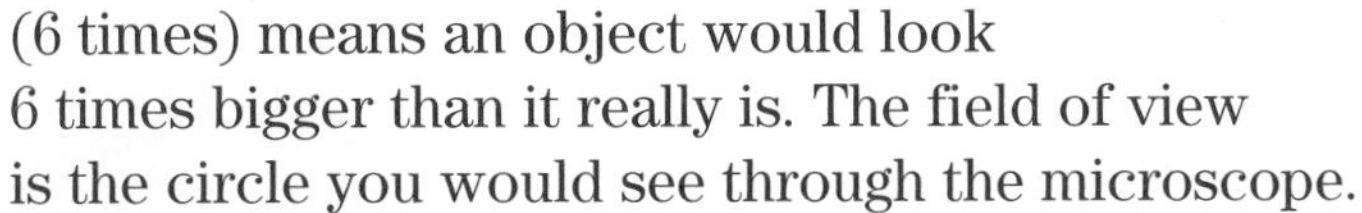

In this activity, each team member will get an opportunity to look through a hand-held microscope. The key question for this activity is:

What are three ways an object looks different through a microscope as its magnification increases?

Record the key question for the activity on your record sheet.

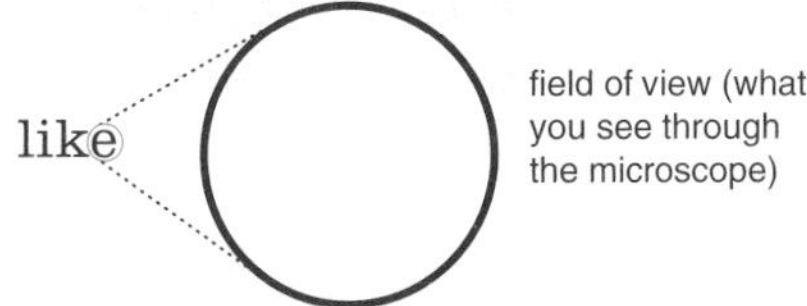

We Think

Suppose you looked through a hand-held microscope at the letter “e” in the word “like.” The letter is about 2 mm high and 2 mm wide.

What do you think you would see in the field of view? Discuss this question with your team.

On your record sheet, draw what you think you would see.

Participate in a class discussion.

Explore Your Ideas

You will do two explorations in this activity. In the first, you will look at the small letter "e" through your microscope. In the second, you will look at a leaf.

Your team will divide up into two pairs. Each pair of partners will work with one hand-held microscope. One partner (*Procedure Specialist*) should read the procedure out loud while the other partner works with the microscope. You should then switch roles.

Your team will need:

- 2 illuminated hand-held microscopes
- 2 leaves

Exploration 1: Looking at the Letter "e"

STEP 1 Begin by becoming familiar with your microscope. Look at the diagram that shows the parts of an illuminated, hand-held microscope. Your microscope may be a different size and have switches and wheels located in different places.

Parts of a Hand-held Microscope

- Find the *focusing wheel* on your hand-held microscope. When you turn this wheel, the *eyepiece* should move up and down.
- Find the *zoom switch,* which changes the magnification of your microscope. *Make sure this switch is set to the lowest magnification, 60x.*
- Find the *transparent cover* and the *bulb.* Some microscope models have a *bulb rotation knob.* Be sure to move this knob so that the tip of the bulb is pointed toward the hole in the transparent cover that is below the eyepiece.

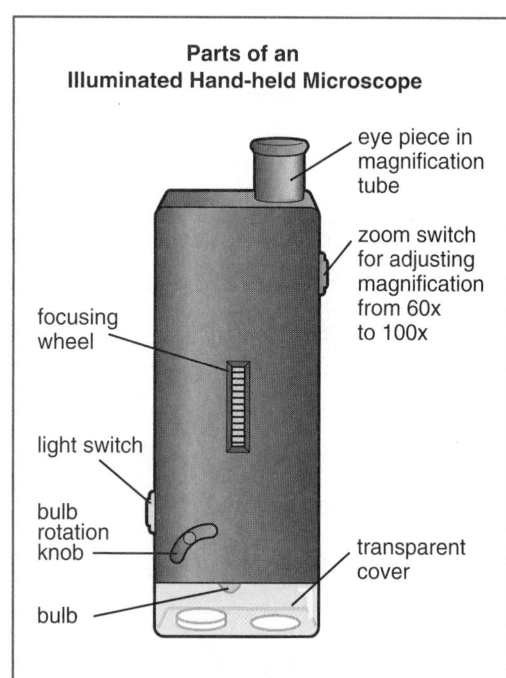

- Find the *light switch.* Turn on the bulb.

STEP 2 Place your book flat on your desk or table. You are going to look at the small letter "e" below. Move the hand-held microscope until the hole in the transparent cover, which is below the eyepiece, is above the letter. Then stand the microscope on this page so that the "e" is about in the middle of the hole.

e

STEP 3 Look through the eyepiece at the letter. *Gently and slowly* move the microscope until you begin to see at least part of a fuzzy letter. Steady the microscope with one hand and use the *focusing wheel* with your other hand until you see the letter clearly.

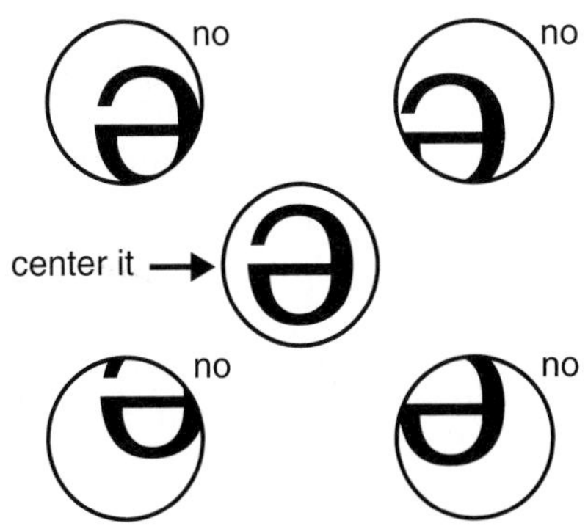

STEP 4 Move the microscope so that the letter is in the *center* of the field of view. Recall that the field of view is the circular area that you see.

Draw what you see in Table 1 on your record sheet. Use a table similar to the one shown. Be sure to draw the object within the circle in its correct size with relation to the size of the field.

STEP 5 *Center* the letter again before you switch to a higher magnification. Hold the microscope *steadily* over the centered letter and use the *zoom switch* to move the eyepiece to the next higher magnification, 80x. Use the *focusing wheel* until you see the letter clearly.

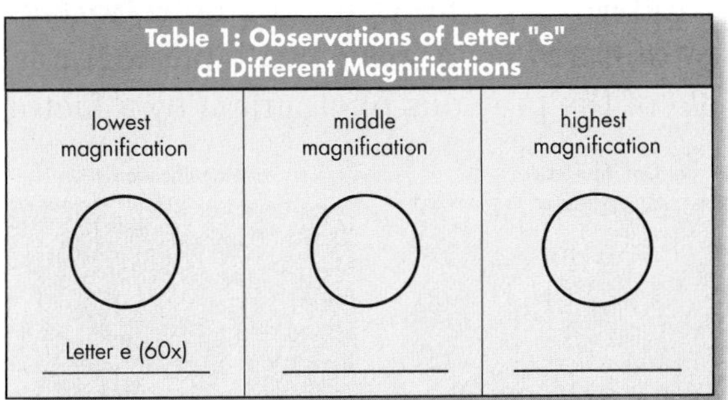

Table 1: Observations of Letter "e" at Different Magnifications		
lowest magnification	middle magnification	highest magnification
Letter e (60x)		

Draw what you see in the second field-of-view circle in the table. Be sure to draw the object within the circle *in its correct size with relation to the size of the field.* Label your drawing with the object (letter "e") and the magnification.

STEP 6 Hold the microscope *steadily* and use the *zoom switch* to move the eyepiece to the highest magnification, 100x. Use the *focusing wheel* until you see the letter clearly.

Draw what you see in the third field-of-view circle in the table. Be sure to draw the object within the circle *in its correct size with relation to the size of the field.* Label your drawing.

STEP 7 Switch roles and repeat Steps 2 through 6.

Exploration 2: Looking at a Leaf

STEP 1 Place the top of the leaf flat on a white piece of paper. Set the microscope magnification at 60x. Move the microscope until the hole in the transparent cover is above a vein in the leaf. Then stand the microscope on the leaf.

STEP 2 Look through the eyepiece and adjust the focusing wheel until you see the surface of the leaf clearly. Change the magnification to 100x. Refocus on the surface of the leaf.

Draw what you see in the field-of-view circle in Table 2 on your record sheet.

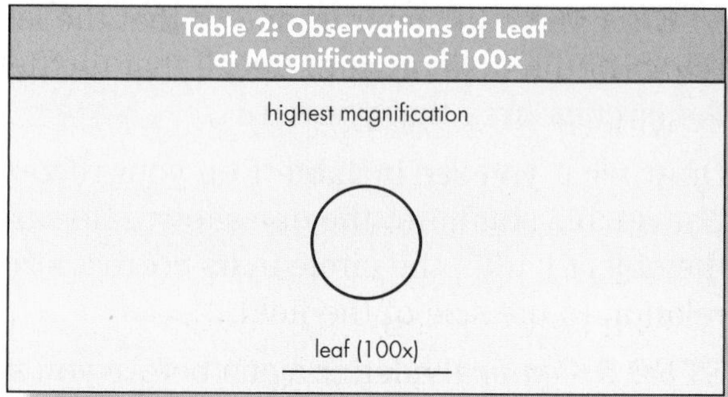

Table 2: Observations of Leaf at Magnification of 100x
highest magnification
leaf (100x)

More about Magnification

Your teacher will show you a series of photographs, each the same size, starting with a view of Earth from space. Each photograph zooms in on the square in the middle of the previous photograph by a factor or "power" of ten.

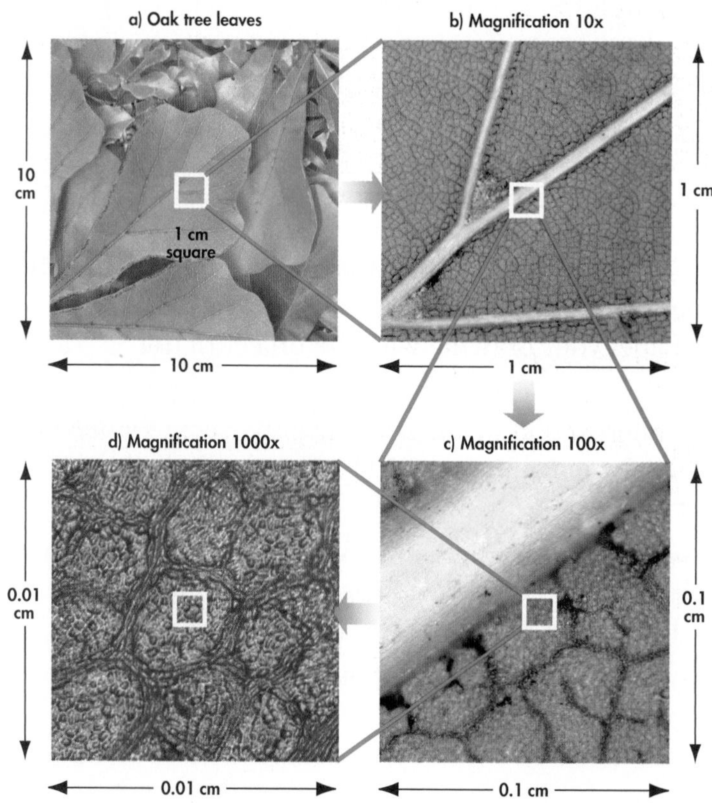

Look at the pictures of the leaf on the previous page. The first photograph (a) represents a 10 cm square view of some oak leaves. The next photograph (b) shows a 1 cm square of a leaf surface that is magnified 10 times (10x). Photograph (c) shows a 0.1 cm square of the leaf surface that is magnified 100 times (100x). The last photograph (d) shows a 0.01 cm square of the leaf surface that is magnified 1000 times (1000x).

Complete the following statements on your record sheet.

1. With each increase in the magnification, you see __________ (*a bigger, the same, a smaller*) part of the leaf.
2. With each increase in the magnification, the part of the leaf you see looks 10 times __________ (*bigger, smaller*).
3. With each increase in the magnification, you can see __________ (*more, the same, fewer*) details of the leaf surface.

Make Sense of Your Ideas

Examine the drawing in your table for the three different magnifications of the letter "e." Discuss the following questions with your team and record your answers on your record sheet.

1. As the magnification of the microscope increases, what happens to the *amount* of the letter you see in the field of view?
2. As the magnification of the microscope increases, what happens to the *thickness* of the lines you see in the field of view?
3. As the magnification of the microscope increases, what happens to the *details* in the ink of the letter that you can see? Describe at least two differences.

Participate in a class discussion to review the team answers.

Our Consensus Ideas

The key question for this activity is:

What are three ways an object looks different through a microscope as its magnification increases?

1. Write your best answer on your record sheet.

Participate in a class discussion.

2. Record the class consensus ideas on your record sheet.

MAKING SENSE OF SCIENTISTS' IDEAS

Activity 4
Chemical Interactions

Purpose

Look at *How To Identify Interaction Types* in the Appendix. Read the second row of rectangles on the map, which briefly describes what is *required* for each interaction to occur. Notice that most interactions are defined in terms of the *types of objects* required and/or the *circumstances* required for the interaction.

Science Words

chemical interaction: any type of interaction that results in at least one new material

physical interaction: any type of interaction that does not result in any new materials

Chemists have discovered that it is more useful to describe interactions of materials (matter) by defining two categories of interactions. These categories are based on the results of the interaction rather than what is required for the interaction. A **chemical interaction** is any type of interaction that results in at least one new material. A **physical interaction** is any type of interaction that does not result in any new materials.

In this activity, you will investigate three different chemical interactions. You will also read about other everyday chemical interactions. The key questions for this activity are:

1. What observations can provide evidence of chemical interactions?
2. What are chemical properties of materials?

Record the key questions for the activity on your record sheet.

You will learn more about physical interactions in the next activity.

Explore Your Ideas

In a chemical interaction, at least one new material is made. What is the evidence that a new material is made during an interaction? You learned in Activity 2 that a material is defined by its unique *set* of properties. You can find out whether a new material is made during an interaction by comparing the properties of the original materials (before the interaction) with the properties of the materials after the interaction. If one or more of the materials after the reaction have a different set of properties, you know that a new material was made and the interaction was chemical.

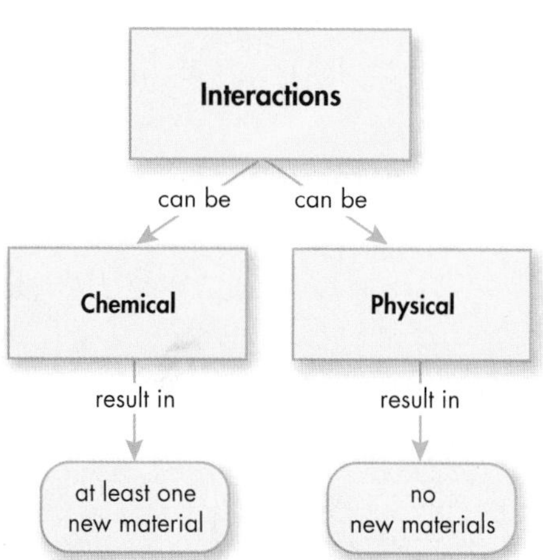

Exploration 1: Crushed Chalk and Vinegar

STEP 1 *Team Manager:* Place about ¼ of a teaspoon of crushed chalk on the wax paper. Then place a few drops of vinegar at a different spot on the wax paper. Observe the properties (what you see, feel, and smell) of each material.

Record the properties of the crushed chalk and vinegar in the first column (Properties of original materials) of Table 1 on your record sheet.

Table 1: Observations of Chemical Interactions

	Properties of original materials (Before the interaction)	Properties of materials left (After the interaction)	Other evidence (During the interaction)
Exploration 1	chalk dust: vinegar:		
Exploration 2	milk: lemon juice:		
Exploration 3	match: air:		

STEP 2 *Supply Master:* Use the beaker to measure about 5 mL of crushed chalk. Pour the crushed chalk into the plastic bag. Then measure 30 mL of vinegar. Pour the vinegar into the bag *and seal the bag quickly.* Observe any changes happening in the bag.

Record your observations in the third column (Other evidence) of your table.

STEP 3 Take turns observing the properties (what you see, feel, and smell) of the materials left in the bag after the interaction has stopped. Does the gas in the bag have an odor? Place a few drops of the liquid left in the bag on your wax paper to observe its properties.

Record these properties in the second column (Properties of materials left) of your table.

Your team will need:

- container of crushed chalk
- container of vinegar
- clean plastic teaspoon
- 50 mL beaker (for measuring)
- piece of wax paper
- plastic re-sealable bag

Wear eye protection.

Clean up spills immediately.

Smell carefully by wafting the gas toward your nostrils with your hand rather than smelling it directly.

Exploration 2: Souring Milk

STEP 1 *Recycling Engineer:* Place a few drops of milk on your wax paper. Then place a few drops of lemon juice on a different spot on your wax paper. Observe the properties (what you see, feel, and smell) of each material.

Record the properties of the milk and lemon juice in the first column of your table.

Your team will need:

- container of whole milk
- container of lemon juice
- clean 50 mL beaker
- eyedropper
- piece of wax paper
- stir stick

STEP 2 *Procedure Specialist:* Pour 10 mL of milk into the beaker. Using the eyedropper, add about 5 mL of lemon juice. Stir the mixture for a few minutes. Observe any changes happening in the beaker.

Record your observations in the third column of your table.

STEP 3 Take turns observing the properties (what you see, feel and smell) of the stuff in the beaker. Swirl the stuff around in the beaker and look at the sides of the beaker.

Record these properties in the second column of your table.

Your team will need:

- 2 small wooden matches in a matchbox
- small aluminum foil pan

Exploration 3: Burning Wood

STEP 1 Observe the properties (what you see, feel, and smell) of a match.

Record the properties of the match in the first column of your table. Also, write down some properties of air.

STEP 2 *Supply Master:* Break a match into as many pieces as possible and place them in the pan. Use the second match to light the pieces of the match in the pan. Observe any changes happening in the pan.

Tie back hair and loose sleeves before lighting the match.

Record your observations in the third column of your table.

STEP 3 When the burning has stopped and the pan has cooled, observe the properties (what you see, feel, and smell) of the material left in the pan.

Record these properties in the second column of your table.

STEP 4 Follow your teacher's directions for cleaning your work area.

Wash hands well after the activity.

Participate in a class discussion about your observations in the three explorations.

Record any observations you missed in your table.

Make Sense of Your Ideas

In a chemical interaction, at least one new material is produced. The evidence of a chemical interaction is that the materials at the end of the interaction have different properties than the original materials (before the interaction).

In Activity 2, for example, you poured vinegar onto baking soda in a flask. Is there any evidence that a chemical interaction occurred between the vinegar and the baking soda—that the interaction resulted in new material(s)?

Table 2 shows the properties of the original vinegar and baking soda (before the interaction), the properties of the materials left after the interaction, and other observations during the interaction.

1. Why is the appearance or disappearance of room temperature gas good evidence that a new material was made?

Participate in a class discussion about your answers.

Discuss and answer the following questions with your team. Use the information in Table 1.

Table 2: Observations of Interaction between Vinegar and Baking Soda

Interaction	Properties of original materials (Before the interaction)	Properties of materials left (After the interaction)	Other evidence (During the interaction)
Pouring vinegar onto baking soda	Vinegar: colorless, transparent liquid with a sour taste and smell	In flask: colorless, transparent liquid with a tangy smell	Bubbling
	Baking soda: white, powdery solid	In baggie: colorless, odorless, transparent gas	Some baking soda disappears

2. What is the evidence that a chemical interaction occurred between the crushed chalk and the vinegar—that the interaction resulted in new material(s)?
3. What is the evidence that a chemical interaction occurred between the milk and the lemon juice—that the interaction produced new material(s)?
4. What is the evidence that a chemical interaction occurred between the wood and the air—that the interaction produced new material(s)?

Participate in a class discussion about your answers.

5. Begin a list like the one shown on your record sheet.

Evidence of a Chemical Interaction

Observations that can provide evidence of a chemical interaction include:

1. __________
2. __________
3. __________
4. __________
5. __________
6. __________

Make Sense of Scientists' Ideas about Chemical Interactions

The first key question for this activity is: *What observations can provide evidence of chemical interactions?*

You began to answer this question by listing observations from your explorations that can provide evidence of the appearance of new materials. To make sure this list is complete, you will consider observations from other everyday chemical interactions.

In Activity 2, you bubbled carbon dioxide gas through the BTB indicator.

1. What is the evidence of a chemical interaction between the carbon dioxide gas and the BTB indicator?

A lot of cooking involves chemical interactions. When you put bread dough in the oven, a gas is produced that causes the bread dough to rise.

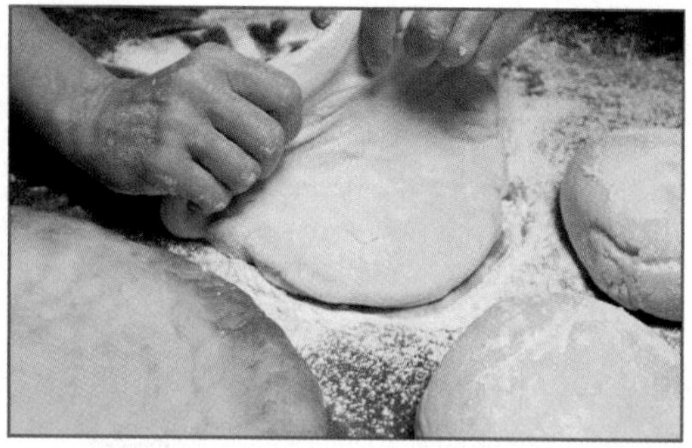

2. What is some other evidence that *baking* cakes and bread is a chemical interaction? (*Hint:* How are the properties of bread different from the properties of the original bread dough?)

Have you ever walked through a forested area and seen or felt rotting wood? Although bacteria are involved, rotting (decaying) is a chemical interaction.

3. What is the evidence that the *rotting of wood* is a chemical interaction?

You learned in your explorations that some interactions are fast. The rotting wood example shows that some chemical interactions can be very slow. Another slow chemical interaction is rusting. You will investigate rusting in Unit 7.

Some interactions can be explosive. For example, when you ride in a car, the gasoline burns explosively and produces new gases that come out of the exhaust. The fast burning of dynamite causes explosions large enough to topple tall buildings.

In fact, you are familiar with many chemical interactions because they occur all around you every day. When you digest food, it reacts with the oxygen you breathe. New materials are made (proteins and fats) that help you move and grow. The changing colors of the leaves in autumn are also evidence of chemical interactions. The green chlorophyll in the leaves breaks down and the yellow, orange, and red colors of the other substances in the leaves appear.

Participate in a class discussion about the observations that provide evidence of chemical interactions in these examples.

4. If appropriate, add to your list of observations.

5. List some common names for chemical interactions from this reading.

Make Sense of Scientists' Ideas about Chemical Properties

The second key question for this activity is: *What are chemical properties of materials?*

The *chemical properties* of a material are descriptions or measurements of how the material interacts *chemically* with other materials. For example, in Activity 2 you tested how the gas carbon dioxide interacts with the flame of a burning splint. You observed that the flame went out right away. So one *chemical property* of carbon dioxide gas is that it does not support burning.

1. What is a chemical property of oxygen when it interacts with the flame of a burning splint? (*Hint*: Look at Table 1 on your record sheet for Activity 2.)

Another chemical property of materials is whether they burn. This property is sometimes called *flammability*. For example, one chemical property of wood is that it burns (is flammable). A chemical property of aluminum is that it does not burn (is not flammable). Have you ever seen the blue flame on a gas stove? The gas that burns in the stove is methane. One chemical property of methane gas is that it burns. Some liquids can also burn. For example, a chemical property of alcohol, kerosene, and gasoline is that they burn. You will learn more about the burning chemical interaction in Unit 7.

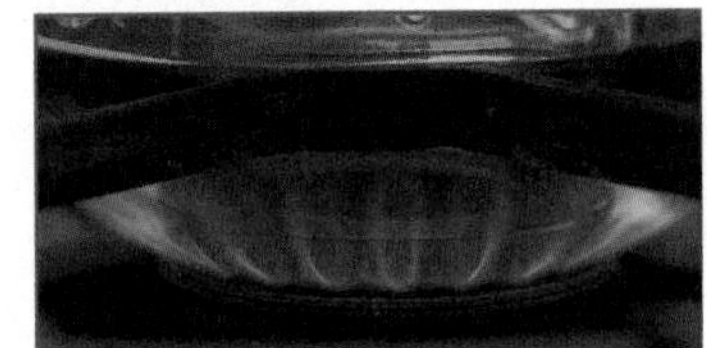

Our Consensus Ideas

Think about the key questions for this activity. Discuss them with your team.

1. What observations can provide evidence of chemical interactions?
2. What are chemical properties of materials?

Look over your list of observations that can provide evidence of a chemical interaction that you made during this activity. Do all team members have the same observations?

1. Write your best answer to the second activity key question on your record sheet.

Participate in a class discussion.

2. Record the class consensus ideas on your record sheet.

MAKING SENSE OF SCIENTISTS' IDEAS

Activity 5
Physical Interactions

Purpose

You learned that chemists find it useful to define two categories of interactions *based on the results of the interaction*, chemical interactions and physical interactions. A chemical interaction is any type of interaction (such as a mechanical, heat-conduction, or electric-current interaction) that results in at least one new material. In this activity, you will explore *physical* interactions. The key question for this activity is:

What are physical interactions and physical properties?

Record the key question for the activity on your record sheet.

Explore and Make Sense of Your Ideas

Science Words

physical property: a description or measurement of what happens to a material during physical interactions

A *physical interaction* is any type of interaction (such as a mechanical, heat-conduction, or electric-current interaction) that *does not result* in any new materials. **Physical properties** of materials are descriptions or measurements of what happens to a material during physical interactions. You are already familiar with many physical interactions and physical properties.

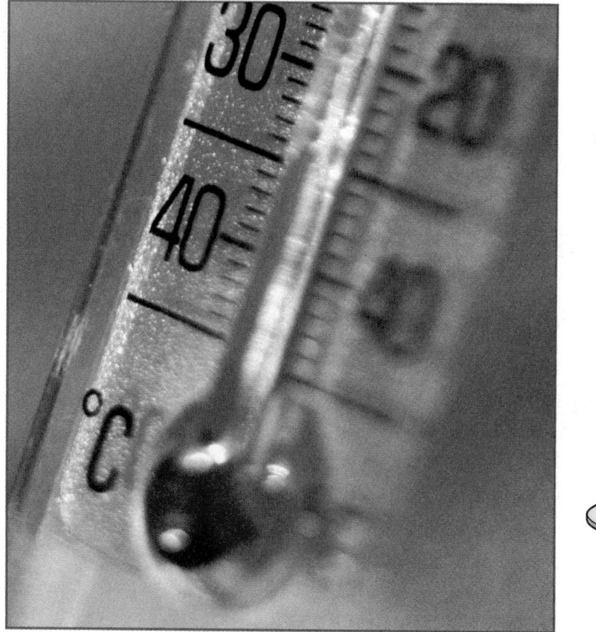

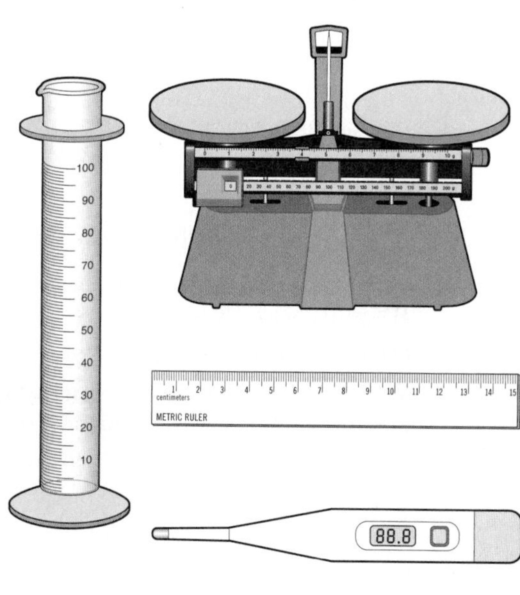

Suppose you were sawing a wood log. The interacting objects are the saw blade and the log. Is this a chemical interaction (results in new material) or a physical interaction (does not result in new material)? To answer this question, you need to compare the properties of the original materials (saw blade and log) and the properties of the materials after the interaction (saw blade and sawdust). These properties are shown in Table 1.

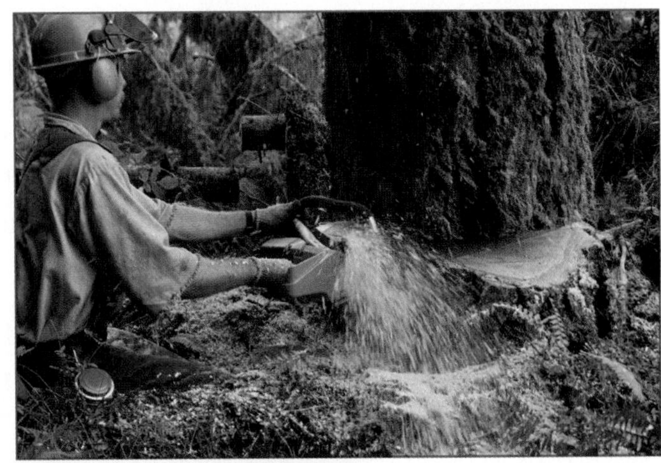

Table 1: Observations of Interaction between a Saw Blade and Log

Interaction	Properties of original materials (Before the interaction)	Properties of materials left (After the interaction)	Other evidence (During the interaction)
Sawing a log of wood	log (without bark): solid; tan color, smooth, woody smell saw blade: solid; metallic; circular shape; many sharp teeth on rim	sawdust: solid, light, small flakes; tan color; woody smell saw blade: hot, solid; metallic; circular shape; many sharp teeth on rim	temperature of saw blade increases

1. What physical property of the saw blade is different before and after the interaction?
2. What physical (sense) properties are different between the log and the sawdust?
3. Imagine that you could determine some *chemical properties* of a log and some pieces of the sawdust made from sawing the log. For example, you could test the flammability (does it burn?) of a log and some pieces of sawdust. You could find out whether the log and the sawdust interact chemically with bromothymol blue (BTB), with alcohol, or with other substances. Do you think the chemical properties of the log and the sawdust would be the same or different? Why or why not?

Participate in a class discussion about your answers.

The evidence of a *physical* interaction is a change in the physical properties of an object (such as its temperature, density, hardness, texture, flexibility, size, or shape) but no change in the chemical properties of the object (such as whether or not it burns or reacts with BTB). In other words, the materials that make up the original object stay the same in a physical interaction.

4. Imagine you tear a paper sheet in two pieces. Tearing is an applied mechanical interaction between your hands and the paper. Do the physical properties of the original sheet change? Do the chemical properties of the original sheet and two pieces (after the interaction) change?

5. Imagine that you stretched the rubber band of a slingshot with your hands. This is a mechanical, elastic interaction. What is the evidence that this is a physical interaction?

6. Tearing and stretching are two examples of everyday words for physical interactions that change the size and/or shape of an object. Discuss this with your team and list some more words for physical interactions.

Participate in a class discussion about your answers.

7. What is the evidence that ironing a piece of clothing (heat-conduction interaction) is a physical interaction?

8. What is the evidence that warming your hand near a cup of hot chocolate (infrared interaction) is a physical interaction?

9. A hot iron is left on a piece of clothing for a few minutes. There is a burning smell. The part of the shirt under the iron is dark brown and brittle. Is this heat-conduction interaction a physical interaction or a chemical interaction? What is the evidence?

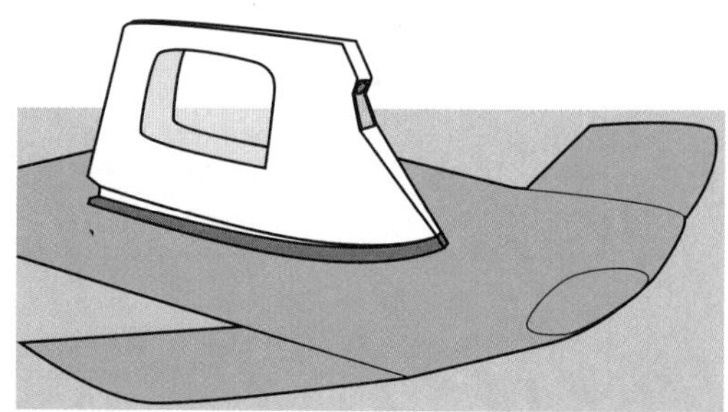

Sometimes, heat-conduction and infrared-interactions result in a phase change (substance changes between its solid, liquid, or gas forms). There are many examples of phase changes in your everyday lives.

10. List four heat-conduction and infrared interactions that result in a phase change.

Interactions that result in the phase change of a substance are physical interactions because no new materials are produced. For example, water does *not* turn into a different material, such as alcohol or gasoline, during a phase change.

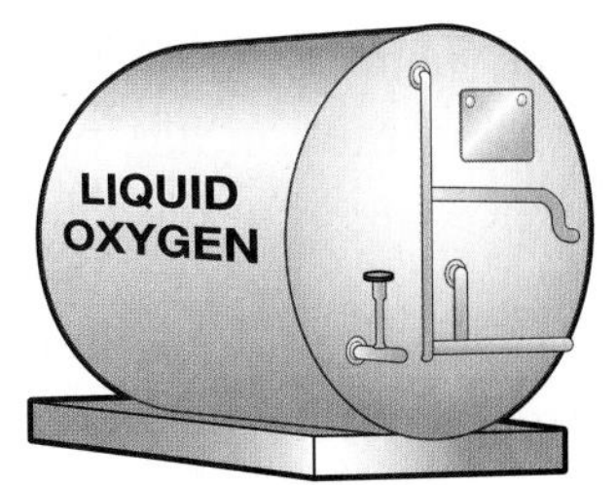

11. What is the evidence that cooling oxygen gas until it condenses into a liquid (heat-conduction interaction) is a physical change?

In Activity 4, you listed several changes that can be evidence of a chemical interaction. However, these clues can fool you. Sometimes, the evidence for a physical interaction and a chemical interaction are the same. Look at what Rebecca has to say about the freezing of water.

12. You be the teacher and explain why this student's conclusion is wrong.

Ice is often a different color than water and is not transparent. This means that freezing is an example of a chemical interaction.

Rebecca

Have you ever added sugar to a drink or sprinkled salt onto your food? Sugar and salt dissolve in watery liquids. Compare what happens when you dissolve salt in water and dissolve an Alka-Seltzer™ tablet in water.

What is the evidence of an interaction between the Alka-Seltzer™ tablet and water? First, the *solid tablet disappears*. Second, the mixture bubbles, *producing a gas*. When you tested this gas in Activity 2, you found that it is carbon dioxide. Finally, the *liquid left in the flask has a tangy odor*. Dissolving an Alka-Seltzer™ tablet in water is a chemical interaction because new substances are made—carbon dioxide gas and a liquid with a tangy odor. You know these are new substances because they have different physical and chemical properties from the original Alka-Seltzer™ and water.

13. When salt is dissolved in water, what are *two* pieces of evidence that an interaction has occurred? (*Hints*: What happens to the solid salt? Does the liquid left in the flask have the same properties as water?)

14. Do you think dissolving salt in water is a chemical interaction or a physical interaction? Why?

Participate in a class discussion about your answers to these questions.

Complete a table, like the one shown below, of some reasons for thinking dissolving salt in water is a physical interaction and some reasons for thinking it is a chemical interaction.

Table 2: Reasons for Classifying Dissolving

Dissolving Salt in Water is a Physical Interaction	Dissolving Salt in Water is a Chemical Interaction

Sometimes, it is difficult to decide whether an interaction is chemical or physical. Some important interactions, such as dissolving salt or sugar in water, might be regarded as a chemical interaction, a physical interaction, or both. In the Practice for Activity 7, you will learn another piece of evidence that chemists use to decide whether a particular case of dissolving is a chemical or physical interaction.

Our Consensus Ideas

The key question for this activity is:

What are physical interactions and physical properties?

1. Answer this key question on your record sheet.

Participate in the class discussion about the key question.

2. Record the class consensus ideas on your record sheet.

DEVELOPING OUR IDEAS

Activity 6
Mixtures

Purpose

You learned in Activity 1 that there is no right or wrong way to classify materials using their properties. However, scientists have learned that properties that tell us something about the *make-up or composition of the material* are the most useful. Therefore, chemists classify materials into two groups: single substances and mixtures of two or more single substances.

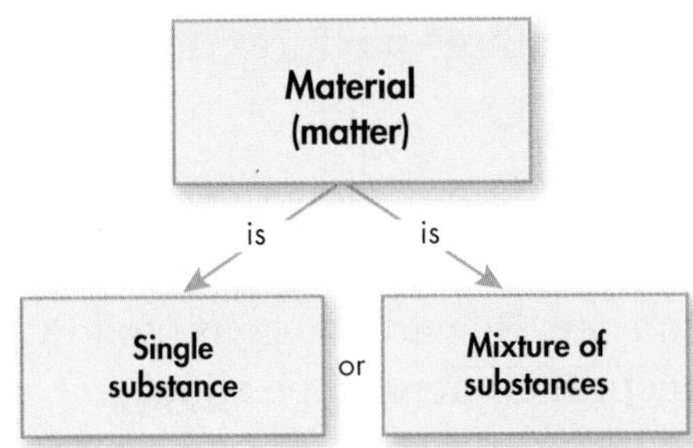

A mixture is a combination of two or more substances in any amounts. Many objects are mixtures. Suppose you mixed some salt in water and stirred. The salt dissolves in the water and seems to disappear. Salt water is clearly a mixture, but does it have tiny specks of salt that are too small to see with your eyes?

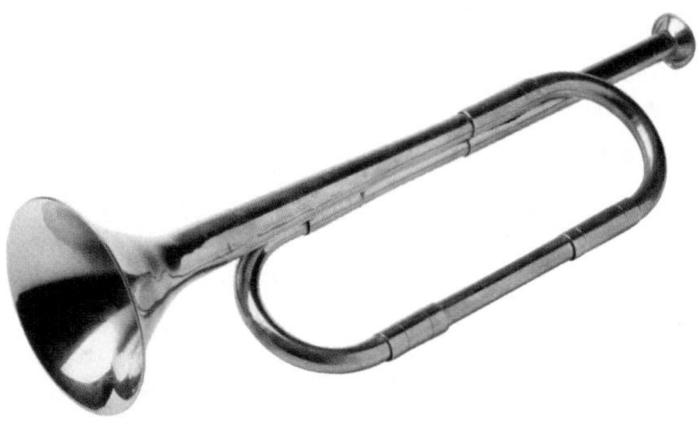

Brass is a mixture of zinc and copper. First, zinc is melted at a very high temperature. Then some liquid copper is mixed with the liquid zinc. As the mixture cools, it hardens into a solid.

A chocolate chip cookie is a mixture of at least two substances.

In this activity, you will use the hand-held microscope to look for tiny specks of the different substances in mixtures. The key questions for this activity are:

1. How big are the smallest pieces in a mixture that you can see with a visible-light microscope?

2. What is one property that defines two types of mixtures?

Record the key questions for the activity on your record sheet.

Clean up spills immediately.

Wash hands well after the activity.

We Think

Think about four of the mixtures from Activity 1: salt water, Tabasco® sauce, brass, and Styrofoam™. Discuss the following question with your team and record your team's answer. You will look at each of these mixtures through your microscope.

In which of these mixtures do you think you will see distinct specks or pieces of the different substances in the mixture? Why?

Participate in the class discussion about the question.

Explore Your Ideas

In the first exploration, you will examine two mixtures that pour. In the second exploration, you will examine three mixtures that do not pour.

Your team will need:

- 2 hand-held microscopes
- 4 microscope slides
- 4 wooden blocks, same size
- small container of salt water
- small container of Tabasco® sauce
- paper towels
- access to soap and water

Exploration 1: Mixtures That Pour

You will look through a microscope at two mixtures that pour, salt water and Tabasco® sauce. You will decide whether these mixtures have tiny specks of different substances that you can see. Be careful as you follow the directions or you will not see anything except the microscope slide!

Each pair of partners will work with one hand-held microscope. One partner (*Procedure Specialist*) should read the procedure out loud while the other partner works with the microscope. They should then switch roles.

STEP 1 Clean the microscope slides with soap and water. Dry the slides and hold them by the edges to avoid making fingerprints.

STEP 2 Place a slide across two blocks, as shown in the diagram. Set the magnification to 60x. Turn on the microscope bulb and gently place the microscope on top of the slide.

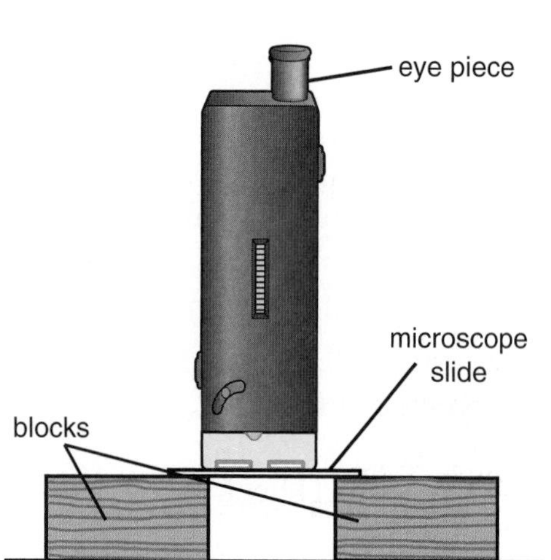

Look through the eyepiece and focus. You will see specks of light and bigger scratches and chips on the microscope slide.

Take turns looking at the slide. Increase the magnification to 100x and take turns looking at the slide again. *Remember what the slide looks like.*

STEP 3 Remove the microscope, turn off the light, and return the magnification to 60x. Make sure the lid of the Tabasco® sauce bottle is on tight and shake the bottle. Open the bottle and carefully place a drop of the sauce in the center of the slide.

If the sauce does not spread out evenly, the slide is dirty. In that case, repeat Step 1.

STEP 4 Put the slide with the drop on top of the blocks again. Identify the hole in the transparent cover of the microscope that is directly below the eyepiece. Center this hole on the drop. Gently lower the microscope until it rests on the slide.

Be sure that some of the drop is inside the hole below the eyepiece.

STEP 5 Take turns focusing and looking at the drop through the microscope.

Be sure to refocus on the liquid, not on the slide. Remember what the slide looks like. If you see only specks of light and scratches, you need to refocus on the liquid.

When you are focused on the Tabasco® sauce, you should see small red specks in the field of view. If you do not see the red specks, move the microscope slowly and gently over the drop until you see the red specks.

STEP 6 Change the magnification to 100x and refocus. Take turns looking at the drop through the microscope.

Draw a picture of what you see in a field-of-view circle in Table 1 on your record sheet. Be sure to draw any specks in their correct size and approximate location with relation to the field of view of the microscope.

Write a brief description of what you see through the microscope.

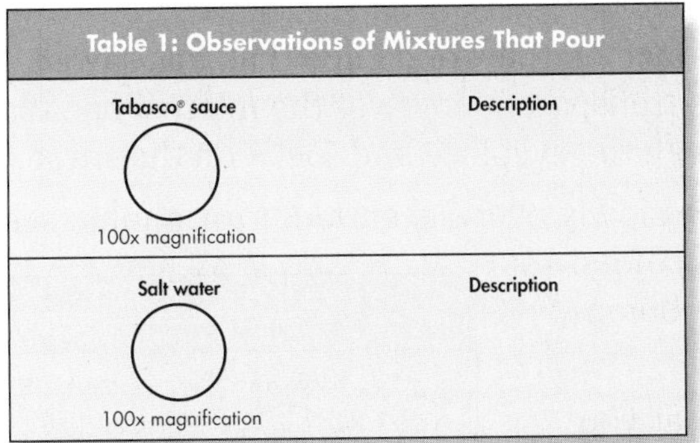

Table 1: Observations of Mixtures That Pour	
Tabasco® sauce 100x magnification	Description
Salt water 100x magnification	Description

Clean up spills immediately.

STEP 7 Dampen a paper towel and wipe the Tabasco® sauce off the holes in the transparent cover of the microscope. Clean the microscope slide also.

STEP 8 Place a drop of salt water on the slide and repeat Steps 4 through 7, recording your observations on your record sheet.

It is very difficult to focus on the salt water because it is transparent like the glass slide. Remember what the slide looks like. If you see only specks of light and scratches, you need to refocus on the liquid.

You may see dark specks of dust or hair floating in the salt water. Do you see white chunks of salt floating in the water?

Your team will need:

- 2 hand-held microscopes
- Styrofoam™ ball
- wood cube
- brass cube
- access to a copper cube, a brass and copper cleaner, and a soft cloth

Exploration 2: Mixtures That Do Not Pour

You will look through your microscope at three mixtures that do not pour: Styrofoam, wood, and brass.

Each pair of partners will work with one microscope. For each step, one partner (*Procedure Specialist*) should read the procedure out loud while the other partner works with the microscope. They should then switch roles.

STEP 1 Return to a magnification of 60x. Hold the Styrofoam™ ball under the hole in the transparent cover of the microscope that is below the eyepiece. Look through the eyepiece and focus on the surface of the ball. You should see compartments (bubbles or pockets).

Turn the ball a few times to see different parts of the surface.

Refocus several times to see the walls of compartments that are deeper in the Styrofoam™ and are not broken.

Draw a picture of what you see in a field-of-view circle in Table 2 on your record sheet. Be sure to draw some unbroken compartments (bubbles or pockets) in their correct size and approximate position with relation to the field of view of the microscope. Label any parts you think have different properties.

Write a brief description of what you see through the microscope.

STEP 2 Find the surface on the wood cube that was sawed. Hold this surface under the hole in the transparent cover of the microscope that is below the eyepiece. Look through the eyepiece and focus on the surface of the wood.

Move the microscope slowly over the surface. You should see distinctly different shapes and colors. Do you see any dips, bumps, and shadows?

Draw a picture of what you see in a field-of-view circle in Table 2. Label any parts you think have different properties.

Write a brief description of what you see through the microscope.

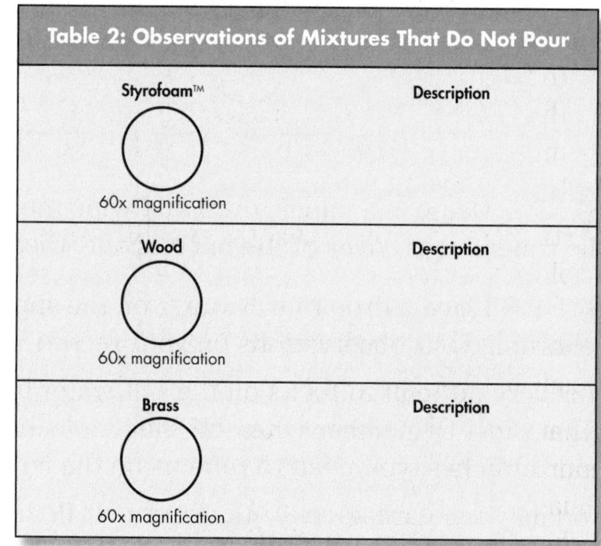

Table 2: Observations of Mixtures That Do Not Pour	
Styrofoam™ 60x magnification	Description
Wood 60x magnification	Description
Brass 60x magnification	Description

STEP 3 Clean one surface of the brass cube with the brass and copper cleaner and a soft cloth. Hold the brass cube by the sides to avoid making fingerprints.

STEP 4 Hold the cleaned surface under the hole in the transparent cover of the microscope that is below the eyepiece. Look through the eyepiece and focus on the surface of the brass.

Move the microscope slowly over the surface of the brass, refocusing several times. You will probably see some long scratches, chips, and bumps. Try to refocus to look inside the scratches, chips, or bumps.

Do you see any compartments or bubbles? Brass is a mixture of the metals copper and zinc. Copper is a bright, reddish brown color and zinc is a silvery gray color. Do you see any distinct chunks of zinc (silvery gray) mixed in with the copper (bright, reddish brown)?

Draw a picture of what you see in a field-of-view circle in Table 2. Label any parts with different properties.

Write a brief description of what you see through the microscope.

STEP 5 Look at the single substance, copper, for a comparison. First, clean one surface of the copper cube (follow Step 3). Look at the cleaned surface through your microscope. Move the microscope slowly over the surface of the copper, refocusing several times.

1. What is similar about the appearance of the copper (single substance) and the brass (mixture)?

Discuss the observations and drawings you made in the explorations with your team before answering the next two questions.

2. In which of the five mixtures did you distinctly see one or both of the different substances in the mixture? What is your evidence?

3. In which of the five mixtures did you *not* see distinctly one or both of the different substances in the mixture? (Do not count the scratches on the microscope slide or the shadows of the scratches and bumps on the solid brass.) What is your evidence?

Participate in a class discussion about your observations in the two explorations.

Make Sense of Your Ideas

Read through the question. Discuss your answer with your team.

What two substances do you think are mixed together in Styrofoam™? Justify your answer. Remember: Unlike most solids, Styrofoam™ can be easily squashed between your fingers. Why? What do you think is inside the *unbroken* compartments of the Styrofoam?

When you looked through your hand-held microscope (magnification 60x) deeper inside Styrofoam™, you saw pieces with different properties. You saw *unbroken* compartments (bubbles or pockets) with solid walls. Air is trapped inside these compartments. In Styrofoam™, you saw one of the two substances in the mixture—the plastic walls of the compartments and bubbles.

When you looked at Tabasco® sauce and wood, you also saw pieces with different properties. In Tabasco® sauce, you saw tiny specks of solid red pepper floating in a liquid. Wood has solid pieces with different colors and distinct shapes. So in Tabasco® sauce and wood, you saw the different substances in the mixtures.

It is harder to decide whether salt water and brass have pieces of different substances that you can see. If your slide was not clean, you may have seen specks of dust floating in the salt water. Sometimes, it is hard to see the surface of the brass clearly with all the bumps and scratches. You may have noticed small bumps with a dark brown color on the brass surface. On the other hand, the copper (single substance) had similar, dark brown bumps as the brass (mixture).

Discuss the following question with your team. Be prepared to share your team's answer and reasons with your class.

- Suppose you looked at salt water and brass through a microscope with a higher magnification than 100x. Do you think you would see the white salt in the salt water and silver-gray zinc in the reddish-brown copper?

Participate in a class discussion about your answers.

Make Sense of Scientists' Ideas about Mixtures

Limit to Maximum Magnification

Your eyes limit the details you can see through a microscope. That is why the magnification of good laboratory microscopes is usually about 1300x (1300 times). A higher magnification than 1300x makes the image bigger, but you can't see any more details.

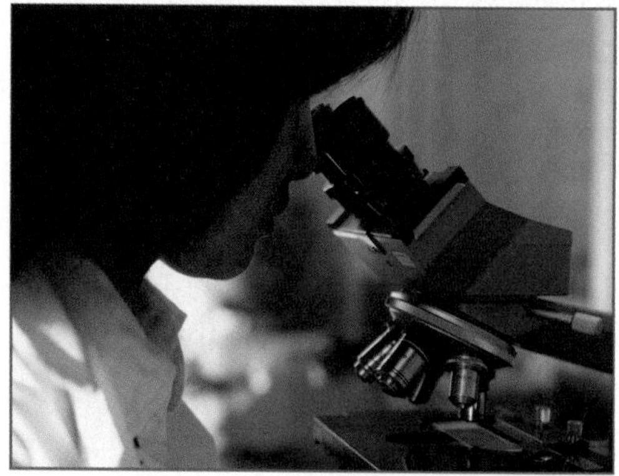

Scientists and engineers have designed special visible-light microscopes that have magnifications up to about 500,000x (500,000 times). You do not look through these microscopes. They use special techniques, digital cameras, and computer software to make photographs of materials. Your teacher will show you some photographs of an oak leaf taken through one of these special microscopes.

Suspensions and Solutions

Chemists often use the limit of what you can see with a light microscope to define two types of mixtures. One type of mixture is called a **suspension**. A suspension has visible pieces of at least one of the different substances in the mixture. This means that the pieces must be larger than 0.2 micrometers (abbreviated μm). (A micrometer is one-millionth of a meter.) The other type of mixture is called a **solution**. A solution is a mixture that does not have any visible pieces of the different substances in the mixture, even when viewed through the best research microscopes.

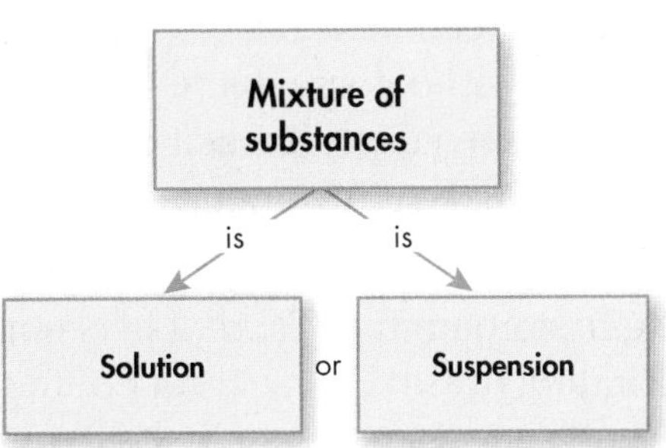

Science Words

suspension: a mixture that has visible pieces of at least one of the substances. The pieces must be larger than 0.2 μm.

solution: a mixture that does not have any visible pieces of the different substances

In a suspension, one (or more) material floats in a different supporting material. You are familiar with many different types of suspensions. Tabasco® sauce is a suspension of solid red pepper floating in a supporting liquid. Orange juice is a suspension of solid fragments of orange floating in the supporting juice.

Styrofoam™ and puffed cereals are suspensions with small pockets or bubbles of air "floating" in a solid. Tiny bubbles of air can also float in a supporting liquid. Soapsuds, shaving cream, and hair mousses are examples of suspensions with tiny air bubbles floating in a supporting liquid.

Some suspensions have tiny solid specks or liquid droplets floating in a supporting gas. For example, small specks of dust can float in the air. Steam and fog are tiny droplets of water floating in air. Hair spray is also liquid droplets floating in air.

Some suspensions have small solid pieces "floating" in another solid. Many minerals, like granite, have small pieces of feldspar and mica floating in quartz.

In a solution, one (or more) substance is dissolved in another substance. Many *liquid* solutions are found in and around the home. Syrups are made by dissolving sugar, extracts of fruit, or tree sap in water. Many juices, such as clear apple juice and grape juice, are solutions. Some medicines are also solutions. Tincture of iodine (an antiseptic) is a mixture of iodine dissolved in alcohol. Some eyewashes contain boric acid dissolved in water.

Gases can also dissolve in liquids. For example, ocean water has oxygen as well as salt dissolved in water. All carbonated fizzy drinks are solutions of carbon dioxide dissolved in a flavored liquid.

A solution can be a gas or a solid, as well as a liquid. For example, the air you breathe (without specks of dust or smoke) is a solution of oxygen and other gases dissolved in nitrogen. *Alloys* are solid solutions made by dissolving one substance (usually a metal) in another metal. For example, stainless steel knives, forks, and spoons are made of nickel and chromium dissolved in iron.

The first column in Table 3 lists some of the mixtures from Activity 1, including brass. The second column describes the observations made through a research microscope up to a magnification of about 500,000x.

Use the results of the observations to classify each mixture as a suspension or a solution. Write your decision in the third column of Table 3 on your record sheet.

Table 3: Suspensions and Solutions

Mixture	Observation through a microscope with a magnification of 500,000x	Suspension or solution?
air	No pieces of any gas or gas mixture are seen through a microscope.	
brass	No pieces with distinctly different properties are seen (when sanded and polished by machines).	solution
muddy river water	Mud is seen floating around in the water.	
salt water	No pieces with distinctly different properties are seen.	
Styrofoam™	Compartments of plastic (that can hold air) are seen.	
Tabasco® sauce	Red pieces are seen floating in a liquid.	

1. Justify your answer for the mixtures you decided were suspensions.
2. Justify your answer for the mixtures you decided were solutions.

Participate in a class discussion about your answers to these questions.

Our Consensus Ideas

The key questions for this activity are:

1. How big are the smallest pieces in a mixture that you can see with a visible-light microscope?

2. What is one property that defines two types of mixtures?

1. Use the information in the Make Sense section to answer the two key questions for this activity.

Participate in a class discussion about the key questions.

2. Write the class consensus ideas on your record sheet.

MAKING SENSE OF SCIENTISTS' IDEAS

Activity 7
Solutions and Chemicals

Purpose

Chemists classify materials as mixtures or single substances. You learned in the last activity about the two kinds of mixtures: suspensions and solutions. Chemists call single substances chemicals.

Solutions and chemicals (single substances) have one similarity. They do not have pieces with different physical properties that can be seen through light microscopes. You cannot always tell by looking whether a material is a solution or a chemical (single substance). For example, consider gases such as helium and carbon dioxide, liquids like vinegar and alcohol, and solids like steel and aluminum. They do not have any visible pieces with different properties. Are any of these materials solutions of two or more substances or are they all chemicals (single substances)?

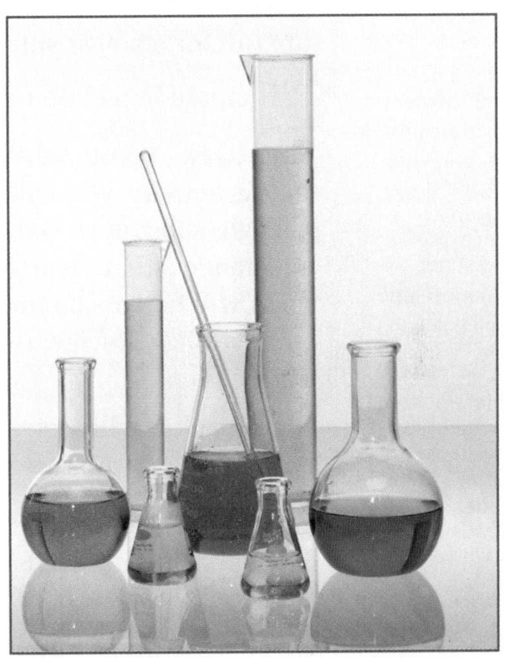

Can you tell which of these transparent liquids is a solution (mixture) and which is a chemical (single substance)?

This activity has two parts. In the first part, you will investigate a property that chemists use to classify materials as either solutions or chemicals. In the second part of the activity, you will consider some methods of separating the different substances in the two types of mixtures: solutions and suspensions.

Here are the key questions for this activity:

1. **What is a property that helps you decide whether a substance is a solution (mixture) or a single substance (chemical)?**
2. **What is similar about the methods of separating the different substances in a solution and in a suspension?**

Record the key questions for the activity on your record sheet.

We Think

Discuss the following question with your team. Write your team's answer on your record sheet.

Suppose you mix salt and water until the salt dissolves and the liquid is clear. You *know* the result is a mixture because you made it. Imagine that someone gives you a clear liquid at room temperature. How could you find out if it is a mixture or a single substance? (You are not allowed to taste it or smell it!)

Participate in a class discussion.

Your team will need:

- butane or other burner
- boiling chips (for test tube)
- ringstand
- test-tube clamp (for ringstand)
- large test tube
- centimeter ruler
- 10 mL of liquid assigned to your team
- test-tube holder (hand-held)

Explore Your Ideas about Solutions

In this activity, you will explore what happens when you boil a solution and a single substance. Half of the class will boil pure water. Pure water is a single substance—there is nothing else mixed in the water. The other half of the class will boil a salt-and-water solution. You will observe what happens (if anything). Review the directions in *How To Follow Safety Rules during an Experiment.*

Your teacher will assign your team either pure water (single substance) or a salt-and-water mixture.

STEP 1 *Procedure Specialist:* Set up the equipment as shown in the diagram. Place five boiling chips in the test tube. Then pour 10 mL of your liquid into the test tube.

If you are using a butane burner, make sure that the bottom of the test tube is 6 cm above the top of the burner.

Wear eye protection.

Be careful with the alcohol lamp so that no alcohol spills.

Do not leave the flame unattended.

Tie back long hair and loose sleeves before lighting the burner.

Use a test-tube holder or tongs if you must handle the hot test tube—remember that hot glass looks the same as cool glass.

If you are using an alcohol burner, make sure that the bottom of the test tube is 1-2 cm above the top of the burner.

STEP 2 *Team Manager:* Light the burner and bring the liquid to a boil. Be very careful with the hot glassware.

STEP 3 Continue boiling until the liquid has almost disappeared. Then turn off the burner.

STEP 4 After the test tube has cooled a little, observe the test tube.

Is there anything left in the test tube? If so, what does it look like? Record your observations in Table 1 on your record sheet.

STEP 5. Use the hand-held test-tube holder to hold the test tube in case it is still warm.

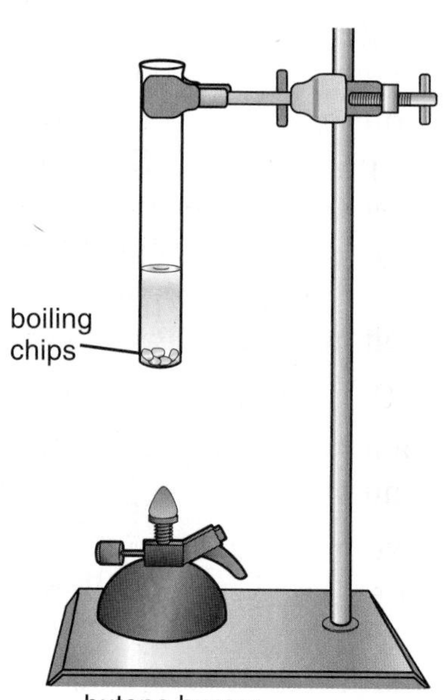

butane burner

Exchange your test tube with a team that boiled the other liquid. Observe their test tube and consider if the result of their exploration was the same or different from your team's result.

Participate in the class discussion about your observations and the observations of the other teams.

Record the observations of the other teams in Table 1 on your record sheet.

Table 1: Boiling Pure Water and Salt Water		
	Pure water (Single substance)	Salt water (Solution)
Is there anything left in the test tube after most of the liquid has boiled away? If so, describe what is left.		

What is *different* about what is left after the pure water (single substance) and the salt-water solution (mixture) boil until the liquid is gone?

Participate in a class discussion about your answer.

Make Sense of Scientists' Ideas about Solutions

In your exploration with the salt-and-water solution, you separated the salt from water by heating the solution until most of the liquid water boiled away and the solid salt was left. *All* solutions (solids and gases as well as liquids) can be separated into different substances during a physical interaction that results in a phase change. Separation of the different substances in a solution occurs not just during boiling (liquid to gas) but also in the other three phase changes (melting, condensing, and freezing).

1. Complete the following statement on your record sheet.

If a material separates into different substances during a phase change, it must be a ______________________________.

In your exploration with pure water (single substance or chemical), all the water boiled away and nothing (no solid) was left. The water did not separate into different substances. All single substances behave this way during any kind of phase change (melting, boiling, condensing, or freezing).

2. Complete the following statement on your record sheet.

If a material does *not* separate into different substances during a phase change, it must be a ________________.

The first column in Table 2 lists the nine materials from Activity 1 that do not have visible pieces (larger than 0.2 µm) of different substances. These materials could be solutions (mixtures) or single substances (chemicals). The second column describes what happens to each material during a heat-conduction interaction that results in a phase change.

Classify each material as a single substance *(ss)* or a solution *(sol)*. Write your decision in the third column of the table on your record sheet.

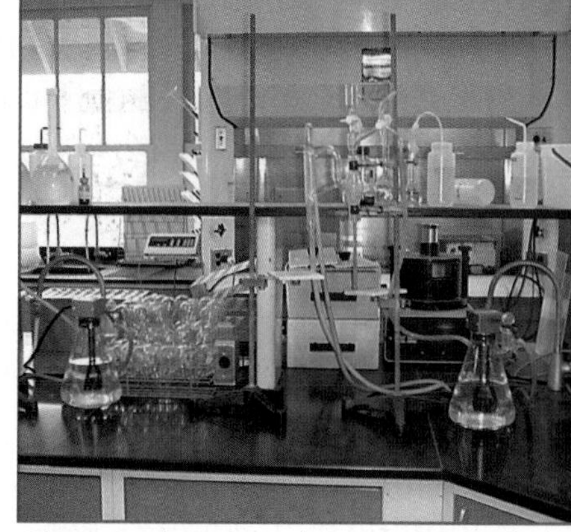

Chemists use special equipment called *distillation* apparatus to separate the substances in gas and liquid solutions.

Table 2: Solutions and Single Substances

Material	What happens to material during a heat-conduction interaction	Single substance or solution?
pure water	Water does not separate into different substances during a phase change.	ss
salt water	Salt water separates into salt and water when it is boiled.	sol
air	Air separates into different gases (mostly nitrogen and oxygen) when it is cooled and liquefies.	
aluminum	Aluminum does not separate into different substances during a phase change.	
brass	Brass separates into different metals (copper and zinc) during a phase change.	
carbon dioxide	Carbon dioxide does not separate into different substances during phase change.	
charcoal	Charcoal does not separate into different substances during a phase change.	
helium	Helium does not separate into different substances during a phase change.	
sugar	Sugar does not separate into different substances during a phase change.	

3. Justify your answer for the materials that you decided were single substances.
4. Justify your answer for the materials that you decided were solutions.

Participate in the class discussion about your classifications.

Explore and Make Sense of Your Ideas about Separating the Substances in Solutions and Suspensions

The different substances in a solution can be separated by heat-conduction and/or infrared interactions that result in a change of phase of the substances. This method works because different substances change phase (melt/freeze and boil/condense) at different temperatures.

You learned in Activities 4 and 5 that chemists classify all types of interactions as either chemical or physical. Chemical interactions result in new substances. Physical interactions do not result in new substances.

Discuss the following question with your team and write your answer on your record sheet.

1. What category of interactions, chemical or physical, separates the invisible substances in a solution? Justify your answer.

Participate in a class discussion about your answer.

Think about different methods for separating the different substances in a suspension. The visible pieces in suspensions can range in size from large to small to microscopic (can only be seen through a microscope). What types of interactions can be used to separate the large, small, or microscopic substances from the supporting material in a suspension?

You can use common mechanical devices to separate *large chunks* in a suspension from the other substance(s) in the suspension. For example, you can use your fingers or a kitchen knife to separate the chocolate chips from the other materials in a chocolate chip cookie.

2. What are some other devices you can find at home that are used to separate suspensions into different parts?

Participate in a class discussion about your answer.

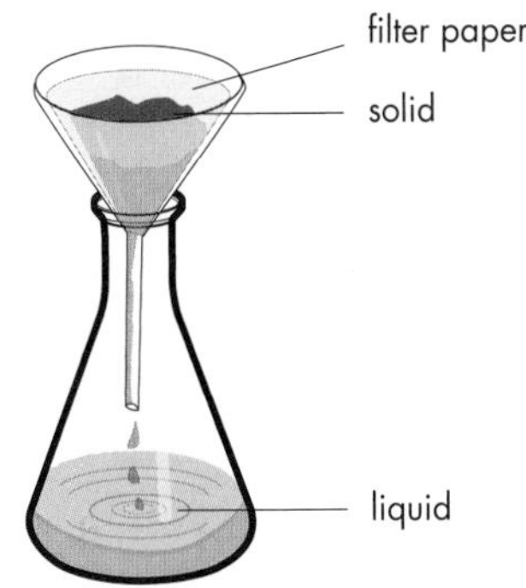

Chemists use a variety of devices to separate suspensions. They often pour suspensions through strainers with different size holes to separate *large* pieces of solid from the supporting material. Chemists also use filter paper and a funnel to separate *small* solid parts from the supporting liquid. Filter paper is like a strainer with very small holes.

3. If all of a liquid material pours through filter paper, can you tell whether the material is either a solution or a single substance? Explain your answer.

Participate in a class discussion about your answer.

Chemists also use a mechanical device called a *centrifuge* to separate tiny, *microscopic parts* in suspensions. Tubes of a suspension are placed in the center of the centrifuge. The tubes are spun at high speeds. The heavier (denser) parts of the suspension move to the bottom of the tubes. The lighter (less dense) parts move to the top. The different parts can then be observed separately.

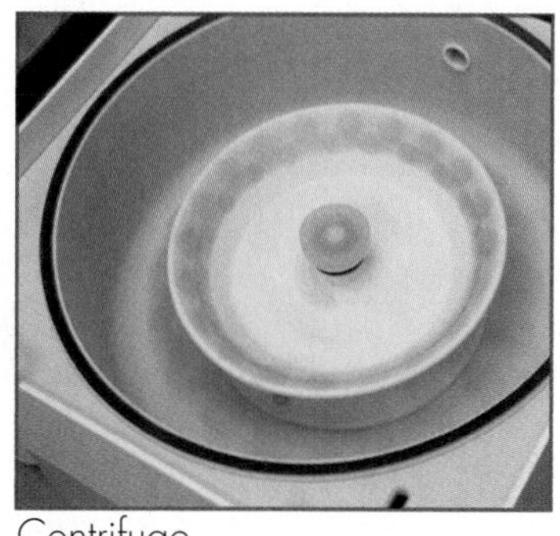
Centrifuge

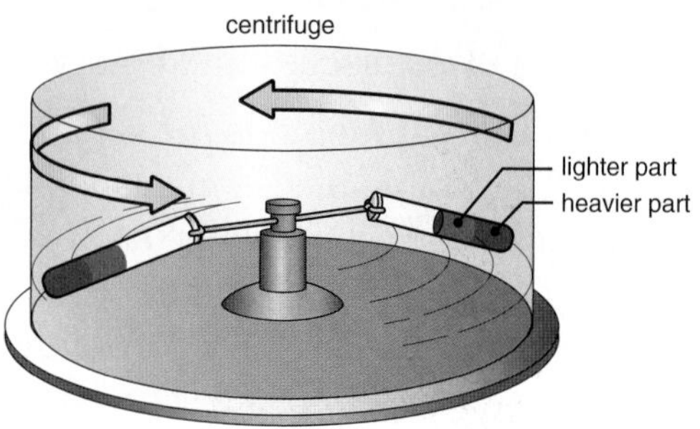

Discuss the following question with your team and record your answer on your record sheet.

4. What category of interactions, *chemical* or *physical*, can separate the visible (larger than 0.2 µm) substances in a suspension? Justify your answer.

Participate in a class discussion about your answers.

Our Consensus Ideas

Recall the key questions for the activity:

1. What is a property that helps you decide whether a substance is a solution (mixture) or a single substance (chemical)?

2. What is similar about the methods of separating the different substances in a solution and in a suspension?

Discuss the key questions with your team.

1. Write your answers on your record sheet.

Participate in a class discussion about the key questions.

2. Write the class consensus ideas on your record sheet.

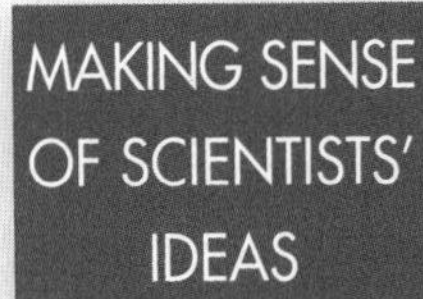

Activity 8
Classifying Chemicals

Purpose

In Activities 6 and 7, you learned that there are two types of mixtures: suspensions and solutions. Different substances in both types of mixtures can be separated by *physical interactions*. Single substances, which are called *chemicals*, cannot be separated into different substances during physical interactions.

Chemists classify single substances into two groups: elements and compounds. This activity has two parts. In the first part, you will do an exploration with electrical-energy transfer to the chemical water. The key question for this part of the activity is:

What is a property that helps you decide whether a chemical (single substance) is an element or a compound?

Record the key question for the first part of the activity on your record sheet.

In the second part of the activity, you will be introduced to the Periodic Table of the Elements.

Explore Your Ideas

Your team will need:

- small beaker
- distilled water
- 1.5 mL (about ¼ tsp.) Epsom salt
- teaspoon
- 9-V battery
- 2 connecting wires (30 cm or more) with alligator clips
- 2 pencils, sharpened at both ends
- 2 magnifying glasses

STEP 1 *Supply Master:* Fill the beaker about one-third full with the distilled water. Add about 1.5 mL (¼ tsp.) Epsom salt to the water. Be sure to stir the Epsom salt in the water until it is *completely* dissolved.

Note: The dissolved Epsom salt just helps the water conduct electricity better. Chemists have found in many experiments that the salt does not change or take part in the interaction.

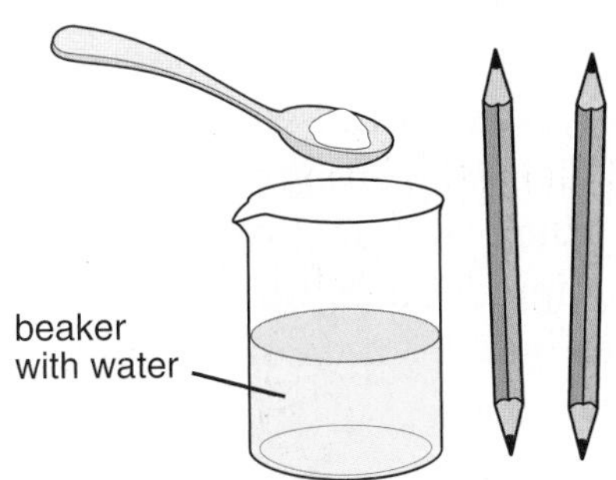

STEP 2 *Procedure Specialist:* Attach one end of the wire to one end of the pencil. Connect the other end of the wire to one of the terminals of the 9-V battery.

Wear eye protection.

Clean up spills immediately.

Do not leave the battery unattended when it is connected.

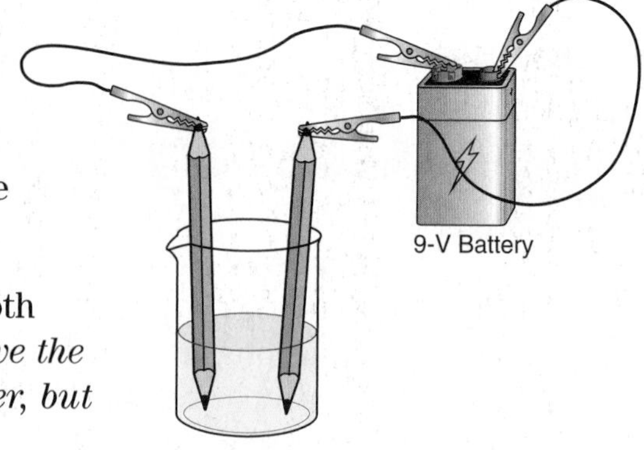

STEP 3 *Recycling Engineer:* Attach one end of the other wire to one end of the pencil. Connect the other end of the wire to the other terminal of the 9-V battery, as shown in the diagram.

STEP 4 *Manager:* Place the tips of both pencils in the water in the beaker. *Move the pencils until the tips are close together, but not touching each other.*

STEP 5 Use a magnifying glass to observe what happens near the pencil connected to the *negative* terminal of the battery.

In Table 1 on your record sheet, draw a picture and describe what you see happening in the liquid near this (negative) pencil.

Table 1: What Happens Near the Negative (–) and Positive (+) Battery Terminals	
Draw a picture and describe what happens near the pencil attached to the **negative** (–) end of the battery.	Draw a picture and describe what happens near the pencil attached to the **positive** (+) end of the battery.

STEP 6 Use a magnifying glass to observe what happens near the pencil connected to the *positive* terminal of the battery.

In the table on your record sheet, draw a picture and describe what you see happening in the liquid near this (positive) pencil. (You may need to watch closely for a longer time.)

Discuss the following question with your team and then write your best answer.

1. What is the evidence that an interaction occurred in your exploration?

Participate in a brief class discussion about the team's answers.

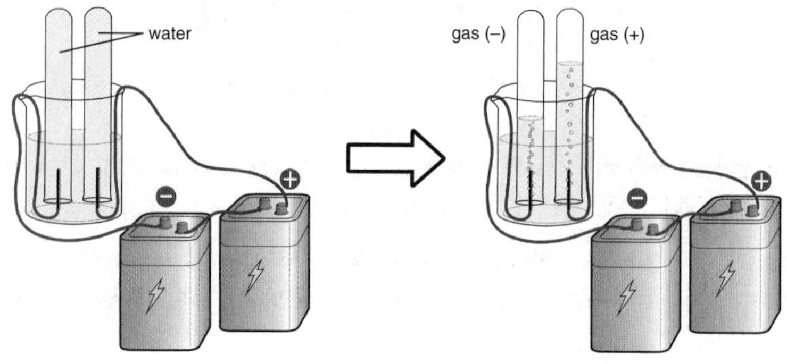

STEP 7 While you were completing your exploration, your teacher used special equipment, called electrolysis equipment, to collect the gases. Instead of pencils, this equipment uses thin, steel rods. The gas bubbles are collected in two test tubes.

2. Do you think that the gases collected near the two battery terminals are the same or different gases? Why?
3. What gas or gases do you think are made near each terminal? Explain your reasons.

STEP 8 Your teacher will show you a video or demonstrate the flame test for the two gases.

Record the results of these tests in Table 2 on your record sheet.

Table 2: Flame Test of Two Gases	
Flame test of gas collected from the **negative** (–) battery terminal.	Flame test of gas collected from the **positive** (+) battery terminal.

Discuss the following questions with your team and write the answers on your record sheet.

4. What gas do you think is made near the steel rod attached to the *positive* terminal of the battery? What is the evidence? (*Hint:* Look at Table 1 for Activity 2 on your record sheets.)
5. Is the gas made near the steel rod attached to the *negative* battery terminal the same gas or a different gas? Do you think this gas is the same as any of the gases you tested in Activity 2? What is your evidence?
6. Chemists have done further tests of the gas made at the negative battery terminal. This gas has a boiling point of –252°C. The mass of one liter (density) of this gas is 0.08 g. Look at the *Table of Densities* and *Table of Melting and Boiling Points* in the Appendix. What is this gas?
7. The gas collected at the positive terminal has a boiling point of –183°C. The mass on one liter of the gas is 1.33 g. What is the gas?
8. Is the electrolysis of water a physical interaction or a chemical interaction? What is your evidence?

Participate in a brief class discussion about the answers.

Make Sense of Scientists' Ideas about Classifying Chemicals

Elements and Compounds

When you transfer electrical energy to pure water (single substance), a chemical interaction occurs. Two gases are produced in the chemical interaction. Chemists have found that most single substances can be broken down into other substances during chemical interactions.

Chemists have also found a small set of single substances, like nitrogen, carbon, silver, and lead, that *cannot* be broken down into other substances during chemical interactions.

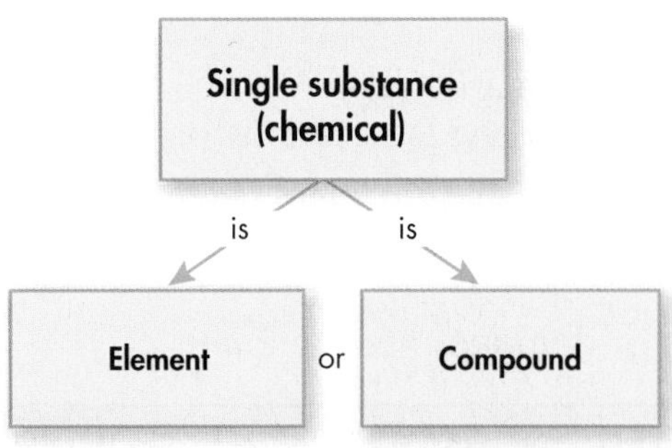

Chemists classify single substances (chemicals) into two groups, **elements** and **compounds**. Elements do not break down into simpler substances during chemical interactions. Compounds are chemicals that break down into elements during chemical interactions.

Scientists have discovered 92 different elements in materials found on Earth. (They have made other elements in the laboratory.) Most of the 92 elements chemically interact with other elements to make compounds. Very few elements are found in their pure state in nature. These include carbon, nitrogen, oxygen, sulfur, helium, neon, argon, krypton, xenon, and radon.

Science Words:

element: a single substance (chemical) that does not break down into simpler substances during chemical interactions

compound: a single substance (chemical) that breaks down into elements during chemical interactions

Suspensions and solutions are mixtures of compounds and/or elements. Compounds consist of two or more elements. *All* materials found on Earth are made up of some combination of the 92 elements. For this reason, elements are called the "basic ingredients" of all materials (matter).

1. Complete the statement below on your record sheet.

If a single substance (chemical) breaks down into elements during chemical interactions, it must be a(n) _______________.

2. Complete the statement below on your record sheet.

If a single substance (chemical) does not break down into simpler substances during chemical interactions, it must be a(n) _______________.

Table 3: Compounds and Elements

Single Substance	What happens during chemical interactions	Compound or Element?
pure water	Water breaks down into two elements (hydrogen and oxygen) during a chemical interaction (electrolysis).	c
aluminum	Aluminum does not break down into simpler substances.	
carbon dioxide	Carbon dioxide breaks down into two elements (carbon and oxygen).	
charcoal	Charcoal does not break down into simpler substances.	
helium	Helium does not break down into simpler substances.	
sugar	Sugar breaks down into three elements (carbon, hydrogen and oxygen).	

The first column in Table 3 lists the single substances (chemicals) you classified in Activity 1. The second column describes what happens to these substances during chemical interactions.

Use the results to classify each material as a compound *(c)* or an element *(e)*. Write your decision in the third column of Table 3 on your record sheet.

3. Justify your answer for the materials that you decided were elements.

4. Justify your answer for the materials that you decided were compounds.

Participate in a class discussion of the answer to these questions.

More about Elements

Before the battery was invented by Alessandro Volta in 1800, chemists had a much shorter list of elements. It included the metals copper, silver, tin, gold, and lead, other solids such as sulfur, arsenic, phosphorus, and boron, the liquid mercury, and the gases hydrogen, nitrogen, and oxygen. Electrolysis helped chemists discover many new elements. It also helped chemists discover that some substances they thought were elements were really compounds because they could be broken down into two or more elements.

Introducing the Periodic Table of the Elements

By about 1850, several chemists had discovered that there were families of elements with similar properties. These were grouped together to construct the table called the *Periodic Table of the Elements*. Look at the modern version of the Periodic Table on the inside back cover of this book. Columns of the Periodic Table are numbered 1 through 18 and rows are labeled 1 through 7. The two rows of elements, 57–71 and 89–103, actually fit into the 6th and 7th rows. They are separated from the rest of the table to make the table fit on a page. The lighter (less dense) elements are listed toward the top of the table and the heavier (more dense) elements are near the bottom.

At the bottom left hand corner of the Periodic Table, you will see a Key. The Key shows the signs for elements that are solids, liquids, and gases at room temperature. The sign is for elements that are made in the laboratory (artificially made).

5. How many elements are made in the laboratory (artificially made)?

6. How many elements are liquids at room temperature?

Notice that each element in the Periodic Table has a different symbol (one, two, or three letters). These symbols have the same meaning in any language.

7. The 8th element on the Periodic Table is oxygen. What is the chemical symbol for oxygen?

8. The 26th element on the Periodic Table is iron. What is the chemical symbol for iron?

9. Many elements listed in the Periodic Table of the Elements are scarce. Either the Earth contains very little of the element or it can only be made in the laboratory. Make a list of at least *five* elements that are unfamiliar to you.

Many symbols on the Periodic Table are taken from the element's name. For example, the symbol for hydrogen is H. The symbol for argon is Ar. Some symbols seem to be unrelated to the element's name. For example, Pb is the symbol for lead and Sn is the symbol for tin. These symbols come from the Latin names for these elements. Latin was the language of science when these substances were identified as elements. So Pb stands for plumbum, which is the Latin name for lead. Sn stands for stannum, which is the Latin name for tin.

Chemists named some of the recently discovered elements after famous people or places. The 101st element, Md, is named mendelevium after the Russian chemist Dimitri Mendeleev (1834 – 1907). He was the first chemist to make a table of elements grouped together by similar properties. The 98th element, Cf, is named californium after California, the state in which it was discovered.

10. What are two elements named after a place (not including californium)?

11. Find another element named after a famous scientist (not including mendelevium)?

Participate in a class discussion about the answers to these questions.

Our Consensus Ideas

The key question for this activity is:

What is a property that helps you decide whether a chemical (single substance) is an element or a compound?

Discuss the key question with your team.

1. Write your answer on your record sheet.

Participate in the class discussion about the key question.

2. Write the class consensus ideas on your record sheet.

MAKING SENSE OF SCIENTISTS' IDEAS

Activity 9
Metals, Nonmetals, and the Periodic Table

Purpose

In Activity 8, you learned that chemicals (single substances) are classified as elements or compounds. Families of elements with similar properties are grouped together in the Periodic Table of the Elements. Three of the families are called metals, nonmetals, and the noble gases. In this activity, you will investigate where the metal, nonmetal, and noble gas elements are located on the Periodic Table.

The key question for this activity is:

How are metals, nonmetals, and noble gases organized on the Periodic Table of the Elements?

Record the key question for the activity on your record sheet.

Explore Your Ideas

You will need:
- envelope of Element Cards

You will do three short explorations. First, you will classify 20 elements as metals or nonmetals based on their properties. Second, you will locate these metals and nonmetals on the Periodic Table and look for a pattern in their location. Finally, you will discover the similar property of the noble gases elements in column 18 of the Periodic Table.

Exploration 1: Sorting Metal and Nonmetal Elements

The table on the next page summarizes the trends in the physical properties of metal and nonmetal elements. In this exploration, you will classify 20 elements based on their properties.

STEP 1 *Team Manager:* Divide the Element Cards so that each member of the team has about the same number of cards. Team members place their cards face down.

STEP 2 *Recycling Engineer:* Turn over your first Element Card and read it aloud to the team. Discuss whether the element is a metal or nonmetal. Use the information in the table to the right.

Place the Element Card face up in one of two groups in the middle of the table, one for metal elements and one for nonmetal elements.

Table: Physical Properties of Metal and Nonmetal Elements

Physical Property	Description of Property	Metals	Nonmetals
Phase	gas, liquid or solid at room temperature	solid	gas, liquid or solid
Color	interaction with eyes	most are gray	solids have different colors
Shiny or dull	how well it reflects light	bright, shiny	dull
Melting point	temperature at which it melts	medium to high	low
Density	mass of 1cm cube	medium to high	low
Malleable or brittle	how well it can be hammered into thin sheets	malleable	brittle
Heat conductivity	how well it conducts heat	good conductors	nonconductors (insulators)
Electrical conductivity	how well it conducts electric current	good conductors	nonconductor

STEP 3 Take turns repeating Step 2 until all the Element Cards have been sorted.

1. Record the elements you classified as metals and nonmetals.

Participate in a class discussion about your classification.

Exploration 2: Metals, Nonmetals, and the Periodic Table

STEP 1 Find the location of the metals and nonmetals on the Periodic Table on the inside back cover of the book.

STEP 2 Earlier in this activity, you classified 20 elements (Element Cards) as metals and nonmetals. Find these elements on the Periodic Table.

2. Was your classification accurate? If it was not correct, try to explain how or why you made the error.

3. Describe the pattern of metal and nonmetal elements on the Periodic Table.

Exploration 3: A Special Group of Nonmetals

STEP 1 Locate and read your Element Cards for the nonmetals in *Column 18* on the Periodic Table, as shown in the diagram on the next page. These elements are called the *noble gases.* Compare the properties of these noble gases with the properties of some of the other nonmetals.

4. What *chemical property* do the noble gases have in common that is different from the other nonmetals?

Participate in a class discussion about your answer to this question.

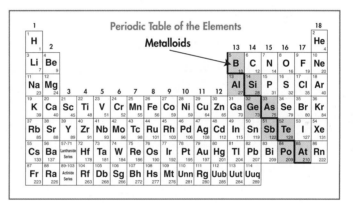

Periodic Table of the Elements

Noble gases (column 18)

1	2	3	4	5	6	7	8	9	10	11	12	13	14	15	16	17	18
1 H 1																	2 He 4
3 Li 7	4 Be 9											5 B 10	6 C 12	7 N 14	8 O 16	9 F 19	10 Ne 20
11 Na 23	12 Mg 24											13 Al 27	14 Si 28	15 P 31	16 S 32	17 Cl 35	18 Ar 40
19 K 39	20 Ca 40	21 Sc 45	22 Ti 48	23 V 51	24 Cr 52	25 Mn 55	26 Fe 56	27 Co 59	28 Ni 59	29 Cu 64	30 Zn 65	31 Ga 70	32 Ge 73	33 As 75	34 Se 79	35 Br 80	36 Kr 84
37 Rb 85	38 Sr 88	39 Y 89	40 Zr 91	41 Nb 93	42 Mo 96	43 Tc 98	44 Ru 101	45 Rh 103	46 Pd 106	47 Ag 108	48 Cd 112	49 In 115	50 Sn 119	51 Sb 122	52 Te 128	53 I 127	54 Xe 131
55 Cs 133	56 Ba 137	57-71 Lanthanide Series	72 Hf 178	73 Ta 181	74 W 184	75 Re 186	76 Os 190	77 Ir 192	78 Pt 195	79 Au 197	80 Hg 201	81 Tl 204	82 Pb 207	83 Bi 209	84 Po 209	85 At 210	86 Rn 222
87 Fr 223	88 Ra 226	89-103 Actinide Series	104 Rf 263	105 Db 268	106 Sg 266	107 Bh 272	108 Hs 277	109 Mt 276	110 Unn 281	111 Rg 280	112 Uub 285	113 Uut 284	114 Uuq 289				

Make Sense of Scientists' Ideas

You may have noticed in Exploration 2 that the properties of metals and nonmetals are not *clearly* defined for all elements. For example, look at the properties of the elements along the zigzag line, starting with boron and moving down and to the right.

The elements that border the zigzag line are called **metalloids** or **semi-metals**. Metalloids have properties of both metals and nonmetals. All metalloids are shiny solids, but they are not as shiny as the metals. Most metalloids are brittle or powdery.

Periodic Table of the Elements

Metalloids

1	2	3	4	5	6	7	8	9	10	11	12	13	14	15	16	17	18
1 H 1																	2 He 4
3 Li 7	4 Be 9											5 B 10	6 C 12	7 N 14	8 O 16	9 F 19	10 Ne 20
11 Na 23	12 Mg 24											13 Al 27	14 Si 28	15 P 31	16 S 32	17 Cl 35	18 Ar 40
19 K 39	20 Ca 40	21 Sc 45	22 Ti 48	23 V 51	24 Cr 52	25 Mn 55	26 Fe 56	27 Co 59	28 Ni 59	29 Cu 64	30 Zn 65	31 Ga 70	32 Ge 73	33 As 75	34 Se 79	35 Br 80	36 Kr 84
37 Rb 85	38 Sr 88	39 Y 89	40 Zr 91	41 Nb 93	42 Mo 96	43 Tc 98	44 Ru 101	45 Rh 103	46 Pd 106	47 Ag 108	48 Cd 112	49 In 115	50 Sn 119	51 Sb 122	52 Te 128	53 I 127	54 Xe 131
55 Cs 133	56 Ba 137	57-71 Lanthanide Series	72 Hf 178	73 Ta 181	74 W 184	75 Re 186	76 Os 190	77 Ir 192	78 Pt 195	79 Au 197	80 Hg 201	81 Tl 204	82 Pb 207	83 Bi 209	84 Po 209	85 At 210	86 Rn 222
87 Fr 223	88 Ra 226	89-103 Actinide Series	104 Rf 263	105 Db 268	106 Sg 266	107 Bh 272	108 Hs 277	109 Mt 276	110 Unn 281	111 Rg 280	112 Uub 285	113 Uut 284	114 Uuq 289				

Science Words

metalloids (semi-metals): elements that have properties of both metals and nonmetals

Most metalloids are poor conductors of heat and electric current. However, the poor electrical conductivity of metalloids like silicon and germanium make them useful for constructing semiconductors. Semiconductors are used in the integrated circuits of electronic devices, such as computers and video games.

1. On your record sheet, list some properties of metalloids.

Another element that does not fit well in the metal-nonmetal classification is hydrogen (H). Hydrogen is a gas, not a solid. In many chemical interactions, it behaves like a metal, but in others, it behaves like a nonmetal.

18
2 He 4
10 Ne 20
18 Ar 39
36 Kr 84
54 Xe 131
86 Rn 222

The nonmetal elements on the far right of the Periodic Table (Column 18) are called the *noble gases.* These elements rarely react chemically with other elements (like nobility refusing to associate with common people). Chemists say they are *stable* or *chemically non-reactive.* Noble gases are so stable that they all exist in the Earth's atmosphere.

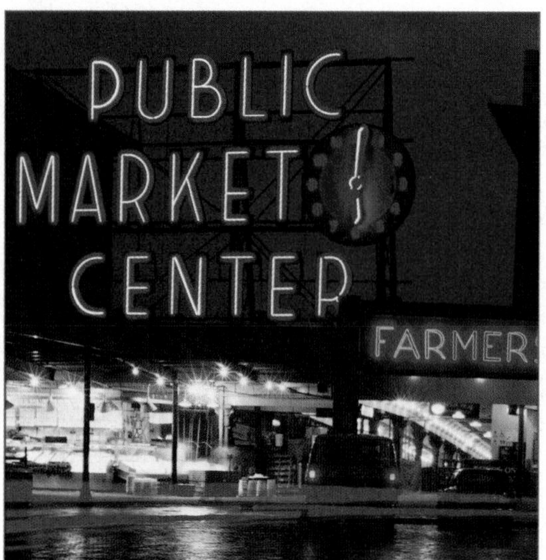

Tubes filled with different noble gases display different colors when an electric current is passed through them.

Since helium is less dense than air and non-reactive, it is used to fill balloons and blimps.

The six noble gases are all colorless and odorless. Neon (Ne), argon (Ar), krypton (Kr), and xenon (Xe) can be removed from the air. You have probably seen neon and other noble gases in advertising signs. Helium is the second lightest gas.

2. On your record sheet, list the properties of the nonmetal noble gases.

An important idea is that the elements right next to each other, either in a row or a column, have the most *similar* properties. For example, consider the elements iron (Fe), cobalt (Co), and nickel (Ni). These are the only elements that are *magnetic.* Notice that these metals are next to each other, in a row.

8	9	10	11
26 Fe 56	27 Co 59	28 Ni 59	29 Cu 64
44 Ru 101	45 Rh 103	46 Pd 106	47 Ag 108
76 Os 190	77 Ir 192	78 Pt 195	79 Au 197

The naturally-occurring metals in Column 11 are copper (Cu), silver (Ag), and gold (Au). One common property of these metals is that they are very good conductors of electric current. Another property is that they are *ductile.*

The property ductility is a description of how easily a substance can be stretched or drawn out into thin wire in special machines. Copper and gold especially have high ductility.

copper wire

3. Why are copper and gold used to make wires for electric circuits?

The metal elements in Column 11 have an important chemical property in common. They resist corrosion. Corrosion is the *gradual* wearing away of metals when they interact chemically with other substances in the environment. The Column 11 elements are often called the *coinage metals* because they have been used for thousands of years to make coins.

4. Why is the chemical property of resistance to corrosion so important for making coins?

Participate in a class discussion about your answers to Questions 1 through 4.

Our Consensus Ideas

The key question for this activity is:

How are metals, nonmetals, and noble gases organized on the Periodic Table of the Elements?

1. Answer this question as a team. To do this, describe the pattern of the location of metal and nonmetal elements on the Periodic Table.

Participate in a class discussion about the pattern in the location of metals, nonmetals, and noble gases on the Periodic Table.

2. Write the class consensus ideas on your record sheet.

Activity 10
Describing and Classifying Materials

Comparing Consensus Ideas

Recall the first key question for this unit.

1. What are physical and chemical interactions?

You learned in this unit that chemists find it useful to define two categories of interactions, chemical interactions and physical interactions.

Your teacher will review with you all of the ideas that the class developed during this unit to help you answer this key question. Be prepared to contribute to this discussion.

Your teacher will distribute copies of *Scientists' Consensus Ideas: Interactions and Properties.* Read the scientists' ideas and compare them with the ideas your class developed during this unit.

Write the examples where requested on the *Scientists' Consensus Ideas* form.

Your teacher will lead a whole-class discussion.

The second key question for this unit is:

2. How do chemists describe and classify materials?

You learned that all materials are either single substances or mixtures of two or more single substances. The two types of single substances are compounds and elements. The two types of mixtures are suspensions and solutions. So all of the millions of different materials in the world can be classified as a *suspension, solution, element,* or *compound.*

You used three diagrams in this unit to classify matter.

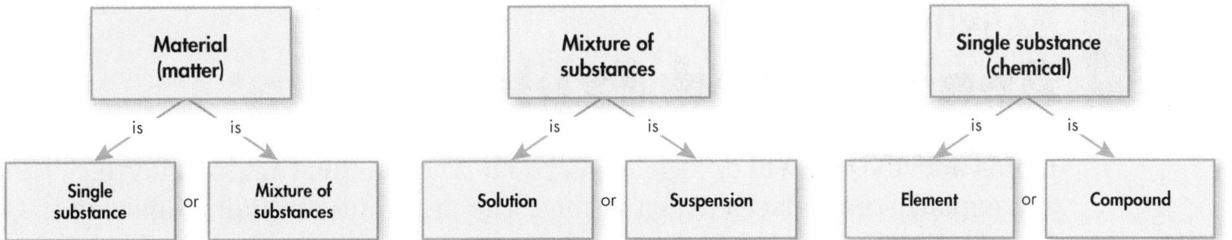

Chemists often put the three diagrams together into a single classification map of materials.

To help you answer the second key question, your teacher will work with the class to combine the three diagrams into one. Be prepared to contribute to this exercise.

Your teacher will distribute copies of *Scientists' Consensus Ideas: Describing and Classifying Materials*, which shows a classification map.

Write the examples from this unit where requested on the *Scientists' Consensus Ideas* form.

Your teacher will lead a whole-class discussion.

I'm confused! Brass is made up of zinc and copper and is a mixture (solution). But water is made up of oxygen and hydrogen and is a compound. Why aren't they **both** mixtures?
Chantel

Chantel is confused. See if you can help her. To help her understand, answer this question:

What is one difference between a *mixture* (such as brass or muddy water) and a *compound* (such as water or rust)? (*Hint:* Think about how a mixture and compound are separated into their different parts.)

Participate in a class discussion about this question.

IDEA POWER

Activity 11
Analyze and Explain

In this activity, you will use the ideas you learned in this Unit to analyze and explain some interesting situations. The procedure for doing this is in *How To Write an Analysis and Explanation* in the Appendix. Use the ideas in your two *Scientists' Consensus Ideas* handouts, *Interactions and Properties* and *Describing and Classifying Materials.*

The analyses for these situations can include identifying the type of interaction or the most useful properties to answer the question. You do not need to draw force or energy diagrams. You will need to use the Periodic Table of the Elements.

Be sure that what you write would get a good evaluation using *How To Evaluate an Analysis and Explanation.*

Chemical or Physical?

1. Analyze and explain why each of the following situations is a chemical interaction, a physical interaction, or there is not enough information to tell whether it is chemical or physical. An example has been done for you.

Example: A white solid is heated and the room fills with choking fumes.

Analyze and explain: Either the material could have boiled (physical) or a new gas could have been produced (chemical). You can't tell if the interaction was physical or chemical.

a) A liquid is cooled and a solid forms on the bottom of the beaker.

b) Tea turns cloudy and a lighter color when lemon is added.

c) After a few days, a banana turns brown and black spots appear.

d) A nail is bent when hammered.

e) During the exploration, the temperature of the liquid increased 15°C.

Participate in a class discussion about your answers.

Cleaning Aluminum

Here's a practical use for an acid-base chemical interaction.

All aluminum pans have a thin coating of aluminum oxide, a base. It is this coating that gets stained when certain foods are cooked in aluminum pans. You can remove the stain by putting something acidic, such as vinegar, on the basic aluminum oxide.

2. Explain how this acid-base interaction is used to remove stains from an aluminum pan. Use the terms *vinegar*, *aluminum oxide*, *new substance*, and *chemical interaction* in your explanation.

Participate in a class discussion about your answer.

The Case of the Poisonous Gas

A chemist working in a government laboratory had to quickly identify an unknown, poisonous gas. The gas separated into three different gases when it was slowly cooled to very low temperatures. Only one of the three gases was poisonous.

3. Is the unknown gas a *suspension*, *a solution*, *a compound*, or *an element*? Analyze and explain.

Participate in a class discussion about your answer.

The Case of the White Powder

When the forensic scientists arrived at the scene of the robbery, they found a white powder on the bedroom dressing table. Back at the lab, no pieces of a different substance were seen in the white powder under a powerful light microscope. The white powder did not separate into different substances when it was melted. It did separate into two elements in a series of chemical interactions.

4. Is the unknown white powder a suspension, a solution, a compound, or an element? Analyze and explain.

Participate in a class discussion about your answer.

The Hurricane Disaster

Recently, a hurricane isolated a town when the roads and bridges into the town were closed. A major factory in this town uses cadmium (Cd) in a process that is important to the survival of the town. As a result of the hurricane, all sources of cadmium were cut off.

5. Use your Periodic Table of the Elements to suggest two elements that might replace cadmium in the factory's process. Explain your suggestions.

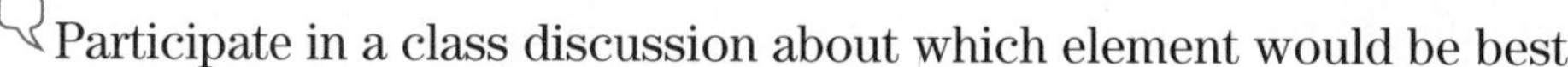

Participate in a class discussion about which element would be best.

Salt, Salt, and More Salt!

Your teacher will show you a video of the chemical reaction between chlorine gas and sodium metal. The new substance made in this reaction is sodium chloride, or table salt.

6. What other elements, in addition to sodium, would you expect to react with chlorine? Why?

Participate in a class discussion about which element would be best.

Challenge Problem: Yellow Goop

Your teacher will show you a video of the chemical reaction between a potassium iodide solution and a lead nitrate solution.

7. What would you expect to happen if a *potassium bromide* solution were added to the lead nitrate solution?

8. What other potassium compound(s) should behave in the same way? Why?

Participate in a class discussion about your answers.

LEARNING ABOUT OTHER IDEAS

Activity 12
Acid, Base, or Neutral?

Purpose

You learned in Activity 1 that there are many different ways to classify materials. For example, all substances can be classified as an acid, a base, or neutral (neither an acid nor a base). Knowing whether a substance is an acid, base, or neutral helps scientists to predict how the substance will interact with other materials. In this activity, you will learn *how* scientists classify compounds as acidic, neutral, or basic. At the end of this activity, you will be able to answer the following key questions:

1. How can you determine whether a substance is an acid, a base, or neutral?

2. How can you determine the strength of an acid or a base?

Record the key questions for the activity on your record sheet.

Learning the Ideas

Acid-Base Indicators

Indicators are used to determine whether a substance is an acid, base, or neutral. An indicator is a compound (dye) that changes color in an acid, in a base, or in both. You used an acid-base indicator in Activity 2. You bubbled carbon dioxide gas through a bromothymol blue (BTB) indicator. Some of the carbon dioxide interacted chemically with water in the solution to make weak carbonic acid. The BTB indicator changed color from blue to yellow.

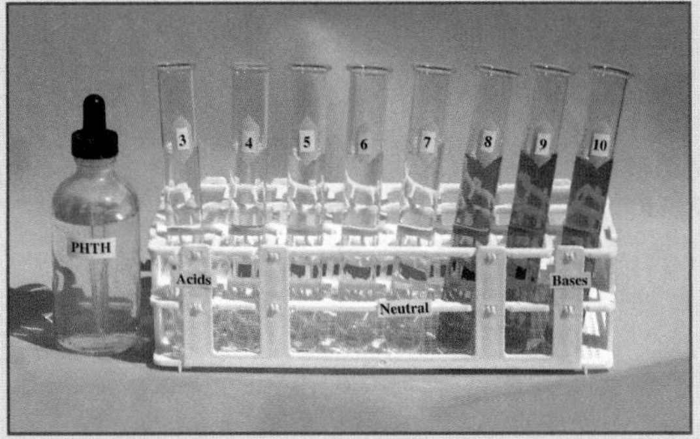

Phenolphthalein Indicator.

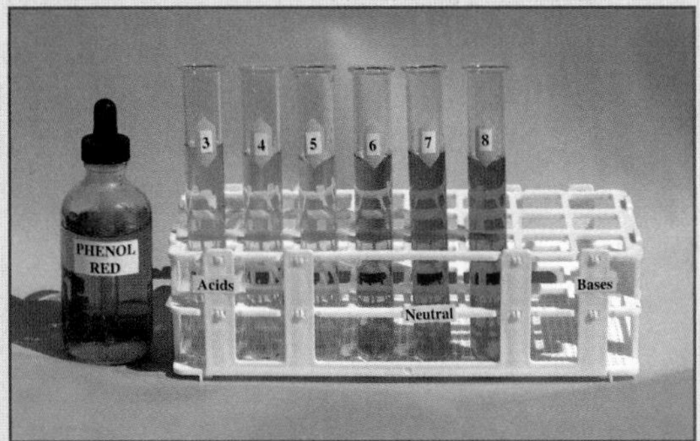

Phenol Red Indicator.

Chemists use many different indicators with a wide range of color changes, as shown in the photographs on the previous page. For example, the indicator phenolphthalein stays clear (no color change) with acids and neutral substances but turns pink in bases. The indicator phenol red turns orange or yellow with acids, stays red (no color change) with neutral substances, and turns pink with bases.

1. What is an acid-base indicator?
2. When an unknown substance is mixed with phenol red, the color changes from red to yellow. Is the unknown substance *an acid, neutral, a base*, or *can't you tell*? Explain your reasoning.
3. When an unknown substance is mixed with phenolphthalein, it does not change color. Is the unknown substance *an acid, neutral, a base*, or *can't you tell*? Explain your reasoning.

Your teacher will review the answers to these questions with the class.

Many plants contain indicators. Some flower petals, such as red poppy, bluebottle, and hydrangea petals, contain indicators that show different colors depending on the soil the plant is grown in. Fruits and vegetables often contain indicators. When some fruits ripen, they produce an acid. The indicator in the fruit changes color as the acid is produced. For example, cherries, blueberries, grapes, and mulberries contain an indicator that turns a red-purplish or a violet color as the fruit ripens and its acid content decreases. You can even make an indicator by boiling red cabbage.

The indicator in the grapes changes color as the fruit ripens.

The Strength of Acids and Bases

Some acids and bases are *weak* and harmless to the touch. Other acids and bases are *strong*, poisonous, and can cause serious skin burns. Scientists measure the strength of acids and bases on a scale called the **pH** scale. The pH scale is a series of numbers from 0 to 14. A substance with a pH number lower than 7 is an acid. A substance with a pH number larger than 7 is a base. A substance with a pH number near 7 is neutral.

Science Words

pH: a quantity used to represent how acidic a solution is

Acids with a pH lower than 2.0 are called *corrosive* because they dissolve or destroy other substances, especially metals, during chemical interactions. Bases with a pH above about 13 are also corrosive because they can dissolve hair and grease.

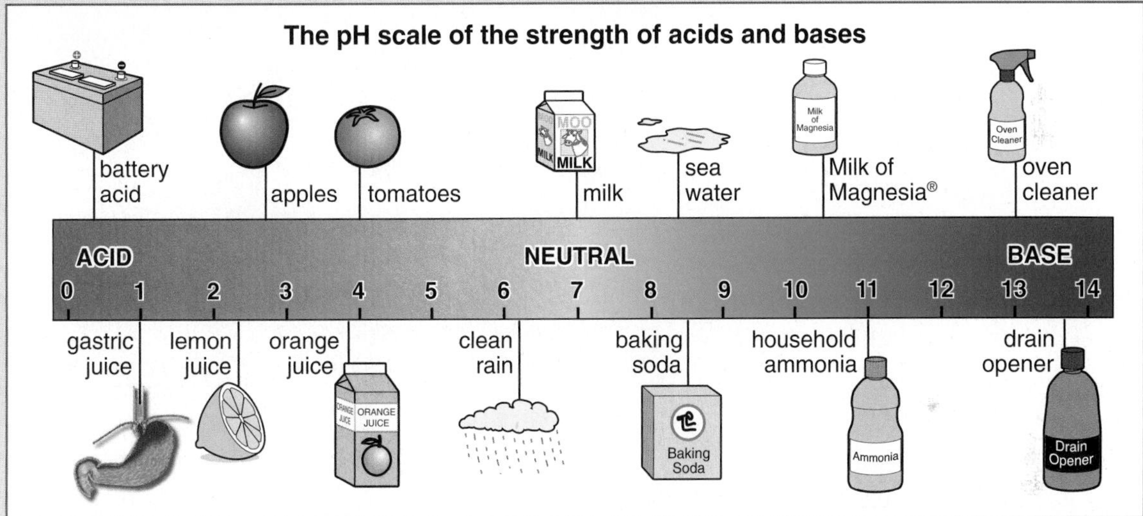

Use the pH scale to answer the following questions.

4. Which two acids shown on the pH scale are corrosive? Explain your answer.

5. Which two bases shown on the pH scale are corrosive? Explain your answer.

6. A brand of hair remover has a pH number of 12.3. Is this brand of hair remover *acidic*, *neutral*, or *basic*? Explain your answer.

7. A brand of cola has a pH number of 2.4. Is this brand of cola *acidic*, *neutral*, or *basic*? Explain your answer.

8. Which acid is stronger, the acid in apples or the acid in orange juice? Explain your answer.

Your teacher will review the answers to these questions with the class.

Measuring the pH Numbers of Substances

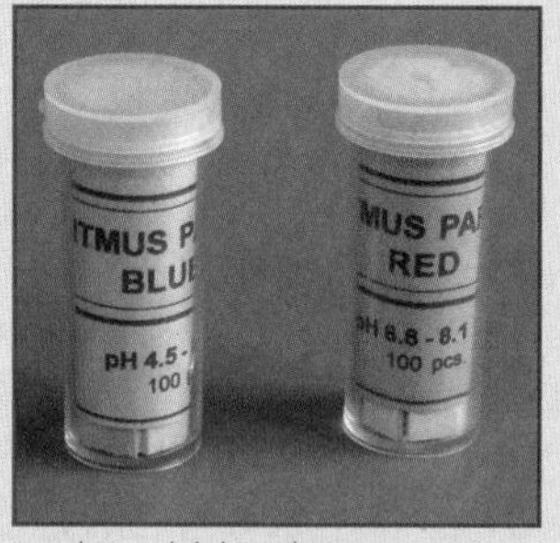

Red and blue litmus paper.

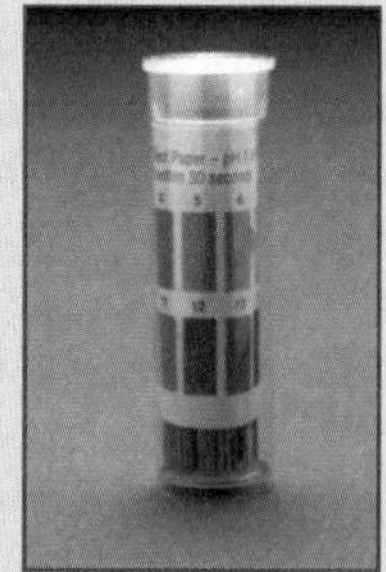
Universal-indicator paper.

The most well-known acid-base indicators are red and blue litmus paper. Acids turn blue litmus paper red. A base turns red litmus paper blue. However, litmus paper cannot give an indication of the exact pH number.

Universal-indicator paper is more useful than litmus paper. Universal indicator is red at pH 1. It goes through a series of colors from red through orange, yellow, green, and finally blue for strong bases.

A laboratory pH meter.

The best way to measure pH is with an electronic pH meter. There are many brands of pH meters. Some are hand-held and are useful to biologists working in the field. Some pH meters are designed for use in chemical laboratories.

What We Have Learned

Recall the key questions for this activity.

1. **How can you determine whether a substance is an acid, a base, or neutral?**
2. **How can you determine the strength of an acid or a base?**

Participate in a class discussion to review the answers to these key questions.

Write the answers to the key questions on your record sheet.

LEARNING ABOUT OTHER IDEAS

Activity 13

Groups and the Periodic Table of the Elements

Purpose

You learned in Activities 8 and 9 that elements are organized into families on the Periodic Table of the Elements. Consider another type of periodic table that you are familiar with, the calendar. You can think of a calendar as a grid made up of horizontal rows and vertical columns. Each row represents a different week. Each column represents a different day of the week. Each day is repeated in every row (week) through the whole month. This arrangement of the days in the month is **periodic**. Things that are periodic have a regular, repeating pattern.

Science Words

periodic: having a regular, repeating pattern

NOVEMBER						
S	M	T	W	T	F	S
	1	2	3	4	5	6
7	8	9	10	11	12	13
14	15	16	17	18	19	20
21	22	23	24	25	26	27
28	29	30				

In this activity, you will learn about what is periodic in the Periodic Table of the Elements. The key questions for the activity are:

1. What is the repeating pattern in the Periodic Table of the Elements?
2. What are some common groups of metal and nonmetal elements?

Record the key questions for the activity on your record sheet.

Learning the Ideas

What is Periodic about the Periodic Table of the Elements?

Like the days of the month on a calendar, elements are arranged on the Periodic Table in a way that shows a repeating, or periodic, pattern of *chemical properties*. As you read across the rows of the calendar, the days repeat. Across the Periodic Table, the c*hemical (and some physical) properties* of the elements repeat.

18
2 He 4
10 Ne 20
18 Ar 39
36 Kr 84
54 Xe 131
86 Rn 222

In a calendar, days with the same name are organized into columns. In the Periodic Table, elements with similar *chemical (and some physical) properties* are organized into columns called groups. The numbers 1 through 18 across the top of the columns of the Periodic Table identify each element's group. For example, iron (Fe) is in Group 8 and carbon (C) is in Group 14.

In the next sections, you will learn about three groups of elements: the noble gases (Group 18), the highly reactive metals (Groups 1 and 2), and the highly reactive nonmetals (Group 17).

Noble Gases (Group 18)

You learned in Activity 9 about the nonmetal noble gases, Group 18, located at the far right of the Periodic Table. Helium (He), neon (Ne), argon (Ar), krypton (Kr), xenon (Xe) and radon (Rn) are all colorless and odorless. Noble gases are so stable that they all exist in the Earth's atmosphere. Noble gases rarely react chemically with other elements. They are stable or chemically non-reactive.

1. What chemical property do the noble gases have in common?

Highly Reactive Metals (Groups 1 and 2)

Like other metals, the metals in Groups 1 and 2 are a shiny, silvery color and are malleable. They are good conductors of heat and electric current, but they are also highly chemically reactive with other substances. These metals are so reactive that they are never found in nature by themselves. Many burn spontaneously in air, so they have to be stored in oil. The metals in Group 1 are a little more reactive than the metals in Group 2.

The metals in Group 1 are called *alkali metals* and the metals in Group 2 are called *alkaline earth metals*. Alkali is another word for *base*. One chemical property of the highly reactive metals is that they react with water to make bases.

Group 1:	Group 2:
lithium	beryllium
sodium	magnesium
potassium	calcium
rubidium	strontium
cesium	barium
francium	radium

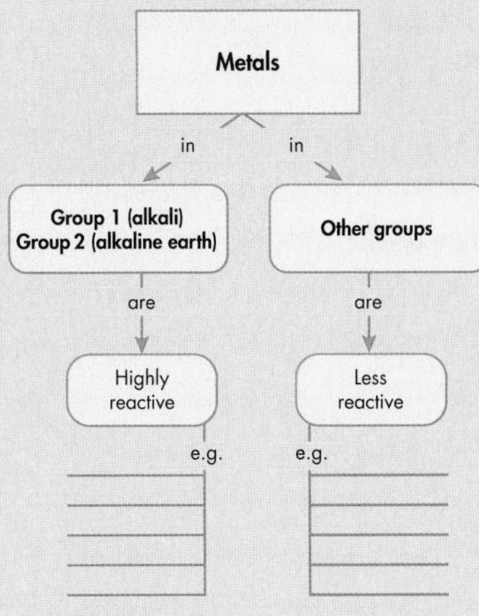

2. What are the metals in Group 1 called? What are the metals in Group 2 called?
3. What *general* chemical property do Group 1 and 2 metals have in common?
4. List two *specific* chemical properties of Group 1 and Group 2 metals.
5. On your record sheet, create a concept map like the one shown here. Write four examples of highly reactive metals on your concept map.
6. Write four examples of common metals in the Other Groups column in your concept map.

Your teacher will review the answers to these questions with the class.

Highly Reactive Nonmetals (Group 17)

The nonmetals in Group 17, fluorine (F), chlorine (Cl), bromine (Br), iodine (I), and astatine (At), are the most chemically reactive nonmetals. They are so chemically reactive that they are never found in nature by themselves. They are always combined in compounds of all the other naturally-occurring elements (except for the noble gases). For example, a chemical property of this group of elements is that they all react with water to make acids.

Group 17 nonmetals are called *halogens*, which means "salt-former." They react with metals to form compounds called salts (halides). You are most familiar with table salt (sodium chloride). Your body needs sodium chloride to conduct nerve impulses. Your teacher will show you a short video of the reaction of sodium and chlorine.

Fluorine (F) is the most reactive of all nonmetals. It comes from the mineral called fluorspar. Fluoride toothpaste is made with fluorine. You are probably familiar with the non-stick coating on pans. This coating is made of fluorine combined with other nonmetals.

Group 17:
fluorine
chlorine
bromine
iodine
astatine

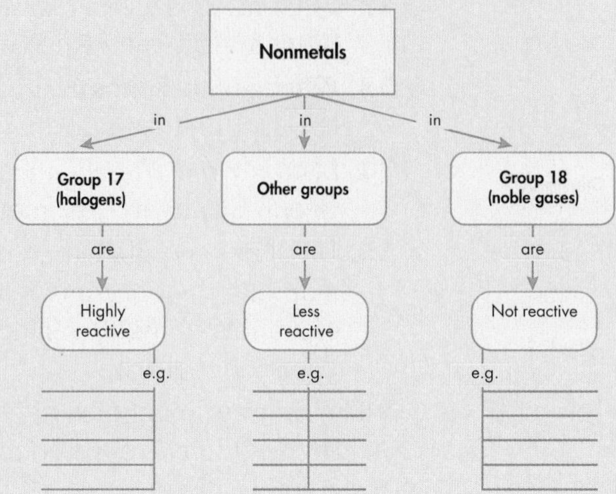

You may also be familiar with halogen lamps. In a halogen light bulb, small amounts of iodine, and sometimes the other halogens, are added to the filament. This produces a brighter light than ordinary (tungsten) filaments.

7. What are the nonmetals in Group 17 called?
8. What *general* chemical property do Group 17 nonmetals have in common?
9. List two *specific* chemical properties of Group 17 nonmetals.
10. On your record sheet, create a concept map like the one above. Write four examples of Group 17 nonmetals on your concept map.
11. Write four examples of common nonmetals in the other groups in your concept map.
12. Write four examples of Group 18 nonmetals (noble gases) in your concept map.

Your teacher will review the answers to these questions with the class.

What We Have Learned

Recall the key questions for this activity.

1. What is the repeating pattern in the Periodic Table of the Elements?

2. What are some common groups of metal and nonmetal elements?

Participate in a class discussion to review the answers to these key questions.

Write the answers to the key questions on your record sheet.

UNIT 6

PHYSICAL INTERACTIONS AND PHASES

INTRODUCTION

What will you learn about in Unit 6?

You learned in Unit 5 that scientists classify all matter as:

- suspensions
- solutions
- elements
- compounds.

You also were introduced to the Periodic Table of the Elements.

In Unit 6, you will observe and build many kinds of models of what gases, liquids, and solids are like on a scale too small to see, even with the best light microscopes. You will learn how the scientists' theory of materials helps you make sense of some physical interactions and the physical properties of solids, liquids, and gases.

A model is a **simplified imitation** of something that we hope can help us understand it better. Scientists often use physical models in experiments to help them figure out how something works.

Physical Interactions and Phases

The key question for this cycle of learning is:

What is the scientists' theory of what gases, liquids, and solids are like on a scale too small to see?

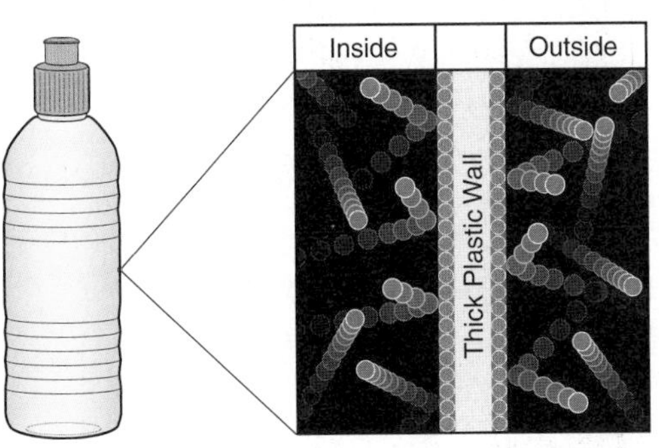

UNIT 6:
PHYSICAL INTERACTIONS AND PHASES

Key Question

What is the scientists' theory of what gases, liquids, and solids are like on a scale too small to see?

DEVELOPING OUR IDEAS

Activity 1
What's Inside Air?

Purpose

In Units 3, 4, and 5, you did explorations that provided evidence that gases are one phase of matter. You showed that gases have mass, can push on things, interact with a flame, and interact with other substances like BTB indicator. But you cannot see a gas, even when you look through the best microscopes. In this activity, you will think about and share your initial idea of what is inside a gas like air. The key question for this activity is:

What is air like on a scale smaller than you can see through light microscopes?

Record the key question for the activity on your record sheet.

We Think

Imagine you have three hand pumps, all sealed. One contains air, one contains water, and another one contains sand. You push down on the plungers of the three pumps. Discuss the following question with your team.

- What do you think will happen to the contents of the pumps?

Participate in a class discussion.

Explore Your Ideas

A good idea helps you make sense of your observations. In this activity, you will first observe what happens when you try to *compress* air into a smaller volume (space), and try to *expand* air into a larger volume (space). Then you will construct an idea of what's inside air that helps you make sense of your observations.

Part A: Expanding and Compressing Air

Each pair in your team will work with one of the air pumps and *take it in turns*, so each student gets to try the compression and expansion. Your teacher will pass around the class a sealed pump with sand in it, and a sealed pump with water in it.

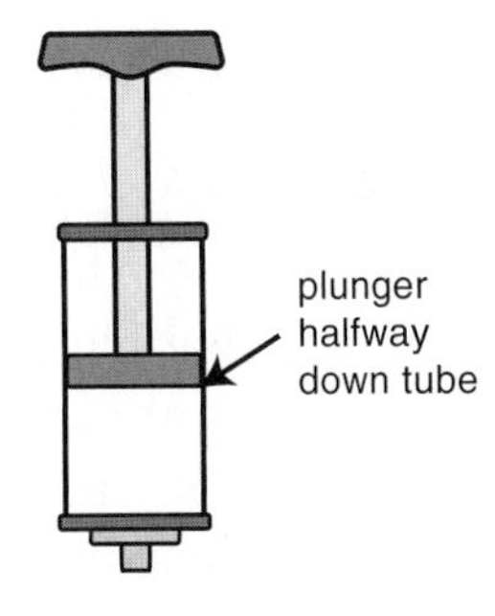

Your team will need:

- 2 plastic air pumps

Your class will need:

- sealed pump with water
- sealed pump with sand

STEP 1 Start with the plunger about halfway down the tube, as shown in the diagram.

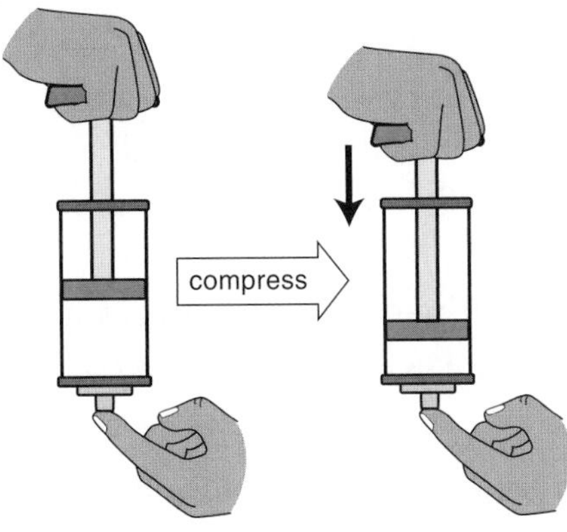

STEP 2 Block off the bottom of the pump with a finger to prevent any air escaping from inside the pump.

STEP 3 Push down on the plunger to compress the air into a smaller volume. Push it down until it is about a half of the original volume (space).

Use the pumps as directed by your teacher.

Wear eye protection.

1. What happens when you let go of the plunger?

STEP 4 Start again with the plunger of the pump about halfway down the tube. Block off the bottom of the pump with a finger.

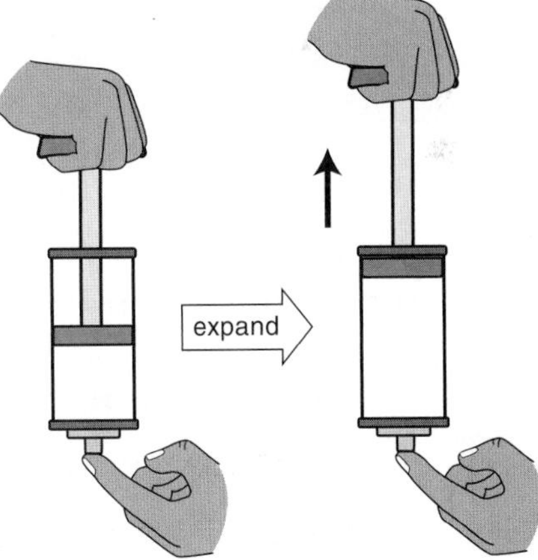

STEP 5 Pull up on the plunger to expand the air into a larger volume. Expand it to about twice the original volume (space).

2. What happens when you let go of the plunger?

STEP 6 When the water-filled pump gets to your team, take turns trying to compress the water.

3. Write a sentence or two comparing how easy it is to compress the air and the water.

STEP 7 When the sand-filled pump gets to your team, take turns trying to compress the sand.

4. Write a sentence or two comparing how easy it is to compress the air and the sand.

Magnification 100,000,000x

Ultrascope

field of view

10 nm

Part B: Constructing an Idea

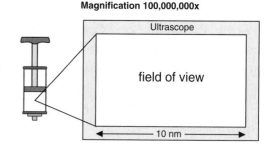

You learned in Unit 5 that that there is a limit to what you can see with a light microscope. The smallest pieces you can see that have distinctly different properties are about two-tenths of a millionth of a meter wide (0.2 µm). This is a magnification of about 500,000x (five hundred thousand times). Even at this magnification, you cannot see anything of a pure gas (that is, a gas with no smoke or specks of dust).

Imagine that you have a magic microscope, called an *Ultrascope*, which allows you to see a gas. The prefix *Ultra* means "even more." This magic Ultrascope has a rectangular field of view and magnifies about 100 times more than the best research microscope.

You would be looking at a part of the gas only about one ten-billionths of a meter wide (0.000,000,010 m). A billionth of a meter is called a *nanometer* (symbol is nm).

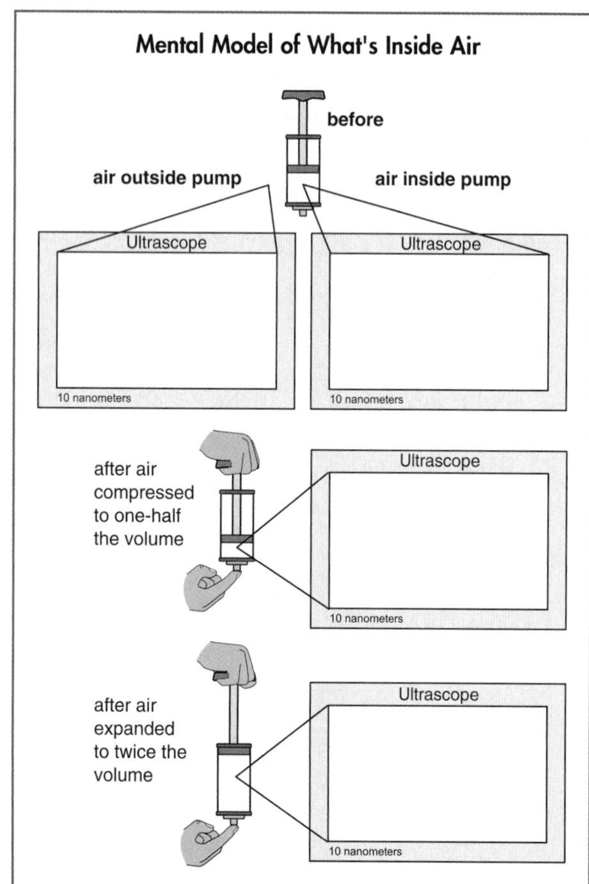
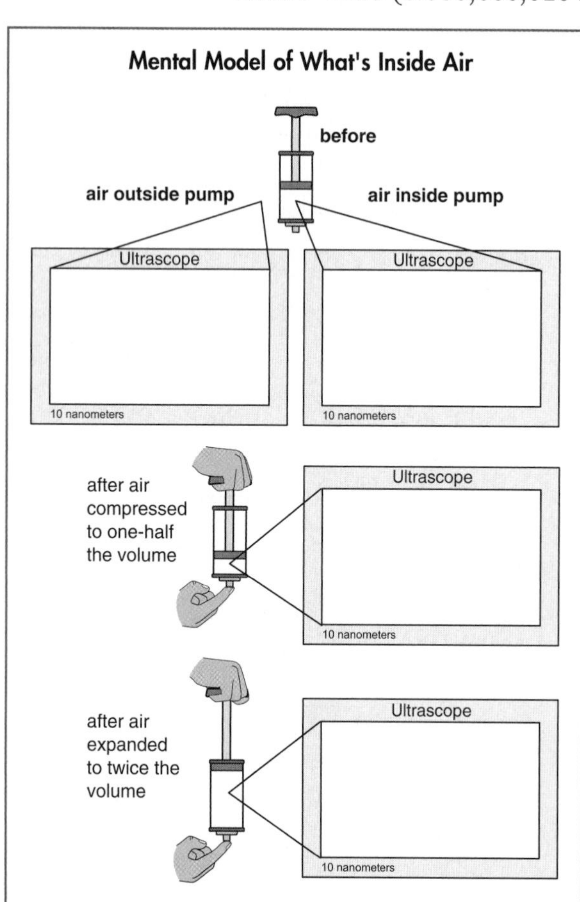

STEP 1 Without consulting your teammates, draw on your record sheet your idea of what is inside air.

- Draw what you think the air *outside* the pump (the air in the room) would look like through the Ultrascope.
- Draw what you think the air *inside* the pump would look like through the Ultrascope before anything is done to the air (when the pump is just sitting there).
- Draw what you think the air that is *inside* the pump would look like through the Ultrascope after the air has been compressed into a smaller volume (space).
- Draw what you think the air that is *inside* the pump would look like through the Ultrascope after the air has been expanded to a larger volume (space).

5. Explain your thinking. How does your idea help make sense of your observations of compressing and expanding a gas?

STEP 2 Share your *idea* (your Ultrascope drawings) with other members of your team. Listen to your team members as they show and explain their ideas. Practice the skill of respecting other team members and their ideas.

STEP 3 As a team, decide on an idea that best represents the thinking of your team members. Be prepared to share the reasoning for each idea in a class discussion.

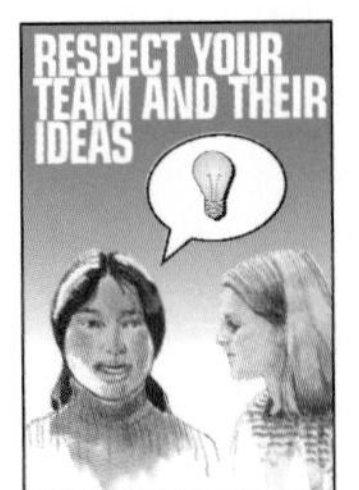

STEP 4 Prepare a Presentation Board. Divide your presentation board into three sections. *Recycling Engineer:* Draw four Ultrascope views, as shown in the diagram of the presentation board below.

In the first Ultrascope view, draw your idea of what you think the air *outside* the pump would look like *before* anything is done to the air.

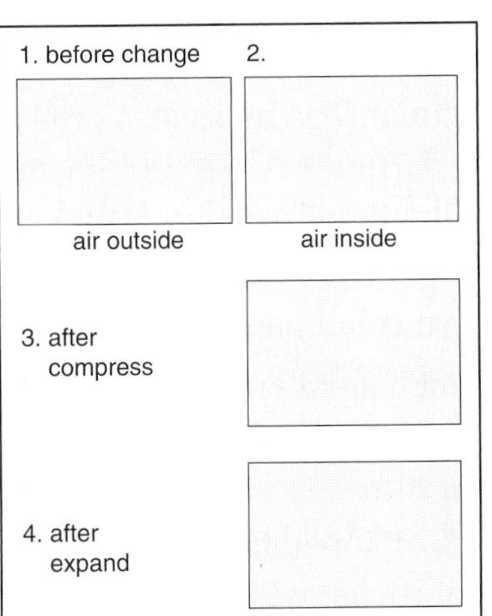

In the second Ultrascope view, draw what you think the air *inside* the pump would look like *before* anything is done to the air.

In the third Ultrascope view, draw what you think the air would look like *after* it has been compressed.

In the fourth Ultrascope view, draw what you think the air would look like *after* it has been expanded.

Participate in the class discussion to summarize the *class ideas* of air.

Our Consensus Ideas

The key question for this activity is:

What is air like on a scale smaller than you can see through light microscopes?

Participate in the class discussion about which idea(s) seems to be the most useful for explaining what happens when air is compressed and expanded.

Record drawings of your class idea (or ideas, if your class has more than one) on your record sheet.

MAKING SENSE OF SCIENTISTS' IDEAS

Activity 2
Scientists' Theory

Purpose

In Activity 1 you saw that it is easy to compress air into a smaller space (volume) when you push down on the plunger of a closed pump. No air was allowed into or out of the pump. You know from the *Conservation of Mass Law* that the mass of the air in the pump stayed the same. As soon as you let go of the plunger, it moved back to its original position. You developed ideas to make sense of these observations.

One idea that makes the observations easy to explain is that the air is made up of very *small particles*. The particles are far apart, so there is a lot of space between the particles. Pushing down on the plunger forces the air particles to move closer together into a smaller volume. The mass stays the same because there is still the same number of air particles. The particles just have less space between them. Letting go of the plunger allows the particles to spread out again.

There are many questions you could ask about these small particles, such as:

1. How tiny are these small particles? How much smaller are they than the smallest pieces of substances you can see through a light microscope?
2. What are the characteristics of the particles of substances? For example, is there an unlimited number of particles with different sizes, shapes, textures, and masses?
3. What is in the space between the particles?
4. Do the particles of a substance move continuously and endlessly? Or do the particles only move when the substance changes, then they gradually slow down and stop? For example, do particles move while you pour a liquid, then slow down and stop moving?
5. Do particles interact with each other? If so, what are the types of interactions between particles?

Scientists put together all of their ideas about the particles of matter into a *Small-Particle Theory* (SPT) of matter. In this activity you will learn the scientists' answers to the first *four* questions above. The answers to these questions are the first four ideas of the Small-Particle Theory.

The key question for this activity is:

What are the first four ideas (mental models) in the Small-Particle Theory of Matter?

Record the key question for the activity on your record sheet.

Explore Scientists' Ideas

What is a Scientific Theory?

Many people think that a theory is "just a guess" or that one theory is just as good as another theory. However, a **scientific theory** is different from how you use the word in your everyday lives. First, a scientific theory is *not* a single idea. A scientific theory is a consistent set of related scientific ideas.

Science Words

scientific theory: a consistent set of related scientific ideas

Criteria for a Good Scientific Theory

Scientific Theory
(consistent set of related scientific ideas)
Idea 1. All matter is made up of particles.
Idea 2. . . . .
Idea 3. . . . .

Criterion #1
a) The ideas in the theory are testable because they allow you to make predictions.
b) So over time, convincing experimental evidence is gathered that supports each idea.

Criterion #2
The theory helps you make sense of and explain everyday objects and events, as well as the conclusions of experiments.

Criterion #3
The ideas in the theory fit well with (do not contradict) each other, and fit with other scientific ideas, laws, and theories.

Criterion #4
The theory is economical—it ties together and explains a wide range of objects and events that seemed disconnected.

Criterion #5
The theory is testable because it allows you to predict new objects or events that you have never observed (because you did not know where or how to look before the theory).

Second, a theory has to meet five very strict criteria to be considered a good scientific theory. These criteria are shown above.

How Tiny are the Small Particles?

Suppose you took a piece of paper and tore into small pieces. Then you took one of the small pieces of paper and cut it into even smaller pieces. Then you took the tiniest piece of paper you could see, looked at it through a microscope, and cut and divided it again into the smallest pieces you could see. If you had a very powerful microscope, you could take your tiniest pieces and keep cutting them, dividing, and looking again. Would this process have an end to it? Would you ever come to a tiniest particle that either you could not cut, or if you did cut it, it would not be paper any more?

The first idea of the Small-Particle Theory (SPT) is:

Idea 1 All single substances, solids, liquids, and gases are made up of tiny particles. A particle is the smallest piece of a substance.

Some early Greek philosophers thought that the small particles were about the same size as small specks of dust. Of course, they did not have modern microscopes. A small speck of dust is about one-thousandth of a meter wide (0.001 m, or 1 mm).

Recall in Unit 5 you looked at the nucleus of a cell of a leaf at the magnification of 100,000x. The nucleus is approximately one millionth of a meter wide (0.000,001 m). A millionth of a meter is called a micrometer (symbol is μm). So the nucleus of a plant cell is a thousand times smaller than a speck of dust! The photograph of the cell nucleus is close to the limit of what you can see with a light microscope.

The second idea of the Small-Particle Theory (SPT) is:

Idea 2 The particles of substances are too small to see through visible-light microscopes.

Just how tiny are the small particles of substances? Your teacher will show you some *visual models* of what is inside a plant nucleus. The models show what scientists believe you *might* see at higher magnifications.

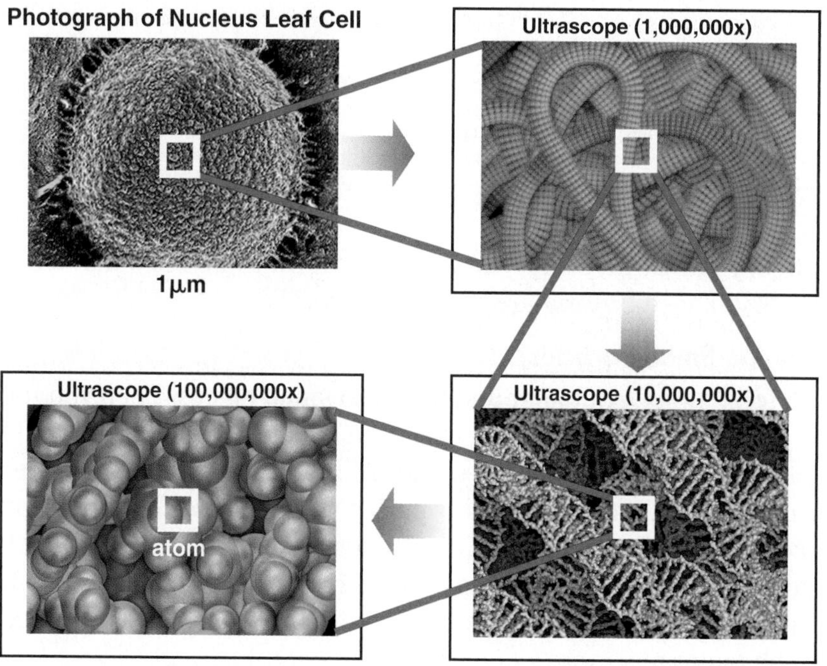

You can see that what appear to be bumps on the photograph of the cell nucleus are actually strings of stuff. The next model (magnification of ten million times) shows that the strings are made up of many twisted strands. (These are DNA strands, which make up the genes of the leaf and influence the leaf's traits.) The strands themselves are made up of connected spheres.

The next picture (magnification of a hundred million times) shows the scientists' model of these spheres. The spheres represent the smallest particles of matter, called atoms. Each atom is about one-tenth of a nanometer wide (0.000,000,000,1 m). This is ten thousand times smaller than the cell nucleus, and ten million times smaller than a speck of dust!

If the particles of substances are so tiny, then everyday objects must be made up of an enormous number of particles. It is very difficult to even imagine the extremely tiny size and huge numbers of particles that make up objects.

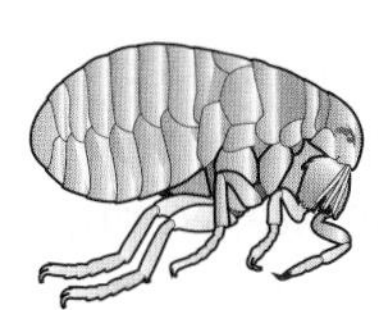

For example, think of how small a flea is. Suppose you magnified a flea to the size of a 20-story building. You would still not be able to see the particles that make up the flea!

Another way to think about the size of the particles is to think about the air that surrounds you. Each liter (lung full) of the air you breathe contains about 30 billion trillion particles,

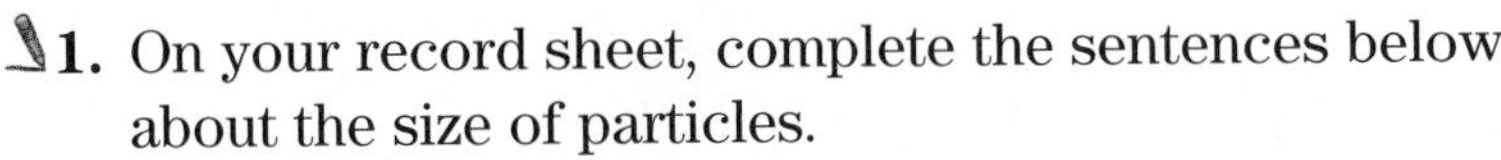

30,000,000,000,000,000,000,000 particles

even though a liter of air only weighs about one gram.

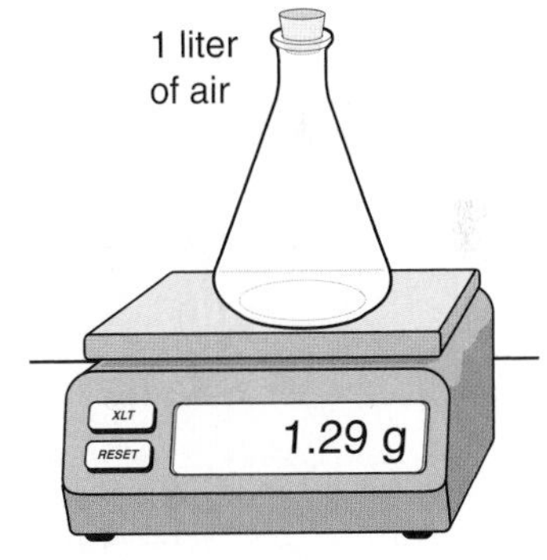

The extremely small size and huge numbers of particles means that each particle has a very tiny mass. For example, an oxygen particle has a mass of about 54 millionth of a billionth of a billionth of a gram, or

0.000,000,000,000,000,000,000,054 g.

1. On your record sheet, complete the sentences below about the size of particles.

   a) Atoms have an average width of about ________ nm (nanometers).

   b) There are millions of trillions of particles in everyday objects. For example, a liter of air contains about __________ particles.

   c) The mass of particles is very small. For example, an oxygen particle has a mass of about ___________ g (grams).

What are the Characteristics of the Small Particles?

The early Greek philosophers proposed that everything was made up of an unlimited number of *atoms* with different sizes, shapes, textures, and masses, as shown in the diagram. Atom in Greek (*atmos*) means unbreakable. The Greeks thought that each of the thousands of millions of different materials in the world had a different atom.

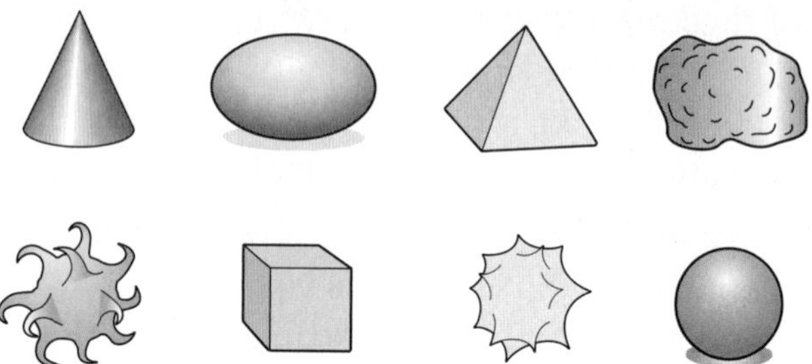

The Greeks did not know about compounds and elements. As you learned in Unit 5, elements are substances that *cannot* be broken down into simpler substances during chemical interactions. Elements are the "basic ingredients" out of which everything is made. The *Periodic Table of the Elements* represents all the elements discovered by scientists so far. Look at the *Periodic Table of the Elements* and answer the following question.

2. How many elements have been discovered (including those made in the laboratory)?

According to the Small-Particle Theory (SPT):

Idea 1 All single substances, solids, liquids, and gases are made up of tiny particles. A particle is the smallest piece of a substance.

A. Particles (smallest pieces) of a substance can be *atoms* or *molecules*.

B. Elements consist of one kind of atom.

1. Atoms of the same element are usually the same mass, and/or size. Atoms of different elements have different masses and/or sizes.
2. The metal elements and the noble gases consist of single atoms. For some nonmetal elements, two identical atoms are joined together in a structure called a molecule.

To make sense of these ideas, consider the air you breathe. Air is a solution of different gases. Some of the gases in air are nonmetal elements, such as helium, argon, oxygen, nitrogen, and hydrogen.

The particles (smallest pieces) of all the noble gases are atoms. A helium atom (He) is small, round, and light. The argon atom (Ar) is about twice as wide.

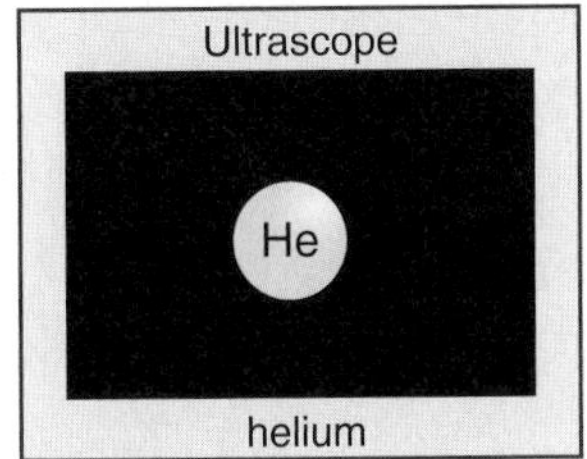

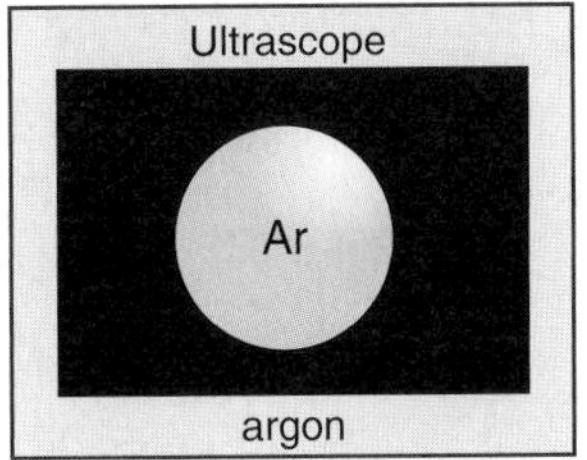

Suppose you had a one-liter balloon of helium. Now imagine that you could see individual atoms of helium through the Ultrascope. The approximately 30 billion trillion helium atoms are the same mass, shape, and size. But the helium atoms are different in size and/or mass from all the atoms of other elements.

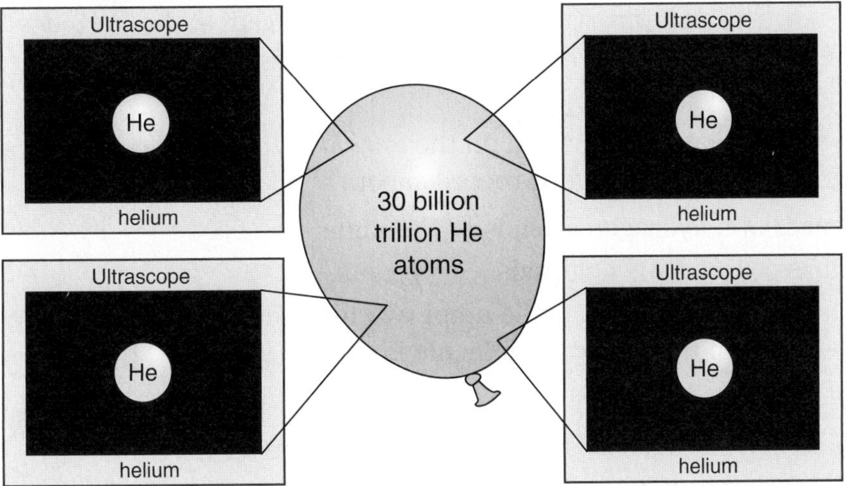

Unlike the helium and argon particles, which exist as single atoms, a hydrogen particle actually consists of two hydrogen atoms (H) joined together in a structure called a *molecule*. Hydrogen molecule are the same mass, size, and shape. The hydrogen atom is the smallest and lightest atom.

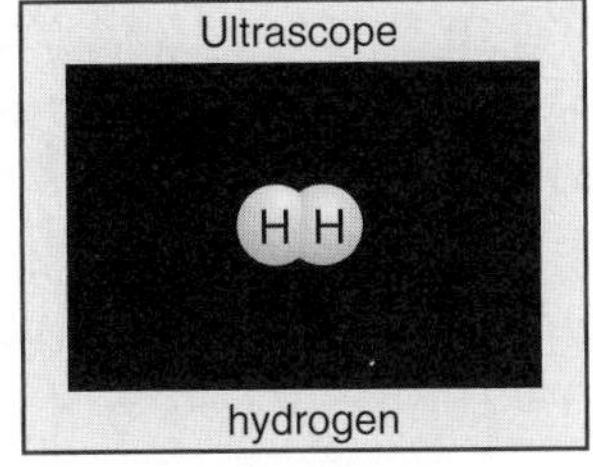

Oxygen (O) and nitrogen (N) particles are also molecules, consisting of two identical atoms joined together in molecules.

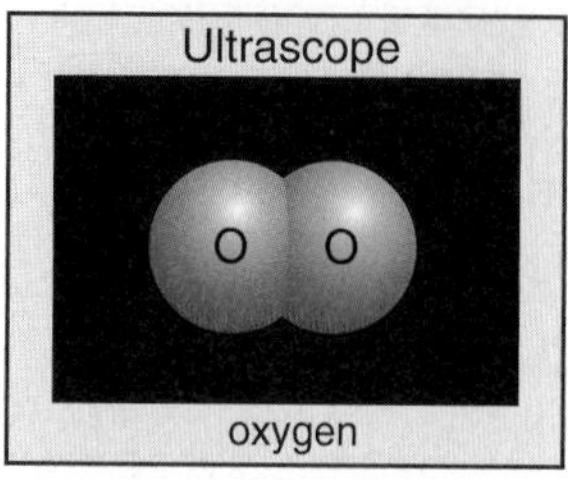

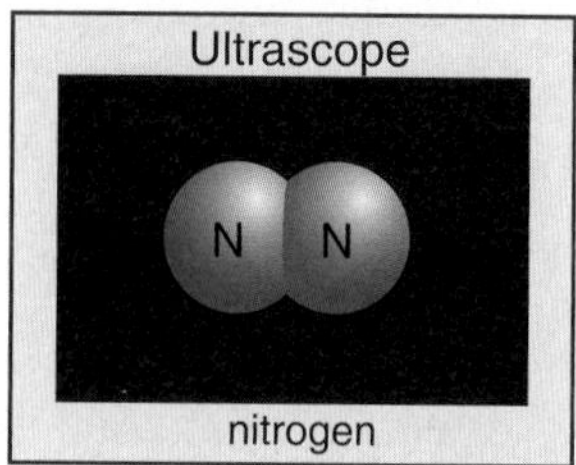

You know that the mass of particles is extremely tiny. For example, a nitrogen atom has a mass of about 23 millionths of a billionth of a billionth of a gram, or

0.000,000,000,000,000,000,000,023 g.

These numbers are too small to record conveniently on the *Periodic Table of the Elements*. Chemists often use a measurement called the **a**tomic **m**ass **u**nit (amu). This number is the mass of the element's atom compared to the mass of a hydrogen atom, which is set to 1. Look again at the *Periodic Table of the Elements*. Below the name of each element is the mass of the element's atom in atomic mass units.

Use the atomic mass information on the *Periodic Table of the Elements* to complete the following on your record sheet.

3. The mass of a hydrogen atom is ______ amu.

4. The mass of a nitrogen atom is ______ amu.

5. The nitrogen *molecule* is made up of two nitrogen atoms joined together. So the mass of the nitrogen molecule is:

______ amu + ______ amu = ______ amu

Participate in a class discussion about your answers.

You learned in Unit 5 that compounds can be broken down into elements during chemical interactions.

According to the Small-Particle Theory (SPT):

Idea 1 All single substances are made up of tiny particles. A particle is the smallest piece of a substance.

C. Compounds consist of two or more atoms of different elements. The atoms of some compounds are joined together in a molecule. Molecules of the same compound are usually the same mass, shape, and size. Molecules of different compounds have different masses, shapes, and/or size.

For example, the Mickey-Mouse shaped water molecule consists of three atoms. Two identical hydrogen atoms are joined to a central oxygen atom. The water molecule is about 18 times heavier than a hydrogen atom.

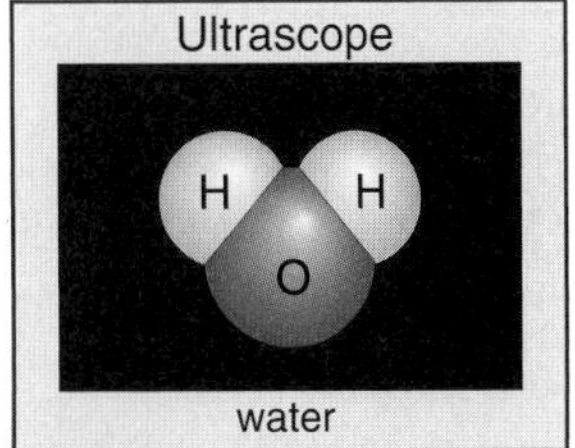

Suppose you could look through the imaginary Ultrascope at some individual molecules of a drop of water.

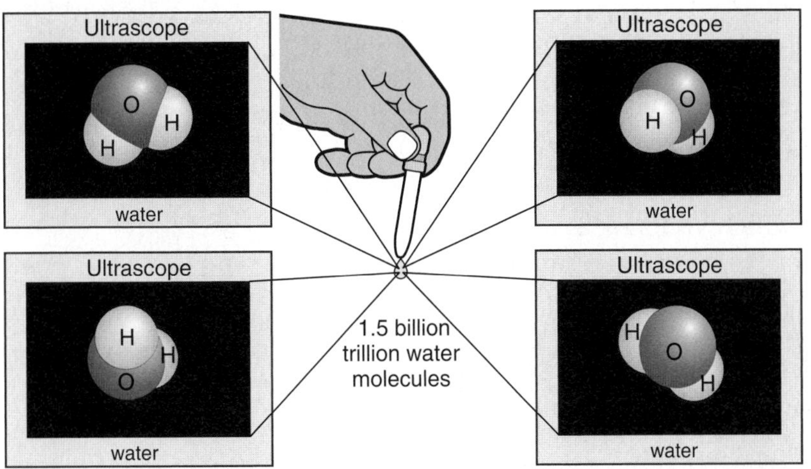

This diagram shows each water molecule from a different view. But *all* of the approximately 1.5 billion trillion water molecules in the drop have the same Mickey-Mouse shape, and are usually the same size and mass.

Water molecules are the smallest particles of water. When you break the water molecules apart during a chemical interaction, then you do not have water any more. As you know, you get the elements hydrogen and oxygen. You will learn more about breaking apart molecules during chemical interactions in Unit 7.

Different compounds are made of different combinations of atoms. For example, the *carbon dioxide molecule* consists of three atoms in a line. Two oxygen atoms are joined to one carbon atom in the middle.

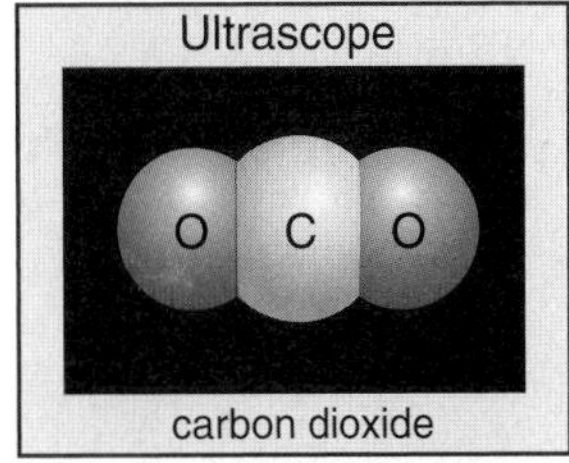

6. An ammonia molecule consists of three hydrogen atoms joined to a nitrogen atom in a pyramid shape. Find the symbols for nitrogen and hydrogen on the *Periodic Table*. On your record sheet, write the symbols on the drawing of the ammonia molecule.

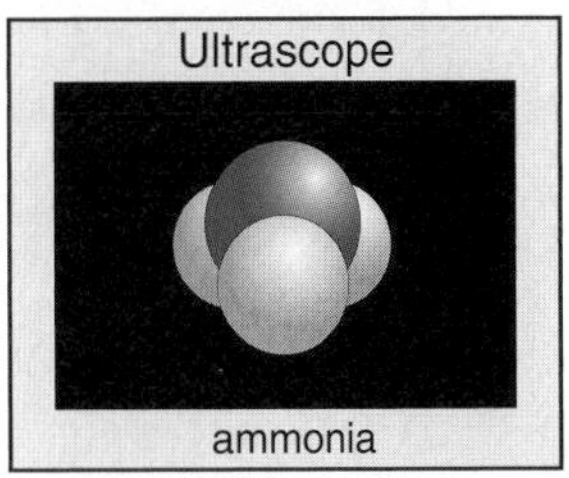

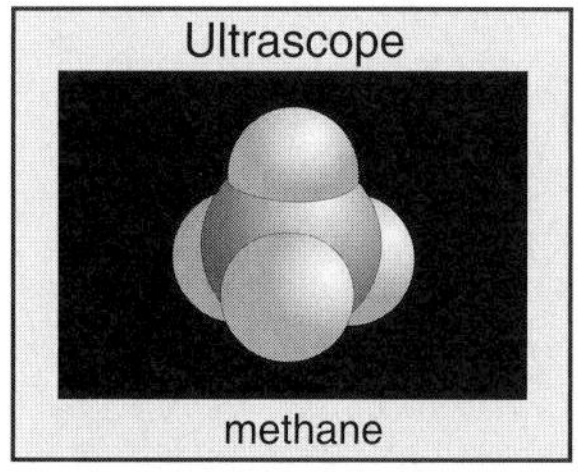

methane

7. A methane molecule consists of four hydrogen atoms joined to a central carbon atom. This shape is called a tetrahedral. Find the symbols for carbon and hydrogen on the *Periodic Table*. On your record sheet, write the symbols on the drawing of the methane molecule.

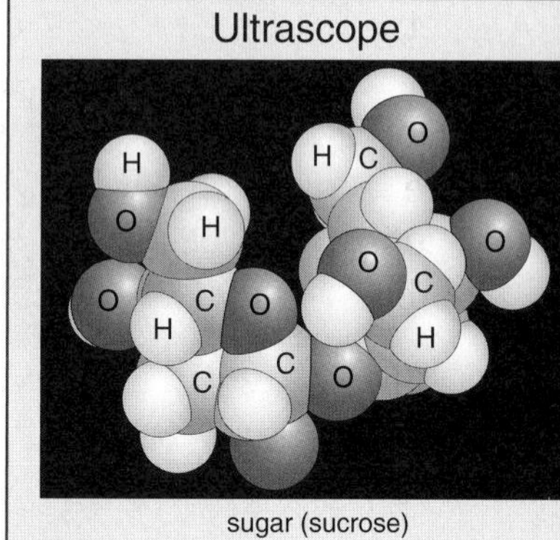

sugar (sucrose)

The number of atoms in a molecule can range from two atoms to thousands of atoms. For example, the sugar molecule shown at the right is made up of 44 atoms: 11 carbon (C) atoms, 11 oxygen (O) atoms, and 22 hydrogen (H) atoms. The giant DNA molecule, which you saw in the *Power of Ten* photographs, is made up of millions of atoms.

What's Between the Particles of Matter?

The third idea in the Small-Particle Theory (SPT) is:

Idea 3 There is nothing between the particles of matter. (The space between the particles is empty of stuff.)

The idea that there is nothing (empty space) between the particles took scientists a long time to believe. It is difficult to imagine "nothingness."

For many years scientists thought there was a unique kind of stuff between the particles, which they called the *ether*. The ether had to have very special properties. For example, to explain why air in a pump is so compressible, the ether had to be very compressible. To explain why the plunger of the pump springs back to its original position when you let go, the ether had to be very "springy."

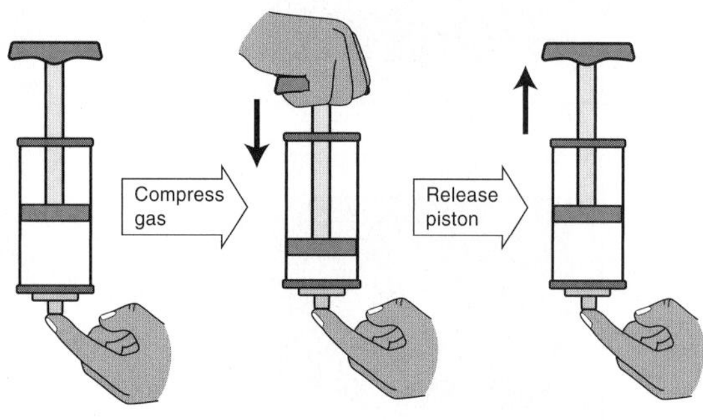

In fact, to explain everything you observe, the ether acquired many very peculiar properties. Scientists finally rejected the idea of a special kind of stuff (ether) between the particles because:

- after many experiments, they could find no evidence that the ether exists.
- the ether with its special, peculiar properties is *not needed* to make sense of everyday experiences and the conclusions of experiments.

Today we have more indirect experience with *nothingness*. Astronauts can travel to the "empty space" above the Earth. The Earth is surrounded by a thin layer of air, called the Earth's atmosphere. This layer of air is surrounded by the blackness of empty space. (Actually, scientists estimate that the space between the Earth's atmosphere and the Moon contains about one particle in each one-centimeter cube of space. But this is empty compared to the approximately 30 million billion particles (30,000,000,000,000,000,000) in each one-centimeter cube of the air you breathe.)

STEP 1 Your teacher will do a class demonstration involving your sense of smell to help you make sense of the idea of empty space between the particles.

Stand around your teacher in the middle of the room. Your teacher will open a bottle of a substance with a very strong smell. Raise your hand when you smell the substance.

8. How did the molecules of the smelly substance get from the bottle to your nose?

Participate in a class discussion. What idea needs to be added to the Small-Particle Theory to explain how the molecules of the smelly substance get from the bottle to your noses?

Do the Particles of Matter Move?

The fourth idea of the Small-Particle Theory (SPT) is:

Idea 4 The particles of substances move continually (all the time) in all directions.

Ideas 3 and 4 help you make sense of how you can smell things at a distance. The particles of the smelly substance move continually in different directions (Idea 4). They spread into the empty spaces (Idea 3) between the air particles until they eventually reach your nose. The continual-motion idea also helps to explain why you can breathe air, even on top of a high mountain. If the air particles slowed down and stopped, then they would sink to the surface of the Earth and you would not be able to breathe!

The idea of continual motion seems strange at first, because in your everyday life there is always *friction* and *drag* to slow things down. But you do have indirect experience with continual motion. You may have seen movies or videos of astronauts working in space taken by NASA (National Aeronautics and Space Administration). When astronauts work outside the space shuttle, they have to be *very* careful about how they move. They must be tied to the shuttle or the satellite they are fixing, like the astronaut shown in the picture. Even the slightest push against the shuttle or satellite sends them off in space at a constant speed. They never slow down or stop (unless they collide with something else or catch hold of something) *because there is no air in space to slow them down.*

Scientists believe that a similar thing happens with the millions of trillions of air particles in each liter of air here on Earth. Since there is nothing between the particles (the space is empty of stuff), there is nothing to slow them down. So, the particles keep moving continually.

Make Sense of Scientists' Ideas

A teacher did a demonstration. She attached a pump to a closed flask and removed about one-half of the air from the flask.

Two students, Isabel and Mike, were discussing what happens when air is pumped out of a flask.

Discuss the following question with your team. Then write your answer on your record sheet. Be prepared to share your team's answer with the class.

Which student's ideas, Isabel's or Mike's, match scientists' ideas? Justify your answer.

Participate in a class discussion about the answer to this question.

I think air contains particles of oxygen, nitrogen, carbon dioxide, and some other particles. The space between the particles is filled with air. When you pump some air out of the flask, the particles left in the flask move around until they are evenly spaced inside the air left in the flask. Then they gradually slow down and stop moving.
Isabel

I think air is made up of particles of oxygen, nitrogen, carbon dioxide, and some other particles. The space between the particles is empty. When you pump some air particles out of the flask, the particles left in the flask move around so they fill all the space inside the flask. But they keep moving. They do not slow down and stop.
Mike

Our Consensus Ideas

Recall the key question for this activity:

What are the first four ideas (mental models) in the Small-Particle Theory of Matter?

To answer this question, complete the following on your record sheet.

Idea 1 All single substances, gases, liquids, and solids, are made up of tiny particles. A particle is the smallest piece of a substance.

A. Particles (smallest pieces) of a substance can be *atoms* or *molecules*.

B. Elements consist of one kind of atom. All atoms of the same element are ______________ in mass, shape, and size. But the atoms of ______________ elements have different masses and/or size.

1. Atoms of the same element are usually ______________ mass and/or size. Atoms of ______________ elements have different masses and/or sizes.
2. The metal elements and the noble gases consist of ______________ atoms. For some nonmetal elements, two identical atoms are joined together in a structure called a ______________.

C. Compounds consist of two or more atoms of different elements.

1. The atoms of some compounds are joined together in a ______________. ______________ of the same compound are usually the same mass, shape, and size. ______________ of different compounds have different masses, shapes, and/or size.

Idea 2 The particles of substances are __________ to be seen through light microscopes.

Idea 3 There is ___________ between the particles of matter (the space between the particles is ______________).

Idea 4 The particles of substances move ______________ in all directions.

Participate in a class discussion about the key question.

DEVELOPING OUR IDEAS

Activity 3
Properties of Gases, Liquids, and Solids

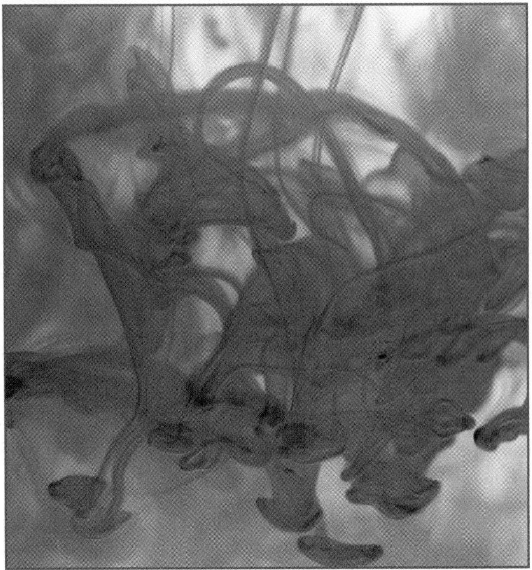

Purpose

One criterion of a good scientific theory is that the ideas in the theory must help you make sense of a wide range of everyday experiences with objects and events. One everyday experience you have is with gases, liquids, and solids. So, before you further explore the Small-Particle Theory (SPT), you will investigate some similarities and differences in the physical properties of gases, liquids, and solids. You will begin by investigating a diffusion interaction. **Diffusion** is the spreading of one substance into another substance in the same phase. Then you will examine the density of solids, liquids, and gases. The two key questions for this activity are:

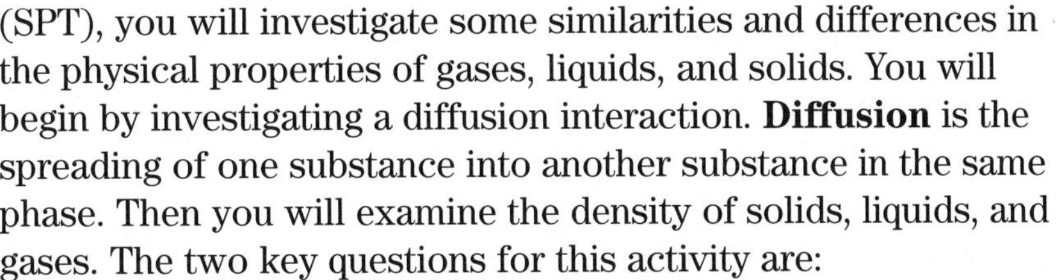

Science Words

diffusion: the spreading of one substance into another substance in the same phase

1. What are the similarities and differences in the diffusion properties of solids, liquids, and gases?

2. What are the similarities and differences in the density properties of solids, liquids, and gases?

Record the key questions for the activity on your record sheet.

We Think

You learned in Activity 2 that a gas *spreads out* into another gas very quickly. This interaction is called diffusion. Any gas will spread (diffuse) about 2 cm into another gas in less than a second.

How fast do you think a liquid will spread 2 cm into another liquid?

Discuss this question with your team. Write your team's best answer on your record sheet.

Participate in a class discussion.

Explore Your Ideas about Diffusion

You will do a short exploration to answer this question. Liquid-liquid diffusion can be seen in something as simple as a drop of food coloring spreading out in a small dish of water.

Your team will need:

- container of room-temperature water
- red food coloring
- 2 Petri dishes (clear, shallow dishes)
- 50 mL beaker or graduated cylinder
- 2 eyedroppers
- 2 bull's-eye patterns with evenly spaced rings
- 2 stopwatches or clock with a second hand
- paper towels and/or newspapers
- container or bucket for waste colored water

STEP 1 *Read the directions all the way through before you start.* You will work in pairs for this exploration.

STEP 2 Pour about 25 mL of room-temperature water slowly and gently into the Petri dish so that it is about halfway full.

STEP 3 Select a location in your workspace where you will not knock or disturb the water in the Petri dish. Then center the dish of water over the center of the bull's-eye pattern.

STEP 4 Take turns practicing your eyedropper skills. Start with some food coloring in your eyedropper. Then carefully place the tip of the eyedropper just under the water and above the center of the bull's-eye pattern. *Gently* squeeze a drop of food coloring into the water and remove the eyedropper.

If you squeeze too hard, the food coloring swirls and spreads out right away. Practice until you can get a drop of food coloring that does not swirl too much.

STEP 5 Empty the Petri dish and repeat Steps 2 and 3.

STEP 6 Place a drop of food coloring *above the black center of the bull's-eye* pattern. Start your stopwatch or note the time on a clock.

When the food coloring has spread to the second circle (2 cm), stop the stopwatch or note the time on the clock.

1. Record the time it took the food coloring to spread 2 cm into water.

Food coloring can stain your clothes, hands, and tables.

(1) Be sure to wear gloves when handling the food coloring.

(2) Cover your workspace with paper towels and/or newspapers.

(3) Wear eye protection.

STEP 7 Clean up your work area. Add your team data to a class data table. Calculate the class average time.

Participate in a class discussion about your observations.

2. Record the class average time it takes food coloring to spread 2 cm into water.

Make Sense of Your Ideas about Diffusion

You know that gas-gas diffusion is very fast. For example, a smelly gas will spread 2 cm into the air in less than a second. This result is shown in *Table1: Summary of Similarities and Differences in Some Physical Properties of Gases, Liquids, and Solids.*

Table 1: Summary of Similarities and Differences in Some Physical Properties of Gases, Liquids, and Solids

Property	Gas	Liquid	Solid
1. Shape Does it have a definite shape?	**no** (no surfaces or boundaries)	**no** (takes shape of container; has surfaces)	**yes** (has surfaces)
2. Volume Does it have a definite volume (# of unit cubes?)	**no** (fills all space inside and outside of a container)	**yes** (can count number of unit cubes)	**yes** (can count number of unit cubes)
3. Diffusion How fast does a substance spread 2 cm into another substance in the same phase?	**very fast** (takes less than a second)		
4. Density How do *equal volumes* (1 L) of substance interact with mass balance?	**very small** (1 L has mass of only a few grams)		
5. Compressive Strength			
6. Drag			
7. Tensile Strength			
8. Thermal Expansion			

1. Does it take seconds, minutes, hours, days, months, or years for a liquid to spread 2 cm into another liquid? Write your answer in the third row of Table 1 on your record sheet.

Given enough time, a solid will also diffuse into another solid. For example, suppose a gold brick is placed on top of a lead brick. About once a month, a microscopic photograph is made of the edge between the gold and lead bricks. After several months, some gold is in the lead brick and some lead is in the gold brick.

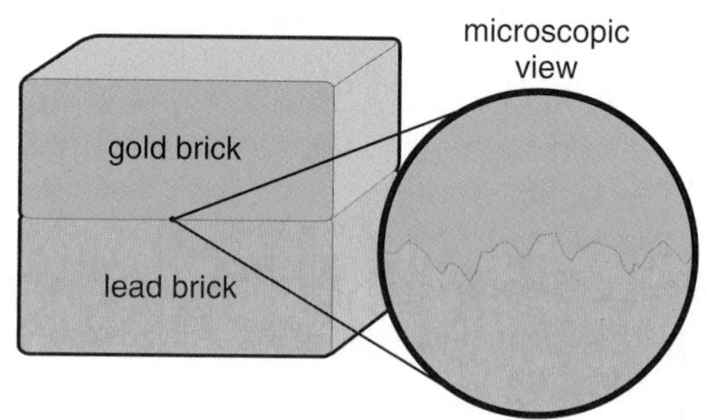

2. Does it take seconds, minutes, hours, days, months, or years for a solid to spread 2 cm into another solid? Write your answer in the third row of Table 1 on your record sheet.

Participate in a class discussion about the answers to these questions.

Explore Your Ideas about Density

The second key question for this activity is: What are the similarities and differences in the densities of solids, liquids, and gases? To answer this question, you will use the *Table of Densities* in the Appendix.

Density is the mass of a standard-unit volume. The standard volume in the *Table of Densities* is one liter. You will determine the range of densities (least to greatest) for gases, liquids, and solids, and record the ranges in *Table 2: Range of Masses of One Liter of Gases, Liquids, and Solids*.

STEP 1 Look at the densities of different gases in the *Table of Densities* in the Appendix. Notice that one liter of different gases ranges in mass from about one-tenth of a gram (for hydrogen) to a little over three grams (for chlorine). These masses are recorded in the first row of the range of masses table.

Table 2: Range of Masses of One Liter of Gases, Liquids, and Solids

Phase	Range (g)	
	Lowest	Greatest
1. Gases	From: **0.08**	To: **3.21**
2. Liquids (excluding mercury)	From:	To:
3. Solids	From:	To:

STEP 2 Now look at the densities of some room-temperature liquids. Notice that mercury is different from the other liquids. One liter of mercury has a mass of over one kilogram (13,000 g). But one-liter volumes of most liquids have much less mass than mercury.

Not counting mercury, which liquid has the *least* mass of one liter? The *greatest*? Write these masses in the second row of Table 2 on your record sheet.

STEP 3 Finally, look at the densities of some room-temperature solids (not powders). There is a wider range of masses of one liter of different solids.

Which solid has the *least* mass of one liter? The *greatest*? Write these masses in the third row of the table.

Participate in a class discussion about your data.

Make Sense of Your Ideas about Density

Use the information in Table 2 to answer the following questions.

1. How do the masses of one liter of most liquids compare with those of most gases?
2. How do the masses of one liter of most solids compare with those of most liquids?
3. How do the masses of one liter of most solids compare with those of most gases?

Participate in a class discussion about the answers to these questions.

4. Go back to *Table 1: Summary of Similarities and Differences.* With your teacher's guidance, complete the fourth row (Density).

Our Consensus Ideas

The key questions for this activity are:

1. What are the similarities and differences in the diffusion properties of solids, liquids, and gases?

2. What are the similarities and differences in the density properties of solids, liquids, and gases?

1. Write your answers on your record sheet.

Participate in the class discussion about the key questions.

2. Write the class consensus ideas on your record sheet, if they are different from your answers.

MAKING SENSE OF SCIENTISTS' IDEAS

Activity 4
Small-Particle Theory of Gases

Purpose

In Activity 2 you learned the first four ideas of the Small-Particle Theory of Matter. In this activity you will see two computer representations of the Small-Particle Theory (SPT) of gases. The key questions for this activity are:

1. What are two ways that the small particles of matter can interact with each other?

2. What are the SPT ideas of the between-particle interaction strength, the motion, and the average spacing of the particles of a gas?

Record the key questions for the activity on your record sheet.

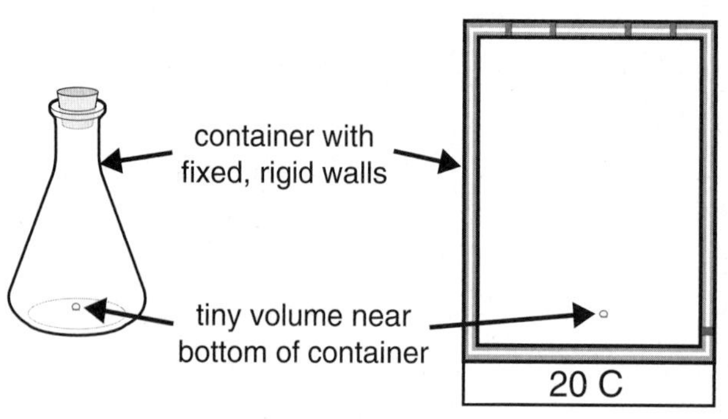

In this activity you will use an Ultrascope simulator and animation. The computer simulator has a container with fixed, rigid walls, as shown in the diagram. The computer container represents any real container with fixed, rigid walls, like a flask with a stopper. The simulator also uses the imaginary microscope, called the Ultrascope.

The Ultrascope field of view is only one molecule thick, as shown in the diagram. The total volume in the field of view of the imaginary Ultrascope is only 8 million, million, billionths of a one-centimeter cube:

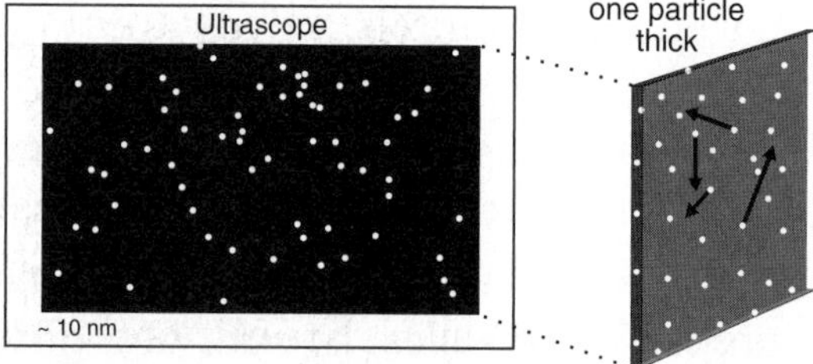

0.000,000,000,000,000,000,008 cm^3 (cubic centimeters)!

The background is black to show that there is nothing (no stuff) between the molecules.

Explore the SPT Ideas about the Motion and Spacing of Gas Molecules

Your team will need:

- 4 cards with a square hole in the middle

You will first take part in a demonstration. Then you will look at two representations of the motion and spacing of gas particles, the Ultrascope (Ideal Gas) simulator and a computer animation.

STEP 1 The card with the hole represents the field of view of the Ultrascope. Hold the card with the hole about 30 cm in front of your eyes. Watch some students moving around in the front of the room.

1. Describe the motion of the students as seen in your *field of view.* Do you see all of the students in the front of the room? Do some students move into your field of view while other students move out of your field of view?

STEP 2 Your teacher will show you the Ultrascope simulation of gas particles near the bottom of the flask. Watch the particles in the field of view of the Ultrascope. You will see particles continually moving into the field of view and out of the field of view.

As you watch, think about these four questions about the motion and average spacing of the particles:

- Do all the gas particles move at the same speed? Different speeds?
- Do all the particles move in the same direction? Different directions?
- What causes a particle to change direction?
- On the average, are the particles in a gas close together or far apart *compared to the size of the particles*?

Participate in a class discussion about the answers to these questions.

STEP 3 Imagine that you could take "snapshot" pictures of the gas particles seen in the Ultrascope. The three white arrows on the Ultrascope picture represent the speed of the three particles the instant the snapshot was taken. The length of each arrow tells you how fast the particle is moving. The long arrows indicate fast speeds and shorter arrows indicate slower speeds. The direction of the arrow tells you the direction the particle was moving at the instant the snapshot was taken.

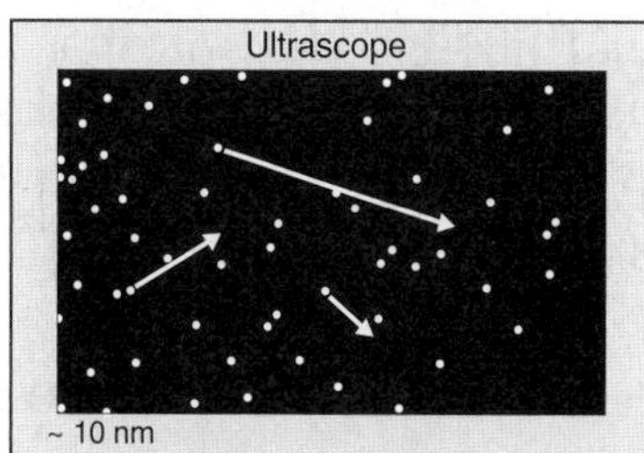

The Ultrascope computer simulation calculates the speed of *each* particle at one instant or snapshot. Then the simulation calculates the average speed of all the particles in the Ultrascope view. The average speed is the sum of the speeds of all the particles divided by the number of particles (represented by the symbol N in the equation on the next page).

$$\text{Average Speed} = \frac{\text{speed of particle 1} + \text{speed of particle 2} + \text{speed of particle 3}\ldots + \text{speed of particle } N}{N}$$

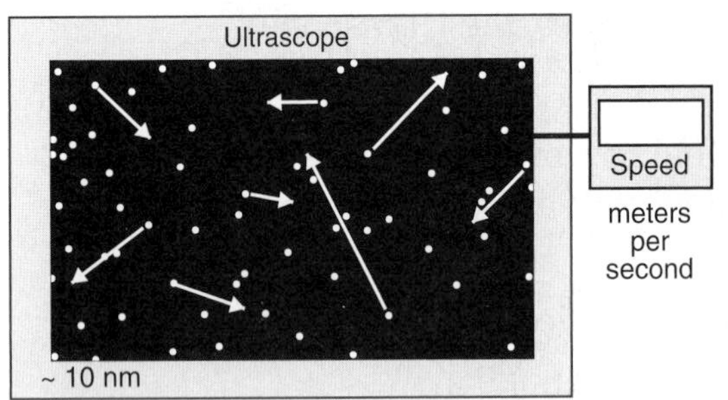

The meter attached to the Ultrascope shows this *average* speed of the particles when the snapshot was taken. The Ultrascope calculates the average speed of the particles about 10 times each second.

The fastest athletes in the world can run 100 m in just under 10 s. This is an average speed of about 10 m/s, or 23 mph.

2. Do you think the average speed of room-temperature gas particles is *greater than, less than*, or *about the same as* the average speed of our best athletes (about 10 m/s)? If greater or less, *how much* greater or less? Write your prediction of the average speed in meters per second.

Participate in a class vote.

STEP 4 Watch the Ultrascope computer model of the Small-Particle Theory to test your predictions. Your teacher will stop the Ultrascope about five times over a 1-2 min time interval. On your record sheet, record the average speed each time the Ultrascope is stopped. Calculate the average of all the average speeds and enter it in the last row.

3. Did your prediction of the average speed of gas particles match the average speed shown in the Ultrascope model?

The Ultrascope model of molecular motion is like watching a motion picture that has been slowed down a lot. If the particles on the screen moved as fast as real gas particles, then your eyes would not be able to see the motion of the particles. So the screen shows the particles moving much more slowly than real gas particles. But the Ultrascope speed meter shows the actual average speeds of real gas particles.

Table 1: Average Speed of Particles

	Average Speed
Time 1	
Time 2	
Time 3	
Time 4	
Time 5	
Average	

STEP 5 Your teacher will show you another kind of computer representation of a gas, an *animation*. This is an animation of pure water vapor. That is, water vapor that is not mixed with air.

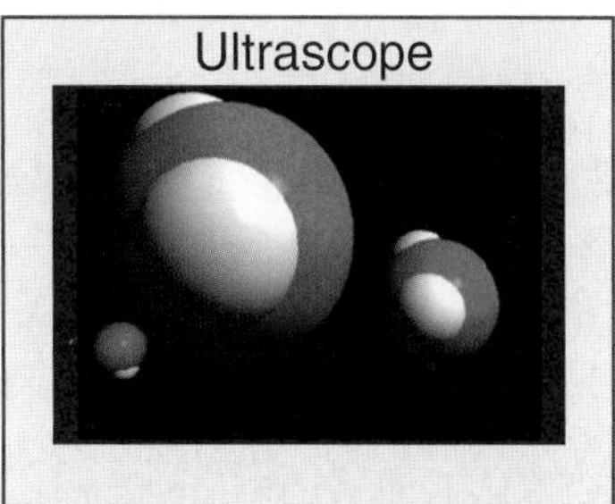

Like the Ultrascope simulation the animation is a slow-motion picture. In real life the molecules are moving much faster (1050 miles per hour on average). The animation is different from the Ultrascope simulation in two ways.

First, the magnification of the animation is much larger than with the Ultrascope. The particles appear bigger. You can see the characteristic Mickey-Mouse shape of the water molecules. Second, the animation shows a field of view that is deeper (many particles thick) instead of just one molecule thick. You saw some water particles that are moving further away, and some that are moving closer. Although the ones that are further away appear smaller and dimmer, are they really smaller?

The photograph shows a crowd of people at a sports stadium. Even though the people farther away appear to be smaller and dimmer than the people closer to the camera, you know that they are not really smaller. In the same way, when you look at the computer animation, the dimmer and smaller water molecules are really the same size as ones that are closer. They are just farther away.

STEP 6 Watch the animation again, this time focusing your attention on the motion of the molecules before and after they hit each other.

4. When the particles are not colliding, do they appear to be *moving at a constant speed*, *speeding up*, or *slowing down*?

Participate in a class discussion about your answer.

Explore the SPT Ideas about the Interactions between Particles

In this section, you will explore the first key question: What are two ways that the small particles of matter can interact with each other?

STEP 1 Start by reviewing what you know about interactions. Look at your *Interaction Type* wall chart.

1. Which interaction types require that the objects be touching each other (or connected)?
2. Which interaction types require that the objects be near each other (not necessarily touching)?

Participate in a class discussion about your answers.

STEP 2 According to the Small-Particle Theory (SPT) Idea 3, there is nothing between the particles. The space between the particles is empty of stuff. This means that there are no drag or friction interactions between particles. In the Ultrascope simulation and animation movie models, the particles seem to touch when they collide with (hit) each other and bounce apart. This collision interaction is similar to an applied mechanical interaction between a bowling ball and pin.

Idea 6 Collision Interaction – Particles Bounce Apart

This interaction occurs when a moving particle (atom or molecule) of a substance collides with (hits) another particle and the particles bounce apart.

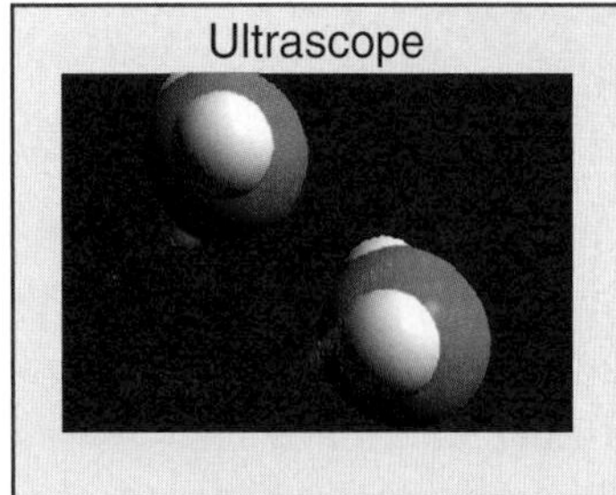

before collision

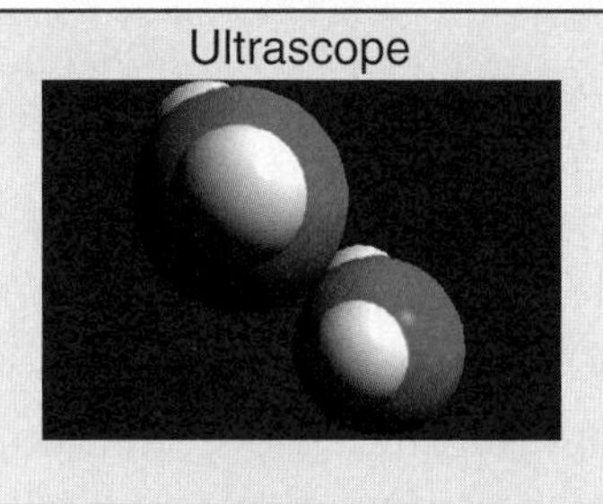

during collision

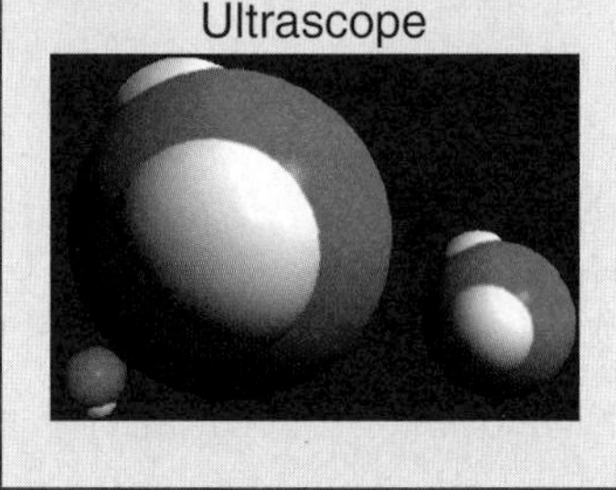

after collision

STEP 3 The second category of interactions between objects can occur when objects are just near each other (not necessarily touching).

Is there an interaction between the particles (atoms or molecules) of gases similar to an electric-charge or magnetic interaction? To answer this question, first think about what you learned in Unit 1 about the electric-charge interaction.

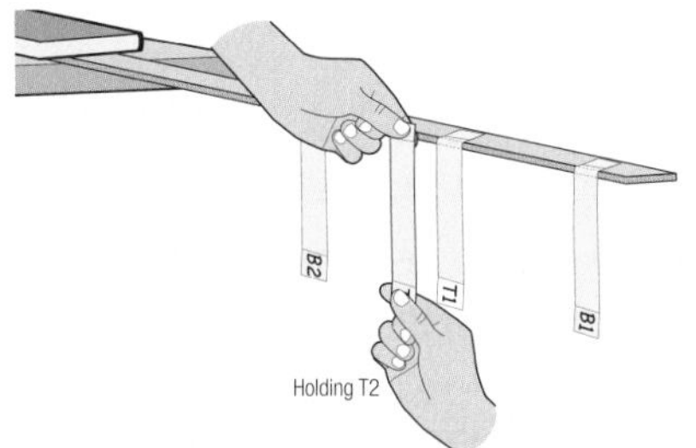

3. What are the two things that can happen when you bring one charged object near another charged object?

Participate in a class discussion about the answer.

STEP 4 Imagine Ultrascope snapshots of *gas* particles taken at two different times (Now and Some time later).

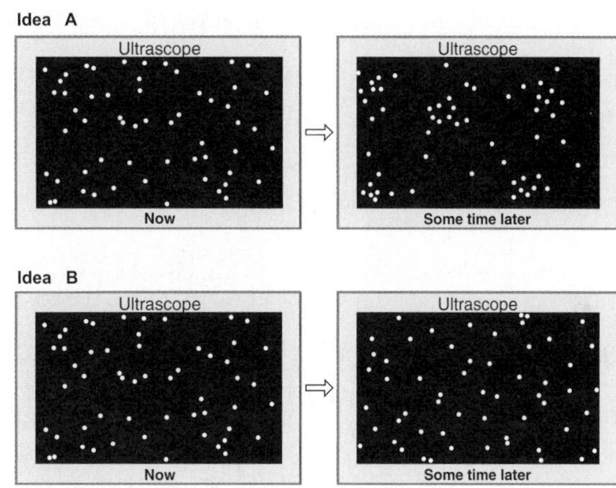

4. Which idea, Idea A or Idea B, more likely represents particles that are continually attracting each other? Explain your reasoning.

5. Which model more likely represents particles that are not interacting with each other? Explain your reasoning.

STEP 5 Your teacher will show you the scientists' Ultrascope simulation again. *Watch the motion of the gas particles carefully for evidence of Idea A or Idea B.*

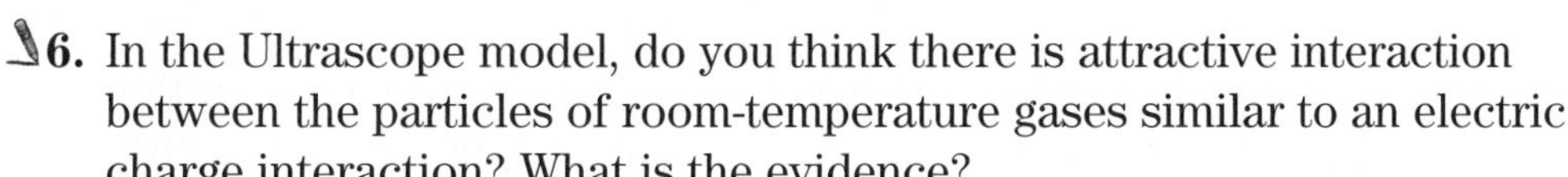

6. In the Ultrascope model, do you think there is attractive interaction between the particles of room-temperature gases similar to an electric charge interaction? What is the evidence?

Participate in a class discussion about your answers.

STEP 6 A model is a simplified imitation of something that we hope can help us understand it better. All models of an object or event have some characteristics that are similar to the object or event, and some characteristics that are different. For example, the model of a glider in a wind tunnel has the same shape as real gliders, but is very different in size.

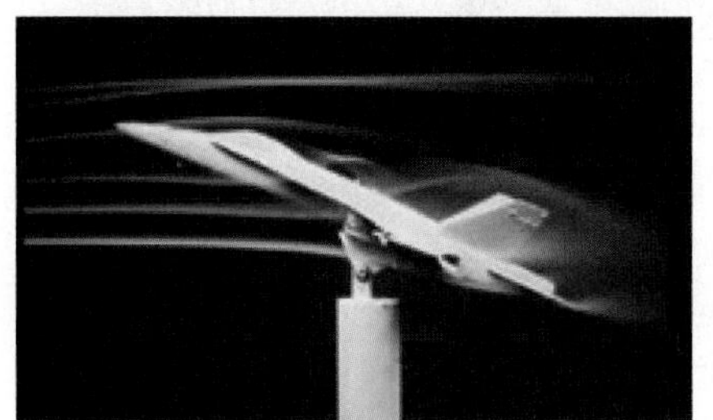

Science Words

cohesion: sticking or holding together to form a whole

The Ultrascope simulation is a simplified imitation (model) of the behavior of the particles of a gas. One major difference between the Small-Particle Theory ideas and the Ultrascope model is that there is, in fact, *a continual attraction between the particles of all substances, even when they are not touching.* The attraction between particles of a substance is often called the **cohesion** interaction because cohesion means holding together to form a whole.

Idea 7 Cohesion Interaction between Particles of a Substance

The particles (atoms or molecules) of an element or compound are continually attracted to each other, even when they are not touching. This is called the cohesion interaction. Cohesion means sticking or holding together.

A. The cohesion interaction is related to different types of the electric-charge interaction.

B. The strength of the cohesion attraction between particles is different for different substances near room temperature.

C. *Very Weak Attraction* When the cohesion attraction between particles is in the very weak range (almost none), then the substance is usually a gas near room temperature. For most gas substances, the attraction is so weak that it can be ignored in many situations.

The Ultrascope model ignores the very weak attraction between the particles of a gas. This interaction is *not* magnetic. It is a special type of electric-charge interaction. You will learn more about the interaction between the particles of substances in later activities.

Table 2: Differences between SPT and Ultrascope Model

SPT Ideas	Ultrascope Model of Gases
Idea 7C When the cohesion attraction between particles is in the very weak range (almost none), the substance is usually a gas near room temperature.	
Idea 7C2 At room temperature the average speed of gas particles between collisions is about 480 m/s (~1050 mph).	
Idea 4 The particles of substances move continually in all three directions.	

STEP 7 As you know, there are two additional differences between scientists' Small-Particle Theory (SPT) ideas and the Ultrascope model of the particles of gases. First, watching the Ultrascope model of molecular motion is like watching a motion picture that has been slowed down a lot. At room temperature, the average speed of the particles between collisions is about 480 m/s (about 1050 miles per hour).

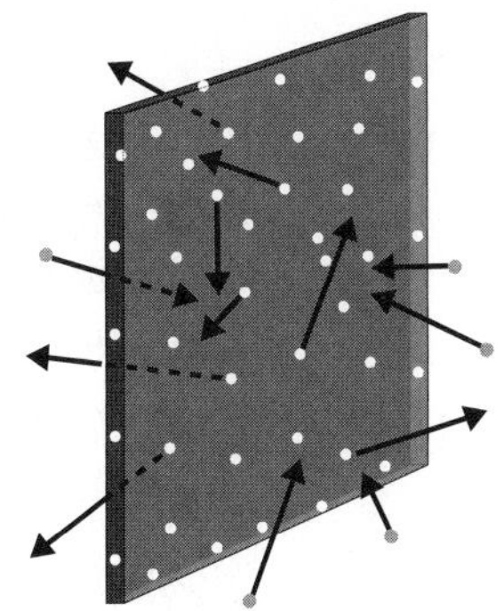

This is close to the speed of a jet fighter, and 48 times faster than our best athletes can sprint! The Ultrascope shows the particles moving at a much slower average speed.

Another difference between the SPT ideas and the Ultrascope model is the directions of motion of the gas particles. The Ultrascope shows a field of view only one particle thick. So, the particles can only move in two directions, up-down and left-right. But gas particles move in *all three* directions.

Summarize the differences between three Small-Particle Theory (SPT) ideas and the Ultrascope Model of gases in Table 2 on your record sheet.

Participate in a class discussion about your answers.

The SPT Ideas

The first key question for this activity is: What are two ways that the small particles of matter can interact with each other?

The answer to this question is summarized in the Small-Particle Theory (SPT) Idea 5.

Idea 5 Small-Particle Interactions

There are only two ways the small particles of matter (atoms or molecules) can interact.

A. The particles continually *attract and/or repel* each other, even when they are not touching. This category of interaction requires that the particles be near each other *or* touching.

B. The particles can interact when they *collide* with each other. This category of interaction requires that one moving particle hits another particle.

1. Complete the classification map of small particle interactions, like the one shown, on your record sheet.

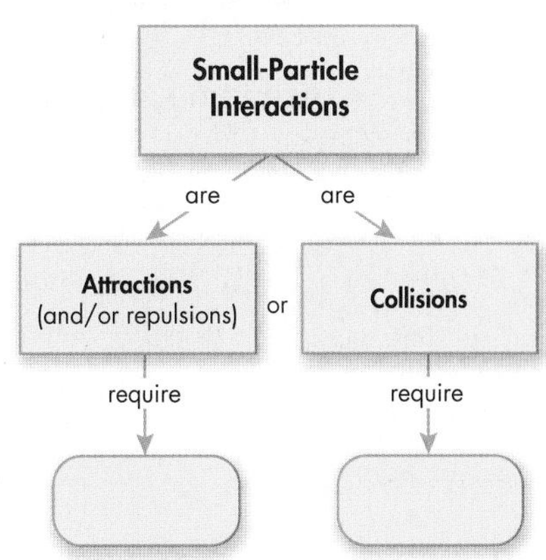

The second key question for this activity is: What are the SPT ideas of the between-particle interaction strength, the motion, and the average spacing of the particles of a gas?

This question asks about three characteristics of the particles of room-temperature gases: the strength of their between-particle attraction, their spacing (average distance apart compared to the size of the particles), and their motion.

With your teacher's guidance, you will summarize your answer to this key question by completing the following statements on your record sheet. Your teacher will keep the Ultrascope model running so you can check your answers.

Idea 7 Cohesion Interaction between Particles of a Substance

C. When the attraction between particles is __________ (almost none), the substance is usually a gas near room temperature.

Consider whether, on the average, the particles of a gas are close together or far apart compared to the size of the particles, and write a sentence about the particle spacing of gases.

2. Particle Spacing ____.

Before writing about the motion of gas particles, consider these questions:

- Do all gas particles move at the same speed?
- Do all the particles move in the same direction?
- Do they speed up, slow down, or move at a constant speed between collisions?
- About how fast do the gas particles move on the average?

Write the average room-temperature speed of the particles in the Ultrascope meter (from the last row of Table 1).

3. Particle Motion ____.

Our Consensus Ideas

Think again about the key questions for this activity.

1. What are two ways that the small particles of matter can interact with each other?

2. What are the SPT ideas of the between-particle interaction strength, the motion, and the average spacing of the particles of a gas?

Are you satisfied now that you are able to answer the key questions?

Participate in a class discussion.

MAKING SENSE OF SCIENTISTS' IDEAS

Activity 5
Small-Particle Theory of Liquids

Purpose

In Activity 4 you saw two computer representations of how the collision interaction and the very weak cohesion interaction influence the average spacing and motion of the particles of gases near room temperature. In this activity, you will explore the following question:

What is the SPT of the between-particle interaction strength, the motion, and the average spacing of the molecules of a liquid?

Record the key question for the activity on your record sheet.

The activity has two parts. First, you will explore how easy substances are to pull apart into smaller pieces. In the second part of the activity, you will explore different ideas of the motion and average spacing of the molecules of a liquid.

Explore the SPT Ideas of Interaction Strength

Strength is a mechanical property of materials. But what does it mean when you say a material, like plastic, is strong? There are different kinds of strength, depending on how well the material holds up to whatever you are trying to do to it. Here are some different kinds of strength tests.

- Try to bend (or flex) it—flexural strength.
- Try to twist it—torsional strength.
- Try to compress it—compressive strength.
- Try to hit it sharply and suddenly (as with a hammer)—impact strength.
- Try to pull it apart—tensile strength.

Typical solids, like rocks, metals, and plastic, have a high tensile strength. They are difficult to pull apart (unless they are very small or thin).

It is impossible to pull a gas apart into smaller pieces because gases have no edges or boundaries. Gases have no tensile strength. In the first part of this exploration, you will investigate the tensile strength of liquid water. How easy is it to pull apart a drop of water into smaller droplets?

Then you will investigate two *marble models* of the strength of the cohesion interaction between water molecules to see which model is most similar to real water drops.

You have a limited time to complete these explorations, so *make sure that the Procedure Specialist reads the directions out loud to the team as you do each exploration.*

Your team will need:

- sheets of wax paper
- paper towels
- container of water
- water drop form (half sheet)
- 4 eyedroppers

Exploration 1: How easy is it to pull apart a water drop?

In this exploration, you will first determine how big a drop you can move 13 cm. Then you will see how far you can pull (stretch) a large drop before it breaks into droplets.

Record your measurements on your record sheet.

Before you start, your teacher will demonstrate how to estimate the size (area) of a water drop.

STEP 1 Hold a piece of wax paper over the Water Drop Move and Stretch on your record sheet. Fill your eyedropper with water and make one *big* drop of water in the Start box.

Estimate the size (area) of the drop by counting the number of squares you can see through the drop when it is at the Start position. You must be looking straight down at the top of the drop.

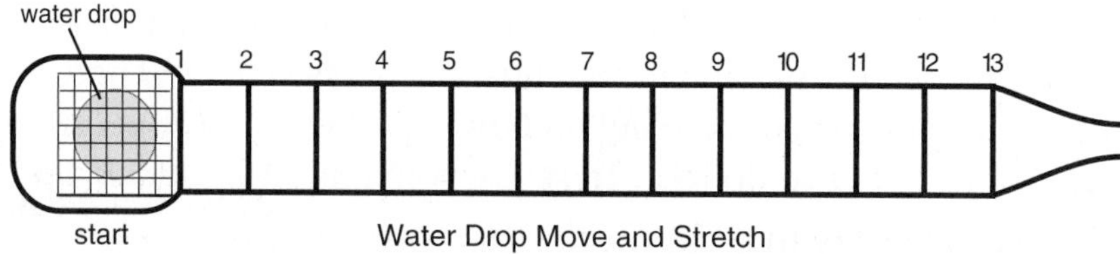

STEP 2 Empty the eyedropper. Use the tip of the eyedropper to move the drop to the right along the scale. If it does not reach the 13-cm mark before breaking up, wipe off the water with paper towels.

Repeat with smaller drops, until you have identified the biggest-size drop you can move 13 cm before it breaks up.

1. On your record sheet, record your measurement of the size (# squares) of the *largest* drop you can move 13 centimeters.

STEP 3 Make a big drop of water in the Start circle of the Water Drop Move and Stretch. *The drop should be larger than the largest drop you could move 13 centimeters.*

Hold the eyedropper almost horizontal and place the tip of the eyedropper in the middle of the drop. How far can you stretch the drop with the eyedropper before it breaks up into smaller droplets? Measure the stretching distance to the nearest ¼ (0.25) of a centimeter.

2. Record your measurement on your record sheet.

Find the *Summary Table of Similarities and Differences in Some Physical Properties of Gases, Liquids, and Solids* from the record sheet for Activity 3.

Compared to a solid, how easy is it to separate a drop of water into pieces (smaller droplets) by pulling on it? Write your conclusion in row 6 of the *Summary Table of Similarities and Differences.*

Your team will need:

- set of regular marbles
- set of attracting marbles
- plastic or cardboard tray (lined with black paper)

Exploration 2: How does the behavior of two physical water-drop models compare with the behavior of real water drops?

You will compare what happens when you move and stretch water drops with what happens when you move and stretch two different models of the molecules of a water drop. These models will help you figure out whether the interaction between liquid molecules is the same strength as the very weak interaction between gas molecules, or stronger.

The physical models use attracting marbles to represent a weak interaction between the molecules and regular marbles to represent no interaction between the molecules. Your fingers will represent the eyedropper you used to move and stretch water drops.

You will record your observations in the table on your record sheet.

STEP 1 *Team Manager:* Place the attracting marbles in the middle of the tray in a heap like a water drop. See how big an attracting-marble drop you can move around the tray. Test what happens when you try to stretch the marble drop.

Tensile strength is the maximum stretching force that a material, for example a wire, can withstand before breaking.

Think about what you observed when you tried to move and stretch the real water drop.

3. Does a similar thing happen when you try to move and stretch the attracting-marble drop as when you move and stretch a real water drop? Describe your observations on your record sheet.

STEP 2 *Supply Master:* Place the regular marbles on the middle of the tray in a heap like a water drop. Take turns observing what happens when you try to move the regular-marble drop. Test what happens when you try to stretch the regular-marble drop.

4. Does a similar thing happen when you try to move and stretch the regular-marbles drop as when you move and stretch a real water drop?

Describe your observations in the Summary Table of Similarities and Differences.

5. Which model of a liquid drop, *attracting marbles* or *non-attracting marbles*, is more similar to the way water drops behave when they are moved or stretched? Justify your answer.

6. Do you think the cohesion interaction between the molecules of a liquid is the same strength as the interaction between the molecules of a gas, or stronger? What is your evidence?

Participate in a class discussion of your observations and conclusions.

You learned in Activity 4 that the cohesion attraction between molecules is not magnetic. Instead, the attraction between the molecules of a substance is related to the electric-charge interaction. In Unit 3, you learned that there are many different types of mechanical interactions (applied, friction, drag, elastic, wave). Similarly, there are many different types of electric-charge interactions. *All of the attractive interactions between atoms and molecules are related to different types of the electric-charge interaction.*

Make Sense of the SPT Ideas of Liquid Molecules

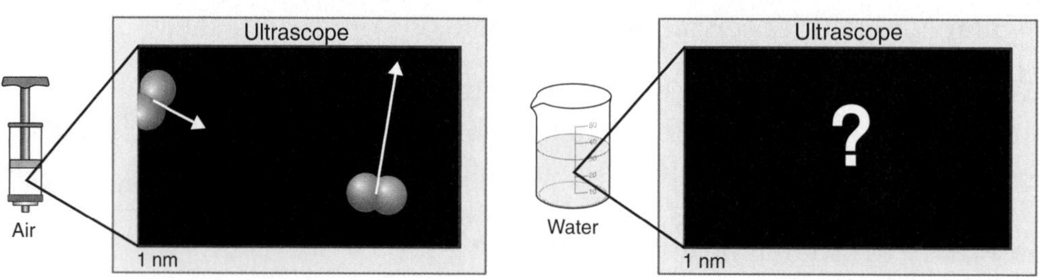

Imagine that teams in another class were asked to draw and describe what they think they would observe if they looked at liquid water through the Ultrascope. How would it be different from looking at the same volume of a gas?

The class came up with two different ideas (Idea A and Idea B) about the average spacing and motion of the molecules of water.

We think that the water molecules are closer together than air molecules because the attraction between the water molecules is stronger. But they do not touch because we can move through liquids. Therefore, there must be some space between the molecules. The molecules move continually in all directions.

Idea A

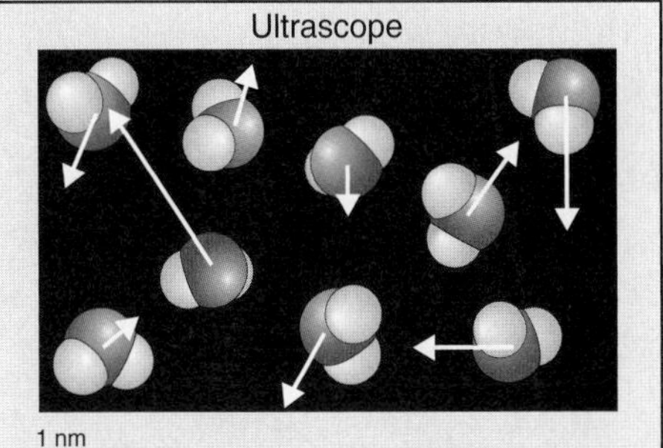

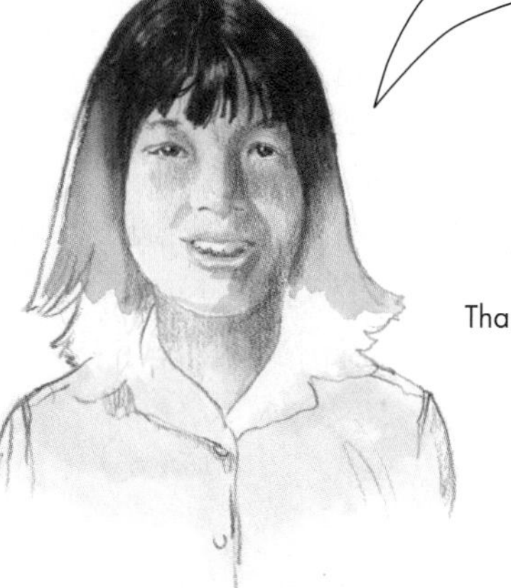

Than

Idea B

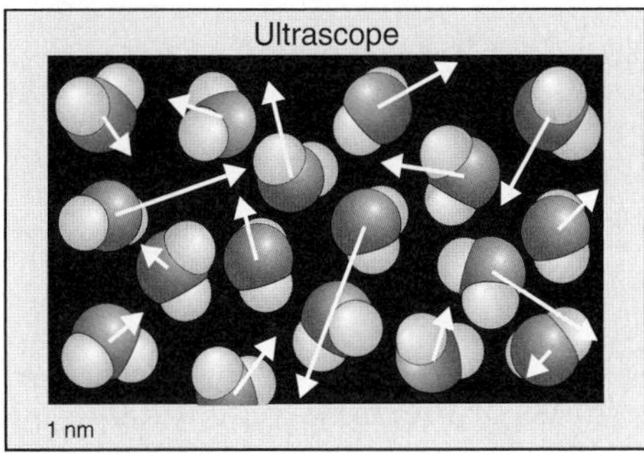

We think that the attraction between water molecules is strong enough to pull them very close together. This explains why liquids cannot be compressed. Liquids are harder to move through than air because there are more molecules to push aside than in air. The molecules move continually in all directions. But they hit each other and change direction more often than in gases. This explains why diffusion is slower in a liquid.

Nguyen

Good ideas of the molecules of liquids should meet these criteria:

- The idea should help to make sense of the properties of liquids.
- The idea should fit well with *and not contradict* the other Small-Particle Theory ideas (Ideas 1 – 7C).

Summary of SPT Ideas 1 – 7C

Idea 1 All matter is made up of particles (called atoms and molecules).

Idea 2 The particles of matter are too small to see through a visible-light microscope.

Idea 3 There is nothing between the particles of matter (the space between the particles is empty of stuff).

Idea 4 The particles of matter move continually in all directions.

Idea 5 There are only two ways that particles can interact with each other: collisions and continual attraction or repulsion.

Idea 6 Collision Interaction—Particles Bounce Apart
This interaction occurs when a moving particle (atom or molecule) of a substance collides with (hits) another particle, then the particles bounce apart.

Idea 7 Cohesion Interaction between the Particles of a Substance

C. *Very Weak Attraction* When the cohesion attraction between the particles of a substance is very weak (almost none), the substance is usually a gas near room temperature.

To help you decide which idea is better, the *Team Manager* should lead a discussion about the first question. Then the *Supply Master* should lead a discussion about the second question.

- Which idea, as presented, explains the most properties? Which idea helps you make sense of the drag property of liquids better, or do they both explain the drag property equally well? Try to explain the compression and diffusion properties with Idea A. Which idea (A or B), helps you make sense of these properties of liquids better, or do they both explain the properties equally well? See the *Summary Table of Similarities and Differences in Some Physical Properties of Gases, Liquids, and Solids* on your record sheet for Activity 3.
- Look at the summary of SPT Ideas 1 – 7C. Does each model (A and B) fit well with SPT Ideas 3 through 7C? If the answer is no, which SPT idea(s) does the Idea A or Idea B contradict?

Participate in the class discussion about which mental model seems to be more useful for explaining some of the properties of liquids.

1. On your record sheet, write the idea that your class decided was better (A or B), and the reasons for the decision.

Your teacher will show you a computer animation of the molecules of liquid water. The animation of gas molecules will also be shown for comparison. Remember that these animations are like slow-motion movies.

As you watch the animations, consider the following questions about the average spacing and motion of the molecules of a liquid.

2. On the average, are the molecules of liquid water close together or far apart *compared to the size of the molecules?* Are they closer together or farther apart than the molecules of water vapor (gas)?
3. Does this model (computer animation) of liquid water fit well with SPT Idea 3 or contradict this idea?
4. Does this model of liquid water fit well with SPT Idea 4 or contradict this idea?
5. Does this model of liquid water fit well with SPT Idea 6 or contradict this idea?
6. How is the motion of the molecules of a liquid *different from* the motion of the molecules of a gas?

Participate in a class discussion about your answers to these questions.

Our Consensus Ideas

The key question for this activity is:

What is the SPT of the between-particle interaction strength, the motion, and the average spacing of the molecules of a liquid?

With your teacher's guidance, summarize on your record sheet the scientists' ideas about the average spacing and motion of the molecules of a liquid. With a white pencil or gel pen, draw arrows on at least five Ultrascope molecules to show a range of speeds and direction of motion.

Idea 7D When the cohesion attraction between the molecules of a substance is weak (but much stronger than for gases), the substance is usually a liquid near room temperatures.

1. *Molecular Spacing* ________________________________.
2. *Molecular Motion* ________________________________.

MAKING SENSE OF SCIENTISTS' IDEAS

Activity 6
Small-Particle Theory of Solids

Purpose

So far you have learned the Small-Particle Theory (SPT) ideas about molecular gases and liquids at room temperature. In this activity about the Small-Particle Theory of solids, you will explore the following questions.

1. What are the SPT ideas about the strength of the cohesion interaction, the average spacing, and the motion of the molecules of solids?

2. What is the SPT of the strength of interaction between neighboring atoms of some solids?

This activity has three parts. First, you will investigate an important property of solids, their crystal shapes. Then you will explore your ideas about the strength of the cohesion attraction, average spacing, and motion of molecules of a solid. Some solids are made up of atoms instead of molecules. You will explore the interaction of the neighboring atoms of these solids in the last part of the activity.

Explore Your Ideas about the Shape of Solids

Solid elements and compounds are crystals when they grow naturally. Crystals of solids like salt, quartz, and snowflakes have different geometrical shapes called their *crystal structure*. For example, snowflakes have a six-sided crystal structure of solid water. Of course, most crystals are not perfect because pieces get broken off.

Rock candy is formed as sugar crystals grow on a wooden stick.

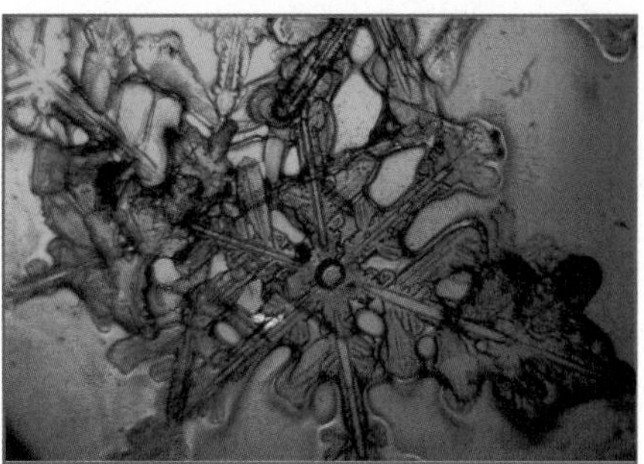

In this exploration, you will examine the crystal structure of table salt and Epsom salt. Each pair of partners will work with one pocket microscope. *One partner should read the procedure out loud while the other partner works with the microscope.* Then switch roles.

Be careful with solids so they do not get in your eyes.

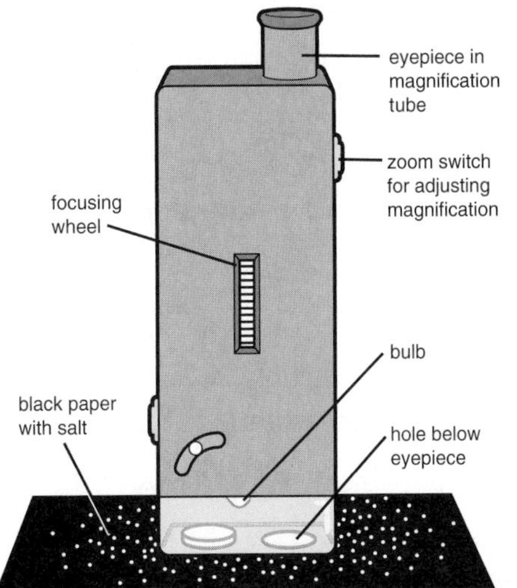

STEP 1 Sprinkle a pinch of table salt on a square of black construction paper. *Do not put too many grains of salt on the paper!* Use the toothpick to spread the grains out. You should have only one layer of grains close together.

Table salt crystals.

STEP 2 Turn on the bulb of the hand-held microscope. Use the bulb-rotation knob to make sure the tip of the bulb is pointed to the hole in the transparent cover that is below the eyepiece.

Your team will need:

- pinch of granulated salt
- pinch of Epsom salt
- 2 hand-held microscopes
- 4 small squares of black construction paper
- toothpick

STEP 3 Place the black paper with salt flat on your desk or table. Set the magnification on the microscope at 60x.

Gently and *slowly* move the microscope while looking through the eyepiece until you see *several* grains of salt in the middle of the field of view. *Focus.* The grains should look like the ones in the photograph. Move the microscope around to see many different salt crystals.

STEP 4 Find a good example of a salt crystal and center the image. Hold the microscope steady and use the zoom switch to move the eyepiece to the highest magnification, 100x.

Draw one salt crystal in the 100x field-of-view circle on your record sheet. Be sure to draw the crystal within the circle *in the correct size with relation to the size of the field of view.* Also, describe the crystal in terms of its properties.

STEP 5 Switch roles and repeat Steps 3 and 4.

Table 1: Microscope Observations of Table Salt and Epsom Salts

One grain of table salt	Description
○ 100x magnification	
One grain of Epsom salts	**Description**
○ 100x magnification	

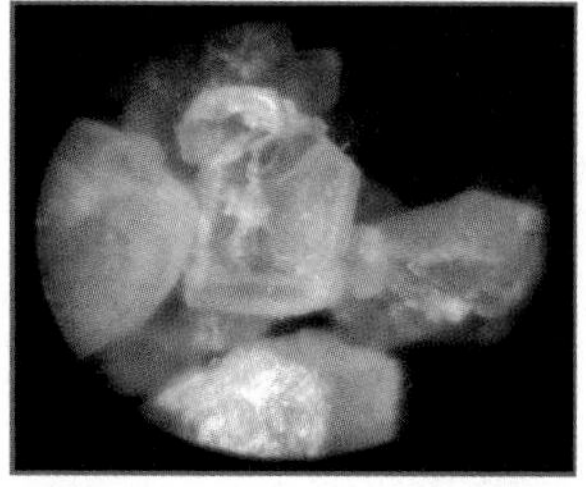

Epsom salt crystal.

STEP 6 Sprinkle a pinch of Epsom salt on a new piece of black construction paper. *Do not put too many grains of Epsom salt on the paper!* Use the toothpick to spread the grains out. You should have only one layer of grains close together.

STEP 7 Repeat Steps 3 through 5 with the Epsom salt. Look for a crystal like the one shown in the photograph.

Participate in a brief class discussion about the crystal structure of table salt compared to Epsom salt.

Your teacher will show you some photographs of other mineral crystal structures. Rock collectors can often identify the elements and compounds in a crystal by its color and structure.

Make Sense of the SPT of Molecular Solids

SPT Ideas about Attraction and Average Molecular Spacing

Imagine that you could look through the Ultrascope at a tiny volume of water and a tiny volume of sugar, as shown in the diagram.

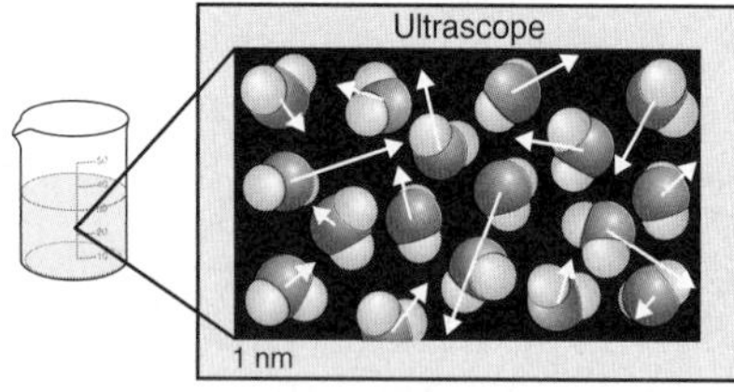

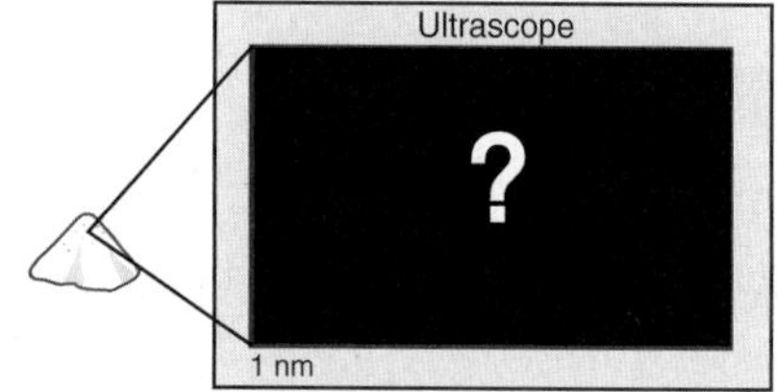

Read the two ideas (Ideas C and D) about the strength of the cohesion *attraction* and the *spacing* of the molecules of sugar. The students drew spheres for the sugar molecule because the molecules took too long to draw.

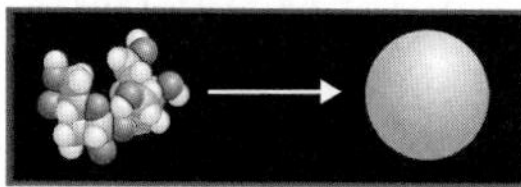

We think that the molecules of sugar are close together like in water because solids cannot be compressed. But we think the attraction between the sugar molecules is a lot stronger. This explains why we can't push the molecules aside when we try to walk through a solid like ice or wood.

Rebecca

Idea C

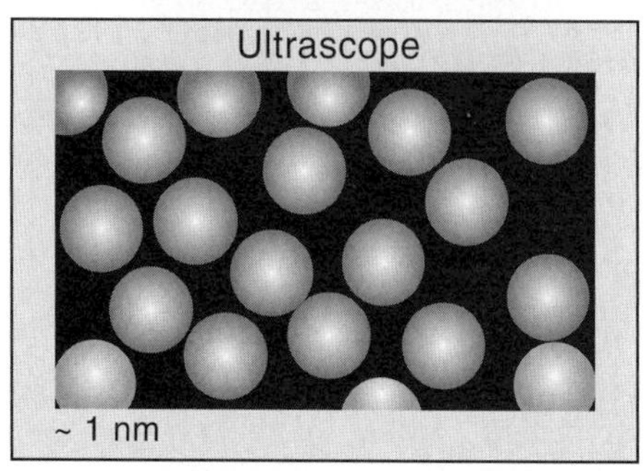

Idea D

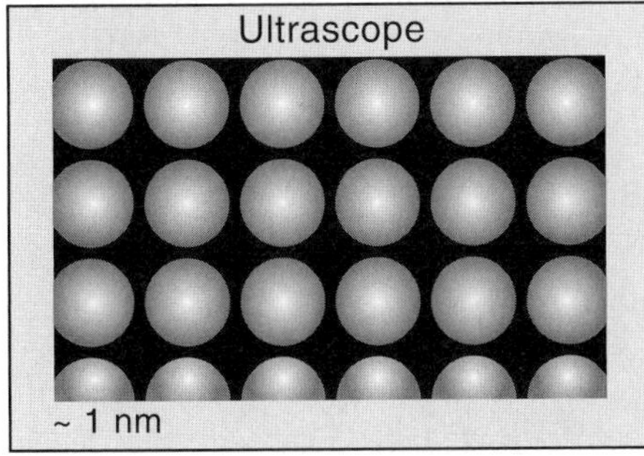

Ultrascope
~ 1 nm

We also think that the molecules are close together and the between-molecule attraction is stronger than in liquid water. But the molecules of a solid must be arranged in a regular, repeating pattern. This would explain why solids have a crystal structure.
Otis

Which idea do you think is *better*? Why?

Good ideas of the molecules of solids should meet these criteria:

- The idea should help to make sense of the properties of liquids.
- The idea should fit well with *and not contradict* the other Small-Particle Theory ideas (Ideas 1 – 7D).

To help you decide which idea is better, the *Procedure Specialist* should lead a discussion about the first question. Then the *Recycling Engineer* should lead a discussion about the second question.

- How do the Ideas (C and D), explain the properties of compressive strength and drag? Try to explain the shape and tensile strength properties using both ideas. Which idea do you think helps you make sense of these properties of solids best, or do they both explain the properties equally well? (See the *Summary Table of Similarities and Differences in Some Physical Properties of Gases, Liquids, and Solids* on your record sheet for Activity 3.)
- Review SPT Ideas 1 – 3 and 5 – 7D (from the previous activity). Does the model of the between-molecule attraction and spacing fit well with these ideas? If the answer is no, which SPT idea(s) does the Idea C or Idea D contradict?

Participate in the class discussion about which model of between-molecule attraction and spacing seems to be more useful for explaining the properties of solids.

1. On your record sheet, write the idea (C or D) of spacing of the molecules of a solid that your class decided was better and the reasons for the decision.

SPT Ideas about Molecular Motion in Solids

Read the conversation below about the motion of the molecules of solids.

I think the molecules of solids do not move. Well, they might shake a little when the whole object is moved, but then they slow down and stop.
Well, I'm just confused. The idea that solid molecules do not move kind of contradicts Idea 4 of the Small-Particle Theory. This is the idea that the particles of substances move continually. Why should the molecules of gases and liquids move continually, but the molecules of solid sugar not move at all?
Yeah! And we know that when a gold brick stays on top of a lead brick for several months, then some gold is found in the lead and some lead is found in the gold. How could this happen if the gold and lead atoms don't move? But the particles must be locked in place because of the crystal structure.
Carlos
Xuan
Jason

If you are also confused about the motion of the molecules of a solid, then you are in good company. Scientists also had a difficult time with this idea. Right now, you do not have enough convincing evidence to decide one way or the other.

Your teacher will show you the scientists' ideas in a computer animation of the molecules of solid water (ice). The computer animation of liquid water will also be shown for comparison. As you watch the animation of ice, try to answer the questions below. Remember that these animations are like very slow motion movies.

2. On the average, are the molecules of ice close together or far apart *compared to the size of the molecules*?
3. How is the motion of the molecules of a solid different from the motion of the molecules of a liquid?

4. In Unit 1 Cycle 2, Activity 5, you flicked the free end of a ruler and listened to the sound produced, as shown in this diagram. Just after you flicked it, how is the motion of the free end of the ruler like the motion of the molecules of a solid? How would you describe this motion?

Participate in a class discussion about your answers to these questions.

Consensus SPT Ideas about Molecular Solids

The first key question for this activity is:

1. What are the SPT ideas about the strength of the cohesion interaction, the average spacing, and the motion of the molecules of solids?

1. With your teacher's guidance, summarize the scientists' ideas about the average spacing and motion of the molecules of a solid on your record sheet, as shown below. With a pencil or white gel pen, draw arrows on at least five Ultrascope molecules to show a range of speeds and direction of motion.

Idea 7E For substances that are molecular solids at room temperature, the cohesion attraction is __________ than the attraction between the molecules of a liquid.

1. *Molecular Spacing* ____________________.

2. *Molecular Motion* ____________________.

Make Sense of the SPT of the Interaction Between Neighboring Atoms of Some Solids

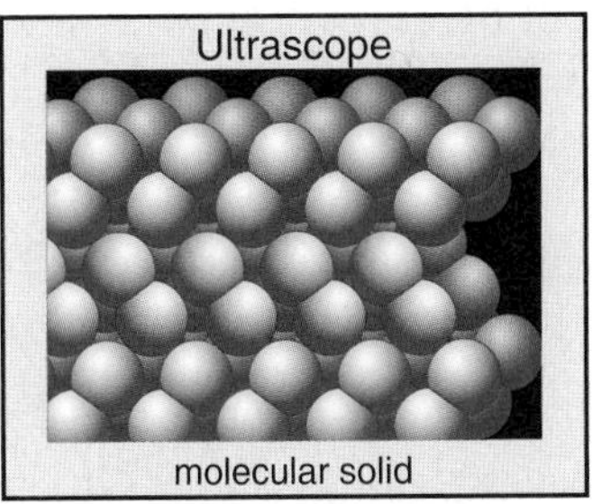

So far, the focus of your explorations has been on the interaction, motion, and average spacing of single substances that consist of molecules, like water, sugar, and the iodine shown in this Ultrascope model. But you know that the particles of some solid substances are atoms instead of molecules.

For example, all solid metal elements, like aluminum, iron, copper, and gold, consist of *atoms*. The identical atoms of a metal element are locked in a pattern, as shown in this Ultrascope model. The attraction between the atoms in a solid metal is a different, *stronger* type of electric-charge interaction than the attraction between the molecules of room temperature solids like sugar.

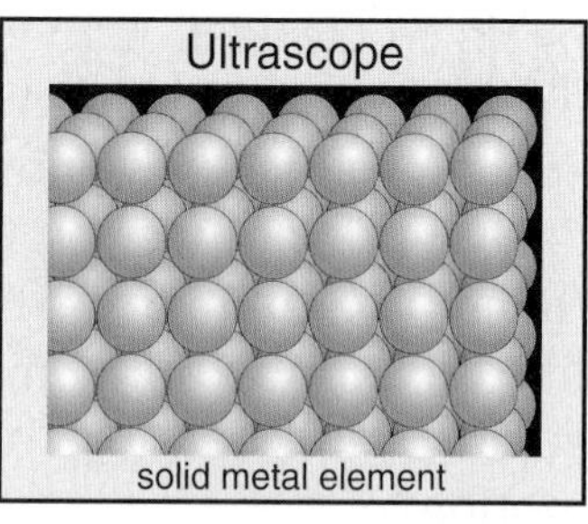

solid metal element

Science Words

ion: a single atom or group of atoms that have either a positive or negative charge

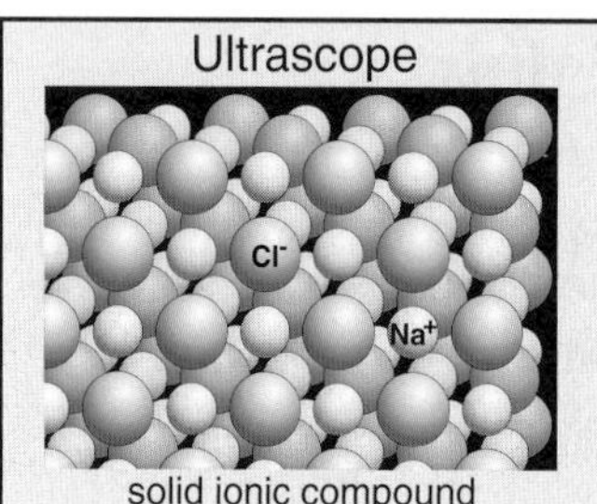

solid ionic compound

In some solid compounds, the atoms of two different elements are oppositely charged. One atom is positive and one atom is negative. Scientists call charged atoms **ions**. The salts you looked at through the microscope, table salt and Epsom salt, are made up of ions. Solid compounds made up of ions are called *ionic* compounds. The attraction between ions is another different, *stronger* type of electric-charge interaction than the attraction between the molecules of molecular solids.

1. Is the cohesion attraction between the molecules of sugar (molecular compound), *greater than*, *about the same as*, or *smaller than* the attraction between the atoms of the metal gold?

2. Is the cohesion attraction between the molecules of ice (molecular compound), *greater than*, *about the same as*, or *smaller than* the attraction between the ions of table salt?

Participate in a class discussion about your answers to these questions.

Your teacher will show you a computer animation of table salt (a solid ionic compound). Then a computer animation of ice (a solid molecular compound) will be shown again beside the model of the salt, so you can compare the two models. As you watch the models, try to answer the following questions.

a) How are the *patterns* of the particles of salt (ions) and ice (molecules) the same? How are they different?

b) How is the *motion* of the particles of salt (ions) and ice (molecules) the same? How is the motion different?

Participate in a class discussion about your answers to these questions.

You probably noticed that for ionic compounds like table salt (sodium chloride), the ions of the different elements form a repeating pattern. What is special about the pattern is that the closest neighbors for any ion are always ions of the other element with the opposite charge. In table salt, for example, six negatively charged chlorine ions surround each positively charged sodium ion, as shown in the diagram below. Similarly, six positively charged sodium ions surround each negatively charged chlorine ion.

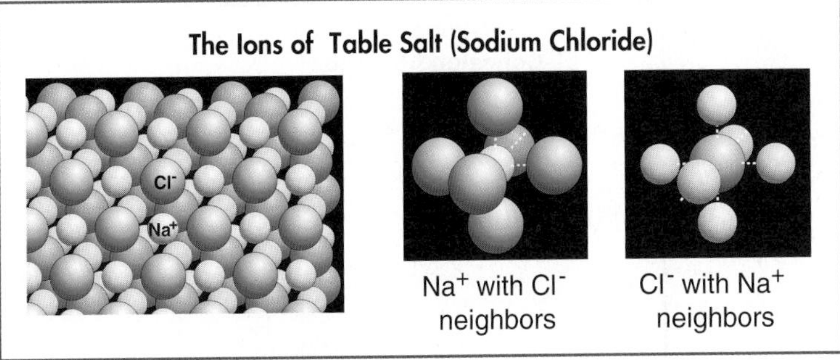

Consensus SPT Ideas about Metals and Ionic Compounds

The second question for this activity is:

2. What is the SPT of the strength of interaction between neighboring atoms of some solids?

With your teacher's guidance, summarize briefly the answer to this key question.

Putting It All Together (Activities 4 – 6)

In Activities 4 and 5 and in this activity, you explored the relationship between the strength of attraction between the particles of a substance, the phase of the substance, and the type of particle of the substance. You used words like *very weak* and *stronger* to compare the cohesion strength of attraction between the particles of a substance. But different gases have different strengths of attraction. So *very weak* refers to a range of attraction strengths on a continuous scale. The scale (arrow) is shown in the diagram on the next page.

Within this very weak (almost none) range of attraction strengths for gases, hydrogen and helium have the least cohesion attraction. Oxygen and nitrogen have a slightly greater attraction.

Liquids have a *weak* range of attraction strengths. Within this range, the molecules of ammonia and rubbing alcohol have a lower strength of cohesion attraction (forces) than water molecules.

Molecular solids have a *medium* range of attraction strengths. Within this range, the molecules of sulfur have a slightly lower strength of attraction (forces) than sugar molecules.

For solids that are made up of atoms or ions (charged atoms), the strength of attraction ranges from *strong to very strong*. For example, the strength of attraction between the ions of Epsom salt is much lower than the strength of attraction between the atoms of iron.

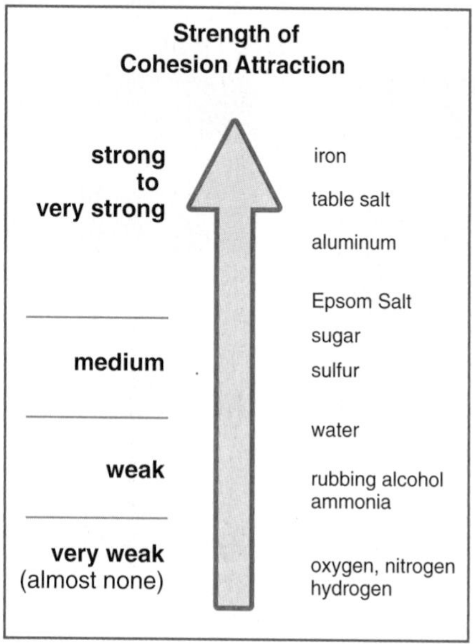

These ideas are outlined on the next page in the *Table: Summary of Between-Particle Attraction Strengths, Phase of Substance and Type of Particle Substance*. The first column of the table shows the four ranges of attraction strengths between the particles of substances—very weak, weak, medium, and strong to very strong.

The second column lists the phase of the substance for each range of the strength of attraction between particles. The third column shows Ultrascope models of the type of particles (molecules, atoms, or ions) in the different ranges of the strength of attraction between particles.

1. On your record sheet, fill in the blanks in the second column for the phase of the substance in the different ranges of the strength of attraction between particles. Use the words *gas*, *liquid*, or *solid*.

2. On your record sheet, fill in the blanks in the third column of the table (next to the Ultrascope models) for the type of particle in the different ranges of the strength of attraction between particles. Use the words *molecules*, *atoms*, or *ions*.

Participate in a class discussion about your answers.

Table: Summary of Between-Particle Attraction Strengths, Phase of Substance, and Type of Particle Substance			
Strength of cohesion attraction between particles of a substance (very weak to very strong)	**Phase of substance near room temperature** (gas, liquid, or solid)	**Type of particle** (molecules, atoms, or ions)	
strong to very strong	________	Ultrascope 1 nm	________
	________	Ultrascope 1 nm	________
medium	________	Ultrascope 1 nm	molecules
weak	liquid	Ultrascope 1 nm	________
very weak (almost none)	________	Ultrascope 1 nm	molecules or atoms of noble gases

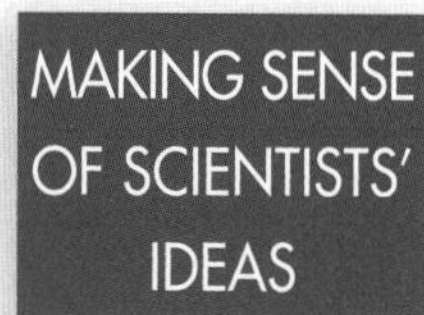

Activity 7
Small-Particle Theory of Thermal Expansion

Purpose

In Unit 4 Cycle 2, you investigated the heat-conduction interaction. A heat-conduction interaction occurs when two objects of different temperatures touch each other. Heat energy is transferred from the warmer object to the colder object. For example, consider holding a piece of ice in your hand.

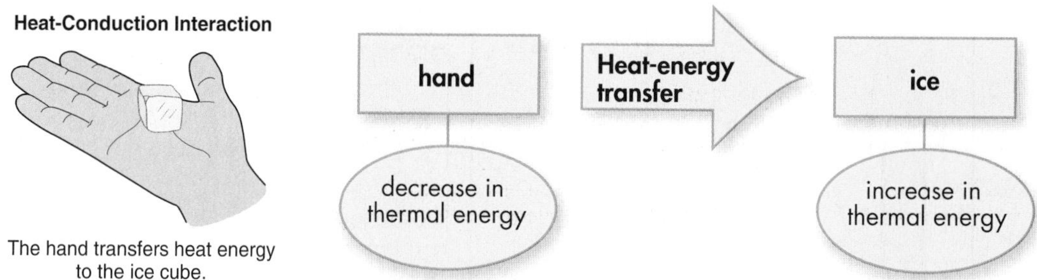

The hand transfers heat energy to the ice cube.

During this interaction, heat energy is transferred from your warmer hand to the cold ice. The evidence of the interaction is that the temperature of your hand decreases while the temperature of the ice increases. The thermal energy of your hand decreases while the thermal energy of the ice increases.

In Unit 4, you measured the temperatures of solids and liquids during a heat-conduction interaction. This activity focuses on what happens to gases during a heat-conduction interaction. The key questions for this activity are:

1. What is the Small-Particle Theory (SPT) of thermal energy?

2. What happens to the motion, average motion energy, and average spacing of the molecules of a gas during a heat-conduction interaction?

Record the key questions for the activity on your record sheet.

Explore Your Ideas about a Heat-Conduction Interaction with a Gas

Demonstration: What happens to the volume (space occupied) of a gas during a heat-conduction interaction?

STEP 1 Your teacher will set up the demonstration as shown in the diagram.

STEP 2 Observe carefully as your teacher begins to heat the air in the flask with a burner.

Record your observations in Table 1 on your record sheet.

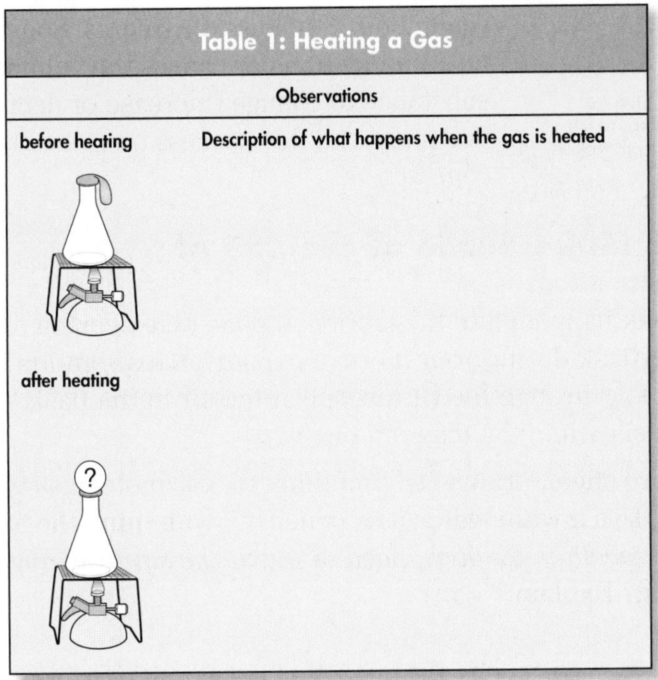

Table 1: Heating a Gas	
Observations	
before heating	Description of what happens when the gas is heated
after heating	

1. Is the flask-balloon-air system an open or closed mass system? Justify your answer. Remember: A *closed mass system* is a system in which no mass is input (added) to the system or output (removed) from a system during interactions. A system that is not a closed mass system is called an open mass system. In an *open mass system*, mass may be input (added) to the system or output (removed) from the system.

2. Does the mass of the balloon-flask-air system change during the heat-conduction interaction? Justify your answer. Remember the *Conservation of Mass Law*. In a closed mass system (a system with no mass inputs and no mass outputs), the mass of the system does not change during interactions. In an open mass system (a system with mass inputs or mass outputs), the mass of the system may change during interactions.
3. Does the volume of the air in the balloon-flask-air system change during the heat conduction interaction? What is your evidence?
4. Scientists have found that all gases behave like the air in the demonstration when they are heated. What is the relationship between the temperature and the volume of a gas?

Participate in a class discussion about your observations and conclusions.

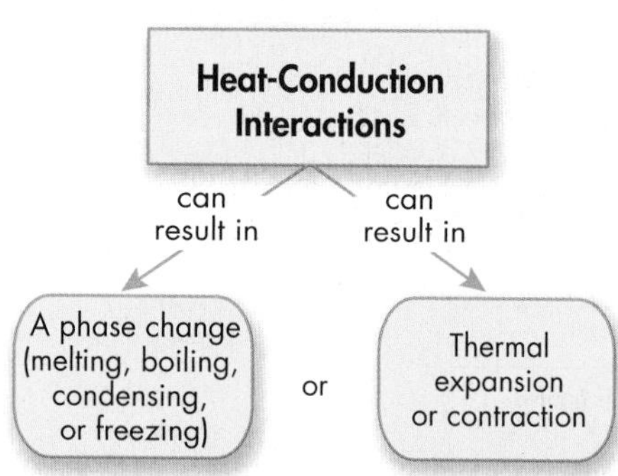

Different things can happen to the objects during a heat-conduction interaction. For example, sometimes the transfer of heat into or out of a substance causes the substance to change phase (melt, boil, condense, or freeze). Sometimes a heat-energy transfer causes the volume of the substance to change (increase or decrease). Scientists call this process *thermal expansion or contraction.*

Explore and Make Sense of the SPT of Thermal Expansion

What do you think happened to the motion and average spacing of the air molecules in the flask during your thermal expansion exploration? *Imagine* that you could look through the Ultrascope at the air in the flask before and after heating, as shown in the diagram below.

1. On your record sheet, draw what you think the air molecules will look like after heating. Use a white gel pen or pencil. Do you think the molecules will be *closer together*, *farther apart*, or *about the same* average spacing as before heating? Explain.

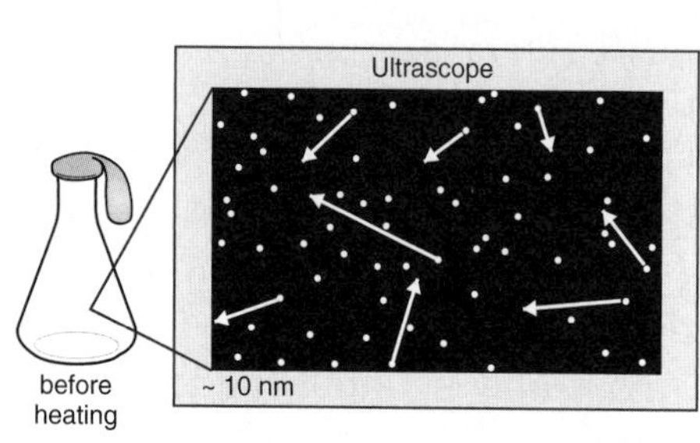

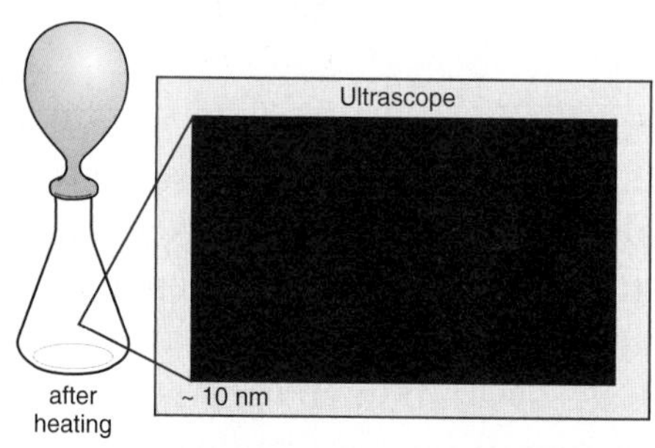

2. Draw arrows on at least five molecules to show the range of speeds of the molecules *after* heating.
3. Do you think the average speed of the molecules will be *higher*, *lower*, or *about the same* as before heating? Explain. Remember: The faster the speed of an object, then the greater its motion energy.

Participate in a class discussion about your predictions.

You will now use the Ideal Gas Simulator to test your predictions about what happens to the average spacing, average speed, and average motion energy of gas molecules during thermal expansion.

Record your simulator results on your record sheet in Table 2.

STEP 1 Open *Unit 6 Act 7 Setup 1* and wait for it to load.

STEP 2 When you open the file, you should see the setup shown in the picture.

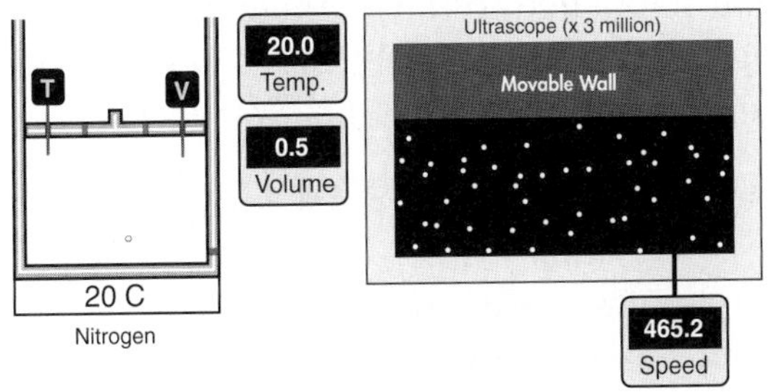

The container holds a gas that is warmed by the heater located below it. Inside the container is a movable wall. This wall represents the balloon on top of the flask. One instrument measures the temperature (degrees Celsius) of the gas in the container (flask and balloon). Another instrument measures the volume of the container (in liters).

The Ultrascope shows a field of view that includes the movable wall of the container. The Ultrascope also has a meter that measures the average motion energy of the gas molecules in the convenient units called million electron volts (MeV).

In Table 2, record the before-heating temperature, volume, and average speed of the gas molecules.

Table 2: Volume, Temperature, Speed, and Motion Energy

Gas	Temperature (°C)	Volume (L)	Average Speed (m/s)	Average Motion Energy (MeV)
After heating				
Before heating				
Change				

STEP 3 Run the simulator until the movable wall reaches the top of the container, then PAUSE (‖) just as the movable wall reaches the top of the container. *While the simulator is running, watch what happens to the average spacing of the gas molecules.*

If you are not sure what happened to the average spacing, look carefully at the particles (after heating). Then select STOP (■) to reset the simulator back to what the particles looked like before heating. Repeat as many times as necessary.

1. Was your prediction correct about what the molecules look like after heating? Are the gas molecules *closer together, farther apart,* or *about the same* average spacing as before heating?

STEP 4 Run the simulator again, and PAUSE (‖) just as the movable wall reaches the top of the container. *While the simulator is running, watch what happens to the temperature and average speed of the gas molecules.*

If you are not sure what happened to the average speed, look carefully at the particles (after heating). Then select STOP (■) to reset the simulator. Repeat as many times as necessary.

On Table 2, record the after-heating temperature, volume, and average speed of the gas molecules.

Calculate the change in temperature, volume, and average speed of the molecules. Record these changes in your table.

Use the information in Table 2 to help you answer this question.

2. Was your prediction correct about what happens to the average speed of the gas molecules after heating? Is the average speed of the molecules *higher, lower,* or *about the same as* before heating? What is the evidence from the simulator?

STEP 5 Select the Ultrascope speed meter. When you double-click on the meter, a menu opens called *Properties of Speed/Energy.* Select *Average Energy* on the menu and click on the OK button.

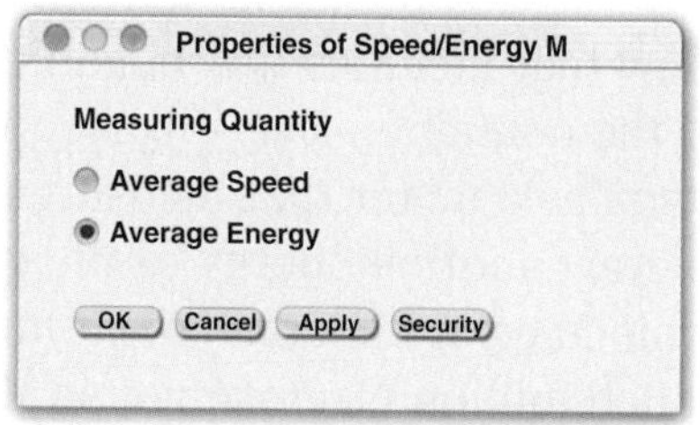

The Ultrascope meter now measures the average motion energy of the gas molecules in the convenient units called million electron volts (MeV).

Record the before-heating average motion energy of the gas molecules in your table.

STEP 6 Run the simulator again, and PAUSE (❚❚) just as the movable wall reaches the top of the container. While the simulator is running, watch what happens to the temperature and average motion energy of the gas molecules.

If you are not sure what happened to the average motion energy, look carefully at the particles (after heating). Then select STOP (■) to reset the simulator. Repeat as many times as necessary.

Record the after-heating average motion energy of the gas molecules.

Calculate and record the change in the average motion energy in your table.

Use the information in Table 2 to help you answer these questions.

3. Was your prediction correct about what happens to the average motion energy of the gas molecules after heating? Is the average motion energy of the molecules *higher*, *lower*, or *about the same as* before heating? What is the evidence from the simulator?

4. In the SPT theory, what is the relationship between the thermal energy of the gas and the average motion energy of the gas particles?

Participate in a class discussion about your simulator observations and conclusions.

Our Consensus Ideas

The first key question for this activity is:

1. What is the Small-Particle Theory (SPT) of thermal energy?

Complete the following Small-Particle Theory (SPT) idea about thermal energy on your record sheet.

Idea 8A The thermal energy of a substance is related to the average motion energy of the particles of the substance.

The higher the temperature and thermal energy of a substance, then

___________________________________.

The second key question for this activity is:

2. What happens to the motion, average motion energy, and average spacing of the molecules of a gas during a heat-conduction interaction?

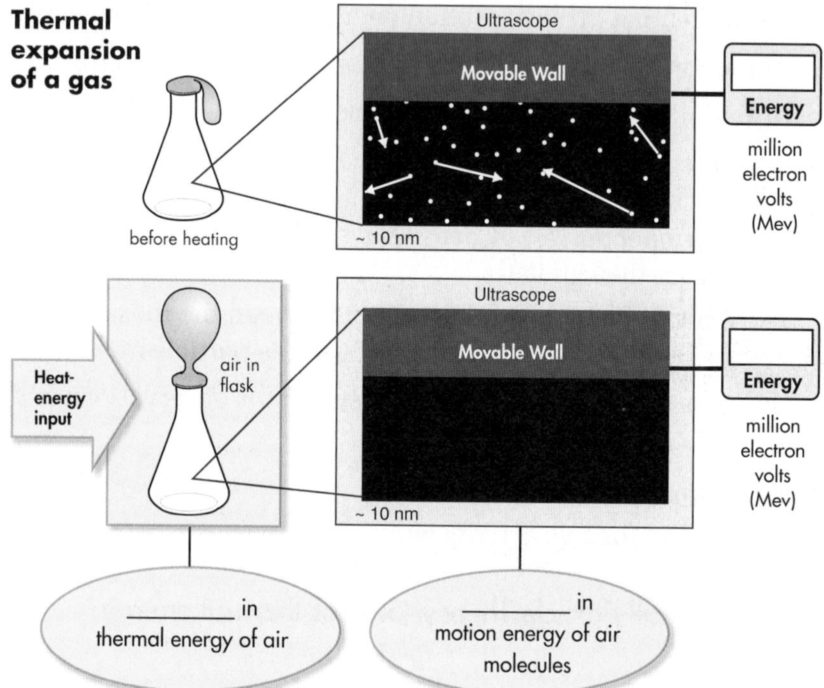

With the guidance of your teacher, summarize the answer to Question 2 on a diagram, like the one shown. Make sure to do the following:

- Complete the energy diagram for heating a gas in the flask with a balloon (movable wall).
- In the Ultrascope meters, write the average motion energy of the air molecules *before heating* and *after heating*. Fill in the blank in the energy oval of the Ultrascope.
- Draw what the molecules look like through the Ultrascope after heating. Use a white gel pen or pencil. Then draw arrows on at least five of your Ultrascope molecules to show the range of speeds of the molecules after heating.

MAKING SENSE OF SCIENTISTS' IDEAS

Activity 8
Small-Particle Theory of Stored Volume Energy

Purpose

When heat energy is transferred into gas, liquid, and solid substances and they expand, then the thermal energy of the substances increases, as shown in the energy diagrams. But when heat energy is transferred to a liquid or solid, then there is also a small increase in another kind of energy—stored volume energy.

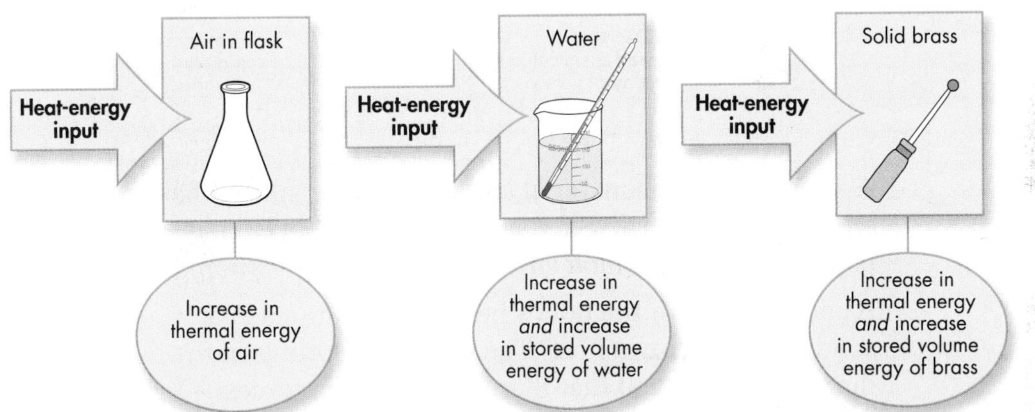

You learned in Activity 7 that thermal energy is related to the average motion energy of the billions of trillions of particles of a substance. In this activity, you will explore the Small-Particle Theory (SPT) of stored volume energy. The key question for the activity is:

What is the Small-Particle Theory (SPT) of stored volume energy?

Record the key question for the activity on your record sheet.

Explore SPT Ideas

Chemists call the attraction between particles (atoms, ions, or molecules) a **bond**. A useful analogy of bonds (attractions between particles) is to think about them as similar to *stretched rubber bands* or *stretched/compressed springs*.

Science Words

bond: the attraction between particles (atoms, ions, or molecules)

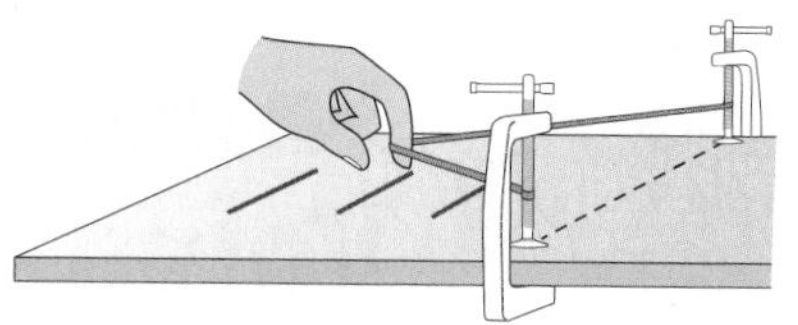

In Unit 2, you investigated the elastic interaction and stored elastic energy. For example, consider the stretching of a rubber band, as shown in the diagram. The transfer of mechanical energy to the rubber band causes an increase in the stored elastic energy of the rubber band, as shown in the energy diagram.

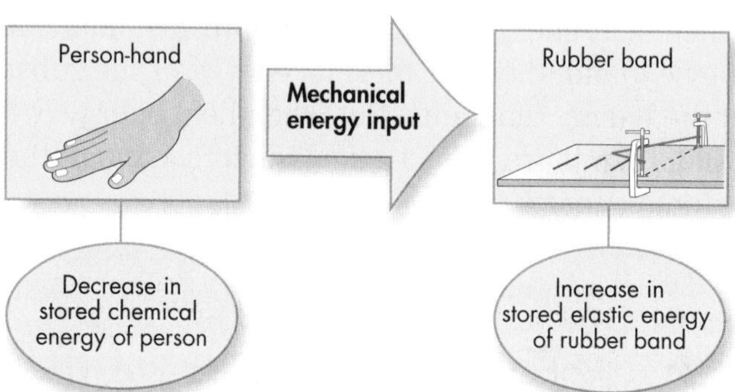

A stretched elastic material (rubber band or spring) has stored elastic energy. Similarly, a system of attracting particles (atoms or molecules) has *stored bond energy.*

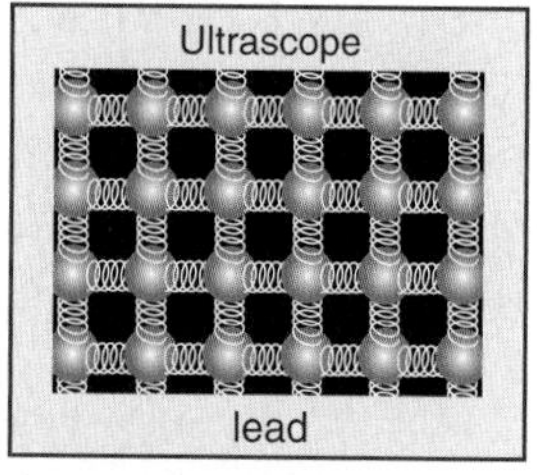

A spring image of the bonds between the atoms in a solid metal is shown here. The stretched springs in the diagram remind you that there are cohesion forces between the atoms. It also reminds you that *energy is stored "in the bonds."*

Exploration: What is the relationship between the strength of the attraction between two particles and the amount of stored cohesive bond energy?

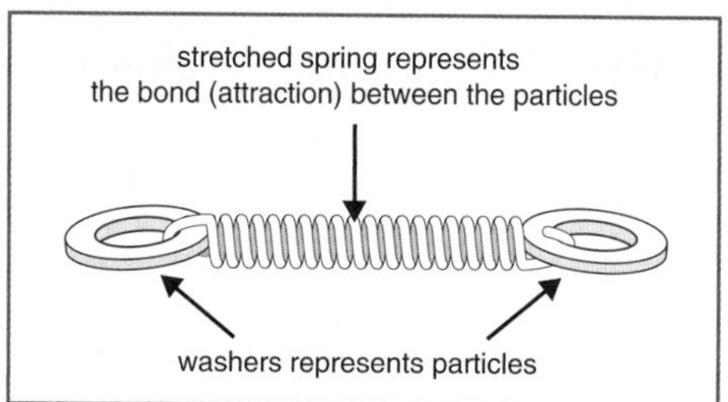

To answer this question, you will use two washers to represent two particles, and a *stretched* spring attached to the washers to represent the bond (attraction) between the particles.

When you transfer mechanical energy to a spring, that energy is stored in the stretched spring. So the *force* you need to stretch the spring a fixed distance (transfer mechanical energy to the spring) represents the amount of elastic energy stored in the spring. The more force it takes you to stretch the spring over a fixed distance, the more elastic energy you have stored in the spring.

The three strengths of attraction in this model are:

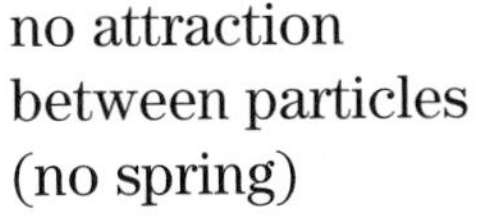

no attraction between particles (no spring)

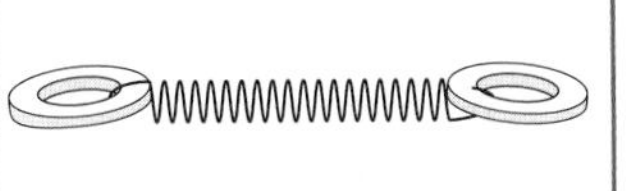

weak attraction between particles (thin spring)

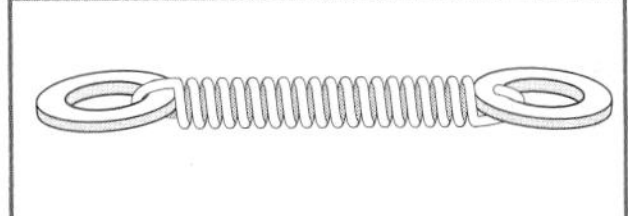

strong attraction between particles (thicker spring)

Your team will need:

- weak spring-washer system
- strong spring-washer system
- ruler

One pair of partners starts with the weak spring-washer system, while the other pair of partners starts with the strong (thicker) spring-washer system. *One partner reads the procedure out loud while the other partner works with the spring-washer system. Then switch roles.*

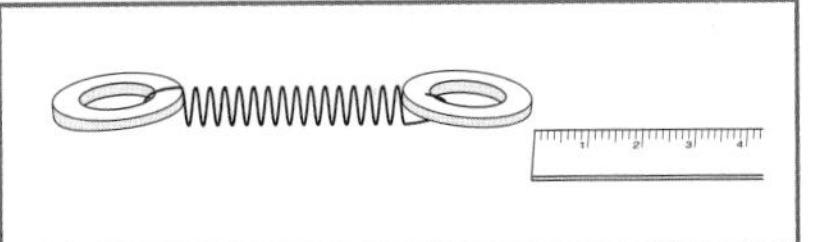

STEP 1 Hold the right washer over the end of a ruler (the zero mark). Hold the left washer so that the weak spring is taut, but not stretched.

Wear eye protection.

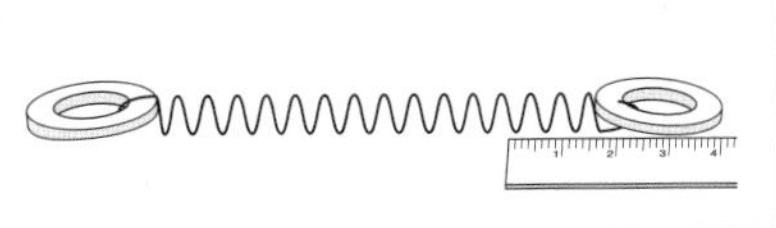

STEP 2 Pull the right washer until it is *about 1 cm* from the end of the ruler. Hold for a few seconds.

STEP 3 Repeat Steps 1 and 2 with the strong spring-washer system.

Which spring-washer system (weak or strong) took the most force to hold with the spring stretched about 1 cm? Record your observations in Table 1.

Table 1: Strength of Attraction and Stored Elastic Energy

Strength of attraction between washers	Force need to stretch spring a distance of 1 cm (stored elastic energy)
none (no spring)	none
weak (thin spring)	
strong (thicker spring)	

For this physical model, the larger the effort it takes you to stretch the spring, the more elastic energy you stored in the spring. Use the results in Table 1 to answer the following questions.

1. What is the relationship between the strength of the spring and the amount of stored elastic energy in the spring-washer system?
2. What do you think is the relationship between the strength of attraction between two particles of matter and the amount of stored-bond energy? Explain your reasoning.

Make Sense of SPT Ideas

1. Complete the following Small-Particle Theory (SPT) ideas on your record sheet. Use the information in this activity, and the results of your Exploration.

Idea 8B Stored Volume Energy and Cohesive Bonds

1. Chemists use the word________________ for the attractions (forces) between neighboring particles. A useful model of ________________ (attractions) is to think about them as similar to stretched or compressed springs. A stretched or compressed spring has stored elastic energy. Similarly, two attracting particles of a substance have ________________.
2. The bonds (attractions) between the particles of a substance are called cohesive bonds. The amount of energy stored in cohesive bonds depends on the strength of the attraction between the particles. The stronger the attraction between the particles, then ______________________________.

Our Consensus Ideas

The key question for this activity is:

What is the Small-Particle Theory (SPT) of stored volume energy?

Each particle of a substance is attracted to its neighboring particles. So there are billions of trillions of cohesive bonds (attractions) between the particles. A tiny amount of energy is stored in each of these cohesive bonds. The sum of the energies stored in the billions of trillions of cohesive bonds between the particles of a substance is called the *stored volume energy.*

Complete the SPT idea below on your record sheet.

The stored volume energy of a substance is __________________________.

Participate in a class discussion about the consensus ideas.

Activity 9
The Small-Particle Theory of Matter

Comparing Consensus Ideas

Recall the key question for this cycle of learning:

What is the scientists' theory of what gases, liquids, and solids are like on a scale too small to see?

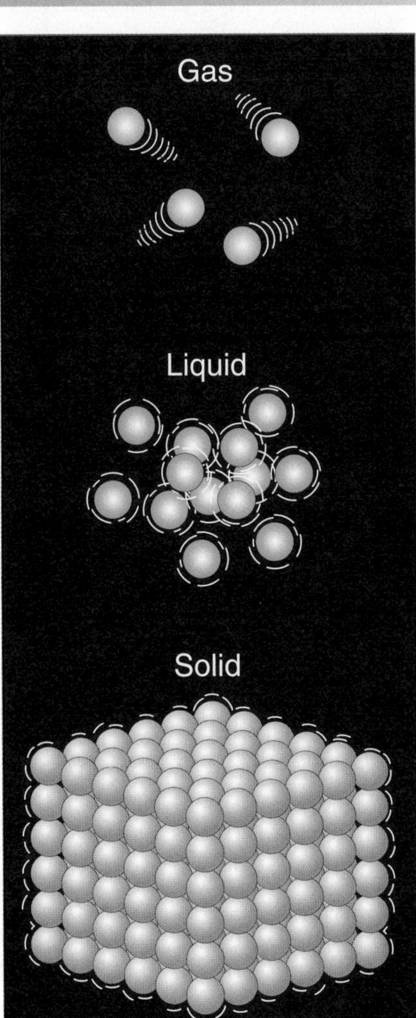

To answer this question, you explored some of the ideas of scientists' Small-Particle Theory (SPT) of matter. You used different kinds of models (marbles, springs) and computer simulatons and animations to explore these ideas.

Your teacher will pass out the sheet titled *Scientists' Consensus Ideas: The Small-Particle Theory of Matter Ideas 1-8.*

Compare the *Consensus SPT Ideas* sections in Activities 2, 4, 5, 6, 7 and 8 with the Scientists' Consensus Ideas.

Participate in a class discussion about these ideas.

On the next page is a concept map that summarizes the ideas about particle interactions. The boxes in light gray will be added when you finish Unit 7.

With your teacher's guidance, read through the map. Fill in the blanks in the map on your record sheet.

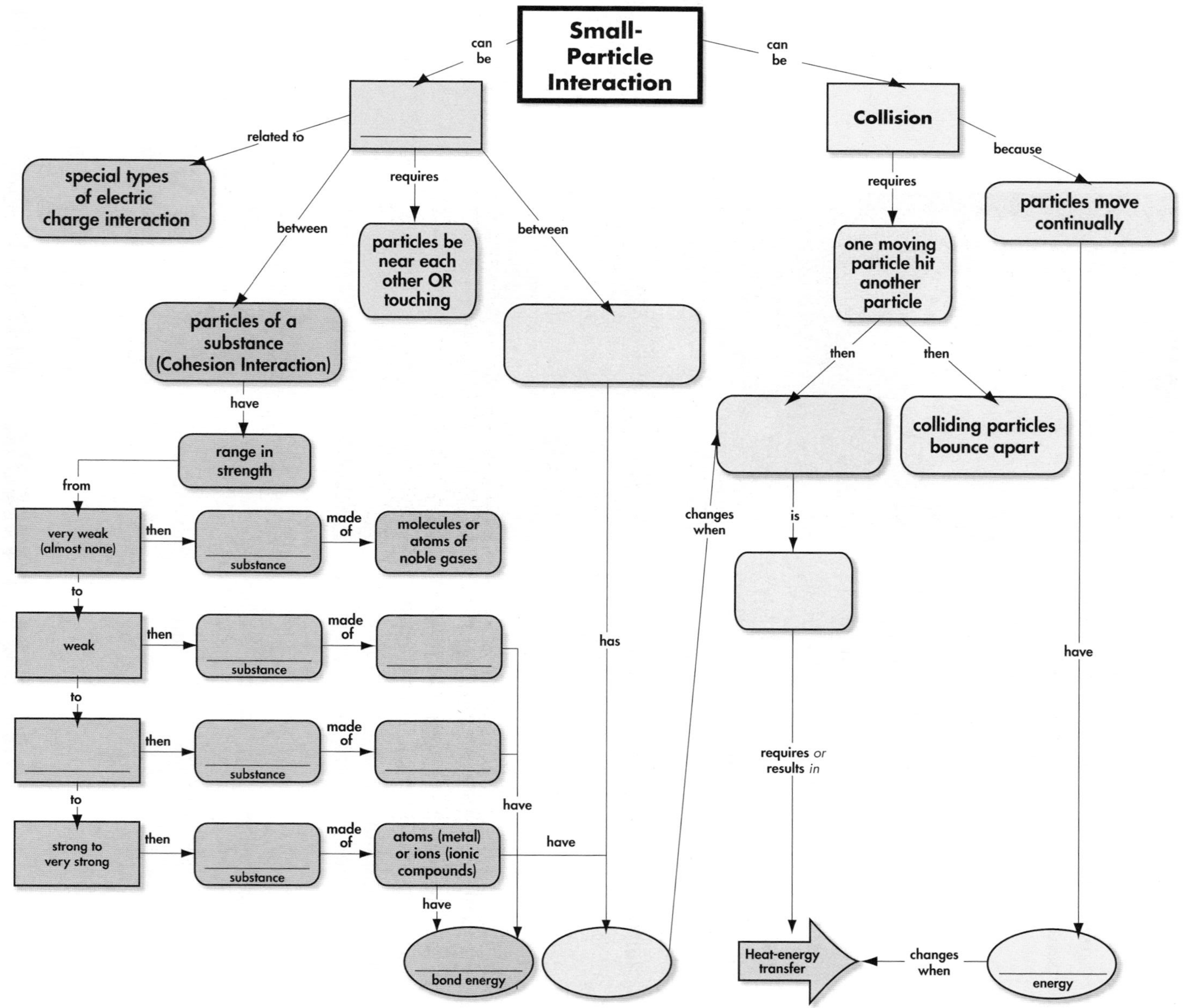

Small-Particle Interaction
can be
can be
related to
special types of electric charge interaction
between
requires
particles be near each other OR touching
between
Collision
because
requires
particles move continually
one moving particle hit another particle
then
then
colliding particles bounce apart
particles of a substance (Cohesion Interaction)
have
range in strength
from
very weak (almost none)
then
substance
made of
molecules or atoms of noble gases
to
weak
then
substance
made of
to
then
substance
made of
to
strong to very strong
then
substance
made of
atoms (metal) or ions (ionic compounds)
have
have
have
bond energy
has
changes when
is
requires or results in
Heat-energy transfer
changes when
have
energy

IDEA POWER

Activity 10
Explaining Properties

Properties of Gases, Liquids, and Solids

Look at the *Table: Summary of Similarities and Differences in Some Properties of Gases, Liquids, and Solids*. You collected information to complete this data table in Activities 3, 5, 6, and 7. Now you can use the Small-Particle Theory (SPT) to explain (make sense of) these properties.

Table: Summary of Similarities and Differences in Some Physical Properties of Gases, Liquids, and Solids

Physical Property	Gas	Liquid	Solid
1. Shape Does substance have a definite shape?	**no** (no surfaces or boundaries)	**no** (takes shape of container; has surfaces)	**yes** (has surfaces)
2. Volume Does substance have a definite volume (# of unit cubes?)	**no** (fills all space inside and outside of a container)	**yes** (can count number of unit cubes)	**yes** (can count number of unit cubes)
3. Diffusion How fast does a substance spread 2 cm into another substance in the same phase?	**very fast** (takes less than a second)	**medium fast** (takes a few minutes)	**very slow** (takes several years)
4. Density How do *equal volumes* (1 L) of substance interact with a mass balance?	**very small** (1 L has mass of only a few grams)	**large** (1 L has mass of hundreds to thousands grams)	**large** (1 L has mass of hundreds to thousands of grams)
5. Compressive Strength How easy is it to compress a substance into a smaller volume (space)?	**low** (easy to compress)	**very high** (very difficult to compress)	**very high** (very difficult to compress)
6. Drag How easy is it to move an object through the substance?	**very easy**	**medium** (more difficult than moving through air, but not impossible)	**impossible**
7. Tensile Strength How easy is it to seperate the substance into pieces by pulling on it?	**none** (cannot separate into pieces)	**low** (very easy to pull apart)	**high** (very difficult to pull apart)
8. Thermal Expansion How large is the change in volume of the substance?	**large** (1 L expands 34 mL for a temperature increase of 10°C)	**small** (1 L expands 1-10 mL for a temperature increase of 10°C)	**small** (1 L expands less than 1 mL for a temperature increase of 10°C)

In Unit 2 Cycle 2, Activity 6 you identified some mechanical interactions involved in a skateboarding situation. This situation also shows many differences in the properties of gases, liquids, and solids.

Read through the following two example questions and explanations.

Example 1 Why is it so difficult to separate solids, like the rope and tree branch, by pulling on them? Explain using the Small-Particle Theory.

Analysis and Explanation

The attraction between the particles of most room-temperature solids is medium to very strong (SPT Idea 7B).

The attraction is strong enough to lock the particles together in fixed positions. (SPT Idea 7E and 7F).

So, it is very difficult to pull the particles apart.

Example 2 Why is it so easy for you to move through air? Explain using the Small-Particle Theory.

Analysis and Explanation

The attraction between the molecules of room-temperature gases is very weak (SPT Idea 7C).

The attraction is so weak that the molecules are essentially free to move in all directions through empty space and stay far apart on the average. (SPT Idea 7C1 and 7C2).

So, it is easy for my body to push the particles aside into the empty spaces between the particles when I walk through air.

You have used interaction ideas to explain situations and events in all of the previous units. Therefore, it should not be surprising that the *interactions between particles* are the key ideas for Small-Particle Theory (SPT) explanations of a property.

Analyze and Explain Properties of Gases, Liquids, and Solids

When you have identified the strength of the interaction for the gas, liquid, or solid, then you can explain the property by stating:

- the effect of the interaction strength on the motion or spacing of the particles; and
- how the effect relates to the property.

Use this general procedure to answer the following questions. Be prepared to discuss your solutions with the class.

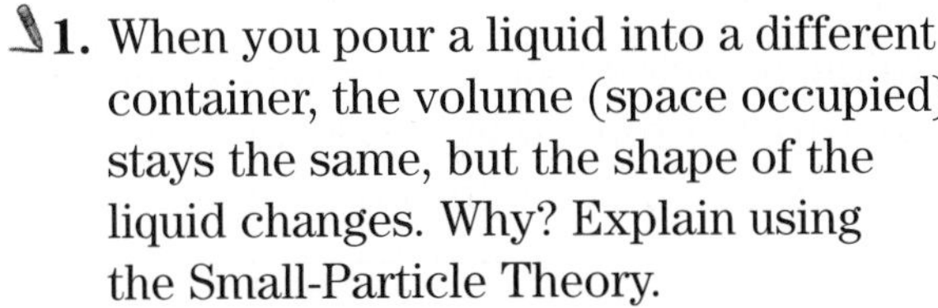

1. When you pour a liquid into a different container, the volume (space occupied) stays the same, but the shape of the liquid changes. Why? Explain using the Small-Particle Theory.

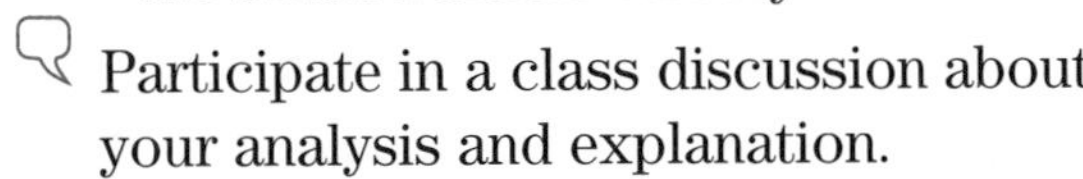

Participate in a class discussion about your analysis and explanation.

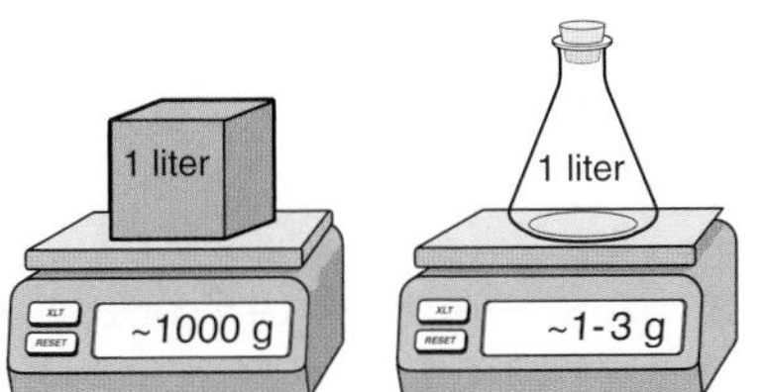

2. One liter of a solid has a mass ranging from hundreds to thousands of grams. But one liter of a gas has a mass of only a few grams. Why are the densities of solids so large? Explain using the Small-Particle Theory.

Participate in a class discussion about your solution.

Use the same analyze-and-explain method to answer Questions 3 and 4.

3. Jason walked into the kitchen and immediately noticed the smell of the soup cooking on the stove. Why does one gas, like a smell, spread out (diffuse) quickly into another gas like air? Explain using the Small-Particle Theory.
4. In Activity 5, you experienced how easy it is to separate a large drop of water into smaller droplets. Why is it so easy? Explain using the Small-Particle Theory.

Air Pressure

Have you ever listened to a weather forecast on TV or radio? Then you have heard of air pressure. Scientists' define air pressure as the *push (force) of the air on each unit square area of a surface.* For example, suppose you looked at four square inches of the surface of an "empty" water bottle.

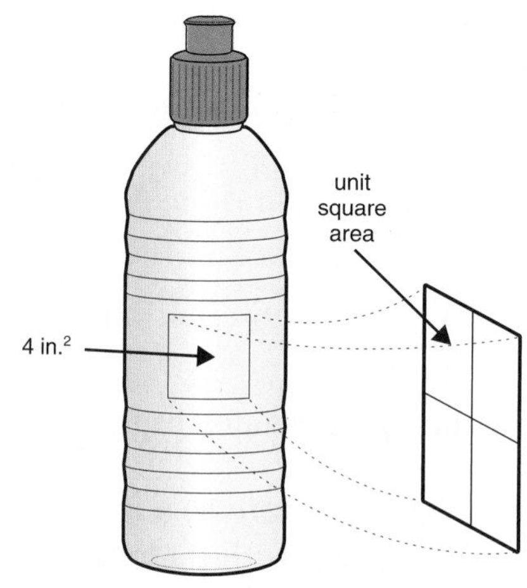

Both the air *inside* the bottle and the air *outside* the bottle push on each unit square area. A force diagram of air pressure is shown. Each force arrow represents the push of the air on a unit square area.

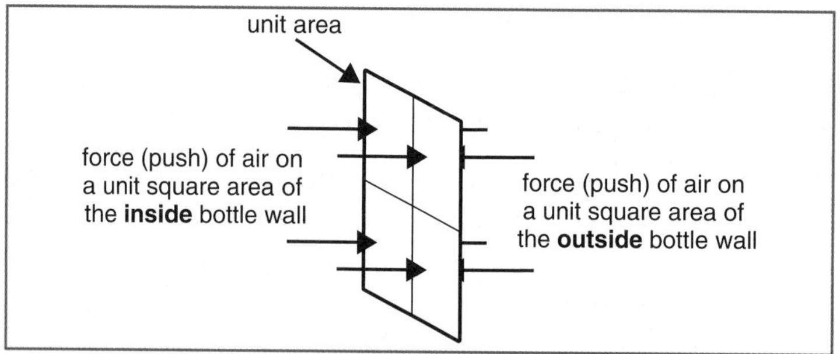

The force (push) of air on a unit square area is very large—about 14.7 pounds of force on each square inch. The average human has a surface area of about 2650 square inches (in^2). So the total force of the air on your body is about 39,000 pounds!

How can air, which is something we cannot see or feel, push that much? The Small-Particle Theory (SPT) is especially helpful in making sense of air pressure. What do you know about the SPT of air?

- You know that air is made up of a huge number of tiny, tiny particles (Ideas 1 and 2).
- There is nothing (empty space) between the particles (Idea 3).
- With nothing to slow them down, the particles of air move continually in different directions and with a range of speeds from almost stopped to very fast (Idea 4).

- There are only two ways particles interact with each other (Idea 5).
- They are continually attracted to each other (Idea 7A).
- They can collide with each other and bounce apart (Idea 6).
- For room-temperature gases like air, the strength of cohesion attraction is so small that it can usually be ignored (Idea 7C). So, the particles of air are, on the average, far apart (Idea 7C1). They move in all directions, occasionally colliding with each other (Idea 7C2).

Analyze and Explain Air Pressure

The surface of a solid is also made up of particles. Your teacher will show you a short computer animation of the air particles near the surface of a wall. Watch what happens to the particles of air near a solid wall.

1. When the particles hit the wall, do they exert a force on the wall?

Participate in a class demonstration and discussion of your answer to this question.

The Ultrascope diagram shows some air-particle collisions with the particles of the walls of a plastic bottle. For convenience, the thick wall of the bottle has been reduced in size to fit on the diagram. (At this very large magnification, the bottle wall would be several feet thick!)

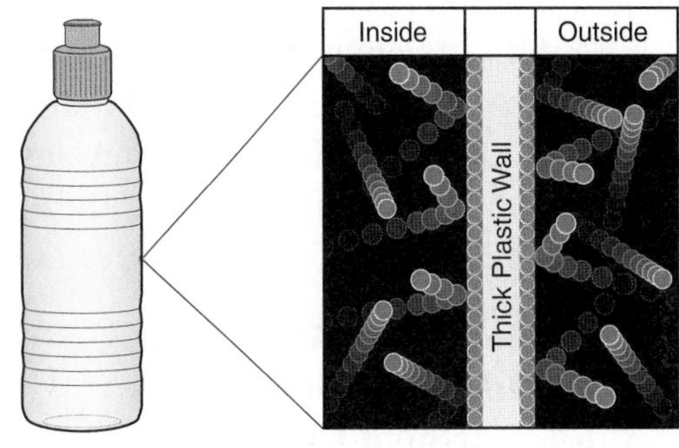

2. Write an SPT explanation of air pressure.

Participate in a class discussion about your explanation.

The total force of the air on the approximately 2650 in^2 of your skin is about 39,000 pounds. This is an enormously large force! So why doesn't this huge force of the air crush us? You are not crushed because the push of the air on each square area of your skin is balanced by the push of your blood on each square area of your skin (your blood pressure), as shown in the diagram.

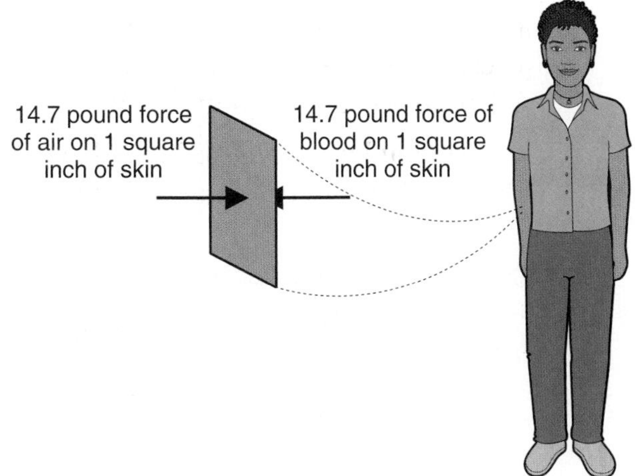

IDEA POWER

Activity 11
Classification Problems

Explore and Make Sense of the SPT of Alloys

You have learned the Small-Particle Theory (SPT) of elements and compounds. But most of the millions of different materials in the world are mixtures. In this activity, you will first explore the SPT of solid solutions (alloys). Then you will solve some classification problems.

In Unit 5 you learned that an alloy is a solution (mixture) of two metals. For example, brass is a mixture of zinc and copper.

Your team will need:

- container of cubes of two different colors
- empty container

A defining characteristic of a mixture is that equal masses of different samples can have different masses of the substances mixed together. For example, consider 100 g samples of brass made at different places. The mass of copper and zinc in the different samples can range from 20 g of copper and 80 g of zinc, to 50 g of copper and 50 g of zinc.

Exploration: What would brass look like through the Ultrascope?

To answer this question, you will do a short exploration. You will use 1-cm plastic cubes to represent the copper and zinc atoms.

Each team will build a model of brass from 27 cubes.

Table 1: Team Models of Brass

Team number	Number of Copper Cubes	Number of Zinc Cubes	Total number of cubes
Team 1	5	22	27
Team 2	7	20	27
Team 3	9	18	27
Team 4	11	16	27
Team 5	13	14	27
Team 6	12	15	27
Team 7	10	17	27
Team 8	8	19	27
Team 9	6	21	27
Team 10	4	23	27

You will record your observations of your model and at least five models of other teams on a data table like the one shown below.

STEP 1 Decide which color of cube will be copper atoms, and which will be zinc atoms in your model.

container with the correct number of copper cubes and zinc cubes

STEP 2 *Supply Master:* Place the correct number of copper cubes in the empty container (See Table 1: Team Models of Brass.)

Procedure Specialist: Place the correct number of zinc cubes in the same container.

STEP 3 *Team Manager:* Mix up the cubes of copper and zinc atoms. Without looking, take one cube at a time out of the container and give it to the *Recycling Engineer.*

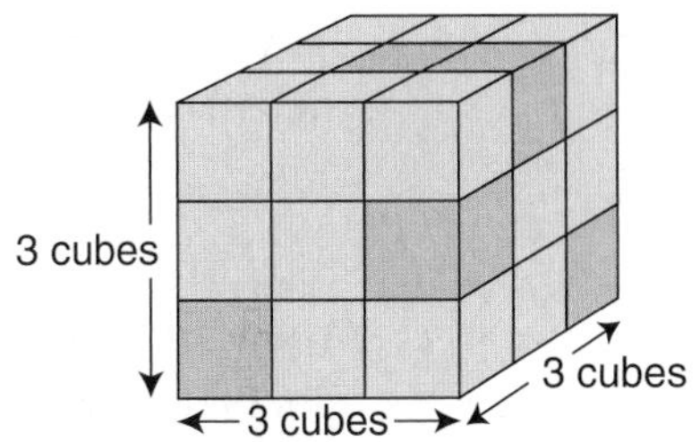

Recycling Engineer: Make a sample of brass that is 3 cubes high, 3 cubes wide, and 3 cubes deep.

Hold the model so you can see three sides, as shown in the diagram.

Shade the copper cubes (atoms) *that you can see* in your data table. For example, the drawing above shows 5 copper cubes. (Your drawing will not be the same as this example.)

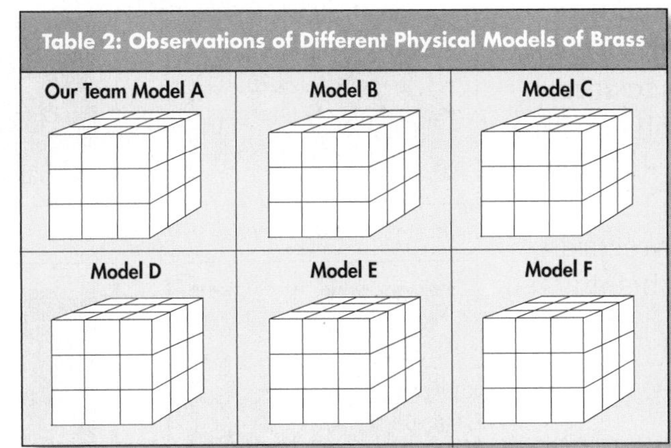

Table 2: Observations of Different Physical Models of Brass

Our Team Model A	Model B	Model C
Model D	**Model E**	**Model F**

STEP 4 Exchange your team model of a brass sample with at least five other teams.

In your data table, for each different model, shade the copper cubes (atoms) *that you can see.*

Do all the models of brass look the same or different? Describe what the models look like.

Participate in a class discussion about your observations and conclusion.

Analyze and Explain

Using SPT Ideas to Solve Classification Problems

Here are ten Ultrascope diagrams of the different classifications of matter. These include suspensions, solutions, and single substances, both elements and compounds. You will need to determine which diagrams represent each classification of matter.

Your team will need:

- container of cubes of at least 3 different colors
- 2 small rectangles of black construction paper

Example: Which diagram, (a through j), represents the particles of a *liquid solution*? Justify your answer.

You will follow the general steps in the familiar procedure for writing an analysis and explanations. You will also build a model to help you apply your ideas.

Your teacher will show you how to use the *Classification of Matter and Small-Particle Theory* consensus ideas to solve this example problem. As you discuss the solution procedure, complete the solution on your record sheet, as shown below.

Small Particle Diagrams

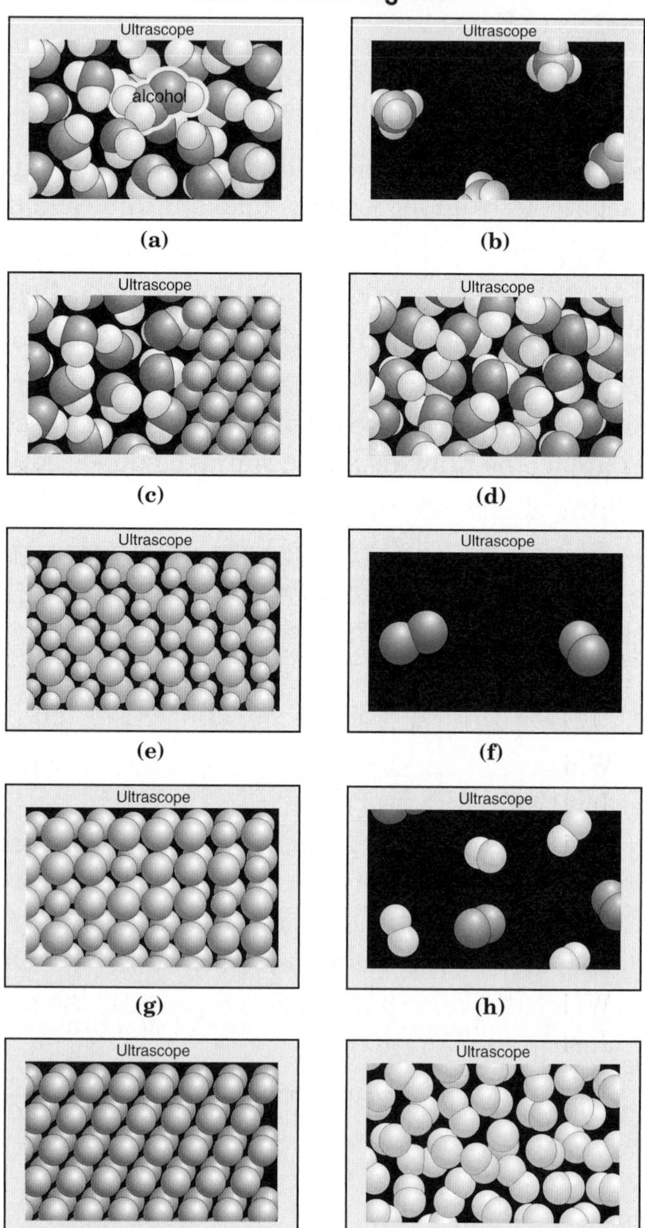

Solution

Analysis

Use the *Classification of Materials* consensus ideas to identify the phase(s), number of substances, and/or kind of substance(s) in the material.

A liquid solution ______________________.

If necessary, use the Small-Particle Theory (SPT) consensus ideas to identify the different diagrams that show the particles in the right phase(s) of the material.

The particles of a liquid ____________________________

Only diagram(s) _____________ show(s) all of the particles in the liquid phase.

Explanation

Build a model with your cubes to help you apply the scientific ideas. Use complete sentences to write your explanation.

Discuss the following problems with your team. Work in pairs to build the models. Be prepared to discuss your answers with the class.

1. Which diagram (b through j) represents the particles of a *gaseous element*? Justify your answer.

2. Which diagram (b through j) represents the particles of a *solid compound*? Justify your answer.

Participate in a class discussion about your solutions.

3. Which diagram (b through j) represents the particles of a *liquid compound*? Justify your answer.

4. Which diagram (b through j) represents the particles of a *solid solution* (alloy)? Justify your answer.

5. Which diagram (b through j) represents the particles of a *gaseous compound*? Justify your answer.

Participate in a class discussion about your solutions.

6. Which diagram (b through j) represents the particles of a *solid element*? Justify your answer.

7. Which diagram (b through j) represents the particles of a *gaseous solution*? Justify your answer.

8. Which diagram (b through j) represents the particles of a *suspension* of a solid in a liquid? Justify your answer.

9. Which diagram (b through j) represents the particles of a *liquid element*? Justify your answer.

Participate in a class discussion about your problem solutions.

LEARNING ABOUT OTHER IDEAS

Activity 12
Small-Particle Theory of Phase Changes

Purpose

In Unit 4 Cycle 2, and Unit 5, you learned about phase changes (melting, boiling, condensing, and freezing). *All* substances can be a solid, a liquid, or a gas if they are cooled or heated enough!

- During the transition from one phase to another phase, the *temperature* and *thermal energy* of a substance does not change (even though you continue to transfer heat energy into the substance).

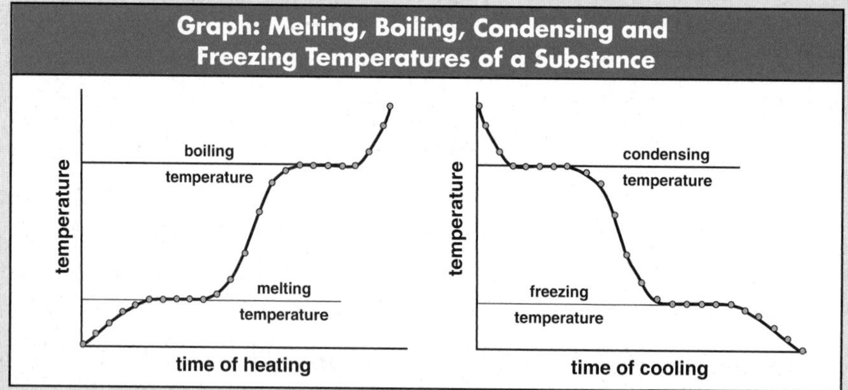

- When a substance changes phase, its stored phase energy either increases (solid to liquid, liquid to gas) or decreases (gas to liquid, liquid to solid). The thermal energy of the substance does not change.

In this activity, you will answer the following key question:

What happens to the particles of a substance during a melting and boiling phase change?

Record the key question for the activity on your record sheet.

Learning the Ideas

Melting

In this activity, you will explore what happens to the molecules when water is heated and changes phase.

Your teacher will show you a scientist's computer animation of particles of water in solid ice *as the ice is heated until it starts to melt.* Watch the animation a few times. Your teacher will set the animation on a continuous loop while you answer the following questions.

1. As the solid ice is heated (but not melting), what happens to the average speed and motion energy of the molecules? (Note: At this Ultrascope magnification, the change in the average spacing of the molecules is so tiny it cannot be noticed.)
2. When the temperature reaches the melting temperature (0°C), what happens to some of the ice molecules?
3. As heating continues, the temperature remains constant. Where does this heat energy go?

Your teacher will review the answers to these questions with the class.

As the temperature reaches the melting temperature, the average motion energy of the molecules is large enough so that some of the faster-moving molecules start to break free from their locked positions and move around. As heating continues the temperature remains constant, so the average motion energy of the molecules (of ice plus water) stays constant. All of the heat energy transferred during melting goes to breaking the molecules free from their locked positions until all the particles are moving around in the liquid state. The liquid water has more stored phase energy than the ice.

4. The melting temperature of the element aluminum (Al) is 660°C. Describe what happens to the atoms of aluminum while it is melting and the temperature remains constant at 660°C.

Your teacher will review the answers to this question with the class.

Boiling

When you boil water large bubbles form on the bottom of the pan or beaker. These bubbles rise through the water and pop when they reach the water surface.

Three students were discussing the question of what's inside the bubbles. Look at their conversation.

I think that oxygen is inside the bubbles. We know that oxygen is dissolved in water. That is how fish breathe. At boiling temperature, the oxygen molecules have enough energy to break free of the water and form bubbles.
Carlos

I disagree. I think that water vapor is in the bubbles. At boiling temperature, the molecules that are close together in water are moving fast enough to break free of the cohesive attraction holding them together. They fly apart into a gas bubble.
Than

Well, I think that the bubbles contain hydrogen and oxygen because water molecules are made of hydrogen and oxygen atoms. At boiling temperature, the water molecules have enough energy to break apart into hydrogen and oxygen molecules.
Isabel

5. Do you agree with Carlos, Than, or Isabel, or do you have a different idea?

Participate in a class discussion about the answer to the question. *Who do you think has the best answer?* Why?

6. How do you know that Isabel's explanation for what is inside the bubbles of boiling water is not right? (*Hint*: What do you need to do to the compound water to get it to break down into the elements oxygen and hydrogen?)

7. How do you know that Carlos' explanation for what is inside the bubbles of boiling water is not right? (*Hint*: When you boil pure water, is there anything left?)

Than had the correct SPT explanation of what is inside the bubbles of boiling water. As you heat water, the motion energy of the water molecules increases and they move faster and faster. At the boiling temperature, molecules that are close together sometimes are moving so fast that the cohesion forces can

no longer hold them together. They fly apart into a bubble of water vapor. The bubbles rise to the surface and pop open, releasing the water vapor into the air. All of the heat energy transferred during boiling goes to breaking the molecules free from loose connections until all the particles are moving randomly around in the gas state.

8. In an Ultrascope diagram like the one shown, draw what you think the molecules inside a bubble of boiling water look like. Use a white gel pen or pencil.

Your teacher will review the answers to these questions with the class.

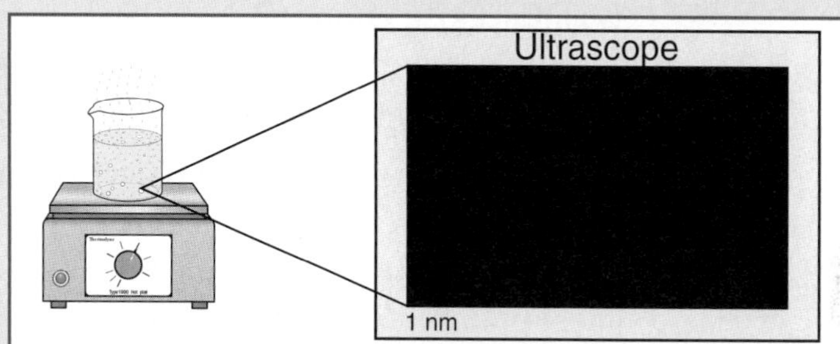

Strength of the Cohesion Attraction and Phase Changes

You know that the higher the temperature and thermal energy of a substance, the greater the motion energy of the particles of the substance (Activity 7). The melting temperature of a substance is related to the strength of the cohesion attraction between the particles of the substance, as shown in the diagram. In general, the higher the cohesion attraction between the particles of a solid substance, the greater the melting temperature of the substance.

For example, when the cohesion attraction is weak, slow-moving particles have enough energy to break free from their locked positions. So, the melting temperature is low.

9. The melting temperature of oxygen is very low (–218°C). What is the SPT explanation for this very low melting temperature?

When the cohesion attraction is strong, the particles have to be moving very fast to have enough energy to break free from their locked positions. So, the melting temperature is high.

10. The melting temperature of iron is very high (1535°C). What is the SPT explanation for this very high melting temperature?

Look at the *Table of Melting and Boiling Points* in the Appendix.

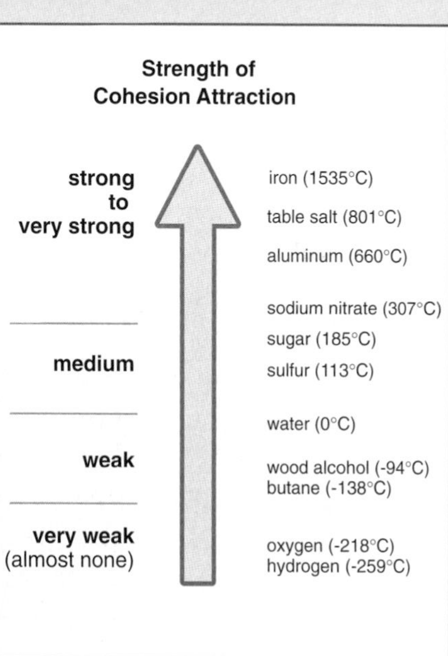

11. Which metal: *aluminum, copper, gold,* or *zinc*, probably has the strongest cohesive attraction between the atoms? Explain your reasoning.

Your teacher will review the answers to these questions with the class.

What We Have Learned

Recall the key question for this activity:

What happens to the particles of a substance during a melting and boiling phase change?

Participate in a class discussion to review the answers to the key question.

1. With your teacher's guidance, write the Small-Particle Theory (SPT) description of melting.
2. With your teacher's guidance, write the Small-Particle Theory (SPT) description of boiling.

LEARNING ABOUT OTHER IDEAS

Activity 13
Atomic Structure

Purpose

John Dalton was a chemist from England. In 1808 he published *New System of Chemical Philosophy*. In it, he presented his theory that each element is made of small atoms. He had the theory that different elements have atoms of different masses. Dalton imagined atoms as tiny, solid balls that are indivisible (cannot be broken into smaller pieces). He backed his theory with some evidence.

In the last two hundred years many scientists, from different parts of the world, collected important data. This data provided evidence to build a new scientific theory. This theory claimed that the atom has structure or parts.

What is the structure of the atom? Atoms are so tiny that they cannot be seen, not even with the most powerful of microscopes. Scientists have relied on indirect evidence to help them develop theories about the structure of the atom. With each new piece of data, the theory of the atom was revised. In some cases, it was replaced by new evidence.

In this activity, you will read about scientists' current theory of the atom. You will examine some important data that is evidence to support this theory. The key question for this activity is:

What is the structure of an atom?

Record the key question for the activity on your record sheet.

Learning the Ideas

Structure of an Atom

You will read about the modern theory of the structure of the atom. As you read, complete the table on your record sheet, like the one shown on the next page. This summarizes the information about the structure of the atom.

Table: Structure of an Atom				
Particle name	**Particle charge**	**Particle mass**	**Particle location**	**Relative number of particles**
Electron (e)				
Proton (p)				
Neutron (n)				

Charge and Mass of Electrons and Protons

According to scientists' current theory, the atom is not an indivisible, solid sphere. In 1897, the British scientist J. J. Thomson discovered that tiny, light, negatively charged particles could be removed from the atom with electrical forces. These particles are called *electrons*. Other experiments also supported the idea that the electrons are very light compared to the mass of the atom.

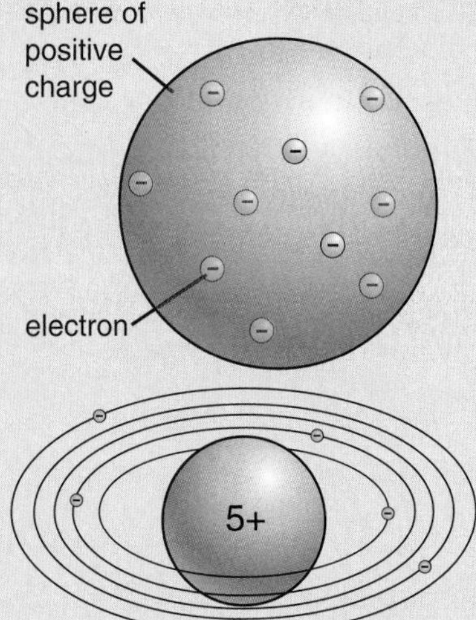

Atoms are usually neutral. If electrons were negatively charged, that meant the rest of the atom must be positively charged. Also, since electrons were very light, the rest of the atom must be much more massive than the electron. But no one knew how electrons, positive charge, and atoms were related. Based on what he knew, J. J. Thomson suggested a model for the atom. He proposed that atoms are balls of positive charge with electrons scattered within the positive charge, like raisins in a muffin.

The Japanese scientist Hantaro Nagaoke proposed, in 1904, a different model of the atom. His model showed the negative electrons revolving around a large, positively charged sphere, like the planets revolving around the Sun.

In 1911, the British scientist Ernest Rutherford designed an experiment that tested these models of the atom. He fired a beam of positively charged particles (called alpha particles) at a very thin sheet of gold foil. He used gold as a target because it can be made into sheets that are only a few hundred atoms thick. Rutherford wanted to see how the paths of the positive alpha particles would change when they hit the gold atom.

The results were surprising. Most of the particles passed straight through the gold foil as if nothing were there. Some changed direction slightly, and a few (about 1 in 20,000) actually bounced back at large angles.

Rutherford drew two conclusions from his experiment. First, the alpha particles go straight through the gold foil *because most of the atom is empty space.* Essentially, all of the mass of the atom is in a tiny, dense center or nucleus (NOO-klee-us) of the atom. Second, the tiny nucleus must be positively charged. Only when a positive alpha particle comes near a massive, positive nucleus is it repelled strongly enough to make it bounce backward.

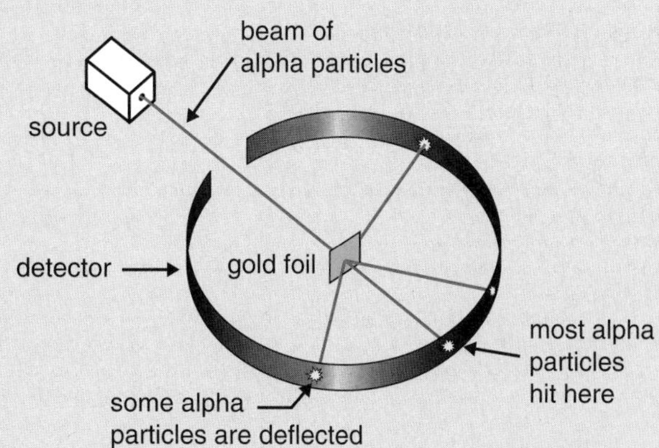

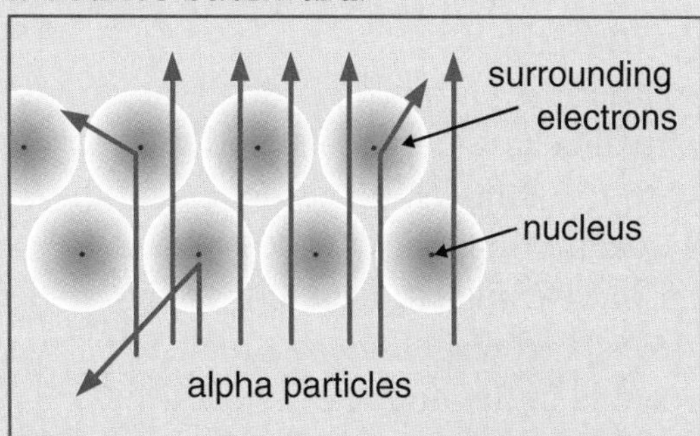

You could compare Rutherford's experiment to what happens when you throw pebbles at a chain-link fence. Most of the pebbles go right through the fence because it is mostly empty space. A few pebbles, however, hit the wire and bounce back.

The positively charged particles in the nucleus are called *protons*. Later, the British scientist James Chadwick discovered that he could remove neutral particles (no charge) from the nuclei of atoms by bombarding them with special particles. These neutral particles are called *neutrons*. The mass of the proton is about the same as the mass of the neutron. But a proton and a neutron are each about 2000 times heavier than an electron.

Complete the second column (Particle charge) in your *Table: Structure of an Atom* on your record sheet.

Complete the third column (Particle mass) of your table.

Your teacher will review the answers to these questions with the class.

Little Particles, Big Spaces

Essentially all of the mass of an atom comes from the protons and neutrons in the nucleus. But an atom's volume is the space around the nucleus in which the electrons move. Rutherford showed that this space is huge compared to the space occupied by the nucleus. To picture the difference, imagine that you were standing in the center of a football stadium holding a small pin, as shown in the diagram on the next page. If the nucleus were the size of the head of your pin, then the space occupied by the electrons would be the size of the entire stadium, including the top row of seats! The width (diameter) of an atom is about one hundred thousand times (100,000) the width of a nucleus.

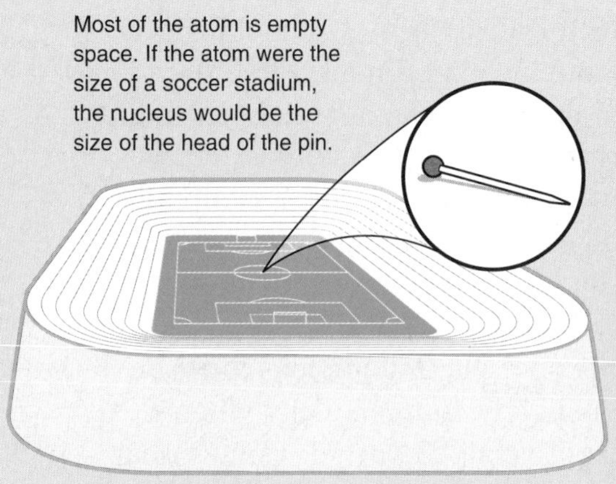

Most of the atom is empty space. If the atom were the size of a soccer stadium, the nucleus would be the size of the head of the pin.

1. Why does the nucleus make up most of an atom's mass? Explain your reasoning

Diagram of the Structure of an Atom

The tiny electrons are constantly moving within the empty space that surrounds the nucleus. The nucleus contains the positively charged protons and the neutral neutrons. You know that objects with the same charge repel each other, while objects with different charges attract each other. So, why doesn't the repulsion between the positively charged protons cause the nucleus to fly apart? And why don't the negatively charged electrons, which are attracted to the protons, slow down and stick to the protons in the nucleus? To answer these and other questions, scientists developed a new theory called *quantum mechanics*. This theory predicts the behavior of atoms and their particles.

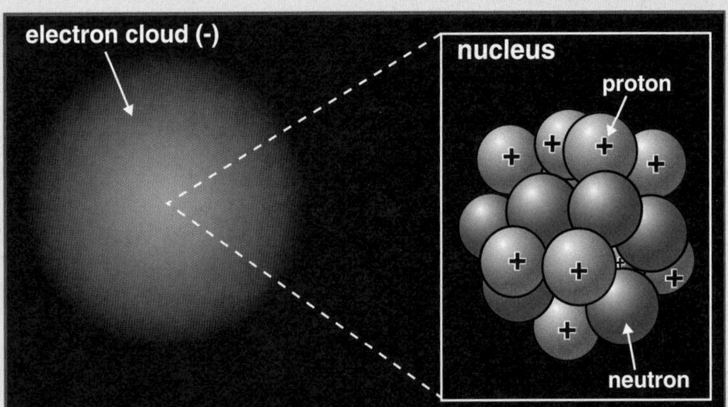

According to the quantum mechanics theory, the electrons do not move like ordinary objects. Sometimes they are close to the nucleus, and sometimes they are far away. Occasionally, they are even inside the nucleus. *But it is impossible to determine exactly where an electron is at any given time.* Scientists represent the location of the electrons in an atom with an *electron (probability) cloud*, as shown in the diagram. The thicker the cloud, the more likely it is to find an electron at that location at any given moment. The thinner the cloud, the less likely it is to find an electron at that location.

Complete the fourth column (Particle location) of your table.

2. Look at the electron cloud in the diagram of the structure of the atom. Is it more likely to find an electron far away from the nucleus or closer to the nucleus? Explain your reasoning.

Your teacher will review the answers to these questions with the class.

Your teacher will show you a sequence of models of the inside of the atom from the *Power of Ten* animation. In each picture, the magnification is increased by 10x.

Atoms and Elements

All atoms of an element have the same number of protons. For example, the atom pictured in the diagram on the previous page showing the structure of the atom is fluorine. If you count the number of protons in the nucleus, you will see that there are nine protons. All fluorine (F) atoms have nine protons. But all carbon (C) atoms have six protons, and all iron (Fe) atoms have 26 protons. The atoms of different elements have different numbers of protons.

3. What is the same about all the atoms of an element?
   What is different about the atoms of *different* elements?

In the atom of any element, the number of protons equals the number of electrons. As a result, the total positive charge and total negative charge balance each other, making the atom neutral. The number of neutrons in an atom may be the same as the number of protons, but it may not.

Complete the fifth column (Relative number of particles) in your table.

4. A fluorine (F) atom has nine protons. How many electrons are in a fluorine atom?
5. An aluminum (Al) atom has 13 protons. How many electrons are in an aluminum atom?

Your teacher will review the answers to these questions with the class.

Review of the Structure of an Atom

Read each evidence statement below and answer the question that follows. It may be useful to refer back to your table to answer these questions.

Evidence: *An unknown scientist discovered that atoms are not attracted to positive or negative charges. That is, there is no charge on the atom itself—it is neutral.*

6. For an atom to be neutral (no charge), what must be true about the number and charge of protons and the number and charge of electrons?

Evidence: *In 1897, the British scientist J. J. Thomson discovered that tiny negative particles could be removed from the atom with electrical forces.*

7. What tiny particle did J. J. Thomson remove from the atom?

Evidence: *In 1911, the British scientist Ernest Rutherford discovered that if special particles were shot toward a piece of gold foil, most of the particles went straight through while a few bounced back in the direction from which they came.*

8. What region of the atom do these special particles have to pass through in order to go straight through? Explain your reasoning.

9. What region of the atom did the particles have to collide with in order to be bounced back in the direction from which they came? Explain your reasoning.

Evidence: *In 1932, the British scientist James Chadwick discovered that he could remove neutral particles from atoms by bombarding them with special particles. The neutral particles are approximately the same mass as the proton.*

10. What particle did Chadwick remove from the atom?

11. About how much mass do the neutrons contribute to the total mass of an atom?

a) about 1/10 the mass of atom

b) about 1/2 the mass of the atom

c) most of the mass of an atom

Explain your reasoning.

Evidence: *The current theory of the atom came from the work of many scientists from the 1920s to the current time.*

12. Describe the modern theory of the location of electrons around the nucleus.

Your teacher will review the answers to these questions with the class.

What We Have Learned

Consider the key question for this activity:

What is the structure of an atom?

Participate in a class discussion about the answer to the key question.

Write the answer to the key question on your record sheet

LEARNING ABOUT OTHER IDEAS

Activity 14
Atomic Structure and the Periodic Table

Purpose

This activity has two different parts. Your class may complete only one part of the activity or both parts of the activity.

In Part A of this activity, you will explore atomic number and atomic mass. The key question for the first part of the activity is:

1. How does atomic structure help you make sense of the Periodic Table of the Elements?

The key question for Part B is:

2. What are isotopes of an element?

Record the key questions for the activity on your record sheet.

Part A: Atomic Structure and the Periodic Table

Learning the Ideas

Atomic Number

One of the triumphs of the quantum mechanics theory is that it explains the order of the elements on the Periodic Table of the Elements. Look at the Periodic Table on the inside back cover of this book.

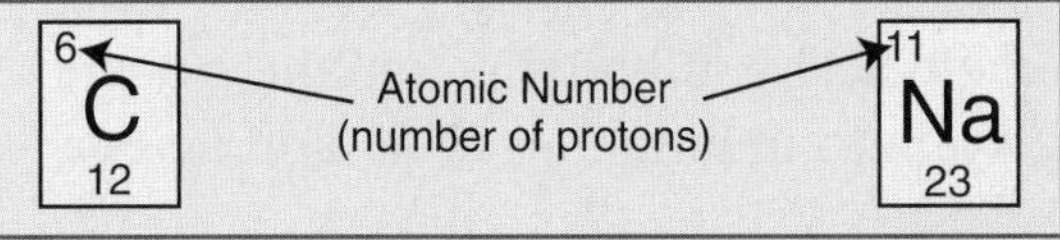

The number at the top left corner of the block for each element is the *atomic number* of the element. The atomic number is the number of protons in an atom of the element. *All atoms of an element have the same number of protons.* For example, the atomic number of carbon (C) is 6, so all atoms of carbon have six protons. The atomic number of sodium (Na) is 11, so all atoms of sodium have 11 protons.

Use your Periodic Table of the Elements to answer the following questions.

1. How many protons do the atoms of iron (Fe) contain?

2. How many protons do the atoms of argon (Ar) contain?

3. How many protons do the atoms of barium (Ba) contain?

Your teacher will review answers to these questions with the class.

The elements in the Periodic Table are arranged from left to right and top to bottom *in order of increasing atomic number* (number of protons). For example, the second row has eight elements, starting with lithium (atomic number 3). The atomic numbers increase across the row—from beryllium (4), to boron (5), to carbon (6), to nitrogen (7), to oxygen (8), to fluorine (9), and finally to neon (10). The next row (row 3) starts with sodium, with an atomic number of 11, and so on.

Atomic Mass Number

You learned in Activity 2 that the number at the bottom of the block for each element is called the atomic *mass number*. This is the number of protons plus the number of neutrons in the atom of the element.

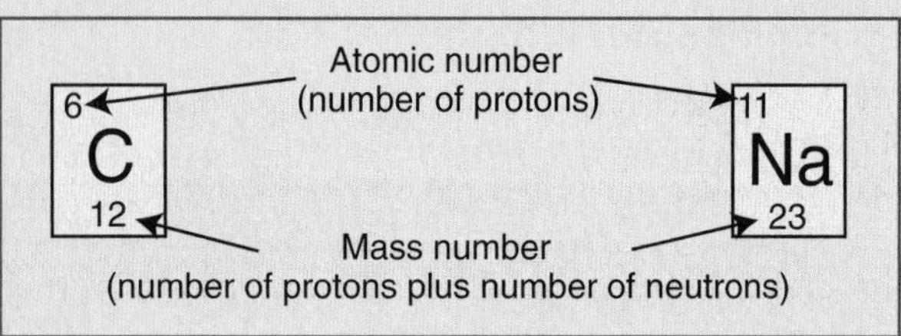

Essentially all of the mass of the element is in the nucleus of the atom, which contains the protons and the neutrons. So, the mass numbers tell you how the masses of the elements compare. For example, sodium (Na) has a mass number of 23 and carbon (C) has a mass number of only 12. The sodium atom is almost twice as heavy as the carbon atom.

You can determine the atomic number and mass number of an atom from the number of protons and neutrons in the atom. For example, suppose a scientist determines that the atoms of an unknown element contain 47 protons and 61 neutrons. The atomic number (number of protons) for this element is 47.

$$\begin{aligned} \text{Mass Number} &= \text{no. of protons} + \text{no. of neutrons} \\ &= 47 + 61 \\ &= 108 \end{aligned}$$

Look at your Periodic Table of the Elements. The unknown element is silver (Ag) because it is the only element with an atomic number of 47 and a mass number of 108.

4. The atoms of an unknown element contain 26 neutrons and 22 protons. What is the atomic number of this element? What is the mass number of this element? What is this unknown element?

5. The atoms of an unknown element contain 50 protons and 69 neutrons. What is the atomic number of this element? What is the mass number of this element? What is this unknown element?

Your teacher will review answers to these questions with the class.

You can also calculate the number of neutrons in an atom of an element by subtracting the number of protons (atomic number) from the total number of protons plus neutrons (the mass number).

$$\begin{aligned}\text{No. of neutrons} &= (\text{no. of protons} + \text{no. of neutrons}) - \text{no. of protons}\\ &= \text{mass number} - \text{atomic number}\end{aligned}$$

For example, the carbon atom (C) has 12 protons-plus-neutrons (mass number) and 6 protons (atomic number).

$$\begin{aligned}\text{No. of neutrons in C atom} &= \text{mass number} - \text{atomic number}\\ &= 12 - 6\\ &= 6\end{aligned}$$

Sodium (Na) has 23 protons-plus-neutrons (mass number) and 11 protons (atomic number), so it must have 12 neutrons.

$$\begin{aligned}\text{No. of neutrons in Na atom} &= \text{mass number} - \text{atomic number}\\ &= 23 - 11\\ &= 12\end{aligned}$$

6. Find the element beryllium (Be) on the *Periodic Table of the Elements*. How many protons are in the atoms of beryllium? How many protons-plus-neutrons are in the atoms of beryllium? How many neutrons are in the atoms of beryllium?
7. Find the element bromine (Br) on the Periodic Table of the Elements. How many protons are in the atoms of bromine? How many protons-plus-neutrons are in the atoms of bromine? How many neutrons are in the atoms of bromine?
8. Find the element gold (Au) on the Periodic Table of the Elements. How many protons are in the atoms of gold? How many protons plus neutrons are in the atoms of gold? How many neutrons are in the atoms of gold?

Your teacher will review answers to these questions with the class.

As the atomic number (number of protons) increases, the mass of the atoms of succeeding elements generally increases, although exceptions exist. Typically, however, the atoms of the elements in the periodic table increase from left to right, and those elements listed in the lower rows have more mass than those in the upper rows.

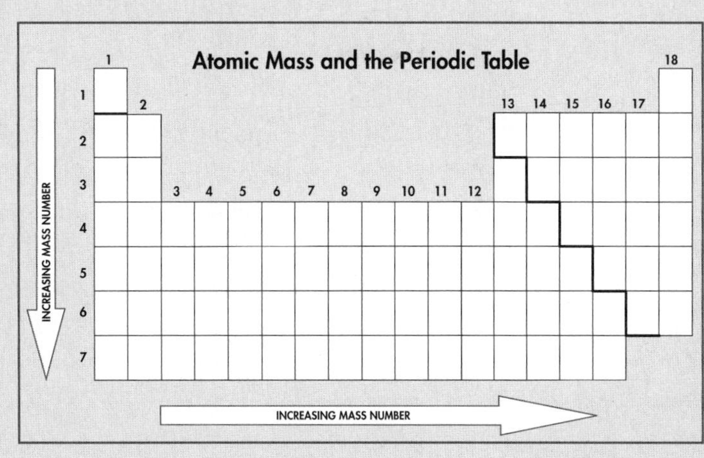

What We Have Learned

Consider the key question for the first part of the activity:

1. How does atomic structure help you make sense of the Periodic Table of the Elements?

Participate in a class discussion about the answer to the key question.

Write the answer to the key question on your record sheet.

Part B: Isotopes of an Element

Learning the Ideas

Your teacher will give each team a copy of a different Periodic Table of the Elements. If you look at this table, you will notice that the atomic mass numbers are not whole numbers. For example, the atomic mass number for hydrogen (H) is 1.008 instead of 1, and the atomic mass number for iron (Fe) is 55.85 instead of 56. How can this be? Is it possible to have fractions of a proton or neutron?

Science Words

isotope: an element that has different numbers of neutrons

It turns out that the atoms of elements do not always have the same number of neutrons (although they always have the same number of protons). An element that has different numbers of neutrons is called an **isotope** (EYE-suh-tohp) of the element.

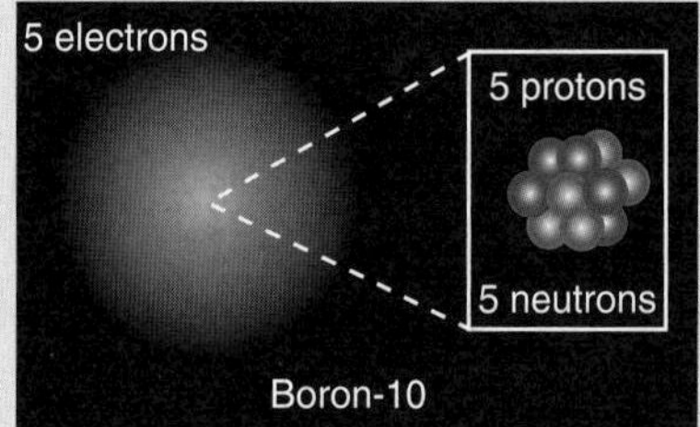

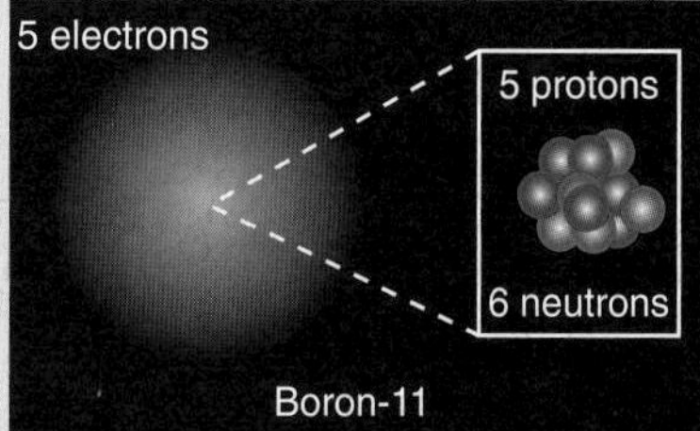

Suppose you have a sample of the element boron. Boron, a metalloid, is usually a gray powder. All boron atoms have 5 protons in the nucleus. But there are two different isotopes of boron with different numbers of neutrons.

One isotope of boron has 5 protons and 5 neutrons in the nucleus, as shown in the diagram. The second isotope of boron has 5 protons and 6 neutrons in the nucleus.

The isotopes are identified by the name of the element followed by the atomic mass number of the isotope. For example, one isotope of boron is called boron-10 because it has an atomic mass number of 10 (5 protons plus 5 neutrons). The other isotope of boron is called boron-11 because it has an atomic mass number of 11 (5 protons plus 6 neutrons).

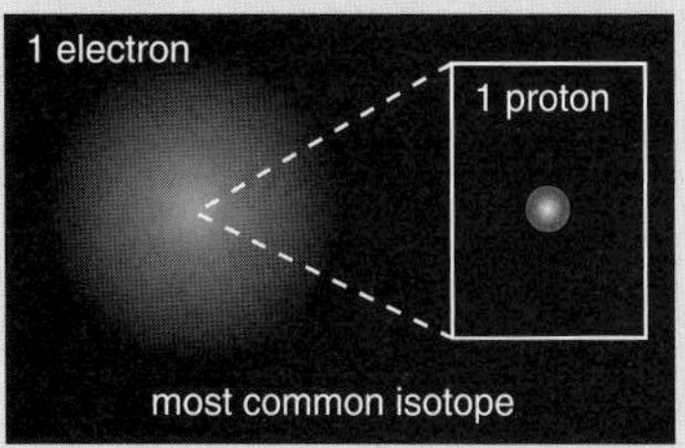

Most elements have more than one isotope. The atomic mass numbers on the Periodic Table are stated as decimal numbers because each number is an average of the mixture of isotopes for each element. For example, about two out of every 10 atoms of boron are boron-10. About 8 out of every 10 atoms are boron-11. The average mass number for boron is 10.81.

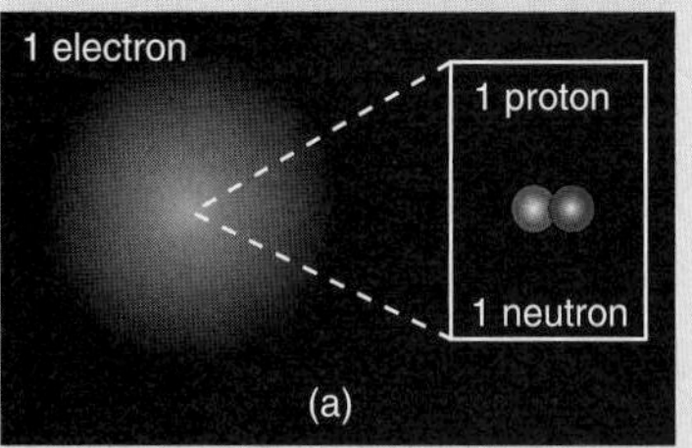

(a)

All hydrogen atoms have one proton in the nucleus. There are three isotopes of hydrogen. The first isotope, hydrogen-1 has one proton and no neutrons. The other isotopes of hydrogen, hydrogen-2 (deuterium) and hydrogen-3 (tritium), are also shown in the diagrams.

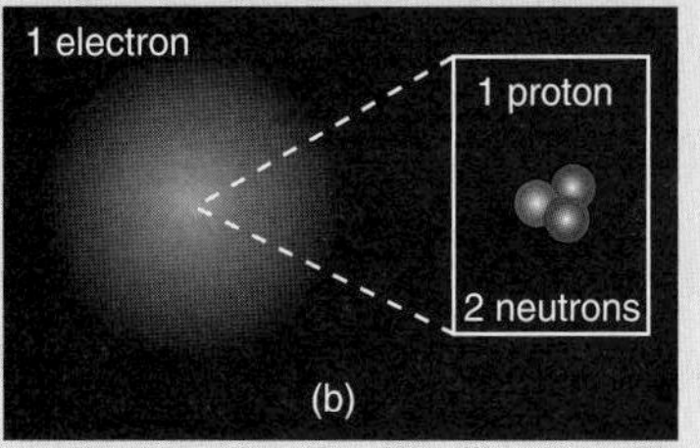

(b)

1. Which diagram, (a) or (b), shows the hydrogen-2 isotope? Explain your reasoning.
2. Which diagram, (a) or (b), shows the hydrogen-3 isotope? Explain your reasoning.

The most abundant isotope of hydrogen is hydrogen-1. Out of thousands of hydrogen atoms, only a few are hydrogen-2 and hydrogen-3. So the average mass number for hydrogen is 1.008.

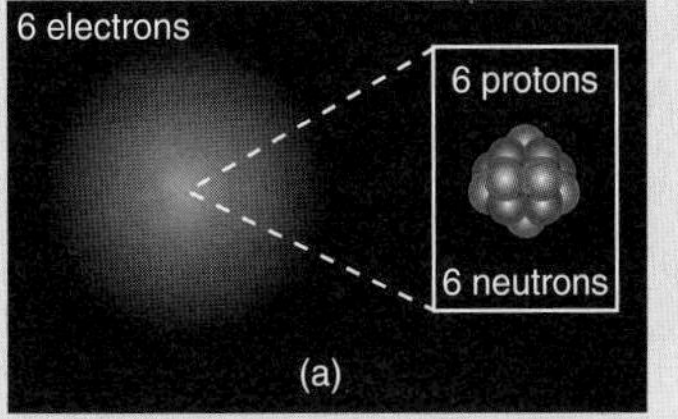

(a)

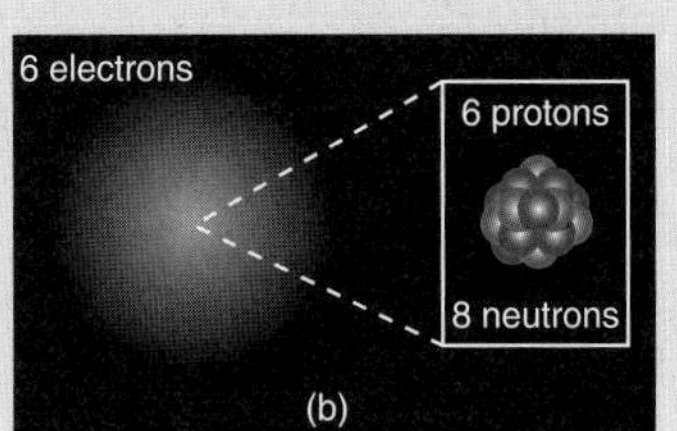

(b)

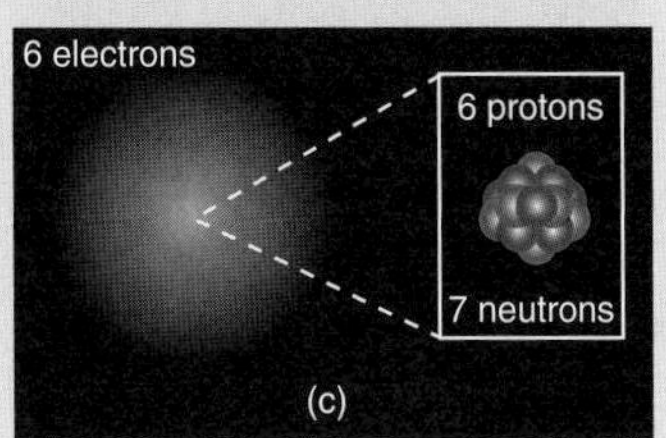

(c)

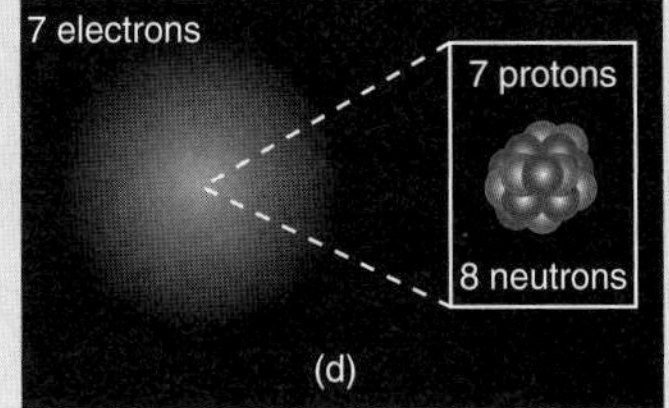

(d)

3. The element carbon has many isotopes. The isotope used in the dating of fossils is carbon-14. Which diagram, (a), (b), (c) or (d), shows the carbon-14 isotope? Explain your reasoning.

What We Have Learned

Think about the key question for the second part of the activity:

2. What are isotopes of an element?

Participate in a class discussion about the answer to the key question.

Write the answer to the key question on your record sheet.

LEARNING ABOUT OTHER IDEAS

Activity 15
Nuclear Interactions

Purpose

More than a thousand years ago, people thought that it was possible to change one substance into a different substance. For hundreds of years chemists tried to make a "philosopher's stone." This was a substance that could convert other metals, like lead, into bright, valuable gold. They heated metals, cooled metals, and added acids to metals. They ground metals into a powder and mixed it with everything they could think of. But nothing worked. Physical and chemical interactions do not convert one element into another element.

Part of painting by Joseph Wright of Derby, 1734-1797: The Alchemist in Search of the Philosopher's Stone Discovers Phosphorus.

Even so, some elements do change into other elements. Atoms of carbon can become atoms of nitrogen. Atoms of iodine can become atoms of antimony. A uranium atom can become a thorium atom. These changes occur as a result of nuclear reactions. Unfortunately, atoms of lead never become atoms of gold! The key question for this activity is:

What are nuclear reactions and radiation?

Record the key question for the activity on your record sheet.

Learning the Ideas

Radioactive Decay

Interactions involving the particles of a nucleus (protons and neutrons) are called *nuclear reactions*. You learned about isotopes in Activity 14. Isotopes are atoms of an element that have the same number of protons but a different number of neutrons. The isotopes of an element have the same atomic number (number of protons) but different atomic masses (number of protons plus neutrons). Some isotopes are unstable. The nucleus of an unstable atom does not hold together well. It breaks apart, sometimes forming atoms with a different number of protons or neutrons. When an unstable atom breaks apart (decays), the nucleus emits fast-moving particles and energy. This process is called *radioactive decay*.

All the isotopes of some elements are radioactive, such as element 43, technetium, or element 86, radon. No stable samples of those elements exist in nature. Element 92, uranium, is another example of an element in which no stable isotopes exist. But uranium decays so slowly that it is still found in the Earth's crust.

There are three types of radioactive decay.

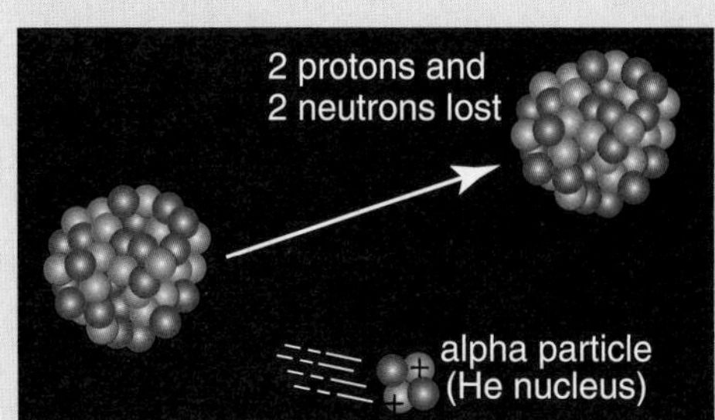

Alpha Decay An alpha particle consists of two protons and two neutrons. It is the same as a helium nucleus. The release of an alpha particle decreases the atomic number by 2 and the atomic mass by 4. Alpha decay can cause a bad skin burn. Alpha particles can be stopped by a sheet of paper, a thin piece of metal foil, and clothing. Uranium, plutonium, thorium, and radon are typical alpha emitters.

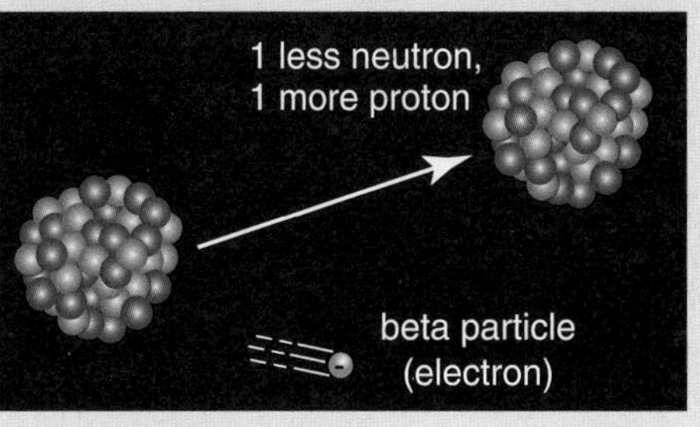

Beta Decay Beta particles are just electrons from the nucleus. The term "beta particle" was used in the early history of radioactivity. An electron is released when a neutron decays into a proton and an electron. The new proton stays inside the nucleus. This means that the nucleus now has one less neutron and one more proton. The mass number stays the same, but the atomic number increases. Cobalt-60 is much used in medical practice. It decays to the more stable nickel-60. Tritium, krypton-85, and the potassium in fertilizers and food are also examples of beta emitters.

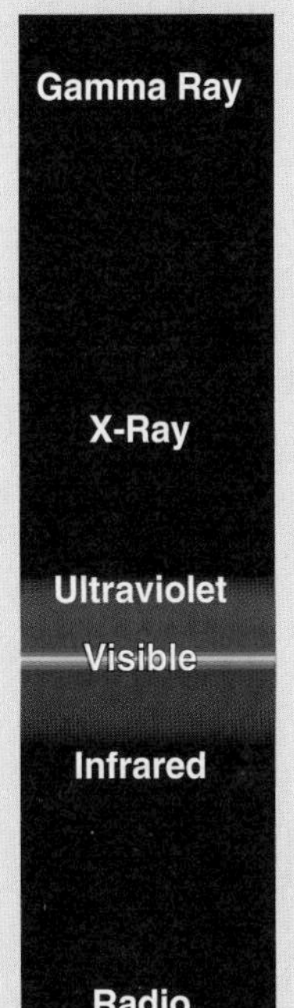

The beta particles (electrons) travel much faster than the alpha particles. They can be stopped by plastic, glass, or a sheet of metal such as aluminum.

Gamma Decay Alpha and beta decay are almost always accompanied by gamma radiation. Gamma radiation is like infrared, visible light, and x-ray radiation, only much higher energy.

no gain or loss of particles

gamma rays

Gamma radiation (also called gamma rays) does not cause a change in either the atomic number or atomic mass of the decaying nucleus. But the energy released is the most penetrating. You would need a piece of lead or steel several centimeters thick, or a concrete wall about a meter thick to stop this type of radiation.

Neutrons in Nuclear Reactions

In addition to alpha, beta, and gamma decay, many nuclear reactions, like fusion and fission, produce neutrons. Fusion and fission reactions occur in stars and release enormous quantities of energy compared to chemical reactions. Nuclear fusion is the combining of two nuclei with low masses to produce one nucleus of larger mass along with some neutrons. In a fission reaction, a neutron collides with a large nucleus (atomic numbers larger than 90) to produce two smaller nuclei and some neutrons.

Neutrons are light and have no electric charge. So, they can penetrate into materials deeper than alpha particles, beta particles, and gamma rays. Neutrons can penetrate even lead. Materials with a high-hydrogen content, for example, water or plastics, provide the most effective shielding against neutrons.

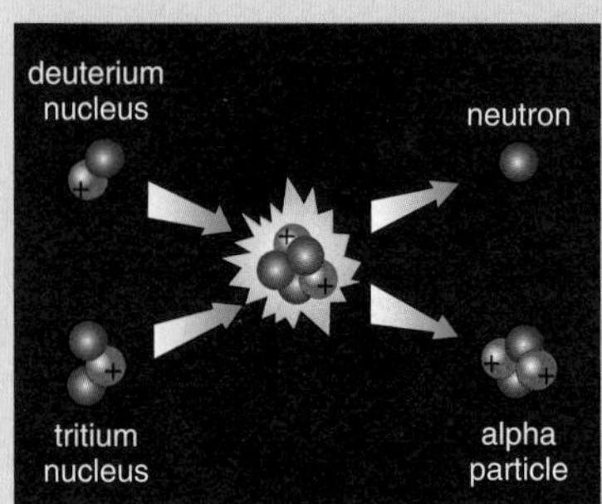

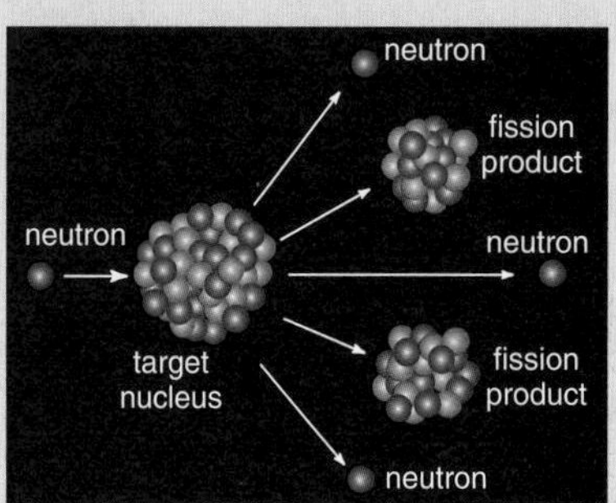

1. What are four forms of nuclear radiation?
2. Which type of radioactive decay (alpha, beta, or gamma) does not result in a different element? Explain your reasoning.
3. Which type of radiation is the most penetrating, alpha particles, beta particles (electrons), gamma rays, or neutrons? (*Hint*: Use the diagram below showing the penetration of radiation in different materials.)
4. **Challenge Problem** Look at the table of radioactive isotopes on the next page. With the help of your Periodic Table of the Elements, predict the new element that forms in each case. Write the name of each element with its atomic mass number in the table on your record sheet.

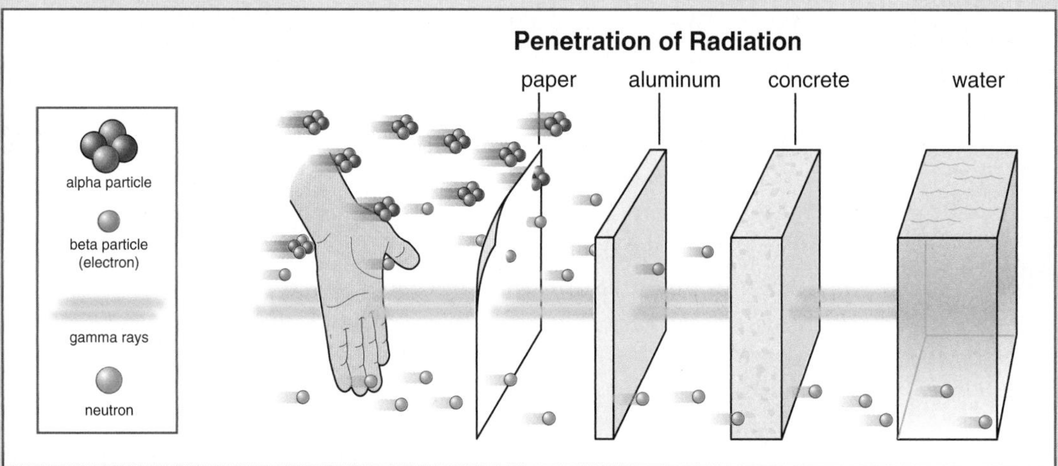

Table: New Element Predictions		
Isotope	Type of Decay	New Element
Uranium - 238	alpha	
Nickel - 63	beta	
Iodine - 131	beta	
Radon - 226	alpha	

Your teacher will review answers to these questions with the class.

Using Radioactive Isotopes

Radioactive isotopes are used by scientists to determine the age of a rock or fossil. Isotopes are also used as *tracers* and as *sources of radiation.*

The radiation given off by isotopes can be detected with special equipment. A radioactive isotope "signals" where it is like a lighthouse flashing in the night. Tracers are radioactive isotopes that can be followed (traced) through the steps of a chemical reaction or industrial process. For example, tracers are used in finding weak spots in metal pipes, especially oil pipelines. When the tracers leak out of a pipe, they can easily be detected.

Tracers behave the same way as the nonradioactive isotopes of an element in chemical reactions. For example, some scientists (neuroscientists) study brain systems and functions. They inject into the bloodsteam of volunteers molecules that are already used by the brain, but with a radioactive tracer attached to the molecule. For example, flourine-18 atoms can be attached to sugar molecules to study the parts of the brain that are most active (burn the most sugar).

A volunteer rests for 30 to 45 min after the tracer is injected into the bloodstream. Then the volunteer is placed inside a large tube with an array of detectors. As the flourine-18 decays, the gamma radiation is detected, analyzed, and a computer produces a final image of the brain.

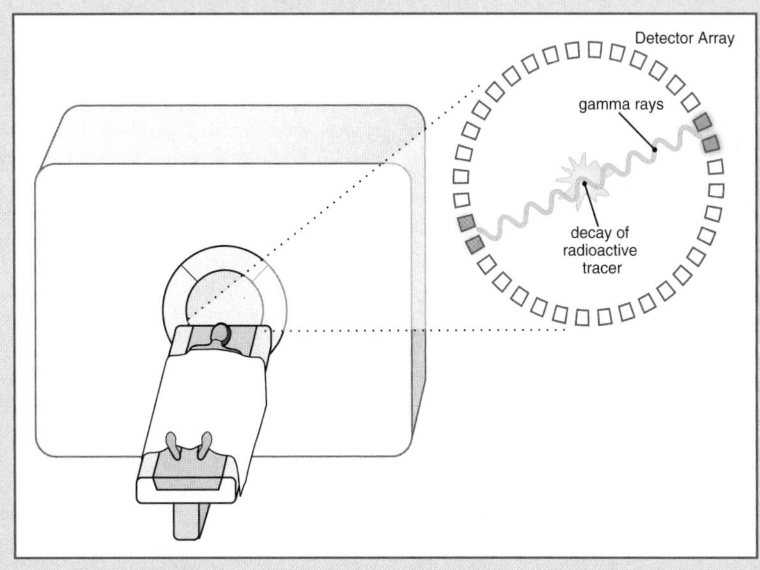

5. Why do you think volunteers have to wait 30 to 45 min to be scanned after injections of the radioactive tracer?

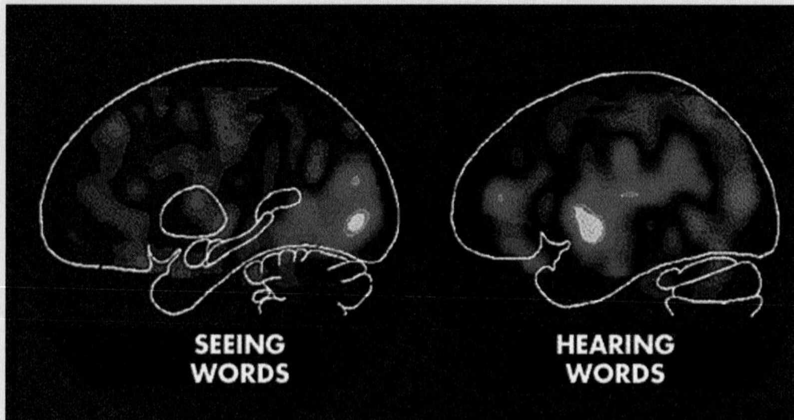

The image on the left shows two areas of the brain (red and yellow) that become particularly active when volunteers read words on a video screen. The image on the right shows two areas of the brain (red and yellow) that become active when volunteers hear words through ear phones.

6. Look at the two images of the brain and read the caption. What would you conclude about the parts of the brain you use when you read language and see language?

Radioactive tracers are designed to deliver a minimum amount of radiation to a volunteer, often less than would normally be received in an x-ray. Because they are safe, tracers are used by doctors to diagnose diseases.

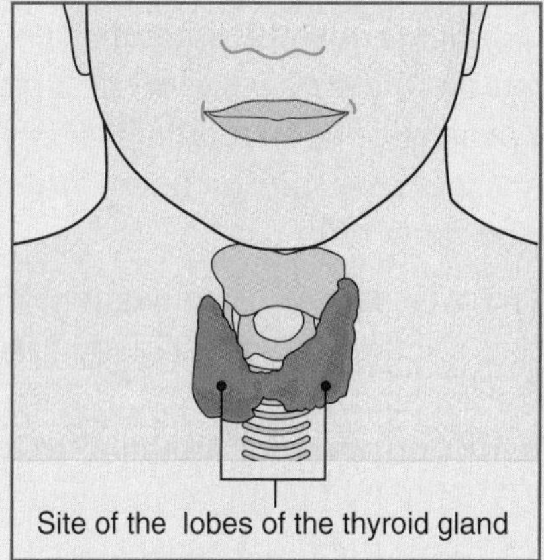

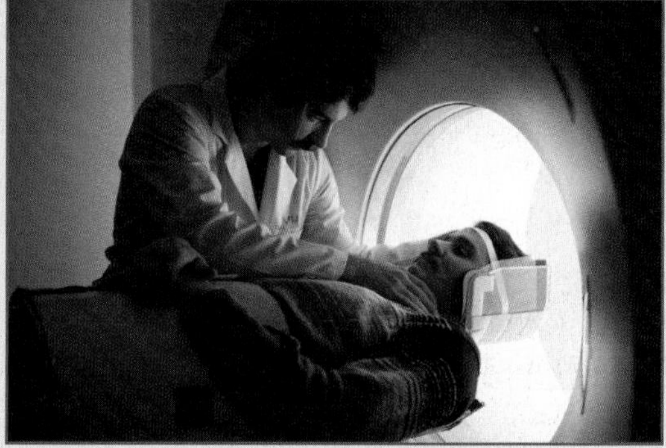

A scanner making an image of this patient's thyroid gland.

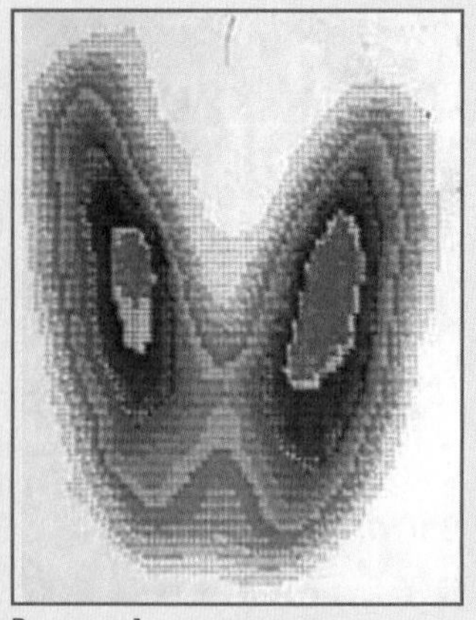

Patient 1

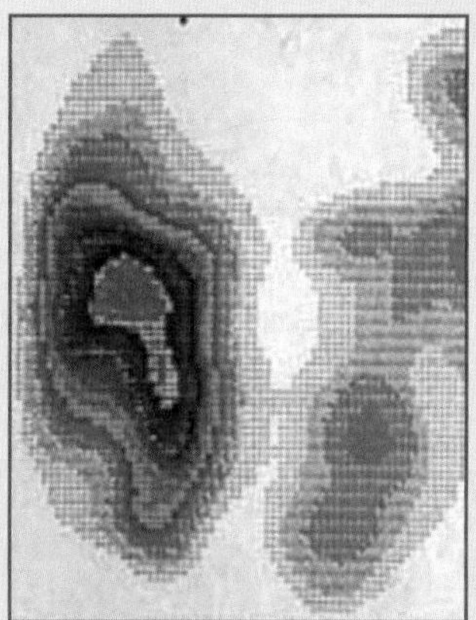

Patient 2

For example, the thyroid gland helps regulate growth. Cells of the thyroid normally absorb iodine from the blood stream (obtained from foods you eat) and use it to make thyroid hormone. If a patient is suspected of having a thyroid disease, he or she may be given a tiny amount of tracer like iodine-131.

7. Look at the images on the left of the thyroid gland of two different patients. The red areas show the most active processing of iodine. Which patient do you think has a normal thyroid and which patient has a diseased thyroid? Why?

Using Sources of Radioactivity

Nuclear reactions release enormous quantities of energy compared to chemical reactions. For this reason, some power plants use radioactive isotopes as fuel, most often uranium-235. Carefully controlled reactions provide electric power in many parts of the world.

Radioactive sources are also used in medicine to destroy cancer cells. In radiation therapy, beams of radiation are focused on the area with cancerous cells. Over time (5 - 7 weeks), this focused radiation damages cells that are in the path of its beam—normal cells as well as cancer cells. But cancer cells are more easily destroyed by radiation, while healthy, normal cells repair themselves and survive.

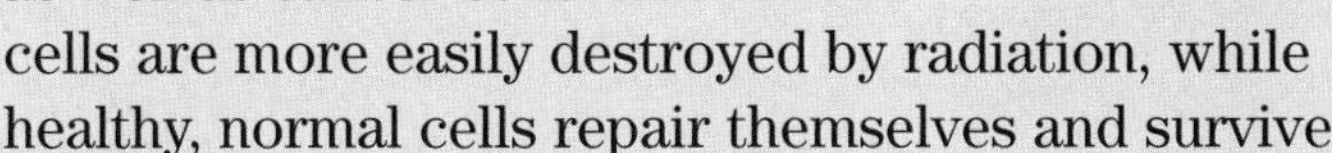

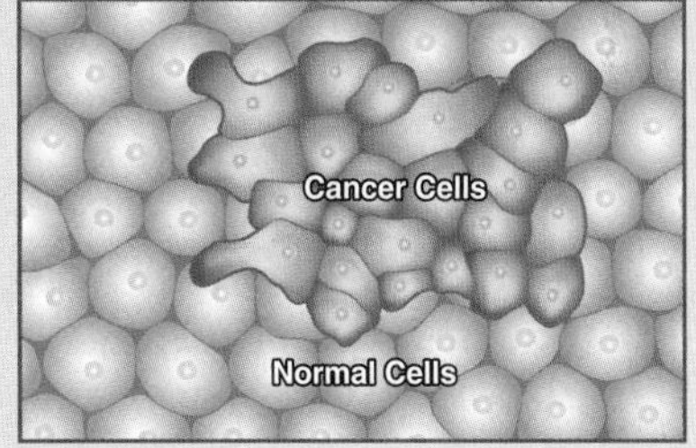

Beams of radiation are used to treat localized solid tumors, such as cancers of the skin, tongue, larynx, brain, and breast. For some cancers, a large dose of external radiation is directed at the tumor and surrounding tissue during surgery. For other types of cancers, a radioactive source is planted in or near a tumor.

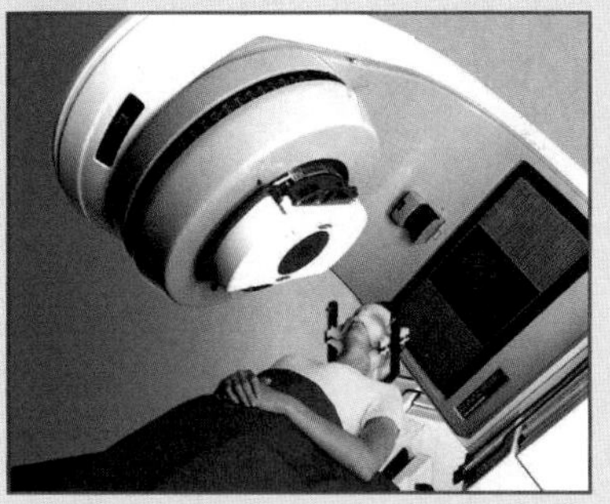

8. Give two examples of how tracers are used. Why do radioactive isotopes work as tracers?

9. Give two examples of how radioactive sources are used.

Your teacher will review answers to Questions 5 through 9 with the class.

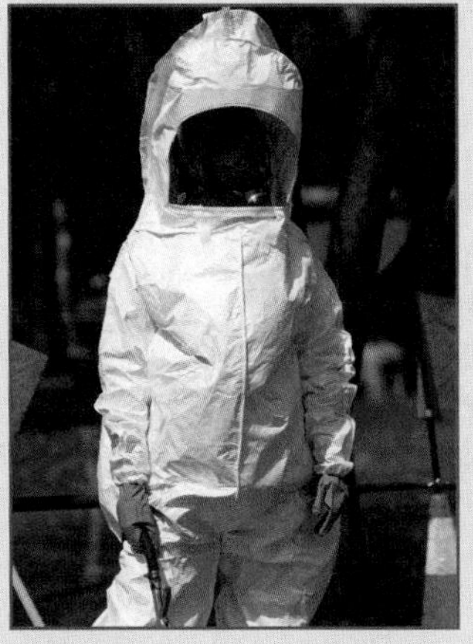

Safe Use of Radioactive Materials

Nuclear reactions have many useful applications. But the radiation from these reactions can be dangerous. Radiation penetrates and damages living cells. Illness, disease, and even death can result from an overexposure to radiation. People who work with radioactive materials must wear protective clothing and use insulating shields.

Radioactive wastes cannot just be thrown away. For example, after radiation therapy, contaminated equipment and clothing can still be hazardous. These items must be disposed of properly. Materials with low levels of radiation may be buried in landfills that are carefully monitored to prevent contamination of the environment.

Radioactive isotopes that decay over hundreds or even thousands of years are the most hazardous. Nuclear power plants produce these kinds of materials. The government is developing procedures for disposing of these materials in specially designed containers buried in underground tunnels about 2000 feet below the Earth's surface.

Radioactive materials are stored in secure containers and buried in a landfill.

Waste Isolation Pilot Plant (WIPP) is a site in New Mexico where the United States government is developing safe storage for radioactive wastes 2150 feet (almost one-half mile) underground. Large rooms house the waste in secure barrels.

What We Have Learned

Think about the key question for this activity:

What are nuclear reactions and radiation?

Participate in a class discussion about the answer to the key question.

Write the answer to the key question on your record sheet.

CHEMICAL INTERACTIONS

What will you learn about in Unit 7?

In Unit 5, you learned about chemical interactions. You learned that a chemical interaction occurs when at least one new substance is made. You know the substance(s) are new because they have a different set of characteristic properties from the original substances.

In Unit 6 you learned about the criteria for a good theory. You recognized that a good theory helps you make sense of (explain) everyday objects and events. You also learned that the ideas in a good theory fit well with (do not contradict) each other and with other scientific laws and theories. You then applied these criteria to the Small-Particle Theory (SPT) of physical interaction.

In this unit, you will continue to apply the criteria of a good theory, but to chemical interactions. You will observe and build different models of what happens during chemical interactions on a scale too small to see, even with the best light microscopes. You will learn how chemists use the Conservation of Mass Law and the Conservation of Energy Law to describe chemical interactions.

Chemical Interactions

The key questions for this cycle of learning are:

1. **What is the scientists' theory of chemical interactions on a scale too small to see?**
2. **How do chemists use their theory to describe the Conservation of Mass and the Conservation of Energy during chemical interactions?**

Three Simple Sugars

UNIT 7: CHEMICAL INTERACTIONS

Key Questions

1. **What is the scientists' theory of chemical interactions on a scale too small to see?**
2. **How do chemists use their theory to describe the Conservation of Mass and the Conservation of Energy during chemical interactions?**

OUR FIRST IDEAS

Activity 1
Small-Particle Theory of Chemical Reactions

Purpose

In Unit 5 you learned about chemical interactions. In Unit 6 you learned how the Small-Particle Theory (SPT) helps you make sense of many physical interactions. In this unit, you will continue to apply the criteria of a good theory, but to chemical interactions.

The key questions for this cycle of learning are:

1. What is the scientists' theory of chemical interactions on a scale too small to see?

2. How do chemists use their theory to describe the Conservation of Mass and the Conservation of Energy during chemical interactions?

Record the key questions for the cycle on your record sheet.

In this activity, you will explore your own ideas as well as the class ideas about what happens to the particles (atoms or molecules) during a chemical interaction.

We Think

About four-fifths of the air is nitrogen and about one-fifth is oxygen. The nitrogen in the air does not usually react with other substances. But at the very high temperatures inside a car engine, the nitrogen interacts with oxygen to produce a new substance, nitrogen monoxide. Nitrogen monoxide is a colorless, poisonous gas.

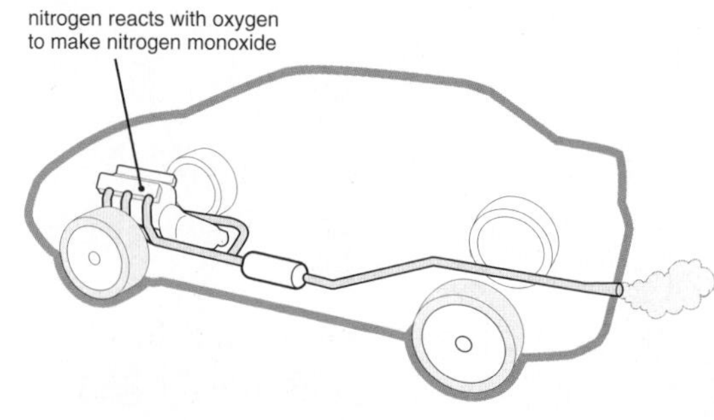
nitrogen reacts with oxygen
to make nitrogen monoxide

Two teams of students came up with different ideas (Idea X and Idea Y) of what happens to the molecules of nitrogen and oxygen when they interact in the car engine. Read the two ideas.

The nitrogen and oxygen molecules are moving very fast in the hot car engine. We think they collide and a new particle of nitrogen monoxide is made. It's as if the two original molecules disappear and change into a completely new particle. But, we know that mass is never created or destroyed, so the mass of the new particle must be the sum of the masses of nitrogen and oxygen molecules.

Idea X

Nadia

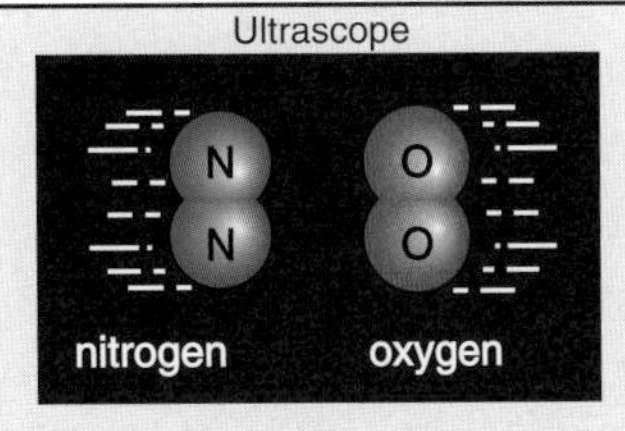

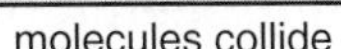

molecules collide

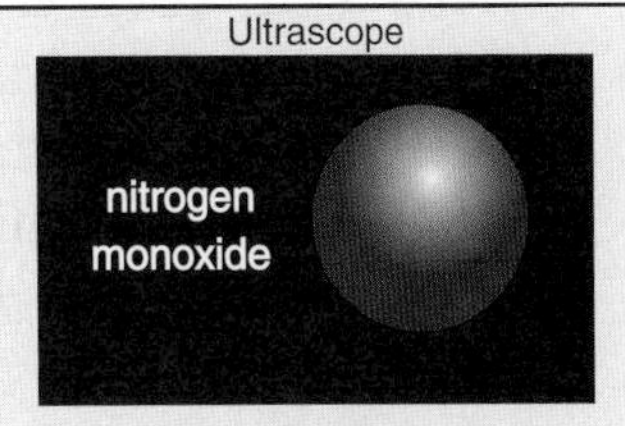

new particle

We also think the nitrogen and oxygen molecules collide. But instead of making a new particle, we think the atoms in the original molecules rearrange into new molecules of nitrogen monoxide. Mass is conserved because the number of atoms of oxygen and nitrogen stays the same.

Idea Y

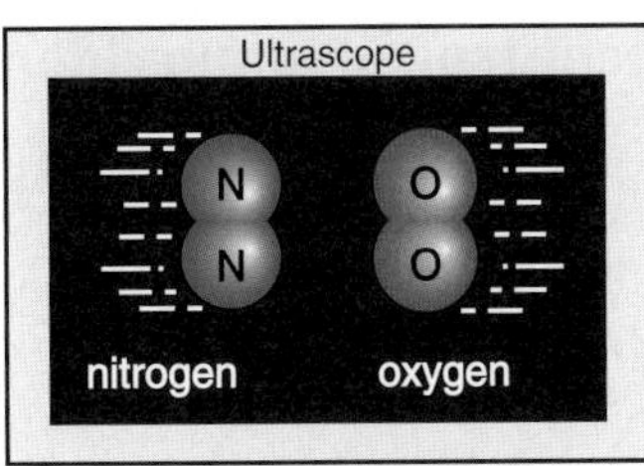

molecules collide

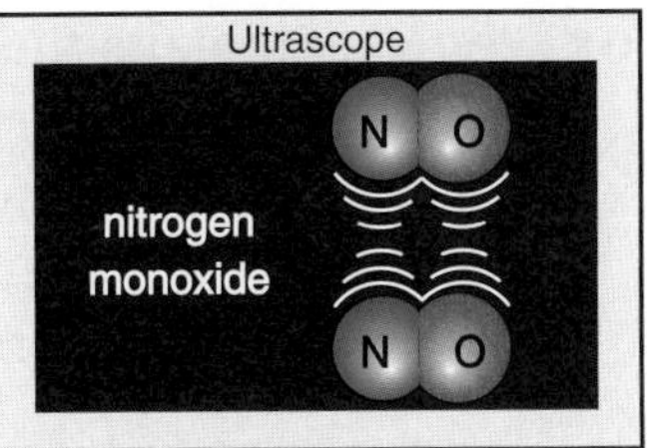

new combination

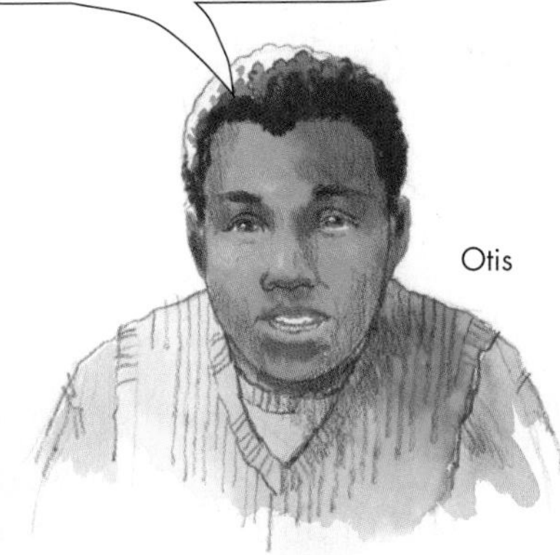

Otis

Discuss the following questions with your team. Be prepared to share your team's answers and reasons in a class discussion.

1. What is the same in both Idea X and Idea Y?
2. What is different in Idea X and Idea Y?
3. Which Small-Particle Theory (SPT) idea, X or Y, do you think is better? Explain your reasoning.

Participate in a class discussion of your answers and reasons.

Our Class Ideas

Record the number of *votes* and *reasons* for each SPT idea of a chemical interaction.

a) _______ students in our class like SPT Idea X better *because* __________.

b) _______ students in our class like SPT Idea Y better *because* _________.

The Language of Chemistry

In a chemical interaction, at least one new substance is made from one or more original substances. Chemists have developed a shorthand language for talking about chemical interactions. Chemists call chemical interactions *reactions*. Instead of using the phrase *interacts chemically*, they just use the word *reacts*. For example, chemists would say: "Nitrogen gas reacts with oxygen gas to make the gas nitrogen monoxide."

Science Words

reactant: original substance in a chemical reaction

product: end (new) substance made in a chemical reaction

Chemists also have special words for the original substances and the new substances produced in a reaction (chemical interaction).

- The original substance(s) are called the *reactant(s)*.
- The end substances (new substances formed) are called the *product(s)*.

For the reaction you considered, the reactants (original substances) are the gases nitrogen and oxygen. The product (new substance) of the reaction is the gas nitrogen monoxide.

Nitrogen gas reacts with **oxygen** gas to make the gas **nitrogen monoxide**.

reactants **product**

Your teacher will show you the video of testing the hydrogen gas from the electrolysis experiment. When the mouth of the hydrogen test tube is brought near a flame, hot hydrogen and oxygen gas mix. The pop you hear is evidence of a reaction between the oxygen and hydrogen, making water (new substance).

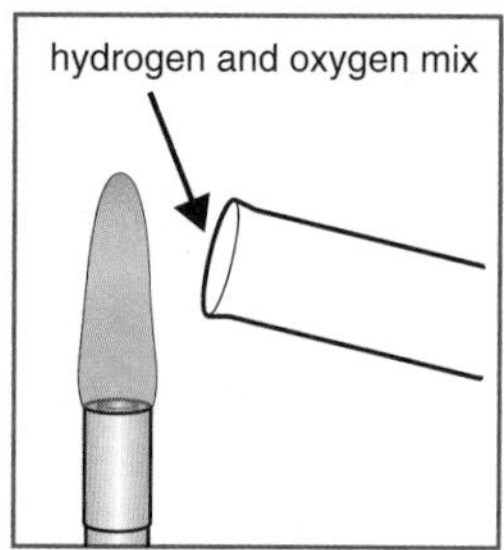
hydrogen and oxygen mix
Hydrogen gas reacts with oxygen gas to make water vapor.

1. What are the reactants in this chemical reaction?

2. What is the product of this chemical reaction?

DEVELOPING OUR IDEAS

Activity 2
What's the Evidence?

Purpose

In the last activity, you learned that chemists call a chemical interaction a chemical reaction. The original substances in a reaction are called the *reactants*. The new substances produced in the reaction are the *products*. You also learned in Unit 5 the kinds of observations that can be evidence of a chemical reaction.

In this activity, you will investigate the chemical reaction called burning. The key question for this activity is:

What are the reactants and products in the burning reaction, and what is the evidence for the reaction?

Record the key question for the activity on your record sheet.

We Think

In Unit 5, Activity 4, you did an exploration with a burning match. In this activity, you will light a birthday candle inside a glass jar, then place a lid on the jar so there is a good seal.

Discuss the following questions with your team.

1. What do you think are the reactants (starting substances) in the burning reaction of a birthday candle? What are your reasons?
2. What do you think are the products (new substances) of this reaction? What are your reasons?
3. Suppose you put the sealed jar with the burning candle on a mass balance. Do you think you would observe a change in mass as the candle burns? Why or why not?

Participate in a class discussion about your predictions and reasons.

Explore Your Ideas

Your team will need:

- birthday candle
- 500 mL (16 oz) specimen jar
- small ball of clay
- container of bromothymol blue (BTB) indicator
- long matches

You will first observe what happens when you burn a candle inside a closed jar. Then you will test to see if a new gas is produced. You will use the same indicator, bromothymol blue (BTB), that you used in Unit 5, Activity 2. Your teacher will also do a demonstration experiment to determine whether the mass of the sealed jar changes while the candle burns inside the jar.

STEP 1 Observe the properties (what you see, feel, and smell) of candle wax.

Record the properties of the candle wax in the "before" column of the data table on your record sheet. Also, write down some properties of air.

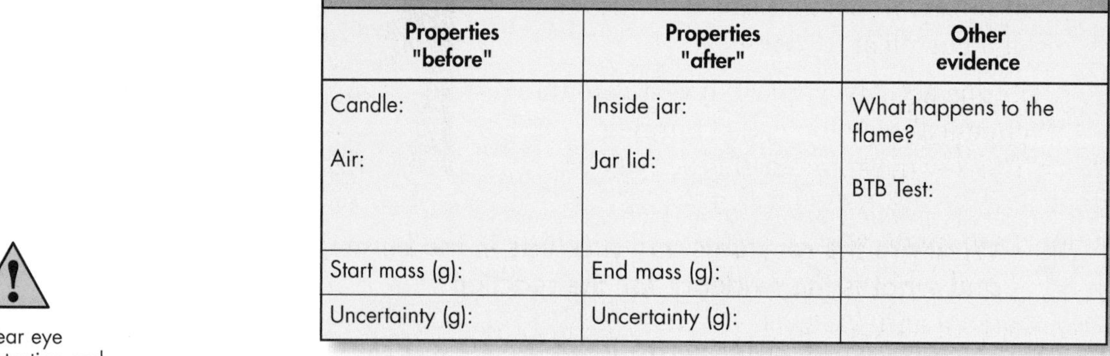

Table: Burning a Candle inside a Closed Jar		
Properties "before"	**Properties "after"**	**Other evidence**
Candle: Air:	Inside jar: Jar lid:	What happens to the flame? BTB Test:
Start mass (g):	End mass (g):	
Uncertainty (g):	Uncertainty (g):	

Wear eye protection and lab apron.

Tie back long hair and loose sleeves before lighting candle.

Alert your teacher immediately if you break any glass.

Allow jar to cool before removing lid.

STEP 2 *Procedure Specialist:* Place the candle in the middle of the ball of clay. Make sure that the candle is firmly in place so it will not tip over. Then place the candle in the middle of the jar.

STEP 3 *Manager:* Light the candle with a long match. Quickly tighten the lid on the jar. Observe any changes happening in the jar. What happens to the flame? What happens to the inside of the jar? You may need to experiment a few times to see what happens to the inside of the jar.

Record your observations in the "Other evidence" column of your data table.

STEP 4 When the burning reaction has stopped, open the jar. Observe the properties (what you see, feel, and smell) of the substances left in the jar. What are the properties of the stuff *on the inside walls of the jar*? What are the properties of the stuff *on the lid of the jar*?

Record these properties in the "after" column of your table.

STEP 5 *Recycling Engineer:* Take the candle out of the jar. Clean the inside of the jar and lid with a paper towel.

STEP 6 *Supply Master:* Measure 10 mL of the BTB indicator. Pour the BTB into the clean jar. Then place the candle inside the jar.

Do not touch, smell, or taste the BTB.

Be careful; BTB can stain skin and clothing.

If you accidentally touch BTB, wash well with water.

Clean up spills immediately.

Dispose of BTB according to teacher's directions.

Wash hands after activity.

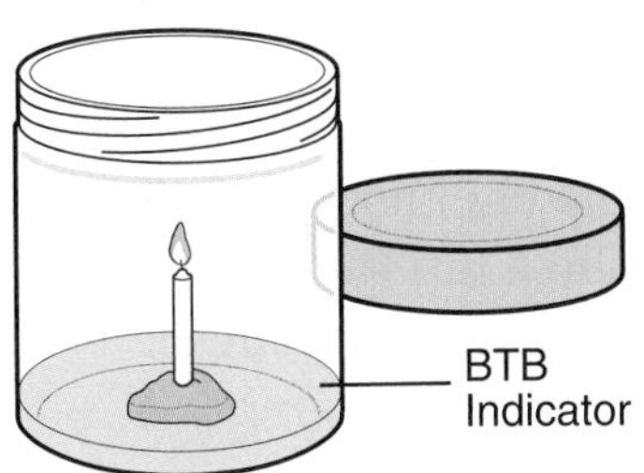

STEP 7 *Team Manager:* Light the candle with a long match. Quickly tighten the lid on the jar.

STEP 8 Wait until the candle has gone out and the jar is cool. *Recycling Engineer:* Hold the jar by the lid. Move jar gently in a circular motion so the BTB swirls around.

Observe what happens to BTB. Does it stay the same color or change color?

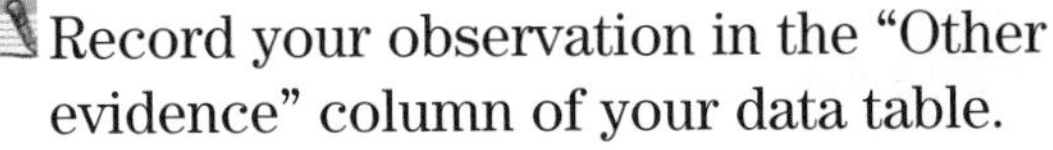

Record your observation in the "Other evidence" column of your data table.

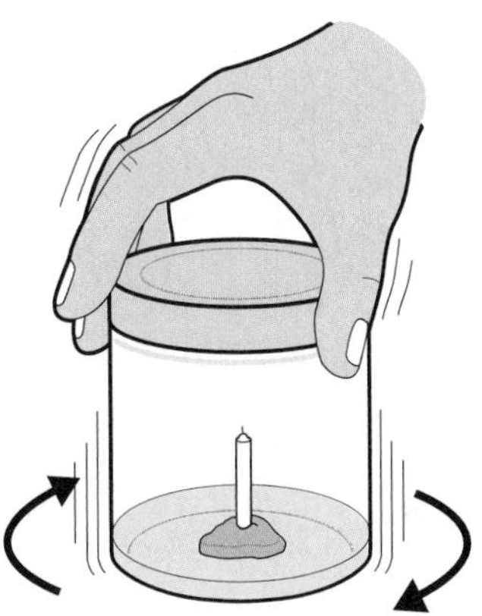

STEP 9 Your teacher will do a demonstration to measure the start mass and end mass of the burning candle in the jar.

Record the start mass and end mass in your data table.

Participate in a class discussion about your observations.

Make Sense of Your Ideas

1. What evidence do you have that an invisible gas in the air is one of the *reactants* in the chemical interaction of burning? What do you think this invisible gas is?

2. What do you think are the products of this reaction? What is your evidence? (*Hint*: What substance do you think collected on the inside walls of the jar? What substance do you think collected on the inside of the lid? Also, look at Table 1 on your record sheet for Unit 5, Activity 2. What gas changes the color of BTB?)

3. Was any energy transferred during the burning interaction? What form of energy? Was it transferred into or out of the burning candle-air system? What is your evidence?

Participate in a class discussion to summarize the answers to these questions.

Make Sense of Scientists' Ideas about Burning

Candle wax and wood are two substances in a large category of fuels called *hydrocarbons*. Hydrocarbons are compounds that are made up of carbon and hydrogen. Candle wax and wood are made up of long chains of carbon atoms with attached hydrogen atoms.

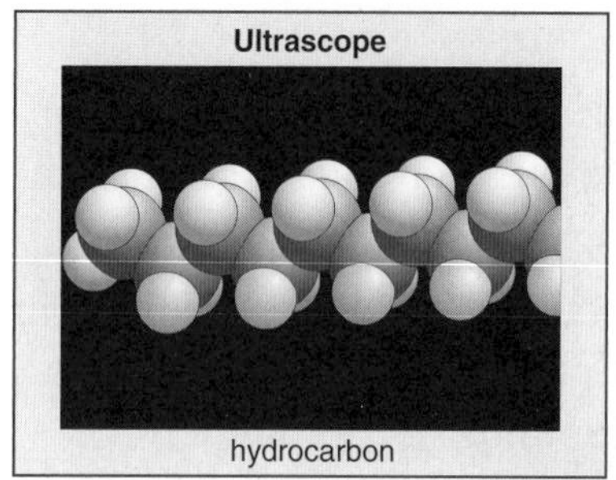
Ultrascope
hydrocarbon

There are many, many reactions occurring in a wood or candle flame, and scientists do not fully understand all the chemistry of a flame. But they do know quite a bit. Here is a simplified summary for a candle flame.

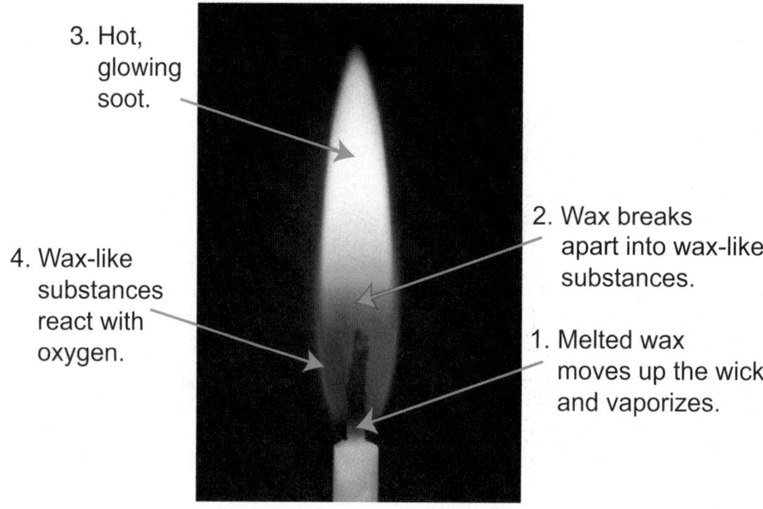
3. Hot, glowing soot.
2. Wax breaks apart into wax-like substances.
4. Wax-like substances react with oxygen.
1. Melted wax moves up the wick and vaporizes.

1. As the wax at the top of the candle melts, it moves up the wick and vaporizes (turns into a gas). The wax gas moves into the area surrounding the wick.
2. There is not much oxygen near the wick, where the flame appears dark red or orange. In most of the reactions, the wax breaks apart into other wax-like substances.
3. One of these substances is soot. This is the black material you saw on the lid of your jar. It is a carbon-rich substance. The hot soot glows in the yellow part of the flame, giving off light like the filament of a bulb. Some of the soot escapes.
4. Near the outside, blue part of the flame, the oxygen can enter from the surrounding air. Some of the wax-like substances react with the oxygen, eventually making water and carbon dioxide.

1. Complete the following description of the burning candle reaction on your record sheet.

Candle wax reacts with ______________ to produce soot, ________________, and ________________.

Not all hydrocarbon fuels are long-chain molecules. The lightest hydrocarbon is *methane* (natural gas). Methane is the fuel burned in the stoves and furnaces of many homes. Methane burns with a blue flame, indicating that it burns *cleanly*. This means that there is no soot produced.

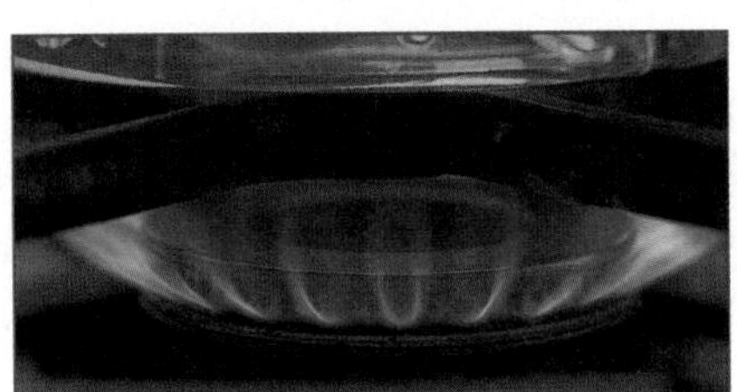

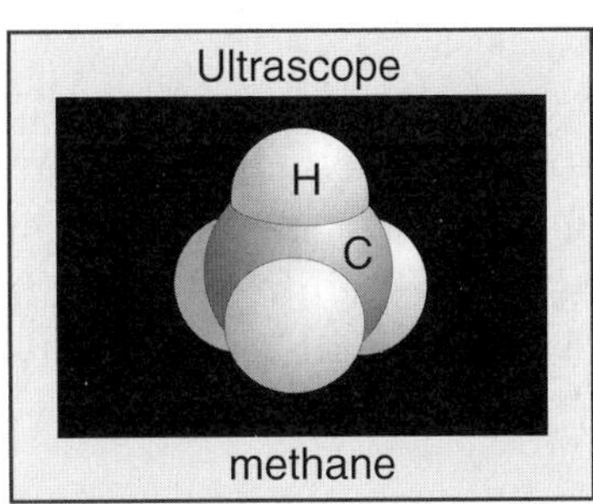

2. Complete the following description of the burning methane reaction on your record sheet.

   Methane reacts with ______________ to produce ______________ and ______________.

Another hydrocarbon gas you may be familiar with is butane. Butane is the gas burned in many camp stoves, laboratory burners, and disposable lighters. An Ultrascope diagram of the butane molecule is shown on the right. Butane also burns cleanly (no soot produced).

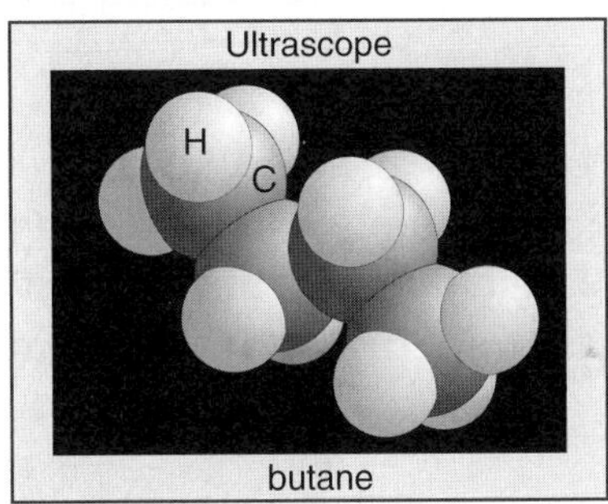

3. Complete the following description of the burning butane reaction.

   Butane reacts with ______________ to produce ______________ and ______________.

Our Consensus Ideas

The key question for this activity is:

What are the reactants and products in the burning reaction, and what is the evidence for the reaction?

1. Write your answer on your record sheet.

Participate in the class discussion about the key question.

2. Write the class consensus ideas on your record sheet.

MAKING SENSE OF SCIENTISTS' IDEAS

Activity 3
Particle Interactions

Purpose

Earlier you considered two ideas about what happens to the particles of the reactants during a chemical reaction (Activity 1). The reactant particles must collide with each other in both ideas. One idea (Idea X) proposes that the two particles then change into a completely new particle, as shown in the diagram. The new particles explain why the product has different properties from the original reactants.

Idea X

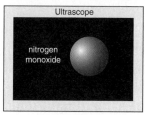

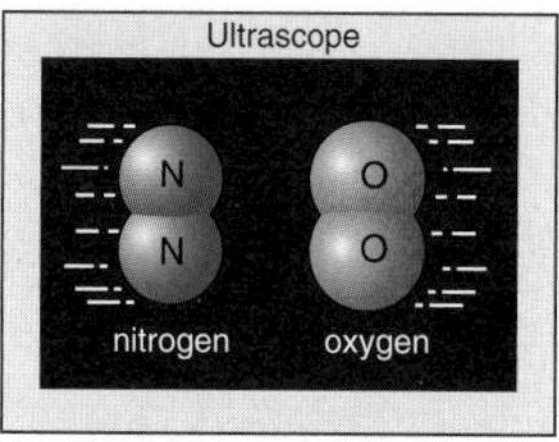

molecules collide

Ultrascope

nitrogen monoxide

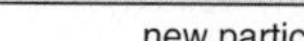

new particle

The second idea (Idea Y) proposes that the atoms rearrange into a new combination (molecule) after the reactant particles collide. The atoms keep their original identity, but the new combination (molecule) accounts for the different properties of the new substance.

Idea Y

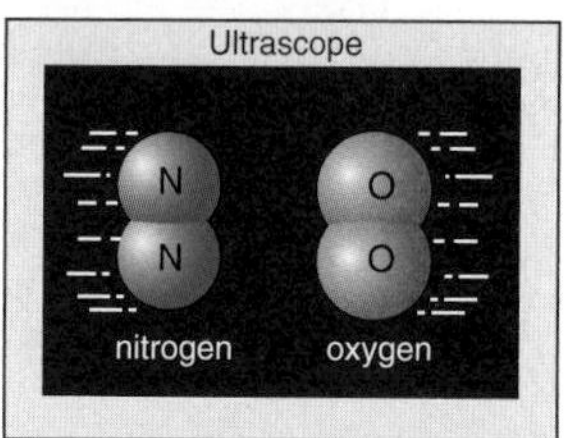

molecules collide

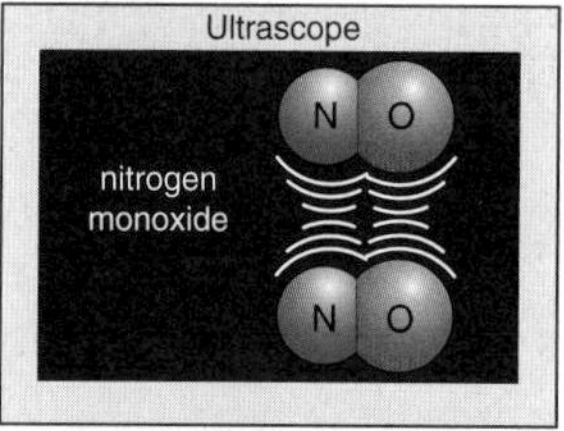

new combination

In this activity, you will learn which idea is better. The key question for this activity is:

What happens to the particles (atoms or molecules) of the reactant(s) during a chemical interaction?

First, you will heat a substance we will call "yellow goop" and observe what happens in the chemical reaction. Then you will use the results of your exploration to make sense of the Small-Particle Theory (SPT) ideas of what happens to the particles of yellow goop (reactant) during the chemical reaction.

Explore Your Ideas

This exploration with yellow goop has two parts. First, you will heat a small amount of yellow goop on a charcoal stick and observe what happens. Then you will heat a small amount of yellow goop by itself.

STEP 1 Take turns looking at the yellow goop in the first test tube with your magnifying glass.

Record your observations of some properties of the yellow goop in the first column of a data table like this one.

Table: Heating Yellow Goop

Properties before heating	Properties after heating
Yellow goop:	With a charcoal stick: By itself:

Your team will need:

- 2 stoppered test tubes, each with pea-sized pieces of yellow goop
- test-tube holder
- matches
- charcoal stick
- alcohol or butane burner
- tweezers
- disposable gloves
- large beaker or test-tube rack
- magnifying glasses

Heating Yellow Goop with Charcoal

STEP 2 *Supply Master:* Pick up the charcoal stick with the tweezers and drop it into the moist yellow goop at the bottom of the test tube. If the yellow goop is too dry to cling to the stick, moisten it with a little distilled water.

STEP 3 *Procedure Specialist:* Light your burner. If you have a butane burner, adjust the burner for a cool flame.

STEP 4 *Manager:* Hold the test tube with the test-tube holder and carefully heat the test tube over the flame until the yellow goop seems to disappear.

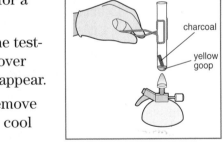

After the yellow goop appears to be gone, remove the test tube from the heat. Let the test tube cool in a large beaker or rack.

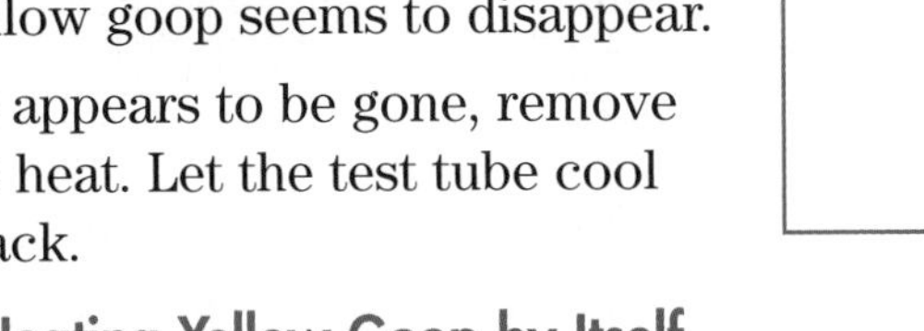

charcoal
yellow goop

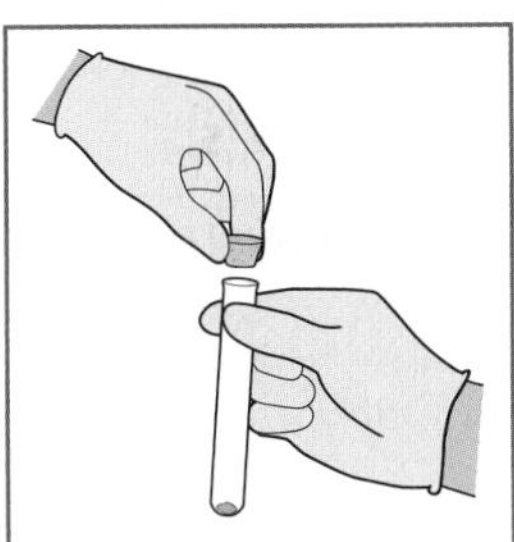

Heating Yellow Goop by Itself

STEP 5 *Recycling Engineer:* Gently place a stopper into the second test tube containing the pea-sized piece of yellow goop. Do *not* press the stopper into the test tube. The stopper should be very loose.

Wear eye protection, disposable gloves and lab apron.

Do not touch, smell, or taste *any* of the materials in this activity.

Alert your teacher immediately if you break any glass or spill any materials.

Tie back long hair and loose sleeves before lighting burner.

Do not leave burner unattended.

Use tongs for warm test tube.

Let test tube cool in rack or beaker.

Do not press the stopper into the test tube tightly.

Carry out in fume hood if available.

Remove tube from heat as soon as purple vapor appears.

Dispose of materials and clean glassware according to teacher's direction.

Do not open the sealed containers of the samples of lead and iodine.

Wash hands after activity.

STEP 6 *Procedure Specialist:* Hold the test tube with the test-tube holder and carefully heat the test tube over the flame. Heat the bottom and sides of the test tube just until you see a purple gas forming inside the test tube. Then immediately remove the test tube from the flame and turn off the burner.

STEP 7 Take turns holding the test tube against a white background, as shown in the diagram.

Describe what you see inside the test tube in your data table. What do you think formed when the yellow goop was heated?

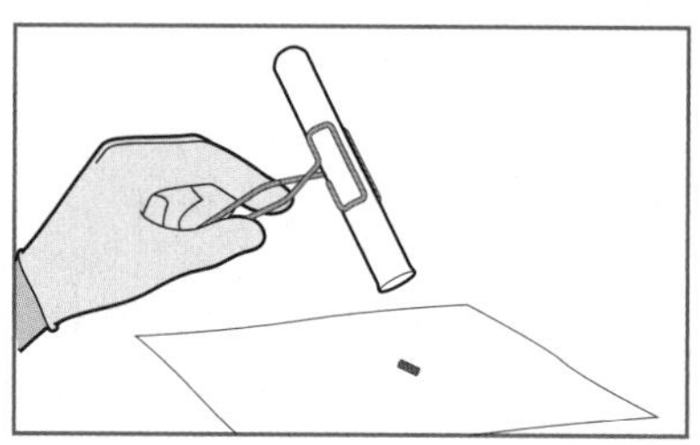

STEP 8 *Supply Master:* When the first test tube is cool, shake the test tube gently until you can see the charcoal stick. Take turns looking at the charcoal stick with your magnifying glass. Then take turns looking at the bottom of both test tubes with the magnifying glass.

In your data table, describe the appearance of the charcoal stick and the bottom of both test tubes. What substance do you think formed on the stick and the bottom of the test tubes?

STEP 9 Your teacher will distribute sealed samples of lead and iodine for you to look at. Lead is a solid metal element (symbol Pb) with a silver color and metallic shine. Iodine is a solid nonmetal element (symbol I). Iodine turns from a solid into a gas (sublimates) when it is heated a little bit.

1. Compare the properties of lead to the properties of the substance on the charcoal stick and the bottom of the test tubes after heating.
2. Compare the properties of iodine to the gas inside the test tube when the goop was heated by itself.

Make Sense of Your Ideas

Examine the results of your exploration with yellow goop. Discuss the following questions with your team and record your answers.

1. Was any energy transferred during the yellow-goop interactions? What form of energy? Was it transferred into or out of the yellow-goop system?
2. What evidence do you have that chemical reactions occurred?
3. What do you think are the products (ending substances) of these reactions? What is your evidence?

Participate in a class discussion about the answers to these questions.

Make Sense of SPT Ideas

The key question for this activity is: *What happens to the particles (atoms or molecules) of the reactant(s) during a chemical interaction?*

You looked at two possible ideas to answer this question. *Both ideas are reasonable. You do not have enough evidence to decide one way or the other at this point.*

It took many years for chemists to gather convincing experimental evidence about which idea is better. The results of their experiments support the idea that atoms are *not* created or destroyed during a chemical interaction. The atoms in the particles of the reactant(s) *recombine* to form the new particles (atoms or molecules) of the products, as in Idea Y.

Idea X

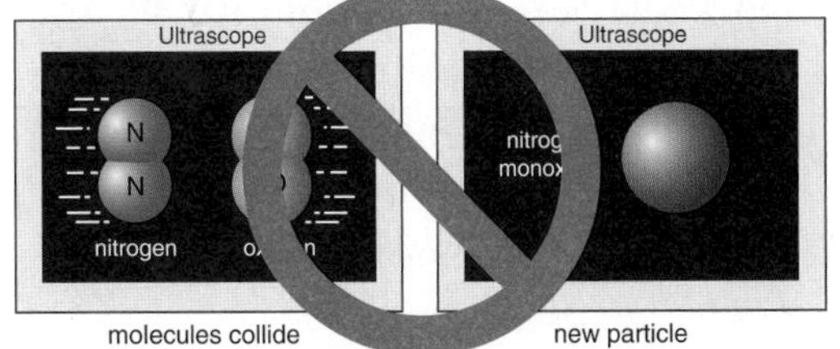

molecules collide

new particle

Idea Y

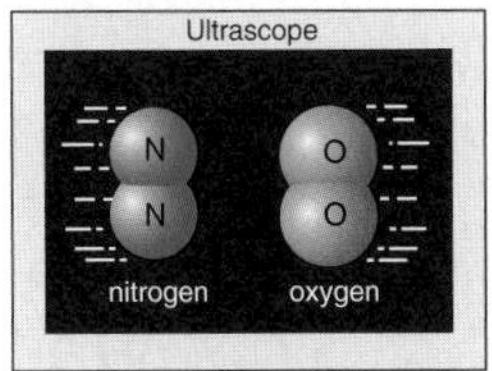

molecules collide

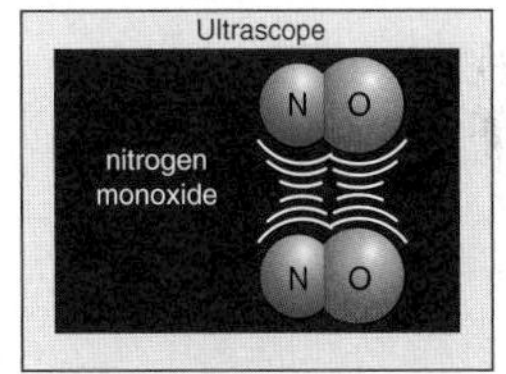

new combination

Let's see how this recombination of atoms works for the yellow-goop reaction.

In Unit 6, you learned about the cohesion attraction between the molecules of substances that are gases, liquids, and solids near room temperatures. You also learned about the different and stronger attraction between the neighboring atoms of a metal element, and the neighboring ions of different elements in ionic compounds (SPT Idea 7F). There is a third type of attraction between neighboring atoms. This is the attraction between the atoms in a molecule. Chemists call these three types of attractions between neighboring atoms *chemical bonds.*

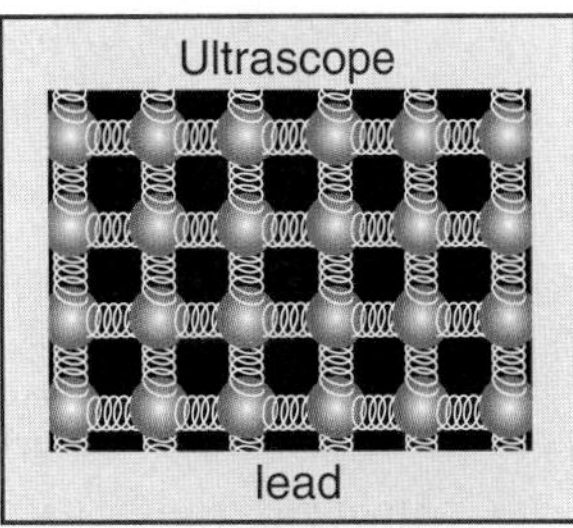

lead

You also learned in Unit 6 that a useful analogy of bonds (attractions) between atoms are stretched/compressed spring. For example, the Ultrascope diagram shows the chemical bonds between the atoms of solid lead.

The chemical bonds between the atoms in a molecule are also represented with stretched springs. The Ultrascope diagram of the iodine molecule shows the chemical bond between the two iodine atoms in an iodine molecule. You will learn more about the chemical bonds between atoms in a molecule in the next activity.

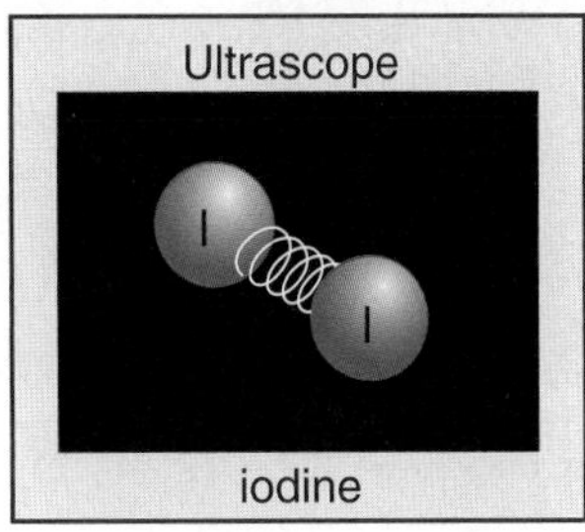

iodine

The chemical name for the yellow goop you used is lead iodide. Two iodine atoms (I) are bonded to a central lead atom (Pb). The chemical bonds between the lead and iodine atoms in a lead iodide molecule are shown in the Ultrascope diagram.

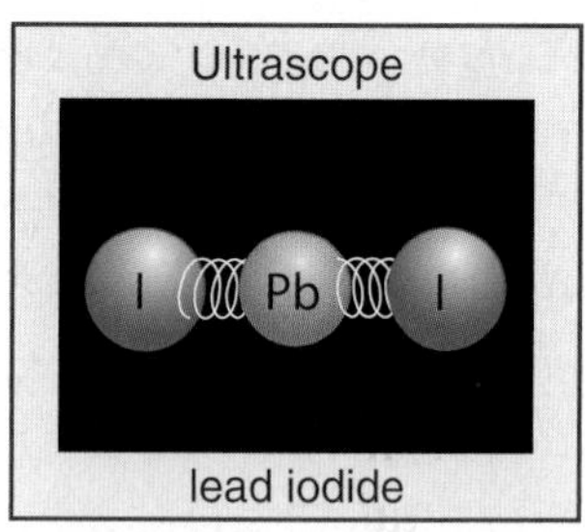

So, how are the chemical bonds in the lead iodide molecule broken and new chemical bonds between the iodine atoms and the lead atoms formed?

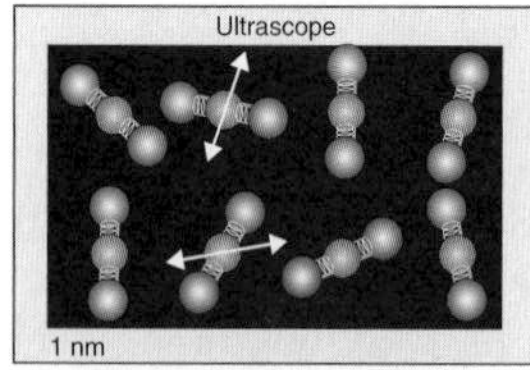

In the solid phase, the lead iodide molecules vibrate back and forth in their locked positions (SPT Idea 7F). The molecules occasionally collide with their neighbors, but nothing happens to them. They just bounce apart.

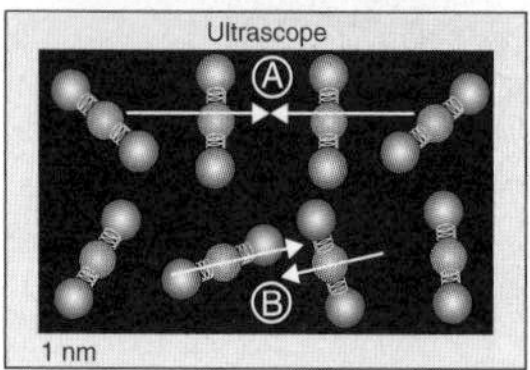

You had to heat the lead iodide in a test tube for a chemical reaction to begin. As the solid is heated, the motion energy and average speed of the molecular vibrations increase (SPT Idea 8A). Because the molecules are moving faster, they collide with their neighbors more often. The Ultrascope picture shows two collisions, A and B.

Study the two Ultrascope diagrams below showing collision A. This results in a new combination of atoms.

(1) Two lead iodide molecules collide.

(2) The lead iodide molecules are moving in the right direction and have enough motion energy when they collide to *break the bonds* (attractions) between the lead atoms and the iodine atoms in both molecules. Chemists describe this process as *breaking the chemical bonds* (attractions) between the atoms of the reactant particles.

Ultrascope Diagram of Collision A: New Combination of Atoms

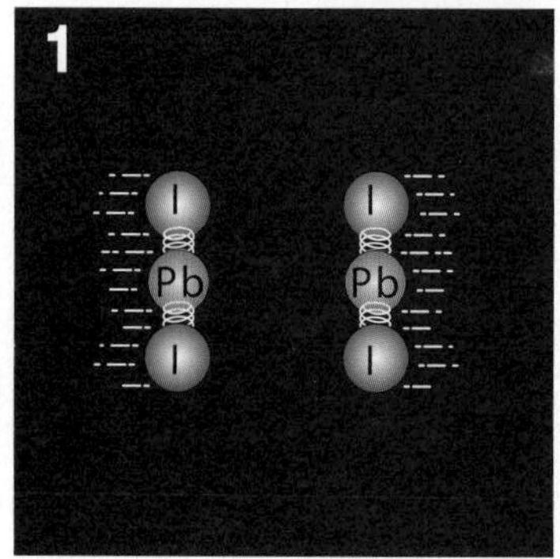

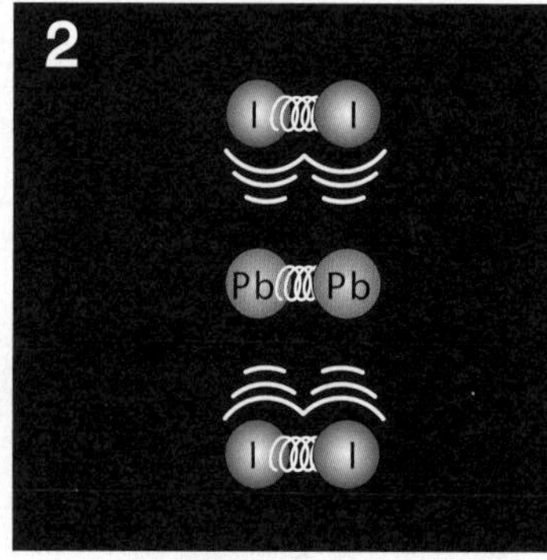

New chemical bonds (attractions) are formed between the iodine atoms. A new chemical bond (attraction) is also formed between the two lead atoms. Chemists call this process *making new chemical bonds* between the atoms in the product particles.

Now imagine that this kind of collision (breaking and making bonds) occurs millions of billions of times as you gently heat the yellow goop in a test tube. Eventually, there are enough iodine molecules in the test tube for you to see the characteristic purple iodine gas. Enough lead atoms collect on the charcoal stick for you to see the characteristic, metallic shine of the metal.

1. How many chemical bonds are broken during the collision of the two reactant molecules of lead iodide?
2. How many new chemical bonds are made in the products (solid lead and iodine gas)?

Look at the diagram showing a collision between nitrogen and oxygen gas molecules. This results in a recombination of the atoms into nitrogen monoxide molecules.

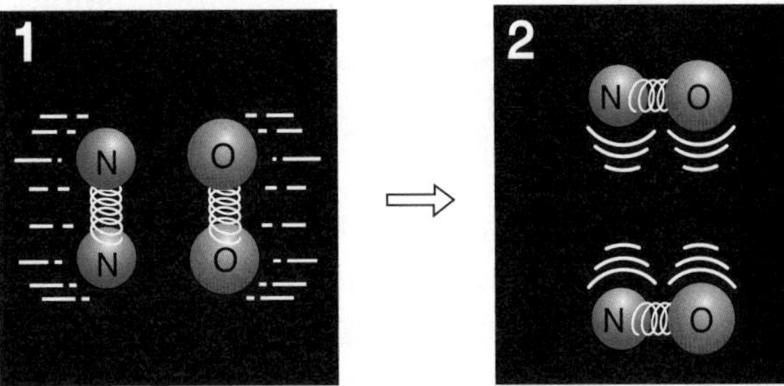

3. Describe the chemical reaction in terms of the energy of motion of the reactant nitrogen and oxygen molecules, the breaking of chemical bonds, and the making of new chemical bonds in the product (nitrogen oxide).

Participate in a class discussion about the answers to these questions.

Of course, not every collision between neighboring lead iodide molecules results in a new combination of atoms. For some collisions, the molecules are not moving in the right direction with enough motion energy to break the bonds (attraction) between the lead and iodine atoms. This situation is shown in the diagram below of collision B. In these cases, the molecules bounce apart. As you continue to heat the lead iodide, more and more molecules have enough energy to break apart when they collide.

Ultrascope Diagram of Collision B: Molecules Bounce Apart

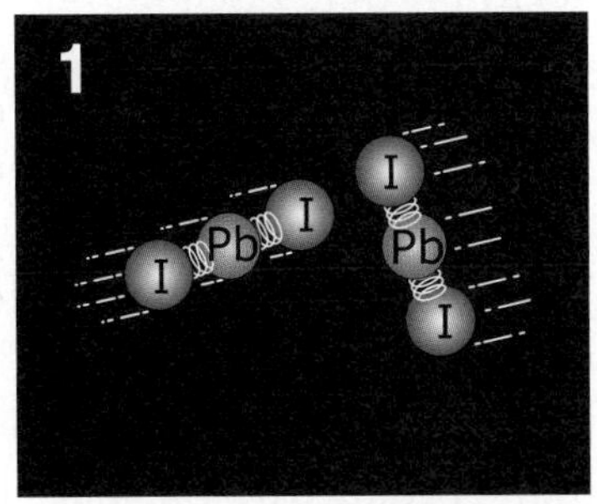

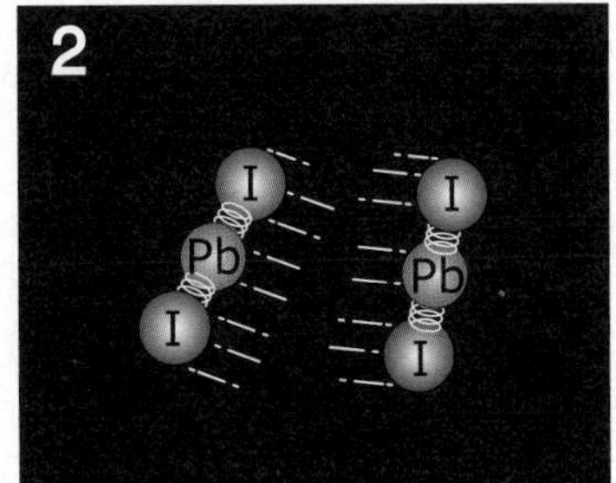

Chemical Formulas

Chemists use atomic symbols to write a shorthand representation of a molecule of a compound. This shorthand representation is called a chemical formula. For example, the chemical formula for lead iodide is shown in the diagram.

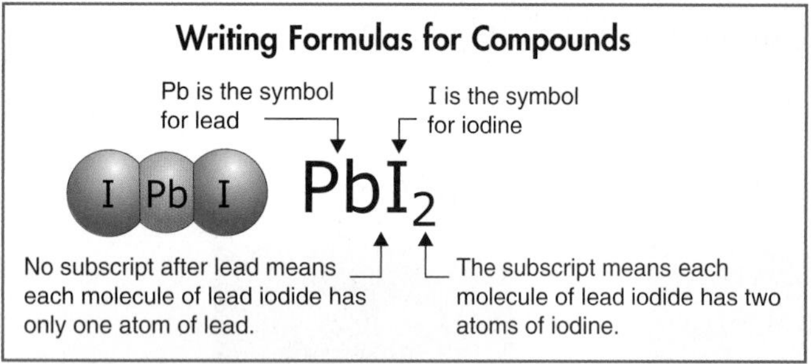
Writing Formulas for Compounds
Pb is the symbol for lead
I is the symbol for iodine
I Pb I
PbI2
No subscript after lead means each molecule of lead iodide has only one atom of lead.
The subscript means each molecule of lead iodide has two atoms of iodine.

The numbers in the formulas, called *subscripts*, show the number of each kind of atom in the compound. When there is more than one atom of an element, a subscript is written to the right and below the element's symbol. If there is only one atom of an element, no subscript is written.

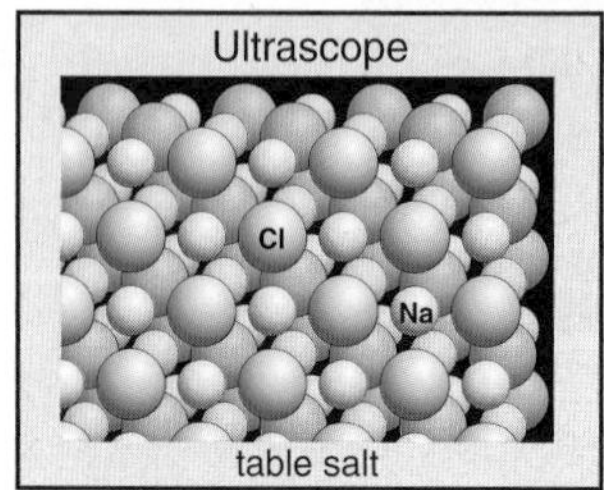
Ultrascope
Cl
Na
table salt

As you know, there are no molecules of ionic compounds. Instead, the charged atoms (ions) are arranged in a large array. An example of the array for table salt (sodium chloride) is shown in the diagram. If there are no molecules, what could be the chemical formula for an ionic compound like table salt?

Chemists use special equipment to determine the number of ions of each element in an ionic crystal. All sodium chloride crystals have the same number of sodium ions as chlorine ions. Chemists write the formula for table salt as NaCl. The formula shows that there is one sodium ion for every chlorine ion.

Our Consensus Ideas

Consider again the key question for this activity:

What happens to the particles (atoms or molecules) of the reactant(s) during a chemical interaction?

To answer this question, complete the following on your record sheet.

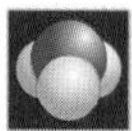

Idea 10 Collision Interaction – Atoms in Particles Recombine

B. For many chemical interactions to occur, at least two reactant particles (atoms, ions, or molecules) must _______________.

1. When the reaction particles collide, the chemical bonds between the atoms of the reactant particles break. Chemists describe this process as __________ between the atoms of the reactant particles.
2. The atoms ____________________ and new bonds are formed to produce the different particles of the product(s). Chemists call this process ____________________ of the product particles.

Participate in the class discussion about the key question.

MAKING SENSE OF SCIENTISTS' IDEAS

Activity 4
Small-Particle Theory of Stored Chemical Energy

Purpose

Look at your *Energy of System* wall chart. In previous *InterActions* units, you explored different types of energy that an object can have: motion energy, stored elastic energy, stored chemical energy, thermal energy, stored volume energy, and stored phase energy. You learned in Unit 6 that thermal energy is related to the average motion energy of the billions of trillions of the atoms or molecules that make up everyday objects. Stored volume energy is the sum of the stored cohesive bond energies of the particles of a substance (gas, liquid, or solid).

In this activity, you will answer the following key questions:

1. What are chemical bonds and chemical-bond energies?

2. What is the Small-Particle Theory (SPT) of the stored chemical energy of a substance?

We Think

You have encountered three types of situations that involve stored chemical energy. Discuss the following situations with your team.

- A student pulls a skateboarder faster and faster.
- A battery and a light bulb are connected in an electric circuit.
- You heat some water with an alcohol burner.

How is stored chemical energy involved in each situation? Draw an energy diagram for each situation.

Participate in a class discussion.

Make Sense of SPT Ideas about Chemical Bonds

You learned in Activity 3 that a *chemical bond* is the name chemists give to the attractions between neighboring atoms of a substance. Chemists use the term *chemical* bonds because these bonds between atoms break apart during chemical reactions. There are three different types of chemical bonds: metallic, ionic, and covalent.

Science Words

metallic bonds: attractions between neighboring metal atoms

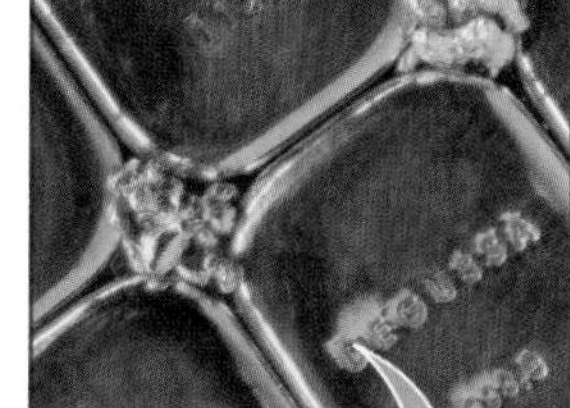

Metallic Bonds

You learned in Unit 6 that the smallest particles of solid metals are atoms. Like all solids, the atoms are arranged in a regular, repeating pattern. The diagram shows an Ultrascope model of gold (Au) atoms.

Metallic bonds are the attractions between neighboring metal atoms. One model of a bond is a stretched spring. The spring model shows some (not all) of the bonds between neighboring gold atoms.

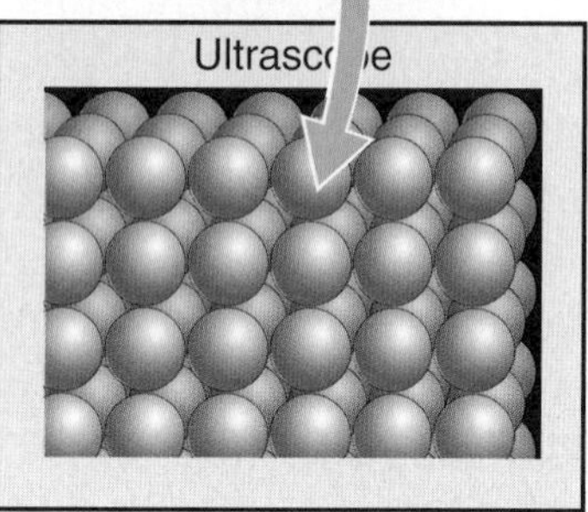

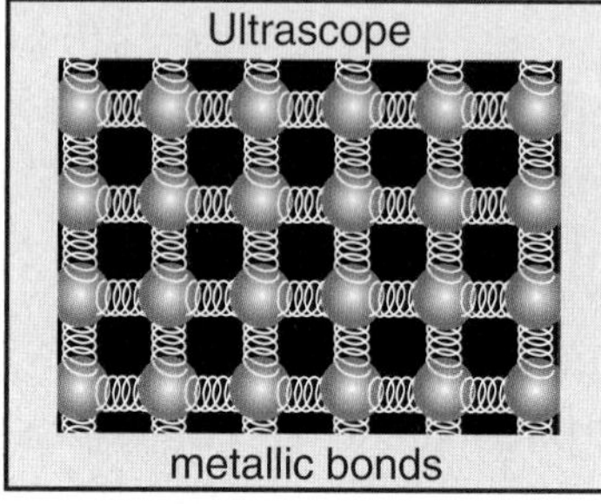

On your record sheet, write a sentence defining metallic chemical bonds in a concept map like the one shown here.

Participate in a class discussion about your answer.

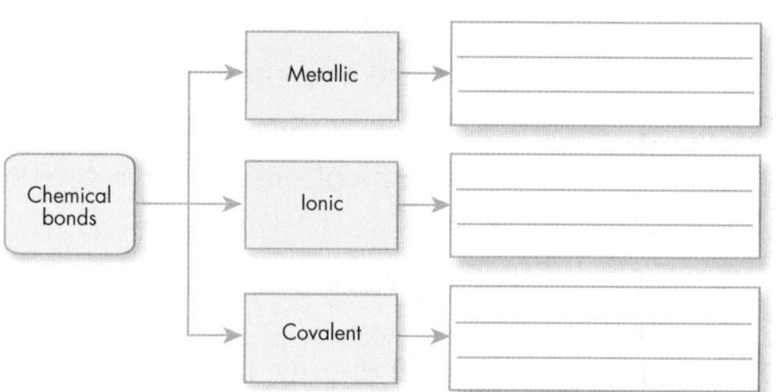

Ionic Bonds

You also learned in Unit 6 that the smallest particles of solid ionic compounds are ions (charged atoms). All ionic compounds consist of a positive metal ion and a negative nonmetal ion.

The ions are arranged in a regular, repeating pattern. Look at the Ultrascope diagram of common table salt, sodium chloride (NaCl). The metal sodium ions (Na^{+}) are positively charged, and the nonmetal chlorine ions (Cl^{-}) are negatively charged.

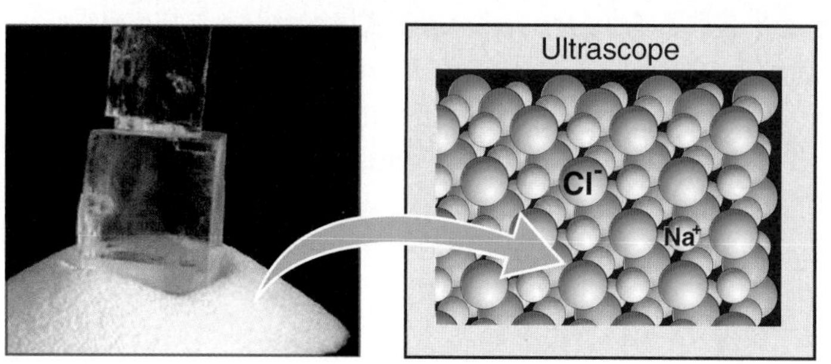

Ionic bonds are the attraction between neighboring nonmetal and metal ions. Notice that the closest neighbors of each ion (Na^+ and Cl^-) are the ions of the opposite charge. The spring image shows only *some* of the ionic chemical bonds (attractions) between neighboring ions.

Science Words

ionic bonds: the attraction between neighboring nonmetal and metal ions

Write a sentence defining ionic chemical bonds in your concept map.

1. What is similar about metallic bonds and ionic bonds? What is different?

Participate in a class discussion about your answers.

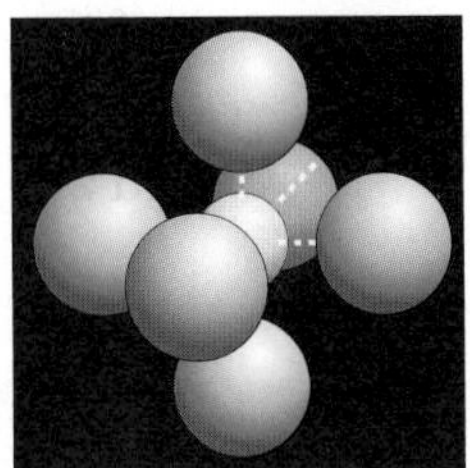

Na^+ with Cl^- neighbors

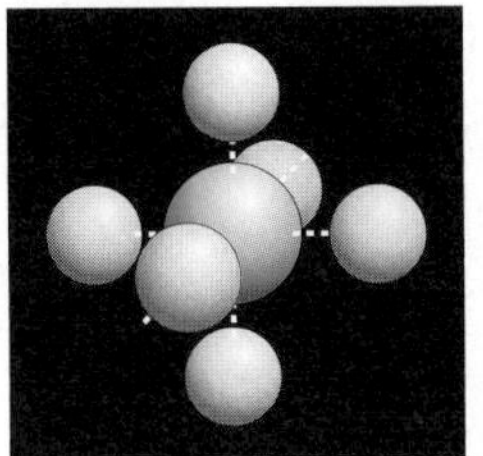

Cl^- with Na^+ neighbors

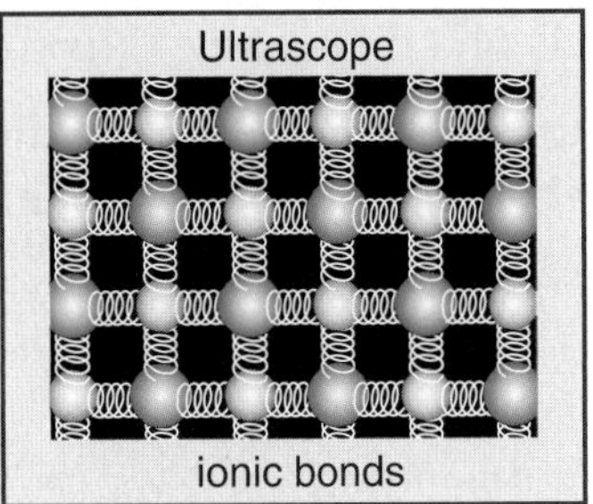

ionic bonds

Covalent Bonds

Molecules are the smallest particles of some elements and compounds. These include oxygen, chlorine, water, sugar, alcohols, oils, and hydrocarbon fuels. Molecular compounds are usually made of nonmetal atoms. A few molecular compounds are formed between metal atoms and nonmetal atoms that are close together on the *Periodic Table.*

In Activity 3, you were introduced to the interaction between some pairs of the atoms in a molecule.

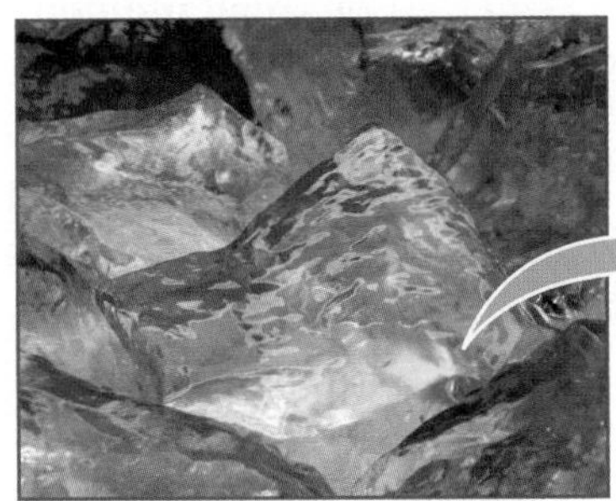

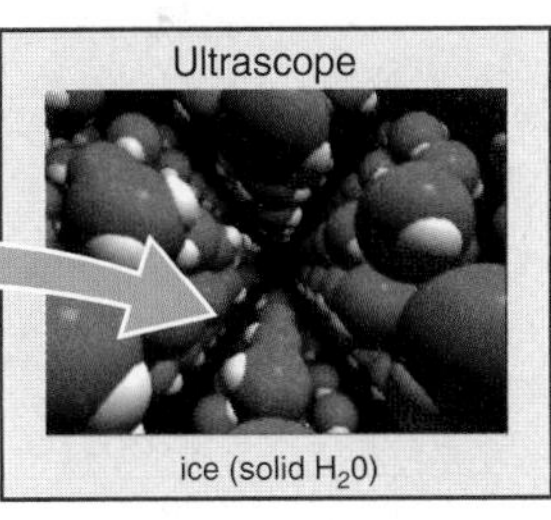

ice (solid H_2O)

Idea 9 Covalent Interaction. Some pairs of atoms in a molecule are continually attracted to each other. This interaction is sometimes called the covalent interaction.

A. The covalent interaction is a unique type of electric-charge interaction.

B. The strength of the covalent attraction between a pair of atoms in a molecule is different for different atom pairs.

Science Words

covalent bond: the attraction between specific pairs of atoms in a molecule

Covalent bonds are the attractions between specific pairs of atoms in a molecule. For example, the spring model of a lead iodide molecule at right shows two identical lead-iodine (Pb-I) covalent bonds. Notice that the two iodine atoms are not bonded to each other.

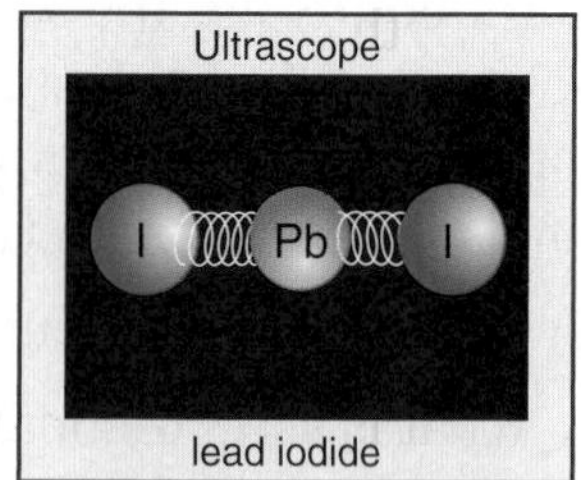

lead iodide

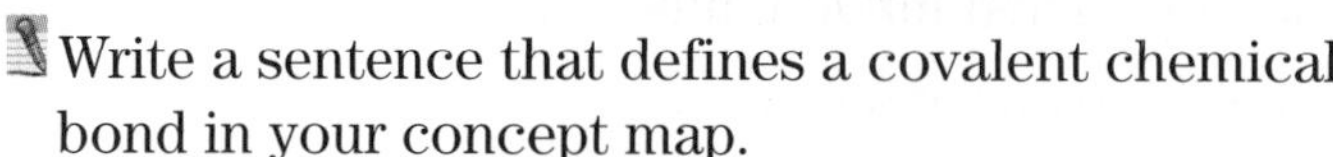
Write a sentence that defines a covalent chemical bond in your concept map.

2. What is one difference between covalent bonds and metallic or ionic bonds?

Look at the diagrams that show the spring images of the covalent bonds in water (H_2O) and methane (CH_4) molecules.

3. How many oxygen-hydrogen (O-H) covalent bonds are there in a water (H_20) molecule?

4. How many carbon-hydrogen (C-H) covalent bonds are there in a methane (CH_4) molecule?

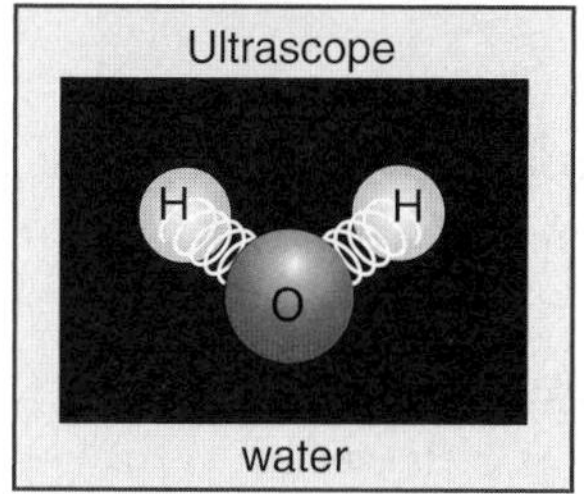

water

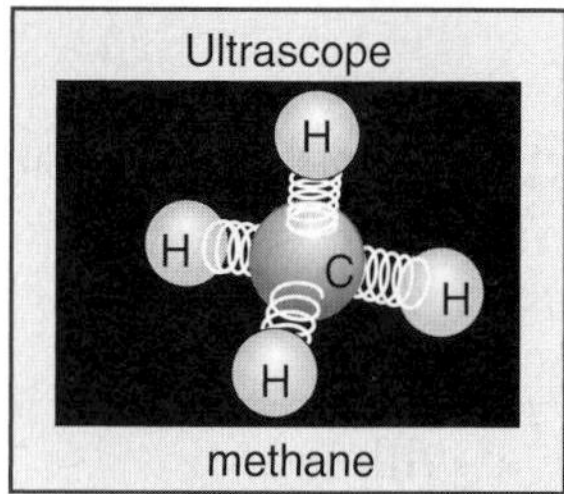

methane

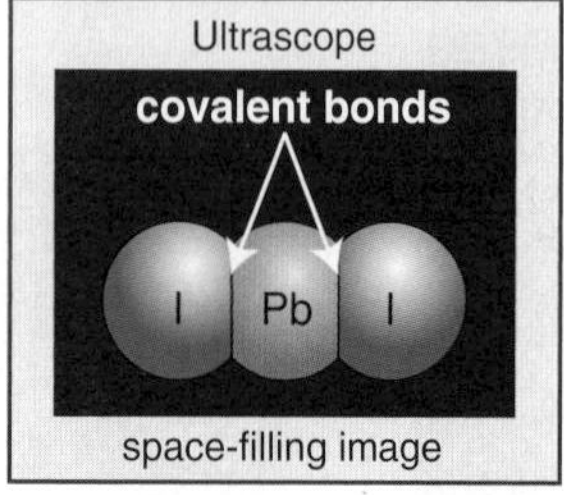

space-filling image

For small molecules, you do not really need the spring image to identify the covalent bonds in a molecule. The images you are familiar with (called space-filling images) show the bonds where the atom spheres are joined together. When two spheres do not connect directly, like the two iodine (I) spheres, then there is no bond between these atoms.

For images of large molecules, like the space-filling image of the sugar molecule, it is difficult to tell which atoms are bonded together. It is also difficult to draw a lot of springs. So, chemists represent covalent bonds with *ball-and-stick* images. The ball-and-stick image for sugar is shown here.

Models of a sugar molecule

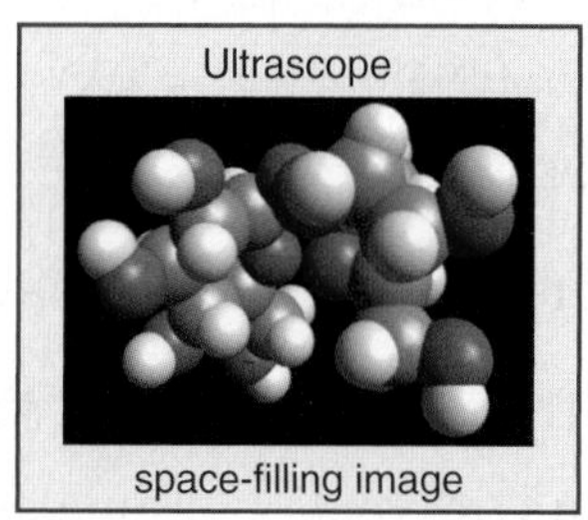

space-filling image

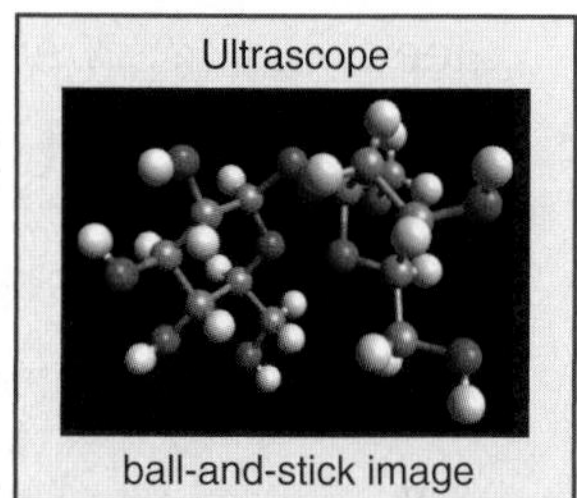

ball-and-stick image

Ultrascope

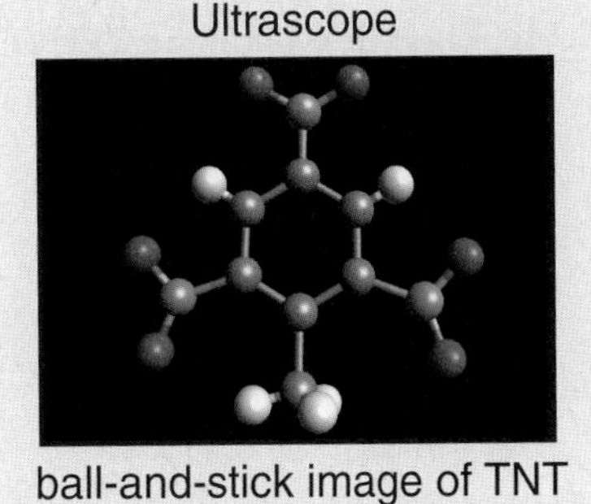

ball-and-stick image of TNT

Look at the ball-and-stick image of a molecule of TNT (trinitrotoluene), part of dynamite. The seven gray atoms are carbon (C), the three blue atoms are nitrogen (N), six red atoms are oxygen (O), and the five white atoms are hydrogen (H).

5. What is the chemical formula of TNT (trinitrotoluene)?

6. How many carbon-carbon (C-C) bonds are there in a TNT molecule?

7. How many carbon-nitrogen (C-N) bonds are there in a TNT molecule?

8. How many carbon-hydrogen (C-H) bonds are there in a TNT molecule?

9. How many nitrogen-oxygen (N-O) bonds are there in a TNT molecule?

Participate in a class discussion about your answers.

Explore SPT Ideas about Chemical-Bond Energy

In a chemical interaction, the particles of the reactant(s) must be moving fast enough to break the chemical bonds (metallic, ionic, or covalent) between the atoms. One model for breaking a bond is stretching a rubber band or spring until it breaks. A more useful model is pulling apart two attracting magnets. You will use magnets to expolore the relationship between breaking a bond and the amount of energy stored in a chemical bond.

One pair of team members will start with the strong magnets while the other pair starts with the weak magnets. Then exchange magnets.

Your team will need:
- 2 strong magnets
- 2 weaker magnets

STEP 1 *Partner 1:* Hold the two magnets (both strong or both weak) so they are stuck together.

STEP 2 *Slowly* increase your pull on both magnets until the magnets are separated—no longer stuck together.

1. As you slowly increase your pull on both magnets, do they move apart slowly or do they suddenly "break" apart? First, write the strength of the magnets (strong or weak) in your record sheet. Then answer the question.

STEP 3 *Partner 2:* Repeat Steps 1 and 2.

STEP 4 Exchange magnets with your other team members.

STEP 5 Repeat Steps 1 through 3 with the second pair of magnets.

2. As you slowly increase your pull on both magnets, do they move apart slowly or do they suddenly "break" apart? First, write the strength of the magnets (strong or weak) in your record sheet, and then answer the question.

3. Compare the effort it took you to pull apart the pair of weak magnets and the pair of strong magnets.

4. The energy stored in a two-magnet system is the amount of energy it takes to "break" the bond. Which pair of magnets stores the most energy?

Make Sense of SPT Ideas about Chemical-Bond Energy

Two attracting magnets stuck together is a useful model for stored chemical bond energy. You use energy as you pull on the two magnets. When you pull on the magnets hard enough, they suddenly break apart. The energy stored in a two-magnet system is the same as the amount of energy you need to pull the magnets until they break apart. Similarly, the amount of energy stored in a chemical bond is the same as the energy needed to break apart the two attracting atoms or ions.

There are different strengths of attraction between two magnets, ranging from weak to strong. You need more energy to pull apart two strong magnets than to pull apart two weak magnets. The energy stored in a two-magnet system depends on the strength of the attraction between the two magnets.

1. What is the relationship between the strength of attraction between two magnets and the amount of energy stored in the two-magnet system?

Participate in a class discussion about your answers.

Just like magnets, there are different strengths of attraction between two neighboring atoms or ions. For example, the metallic bond (forces) between two gold atoms (Au-Au) is stronger than the bond (forces) between two aluminum atoms (Al-Al).

2. Which pair of atoms, gold or aluminum, needs the most energy to break the metallic bond between their atoms. Justify your answer.

3. Which pair of atoms has the most stored chemical-bond energy? Why?

There are also different strengths of covalent bonds (attractions). For example, the covalent bond between two nitrogen atoms (N-N) is stronger than the covalent bond between two oxygen atoms (O-O).

4. Which molecule, nitrogen (N_2) or oxygen (O_2) needs the most energy to break the bond between their atoms? Justify your answer.

5. Which molecule has the most stored chemical–bond energy? Why?

6. What do you think is the relationship between the strength of attraction between two neighboring atoms or ions and the amount of energy stored in the chemical bond?

There are billions of trillions of metallic, ionic, or covalent bonds (attractions) between the atoms of a substance. A tiny amount of energy is stored in each of these bonds. The sum of the energies stored in the billions of trillions of bonds between the atoms is called the *stored chemical energy* of a substance.

7. Imagine that you are holding in your hands a piece of aluminum and a piece of gold. The two pieces have the same number of atoms. Which piece, gold or aluminum, has the greatest stored chemical energy?

Participate in a class discussion about your answers to the last six questions

SPT Consensus Ideas

The two key questions in this activity are:

1. What are chemical bonds and chemical-bond energies?

2. What is the Small-Particle Theory (SPT) of the stored chemical energy of a substance?

To answer these questions, complete the following on your record sheet.

Idea 11 Particle Interactions and Energy

The stored chemical energy of a substance is ______________.

Chemists call the attraction between neighboring atoms chemical bonds because these bonds break apart during chemical interactions.

A. There are three types of chemical bonds, ______, _____, and _____.

1. A ______________ bond is the attraction between the identical neighboring atoms of a solid metal.
2. An ________ bond is the attraction between neighboring positive metal ions and negative nonmetal ions of solid ionic compounds.
3. A ___________ bond is the attraction between specific atoms in a molecule.

B. Stored Chemical Energy and Chemical Bonds

1. Two attracting magnets stuck together is a useful physical model for stored chemicalbond energy. The energy stored in the two-magnet system is ______ as the energy you need to pull the magnets apart. Similarly, the energy stored in a chemical bond is________ as the energy needed to break the bond (pull the atoms apart).
2. The amount of energy stored in a chemical bond depends on the ________ of the attraction between pairs of atoms in the molecules or two neighboring atoms/ions in solidmetals and ionic compounds. The stronger the attraction between the atoms/ions, the________ the stored chemical-bond energy.
3. The stored chemical energy of a substance is ______________ .

Participate in a class discussion about the key questions.

DEVELOPING OUR IDEAS

Activity 5
The Balancing Act

Purpose

The Conservation of Mass Law is:

In a *closed mass system* (a system with no mass inputs and no mass outputs), the mass of the system does not change during interactions.

For example, in Activity 2 you learned that candle wax reacts with the oxygen in the air to produce soot, water, and carbon dioxide. Mass is conserved in this interaction, as shown in the diagram.

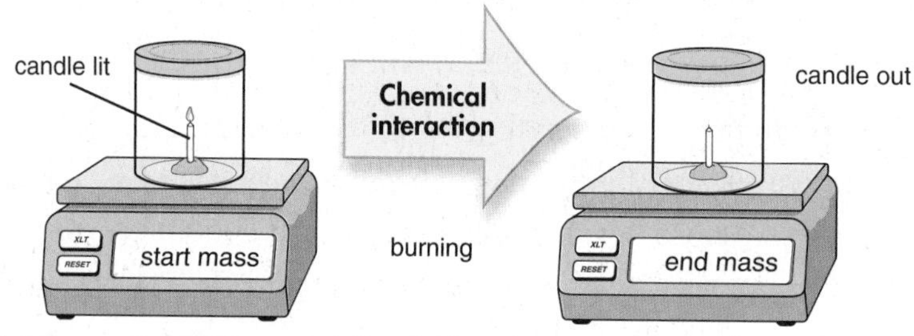

In this activity, you will answer the following question.

What is the Small-Particle Theory (SPT) representation of the Conservation of Mass Law for chemical interactions?

Record the key question for the activity on your record sheet.

To answer this question, you will build models of two chemical reactions, making water and burning methane.

Explore SPT Ideas

Part A: Making Water

Your team will need:

- model kit of plastic interlocking cubes (three colors)
- colored pencils

Your teacher will guide you through this first exploration.

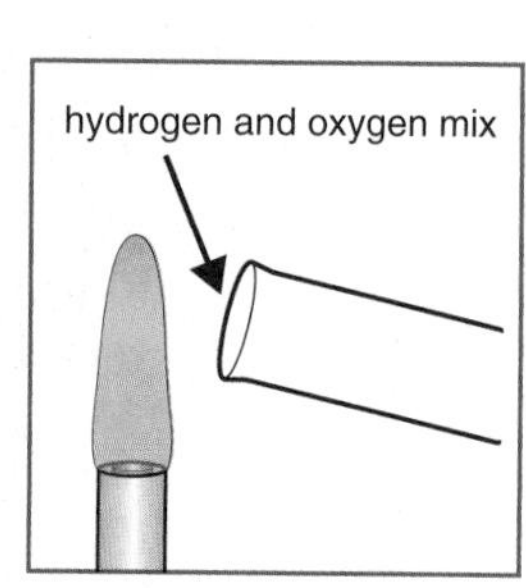

In Activity 1, you identified the reactants and product in the chemical reaction for making water. In this exploration, you will complete *Table 1: Conservation of Mass for the Making Water Reaction.*

STEP 1 The *reactants* are oxygen and hydrogen.

Write a few phrases for the properties of oxygen and hydrogen in row 1 of the reactants column of Table 1.

Table 1: Conservation of Mass for the Making Water Reaction			
	Reactants		Product
1. Common names and descriptions	Hydrogen	Oxygen	Water
2. Models of the particles			
3. Chemical formulas			
4. Model of the reaction			
5. Keeping track of the atoms	Number of _____ atoms:_____ Number of _____ atoms:_____ Total:_____		Number of _____ atoms:_____ Number of _____ atoms:_____ Total:_____

STEP 2 The smallest particle of oxygen is a molecule, as shown in the space-filling image and spring images.

Copy the space-filling image of the oxygen molecule (without the Ultrascope frame) in row 2 of the reactants column of your table.

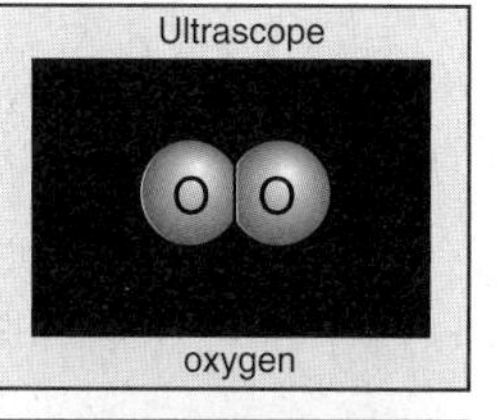

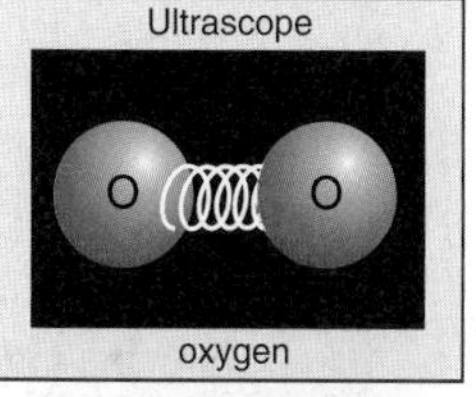

STEP 3 The smallest particle of hydrogen is also a molecule, as shown in the diagrams.

Copy the space-filling image of a hydrogen molecule (without the Ultrascope frame) in row 2 of the reactants column of your table. Use a *different color* than you used for oxygen.

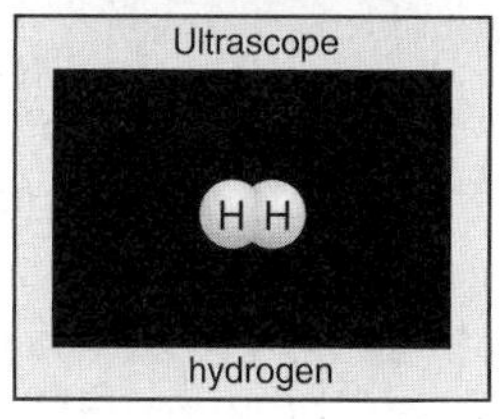

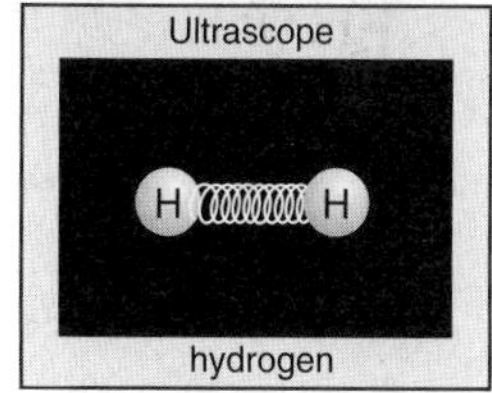

STEP 4 The chemical formula for oxygen is O_2. What is the chemical formula for hydrogen?

Write the chemical formulas for oxygen and hydrogen in row 3 of the reactants column of your table.

STEP 5 A cube in your model kit represents an atom. Make a molecule of oxygen from your kit. Then use cubes of a different color to make a molecule of hydrogen.

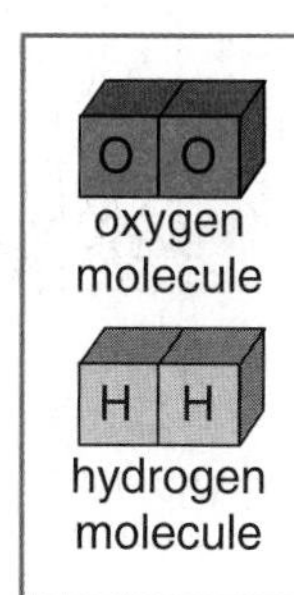

STEP 6 Make one more oxygen and hydrogen molecule in case you need them later. Return all unused pieces to the model kit.

STEP 7 The *product* of this reaction is water.

Briefly describe a few properties of water in row 1 of the products column of your table.

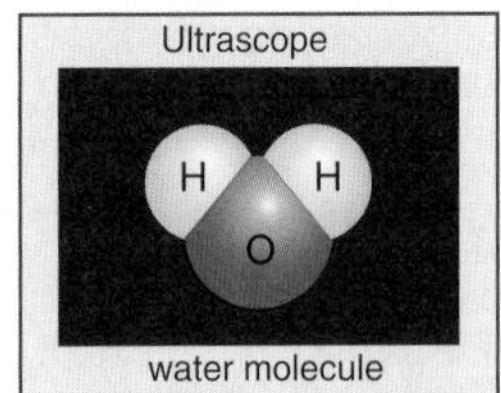

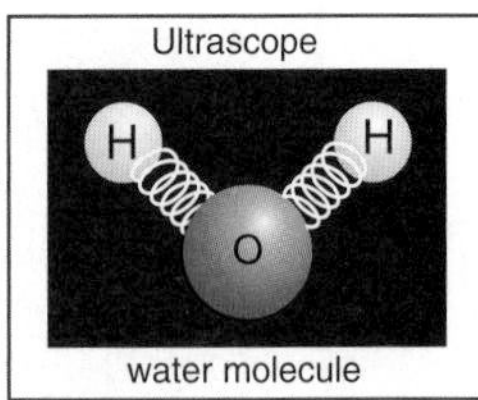

STEP 8 The smallest particle of water is a molecule, as shown in the Ultrascope diagrams.

Copy the space-filling image of a water molecule in row 2 of the products column of your table. Use the same colors for oxygen and hydrogen as you used before.

STEP 9 How many oxygen atoms are there in a water molecule? How many hydrogen atoms are there in a water molecule?

Write the chemical formula for a water molecule in row 3 of the products column of your table.

You learned in Activity 2 that atoms are not created or destroyed. Instead, the atoms in the starting substances (reactants) rearrange into at least one new combination (product). For this to happen, the chemical bonds between the atoms of the reactant particles must break apart and new chemical bonds form in the product particle(s).

You should have models of several hydrogen and oxygen molecules you made from your kit.

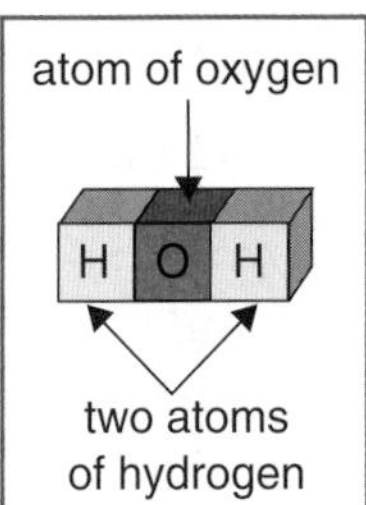

STEP 10 Take an oxygen molecule apart (break the O-O covalent bond). Take a hydrogen molecule apart (break the H-H covalent bond). Grab an oxygen atom and two hydrogen atoms and build a molecule of water (H_2O).

Is anything left over? Yes. There is an oxygen atom left over. *Chemists have found that oxygen gas near room temperature is always made up of molecules, not single atoms.* So what happens? Two molecules of water are formed. Make another molecule of water from the leftover atom of oxygen and another oxygen molecule. Remember that particles of matter are very, very tiny, and oxygen and hydrogen gases are made up of billions of trillions of particles. So there are plenty of oxygen and hydrogen molecules to interact.

Put all the unused oxygen and hydrogen molecules back in your kit. They did not interact in this model.

Draw space-filling images of the two *product* (water) molecules in row 4 of the *products column* of your table. Label your pictures and use the same colors for the atoms as you used before.

STEP 11 Count how many oxygen molecules and hydrogen molecules you used to make two molecules of water. (If you cannot remember, take apart the two water molecules to make the original reactants. Then count them.)

Draw space-filling images of the correct number of *reactant* molecules (hydrogen and oxygen) in row 4 of the reactants column of your table. Label your pictures and use the same colors you used before.

STEP 12 In row 5 of your table, labeled "Keeping track of the atoms," record the names of the atoms of different elements that are present in the reactants and products. Then decide *how many atoms* of each element are present.

On your table, record the number of atoms of each element that are present in the reactants and products. Then calculate and record the total number of reactant and product atoms.

Explain why the total number of atoms in the product particles is equal to the total number of atoms in the reactant particles.

Part B: Burning Methane

It's your turn to keep track of the mass (atoms) for the burning methane chemical interaction. You will complete *Table 2: Conservation of Mass for the Burning Methane Reaction.*

Table 2: Conservation of Mass for the Burning Methane Reaction

	Reactants		Product	
1. Common names and descriptions	Methane *invisible gas*	*invisible gas*	*liquid droplets*	*invisible gas*
2. Models of the particles				
3. Chemical formulas				
4. Model of the reaction				
5. Keeping track of the atoms	Number of _____ atoms:_____ Number of _____ atoms:_____ Number of _____ atoms:_____ Total:_____		Number of _____ atoms:_____ Number of _____ atoms:_____ Number of _____ atoms:_____ Total:_____	

STEP 1 Follow directions carefully. Be sure the *Procedure Specialist* reads the directions out loud, step by step. The two *reactants* are methane and oxygen, as shown in the space-filling images on the next page.

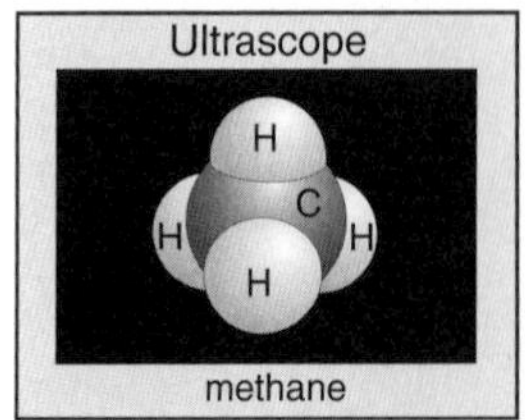

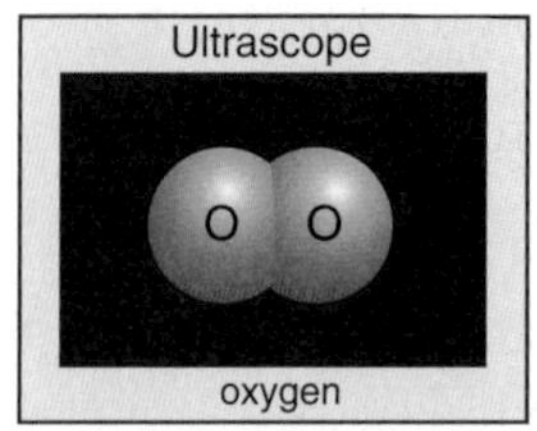

- Write the common names and a brief description of the reactants in row 1 of the reactants column of your table.
- Copy (draw) the space-filling images of methane (CH_4) and oxygen (O_2) molecules in row 2 of your table. Use *different colors* for the carbon, hydrogen, and oxygen atoms. Label the atoms with their symbols.
- Write the formula for each reactant molecule in row 3 of your table.

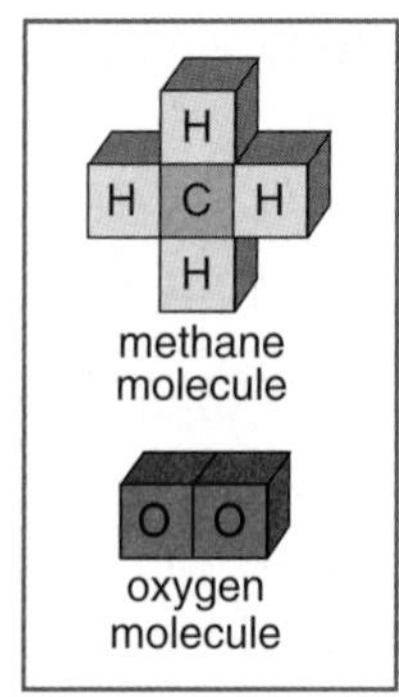

STEP 2 Make a physical model of a methane molecule (CH_4) from your kit. Join (bond) four atoms of hydrogen to one central carbon atom. Use different colors for the two elements.

Then make a physical model of an oxygen molecule. Join (bond) two atoms of oxygen together using a third color.

STEP 3 Make several methane and oxygen molecules in case you need them later. Return all unused pieces to the model kit.

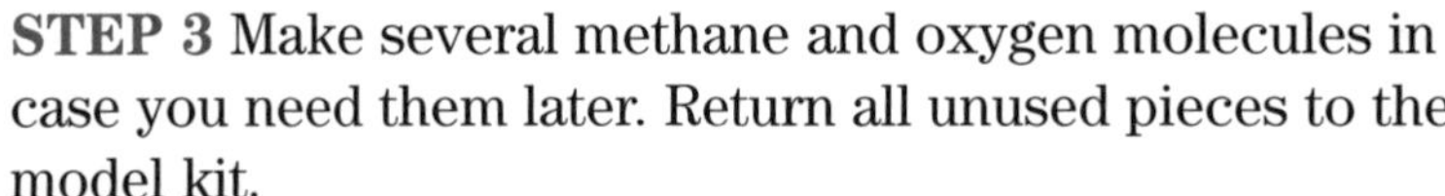

STEP 4 The two products are carbon dioxide and water, as shown in the space-filling images.

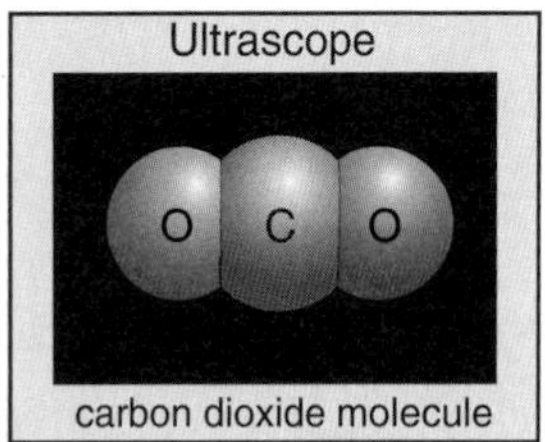

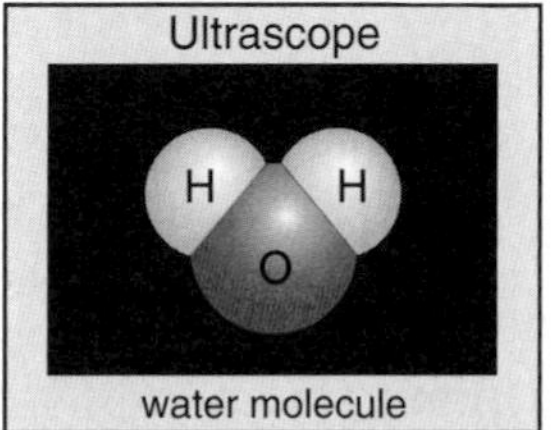

- Write the common names and a brief description of the products in row 1 of the products column of your table.
- Copy (draw) the space-filling images of carbon dioxide (CO_2) and water (H_2O) molecules in row 2 of your table. Use the same colors for the carbon, hydrogen, and oxygen atoms as you used before. Label the atoms with their symbols.
- Write the formula for each product molecule in row 3 of your table.
- Participate in a class discussion about your table.

STEP 5 You should have models of several methane molecules and several oxygen molecules you made from your kit. Take an oxygen molecule (O_2) apart (break the O-O covalent bond). Take a methane molecule (CH_4) apart (break the four C-H covalent bonds).

Grab a carbon atom and two oxygen atoms and build a carbon dioxide molecule (CO_2).

What is left over? Do you have enough oxygen to make the second product, water (H_2O)? *No*. Take apart another oxygen molecule and build another water molecule.

Put all the unused methane and oxygen molecules back in your kit. They did not interact.

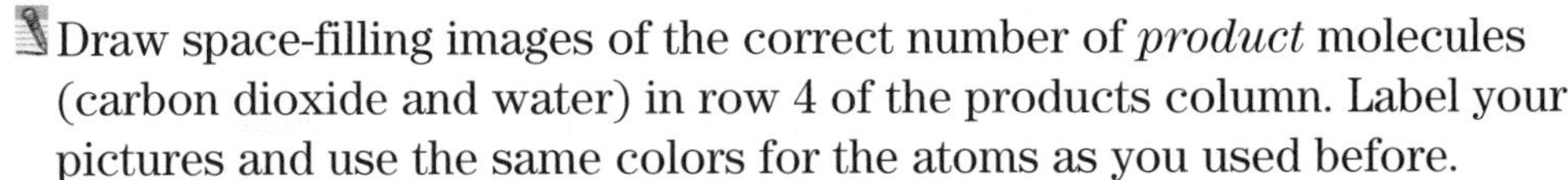

Draw space-filling images of the correct number of *product* molecules (carbon dioxide and water) in row 4 of the products column. Label your pictures and use the same colors for the atoms as you used before.

STEP 6 Count how many methane molecules and oxygen molecules you used to make the products. (If you cannot remember, take apart the carbon dioxide and water molecules to make the original reactants. Then count them.)

Draw space-filling images of the correct number of *reactant* molecules (methane and oxygen) in row 4 of the reactants column. Label your pictures and use the same colors you used before.

STEP 7 In row 5 of your table, record the atoms of different elements that are present in the reactants and products. Then decide *how many atoms* of each element are present. (*Hint: Count* the atoms of the different elements in row 4 labeled "Model of the reaction.")

On your table, record the number of atoms of each element that are present in the reactants and products. Then calculate and record the total number of reactant and product atoms.

Do you have the same number of atoms of each element in the reactants as in the products? If not, then you must have made a mistake because you have not conserved mass.

Participate in a class discussion about your tables.

Make Sense of SPT Ideas

Chemists represent the Conservation of Mass Law for chemical reactions in a form called *balanced equations*. A balanced equation shows the correct number of atoms of each element in the reactant and product particles so that mass is conserved. No atoms are created or destroyed.

Look at row 4 labeled "Model of the reaction" in Table 1. The addition of a plus sign (+) and a yield sign (⇒) turns this model into a *balanced picture equation*, as shown in the diagram.

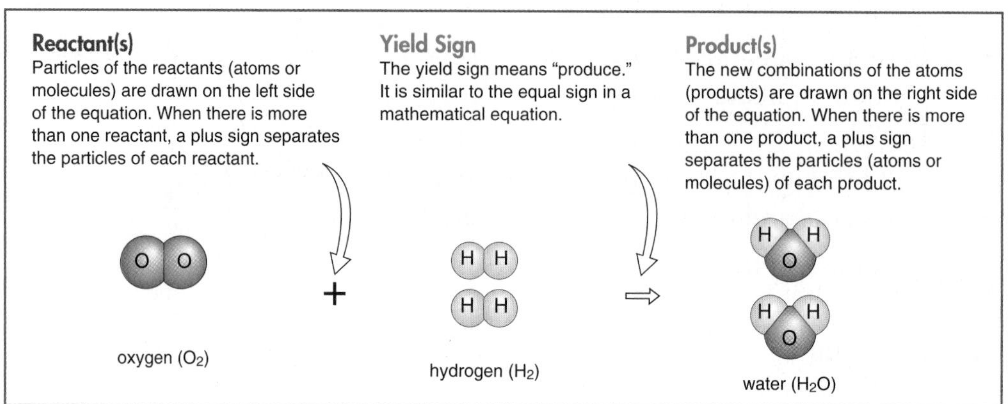

Reactant(s)
Particles of the reactants (atoms or molecules) are drawn on the left side of the equation. When there is more than one reactant, a plus sign separates the particles of each reactant.
Yield Sign
The yield sign means "produce." It is similar to the equal sign in a mathematical equation.
Product(s)
The new combinations of the atoms (products) are drawn on the right side of the equation. When there is more than one product, a plus sign separates the particles (atoms or molecules) of each product.
O O
+
H H
H H
⇒
H H O
H H O
oxygen (O_2)
hydrogen (H_2)
water (H_2O)

Add the plus sign (+) and a yield sign (⇒) to your table.

Look at row 4 in Table 2.

Add the plus sign (+) and a yield sign (⇒) to your table to make the balanced picture equation for the burning methane reaction.

A balanced picture equation shows that the number of atoms of each element in the reactant particles *balances* (is equal to) the number of atoms of each element in the product particles. A balanced picture equation is also shorthand for other information about the chemical interaction. For example, you learned in Activity 3 that the reactant molecules must collide with enough energy to break the chemical bonds. Then the atoms recombine into new molecules.

1. Look at your balanced picture equation for making water. How many reactant molecules are there? How many molecules must collide for the reaction to occur?

2. How many chemical bonds are broken during this reaction?

3. How many product molecules are there?

4. How many new chemical bonds are formed to make the product molecules?

Check your answers by referring to the diagrams on the next page.

Ultrascope Diagram of a Collision That Results in a Chemical Reaction

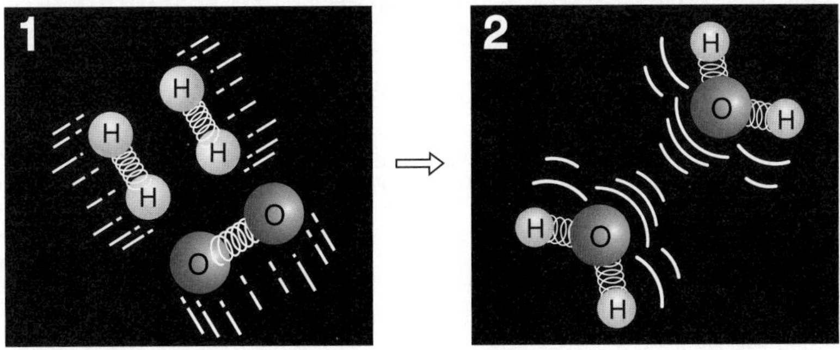

Participate in a class discussion about the answers.

5. Write a paragraph describing what happens to the molecules and the chemical bonds during the making of water reaction.

Suppose you are given a picture equation for a chemical reaction. How can you tell whether the picture equation is balanced? To answer this question, consider the electrolysis exploration you did in Unit 5.

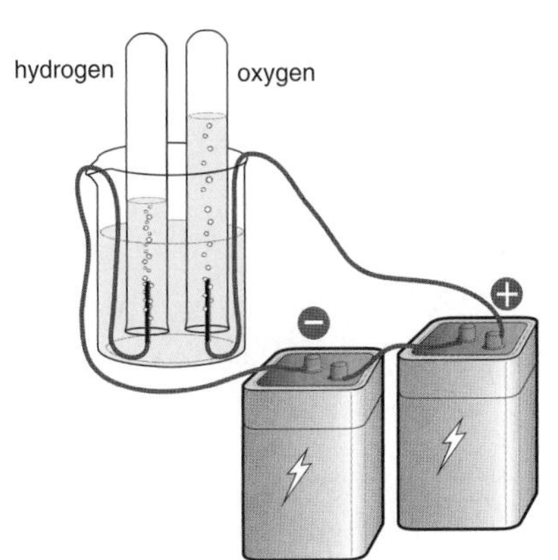

The electrolysis exploration is the opposite of making water. An electrical energy source (battery) causes a chemical reaction in water. The water molecules break apart to form the two gases, hydrogen and oxygen.

Look at the picture equation below drawn by another student.

6. Is this equation balanced? In other words, does this equation correctly represent the Conservation of Mass Law for the electrolysis reaction?

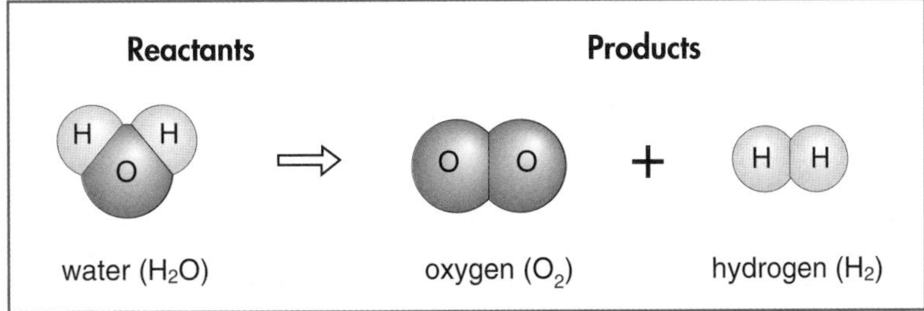

To decide whether a picture equation is balanced, you can use the procedure shown on the next page. This procedure is also included in *How To Decide Whether a Picture Equation of a Chemical Interaction is Correct (Balanced)* in the Appendix.

How To Decide Whether a Picture Equation of a Chemical Interaction is Correct (Balanced)

A. Count the number of atoms *of each element* in the picture of the reactant(s).

B. Count the number of atoms *of each element* in the picture of the product(s).

C. Is the number of atoms of each element the same for the reactant(s) and the products?

Yes —The picture equation is correct. Mass is conserved because the reactant atoms are the same in *type* and *number* as the product atoms.

No — The picture equation is not correct. Mass is not conserved because a type of product atom is incorrect and/or the number of the product atoms is incorrect.

7. The diagram below shows Step A. Complete Step B for the products on your record sheet.

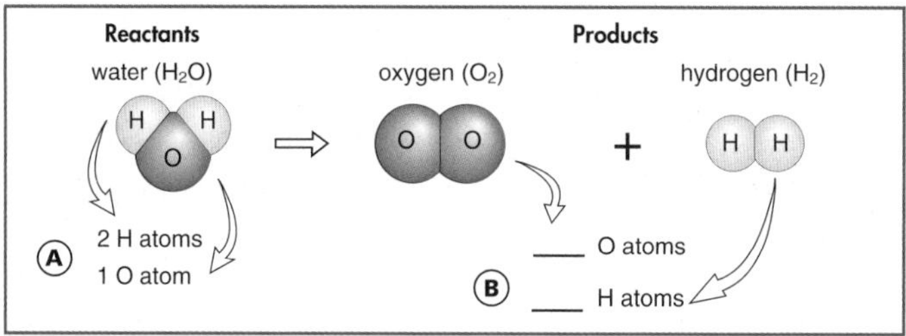

8. Complete Step C. Be sure to explain your reasoning.

Participate in a class discussion about the answers.

Our Consensus Ideas

Recall the key question for this activity.

What is the Small-Particle Theory (SPT) representation of the Conservation of Mass Law for chemical interactions?

1. Write your answer on your record sheet.

Participate in the class discussion about the key question.

2. Write the class consensus ideas on your record sheet.

DEVELOPING OUR IDEAS

Activity 6
The Rusting Reaction

Purpose

You will investigate the chemical reaction called rusting. The key question for this activity is:

What are the reactants and products in the rusting reaction, and what is the evidence for the reaction?

Record the key question for the activity on your record sheet.

We Think

Items made of steel, such as the ones shown in the photograph, tend to rust. Discuss the following questions with your team.

1. Under what conditions does rusting usually occur?
2. Is rusting a chemical reaction? Why do you think so?

Participate in a class discussion.

Explore Your Ideas

You will place some clean steel wool inside a flask, and then seal the flask with a balloon. You will observe the properties of the steel wool before and after the reaction, and what, if anything, happens to the balloon and the temperature of the flask during the rusting reaction. Your teacher will do a demonstration to determine whether the mass changes during the rusting action.

Your team will need:

- piece of steel wool about the size of a Ping-Pong® ball
- beaker or bowl of vinegar (for cleaning the steel wool)
- 2-3 sheets of paper towels
- clean 250-mL Erlenmeyer flask
- balloon
- scissors or long forceps

STEP 1 Your teacher will measure the mass of the steel wool, flask, and balloon.

Record this mass as the "start mass" in the table on your record sheet.

STEP 2 Observe the properties (what you see, feel, and smell) of the steel wool. Steel is a solid solution made by combining small amounts of carbon or other metals with iron.

Table: Rusting Iron Reaction		
Properties "before"	**Properties "after"**	**Other evidence**
Steel wool:	Rust:	Balloon: Flask:
Start mass ________ (g)	End mass ________ (g)	
Uncertainty ________ (g)	Uncertainty ________ (g)	

Steel wool is mostly iron, and it is the iron that is one of the reactants in the rusting interaction.

Record the properties of the steel wool in the "before" column of your data table.

STEP 3 Steel wool comes with a protective coating to keep it from rusting. Before you can start your exploration, you must clean off this protective coating. You will use vinegar to clean the steel wool.

Supply Master: Open up the piece of steel-wool pad a little. Then submerge the steel wool in the vinegar *for at least a minute.* Stir the pad around in the vinegar to make sure that there is good contact between the vinegar and the strands of steel wool.

STEP 4 *Supply Master:* Squeeze the vinegar out of the steel-wool pad. Then open the pad and place it on several layers of paper towels.

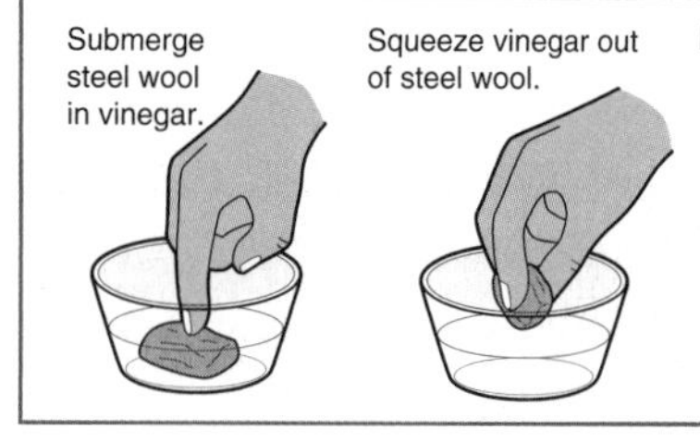
Submerge steel wool in vinegar.
Squeeze vinegar out of steel wool.

Blot the steel wool.

STEP 5 *Procedure Specialist:* Blot the steel wool between a folded sheet of paper towel to remove the vinegar. You may need to use several sheets. Try to get the steel wool as dry as possible.

Wear eye protection, and lab apron.

Clean up spills immediately.

STEP 6 *Manager:* Pull apart the strands, then push the steel wool into the flask.

STEP 7 *Recycling Engineer:* Blow up the balloon a few times to stretch the rubber. *Squeeze the air out of the balloon.* Then stretch the deflated balloon over the top of the flask to make a tight seal. Be sure the opening of the balloon is over the opening of the flask.

STEP 8 Observe any changes happening during the chemical reaction. Take turns feeling the sides of the flask with your hands. What, if anything, happens to the temperature of the flask? What, if anything, happens to the balloon?

Record your observations in the "Other evidence" column of your table.

STEP 9 Wait until you have clear evidence of the rusting reaction. *Manager:* Take the balloon off the flask. Use scissors or another long tool to remove the steel wool from the flask. Place it on a clean paper towel.

Dispose of materials and clean glassware according to teacher's direction.

Wash hands after activity.

Compare this piece of steel wool (after the chemical reaction) with a clean piece of steel wool. Observe the properties (what you see, feel, and smell) of the after-reaction steel wool.

Record these properties in the "after" column of your table.

STEP 10 Your teacher will measure the mass of the demonstration flask after the chemical reaction.

Record this mass as the "end mass" in your table on your record sheet.

Participate in the class discussion about your team's observations.

Make Sense of Your Ideas

Discuss the following questions about the rusting exploration with your team, then record your answers.

1. Iron is one of the reactants in the rusting reaction. What evidence do you have that an invisible gas in the air is the other reactant? (*Hint:* What do you think caused the balloon to be pushed into the flask?)
2. What evidence do you have that a new substance (product) was formed?
3. Was any energy transferred during the rusting interaction? What form of energy? Was it transferred into or out of the steel wool-air system?
4. Was mass conserved in the closed system during the rusting interaction? Compare the start mass with the end mass. Remember that all measurements have some uncertainty.

Participate in the class discussion about your team's answers to these questions.

Our Consensus Ideas

The key question for this activity is:

What are the reactants and products in the rusting reaction, and what is the evidence for the reaction?

1. Write a sentence that describes the rusting reaction.

Participate in the class discussion about the key question.

2. Write the class consensus ideas on your record sheet.

DEVELOPING OUR IDEAS

Activity 7
Balancing the Rusting Reaction

Purpose

In this activity, you will repeat the process of balancing chemical equations that you learned in Activity 5. This time, you will work with the rusting iron reaction. The key question for the activity is:

What is the balanced chemical equation for the rusting iron reaction?

Record the key question for the activity on your record sheet.

Explore SPT Ideas about the Rusting Reaction

Part A: Building Models of the Reactants and Products

To answer the key question, you will build models of the reactant and product particles for the chemical reaction of rusting iron.

Table: Conservation of Mass for Rusting Iron Reaction

	Reactants	Product
1. Common names and descriptions	Iron: Invisible gas:	Rust:
2. Models of the particles		
3. Chemical formulas		
4. Model of the reaction (balanced picture equation)		
5. Keeping track of the atoms (how to check your equation)	Number of ____ atoms:____ Number of ____ atoms:____ Number of ____ atoms:____ Total:____	Number of ____ atoms:____ Number of ____ atoms:____ Number of ____ atoms:____ Total:____

STEP 1 One reactant is the iron in the steel wool. What is the other reactant?

Write the common names and a brief description of the reactants in row 1 of the reactants column of your chart.

STEP 2 Iron is a solid metal element. A particle of iron is a single atom of iron.

Oxygen is a gaseous element. An oxygen molecule is made up of two oxygen atoms bonded together.

Your team will need:

- model kit of plastic cubes (three colors)
- colored pencils

Copy the space-filling images of the iron atom and oxygen molecule in row 2 of the reactants column of your chart. Use a different color for the iron and oxygen atoms. Label the atoms with their symbols.

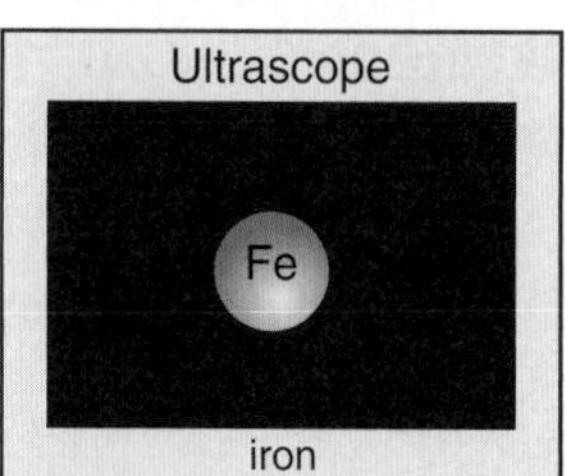

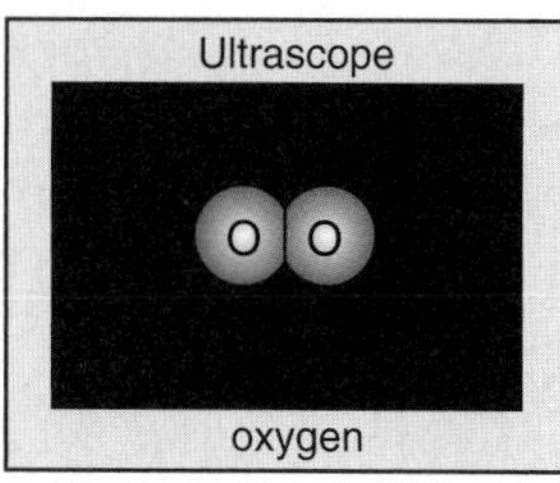

Write the chemical formulas for the iron atom and oxygen molecules in row 3 of the reactants column.

STEP 3 Make a model of an oxygen molecule. Join or bond two atoms of oxygen. Make a model of six iron atoms bonded together, using the other color.

STEP 4 Make several oxygen molecules in case you need them later. Return all unused pieces to the model kit.

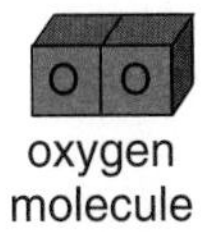

oxygen molecule

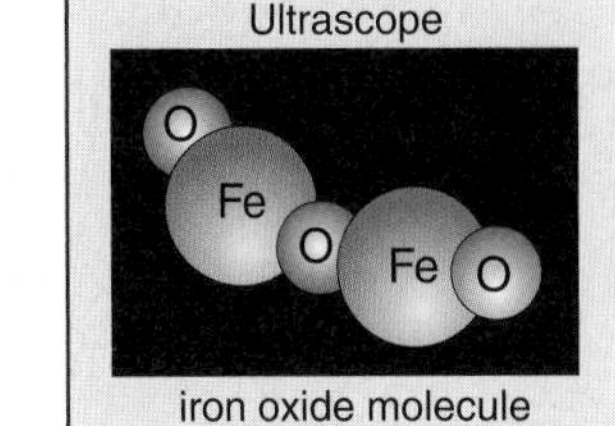

iron oxide molecule

STEP 5 What is the product of this reaction?

Write the common name and a brief description of the product in row 1 of the product column of your chart.

Fe Fe Fe
iron atoms

STEP 6 During the rusting interaction, the oxygen molecules in the air collide with the iron atoms in the solid steel wool. According to the Small-Particle Theory, the chemical bonds (attractions) between the oxygen and iron atoms break. The iron and oxygen atoms recombine to make new bonds in the molecules of rust.

How many atoms of iron are in a rust (iron oxide) molecule? How many atoms of oxygen are in a rust molecule?

Copy the space-filling image of the rust molecule in row 2 of the product column. Use the same colors for the atoms as you used above.

STEP 7 If you said a rust molecule contains *two* atoms of iron and *three* atoms of oxygen, you are correct. The chemical formula for a rust molecule is:

$$Fe_2O_3$$

Write the formula for rust in row 3 of the product column.

Part B: Making a Balanced Equation

You will use your models of the reactant particles to make a balanced picture equation of the rusting iron reaction. You should have models of several oxygen molecules and of six iron atoms bonded together.

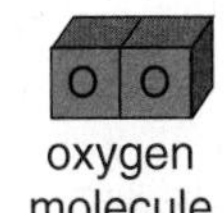

oxygen molecule

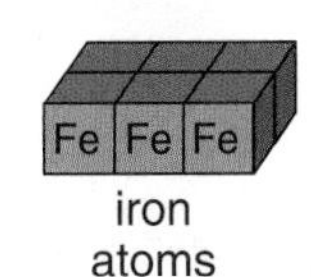

iron atoms

STEP 1 Take an oxygen molecule (O_2) apart (break the covalent bond). Take an iron atom (Fe) from the solid iron (break the metallic bonds). Start building a molecule of rust (iron oxide, Fe_2O_3).

Are one atom of iron and one molecule of oxygen enough? *No*. Take a second molecule of oxygen apart (break the bonds). How many iron atoms do you need?

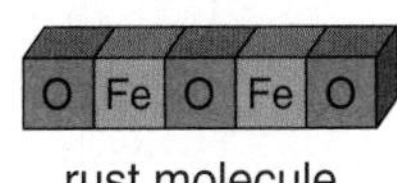

rust molecule

Is anything left over? *Yes.* There is an oxygen atom left over. Oxygen is always made up of molecules, not single atoms. So what happens? *Two molecules of rust are formed.*

STEP 2 Make another molecule of rust from the leftover atom of oxygen and any other iron atoms and oxygen molecules you need.

Put all the unused iron atoms and oxygen molecules back in your kit. They did not interact in this model.

Draw space-filling images of the two product (rust) molecules in row 4 of the product column. Label your pictures and use the same colors for the atoms as you used before.

STEP 3 Count how many iron atoms and oxygen molecules you used to make two molecules of rust. (If you can't remember, take apart the two rust molecules to make the original reactants. Then count them.)

Draw space-filling images of the correct number of reactant particles (iron atoms and oxygen molecules) in row 4 of the reactant column. Label your pictures and use the same colors you used before.

STEP 4 Add plus (+) signs and a yield (⇒) sign to complete your balanced picture equation.

STEP 5 Finally, check to see if your picture equation is balanced. Follow the procedure in *How To Decide Whether a Picture Equation of a Chemical Interaction is Correct (Balanced)* in the Appendix.

Record the answers to Steps A and B in row 5, labeled "Keeping track of the atoms."

Participate in a class discussion about your balanced picture equation.

Make Sense of SPT Ideas

Writing Balanced Equations

Suppose you want to describe to someone the chemical reaction of rusting iron. You could say, "Oxygen gas molecules collide with iron atoms on the surface of the solid strands of steel wool. The covalent bonds between the atoms in the oxygen molecules and the metallic bonds between the surface iron atoms break apart. New bonds form between the atoms to produce rust (iron oxide) molecules."

This is a long description. It also does not describe how mass is conserved during the interaction. Also, the person has to be an English speaker to understand the description.

You know that a balanced picture equation uses chemical and mathematical symbols. The balanced picture equation represents the Conservation of Mass Law at the particle level, and can be understood by scientists all over the world.

But pictures take a long time to draw, so chemists use shorthand "word" and "symbolic" balanced equations. You can use a balanced picture equation to write a balanced word equation. The procedure is in *How To Write a Word Equation of a Chemical Interaction from a Balanced Picture Equation.* This is also included in the Appendix.

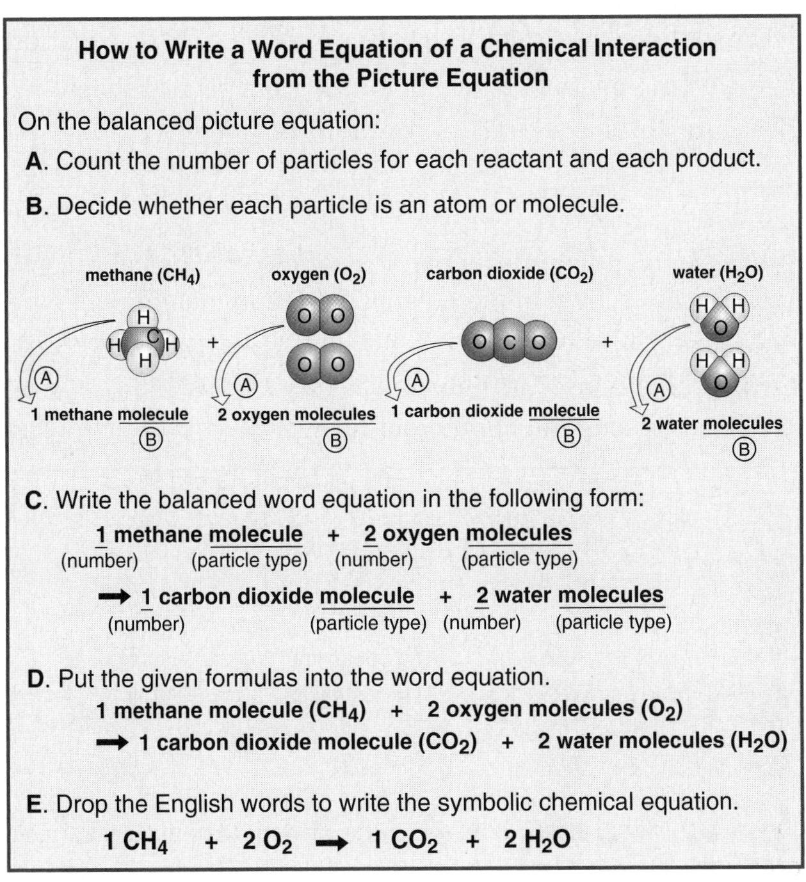

How to Write a Word Equation of a Chemical Interaction from the Picture Equation

On the balanced picture equation:

A. Count the number of particles for each reactant and each product.

B. Decide whether each particle is an atom or molecule.

C. Write the balanced word equation in the following form:

1 methane molecule + 2 oxygen molecules (number) (particle type) (number) (particle type)

→ 1 carbon dioxide molecule + 2 water molecules (number) (particle type) (number) (particle type)

D. Put the given formulas into the word equation.

1 methane molecule (CH_4) + 2 oxygen molecules (O_2)
→ 1 carbon dioxide molecule (CO_2) + 2 water molecules (H_2O)

E. Drop the English words to write the symbolic chemical equation.

$$1\ CH_4 + 2\ O_2 \rightarrow 1\ CO_2 + 2\ H_2O$$

Balanced word equations for the Conservation of Mass still have a few English words. Chemists usually drop the words and write an equation with symbols only. The Balanced symbolic equation for the burning methane reaction is:

$$CH_4 + 2\ O_2 \Rightarrow CO_2 + 2\ H_2O.$$

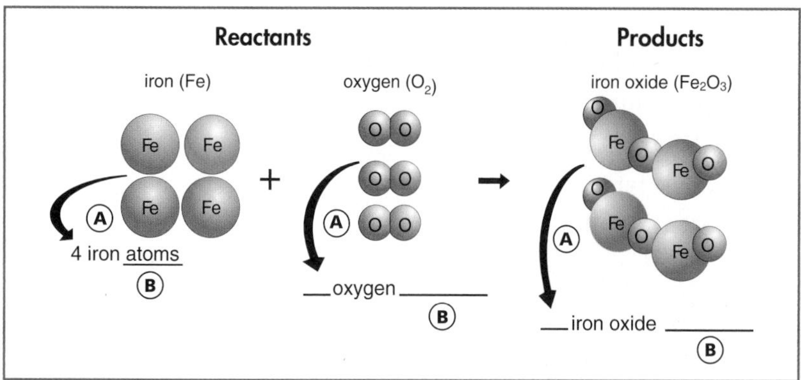

1. The diagram to the left shows Steps A and B for the first reactant (iron). Complete these steps for oxygen and iron oxide.

Then you can substitute the words for the pictures in the equation, putting the given formulas for the particles in parentheses after each part of the equation.

2. Complete the word equation shown below on your record sheet.

4 iron atoms (Fe) + ___ oxygen __________ (   )
⇨ ___ iron oxide __________ (   ).

Participate in a class discussion about your answer.

3. Photosynthesis in plant cells is a complex series of chemical reactions. The overall effect is shown in the balanced picture equation below. Complete Steps A and B for the reactants and products of photosynthesis.

4. Write the word equation for photosynthesis.

Participate in a class discussion about your answer.

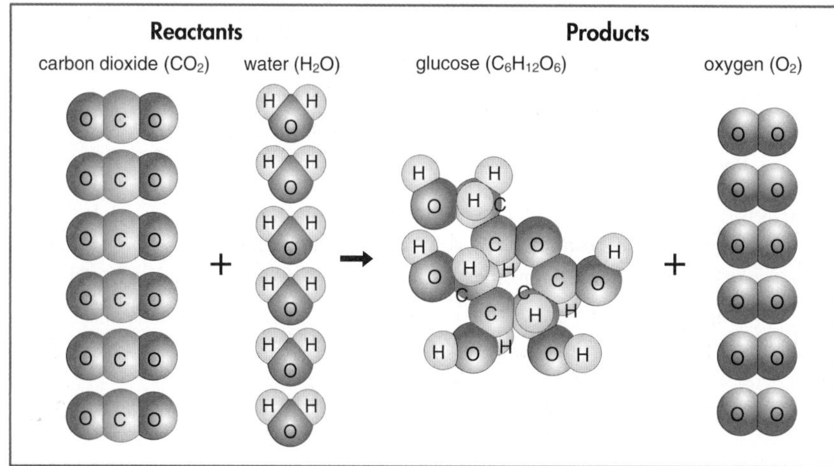

Our Consensus Ideas

Recall the key question for the activity.

What is the balanced chemical equation for the rusting iron reaction?

Are you satisfied that you have answered the key question?

MAKING SENSE OF SCIENTISTS' IDEAS

Activity 8
Energy and Reactions

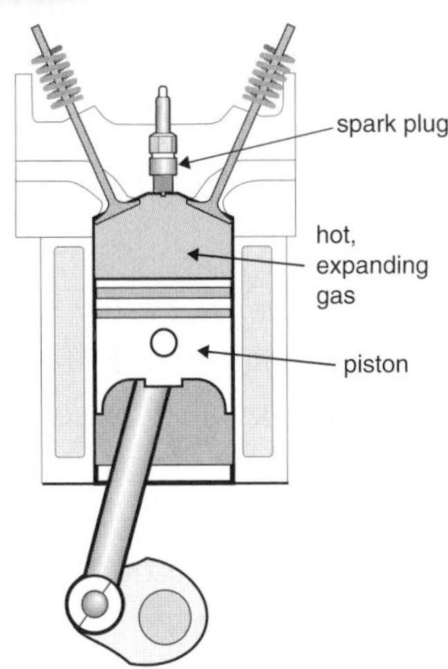

Purpose

You learned in Activity 4 the Small-Particle Theory (SPT) of stored chemical energy. You also know that for some chemical reactions, *energy is transferred out of the system of interacting substances.* For example, the burning gasoline in a car engine produces a hot, expanding gas that pushes on a piston and causes the wheels to turn.

Other chemical reactions *require a continual transfer of energy into the system* for the reaction to occur. For example, in Activity 3 you had to heat lead iodide (PbI_2) to produce solid lead (Pb) and iodine gas (I_2).

In this activity you will explore the following question.

What is the relationship between the transfer of energy into or out of a chemical reaction and the chemical-bond energy of the reactant and product molecules?

Record the key question for the activity on your record sheet.

Explore Your Ideas

In this exploration, you will determine whether the chemical interaction between baking soda ($NaHCO_3$) and the acetic acid in vinegar (CH_3COOH) requires a transfer of heat energy into the system (absorbs heat energy), or if it transfers energy out of the system (produces heat energy).

Your team will need:

- 10 g of baking soda (approximate)
- 25 mL of distilled white vinegar (approximate)
- plastic baggie

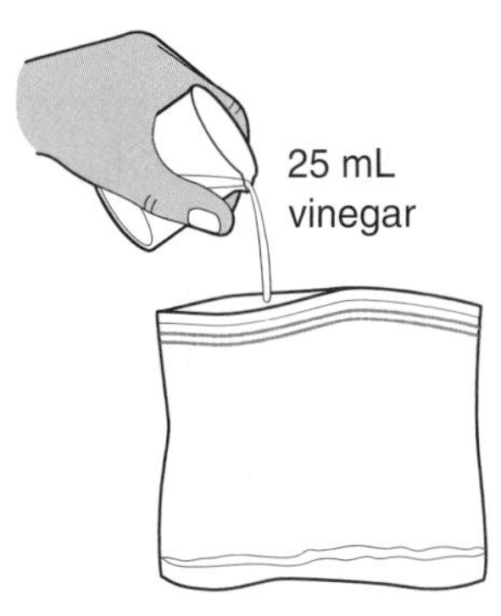

STEP 1 *Team Manager:* Pour about 25 mL of vinegar into the baggie.

Take turns holding the baggie in both hands to feel the temperature of the vinegar.

STEP 2 *Recycling Engineer:* Add about 10 g of baking soda to the baggie.

Wear eye protection and lab apron.

Clean up spills immediately.

Dispose of materials according to teacher's direction.

Wash hands after activity.

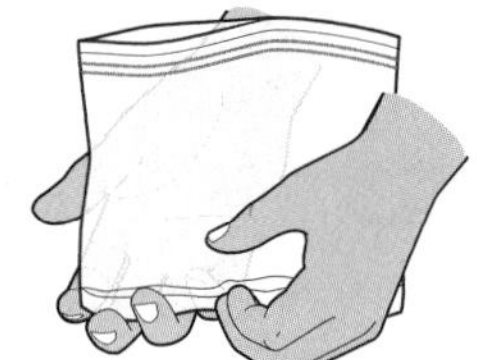

Take turns holding the baggie in both hands during the reaction.

1. Does the baggie feel colder or warmer after the reaction?
2. Is energy transferred into the baggie or out of the baggie during this chemical reaction? Explain your reasoning.

Participate in a class discussion of your observations and conclusion.

Make Sense of Scientists' Ideas

Activation Energy

Science Words

activation energy: the minimum motion energy that reactant particles must have when they collide to break their chemical bonds

All chemical reactions require some energy to get them started. For example, wood does not burst into flames when exposed to the oxygen in the air. The flame of a match provides the "kick-start" energy needed to get the hydrocarbons in wood to react with the oxygen in the air. Similarly, the butane gas in a disposable lighter requires a spark to start the burning reaction. This kick-start energy is called the **activation energy**.

Brainstorm with your team examples of how activation energy is supplied in some chemical reactions.

1. List your examples.

Participate in a class discussion about your answers.

You already know the Small-Particle Theory (SPT) of activation energy. In Activity 3, you learned that reactant particles (atoms or molecules) must collide for a reaction to occur. But not all collisions result in the breaking of bonds and the recombination of the atoms. If the reactant particles do not have enough motion energy when they collide to break the chemical bonds, they simply bounce apart. The motion energy of particles depends on the temperature. The higher the temperature, the greater the motion energy. For example, at room temperature the nitrogen and oxygen molecules in the air do not have enough motion energy to break the very strong N-N bond, as shown in the diagram.

Ultrascope Diagram at Room Temperature
N_2 and O_2 Molecules Collide and Bounce Apart

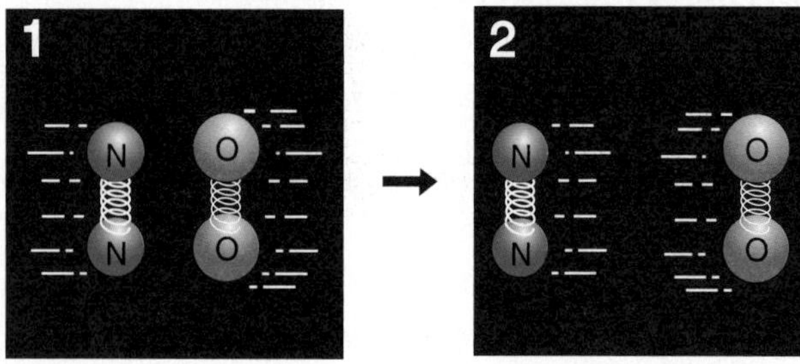

Ultrascope Diagram at High Temperature
N_2 and O_2 Molecules Collide, Atoms Recombine

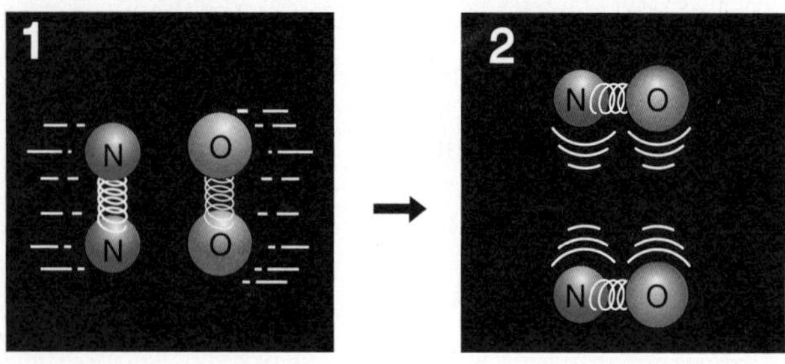

1 molecule N_2 + 1 molecule O_2 → 2 molecules NO

Activation energy is the minimum motion energy that reactant particles must have when they collide to break their chemical bonds. Some reactions have high activation energies. For example, nitrogen and oxygen require the high temperatures inside a car gasoline engine (2800°C) before the motion energy of their molecules is high enough to break the strong N-N bond.

Reactions that happen by themselves at room temperature have a low activation energy. For example, the reaction between vinegar and baking soda has a low activation energy. The reaction between carbon dioxide gas and bromothymol blue (BTB) indicator also has a low activation energy.

You did an exploration in Activity 6 with rusting iron. You are also familiar with the rusting of other things made with iron, like parts of a bicycle or car.

2. Do you think the activation energy for the rusting reaction is *high*, *medium*, or *low*? Explain your reasoning.

Participate in a class discussion about your answer.

Recall the exploration you did in Activity 3 with yellow goop (lead iodide). A diagram and word equation for this reaction is shown below.

3. Do you think the activation energy for this lead iodide reaction is *high*, *medium*, or *low*? Explain your reasoning.

Participate in a class discussion about your answer.

Ultrascope Diagram
PbI_2 Molecules Collide, Atoms Recombine

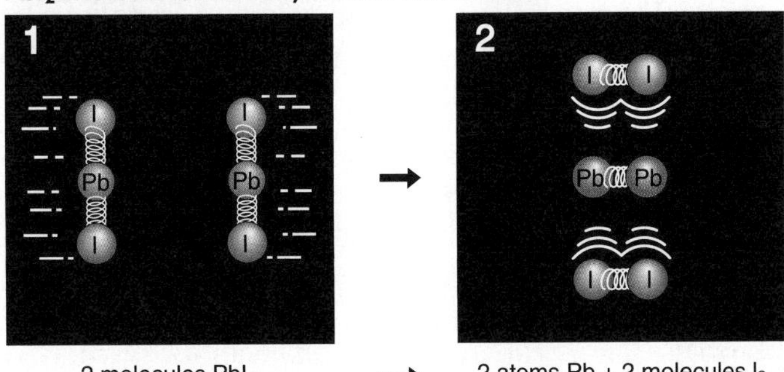

2 molecules PbI_2 → 2 atoms Pb + 2 molecules I_2

4. Complete the following statement about activation energy on your record sheet.

Idea 11C1 The activation energy of a chemical reaction is the minimum ____________ energy that the reactant particles must have to break their chemical bonds. The higher the temperature required to start the reaction, the ______________ the activation energy.

Participate in a class discussion about your answer.

Science Words

exothermic reaction: reactions that produce (release) energy

endothermic reaction: reactions that absorb energy (require continual energy input)

Energy Transfer during Chemical Reactions

Once started, all chemical reactions either:

- produce energy—continually transfer energy out of the system of interacting substances to the surrounding objects; or
- absorb energy—require a continual transfer of energy into the system of interacting substances.

Chemists call reactions that produce (release) energy **exothermic** (ehk-soh-THER-mihk) reactions. *Exo* means "go out" or "exit" and *thermic* means "heat." For example, the explosion of dynamite and the burning of gasoline and wood are exothermic reactions because in both cases heat energy is transferred out of the systems of reacting chemicals.

Chemists call reactions that absorb energy (require continual energy input) **endothermic** (ehn-doh-THER-mihk) reactions. *Endo* means "go in." For example, the lead iodide reaction is endothermic because you had to heat the lead iodide (PbI_2) to produce solid lead (Pb) and iodine gas (I_2).

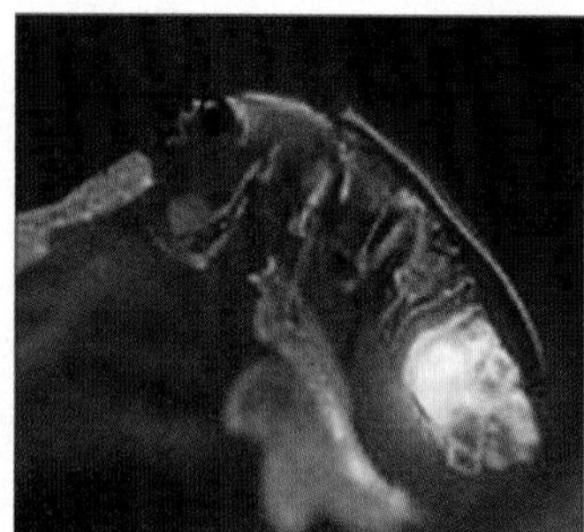

Heat energy is usually transferred into or out of the system during chemical reactions. But other forms of energy transfer can occur, such as light, sound, and electrical energy transfer. For example, the chemical reaction inside a battery results in an electrical energy transfer to a bulb, motor, or other device in a circuit. Fireflies produce light energy through a chemical reaction in their bodies.

Small-Particle Theory (SPT) of Exothermic and Endothermic Reactions

A chemical system consists of both the reactant particles and the product particles. At the start of any reaction, all of the chemical energy of the system is stored in the bonds of the reactant particles. Just as the reaction ends, all the chemical energy of the system is stored in the bonds of the product molecules. Whether a reaction is exothermic (releases energy) or endothermic (absorbs energy) depends on the comparison of the chemical-bond energy of the reactant particles and the chemical-bond energy of the product particles.

Exothermic Reactions

For example, consider again the reaction between nitrogen (N_2) and oxygen (O_2) to make nitrogen monoxide (NO). The chemical-bond energy of the reactant molecules is the sum of the N-N bond energy and the O-O bond energy:

Bond energy of reactant molecules = N-N bond energy + O-O bond energy

1. What is the chemical-bond energy of the product molecules? Be specific.

Bond energy of product molecules = ________________

Participate in a class discussion about your answer.

Chemists have measured the chemical energy stored in the bonds between different atoms. For this reaction, the bond energy of the reactant molecules (N_2 and O_2) is *greater than* the bond energy of the product molecules (NO). The Conservation of Energy Law states that energy is never created or destroyed. It is merely transformed from one kind to another and/or transferred from one system to another.

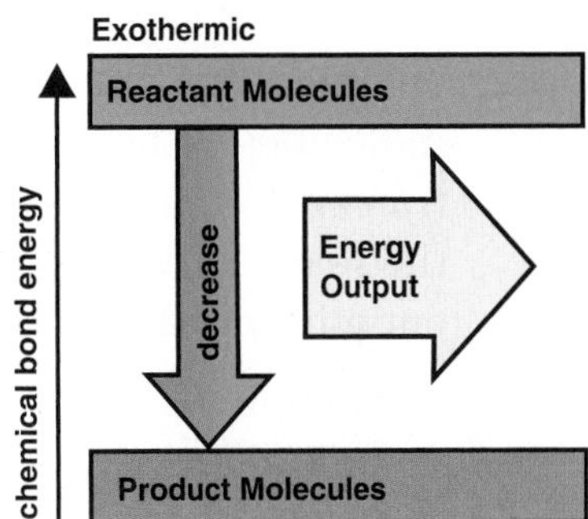

Reactant-product Diagram for an Exothermic Reaction

So what happens to the extra chemical-bond energy of the reactants? Where does this energy go? The extra energy is transferred out of the system, as shown in this reactant-product diagram. The reaction between nitrogen and oxygen is an exothermic reaction.

The stored chemical energy of each substance (reactant or product) is the sum of the billions of billions of chemical-bond energies of the particles of the substance. Below is an incomplete energy diagram for an exothermic reaction.

2. Complete this energy diagram for an exothermic reaction on your record sheet.

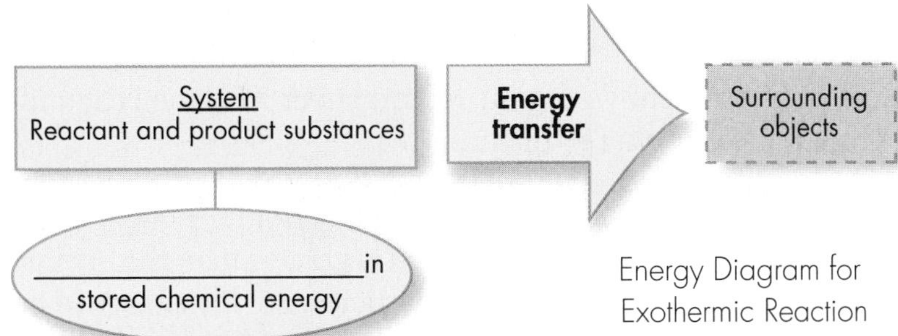

Energy Diagram for Exothermic Reaction

3. Complete the statement below on your record sheet. Use the reactant-product diagram and your energy diagram

> **Idea 11C2** For an exothermic chemical reaction, the chemical bond energy of the ______________ particles is greater than the chemical bond energy of the ______________ particles. So, for energy to be conserved during the reaction, energy is transferred ______________ the system, and the stored chemical energy of the system ______________.

Participate in a class discussion about your answer to these questions about exothermic reactions.

Endothermic Reactions

Consider again the lead iodide (PbI_2) reaction that makes solid lead (Pb) and iodine gas (I_2).

For this reaction, the chemical-bond energy of the *product particles* (2 Pb atoms and 2 I_2 molecules) is greater than the chemical-bond energy of the reactant molecules (2 PbI_2 molecules). Energy must be continually transferred into the system, as shown in the reactant-product diagram for an endothermic reaction.

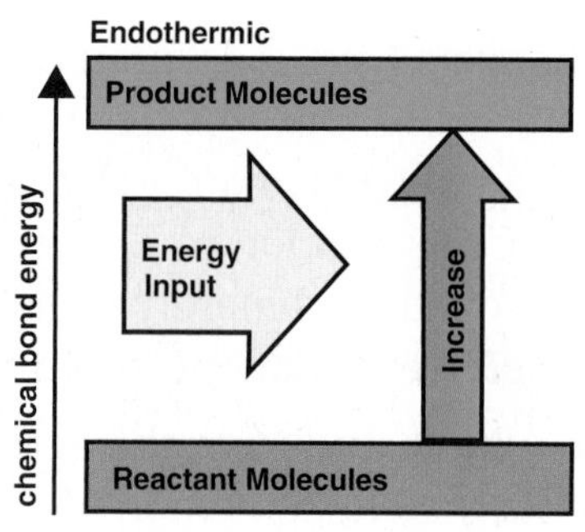

Reactant-product Diagram for an Endothermic Reaction

4. Complete this energy diagram for an endothermic reaction on your record sheet.

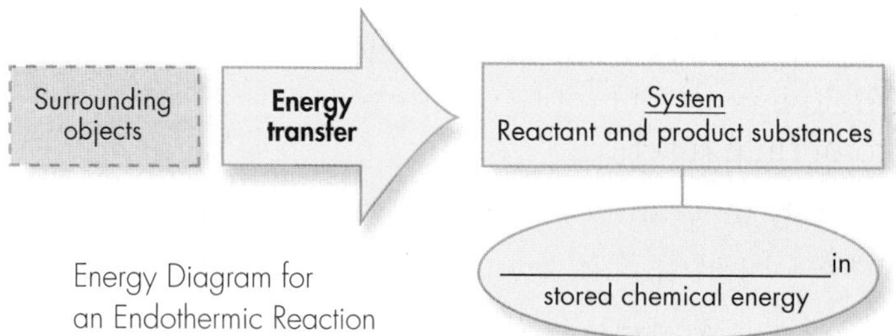

Energy Diagram for an Endothermic Reaction

5. Complete the statement below on your record sheet. Use the reactant-product diagram and your energy diagram

> **Idea 11C3** For an *endothermic* chemical reaction, the chemical bond energy of the ______________ particles is greater than the chemical bond energy of the ______________ particles. So, for energy to be conserved during the reaction, energy is transferred ______________ the system, and the stored chemical energy of the system ______________.

Participate in a class discussion about your answer to these questions about endothermic reactions.

Examples of Reactions

In your exploration, you mixed vinegar with baking soda in a baggie. When you held the baggie, it felt cold. You teacher will work with you as you answer Questions 6 – 9.

6. Is the vinegar-and-baking soda reaction exothermic (continually produces energy) or endothermic (continually absorbs/requires energy)? What is your evidence? (Hint: Look at your answer to the Explore Your Ideas Question 2 on your record sheet.)

Below is an incomplete energy diagram for the vinegar-and-baking soda reaction.

7. On your record sheet, draw and label the energy-transfer arrow (energy input or energy output and the type of energy transfer).
8. Fill in the blank of the energy oval.
9. Which is greater, the chemical-bond energy of the reactant particle ($NaHCO_3$ particle and CH_3COOH particle) or the product particles (CO_2 molecule, H_2O molecule, and CH_3COONa particle). Justify your answer.

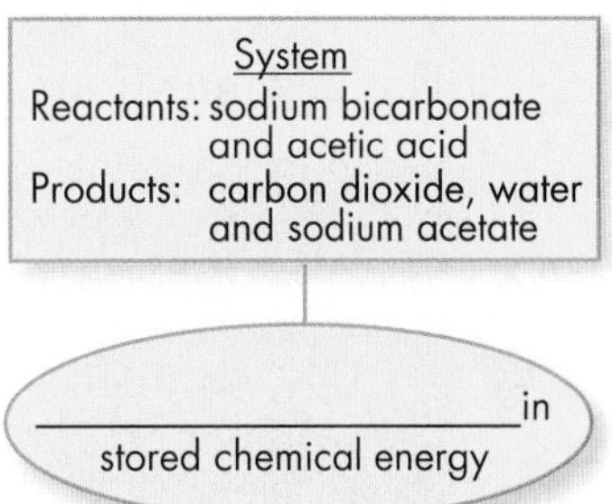

Now it's your turn! In Activity 6, you did a rusting iron exploration. You put clean steel wool in a flask that was sealed with a balloon.

10. Is the rusting iron reaction exothermic (continually produces energy) or endothermic (continually absorbs/requires energy)? What is your evidence? (*Hint:* Look at your answer to the Make Sense of Your Ideas Question 3 on your Activity 6 record sheet.)

Below is an incomplete energy diagram for the rusting iron reaction.

11. On your record sheet, draw and label the energy-transfer arrow (energy input or energy output and the type of energy transfer).
12. Fill in the blank in the energy oval.
13. Which is greater, the chemical-bond energy of the reactant particles (Fe atoms and O_2 molecules) or the product molecules (Fe_2O_3)? Justify your answer.

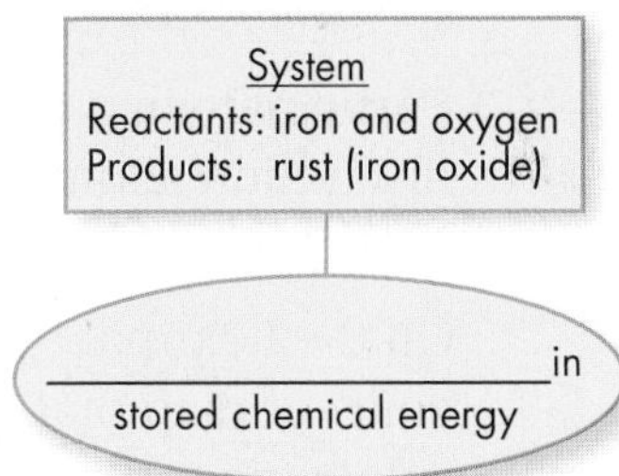

Participate in a class discussion about your answers to these questions.

Our Consensus Ideas

The key question for this activity is:

What is the relationship between the transfer of energy into or out of a chemical reaction and the chemical-bond energy of the reactant and product molecules?

1. Write your answer on your record sheet.

Participate in the class discussion about the key question.

2. Write the class consensus idea on your record sheet, if it is different from yours.

PUTTING IT ALL TOGETHER

Activity 9
Chemical Reactions

Comparing Consensus Ideas

The key questions for this cycle of learning are:

1. What is the scientists' theory of chemical interactions on a scale too small to see?

2. How do chemists use their theory to describe the Conservation of Mass and the Conservation of Energy during chemical interactions?

To answer these questions, you investigated some chemical reactions—the burning reaction, the lead iodide reaction, the rusting reaction, and the vinegar and baking soda reaction. You made sense of several Small-Particle Theory (SPT) ideas to describe these interactions, the conservation of mass, and the conservation of energy.

Your teacher will pass out the sheet titled *Scientists' Consensus Ideas: The Small-Particle Theory of Matter—Ideas 9 –12.*

Compare the Explore and Make Sense of Your Ideas or the Consensus SPT Ideas sections in Activities 3, 4, 5, 7, and 8 with the *Scientists' Consensus Ideas.*

Participate in a class discussion about these ideas.

On the next page is a concept map that summarizes the ideas about small-particle interactions. You completed part of this map towards the end of Unit 6.

With your teacher's guidance, read through the map. Fill in the blanks in the map on your record sheet.

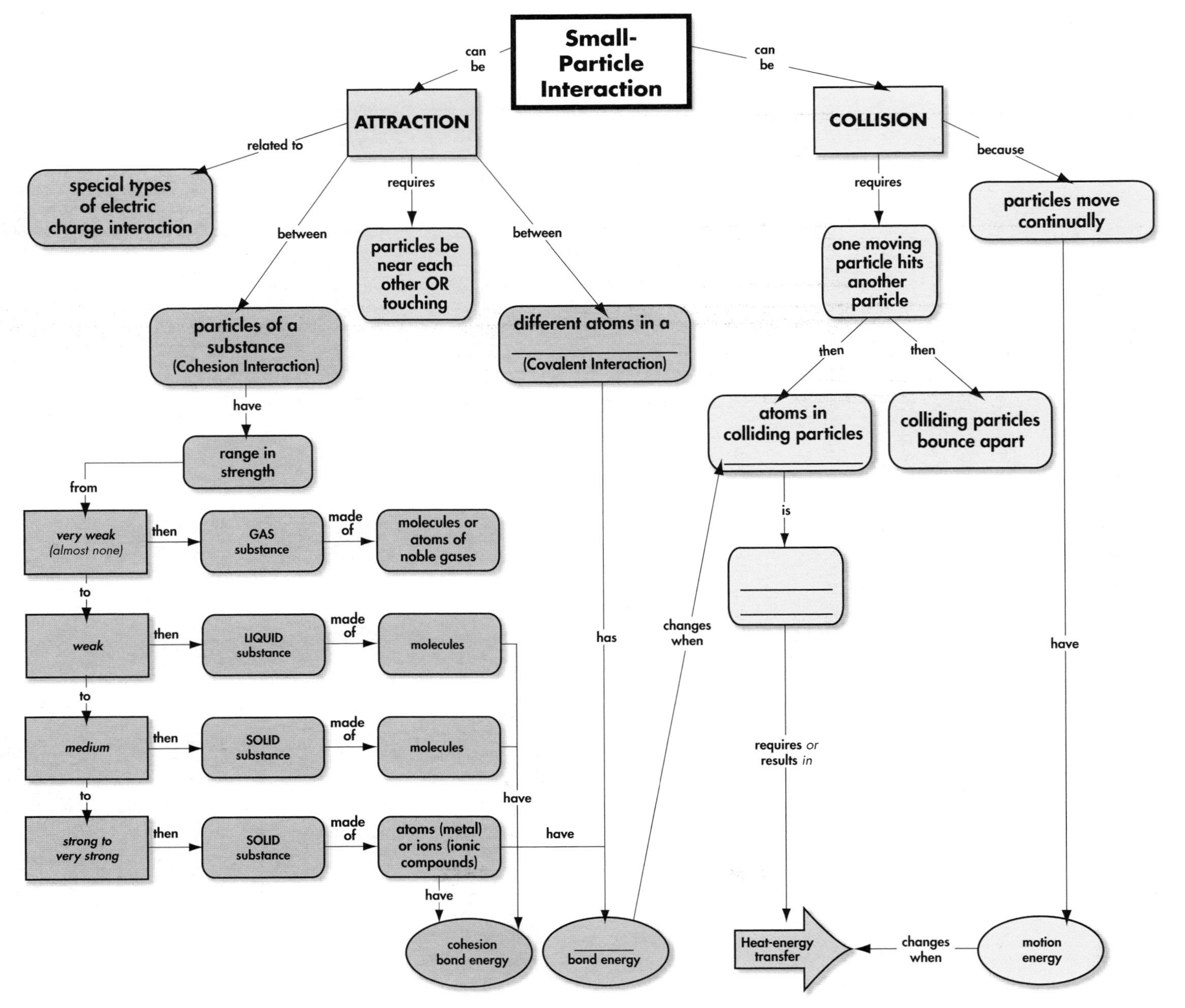

Small-Particle Interaction
can be
ATTRACTION
can be
COLLISION
related to
special types of electric charge interaction
requires
particles be near each other OR touching
between
particles of a substance (Cohesion Interaction)
between
different atoms in a ______ (Covalent Interaction)
have
range in strength
from
very weak (almost none)
then
GAS substance
made of
molecules or atoms of noble gases
to
weak
then
LIQUID substance
made of
molecules
to
medium
then
SOLID substance
made of
molecules
to
strong to very strong
then
SOLID substance
made of
atoms (metal) or ions (ionic compounds)
have
have
have
cohesion bond energy
has
______ bond energy
changes when
requires
one moving particle hits another particle
then
atoms in colliding particles ______
then
colliding particles bounce apart
is
______
requires or results in
Heat-energy transfer
because
particles move continually
have
motion energy
changes when

IDEA POWER

Activity 10
Using the Small-Particle Theory Ideas

In this unit, you learned how to model balanced chemical equations, how to decide if picture equations are correct (balanced) and how to write balanced word equations. Apply these skills in the next problem.

Problem 1

Safety air bags in cars inflate because of a chemical interaction. A head-on crash sets off a detonator cap inside the deflated bag. The activation energy supplied by the detonator starts the chemical reaction. The reactant is solid sodium azide (NaN_3). The products are sodium metal (Na) and nitrogen gas (N_2). The sudden release of nitrogen gas inflates the bag to a volume of 74 L in four-hundredths (0.04) of a second after the crash!

1. Below is a picture equation of this reaction. Is this equation balanced? (Does it show the Conservation of Mass at the small-particle level?) Use the procedure in *How To Decide Whether a Picture Equation of a Chemical Interaction is Correct (Balanced)* in the Appendix.

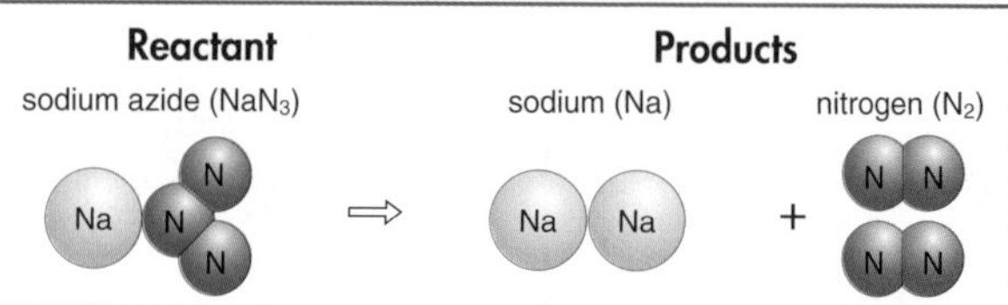

2. If the picture equation is not balanced, then add or cross out pictures of the particles (atoms or molecules) to correct the equation. You may want to use your atom model kit to model the reaction.

3. Write a balanced word equation for this reaction. If necessary, use *How To Write a Word Equation of a Chemical Interaction from the Picture Equation* in the Appendix.

Participate in a class discussion about your answers to these questions.

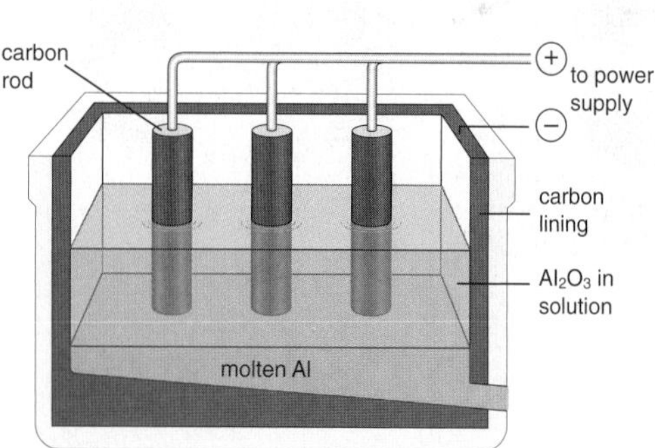

You learned in Activity 8 how to decide whether a reaction is exothermic or endothermic. Apply this skill in the following problems.

Problem 2

Most metal elements on Earth are found in compounds called ores or oxides. Electrolysis is sometimes used to obtain the metal from its ore. For example, aluminum metal is obtained by passing an electric current through molten aluminum ore (aluminum oxide, Al_2O_3), as shown in the diagram.

Large steel pots are lined with solid carbon. Carbon rods hang down into a solution of molten aluminum oxide and a salt called cryolite. The carbon lining and carbon rods are attached to a source of electric current. The aluminum oxide reacts with the carbon in the rods to produce aluminum and carbon dioxide gas. The carbon dioxide bubbles off around the rods. The molten aluminum collects on the bottom of the pot.

4. Is this reaction exothermic (continually produces energy) or endothermic (continually absorbs/requires energy)? What is your evidence? (*Hint*: Is there an energy source required for the aluminum oxide reaction?)

On the right is an incomplete energy diagram for the aluminum oxide reaction.

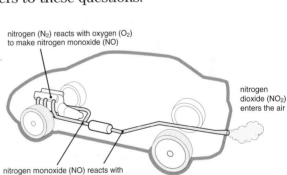

System of Chemicals
Reactants: aluminum oxide and carbon
Products: aluminum and carbon dioxide

____________________ in stored chemical energy

5. On your record sheet, draw and label the energy transfer arrow (energy input or energy output and the type of energy transfer).
6. Fill in the blank of the energy oval.
7. Which is greater, the chemical-bond energy of the reactant particles (Al_2O_3 particles and C atoms) or the product particles (Al atoms and CO_2 molecules)? Justify your answer. (*Hint*: Use the reactant–product diagrams.)

Participate in a class discussion about your answers to these questions.

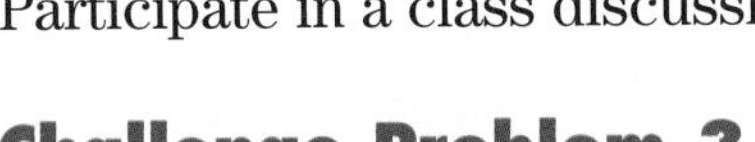

Challenge Problem 3

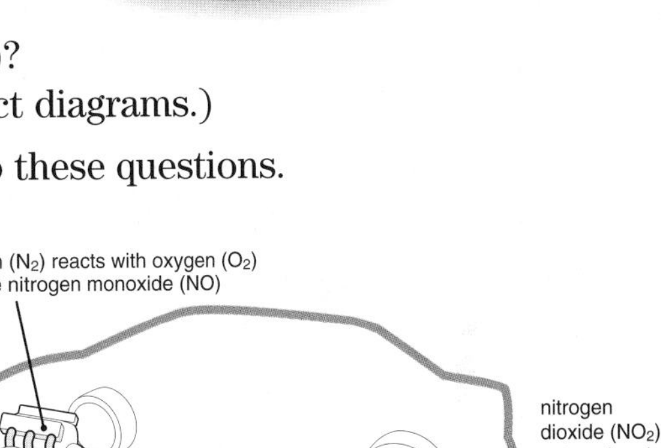

In the exhaust pipe of a car, two nitrogen monoxide molecules (NO) react with an oxygen molecule (O_2) to produce two nitrogen dioxide molecules (NO_2). The two Ultrascope pictures show this collision interaction.

8. Write a balanced word equation for this reaction.

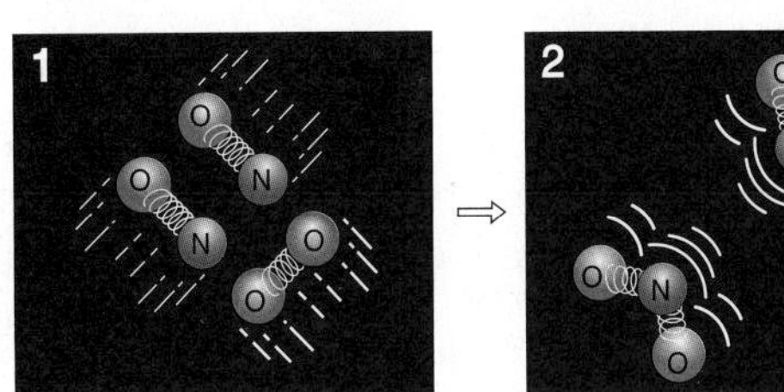

9. Describe the Small-Particle Theory (SPT) of the activation energy for this collision.

10. The chemical-bond energy of the reactant molecules is greater than the chemical-bond energy of the product molecules. Is this an exothermic reaction or an endothermic reaction? Explain your reasoning.

Participate in a class discussion about your answers to these questions.

LEARNING ABOUT OTHER IDEAS

Activity 11
Polymers

Purpose

Look at the photograph of the spider web glistening in the Sun. It was spun overnight from silken fibers produced by the spider's body. These fibers, the plant stems that support the web, and much of the spider itself are made from solid compounds called *polymers* (PAHL-uh-murz).

You learned in Unit 6, Activity 6 that room temperature solids consist of repeating patterns of atoms, ions, or molecules. At the end of the activity, you will be able to answer the key question for this activity.

What are polymers and how are the properties of polymers related to their molecular structure?

Record the key question for the activity on your record sheet.

Learning the Ideas

Synthetic Materials

Have you ever watched a football game? The clothing, helmet, and shoes of the football players are all made of *synthetic* materials. So is the football! Synthetic means that the material was made in laboratories from simpler compounds. The starting compounds came from coal or crude oil. These starting compounds have something in common. They are carbon compounds. Carbon compounds contain carbon atoms bonded to the atoms of other elements.

Carbon's Strings, Rings, and Other Things

Almost every piece of clothing you wear is made of compounds of carbon. So is most of the food you eat and the fuel that keeps you warm. Carbon compounds are put to use in cars, computers, and calculators. The cells of both plants and animals (including you) are made of carbon compounds. In fact, carbon is present in more than 13 million known compounds, and more are being discovered or made every day.

Hydrogen is the most common element found with carbon in its compounds. Other elements include oxygen, nitrogen, phosphorous, sulfur, chlorine, fluorine, bromine, and iodine.

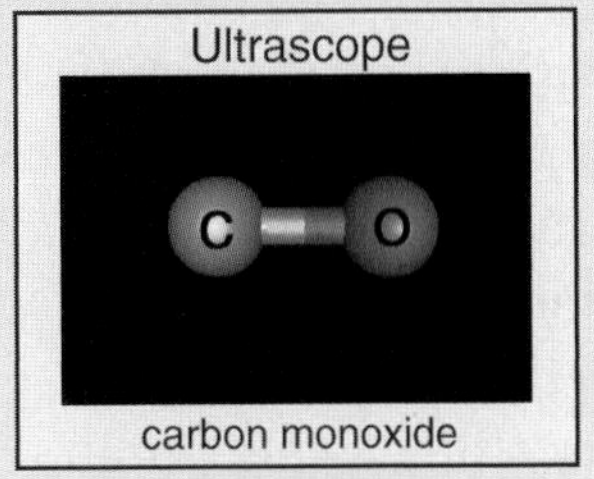

carbon monoxide

one bond

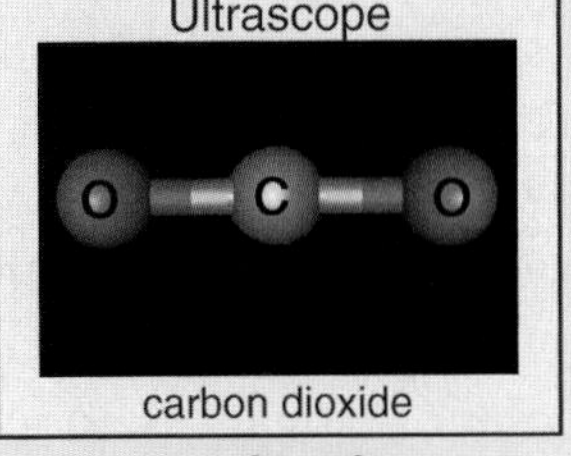

carbon dioxide

two bonds

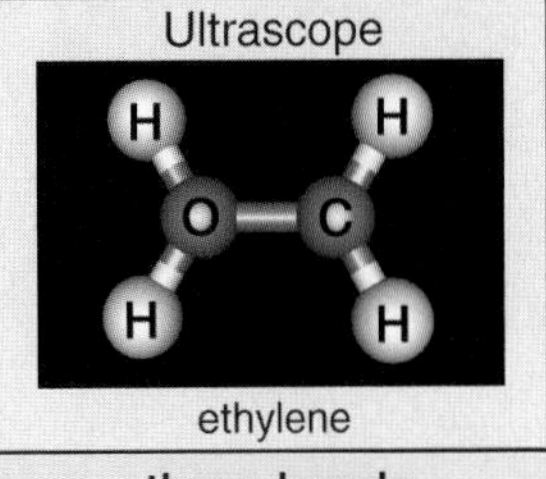

ethylene

three bonds

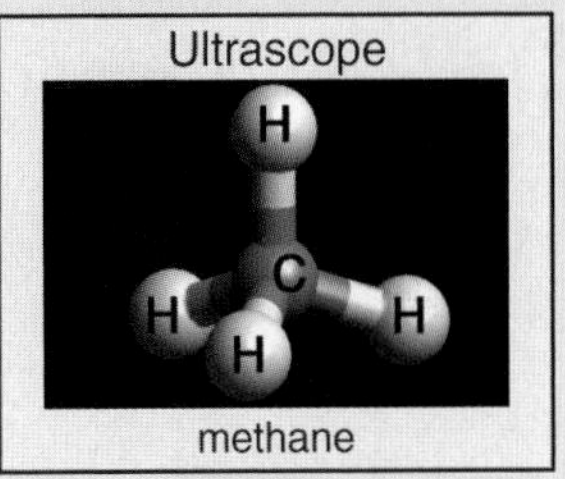

methane

four bonds

Carbon's unique ability to form so many compounds comes from two properties. First, carbon can make from 1 to 4 covalent bonds with itself or with many other elements, as shown in the diagram above. Second, carbon atoms can bond to each other in straight chains, branched chains, and ring-shaped groups. These structures form the "backbones" to which other atoms attach, as shown in the drawing below. In this drawing, the lines represent the covalent bonds that can form between atoms.

straight chain

branched chain

ring

This variety of bonds allows carbon-based molecules to have a wide range of shapes and chemical properties. For example, some carbon-based molecules, like carbon monoxide and carbon dioxide, are linear. The table on the page after the next page shows that some carbon-based molecules, like ethylene and vinyl chloride are planar (all in one plane). Some carbon-based molecules, like methane and styrene, have different 3-dimensional shapes.

1. Why is it possible for carbon to form so many different compounds?

Your teacher will review the answer to this question with the class.

Carbon Compounds Form Polymers

Science Words

polymer: a substance that is a giant molecule made up of many similar small molecules (monomers) linked together in long chains

monomers: the smaller molecules that are linked together to form the giant molecules of polymers

Have you ever made a chain out of paper loops or paper clips? A chain is made up of smaller units (paper loops or metal clips) that are linked together. Similarly, a **polymer** is a giant molecule built of smaller carbon molecules bonded together.

Polymers can be thought of as chains. They are built of smaller parts bonded together.

The smaller molecules or units from which the polymers are built are called **monomers** (MAHN-uh-murz). The term polymer comes from the Greek *poly* (many) and *meros* (part). *Mono* means "one." So polymers are giant molecules made of many parts, each single part being a monomer (small carbon molecule). Polymers are made during chemical reactions that cause the monomers to bond together. Linking only one type of small molecule, or monomer, together, as shown in the diagram below, makes most polymers.

one kind of monomer

You could think of these monomers as linked like the identical cars of a long passenger train. Five monomers (carbon molecules) of different kinds of plastic polymers are shown in the table on the next page.

Some polymers are made with two different types of monomers joined in the same polymer chain. Imagine there are two monomers, called green and purple. These two monomers can be made into a polymer in many different ways, as shown in the diagram. For example, the two monomers can be linked together in random order, or in an alternating pattern. In a block polymer, all of one type of monomer is grouped together, and all of the other is grouped together. When chains of a polymer made of monomer green are joined onto a polymer chain of monomer purple, it becomes a "graft" polymer.

Two-monomer Chains with Different Patterns

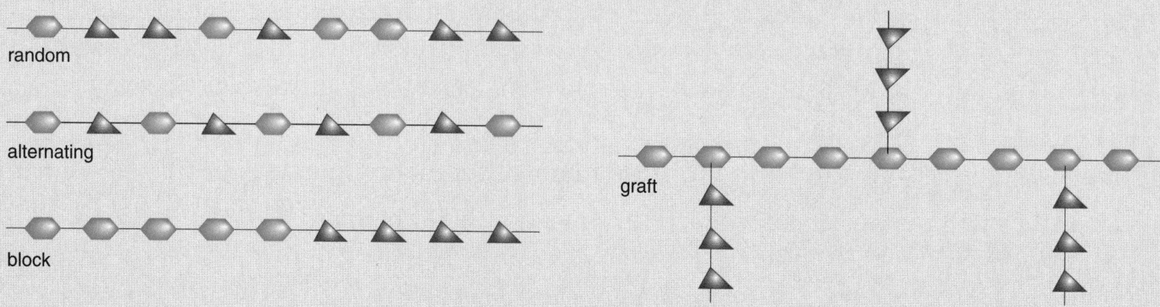

2. How are monomers related to polymers?

3. What are some different ways that monomers bond together to form polymers?

Your teacher will review the answers to these questions with the class.

Table: Monomers and their Plastic Polymers

Monomer	Polymer	Properties	Uses
Ethylene (C_2H4)	Polyethylene (PE)	Low densiity: flexible, soft, melts easily High density: rigid and stronger, higher melting temperatures	Low density: plastic bags, squeeze bottles, electric wire insulation High density: detergent bottles, gas cans, toys, milk jugs
Vinyl chloride (C_2H_3-Cl)	Polyvinyl chloride (PVC)	Tough, flexible	Water pipes, house siding, credit cards, raincoats, shoes
Propylene (C_2H_3-CH_3)	Polypropylene (PP)	Hard; keeps its shape, can be made into fibers	Toys, housewares, carpet backing, medical equipment, electronic components
Styrene (C_2H_3-C_5H_5)	Polystyrene (PE)	Lightweight; can be made into foam	Auto instruments and panels, housing of many applicances, coffee cups, insulation, "peanut" packing
Acrylonitrile (C_2H_3-CN)	Polyacrylonitrile	Used mostly as one of the monomers in polyester fabrics	Orlon, Acrilon, clothing, yarns, wigs

Natural and Synthetic Polymers

Science Words

plastics: synthetic polymers that can be molded or shaped

Natural polymers have been around for as long as life on Earth. Plants, animals, and other living things make many natural materials made of giant polymer molecules. You will learn more about natural polymers in the next activity. Synthetic polymers are made during chemical reactions in a laboratory or factory. The starting materials for synthetic polymers come from coal or oil. **Plastics** are synthetic polymers that can be molded or shaped.

To make things out of plastic, some small plastic pieces are heated until they melt. The melted plastic is then poured, injected, or blown into molded shapes, like drinking glasses and toys. It takes a lot of energy to make plastic shapes. But once made, they stay the same shape. Until it gets hot enough, plastic cannot easily be deformed.

4. What is plastic?

5. Look around your classroom. Name the items made of plastic or plastic parts.

Your teacher will review the answers to these questions with the class.

The table on the previous page shows that different kinds of plastics are made of different monomer molecules. For example, polyethylene is made of the monomer ethylene, as shown in the diagram below. You probably see this plastic most in your daily life. Polyethylene is used to make grocery bags, shampoo bottles, children's toys, and even bulletproof vests.

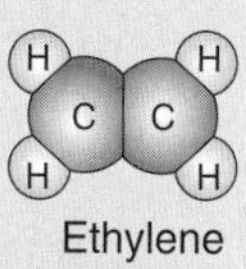

Ethylene monomer

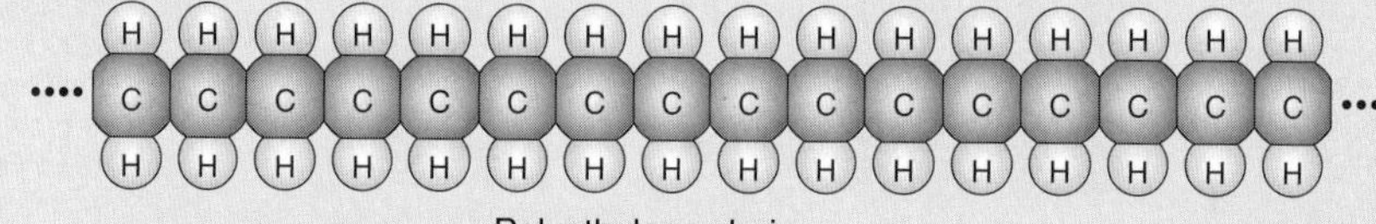

Polyethylene chain

This bag is made of LDPE. The branches in the structure of this polymer makes it flexible and suitable for this use.

The Physical Properties of Polymers

One of the principles of chemistry is that small changes in the *structure* of a molecule can lead to big changes in physical and chemical properties of the material. In this section, you will explore some differences in the structure of polymer molecules that result in changes in their properties.

Chains with or without Branches

Plastic bags are made from **l**ow-**d**ensity **p**oly**e**thylene (LDPE). The process of making the polyethylene produces branches along the polymer chain, as shown in the diagram on the next page.

These branches prevent the polymer chains from getting close to each other. This weakens the cohesion forces between the chains and makes the polyethylene more flexible.

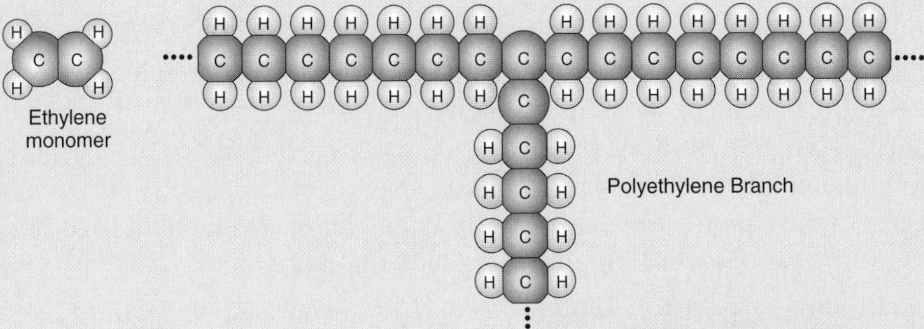

However, a finger-joint implant is made of **h**igh-**d**ensity **p**oly**e**thylene (HDPE). Here the manufacturing conditions result in chains that are without branches. The chains lie side by side to a much larger extent, as shown in the diagram below. This results in stronger cohesion forces and a stronger polymer.

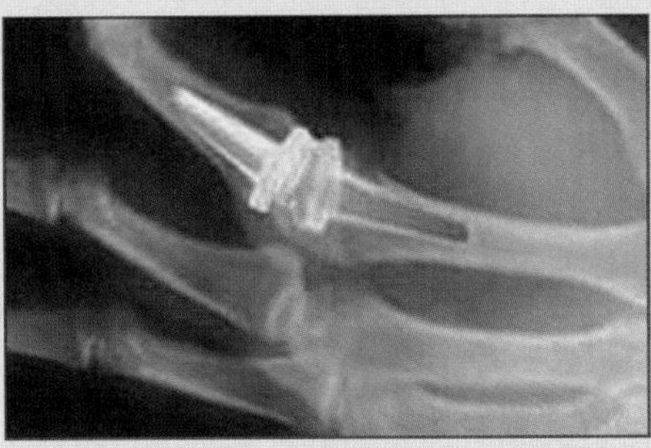

Finger-joint implants are made of HDPE. The chains do not have branches and this results in a stronger polymer.

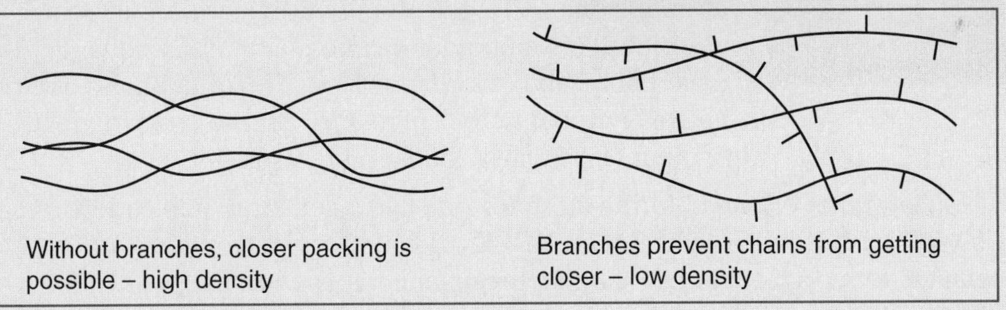

Without branches, closer packing is possible – high density
Branches prevent chains from getting closer – low density

6. Why does the density and mechanical strength of a polymer depend on whether the polymer molecule chain has branches or is without branches?

Your teacher will review the answer to this question with the class.

Chain Lengths

When a polymer is made, the molecular chains are usually different lengths. Two polymer molecules with the same repeating monomer units, but different lengths, are still considered molecules of the same substance. So, chemists refer to the *average chain length* of a polymer. Different polymers can have different average chain lengths, ranging from 500 to 20,000 or more monomer units. The same polymer made with different processes can also have different average chain lengths.

The mechanical strength of a polymer depends on its average chain length. For example, tearing apart a tangled long-chain polymer means breaking the backbone covalent bonds between carbon atoms. These covalent bonds are much stronger than the cohesion forces holding the chains together. Longer chains mean higher mechanical strength. Another property of polymers, the ability to flow when molten (viscosity), also depends on chain length. Long chains are tangled together more, so they flow slower than polymers with shorter chains.

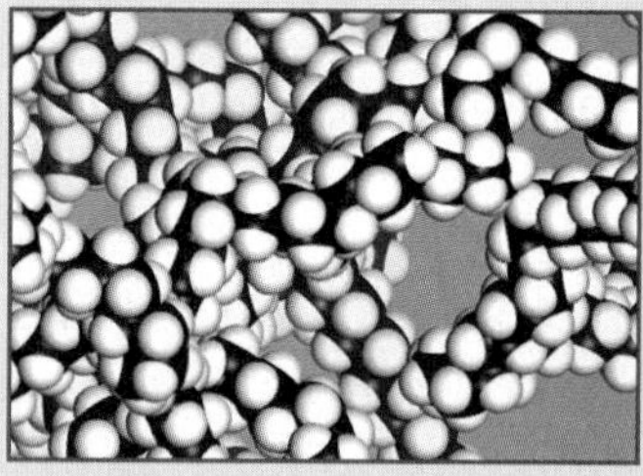
A long-chain polymer.

7. Why does the viscosity (ability to flow) and mechanical strength of a polymer depend on the average length of the polymer chains?

Your teacher will review the answer to this question with the class.

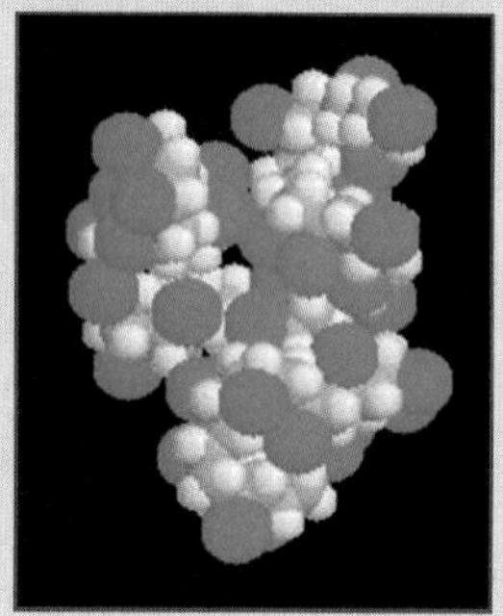
A molecule of PVC.

The Atoms Bonded to the Carbon Backbone

The properties of a polymer depend on the different atoms (elements) bonded to the carbon atoms in the monomer. For example, look at the two monomers, ethylene and vinyl chloride in the Table: Monomers and their Plastic Polymers. If you substitute one chlorine atom for one hydrogen atom in ethylene, you get vinyl chloride. This monomer can be made into a polymer called polyvinyl chloride or PVC. The plumbing in your house is probably PVC pipe. PVC is also used to make linoleum and "vinyl" siding on houses.

The substitution of one chlorine atom for one hydrogen atom makes PVC (polyvinyl chloride) "tougher" than HDPE (high-density polyethylene). The technical term for toughness is impact resistance. It is what happens to the material when you hit it with a hammer sharply and strongly. The cohesion force between the PVC chains is stronger than the cohesion force between the HDPE chains. So, PVC is less brittle (harder to break) than HDPE. The stronger cohesive force between chains also makes thin sheets of PVC more flexible (easier to bend without breaking). This is why credit cards are made out of PVC and not HDPE.

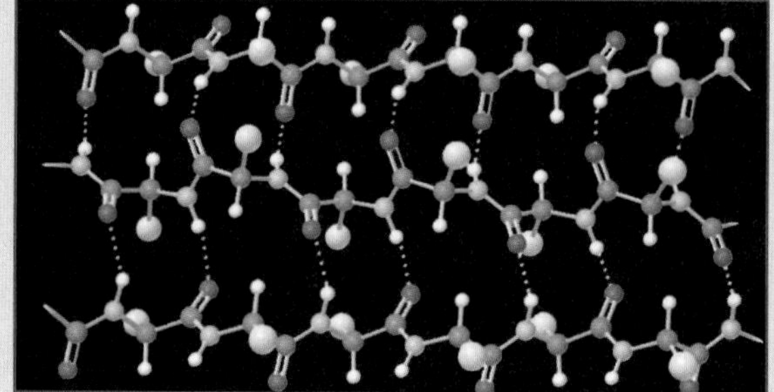
The dotted lines between the chains represent cohesive forces between the chains.

8. Look at the *propylene* monomer in the table. What *group of atoms* replaces one of the hydrogen atoms in ethylene?

Polypropylene is used to make things like dishwasher-safe food containers, because, unlike the more common polyethylene plastic, it melts at a higher temperature so it does not bend or warp in the dishwasher.

9. Look at the styrene monomer in the table. What *group of atoms* replaces one of the hydrogen atoms in ethylene?

Polystyrene is used to make drinking cups. A popular brand of polystyrene foam is called Styrofoam™.

Polystyrene is an inexpensive and hard plastic that is used everywhere. Probably only polyethylene is more common in your everyday life. The outside housing of most computers is made of polystyrene. The housings of things like hairdryers, TVs, and kitchen appliances are also polystyrene. Model cars and airplanes, as well as many other toys, are made from this polymer. So are a lot of the molded parts on the inside of your car, like the radio knobs.

All of these items are made of plastic.

Nylon and other plastic fibers are made from plastic chips that are melted and pressed through narrow openings to make strands. Strands can be packed together to make ropes. Strands can also be twisted together to make threads and yarn from which fabrics are woven.

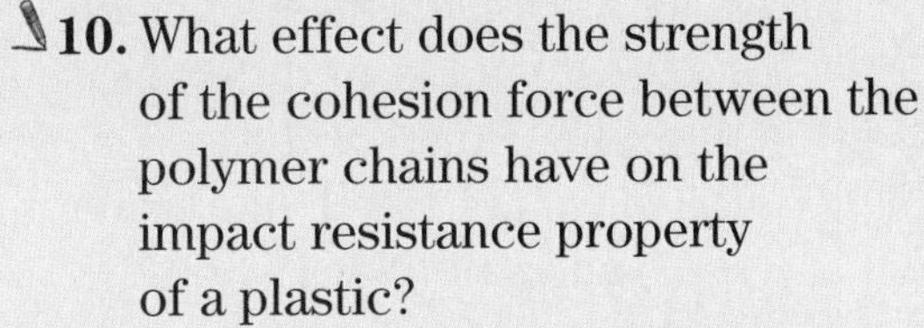

10. What effect does the strength of the cohesion force between the polymer chains have on the impact resistance property of a plastic?

Your teacher will review the answer to this question with the class.

Nylon strands are made from melted plastic.

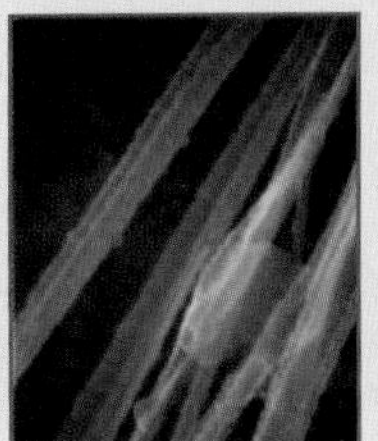

Strands can be twisted together to make yarn.

Strands are packed together to make rope.

What We Have Learned

Recall the key question for this activity:

What are polymers and how are the properties of polymers related to their molecular structure?

Participate in a class discussion to review the answers to the key question.

Write the answer to the key question on your record sheet.

LEARNING ABOUT OTHER IDEAS

Activity 12
The Chemistry of Life – Part 1

Purpose

You learned in Activity 11 that plants, animals, and other living things make thousands of giant polymer molecules from small, carbon-based monomers.

In this activity, you will explore the elements and molecules that are essential for the chemistry of life. The key questions for this activity are:

1. What elements make up most of the mass of living organisms on Earth?

2. What are the major molecules, small and large, that make up organisms and control chemical reactions in the organisms?

Record the key questions for the activity on your record sheet.

Learning the Ideas

The Elements of Life

All living organisms are made of a great variety of different molecules. Some of these molecules are small, like water (H_2O) and salt (NaCl). Some of these molecules are huge polymers, made up of thousand or millions of atoms. But the number of different elements involved is quite small. In fact, most of the mass of all living organisms on Earth is made up of only six elements. These six elements are shown on the periodic table.

1. List the six elements that make up most of the mass of the Earth's living organisms.

Elements in the Human Body	
Oxygen	65.0%
Carbon	18.5%
Hydrogen	9.5%
Nitrogen	3.3%
Calcium	1.5%
Phosphorus	1.0%
Potassium	0.4%
Sulfur	0.3%
Sodium	0.2%
Chlorine	0.2%
Magnesium	0.1%

Of course, some other trace elements are important for plants and animals. For example, the illustration shows the percentage of elements found in the human body.

2. What other elements (besides the six you listed in Question 1) are found in the human body?

Your teacher will review the answers to these questions with the class.

Molecules Important For Life

Water

You know that plants will shrivel up and die without water. You could survive for weeks without food, but would die within a few days without water. In fact, most plants and animals are mostly made of water. For example, your body is about 70 % water. Why is water so essential for life?

The key chemical role for water in plants and animals is to *dissolve* other chemicals of life and transport these chemicals from place to place. Plants and animals are mostly made up of millions to billions of cells, each filled with a watery solution (called cytoplasm). Most of the chemical reactions that allow plants and animals to live occur in this watery solution in cells. Water-based solutions, like blood, allow chemicals to be whisked from place to place as needed.

Small amounts of other elements and vitamins are also essential for life. For example, the sodium (Na) from salt is used to regulate the movement of water into and out of cells. Vitamins are compounds that are necessary for good health, but are not made in our bodies. Each vitamin performs different, specialized tasks.

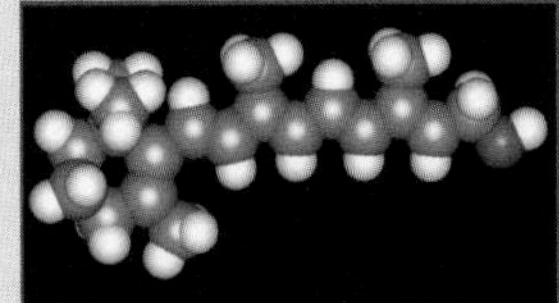

Vitamin A is used in the production of cell walls, color vision, and the proper tooth development and bone growth.

3. What role does water play in the chemistry of life?

Your teacher will review the answer to this question with the class.

Some other categories of molecules important for life are carbohydrates, fats, proteins, and DNA. Explore each of these categories of polymers in the following sections.

Carbohydrates – The Energizers

Carbohydrates are made up of only three elements: carbon, hydrogen, and oxygen. You are probably the most familiar with the carbohydrates sugar and starch. The smallest sugar molecules are fructose (fruit sugar) and glucose (blood sugar). Sucrose, common table sugar, is a combination of fructose and glucose. The molecules of these simple sugars are shown in the diagram.

Glucose is the energizer for both plants and animals. Glucose reacts with oxygen to produce water and carbon dioxide, and energy is released. You use this energy in everything you do from thinking to playing sports. The balanced word equation for "burning" glucose is:

1 glucose molecule ($C_6H_{12}O_6$) + 6 oxygen molecules (O_2)

$\Rightarrow$ 6 carbon dioxide molecules (CO_2) + 6 water molecules (H_2O)

Where does glucose come from? Plants use the light energy from the Sun to produce their own glucose in the complex series of chemical reactions called photosynthesis. The balanced word equation for photosynthesis is:

6 carbon dioxide molecules (CO_2) + 6 water molecules (H_2O)

$\Rightarrow$ 1 glucose molecule ($C_6H_{12}O_6$) + 6 oxygen molecules (O_2)

Plants make more glucose than they need during the day. They use this extra glucose to make two kinds of polymers. One polymer is called starch. Starch molecules are twisted in a tight ball, with many branches and ends. Starch stores thousands of glucose molecules in a small space. When a plant needs some glucose, it breaks down the starch molecules.

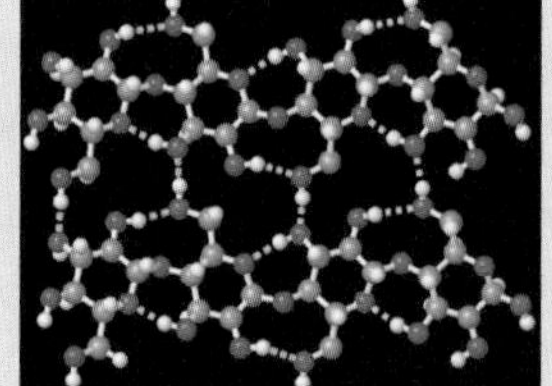

Cellulose (above) and starch (below) are both made of glucose monomers.

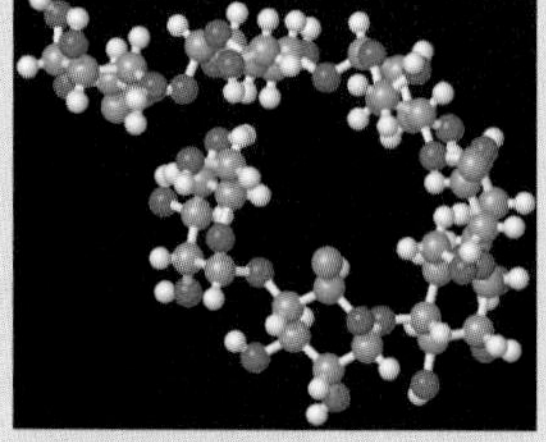

Plants also make a polymer called cellulose. Cellulose and starch are both made of glucose, but the glucose molecules are put together differently, as shown in the diagrams.

Cellulose chains are straight and the chains have strong cohesion bonds between them. That is why plant structures, like cell walls, stems, and wood, are made of cellulose. Plants also use cellulose to make fibers like cotton and hemp. These fibers can be twisted into threads and woven into clothing. Cellulose is the most common natural polymer in the world!

Where do you get the glucose your body needs? From eating plants, of course! The plants in your diet that contain a lot of starch include potatoes, carrots, corn, beans, and grains like rice, wheat, and rye.

Vegetables, especially corn, potatoes, and beans, also contain starch.

Breads and cereals are made from grains, so they contain a lot of starch.

4. What elements make up all carbohydrates?
5. What is the monomer molecule that makes up both starch and cellulose polymers?
6. What kinds of foods do you eat to get sugar and starch?
7. Why is glucose necessary for plants and animals?

Your teacher will review the answers to these questions with the class.

Fats – Stored Energy with a Bad Name

For most people, the word fat means that a person is overweight. But to a chemist, fats are molecules that have important roles in the chemistry of life. Fats are made up of monomer molecules called fatty acids and an alcohol (glycerol) molecule. Fatty acids such as stearic acid and linoleic acid are similar to carbohydrates, but they have a hydrogen (H) atom at one end. At the other end is a group consisting of a carbon atom, two oxygens atoms, and another hydogen atom—the COOH group.

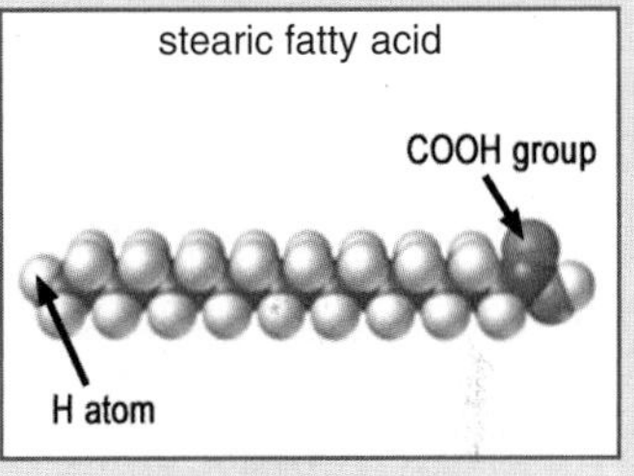

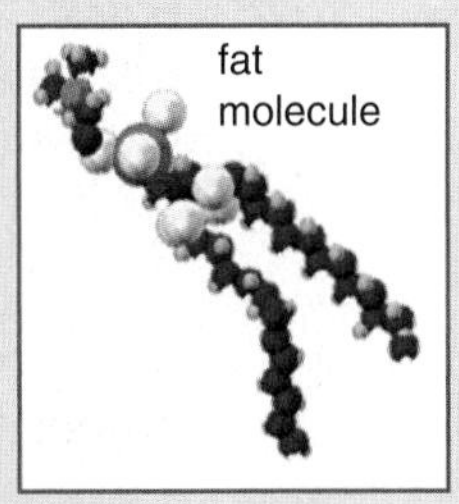

Plants and animals use fat molecules as an efficient means to store energy. Gram for gram, fats store more than twice the chemical energy than starches. Fats do not dissolve in water very well, so they can be stored a long time in the body without washing out. You only use fats as a source of energy when you do not have enough glucose in your blood and cells. Fats form the membranes that surround many of the cell parts. Fat is also a good insulator against cold. Animals that live in cold climates, like polar bears and seals, have very thick layers of fat to keep their organs from freezing.

Fats in plants are called oils because they are liquid at room temperature. Oils are mostly in the seeds of plants. Seeds also contain a lot of starch. Plants use the starch and oils in seeds as a source of energy while they are germinating, before photosynthesis begins. Photosynthesis cannot begin until the growing plant has sprouted leaves above ground. Before that, the plant's growth is

These foods are rich in fat.

powered by the chemical energy stored in the seeds.

You probably eat a lot of fats every day. Hamburgers, ice cream, avocados, butter, bacon, fried foods, potato chips, and many other snack foods are rich in fats. Food labels now have claims about fats and oils, including "no fat" or "low fat" or "low in unsaturated fats." What do these claims mean and why should you care about them?

Something that is "saturated" is full. It is filled to the point where nothing more can be added. Fats like stearic acid are *saturated* because each carbon atom has four covalent bonds. The carbon atoms are *full.* They cannot form any more bonds with hydrogen.

Fat molecules that have some some carbon atoms with only three covalent bonds are unsaturated. The carbon atoms are not full—they can bond with other hydrogen atoms.

carbon atoms with 3 bonds

linoleic fatty acid

Evidence is mounting that too much fat in the diet, especially some saturated fats, can lead to health problems. Saturated fats are associated with the formation of plaques (fatlike material) that can block arteries to the heart or brain. This can result in heart attacks or strokes.

Obesity is also a growing health problem. Obesity is associated with diabetes, high blood pressure, asthma, and arthritis. More than twice the energy is stored in fat molecules than in carbohydrate or protein molecules. That is why it is difficult to "burn off" excess fat.

8. What are three ways animals (including humans) use fats?

9. What are some foods that are rich in fats?

10. How are fats (oils) used in plants?

Your teacher will review the answers to these questions with the class.

What We Have Learned

Recall the key questions for this activity:

1. What elements make up most of the mass of living organisms on Earth?

2. What are the major molecules, small and large, that make up organisms and control chemical reactions in the organisms?

Participate in a class discussion to review the answers to these questions.

LEARNING ABOUT OTHER IDEAS

Activity 13
The Chemistry of Life – Part 2

Purpose

In Activity 12, you learned about the six elements that make up most of the mass of living organisms on Earth. You also learned about some of the molecules, small and large, that play an essential role in the chemistry of life. In this activity, you will learn about two more groups of molecules that also play an essential role. At the end of this activity, you will be able to answer the following question.

What are two major groups of polymers that make up organisms and control chemical reactions in the organisms?

Learning the Ideas

Proteins – The Stuff from Which Living Systems Are Made

Proteins are the primary material of life. The word protein comes from the Greek word *proteios*, which means "of prime importance." Look at a classmate sitting close to you. *Everything* you see is protein—skin, hair, eyeballs, and fingernails. Even some of the clothes your classmate is wearing may be proteins—like wool, leather, and fabrics made with silk. Your bones and organs also contain proteins. So do feathers and the fur, hooves, and horns of other animals.

Proteins are polymers made of 20 different monomers called amino acids. The English alphabet has 26 letters. You can make tens of thousands of words by putting these 26 letters into different sequences with different numbers of letters. In a similar way, the human body makes at least 50,000 different proteins (and possibly twice that many) from only 20 amino acids. Four of the 20 amino-acid molecules are shown in the diagram below. They all contain

Four of the 20 Amino Acids

glycine | serine | cyseine | histidine

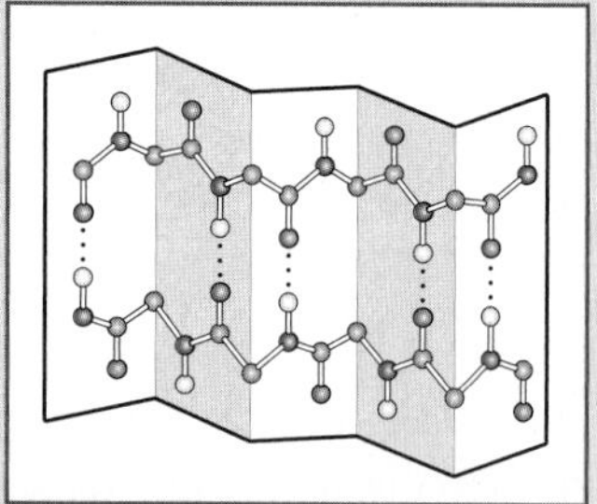
folded or pleated sheets

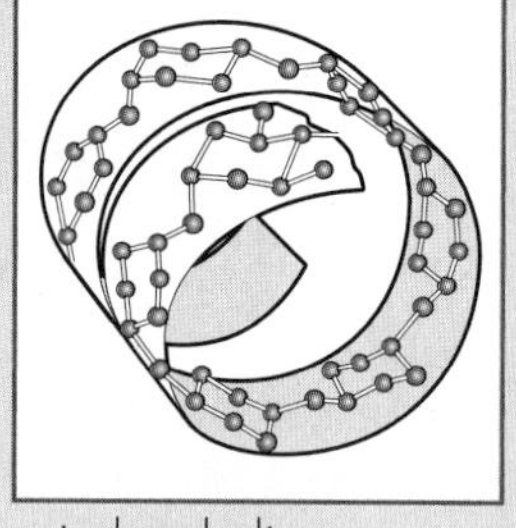
spiral or helix

carbon (grey), hydrogen (white), oxygen (red), and nitrogen (blue) atoms. A few contain sulfur atoms (yellow).

Proteins begin as long, thin chains. Right after a protein is made, different parts of the chain begin to twist or fold into two different shapes. One shape is like a folded sheet. The other shape is a helix, which is like a spiral staircase.

The sheets and helixes fold into their final protein shapes in less than a minute after they are formed. The final shapes may be globular or braided, with distinct corners, bumps, grooves, and planes. It's as if everything in your house were constructed from plastic, pop-together necklaces cleverly combined, wrapped and twisted to make the stairs, windows, furniture, toaster, and other appliances. The shape of the objects in your house relates to their function. For example, you sleep in a flat bed, but sit in a bent chair. A toaster must have slots to hold the toast. Like these objects, each protein has a particular shape that relates to its specific function. The diagram below shows some of the different shapes of proteins in living organisms, including humans.

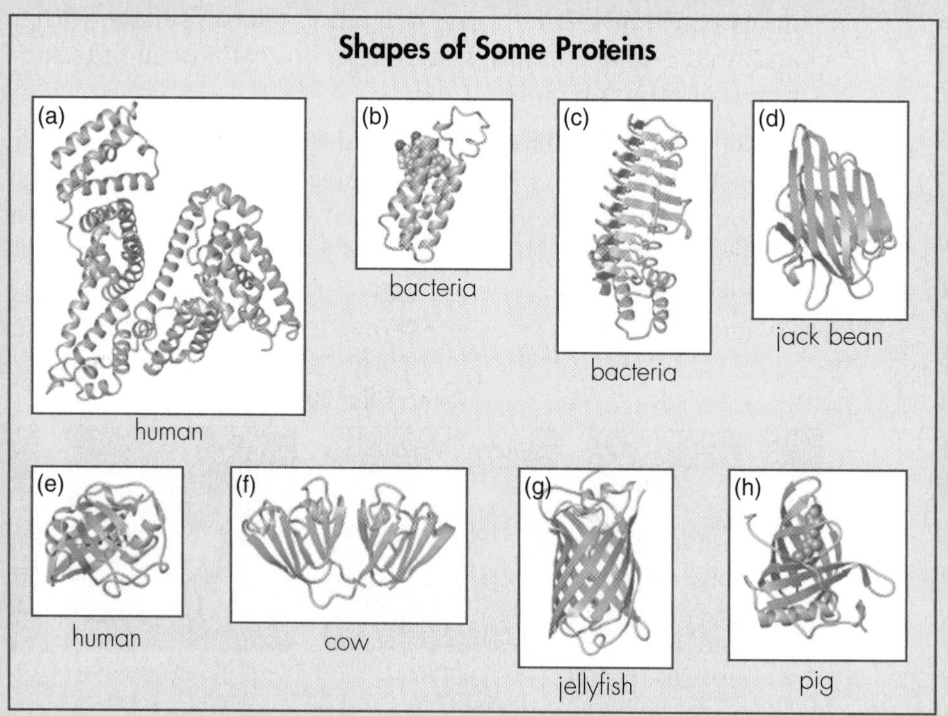

The protein hemoglobin is in the red blood cells in your blood stream. Hemoglobin's molecular shape lets it carry oxygen to all the cells in your body. The most common protein in your body, collagen, is used for support and structure. Collagen is in between all the cells in your body, all around your organs, and even in your teeth and bones. Collagen's molecular shape makes it a good connective tissue.

The average protein survives only two days before it is broken down. Its chemical parts are either recycled or leave the body as waste. So, cells must make proteins around the clock, at a fantastic rate. A cell can make hundreds of proteins each second. Each cell in your body makes tens of millions of protein polymers each day. This requires tens of millions of chemical reactions in each of the several hundred million million cells in your body.

You learned in Activity 2 that for a chemical reaction to occur, reactant molecules must collide with enough motion energy to break their chemical bonds. For the tens of millions of chemical reactions in each cell, this would require an enormous amount of energy. You couldn't eat food fast enough to supply this energy. To solve this problem, your body makes a special group of proteins called *enzymes*. Enzymes are proteins that speed up chemical reactions in a cell, without changing themselves. They make chemical reactions energy efficient.

How Enzymes Help Chemical Reactions

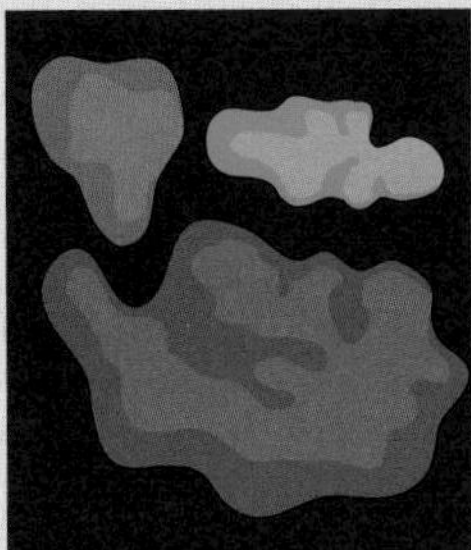

(a) Two molecules approach the dark blue enzyme.

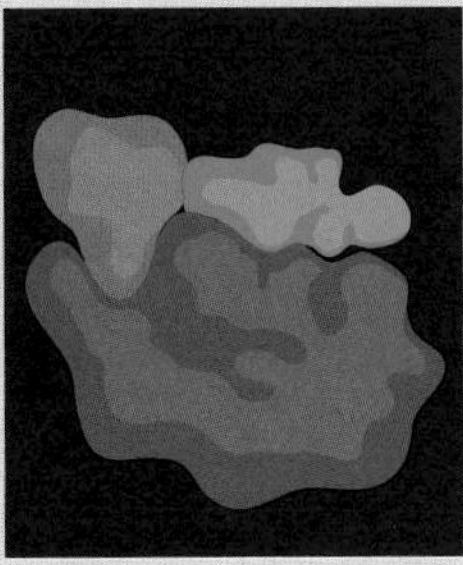

(b) The molecules fit perfectly into the enzyme like a key in a lock.

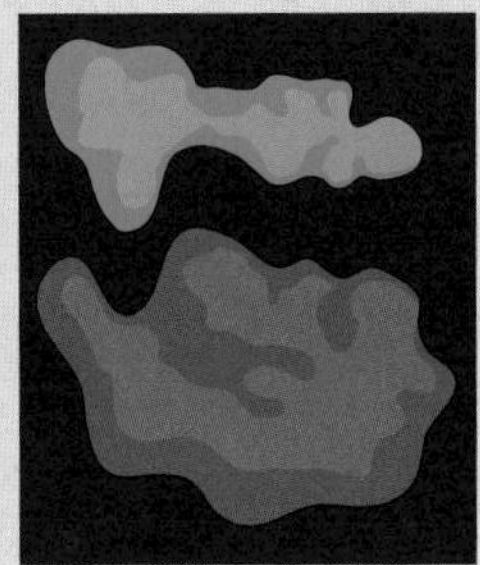

(c) The two molecules react and form a new molecule.

An example of how an enzyme makes a chemical reaction occur is shown in the diagrams above. Some enzymes work the opposite way. They lock in a molecule and help break it apart into two or more product molecules. There are even enzymes for making other enzymes.

1. Why are proteins called the "primary material of life?"
2. Why do proteins have different shapes?
3. What is the role of enzymes in the chemistry of a cell?

Your teacher will review the answers to these questions with the class.

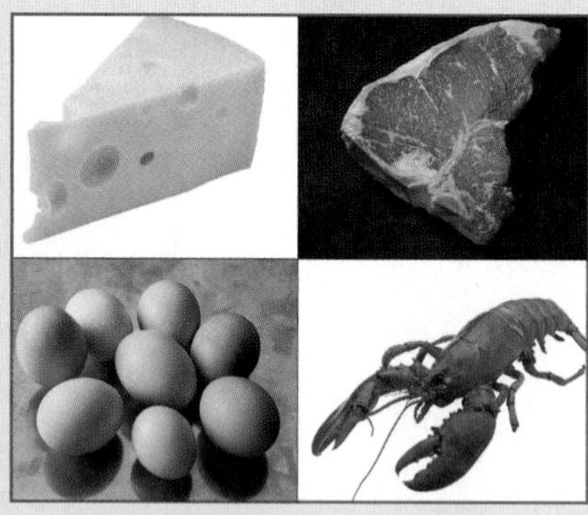
Good sources of protein.

Proteins and Food

The body can make 12 of the 20 amino acids. You have to get the other 8 amino acids from the food you eat. Your body cannot store protein, so a balanced protein diet is required daily. Most of us think of meat and fish when we think of high-protein foods. But many other foods are good sources of protein, including eggs (the white part), milk and cheese, beans, nuts, rice, and other grains.

When you eat protein, enzymes in your stomach and small intestines break the protein down into its amino acids. The freed amino acids travel through your bloodstream first to your liver, then to your cells. There they are built up into the proteins your body needs. If you have eaten more protein than you need, the amino acids react in your liver. Part of the amino acids is made into waste products and leaves your body through your kidneys. The rest of the amino acids are usually converted to storage fat.

Proteins can be used as an energy source, but only in starvation conditions. Carbohydrates and fats are the normal energy sources in the body. The function of proteins is in body building and making enzymes to regulate and control the chemical reactions that provide energy and sustain life.

4. Why do you need to eat proteins in your daily diet?

5. Why does a high-protein diet place an extra burden on your liver and kidneys?

6. What are the differences between the role of proteins and the role of carbohydrates and fat in your body?

Your teacher will review the answers to these questions with the class.

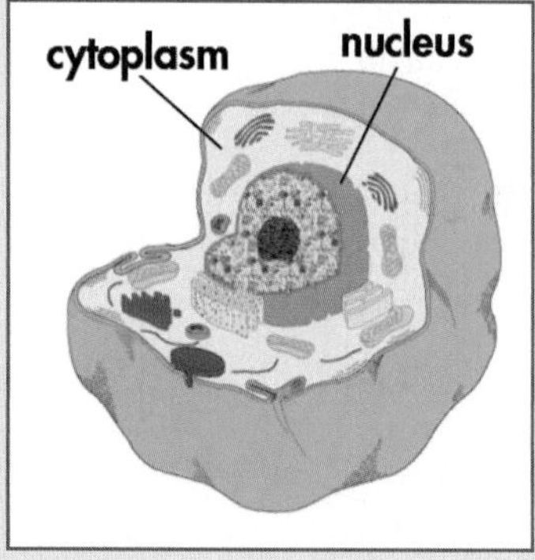

Animal cell.

DNA – The Instruction Molecule for Life

All cells have a similar structure. Proteins are made in the watery solution (cytoplasm) of the cell. But each cell cannot make all of 50,000 (or more) proteins and other compounds needed everywhere in the body. How does a skin cell or a muscle cell know which proteins and compounds to make?

The answer lies in a giant polymer molecule called DNA (deoxyribonucleic acid). DNA is in the nucleus of every cell. When DNA is formed, two long polymer strands twist around each other in opposite directions and bond with each other. The final shape is called a double helix.

The simplified drawing shows the structure of the double helix. The molecules on the outside of the helix form unchanging "backbones." Two bonded monomers called "bases" connect the two backbones together like the rungs of a ladder—but this ladder is twisted. There are four base molecules, as shown in the diagram.

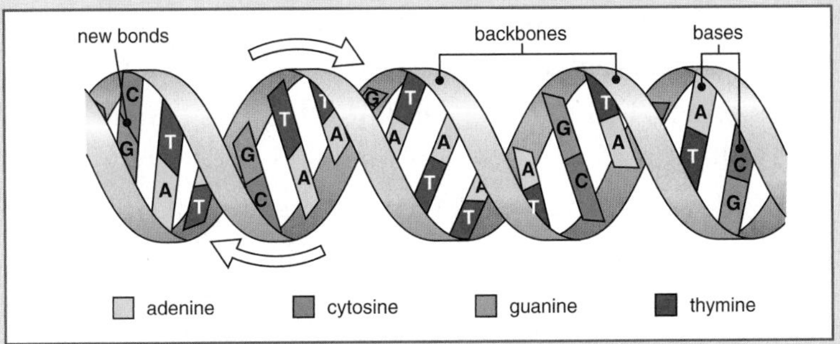

DNA molecules in the cells of both plants and animals are many thousands of bases long. The DNA in the nucleus of a human cell is estimated to be about three billion bases in length. The information for making proteins is stored or "coded" in DNA through the different sequences in which the bases can be arranged. A gene is a long section of DNA that contains a particular sequence of bases able to direct the manufacture of a specific protein. This process is shown in the diagram.

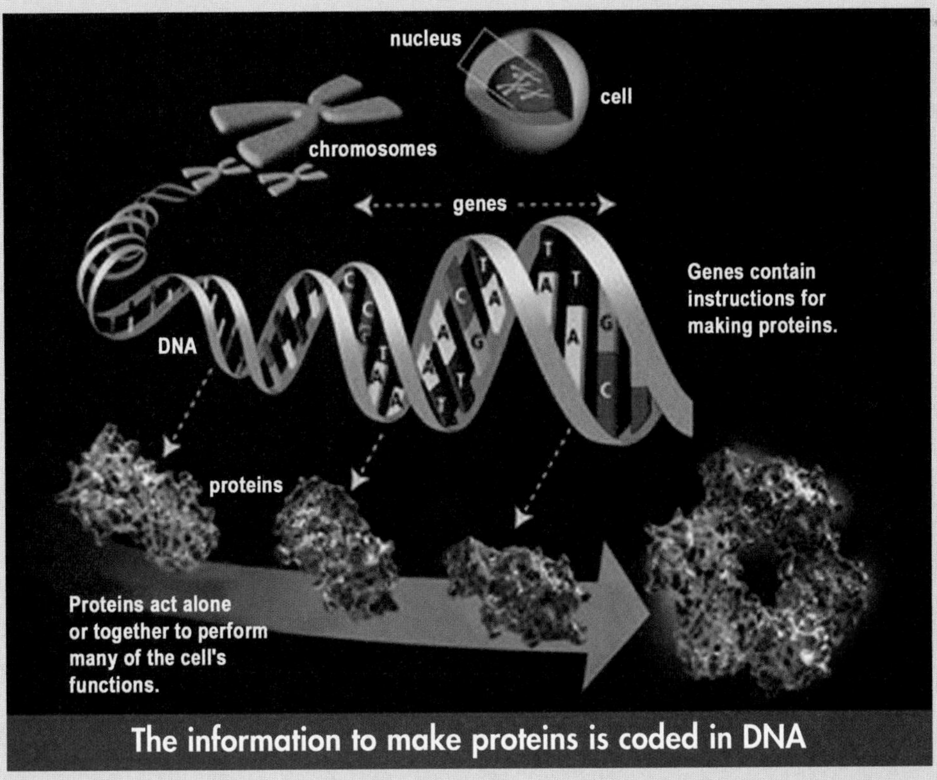

The information to make proteins is coded in DNA

Each cell (except blood cells) has an identical set of DNA in its nucleus. Chemical regulator compounds "turn on" only those genes needed to produce the proteins appropriate for the cell type—skin cell, muscle cell, brain cell, and so on.

7. What does a DNA molecule consist of? Use the terms backbone and bases in your answer.

8. How are proteins made in a cell?

9. How does a skin cell or a muscle cell know which proteins and compounds to make?

Your teacher will review the answers to these questions with the class.

No two people have exactly the same DNA (except identical twins), so the genes of every individual are different. The instructions for characteristics such as how tall you will grow, the shape of your nose, and size of your feet, are coded in the DNA that makes up your genes.

Organization of the Human Body

The diagram on the next page shows the different levels of organization within a person, from the chemicals of life upwards. You know that everything within us is made of small particles—atoms, ions, and molecules. These chemicals have a range of sizes. Some molecules are small, such as water, glucose, vitamins, amino acids, and fats. Some molecules are large polymers, such as proteins and DNA. The chemicals of life operate together in a cell.

Clusters of similar cells form *tissues*, such as smooth muscle tissue, loose connective tissue, and nervous tissue. Different specialized tissues are found within *organs* of the body, such as the stomach, liver, brain, heart, and lungs. A variety of organs work together to form *systems* of the body, such as the muscular system, the nervous system, the circulatory system, and the digestive system.

What We Have Learned

Recall the key question for this activity:

What are two major groups of polymers that make up organisms and control chemical reactions in the organisms?

Participate in a class discussion to review the answers to the key question.

Write the answer to the key question on your record sheet.

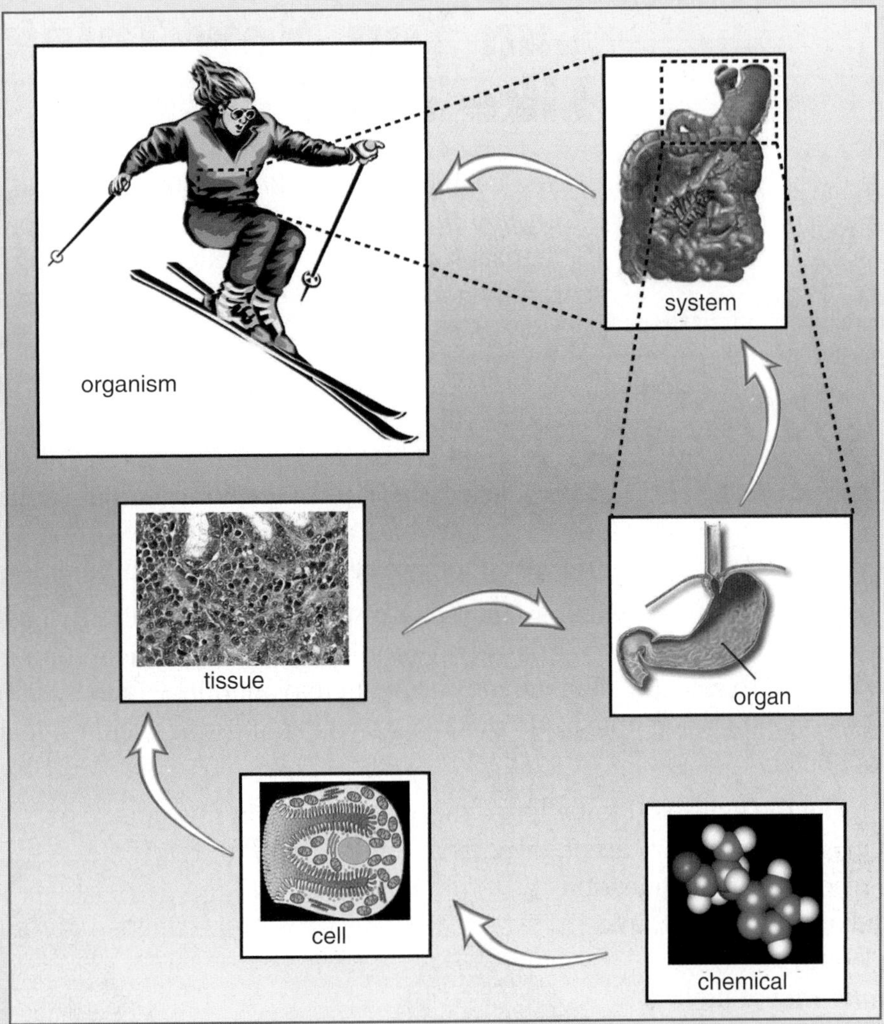
system
organism
tissue
organ
cell
chemical

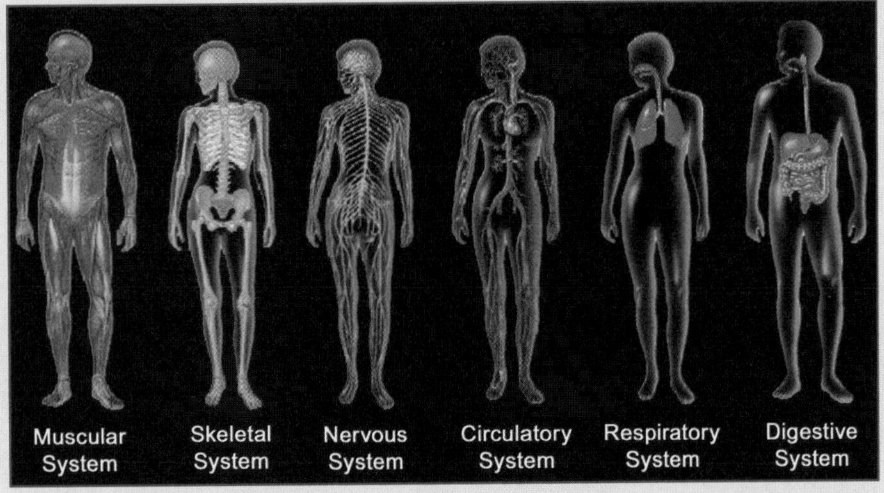
Muscular System
Skeletal System
Nervous System
Circulatory System
Respiratory System
Digestive System

Can You Now Think Like a Scientist?

At the beginning of this book you were asked, *"Can you think like a scientist?"* The answer is **Yes!** Scientists think and develop their ideas using carefully collected evidence – and now you can too.

You have learned how to use evidence to come up with powerful ways to describe how the world around you works. You have learned how to use ideas about energy and interactions to explain everyday situations. You can now explain situations that you may read about in magazines and newspapers, or see on television and the Internet.

You also learned how to evaluate explanations. You can now evaluate the ideas of others to find out if *their* ideas are based on good evidence and logical thinking. This skill is not limited to science. You can also evaluate explanations given by politicians and business people. Being able to do this makes you an informed and critical member of society.

Continue using the skills and knowledge that you learned in this course. Apply them in school, your career, or personal life. Be sure to read, think, do research, think, question others, and always think, think, think. Welcome to the *InterActions* way of life!

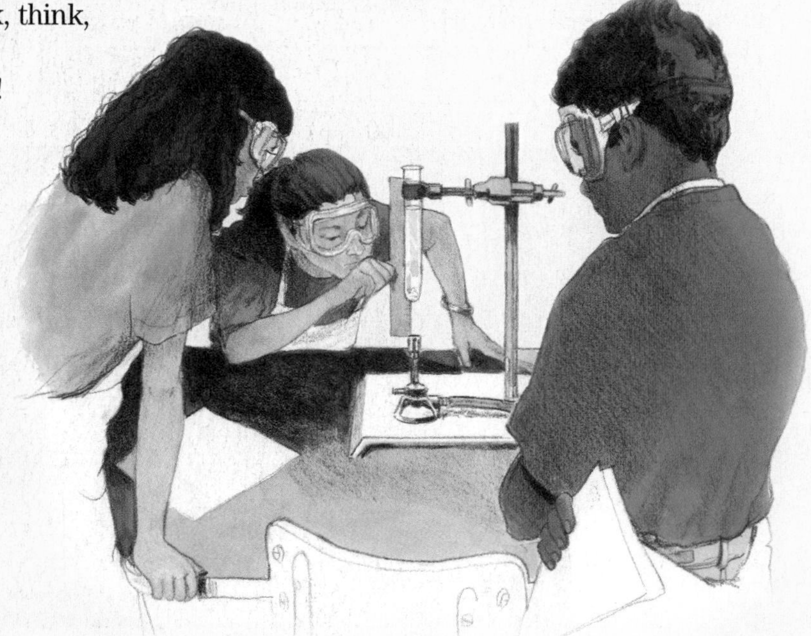

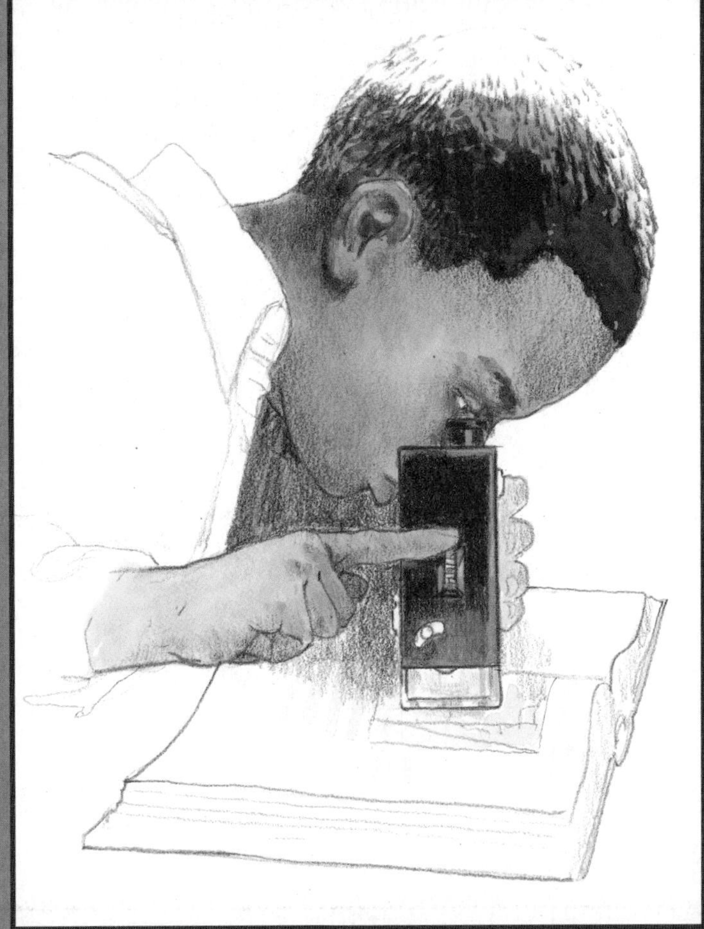

Table of Contents

How To...

How To...

1 Follow Safety Rules during an Experiment

Read over the following safety rules. They will prepare you to work safely in the laboratory. Then, read them a second time. Make sure you understand and follow each rule. Your teacher may wish to add additional rules for your classroom. Ask your teacher to explain any rules you do not understand.

General Rules

1. Always get your teacher's permission to begin any activity in the lab.
2. Read all directions for an experiment before beginning the activity. Carefully follow all written and verbal instructions as directed by your teacher. If you have questions, ask your teacher.
3. Be sure you understand *all* safety symbols included in the procedures. (Read the section on Safety Symbols in this How To.)
4. Use the safety equipment provided for you. Safety goggles and a lab apron must be worn.
5. Do only the experiments assigned or approved by your teacher.
6. Never eat or drink in the lab. Never inhale chemicals. Do not taste any substance or draw any material into your mouth.
7. No playing, running, pushing, or loud talking are permitted during an experiment.

Using Chemicals and Equipment Safely

8. Never use lab glassware as containers for food or drink.
9. If you spill any chemical, quickly wash it off with water. Report the spill to your teacher immediately.
10. Never force glass tubing or a thermometer into a rubber stopper or rubber tubing. Have your teacher insert the glass tubing or thermometer.
11. Keep all materials that can burn away from flames. Never reach across an open flame!
12. When heating a test tube, always point the mouth of the test tube away from yourself and others. Chemicals can splash or boil out of heated test tubes.

13. Never bend over and directly smell a gas. Your teacher will demonstrate a safe technique used by chemists to smell an unknown sample when they suspect an invisible gas may be produced. The technique is called wafting.

14. Do not pick up a container that has been heated until you are sure it has cooled. Hold the back of your hand near the heated container. If you can feel the heat on the back of your hand, then it is too hot to handle. Use an oven mitt.

15. Never heat a liquid in a closed container. The expanding gases could blow the lid off the container, or blow the container apart.

End-of-Experiment Rules

16. Always turn off all burners and disconnect electrical devices.

17. Clean up your work area and return all equipment to its proper place.

18. Dispose of chemicals and other materials as directed by your teacher. Place broken glass and solid substances in the proper containers. To avoid contamination, never return chemicals to their original containers. Never pour untreated chemicals or other substances into the sink or trash containers.

19. Wash your hands after every experiment.

First Aid in the Laboratory	
Injury	Safe Response
Burns	Apply cold water. Call your teacher immediately.
Cuts and bruises	Stop any bleeding by applying direct pressure. Cover cuts with a clean dressing. Apply cold compresses to bruises. Call your teacher immediately.
Fainting	Leave the person lying down. Loosen any tight clothing and keep crowds away. Call your teacher immediately.
Foreign matter in eye	Flush with plenty of water. Use eyewash bottle or fountain. Call your teacher immediately.
Poisoning	Note the suspected poisoning substance and call your teacher immediately.
Any spills	Flush with large amounts of water. Call your teacher immediately.

First Aid

20. Always report all accidents or injuries to your teacher, no matter how minor. Notify your teacher immediately about any fires.

21. Learn what to do in case of specific accidents.

22. Be aware of the location of the first aid kit, but do not use it unless instructed by your teacher.

23. Know the location of the emergency equipment such as fire extinguisher and fire blanket.

24. Know the location of the nearest telephone and whom to contact in an emergency.

SAFETY SYMBOLS		
Dress Safely		
	Safety Goggles	Always wear safety goggles to protect your eyes in any activity involving chemicals, flames, or heating, or the possibility of broken glassware. Wear your goggles any time when there is even the slightest chance that harm could come to your eyes.
	Lab Apron	Always wear a lab apron when you are working with substances that could stain or burn your clothing.
	Tie Back	Always tie back long hair to keep it away from any chemicals, flames, or equipment. Remove or tie back any article of clothing or jewelry that can hang down and touch chemicals, flames, or equipment. Roll up or secure long sleeves.
	Shoes	Do not wear open shoes or sandals.
Heating and Fire Safety		
	Flames	You may be working with flames from a burner, candle, or matches. Before using a burner, make sure you know the proper procedure for lighting and adjusting the burner, as demonstrated by your teacher. Never leave a lighted burner unattended. Never reach across a flame.
	Extreme Temperature	Use an oven mitt when handling hot materials. Before picking up a container that has been heated, hold the back of your hand near it. If you can feel the heat on the back of your hand, it is too hot to handle. Use an oven mitt to pick up a container that has been heated.
Chemical Symbols		
	Toxic	Do not let any poisonous material come in contact with your skin and do not inhale its vapors. Wash your hands when you are finished with the activity.
	Glassware	You are working with materials that could break, such as glass containers and thermometers. Handle breakable materials with care. Do not touch broken glassware. Do not use any glassware that is chipped or cracked.
	Irritant	Always wear gloves when you are working with substances that can irritate the skin or mucus membranes.

How To...

2 Make and Interpret Experiment Measurements

(Unit 1 Cycle 1, Activity 1)

1. When scientists perform an experiment to measure a quantity, they never obtain a true (or exact) value. There is always some uncertainty associated with a measurement. Therefore, scientists always report a best value, not an exact value, for the measurement.
2. The goal of making good measurements is to reduce the amount of uncertainty, so that the best value is as close to the true value as possible. To do this, you need to design an appropriate experimental procedure and follow it very carefully.
3. Each measurement of a quantity may give a value that is either higher or lower than the true value. Therefore, it is common to make many measurements (called multiple trials), and then calculate the average of the measurements. This average is reported as the best value.

   To calculate the average you add up all the measured values, then divide by the number of trials. For example, if there are three trials:

$$\text{Average} = \frac{\text{Trial 1} + \text{Trial 2} + \text{Trial 3}}{3}$$

4. Sometimes one of the measured values is very, very different from the other values. This is called an outlier. If there is a good reason to believe that it is due to a major blunder in measurement, then you should ignore its value when determining the best value.
5. Because there will always be a variation in the values you obtain when making multiple measurements of a quantity, you need a procedure to report the uncertainty associated with your best value. After removing any outliers from your data, you calculate the uncertainty by subtracting your lowest measured value from your highest measured value, and then dividing this difference by two. That is:

$$\text{Uncertainly} = \frac{\text{Highest measured value} - \text{Lowest measured value}}{2}$$

   The value of the uncertainty then tells you by how much the true value may be higher than, or lower than, the best value.

Example:
Imagine a student measured the time for 20 back and forth swings of a pendulum. Her results for four trials, including her values for the best value and uncertainty that she calculated using the above procedure are in the following table. Check the math to make sure the best values and uncertainty calculations are correct.

	Time for 20 Swings (s)
Trial 1	38
Trial 2	42
Trial 3	40
Trial 4	42
Best Value	40.5
Uncertainty	2

She would then report her results as follows:

"The best value is 40.5 s with an uncertainty of 2 s. This means that the true value is probably within the range between 38.5 s and 42.5 s."

How To...

3 Read a Ruler

The metric system is the internationally accepted system of measurement. All of the measuring you will do in science classes will use the metric system. Let's review how to read a metric ruler.

A metric ruler shows markings for millimeters and centimeters. A millimeter (1 mm) is about as wide as the thickness of a dime. A centimeter (1 cm) is about as wide as your fingernail. Some rulers have inches on one side and centimeters (metric system) on the other. Always remember to measure with the metric side of your ruler for science.

Notice the lines on this ruler:

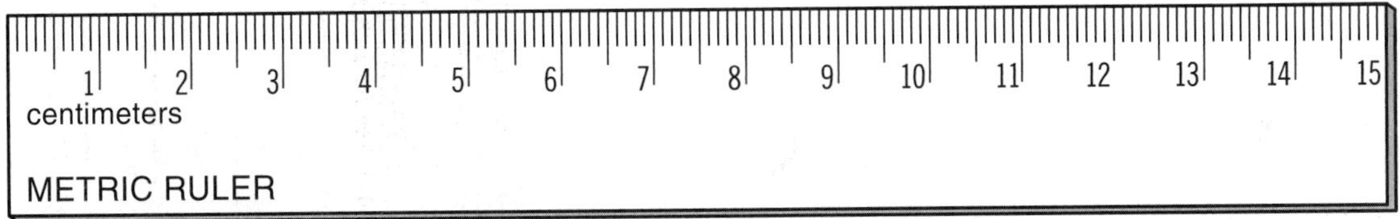

The longest lines are numbered 1, 2, 3... These are the centimeter markings. The shorter lines are not numbered. These are millimeter markings. There are 10 millimeters in each centimeter.

Now take a look at the ruler below. The arrows mark the ends of imaginary lines beginning at the zero end of the ruler.

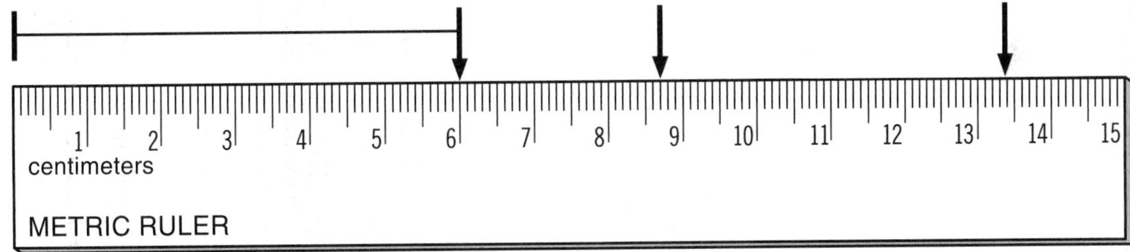

The first arrow on the left points to 6.0 centimeters (6.0 cm). This can also be written as 60 millimeters (60 mm). You do not need to count each millimeter marking. Remember, there are 10 mm in each centimeter. To write centimeters as millimeters, just multiply the centimeter reading by 10.

The arrow in the middle points to 8.7 centimeters (8.7 cm) or 87 millimeters (87 mm).

Sometimes the end of the line you are measuring falls between two millimeter markings (between two short lines). If it does, choose the closest millimeter marking. For example, the arrow on the right falls closest to 13.4 cm or 134 mm. If the arrow falls exactly halfway in between two marks, use the next *highest* millimeter mark.

How To...

4 Answer Multiple-Choice Questions

Multiple-choice questions are used in many different types of student assignments, from homework to exams. They are often the most difficult type of questions to answer. Here are some strategies to help you successfully answer multiple-choice questions.

A. Try to answer the question before looking at the choices.

Read the question and try to answer it *before* looking at the possible choices. After you read the question, you can use the margin to write down your initial ideas before being influenced by the list of choices. For example, think about the answer to the following.

1. A hypothesis is...

B. Read all the choices.

Once you have your own idea of what the answer is, read all of the choices before choosing one. It is very important that you do *not* choose the first choice that sounds correct and ignore the others. There may be two or more choices that seem correct. You need to find the *best* answer to the question.

2. A hypothesis is...

- **a)** an educated guess about what a relationship will be.
- **b)** what your ideas are before you perform an experiment.
- **c)** based on your past experiences.
- **d)** all of the above.

Even though (a) is true, so are (b) and (c), so the correct answer is (d). It is important to read all the choices, and not just take the first correct choice you see. Pay close attention to questions that include "all of the above" or "none of the above."

Often you encounter multiple-choice questions that are difficult or to which you do not know the answer. Here are several strategies you can try for answering those questions.

C. Cross out choices you know are incorrect.

First, read through all the answers and cross out any choices you know are incorrect. You may not know much about the question, but you may know that one or two of the choices are definitely not correct. Once you cross out those choices, you have improved your chances of choosing the correct answer.

Which choices do you know are incorrect in the following question?

3. A *manipulated variable* is...

a) the variable that is not deliberately changed by the experimenter.

b) the variable that is deliberately changed by the experimenter.

c) the variable that is held constant throughout the experiment.

d) the educated guess you have about a relationship before you begin an experiment.

You may not have a clear idea what a *manipulated variable* is but you know that (d) is the definition of *hypothesis;* so you should cross out (d). You might also know that a manipulated variable is not held constant, so you cross out (c). This leaves only two possible choices, (a) and (b).

D. Think about how the choices are different.

Now you have to choose between only two answers. Think carefully about how the two answers differ. If you are still uncertain, choose one and place a star (*) next to the question. This is to alert you to return to this question after you complete the other questions. [The answer to Question 3 is (b).]

Sometimes you may come across a question that appears to have two correct answers. First, read the question to see if you are allowed to choose more than one response.

4. Which of the following variables are *controlled* (held constant) in this experiment? (Circle all that are correct.)

a) size of the paper clips

b) size of the magnets

c) number of paper clips

d) material from which magnets are made

In this question, you can choose more than one answer. Notice the question asks: *Which of the following variables are held constant....* The word "are" tells you that more than one answer is expected. The instruction to *Circle all...* tells you to choose all of the correct answers. [The correct answers to Question 4 are (a) and (d).]

Most of the time, however, you have to select only one correct answer. So what can you do if you come across a question where two choices appear to be correct and you have to decide between them?

E. Rephrase the choice in your own words.

Begin by trying to express or rephrase each answer in your own words. How do they seem to be different from each other? As you are rephrasing the choices, focus on the difference between them.

F. Choose the *best* answer.

You should also consider whether the choices actually answer the question being asked. Even if a choice contains accurate information, it might not answer the question accurately. You may have to look very carefully for clues that help you find the one single choice that is the *best* answer.

In the following example, two of the choices contain accurate information.

5. When taking measurements during experiments...

- **a)** you can obtain an exact value if you are careful enough.
- **b)** the values you obtain are often inexact.
- **c)** you should average all the values obtained to find the best value.

You know that (a) is incorrect, but you may be struggling between (b) and (c). Think about (b). Measurement values are often inexact. In fact, they are *always* inexact. Now consider (c). Scientists don't always average *all* the values they obtain during an experiment. They might notice that one measurement is very different from all the others (an outlier), so they exclude it from the average of the values. So in this case, although (c) is often true, for this question (b) is the better answer.

Strategies for Answering Multiple-Choice Questions

A. Read the question and try to answer it *before* looking at the possible choices.

B. Read all of the possible choices before choosing one.

C. Cross out any choices you know are incorrect.

D. Think carefully about how the choices that seem correct differ.

E. Try to express or rephrase each choice in your own words.

F. Choose the best answer for the question.

How To...

5 Analyze an Experiment Design and Determine if the Experiment is a Fair Test

(Unit 1 Cycle 1, Activity 4)

Analyze the Experiment Design

The *manipulated variable* in an experiment is the variable that is being changed deliberately. The *responding variable* is the one that responds to the change. The question that the experiment is designed to answer is often written in a relationship format similar to one of the following:

> If the *(manipulated variable)* changes, then what happens to the *(responding variable)*?
>
> or
>
> What is the relationship between the *(manipulated variable)* and the *(responding variable)*?

Read the description of the experiment and ask yourself the following questions. They will help you to see what else, apart from the manipulated variable, may be affecting the experiment result.

1. What is the question the experiment is designed to answer? If it's not already written down in a relationship format, try to write it so that it is. (The question is often provided to you in the experiment description, so you just need to copy it. If it is not provided, you need to determine it from the description of the experiment.)
2. What is the *manipulated variable*? What is the *responding variable*? To find out, look at the experiment question, or read the description of the experiment.
3. What are the values (including units) of the manipulated variable?
4. What method is used to measure the responding variable?
5. What variables or conditions are kept the same (*controlled*) during the experiment?

Determine if the Experiment Is a Fair Test

Is there any variable (besides the manipulated and responding variables) that changes in the experiment?

If the answer is "no," then the experiment is a fair test.

If the answer is "yes," then the experiment is not a fair test.

Write the reasons why the experiment is, or is not, a fair test. For example:

The experiment is a fair test because the only variables that changed were the manipulated variable (*write what it is*) and the responding variable (*write what it is*).

The experiment is not a fair test, because as well as the manipulated variable (*write what it is*), another variable (*write what it is*) was also changed. So the responding variable (*write what it is*) may be responding to more than one change.

How To...

6 Evaluate an Experiment Conclusion

(Unit 1 Cycle 1, Activity 5)

STEP 1 Fair Test

Determine whether or not the experiment is a fair test. Respond to the following with a "yes" if it is, and "no" if it is not.

The experiment is a fair test. (See How To Analyze an Experiment Design and Determine if the Experiment is a Fair Test.)

- If "yes," then continue to Step 2.
- If "no," then the conclusion is *not valid*. Stop here and write:

"The conclusion is not valid because the experiment is not a fair test."

STEP 2 Supporting Reasons

If the design of the experiment is a fair test, then evaluate the reasons supporting the conclusion. Read the reasons. Answer "yes" or "no" to indicate whether each of the following criteria is met.

Each supporting reason is based on *evidence*, not opinion.

The supporting reasons use *all the available evidence* (data), not just part of the evidence.

- If you answered "yes" for both criteria, then the supporting reasons are *good* and the conclusion is *valid*. To justify your evaluation, write a statement similar to the following:

"The conclusion is valid because the supporting reasons are *not opinion*, and use *all* of the evidence."

- If you answered "no" for either or both criteria, then the supporting reasons are poor, and the conclusion is *not valid*. To justify your evaluation, describe what is poor about the supporting reasons. The following are examples of statements you could make:

"The conclusion is not valid because the supporting reason ________ (*write the reason*) is based on opinion, not evidence from the experiment."

"The conclusion is not valid because the supporting reason ________ (*write the reason*) uses only part of the available evidence, instead of all the evidence."

How To...

7 Read a Magnetic Compass

People use map compasses to tell geographic direction. Most compasses are magnetic needles that spin freely until aligned with the magnetic field of Earth.

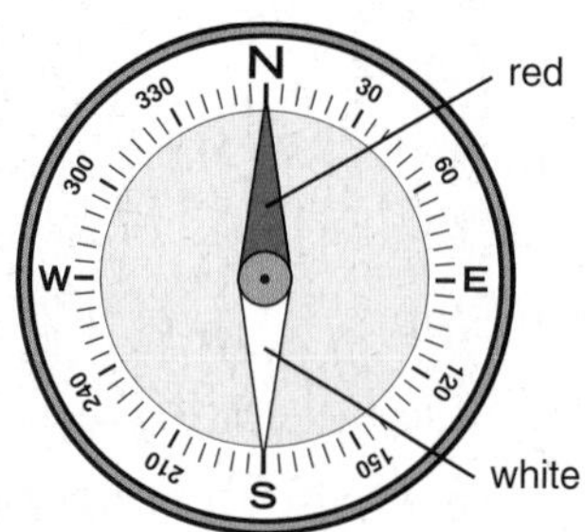

There are several different kinds of map compasses. The compass you will be using in your science class will have a red pointer or arrow. This is called the *compass needle*. On some compasses, the compass needle might be red and white or red and black. Just remember, the *red pointer* is the part you want to use for your measurements.

Notice on the compass in the diagram that the outer ring is marked with the geographic directions North (N), South (S), East (E), and West (W). The compass is divided into 360° (360 degrees), just like any other circle. Between the directional markers, you can see short lines labeled with numbers (30, 60, 120…). These are degree marks. The degrees increase in a clockwise direction. North is always at 0°, so East is at 90°.

Look closely at the lines between North and 30°. How many degrees do you think each line represents? Each of the lines between North and 30° represents 5°. Counting by fives, count each line between North (0°) and 30°. Now count, by fives, all the lines between North (0°) and East. You should get 90°.

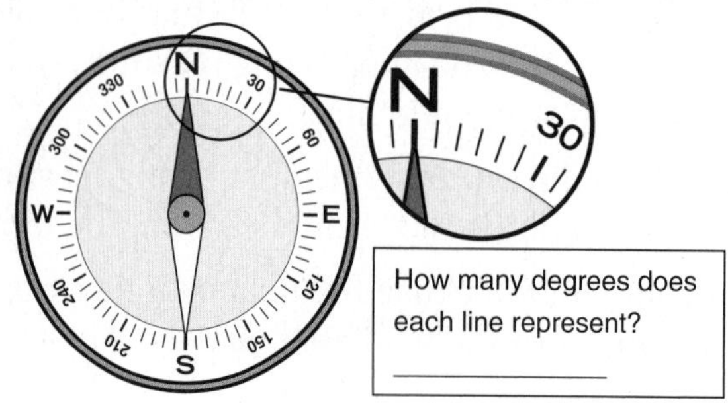

You will use a compass to read how many degrees the compass needle (the red arrow point) moves from 0°. This is called measuring the *deflection* of the needle. In the two compasses shown, assume the needle started at North (0°). How many degrees has the needle moved, or been *deflected*? The needle on the left was deflected 60° and the needle on the right was deflected 40°.

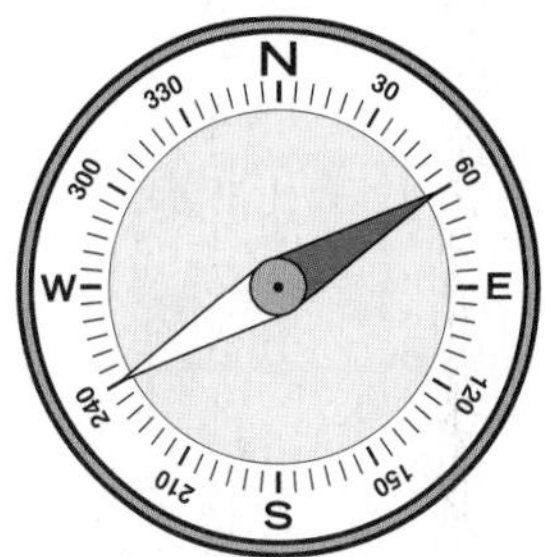

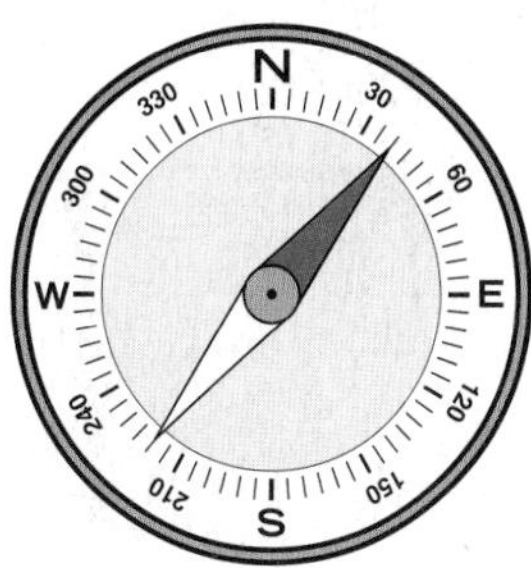

The compass needle can also move, or *deflect*, counterclockwise from North. In the two compasses below, through how many degrees has the needle moved from North (0°)?

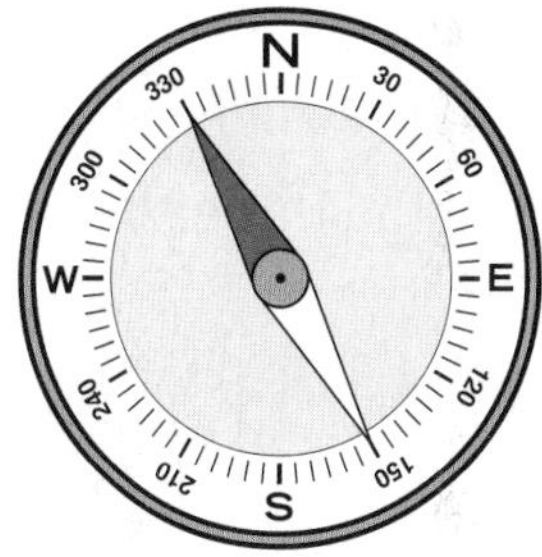

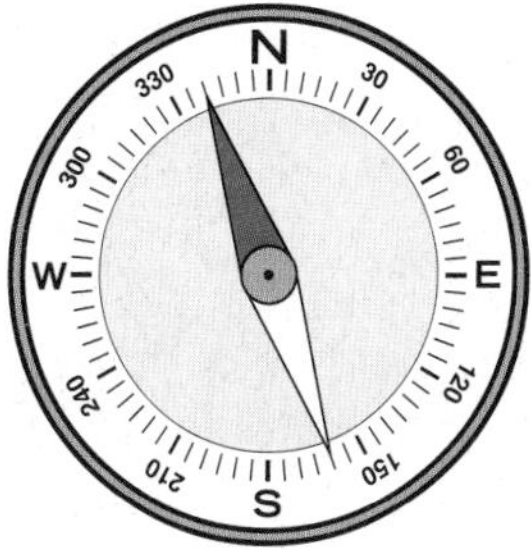

Start at 0 and go counterclockwise. Counting by fives, count each of the short lines. You should get 30; the arrow was deflected 30° from 0. Remember, you are using the compass only to measure how many degrees the needle is deflected from 0. The needle on the compass to the right above was deflected 20° from 0.

How To...

8 Discuss Ideas with Your Team

You will often discuss the answer to questions with your team. Then you will either write the team answer on your record sheet, or present your team's answer to the class. To have a *scientific* discussion of ideas, each team member must practice the three skills shown below.

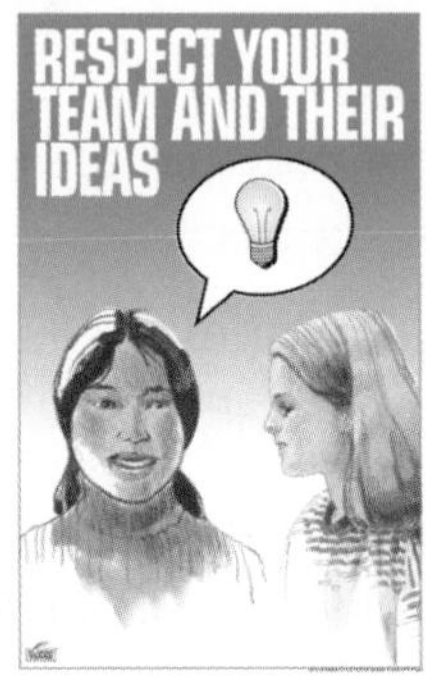

For example, imagine that a team was talking about the answer to the following question.

"Does brass interact with a magnet? What is your evidence?"

Read the team conversation below. Then complete the three statements about this conversation.

Consider how the team could improve their discussion. Carlos and Chantel need to be aware of and monitor their own thinking. Then they could either state their *reasons* for their ideas, or state that they are not sure of their reasons. Chantel and Jason need to respect the ideas of their team members – no put downs. If the ideas of all team members were respected, then Xuan would not be afraid to contribute her ideas and reasons.

Steps for Discussing Ideas with Your Team

STEP 1 A team member states his/her answer to the question. Remember to state both the answer, and the reasons for your thinking. Your reasons should include the evidence from an experiment.

"I think that ____________________ because ____________________ ."

STEP 2 The other team members *take turns* either agreeing with the answer, or suggesting any additions or changes to the answer. Do this by stating your own idea and reason. For example:

"I agree with your answer *because* you covered everything."

or

"I agree with your answer and I think we could add some more evidence from ____________________ ." or

"I disagree with your answer because I think ____________________ ."

STEP 3 After everyone has had a chance to give their answers and reasons, *as a team* choose the answer you like best. Your team might decide that:

- one member's idea is good because the reasons are good;
- a mixture of one or more team members' ideas is best; or
- a totally different answer is needed.

You would record the team answer to the question, along with the supporting reasons (evidence).

STEP 4 Take turns repeating Steps 1 through 3 for each of the questions. Usually the *Team Manager* decides who will go first, second, third, and fourth.

How To...

9 Use the EM Devices Simulator

General Instructions

Step 1 When you open any one of the *InterActions* simulators you will see a box called the setup area, and several palettes surrounding it.

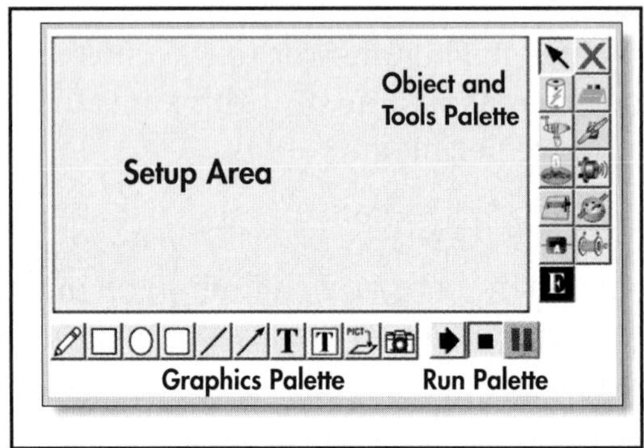

Step 2 To place a battery (or any object) in the setup area:

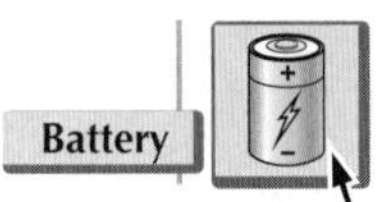

- Click the button that looks like a battery (or the object).
- Move the cursor arrow to the set-up area, and click the mouse button where you want to place the battery.

Step 3 To *move* items:

- Click the **Select** button.
- Click the object you want to move and drag the object to its new location.

Step 4 To *delete* items:

- Click the **Delete** button.
- Click the item you want to delete.

Caution: When you are finished deleting items, click on the **Delete** button again to deselect it. Otherwise, you will continue to delete items, even if you did not want to!

Instructions for Unit 1, Cycle 2, Activity 4

Step 1 Place a battery, ammeter, switch, and bulb (in socket) in the setup area.

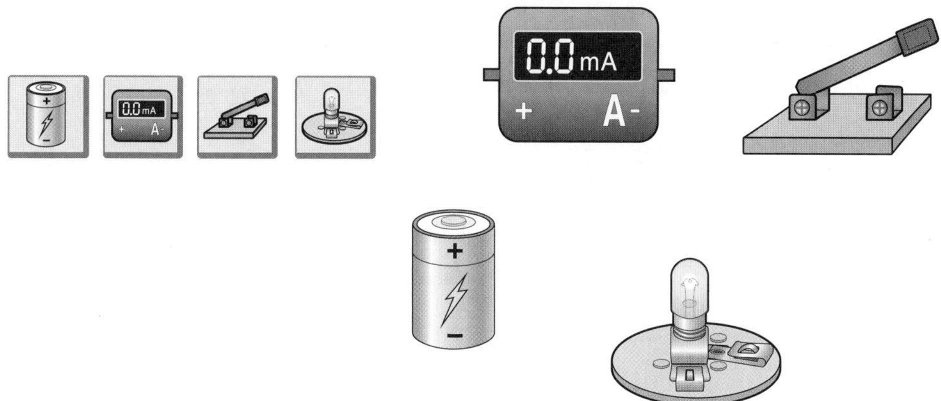

Step 2 Practice opening and closing the switch. Click the **Select** button, then move the cursor to the top of the handle. Drag the handle down to close the switch. Drag it up to open the switch. Leave the switch in the open (handle up) position.

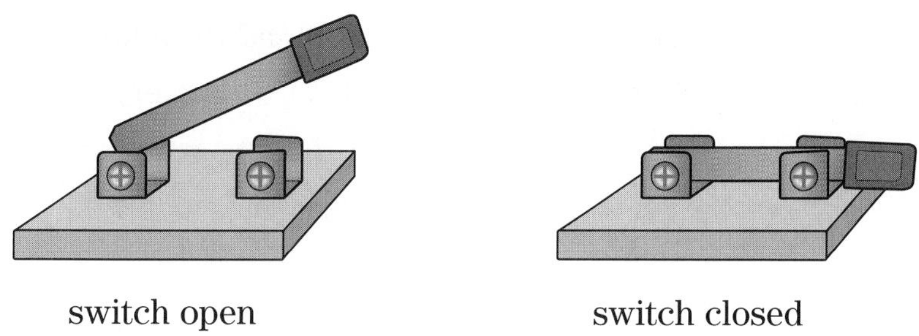

switch open   switch closed

Step 3 For the circuit to work, all the devices must be wired together. To draw a wire between the battery and bulb holder, place the cursor arrow at one end of the battery. Then click and start dragging the cursor towards one end of the bulb holder.

As you get near, the clip will "flash" a small yellow rectangle.

At that point, you can let go of the cursor and the wire will connect itself.

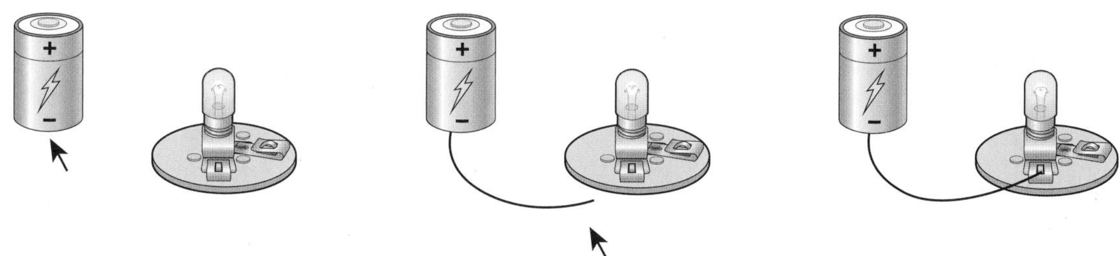

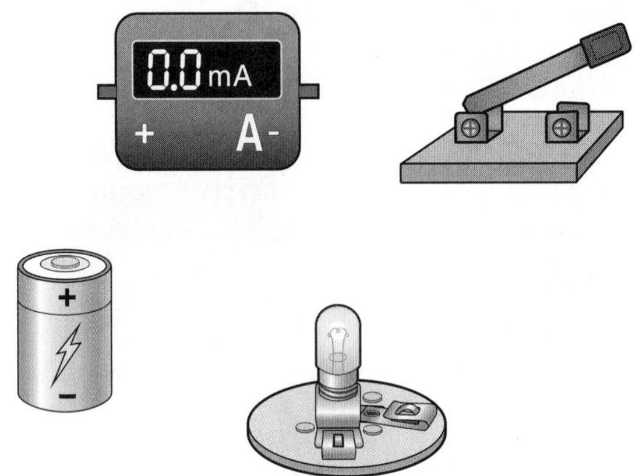

Step 4 Wire together the rest of the circuit.

Step 5 Run the Simulator by clicking the **Run (Play)** button.

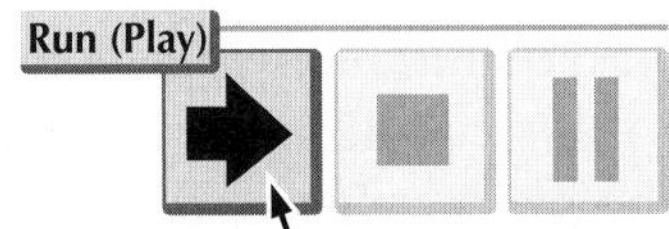

Step 6 Close the switch. The bulb should light and the ammeter should show a reading. Record this ammeter reading.

Step 7 Click **Stop** on the Simulator.

Step 8 Delete the wire between the battery and bulb. Add a second battery. Connect the wires and run the Simulator. Continue with three, then four batteries. Record these ammeter readings.

How To...

10 Use a Graduated Cylinder

The *volume* of a liquid is the amount of space it takes up (occupies). *Graduated cylinders* are used to measure the volume of liquids. The smallest units marked on their scales are milliliters (mL). Graduated cylinders come in several different sizes.

For an accurate reading, the graduated cylinder should be on a level surface. Move your head so that your eye level is at the water level in the graduated cylinder. It may help to hold a white paper behind the cylinder so it is easier to see the top of the liquid.

You will notice that the surface of the water in the cylinder is curved. The curve is called the *meniscus*. The accepted scientific practice is to measure the volume of water by taking the reading at the bottom of the curve (meniscus).

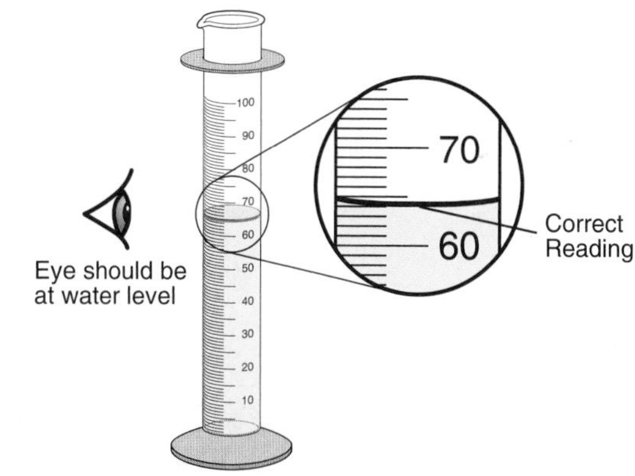

What is the correct volume reading in the picture? _______________

If you count the number of marks from 60 up to 70 on the cylinder, you will count ten marks. Each mark represents 1 mL (milliliter). The bottom of the curve (meniscus) appears to be close to the 64 mark. So, the correct reading is 64 mL, read as "64 milliliters."

Guidelines for Using the Graduated Cylinder

These guidelines will help you avoid measurement mistakes (causing outliers) when using a graduated cylinder.

1. Make sure the graduated cylinder is empty before adding a liquid.
2. To avoid spilling, use a funnel to pour liquids into the graduated cylinder.
3. Try to eliminate any large bubbles that are in the liquid. Tilt the cylinder slowly so that the bubble can burst in the air.
4. Try to "capture" any large drops that stick to the side of the cylinder by tilting it slowly.

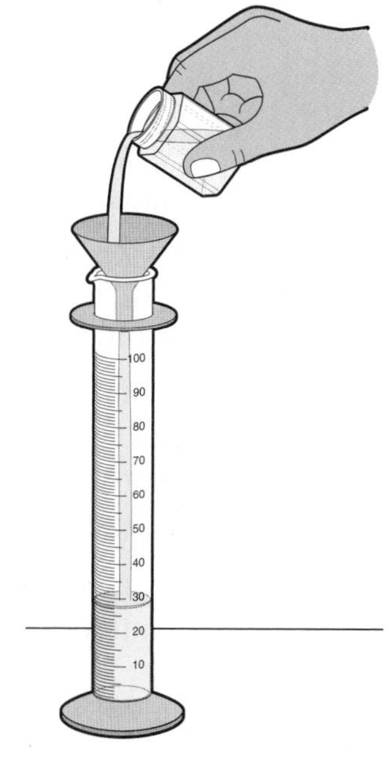

How To...

11 Use Mass Balances

A mass balance is used to measure the mass of different objects. Two common types of mass balances are the triple-beam balance and the equal-arm balance.

Triple-Beam Balance

"Triple beam" refers to the three beams along the side of the object pan. These are called rider beams. Examine the diagram below and notice that the three rider beams are attached to a pointer. When the three beams are balanced, the pointer aligns with a line on the far right of the balance as you see in the diagram.

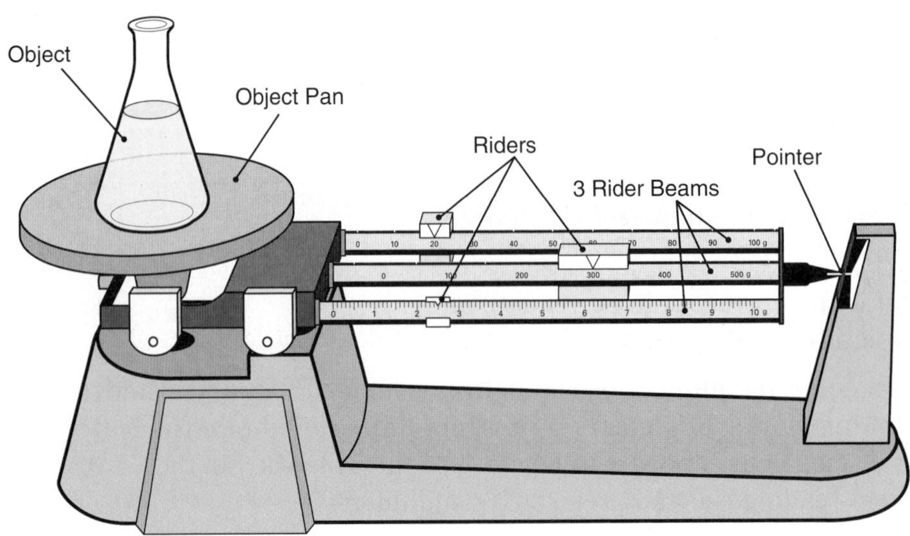

Attached to each rider beam is a rider. The rider is the object that you can move back and forth along the rider beam. The middle rider beam has a large rider and scale of numbers that goes from 0 to 500 g (grams) in 100-g increments. This beam has notches in it. Only when the rider is in a notch, should you read the value from this beam.

The rear rider beam has a scale from 0 to 100 g in 10-g increments. This beam also has notches and the rider should *always* be in a notch before you read the value.

The front beam has a scale from 0 to 10 g and has a small rider that *slides* along the scale. This scale is divided into a hundred small marks, each one-tenth of a gram (0.1 g).

Before you make any mass measurements using the balance, make sure that you calibrate, or zero, the balance. To calibrate the balance, take everything off the pan, and make sure all the riders are exactly on zero. Move your head down so that your eye is at the same level as the pointer. If the pointer does not exactly align with the line, turn the zero knob (usually located under the pan) until the pointer aligns with the line.

How to Read the Triple-Beam Balance

If your object is on the object pan, the riders are in their notches, and the pointer at the end of the three rider beams is lined up with the line at the far right of the balance, then you are ready to read your balance. The procedure is to add up the values of the three rider beams. Always record your answer to the nearest tenth of a gram (0.1 g).

Object mass = Value on center beam + Value on rear beam + Value on forward beam

When reading the balance, always begin with the value on the center beam.

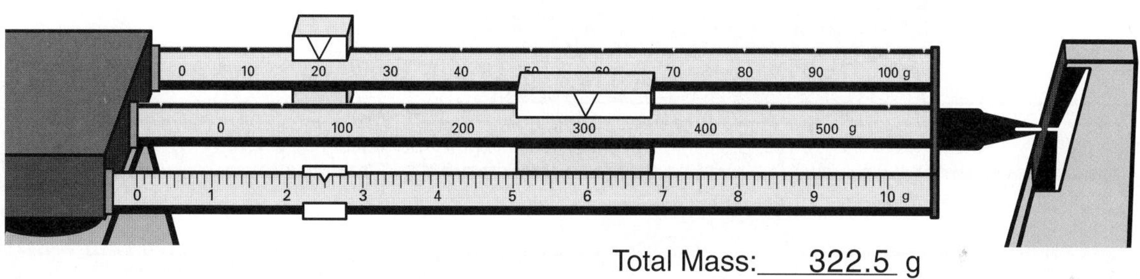

In the diagram, the center beam reads 300 g. Next, add the value on the rear beam. The rear beam in the diagram reads 20 g. Finally, add the value of the forward beam and you have the mass of the object. In the diagram, the value of the forward beam is 2.5 g (notice that each line represents a tenth of a gram). In the diagram:

Object mass = 300 g + 20 g + 2.5 g = 322.5 g

Equal-arm Balance

"Equal-arm" refers to the bar that connects the two pans. The bar appears to have two "arms" of equal length, and the bar balances at its midpoint. Notice that this balance only has two rider beams. If you want to measure objects with a mass greater than 200 g, you must use standard masses, as shown in the diagram.

Examine your balance and notice that the equal arms are attached to the "pointer." If the pans have nothing on them, the riders are in the zero positions, and the balance is properly calibrated or zeroed, the pointer points to zero. When the arms are equally balanced with mass, the pointer again points to zero, and it is then appropriate to read the mass of the object you are balancing.

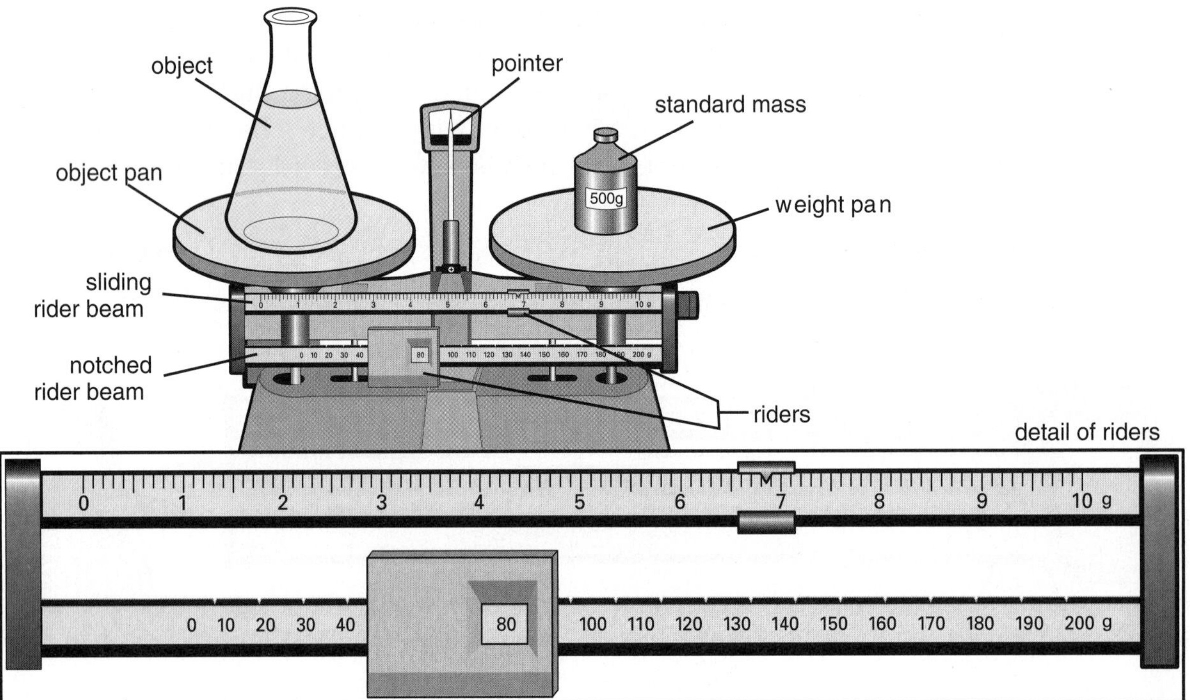

How to Read the Equal-arm Balance

If your object is on the object pan and the pointer is at zero, then you are ready to read the balance. The procedure is to add up the values of the standard mass, the lower beam (the notched rider beam), and the upper beam (the sliding rider beam) in that order. Always measure to the nearest tenth of a gram (0.1 g).

Object mass = Standard masses + value on lower beam + value on upper beam

In the diagram, the standard mass is 500 g, the lower beam reads 80 g, and the upper beam reads 6.9 g (notice that each line represents a tenth of a gram; we rounded up).

Object mass = 500 g + 80 g + 6.9 g = 586.9 g

Rules for Using the Balance

A balance is a sensitive and expensive scientific instrument that can be ruined if it is not treated carefully. Below is a list of rules to follow when using your balance.

1. *Always hold on to both sides of the balance* when you are carrying it. Do not carry the balance by the rider beams.
2. *Always clean the pans* with a damp towel if you have spilled anything on the pans.
3. *Do not lean on or press down* on the object pan.
4. *Always be sure the balance is zeroed (calibrated)* before you make a mass measurement.
5. *Always be sure the riders are in their notches* before you read the mass of your object.
6. *Always move the riders back to the "0" mark* when you are finished using the balance.

How To...

12 Evaluate an Analysis and Explanation

Read the *analysis* and decide "yes," "no," or "not applicable" to indicate whether each of the following criteria is met.

1. The interacting objects and their interaction types are *correctly identified.*
2. The energy diagram *correctly* shows the energy transfers between interacting objects, and *correctly* shows the energy changes within each object.
3. The force arrows diagram *correctly* shows the forces exerted on the object of interest.
4. The system is identified and *correctly* specified as an open or closed system.

- If you decided "yes" or "n/a" for all of the criteria, then the analysis is good.
- If you decided "no" for one or more of the criteria, then the analysis is *poor.* Correct the analysis.

Read the *explanation* and decide "yes" or "no" to indicate whether each of the following criteria is met.

5. The written explanation includes only correct scientific ideas.
6. The written explanation includes all the appropriate scientific ideas; none of the important ones are missing.

- If you decided "yes" for both of the criteria, then the explanation is *good.*
- If you decided "no" for one of the criteria, then the explanation is *poor.* Write a correct explanation.

How To...

13 Write an Analysis and Explanation

1. Analyze the situation.

Perform the analysis needed to answer the question. For example, identify interacting objects and their type of interaction, draw an energy diagram, etc.

In *InterActions in Physical Science* you will often be told what kind of analysis to perform.

2. Write the explanation.

Use your analysis and the science ideas you have developed or learned to help you answer the question. You may refer to this book or to the Scientists' Consensus Ideas forms to review your science ideas.

Write your explanation using complete sentences. Your explanation should be able to pass an evaluation using How To Evaluate an Analysis and Explanation.

How To...

14 Identify Mechanical Interactions

Mechanical Interactions

In a mechanical interaction objects touch each other while pushing or pulling on each other over a distance.

Applied Interaction

An applied interaction occurs when *two non-elastic (i.e., rigid or stiff) objects push or pull on each other.*

Examples include a person rolling a crate and a tugboat towing a barge.

Friction Interaction

A friction interaction occurs when *two surfaces rub against each other.*

Examples include sliding down a slide and engine parts moving against each other.

Drag Interaction

A drag interaction occurs when *an object moves through a gas or liquid. The gas or liquid resists the object's motion.*

Examples include a paper airplane dragging through air, and a boat dragging through water.

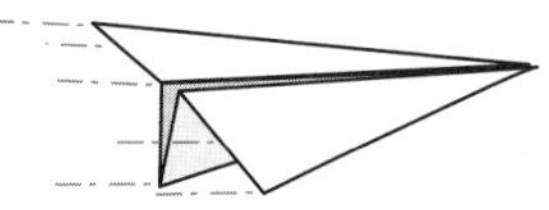

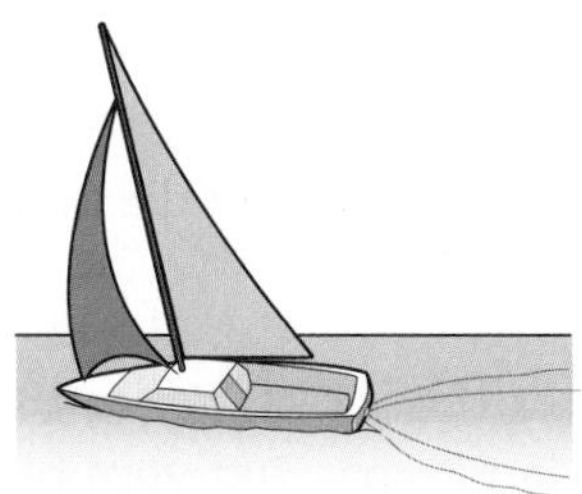

Elastic Interaction

An elastic interaction occurs *when two objects push or pull on each other and at least one of them is stretchy.*

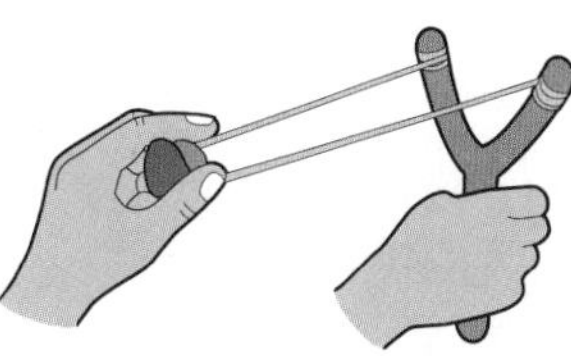

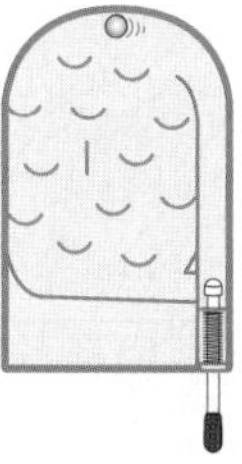

Examples include launching a ball with a slingshot and launching a pinball with a spring.

How To...

15 Do an Experiment as a Team

Most modern scientists and engineers work in research teams. The team members share responsibilities. In this course, you will also carry out specific tasks when you do experiments. Your teacher will assign each member of your team one of the four roles shown in the chart. You will rotate roles for each experiment.

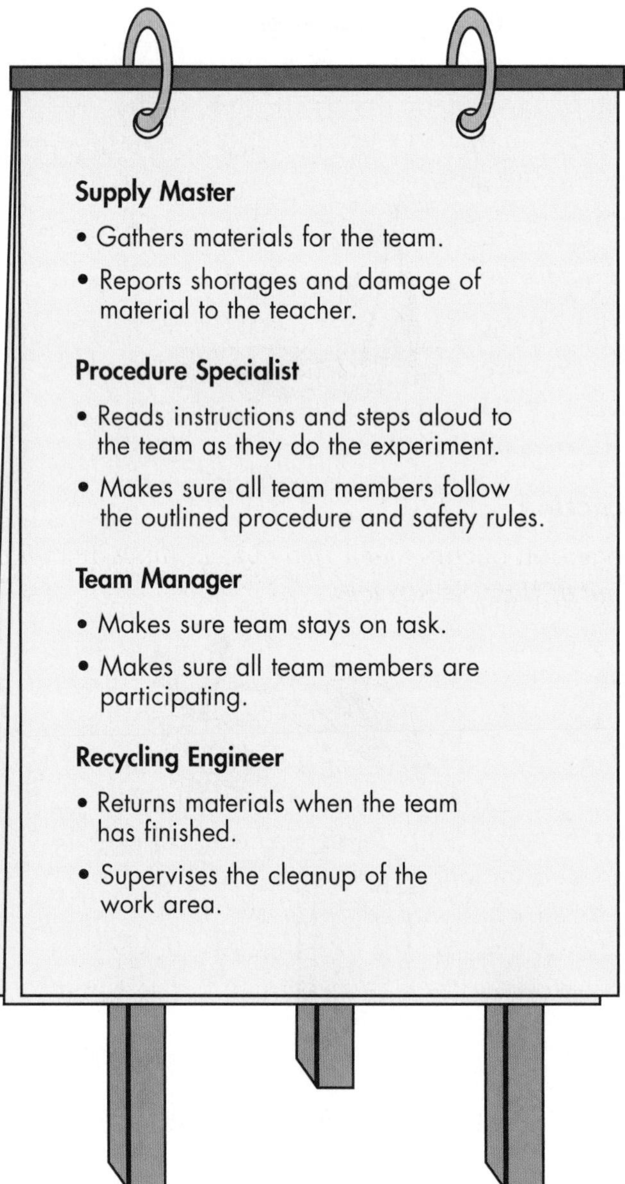

Supply Master
• Gathers materials for the team.
• Reports shortages and damage of material to the teacher.
Procedure Specialist
• Reads instructions and steps aloud to the team as they do the experiment.
• Makes sure all team members follow the outlined procedure and safety rules.
Team Manager
• Makes sure team stays on task.
• Makes sure all team members are participating.
Recycling Engineer
• Returns materials when the team has finished.
• Supervises the cleanup of the work area.

How To...

16 Read a Thermometer

Measuring Temperature

A thermometer is an instrument for measuring temperature. Anything that shows clear evidence of changes in temperature can serve as a thermometer. Most thermometers are narrow glass tubes connected to a hollow bulb of liquid mercury or colored alcohol. The outside of the tube is marked with a numbered *scale*, as shown in the diagram.

In 1714, the German physicist Gabriel Fahrenheit made a mercury thermometer and developed the Fahrenheit temperature scale. On the Fahrenheit scale, the freezing point of water is at 32° above the scale's 0 point, and the boiling point of water is at 212° above 0. The 0 point of the scale stood for the temperature resulting from mixing equal parts by weight of snow and common salt. In the United States people use the Fahrenheit scale for a variety of temperature measurements, including reporting the weather and body temperatures in medicine. For example, a room at 70° F, is comfortable, a room temperature of 90°F is hot. Typically, body temperature is 98.6° F. Scalding hot water is 131° F, and can cause serious injury.

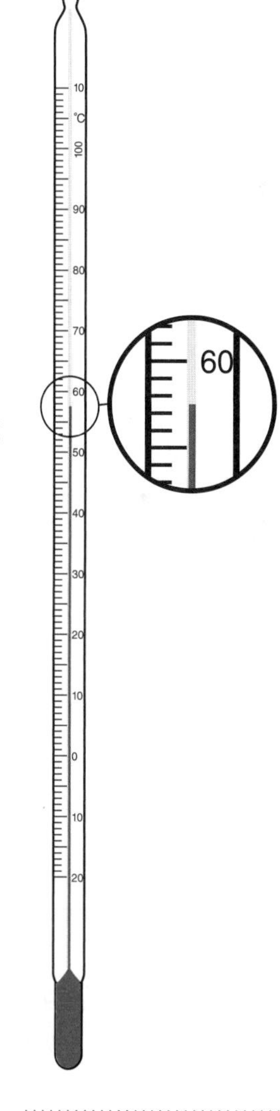

In 1742, the Swedish astronomer and physicist Anders Celsius created a temperature scale on which 0° is the freezing point of water and 100° is the boiling point. This became known as the Celsius scale. Scientists use the Celsius scale. You will also be using the Celsius scale.

Reading a Celsius thermometer is like reading a ruler. There are small marks for each degree, with a larger mark every five degrees. Find the degree mark closest to the level of the alcohol in the tube. Sometimes the level appears to be exactly halfway between two degree marks, rather than closer to one of the marks. In such cases, record the temperature as a decimal. For example, this thermometer reads 57.5° C.

How To...

17 Read an Energy Bars Box

Like an energy diagram, an energy bars box is a way of representing an object's (or system's) energy inputs, outputs, and changes.

The diagrams below shows the relationship between the bars in the energy bars box and the equivalent parts of an energy diagram. The energy bars box displays energy readouts in one of the following ways:

- kilocalories (kC) (in the Freezing and Melting simulator)
- kilojoules (kJ) (in the Interactions and Motion simulator)
- joules (J) (in all other simulators)

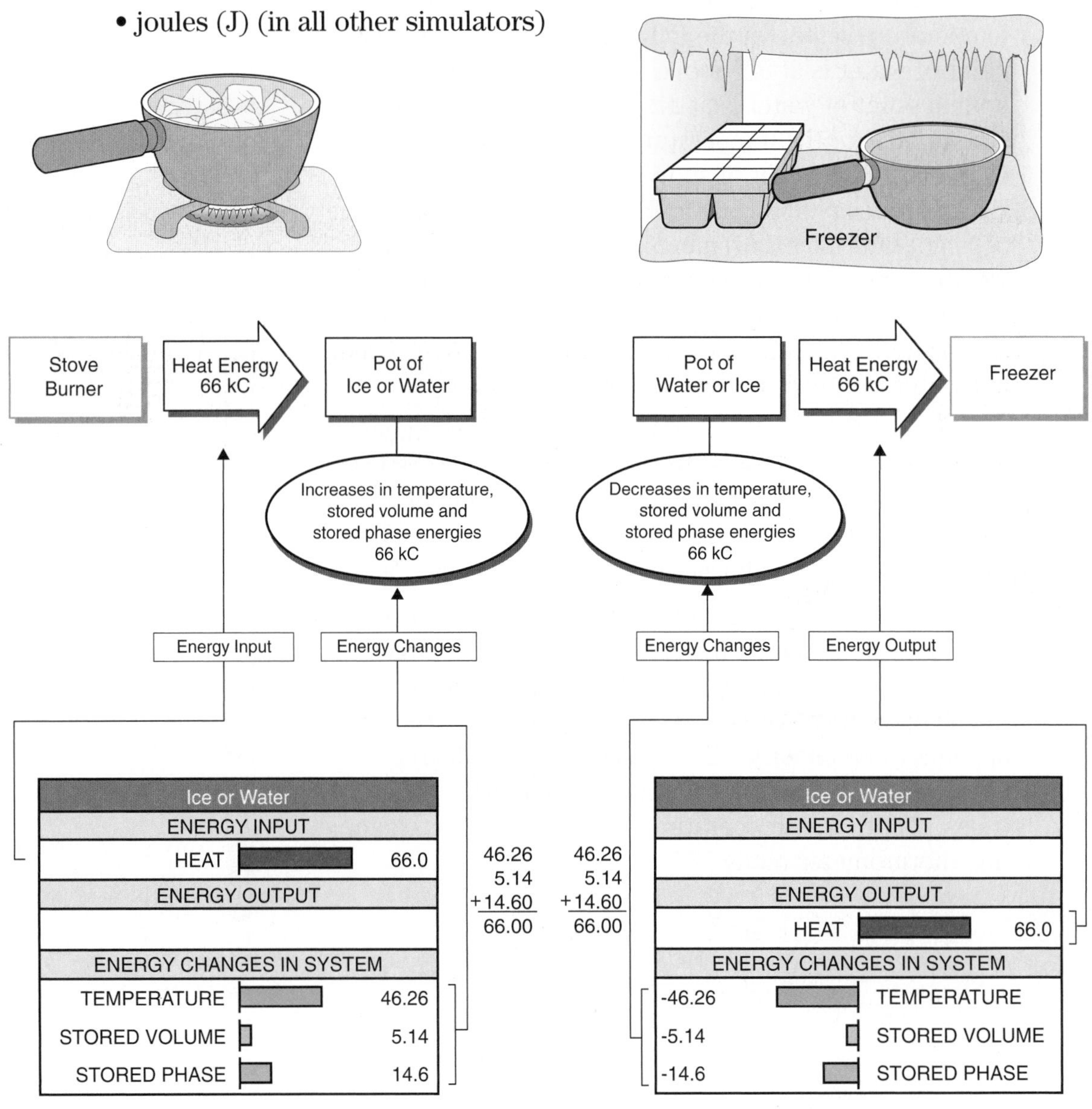

How To...

18 Identify Types of Energy Changes

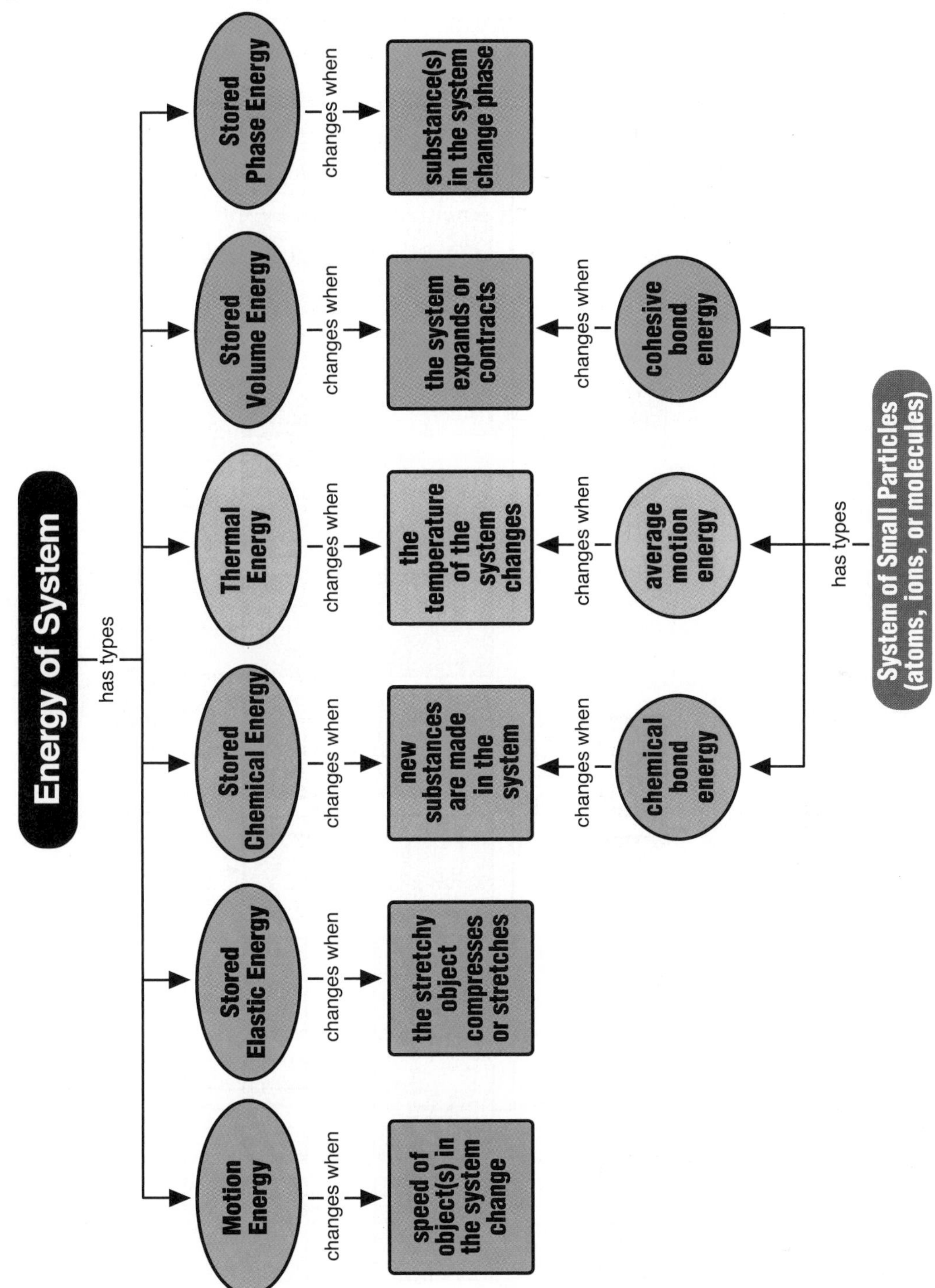

How To... 19 Identify Interaction Types

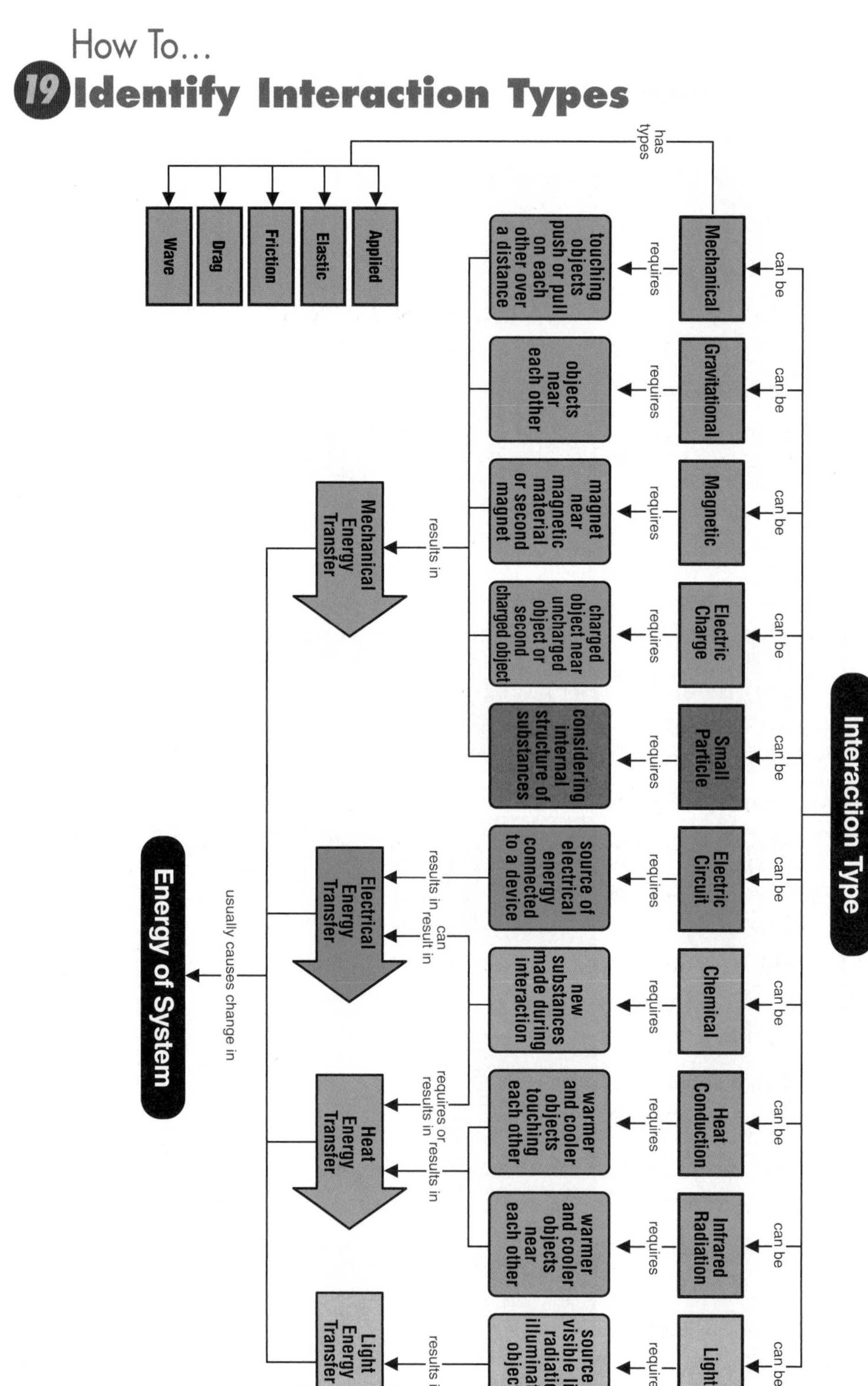

How To...

20 Use the Interactions and Motion Simulator

Here are some simulator functions that you might find useful as you set up the simulator to model the explorations in ***Unit 3, Cycle 2, Activities 5 and 7***.

STEP 1 To ***place an object*** (body) on the screen:

- Click on the button that looks like a cube.
- Move the cursor arrow to the screen area, and click the mouse button where you want to place the object.

STEP 2 To ***move items***:

- Click the **Select** button.
- Click on the object you want to move and drag the object to its new location.

STEP 3 To ***delete items***:

- Click the **Delete** button.
- Click on the item you want to delete.

Caution: When you are finished deleting items, click on another button or the Delete button again to deselect it. Otherwise, you will continue to delete items, even if you did not want to!

STEP 4 To ***rename objects***:

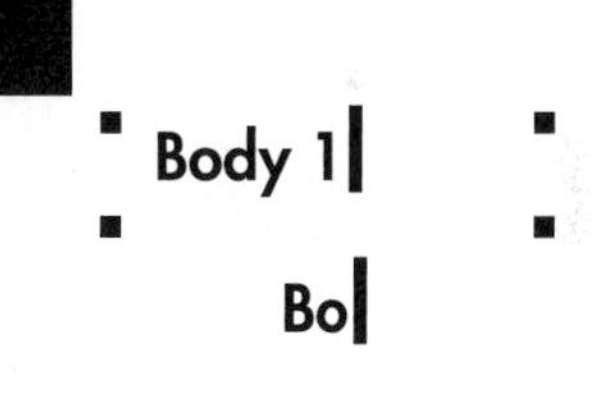

- Click the **Select** button.
- Click ***twice slowly*** on the object's name that you want to change. (If you double-click too fast you will open up the Properties Box for the object.)

- Use the backspace button on the keyboard to erase the old name. Input the new name.

STEP 5 To ***exert a force*** on an object:

- Click on either the **PUSH** or **PULL** buttons.
- Click on the body and drag out a force arrow from the body. The longer the arrow, the larger the force.

STEP 6 To ***change the force strength by changing its length:***

- Click the **Select** button.

- Click on the force arrow to select it.

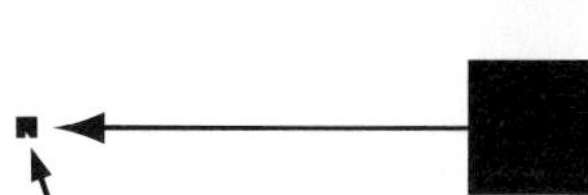

- Click and drag from the end furthest from the object to lengthen or shorten the force.

Step 7 To ***change the force strength by changing its value:***

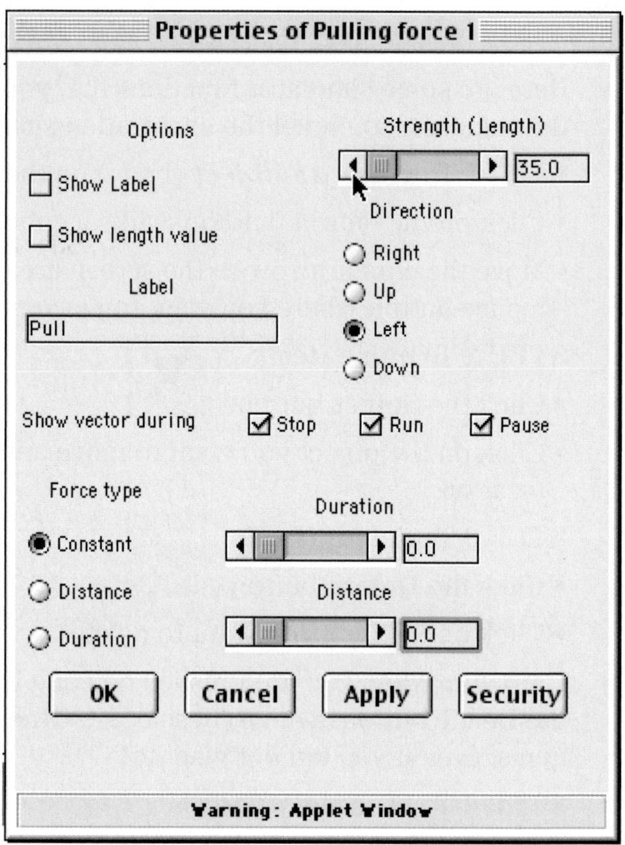

- Click the **Select** button.
- Double-click on the force arrow that you want to change. This will open the force arrow's **Properties** box.
- Set a new value for the force strength (length).
- Click **OK** to leave the **Properties** box.

Step 8 To ***give an initial speed*** to an object:

- Click the **SPEED** button.

- Click on the object and drag out a speed arrow. The longer the arrow, the larger the initial speed.

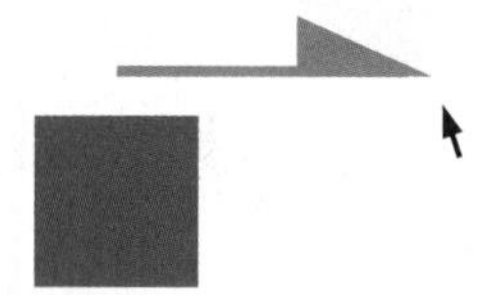

Step 9 To ***run your simulation,*** click on the **RUN** button.

Step 10 To ***stop your simulation,*** click on the **STOP** button.

Step 11 To ***pause your simulation,*** click on the **PAUSE** button.

Step 12 To ***rewind your simulation,*** click on the **REWIND** button.

Step 13 To ***show the speed values*** (for example, a speed value of 40 m/s):

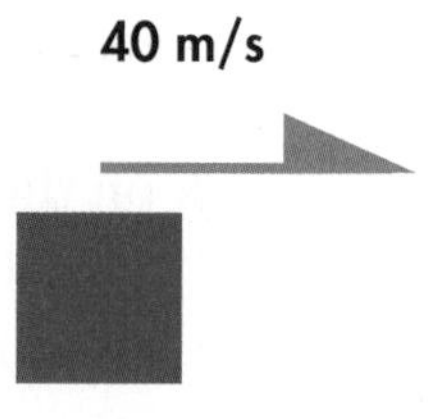

- *Before a run,* double-click the **SPEED** button. *After a run,* or if a speed arrow is shown, double-click on the speed arrow.
- Select "Show length value" in the **Properties** box.

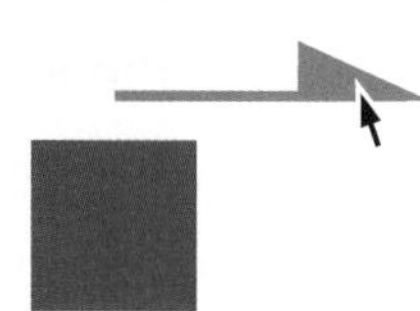

How To...

21 Use a Burner Safely

Two types of burners are usually used in science classrooms: alcohol burners and gas burners. If you are using an alcohol burner, read Part A. Steps for how to use a gas burner are in Part B.

Be sure you know where the fire extinguisher is located before you use a burner. Your teacher may also have special rules on safety and emergencies.

Part A: Alcohol Burner

STEP 1 Make sure that the burner is no more than half full.

STEP 2 Make sure the lid of the burner is on tight.

STEP 3 If the length of your hair is past your earlobes, tie it back or wear a hair net.

STEP 4 Remove the wick cap and light the burner with a match.

STEP 5 Replace the cap on the wick when you are finished. Never return a burner unless the wick is cool.

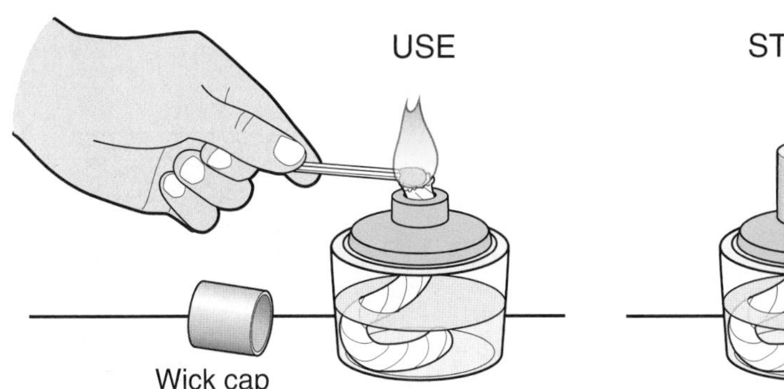

Never use one lighted burner to light another burner. Tipping a burner can cause a spill, and alcohol burns easily.

Step 6 Burners get hot when they are lighted. To avoid burns, handle the burner near its base.

Never carry a burner when it is lighted!

Never use water on an alcohol fire because it spreads the burning alcohol.

STEP 7 If a burner causes a fire, tell your teacher immediately. Put the fire out with a CO_2 (carbon dioxide) or foam extinguisher, sand, or a fire blanket.

CO_2 extinguisher | Foam extinguisher | Fire blanket | Sand

Part B: Gas Burner

STEP 1 If the length of your hair is past your earlobes, tie it back or wear a hair net.

STEP 2 Be sure the gas-supply valve is off before you begin.

STEP 3 Connect the burner to the gas supply with rubber tubing. Check to be sure that both ends of the tubing are connected tightly, so no gas will escape.

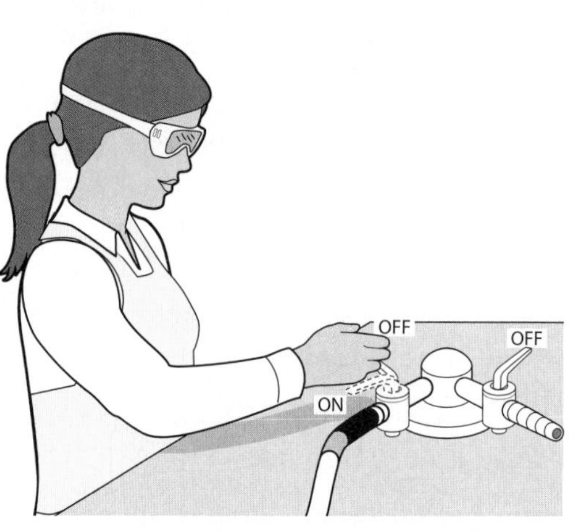

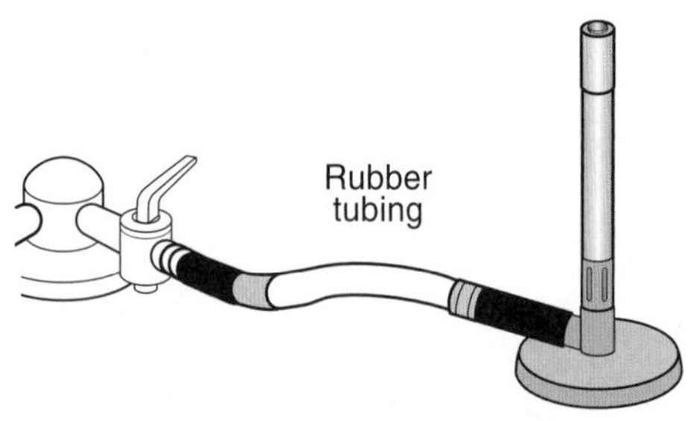

STEP 4 Close the air-mix control on the burner.

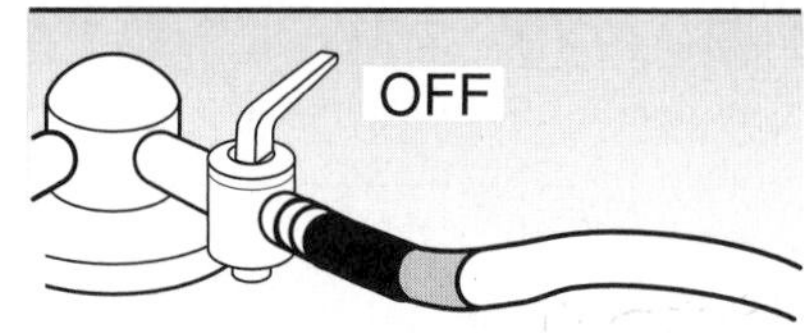

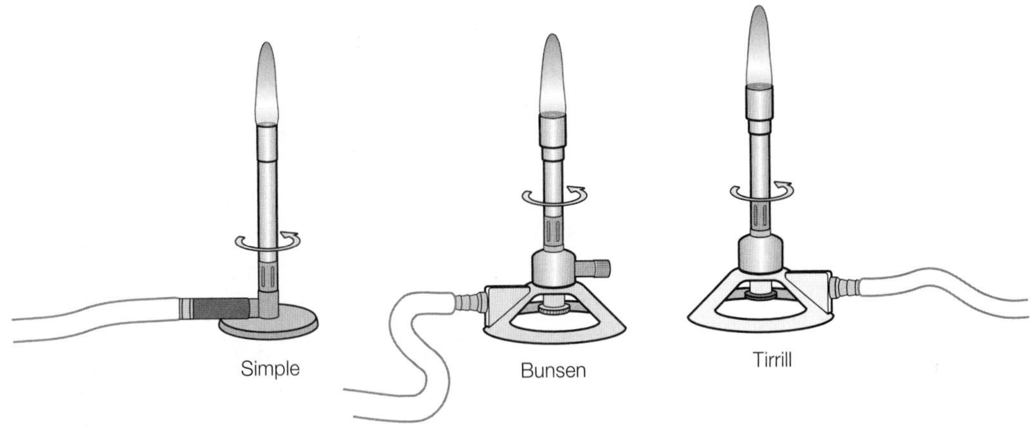

STEP 5 Light a match. Hold the match at the edge of the flame tube (not directly over the tube). Then slowly open the valve on the gas supply.

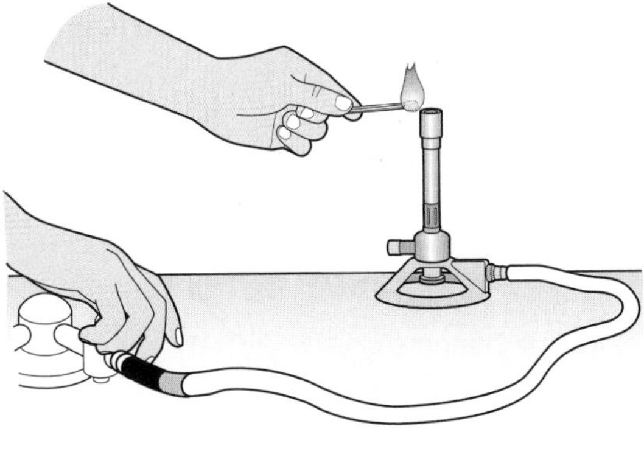

STEP 6 When the burner lights, it usually has a yellow flame. Adjust the air-mix control to get a light blue flame, which is hotter than the yellow flame.

STEP 7 If you get a flame bunting down inside the tube (as shown in the diagram) instead of at the top, then turn the gas supply valve off. Start again, but this time do not open the air-mix control as much.

STEP 8 If you have to move the burner, handle it near the base. The barrel gets hot.

Light blue

Hottest part of flame

Darker blue

STEP 9 To put out the flame, turn the gas supply valve off. Always check that the gas-supply valve is off when you are finished using the burner.

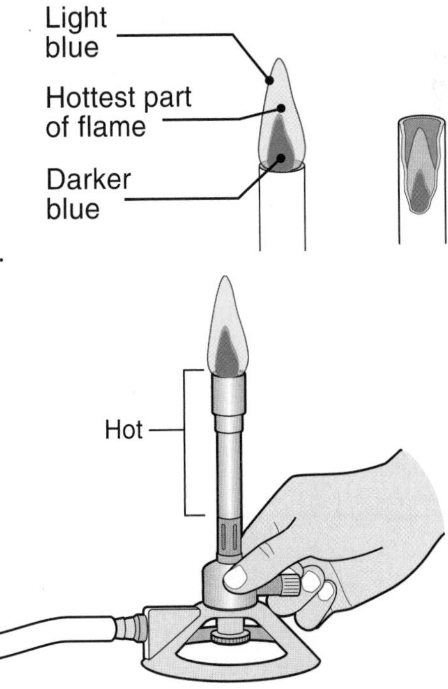

STEP 10 If a fire breaks out, turn off the gas *immediately*, and tell your teacher. Then put out the fire.

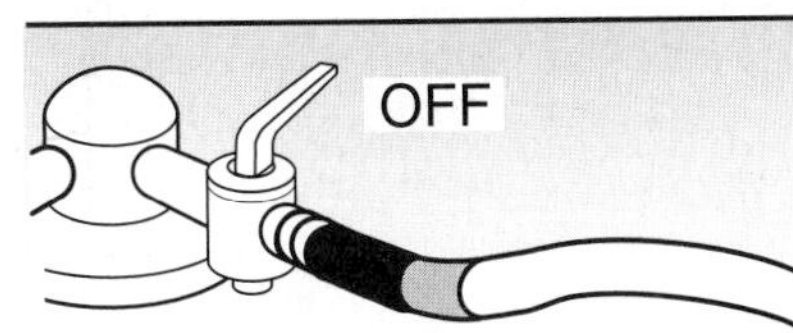

How To...

22 Decide Whether a Picture Equation of a Chemical Interaction Is Correct (Balanced)

(Unit 7, Activity 5)

You are given a picture equation of the Conservation of Mass for a chemical interaction. Follow the steps below to decide if the pictured equation is correct (balanced).

A. Count the number of atoms of each element in the *reactant(s)*.

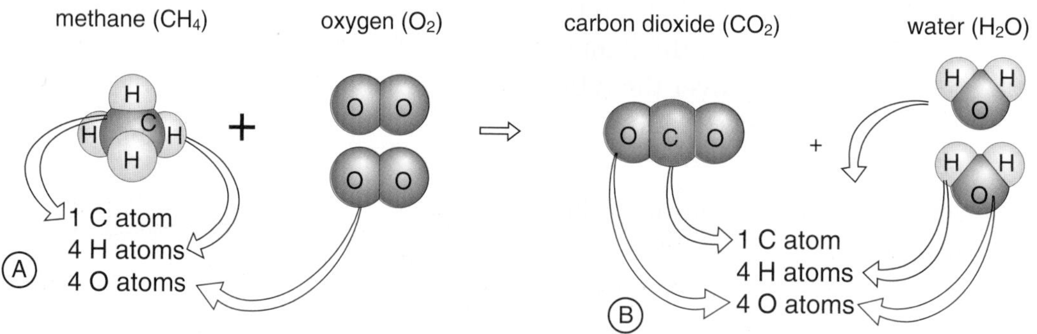

B. Count the number of atoms of each element in the *product(s)*.

C. Is the number of atoms of each element the same for the reactant(s) and the products?

Yes The picture equation is correct. Mass is conserved because the reactant atoms are the same in type and number as the product atoms.

No The picture equation is not correct. Mass is not conserved because a type of product atom is incorrect and/or the number of the product atoms is incorrect.

How To...

23 Write a Word Equation of a Chemical Interaction from the Picture Equation

(Unit 7, Activity 7)

You have a correct balanced-picture equation for a chemical interaction. Follow the steps below to write a balanced-word equation for the interaction.

A. Count the number of *particles* (atoms or molecules) of each reactant and each product in the picture.

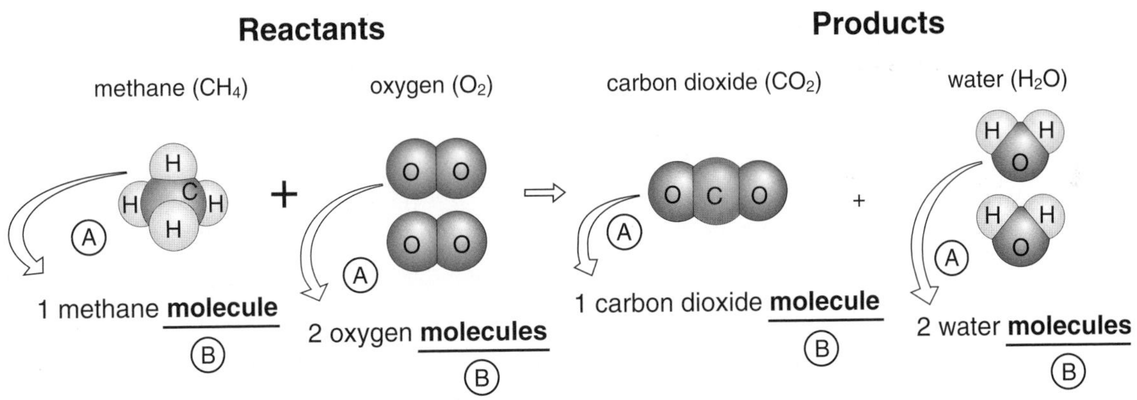

Reactants | **Products**

methane (CH_4) + oxygen (O_2) ⟹ carbon dioxide (CO_2) + water (H_2O)

(A) 1 methane **molecule** (B)

(A) 2 oxygen **molecules** (B)

(A) 1 carbon dioxide **molecule** (B)

(A) 2 water **molecules** (B)

B. Decide whether each particle is an atom or molecule.

C. Write the word equation in the following form:

1 (number) methane molecule (particle type) + 2 (number) oxygen molecules (particle type)

⟹ 1 (number) carbon dioxide molecule (particle type) + 2 (number) water molecules (particle type).

D. Put the given formulas for the particles in your word equation, as shown below.

1 methane molecule (CH_4) + 2 oxygen molecules (O_2) ⟹ 1 carbon dioxide molecule (CO_2) + 2 water molecules (H_2O).

HOW TO...

Table of Densities* (Remember that 1 mL = 1 cm^3)	
Material	**Density**
Solids	
Aluminum	2.7 g/cm^3
Brass (yellow)	8.0 g/cm^3
Copper	8.9 g/cm^3
Oak Wood	0.6–0.9 g/cm^3
Steel	7.6 g/cm^3
Tin (gray)	5.8 g/cm^3
Silver	10.5 g/cm^3
Liquids	
Acetic Acid	1.05 g/mL
Antifreeze	1.11 g/mL
Gasoline	0.74 g/mL
Mercury	13.0 g/mL
Rubbing Alcohol	0.79 g/mL
Salt Water (saturated)	1.20 g/mL
Water	1.00 g/mL
Gases	
Air	0.0012 g/cm^3
Carbon Dioxide	0.0013 g/cm^3
Helium	0.00017 g/cm^3
Hydrogen	0.00008 g/cm^3
Methane	0.00067 g/cm^3
Nitrogen	0.0012 g/cm^3
Oxygen	0.0018 g/cm^3

* *Approximate values at sea level and 20°C.*
IL = 1000 mL

Table of Melting and Boiling Points

Melting Point (M.P.) and Boiling Points (B.P.) are in degrees Celsius (ºC)

	M.P.	B.P.
GAS		
ammonia	–78	–33
butane	–138	–0.5
carbon dioxide	–78†	—
carbon monoxide	–199	–191
chlorine	–101	–35
helium	–272	–269
hydrogen	–259	–252
methane	–182	–164
nitrogen	–210	–253
oxygen	–218	–183
LIQUID		
acetic acid	17	118
acetone	–95	56
antifreeze (ethanethiol)	–144	35
rubbing alcohol (ethanol)	–117	79
water	0	100
wood alcohol (methanol)	–94	65

	M.P.	B.P.
SOLID		
aluminum	660	2467
baking soda (sodium bicarb.)	dec‡	—
calcium oxide	2580	2850
chalk (calcium carbonate)	dec‡	—
charcoal (carbon)	3692†	—
copper	1083	2595
copper sulfate	110	150
Epsom salt (magnes. sulfate)	150	200
gold	1063	2966
iron	1535	3000
potassium chloride	776	1500
sand (silicon dioxide)	1610	2230
sodium nitrate	307	380
sugar (sucrose)	185	dec‡
sulfur	113	445
table salt (sodium chloride)	801	1413
tin	232	2270
zinc	419	907

† Substance turns directly from a solid to a gas or from a gas to a solid (sublimation).
‡ Decomposes when heated or cooled further.

A

acceleration
the change in velocity per unit time

activation energy
the minimum motion energy that reactant particles must have when they collide to break their chemical bonds

ammeter
a device that measures the amount of electric current in a circuit

amplitude
the height of a wave crest. It is related to a wave's energy.

analysis
any procedure that helps you to understand a situation

astronomical unit
a unit of measurement equal to the average distance between the Sun and Earth

atmosphere
the layer or envelope of gases that may surround a planet or moon

average speed
the distance traveled divided by the time taken

B

bond
the attraction between particles (atoms, ions or molecules)

buoyant force
for an object that is placed in a liquid, the force exerted on the object by the liquid

C

characteristic property
a measurement (numbers) that is different for different kinds of materials

chemical interaction
any type of interaction that results in at least one new material

cohesion
sticking or holding together to form a whole

compound
a single substance (chemical) that breaks down into elements during chemical interactions

compound machine (or complex machine)
two or more simple machines working together

compression (longitudinal) wave
a wave in which the motion of the material (medium) is parallel to the direction of the motion of the wave

constant speed
neither speeding up nor slowing down

covalent bond
the attraction between specific pairs of atoms in a molecule

D

density
the mass of a standard-unit volume

diffusion
the spreading of one substance into another substance in the same phase

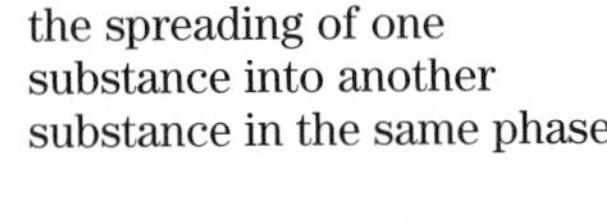

E

earthquake
a sudden motion or shaking of the earth

electric circuit
the path followed by an electric current from a power source through devices that use electricity and back to the source

electrical conductor
a material that allows electric current to exist in it

electric current
a flow of electrons (negative charges) around a circuit

electrical non-conductor
a material that does not allow electric current to exist in it

electromagnet
a temporary magnet consisting of a coil of wire around a core of magnetic material (usually iron); when an electric current flows through the coil, the iron becomes a magnet

element
a single substance (chemical) that does not break down into simpler substances during chemical interactions

endothermic reaction
reactions that absorb energy (require continual energy input)

energy receiver
an object to which the energy is transferred

energy source
an object that is the supplier of energy

evaluation
a judgment of something

evidence
in an experiment, the data collected by the researcher

exothermic reaction
reactions that produce (release) energy

explanation
uses information from the analysis and uses science ideas to answer questions about a situation

F

fair test
an experiment in which only the manipulated and responding variables are allowed to change and all other variables and conditions are kept the same

fault
a fracture in rock, along which the rock masses have moved

fluid
a substance, such as a gas or liquid, that tends to flow and take the shape of its container

force
a push or a pull

frequency
the number of waves produced per unit time

G

gas-giant planet
a large planet with a deep atmosphere that is mostly hydrogen and helium, and a core of icy and rocky material; gas-giant planets include the outer Solar System planets Jupiter, Saturn, Uranus, and Neptune

graduated cylinder
a tall, thin tube, marked off in units, used to measure volume of a liquid

gravitational potential energy
the energy of a system with two objects interacting through gravity. The energy depends on the distance between the object and their masses.

gravity
the force of attraction between two bodies due to their masses

H

hypothesis
a statement that can be proved or disproved by experimental or observational evidence

I

infrared radiation
radiation with energies lower than visible light

input force (or **effort)**
the force exerted on a machine

ion
a single atom or group of atoms that have either a positive or negative charge

ionic bonds
the attraction between neighboring nonmetal and metal ions

isotope
an element that has different numbers of neutrons

K

kinetic energy
the energy an object possesses because of its motion

L

L wave
a seismic wave that travels along the surface of the Earth; it is the last to arrive at a location

length
a measure of distance

light-year
a unit of measurement equivalent to the distance light travels in a year

linear relationship
the relationship between two quantities that, when plotted against each other on a graph, produce a straight line

luminosity: (of a star)
a measure of the star's brightness that does not depend on the distance between the star and the observer

M

magnetic materials
metals that interact with magnets

manipulated variable (also called **independent variable)**
in an experiment, a variable that can be deliberately changed by the scientist and that determines the values of other variables (called responding variables)

mass
the amount of matter that a body contains

mechanical energy
the energy transfer involved in an interaction that causes one or both objects to change position

mechanical interaction
an interaction in which objects touch each other while pushing and pulling each other over a distance

metallic bonds
attractions between neighboring metal atoms

metalloids (semi-metals)
elements that have properties of both metals and nonmetals

monomers
the smaller molecules that are linked together to form the giant molecules of polymers

moon (satellite)
a body that orbits a planet

motion energy
the energy an object has because of its motion

multi-loop circuit (parallel circuit)
a circuit in which two or more single loops connect to the same cell

N

newton
a unit of force

nonlinear relationship
the relationship between two quantities that, when plotted against each other on a graph, do not produce a straight line

nonrenewable energy resource
an energy resource that cannot be replaced

O

opaque
a material through which light cannot travel

orbit
the path that an object takes as it moves through space, usually around another object

outliers
values far from most others in a set of data

output force (or **load**)
the force a machine exerts on an object

P

P wave
a seismic wave that involves motion in the direction in which it is traveling; it is the fastest of the seismic waves

periodic
having a regular, repeating pattern

pH
a quantity used to represent how acidic a solution is

phase change
the conversion of a material from one phase to another phase. Example: solid to liquid, liquid to gas.

physical interaction
any type of interaction that does not result in any new materials

physical property
a description or measurement of what happens to a material during physical interactions

pitch
the quality of a sound dependent mostly on the frequency of the sound wave

plastics
synthetic polymers that can be molded or shaped

polymer
a substance that is a giant molecule made up of many similar small molecules (monomers) linked together in long chains

potential energy
the energy of an object that is dependent on its position

product
end (new) substance made in a chemical reaction

property
a description of how the object interacts with another object

R

reactant
original substance in a chemical reaction

refraction
the change in direction (bending) of a light ray as it passes at an angle from one material to a different material

relationship
an idea about what happens to one variable when a second variable changes

renewable resource
an energy resource that can be replaced in a short period of time

responding variable (also called **dependent variable)**
in an experiment, a variable that responds to the change in the values of the manipulated variables. The scientist cannot set the values of the manipulated variables directly.

S

S wave
a seismic wave that involves vibration perpendicular to the direction the wave is travelling; it arrives later than the P wave

scientific theory
a consistent set of related scientific ideas

seismograph
an instrument that detects seismic waves

simple machine
a simple device that affects the force required to perform a certain task

simulator
a machine or computer program that models a given environment or situation for the purpose of training or research

single-loop circuit (series circuit)
a circuit that has all its parts connected in a single loop

slope
the tilt or slant of a straight line on a graph; the rise divided by the run

solution
a mixture that does not have any visible pieces of the different substances

stored phase energy
the energy associated with the phase of a material

suspension
a mixture that has visible pieces of at least one of the substances. The pieces must be larger than 0.2 μm.

T

terrestrial planet
a small, dense planet similar to Earth that consists mainly of rocky and metallic material; terrestrial planets include the inner Solar System planets Mercury, Venus, Earth, and Mars

transparent
a material through which light can travel

transverse wave
a wave in which the motion of the material (medium) is perpendicular to the motion of the wave

tsunami
a great sea wave produced by an earthquake (or volcanic eruption) on the ocean floor

U

ultrasound
compression waves at much higher frequency than animals or humans can hear

ultraviolet radiation
radiation with energies greater than visible light

V

variable
something in an experiment that changes or can be changed

velocity
how fast an object is moving in a given direction

volume
the measurement of how much space something occupies

W

water displacement
a method used to measure the volume of an object by measuring the amount of water that it displaces

wave
a continuous succession of pulses

wavelength
the distance between identical points along a wave

weight
the force exerted by a planet on an object

A

aceleración (acceleration)
el cambio en la velocidad por unidad de tiempo

amperímetro (ammeter)
un aparato que mide la cantidad de corriente eléctrica en un circuito

amplitud (amplitude)
el desplazamiento máximo de una partícula según pasa por una onda; la altura de la cresta de una onda; está relacionada con la energía de la onda

análisis (analysis)
cualquier procedimiento que te ayude a comprender una situación

años luz (light-year)
una unidad de medida equivalente a la distancia que la luz viaja en un año

atmósfera (atmosphere):
la capa o cubierta de gases que podría rodear un planeta o luna

C

cambio de fase (phase change)
la conversión de un material de una fase a otra fase en una temperatura específica. Ejemplo: sólido a líquido, líquido a gas

cilindro graduado (graduated cylinder)
un tubo delgado y alto con unidades marcadas que se usa para medir el volumen de un líquido

circuito de vuelta sencilla (single-loop circuit [series circuit])
un circuito que tiene todas las piezas conectadas a una vuelta sencilla

circuito de vueltas múltiples (circuito paralelo) (multi-loop circuit [parallel circuit])
un circuito en el cual dos o más vueltas sencillas se conectan a la misma fuente

circuito eléctrico (electric circuit)
el sendero seguido por una corriente eléctrica de una fuente de electricidad a través de instrumentos que usan electricidad y de regreso a la fuente

cohesión (cohesion)
pegados o sostenidos juntos para formar un todo

compuesto (compound)
una sustancia única (química) que se rompe en elementos durante una interacción química

conductor eléctrico (electric conductor)
un material que le permite a la corriente eléctrica existir en este

corriente eléctrica (electric current)
un flujo de electrones (carga negativa) alrededor de un circuito

D

densidad (density)
la masa de una unidad estándar de volumen

desplazamiento de agua (water displacement)
un método usado para medir el volumen de un objeto al medir la cantidad de agua que éste desplaza

difusión (diffusion)
la propagación de una sustancia en otra sustancia en la misma fase

E

electroimán (electromagnet)
un imán temporero consistente de un rollo de alambre alrededor de un núcleo de material magnético (generalmente hierro); cuando una corriente eléctrica fluye a través del alambre, el hierro se convierte en un imán

elemento (element)
una sola sustancia (química) que no se rompe en sustancias simples durante interacciones químicas

energía cinética (Kinetic energy)
la energía que posee un objeto por su movimiento

energía de activación (activation energy)
la energía mínima de movimiento de que partículas reactivas deben tener cuando chocan para romper sus enlaces químicos

energía de fase almacenada (stored phase energy)
la energía asociada con la fase de un material

energía de movimiento (motion energy)
la energía que un objeto tiene debido a su movimiento

energía mecánica (mechanical energy)
la energía transferida envuelta en una interacción que causa que uno o ambos objetos cambien de posición

energía potencial (potential energy)
la energía de un objeto que depende de su posición

energía potencial de gravitación (gravitational potential energy)
la energía de un sistema con dos objetos interactuando a través de la gravedad. La energía depende en la distancia entre los objetos y sus masas.

enlace (bond)
la atracción entre partículas (átomos, iones o moléculas)

enlace covalente (covalent bond)
la atracción entre pares específicos de átomos en una molécula

enlaces iónicos (ionic bonds)
la atracción entre iones de metales y no metales cercanos

enlaces metálicos (metallic bonds)
atracción entre átomos metálicos cercanos

evaluación (evaluation)
un juicio sobre algo

evidencia (evidence)
en un experimento, los datos recogidos por un investigador

explicación (explanation)
usa información de los análisis y usa ideas científicas para contestar preguntas acerca de una situación

F

falla (fault)
una fractura en una roca, a lo largo de la masa de roca que ha movido

fluído (fluid)
una sustancia que tiende a fluir y toma la forma de su envase

frecuencia (frequency)
el número de ondas producido por unidad de tiempo

fuente de energía (energy source)
un objeto que es el suplidor de energía

fuente de energía no renovable (nonrenewable energy resource)
una fuente de energía que no puede ser reemplazada

fuente renovable (renewable resource)
una fuente de energía que puede ser remplazada en un corto período de tiempo

fuerza (force)
un empujón o un halón

fuerza de fluctuación (buoyant force)
para un objeto que se ha colocado en un líquido, la fuerza ejercida en un objeto por el líquido

fuerza o esfuerzo de entrada (input force or effort)
la fuerza ejercida por una máquina

fuerza producida (output force [or load])
la fuerza que una máquina ejerce en un objeto

G

gravedad (gravity)
la fuerza de atracción entre dos cuerpos debido a sus masas

H

hipótesis (hypothesis)
una proposición que puede ser probada o refutada por evidencia experimental o de observación

I

interacción aplicada (applied interaction) (por definir)

interacción física (physical interaction)
cualquier tipo de interacción que no resulte en un material nuevo

interacción mecánica (mechanical interaction)
una interacción en la cual los objetos se tocan unos a los otros mientras se empujan o halan uno al otro sobre una distancia

interacción química (chemical interaction)
cualquier tipo de interacción que tenga como resultado al menos un nuevo material

ión (ion)
un solo átomo o grupo de átomos que tienen ambas cargas positivas o negativas

isótopo (isotope)
un elemento que tiene diferentes números de neutrones

L

lente convergente (converging lens)
los rayos de luz pasando a través de los lentes son traídos a un punto. Están formados para que el medio sea más grueso que sus bordes

longitud (length)
una medida de distancia

longitud de la onda (wavelength)
la distancia entre puntos idénticos a lo largo de una onda

luminosidad de una estrella (Luminosity [of a star])
una medida de la claridad de una estrella que no depende de la distancia entre la estrella y el observador

luna (satélite) (moon [satellite]):
un cuerpo que orbita un planeta

M

máquina compuesta o compleja (compound machine or complex machine)
dos o más máquinas simples trabajando juntas

máquina simple (simple machine)
un aparato simple que afecta la fuerza requerida para desempeñar cierta tarea

masa (mass)
la cantidad de materia que posee un cuerpo

materiales magnéticos (magnetic materials)
metales que interactúan con imanes

metaloides (semi-metales) (metalloids [semi-metals])
elementos que tienen las propiedades de ambos, metales y no metales

monómeros (monomers)
la molécula más pequeña que están unidas para formar las macromoléculas de los polímeros

N

newton (newton)
una unidad de fuerza

no conductor de electricidad (electric non-conductor)
un material que no permite que una corriente eléctrica exista en él

O

onda (wave)
una sucesión continua de pulsos

onda de compresión (longitudinal) (compression [longitudinal] wave)
una onda en la cual el movimiento del medio (material) es paralelo a la dirección del movimiento de la onda

onda L (l wave)
una onda sísmica que viaja a lo largo de la superficie de la Tierra; son las últimas en llegar a un lugar

onda P (p wave)
una onda sísmica que envuelve movimiento en la dirección en la cual está viajando; es la más rápida de las ondas sísmicas

onda S (s wave)
una onda sísmica que envuelve vibración perpendicular en la dirección en la cual la onda está viajando; ésta llega más tarde que la onda P.

onda transversal (transverse wave)
una onda en la cual el movimiento del material (medio) es perpendicular al movimiento de la onda

opaco (opaque)
un material por el cual la luz no puede viajar

órbita (orbit)
el sendero que un objeto toma mientras se mueven a través del espacio, generalmente alrededor de otro objeto

P

pendiente (slope)
la inclinación de una línea recta en una gráfica; la elevación dividida por el escurrimiento

periódico (periodic)
tener un patrón regular repetitivo

peso (weight)
la fuerza ejercida por un planeta en un objeto

pH (pH)
una cantidad usada para representar cuán ácida es una solución

planeta gigante gaseoso (gas-giant planet)
un planeta grande con una atmósfera profunda que es mayormente hidrógeno y helio, y un núcleo de material rocoso y helado; los planetas gigantes de gas incluyen los planetas del exterior del Sistema Solar, Júpiter, Saturno, Uranio y Neptuno

planeta terrestre (terrestrial planet)
un planeta pequeño y denso similar a la Tierra que consiste mayormente de material rocoso y metálico; los planetas terrestres incluyen los planetas internos del Sistema Solar, Mercurio, Venus, Tierra y Marte

plásticos (plastics)
polímeros sintéticos que no se pueden moldear o darle forma

polímero (polymer)
una sustancia que es una macromolécula consistiendo de muchas moléculas pequeñas similares (monómeros) enlazadas en largas cadenas

producto (product)
sustancia final (nueva) hecha en una reacción química

propiedad (property)
una descripción de cómo un objeto interactúa con otro objeto

propiedades características (characteristic property)
una medida (números) que es diferente por sus diferentes tipos de materiales

propiedades físicas (physical property)
una descripción o medida de lo que sucede a un material durante interacciones químicas

prueba justa (fair test)
un experimento en el cual solamente se le permite cambios a las variables de manipulación y de respuesta y donde todas las demás variables y condiciones se mantienen igual

R

radiación infrarroja (infrared radiation)
radiación con energías más bajas que la luz visible

radiación ultravioleta (ultraviolet radiation)
radiación con energías mayores que la luz visible

reacción endotérmica (endothermic reaction)
reacciones que absorben energía (requieren entrada continua de energía)

reacción exotérmica (exothermic reaction)
reacciones que producen (desplazan) energía

reactante (reactant)
sustancia original en un reacción química

receptor de energía (energy receiver)
un objeto al cual la energía es transferida

refracción (refraction)
el cambio en dirección (doblaje) de un rayo de luz cuando pasa en un ángulo de un material a otro material diferente

relación (relationship)
una idea acerca de lo que le sucede a una variable cuando una segunda variable cambia

relación lineal (linear relationship)
la relación entre dos cantidades que cuando son trazadas una con la otra producen una línea recta

relación no lineal (nonlinear relationship)
la relación entre dos cantidades que cuando son trazadas una contra la otra en una gráfica no producen una línea recta

S

simulador (simulator)
una máquina o programa de computadora que modela un ambiente dado o una situación con el propósito de entrenar o investigar

sismógrafo (seismograph)
un instrumento que detecta las ondas sísmicas

solución (solution)
una mezcla que no tiene ninguna pieza visible de las diferentes sustancias

suspensión (suspension)
una mezcla que tiene piezas visibles de al menos una de sus sustancias. Las piezas deben ser más grandes que 0.2 _m

T

teoría científica (scientific theory)
un conjunto de ideas científicas relacionadas

terremoto (earthquake)
un movimiento repentino o sacudida de la tierra

tono (pitch)
la calidad de un sonido dependiente mayormente de una frecuencia de la onda sonora

transparente (transparent)
un material a través del cual la luz puede viajar

tsunami (tsunami)
una gran ola producida por un terremoto submarino (o una erupción volcánica) en el fondo del océano

U

ultrasonido (ultrasound)
ondas de compresión a una frecuencia mayor de la que pueden oir los animales o los humanos

unidad astronómica (astronomical unit)
una unidad de medida igual a la distancia promedio entre el Sol y la Tierra

V

valores extremos (outliers)
valores lejos unos de otros en un conjunto de datos

variable (variable)
algo en un experimento que cambia o puede ser cambiado

variable de manipulación, también llamada **variable independiente (manipulated variable [**also called **independent variable])**
en un experimento, una variable que puede ser cambiada deliberadamente por el científico y que determina los valores de otras variables (llamadas variables de respuesta)

variable de respuesta (responding variable [also called **dependent variable])**
en un experimento, una variable que responde a los cambios en los valores de una variable de manipulación. Los científicos no pueden establecer los valores de las variables de manipulación directamente.

velocidad (velocity)
cuan rápido se nueve un objeto en una dirección dada

velocidad constante (constant speed)
ni aceleración ni reducción de velocidad

velocidad promedio (average speed)
la distancia viajada dividida por el tiempo que toma viajar

volumen (volume)
la medida de cuánto espacio algo ocupa

Index

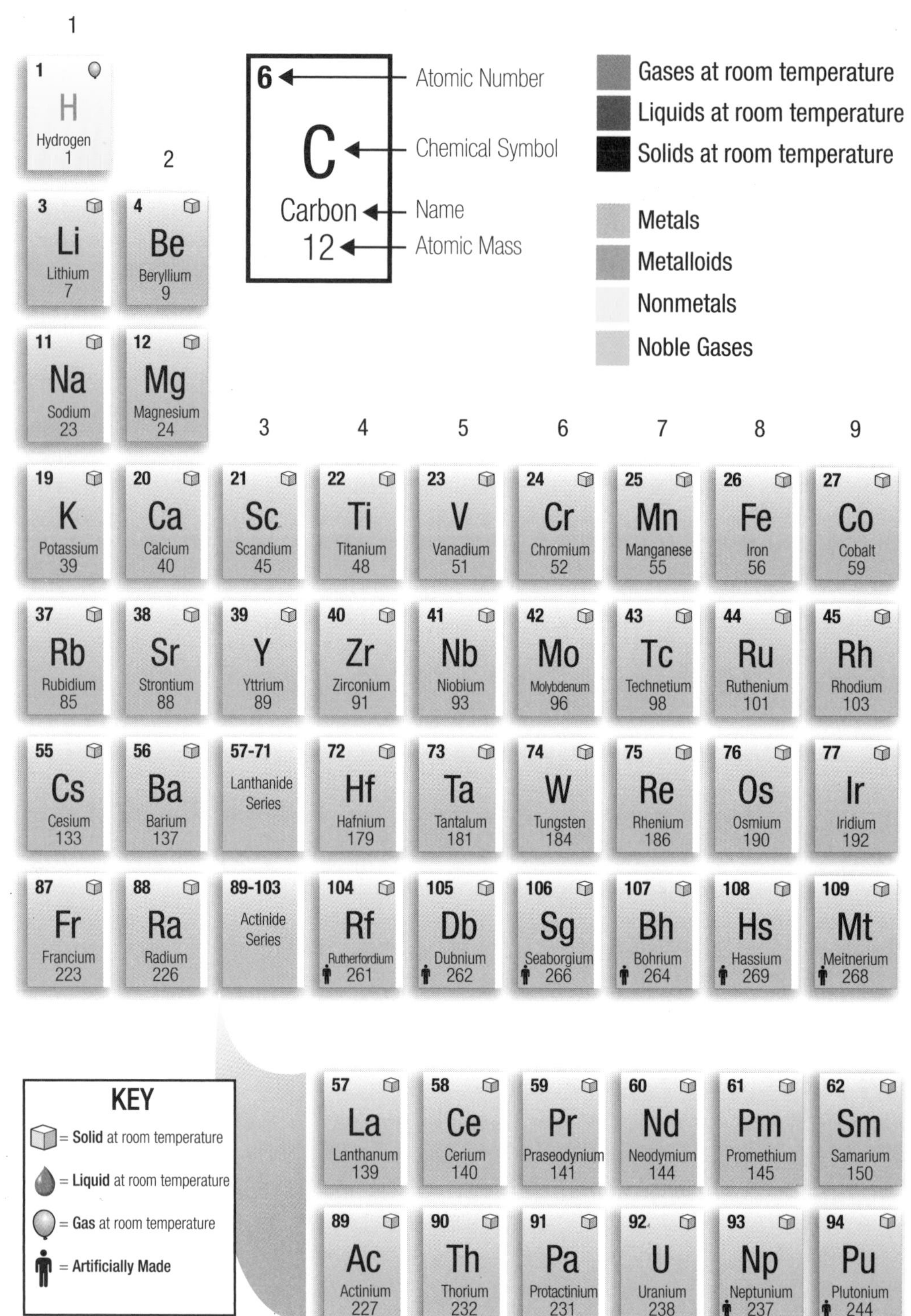

1
2
6 C Carbon 12
Atomic Number
Chemical Symbol
Name
Atomic Mass
Gases at room temperature
Liquids at room temperature
Solids at room temperature
Metals
Metalloids
Nonmetals
Noble Gases
3
4
5
6
7
8
9
1 H Hydrogen 1
3 Li Lithium 7
4 Be Beryllium 9
11 Na Sodium 23
12 Mg Magnesium 24
19 K Potassium 39
20 Ca Calcium 40
21 Sc Scandium 45
22 Ti Titanium 48
23 V Vanadium 51
24 Cr Chromium 52
25 Mn Manganese 55
26 Fe Iron 56
27 Co Cobalt 59
37 Rb Rubidium 85
38 Sr Strontium 88
39 Y Yttrium 89
40 Zr Zirconium 91
41 Nb Niobium 93
42 Mo Molybdenum 96
43 Tc Technetium 98
44 Ru Ruthenium 101
45 Rh Rhodium 103
55 Cs Cesium 133
56 Ba Barium 137
57-71 Lanthanide Series
72 Hf Hafnium 179
73 Ta Tantalum 181
74 W Tungsten 184
75 Re Rhenium 186
76 Os Osmium 190
77 Ir Iridium 192
87 Fr Francium 223
88 Ra Radium 226
89-103 Actinide Series
104 Rf Rutherfordium 261
105 Db Dubnium 262
106 Sg Seaborgium 266
107 Bh Bohrium 264
108 Hs Hassium 269
109 Mt Meitnerium 268
57 La Lanthanum 139
58 Ce Cerium 140
59 Pr Praseodynium 141
60 Nd Neodymium 144
61 Pm Promethium 145
62 Sm Samarium 150
89 Ac Actinium 227
90 Th Thorium 232
91 Pa Protactinium 231
92 U Uranium 238
93 Np Neptunium 237
94 Pu Plutonium 244
KEY
= Solid at room temperature
= Liquid at room temperature
= Gas at room temperature
= Artificially Made